calculus

calculus

lynn loomis *Harvard University*

addison-wesley publishing company
reading, massachusetts · menlo park, california
london · amsterdam · don mills, ontario · sydney

Cover photo by Manolo Guevara, Jr.

Frontispiece courtesy Professor Elmer R. Pearson, Institute of Design, I. I. T.

Chapter opening photographs: Christopher S. Johnson, Chapter 1. The Chimneys Violin Shop, Boiling Springs, Pennsylvania, Chapter 2. Stock, Boston: John T. Urban, Chapter 3; Franklin Wing, Chapters 10 and 20; Clif Garboden, Chapter 12; Frank Siteman, Chapter 15; and Daniel S. Brody, Chapter 19. Manolo Guevara, Jr., Chapter 4. Grant Heilman, Lititz, Pennsylvania, Chapters 5, 7, and 9. DeWys, Inc., New York: David W. Hamilton, Chapter 6; Chapters 13, 16, 17. Bill Finch, Chapters 8 and 11. Erik Hansen, Chapter 14. Hedrich-Blessing, Chicago: Chapter 18.

Third printing, January 1975

ISBN 0-201-04307-6
ABCDEFGHIJ-HA-798765

preface

This book is intended as a text for the standard two- or three-semester calculus course. The choice and the order of topics are traditional. The treatment is intuitive, although questions of estimation receive a fair amount of attention. Answers (or hints) for most of the odd problems are given at the back of the book.

Most of the traditional topics from analytic geometry are covered in Chapter 1 and Appendix 2. Chapter 1 treats lines, circles, translation of axes, and completing the square to simplify quadratic equations, while Appendix 2 develops the conic sections from their standard locus definitions. Some related material involving polar coordinates and parametric equations will be found in Chapters 5 and 11. There are also a few clusters of problems in other chapters that are relevant. For example, in Section 14.6 the optical property of the ellipse is obtained directly from its locus definition by vector differentiation.

The intuitive bias of the book is partly expressed in a free use of variables and "y is a function of x". This old symbolism forms a very useful idiomatic language, and is perfectly respectable when interpreted in terms of our present-day notion of a function. So Leibniz notation and function notation receive about equal time.

In lieu of rigor, there is some emphasis on computation. For the most part this appears as estimation, answering the question, "How good an approximation do I have?" However, Chapter 20 takes up the more basic question, "What must I do to get the accuracy I want?", and in this computational context there is a bit of ϵ, δ reasoning. The two questions are closely related: if a given Δy is estimated from a given Δx by first calculating that $|\Delta y| \leq K |\Delta x|$ for all sufficiently small Δx and Δy, then a simple answer to the ϵ, δ question is ready whenever the question is raised.

The material is organized into chapters in such a way as to allow considerable syllabus flexibility. For example, Chapter 8 (Antidifferentiation) can be taken up any time after Chapter 3 (see page xiii for details), and Chapter 15 (Functions of two variables) any time after Chapter 6, should an early introduction to either of these topics be desired. Possible realignments can be visualized from the following flow chart, which shows the major dependencies between chapters. Any reordering of material that is consistent with the chart should involve only minor problems of accommodation.

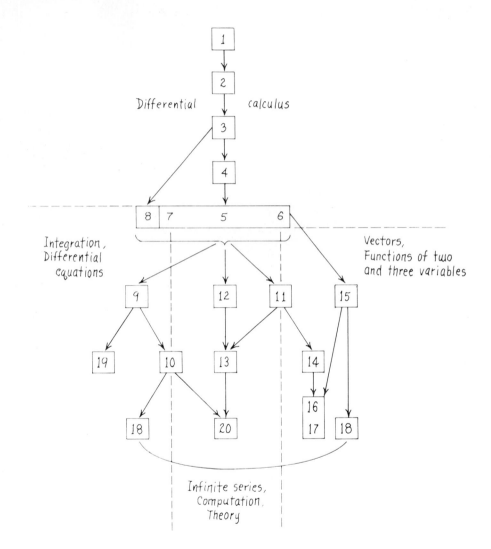

I would like to express my appreciation and gratitude to Nancy Larson for her expert typing, to my editors at Addison-Wesley for their constant help and encouragement, and to Thomas F. Banchoff for his challenging and valuable criticism.

November 1973   L. H. L.

contents

Chapter 3A The Antiderivative

Chapter 8 was originally written to fit in here. A few examples and problems would have to be omitted because they involve functions not yet discussed. See page xiii for details.

Chapter 8   The Antiderivative can directly follow **Chapter 3   The Derivative** by deleting the following examples and problems in Chapter 8 (these examples and problems assume material in Chapters 4 and 7):

Pages	Problems	Pages	Problems
310	19, 20, 21	326	23, 24, 29, 30, 31, 37,
314	10, 11, 12		38, 41, 42, 43
322	Three displayed eqs.	332	18, 20
	at bottom of page	333-6	Omit Section 7
323	Examples 6, 7, 8	337	3, 4
325	6	350	7

Thus the only material that needs to be deleted are three examples; 3 displayed equations (out of too numerous to count); 19 problems; and all of Section 7.

If Chapter 8 is taken up directly after **Chapter 4   Systematic Differentiation** (in which exponential and trigonometric functions are introduced, along with the chain rule), only the following examples and problems in Chapter 8 need to be deleted because they are dependent of material in Chapter 7:

Pages	Problems	Pages	Problems
322	First of three eqs. at bottom of page	323	Example 6
		326	30, 31, 37, 38, 41

introduction

NEWTON, LEIBNIZ, AND THE CALCULUS

The invention of calculus is attributed to two geniuses of the 17th century, Isaac Newton in England, and, independently, Gottfried Leibniz in Germany. However, calculus was not created out of a vacuum by Newton and Leibniz, but rather was the end result of a gradual mathematical evolution extending back over the preceding century and involving many men. The contribution of Newton and Leibniz lay in their realization that a bewildering chaos of ideas could be fitted together to form a harmonious whole, and that the resulting structure could be organized as an algorithmic calculational discipline of enormous power, applicable to all sorts of fundamental questions about the nature of the world.

Although the new calculus obviously worked, Newton and Leibniz did not have a clear idea of *why* it worked. They tried to explain its successes by geometric reasoning, since at that time all mathematical phenomena were viewed in terms of geometry, but their explanations were unsatisfactory. In fact, the logical and philosophical aspects of calculus baffled mathematicians for another century and a half. Some fragmentary progress occurred, and a new point of view gradually formed, based on numbers, variables, and functional relationships between variables. Then, around 1820, the French mathematician, Augustin Cauchy, settled the matter by showing that calculus rests on the properties of the limit operation. This was still not what we today call rigor, and it took another fifty years of deeper probing before ϵ, δ reasoning and the completeness of the real-number system emerged as the ultimate arbiters of analytical correctness.

The failure of Newton and Leibniz to fathom the logical foundations of what they were doing did not prevent them from successfully exploiting their intuitions. Mathematics has more to do with the insights arising out of some sort of intuitive understanding than it does with complete logical rigor; and although the early calculus was crude and blunt compared to its later refinements, it was, nevertheless, an incredibly powerful new tool with which Newton, Leibniz, and their successors were able to do miraculous things.

The chronological development of calculus was marked at several points by leaps in precision and sophistication. Now, three hundred years after Newton and Leibniz, we can start our study of the subject at practically any level we wish.

Since there seems to be little point in repeating the confusions of the first one hundred fifty years, we shall approach calculus at about the level of Cauchy, which is still very intuitive. We can then increase our precision in a natural manner as the subject unfolds, and by the end of the book we will be reasoning with a satisfactory degree of rigor.

chapter 1
graphs

Calculus is about functions, and it is important when approaching calculus to have a reasonably good understanding of what a function is and what its graph can be like. In this preliminary chapter we shall go over the beginnings of analytic geometry, and thus review the graphs of simple equations. Then in Chapter 2 we shall take up the fundamental notion of a *function*.

1. COORDINATES

If a unit of distance is given in the plane or in space, then the distance between any two points is determined. It is the length of the line segment joining the two points, in terms of the given unit. Sometimes it is convenient to use different units for different measurements. For example, we might measure a pipeline and find that it is fifty miles long and one foot in diameter. The possibility of making such measurements of length leads to a systematic scheme for representing points by numbers, called a coordinate system.

The fundamental step is putting coordinates on a geometric line l, and thus making l into a "number line," as pictured below.

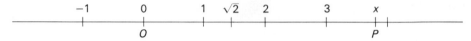

In order to do this, we make l into a *directed line* by choosing one of the two directions along l as the *positive direction*. Then we choose a *unit of distance* (generally by taking some particular line segment as the unit of length). On the line shown below, we have taken the positive direction to the right and have marked off a few points one unit of distance apart.

Definition. *The* signed distance from P to Q *is the distance between P and Q if the direction from P to Q is positive. It is the negative of the distance between P and Q if the direction from P to Q is negative.*

In the figure above, the signed distance from A to B is 2, and the signed distance from C to B is -3. From C to A it is -5, and from B to D it is 4.

We shall represent the signed distance from P to Q by $\overline{PQ}$. Note then that $\overline{QP} = -\overline{PQ}$. If you try various combinations of three points in the figure above you will see that the following **Addition Law** is true.

Addition Law. *If P, Q, and R are any three points on a directed line, then*

$$\overline{PQ} + \overline{QR} = \overline{PR}.$$

For example,

$$\overline{BC} + \overline{CA} = 3 + (-5) = -2 = \overline{BA},$$

$$\overline{AD} + \overline{DC} = 6 + (-1) = 5 = \overline{AC},$$

$$\overline{AE} + \overline{BC} = 2 + 3 = 5 = \overline{AC}, \text{ etc.}$$

Ordinary (unsigned) distances satisfy this identity only when Q is between P and R.

Finally we choose an *origin* point O on l and make the following definition:

Definition. *The coordinate of any point P on l is the signed distance from O to P.*

In this way we assign to each point P on l a uniquely determined numerical coordinate. Conversely, if we are given any real number x, then we obtain a unique point P having x as its coordinate by measuring off the signed distance x from the origin O. Thus, each point P determines a unique number x, and each number x determines a unique point P. This is what we mean when we call the coordinate assignment a *one-to-one correspondence* between the points of l and the real numbers.

Because of the coordinate correspondence, we can think of the real number system $\mathbb{R}$ in its entirety as though it were a geometric line, and we often refer to numbers like 3, π, and $\sqrt{2}$ as being "points" on the "number line."

Coordinates are signed distances from O. Other signed distances are determined from coordinates as follows:

Theorem 1. *The signed distance from the point x to the point t is $t - x$.*

Proof. This is a special case of the Addition Law. Suppose that x is the coordinate of P and that t is the coordinate of Q. That is, $\overline{OP} = x$ and $\overline{OQ} = t$. The identity

$$\overline{OP} + \overline{PQ} = \overline{OQ}$$

then becomes

$$x + \overline{PQ} = t,$$

or

$$\overline{PQ} = t - x. \blacksquare$$

Corollary. *If a new origin O' is chosen at the point with coordinate a, then the point having the old coordinate t has the new coordinate $t - a$.*

Proof. This is just a restatement of the theorem, in view of the definition of coordinates.   ∎

The coordinate plane involves two number lines. We choose a pair of perpendicular lines intersecting in a point *O*, and on each line set up a coordinate system with *O* as the zero point. We call *O* the *origin of coordinates*, or simply the *origin*; and the intersecting number lines are the *coordinate axes*. We choose one of the axes as the "horizontal" axis, and picture it horizontal, with the *positive direction to the right*. The positive direction of the other axis is pictured *upward*.

Given a coordinate system as described above, then each point in the plane determines, and is determined by, an ordered pair of coordinate numbers, as indicated below.

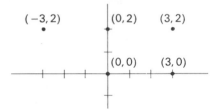

We shall feel free to call (3,2) a point in the coordinate plane, just as we call 3 a point on the number line.

Notice that the point (0,2) is different from the point (2,0). It is crucial that the coordinates form an *ordered* pair. The *first* coordinate is obtained by dropping a perpendicular to the *horizontal* axis, and is the coordinate on the horizontal axis of the foot of this perpendicular. The second coordinate is found similarly on the vertical axis.

The axes divide the plane into four parts called *quadrants*, which are conventionally numbered in the counterclockwise direction, starting from the quadrant that is above the positive half of the horizontal axis.

II		*I*
III		*IV.*

PROBLEMS FOR SECTION 1

For each of the following problems, mark off equally spaced points on a directed line, and then label three of them A, B and C, in such a way that the addition law $\overline{AB} + \overline{BC} = \overline{AC}$ illustrates the given numerical identity.

1. $3 - 2 = 1$
2. $-1 + 4 = 3$
3. $1 - 4 = -3$
4. $-1 - 2 = -3$
5. $5 - 3 = 2$
6. $5 - 6 = -1$

Mark off the letters A through J, in order and equally spaced, on a line l. Suppose that the positive direction is from A to J and that the letters are a unit of distance apart. Verify each of the following instances of the addition law by determining the three signed distances involved:

7. $\overline{AG} + \overline{GB} = \overline{AB}$
8. $\overline{FB} + \overline{BJ} = \overline{FJ}$
9. $\overline{IG} + \overline{GD} = \overline{ID}$
10. $\overline{HC} + \overline{CF} = \overline{HF}$
11. $\overline{DF} + \overline{FC} = \overline{DC}$
12. $\overline{AJ} + \overline{JE} = \overline{AE}$
13. $\overline{DG} + \overline{GA} = \overline{DA}$

14. Again let the points A through J be in order and equally spaced on the line l, but now suppose that $\overline{AC} = -3$. What does this say about the direction of l and the spacing of the letters? Work out Problems 7, 8, and 9 in this new situation.

For each of the following problems, mark two points P and Q on a line and then determine where the origin of coordinates is if P and Q have, respectively, the given coordinates:

15. (-1), (1)
16. (3), (4)
17. (2), (-3)
18. (-1), (-2)

19. Prove that a coordinate system on a line is entirely determined if we know the coordinates of two points.

In the following examples we consider two coordinate systems on a line l, and we use x and y for the two coordinates of a varying point P.

20. Suppose that the two origins are the same and that $y = 2$ when $x = 1$. Show that $y = 2x$ for every point P.

21. Suppose that the two origins are the same and that $y = -1$ when $x = 1$. Show that $y = -x$ for every point P.

22. Suppose that the two origins are the same and that $y = a$ when $x = 1$. Show that $y = ax$ for every point P.

23. Suppose that $y = b$ when $x = 0$ and that $y = b + 1$ when $x = 1$. Show that $y = x + b$ for every point P.

24. Suppose that $y = b$ when $x = 0$ and that $y = b + a$ when $x = 1$. Show that $y = ax + b$ for every point P.

25. Show that in every case there are constants a and b such that

$$y = ax + b$$

for every point P.

26. Suppose that $a < b$ and consider the segment ab on the number line. Show that its midpoint is $(a + b)/2$. [*Hint:* if m is the midpoint, then the signed distance from a to m should equal the signed distance from m to b.]

27. Find the point (number) that is two-thirds of the way from a to b. We say that this point divides the segment ab in the ratio two to one.

28. Find the point (number) that divides the segment ab in the ratio r to s, where r and s are any positive numbers.

29. Draw a plane coordinate system and plot the points $(-2, 1)$, $(2, -1)$, $(-1, 0)$, $(4, 1)$, $(4, -1)$.

30. Draw the collection of all points whose first coordinate is 2 (in a plane coordinate system). Describe this collection geometrically.

31. Draw and describe the collection whose second coordinate is -3.

32. Draw the collection of all points (x, y) for which $y/x = 2$. Is this the same as the collection for which $y = 2x$?

33. Draw the collection of all points (x, y) for which $y = -x$.

34. a) Given the point $P = (3, 4)$, find a point $Q_1 = (x, y)$ such that the perpendicular bisector of the segment PQ_1 is the y-axis.
    b) Now find Q_2 such that the perpendicular bisector of PQ_2 is the x-axis.
    c) Finally, find Q_3 such that the midpoint of the segment PQ_3 is the origin.
    d) What kind of geometric figure is the quadrilateral $PQ_1 Q_3 Q_2$?

35. Answer the same four questions for the point $P = (-2, 1)$.

36. The same for $P = (-1, -3)$.

2. GRAPHS OF EQUATIONS

In order to graph an equation in two variables, say

$$3x + 4y = 6,$$

it is necessary first to label the axes. That is, we decide which variable is going to have its values measured along the horizontal axis. If the variables are x and y, it is conventional to take the horizontal axis as the x-axis. In fact, the horizontal axis is sometimes regarded as being permanently labelled x. But how, then, do we graph the equation

$$3u + 4v = 6 ?$$

Anyway, with the axes labelled—and we shall generally use the standard x, y labelling—we can proceed.

A *solution* of an equation in two variables x and y is a pair of values (x,y) that "satisfies" the equation, i.e., that makes it true. For example, $(x,y) = (3,2)$ is a solution of

$$x^2 + 4y^2 = 25,$$

because it is true that $3^2 + 4(2^2) = 25$. Another solution is $(5,0)$. The pair $(x,y) = (2,3)$ is not a solution because

$$2^2 + 4(3^2) = 4 + 36$$

$$= 40 \neq 25.$$

The collection of all solution pairs is called the *solution set*.

The fact that the ordered pair $(3,2)$ *is* a solution while $(2,3)$ is *not* a solution depends on our stipulation that the first number in an ordered pair is a value of x and the second number a value of y. This goes back to the basic convention that the first coordinate of a point in the plane is the one measured horizontally, and the assumption that the horizontal axis is the x-axis.

The *graph* of the equation is just the geometric representation of its solution set. That is, the graph of an equation is the curve or other configuration in the plane consisting of the points whose coordinates satisfy the equation. The graph of the above equation looks like this:

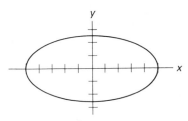

Two equations having the same solution set (i.e., the same graph) are said to be *equivalent*. Here are some other equations equivalent to the equation $x^2 + 4y^2 = 25$:

$$\frac{x^2}{25} + \frac{y^2}{(25/4)} = 1, \qquad \sqrt{x^2 + 4y^2} = 5, \qquad x^2 + 4y^2 - 25 = 0.$$

On the other hand, the two equations

$$x = y \text{ and } x^2 = y^2$$

are *not* equivalent. Why?

If the curve C is the graph of the equation E, then we call E *an* equation of C, because, as we saw above, *other* equations can have the same graph. However, if E has been singled out from various equivalent equations because it has a *standard form*, then we often call it *the* equation of C. For example,

$$\frac{x^2}{a^2} + \frac{y^2}{b^2} = 1$$

is called *the* equation of an ellipse with center at the origin, meaning that this is the standard form for the equation. Sometimes there is more than one standard form. For example, the equations

$$4x - 2y - 8 = 0, \qquad y = 2x - 4, \qquad \frac{x}{2} - \frac{y}{4} = 1$$

are equivalent, and are all standard forms for the equation of a certain straight line.

The *intercepts* of a graph are the points where the graph intersects the axes. An intercept on the x-axis is a point of the form $(a,0)$, but since the second coordinate of an x-intercept is automatically zero, it is customary to say that the x-intercept is a. Any intercept $(a,0)$ of

$$x^2 + 4y^2 = 25$$

must satisfy the equation

$$a^2 + 4(0^2) = 25,$$

giving $a^2 = 25$ and $a = \pm 5$. Thus we obtain the x-intercepts of an equation by setting $y = 0$ and solving for x. Similarly, we get the y-intercepts by setting $x = 0$ and solving for y. The above equation has the y-intercepts $b = \pm 5/2$.

Intercepts can be important "positioning" aids. For example, the equation $3x - 4y - 5 = 0$ has the x-intercept $a = 5/3$ and the y-intercept $b = -5/4$, and once we know that the graph of this equation is a straight line, we can just draw it through these two points.

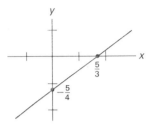

The points whose coordinates satisfy two equations simultaneously, say

$$y = x^2 \qquad \text{and} \qquad y = 2x - 1,$$

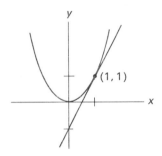

are the points lying on both graphs, i.e., the points of intersection of the graphs. We find these points by the usual techniques for solving simultaneous equations. Here we can eliminate y simply by subtracting the second equation from the first, getting

$$0 = x^2 - 2x + 1 = (x - 1)^2.$$

This gives the solution $x = 1$. The corresponding y value is then obtained from either of the original equations, say $y = x^2 = (1)^2 = 1$, so $(1,1)$ is the only point common to the two curves.

But now suppose that we want to consider how the two graphs are related to each other over all. Suppose, for instance, that we want to compare the points on the two graphs that have the same x-coordinates. To do this we have to use two different y variables, such as in

$$y = x^2 \qquad \text{and} \qquad \bar{y} = 2x - 1,$$

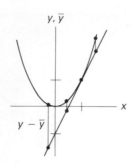

since the two equations will generally determine different y values for a given value of x. Then we can conclude from

$$y - \bar{y} = x^2 - (2x - 1) = (x - 1)^2 \geq 0$$

that the first graph lies everywhere above the second graph, with only the point (1,1) in common.

PROBLEMS FOR SECTION 2

The graph of an equation of the form

$$ax + by + c = 0$$

is always a straight line, provided that a and b are not both zero. The graph of

$$x^2 + y^2 + Ax + By + C = 0$$

is always a circle (or nothing). Assuming these facts, sketch the graphs of the equations in Problems 1 through 9 from their intercepts.

1. $x + y = 1$ 2. $x - y = 2$ 3. $2x + y = 2$

4. $3p + 2q = 6$ (with the p-axis horizontal)

5. $3p + 2q = 6$ (q-axis horizontal)

6. $x^2 + y^2 = 2$ 7. $x^2 + y^2 + 2x - 2y = 0$

8. $x^2 + y^2 - 4x = 0$ (assume the center is on the x-axis)

9. $x^2 + y^2 + 2x - 3 = 0$

10. a) Draw the graphs of Problems 1 and 2 above on the same coordinate system, and estimate their point of intersection.
    b) Compute the point of intersection by solving the equations simultaneously.

11. Same for (1) and (3).   12. Same for (2) and (3).

13. Same for (2) and (6).   14. Same for (1) and (6).

15. Same for (8) and (9).   16. Same for (7) and (8).

17. Graph the circle $x^2 + y^2 = 4$ and the line $x + y = 3$. Do these graphs appear to intersect? Show by algebra that they do not.

18. Graph the equation $y = x^2$. Here we don't know the shape of the graph, and the only way we can start is to plot a few points on the graph. Make up a little table of values of $y = x^2$, when x has the values $-2, -\frac{3}{2}, -1, -\frac{1}{2}, 0, \frac{1}{2}, 1, \frac{3}{2}, 2$. Then plot these points (x, y) and draw a smooth curve through them.

19. Graph the equation $y = x^2 - 2$, proceeding as in the above problem.

20. Graph the equation $y = x^2 - 2x - 2$, using the x-values of Problem 18.

21. Graph the equation $y = 2 - x^2$.

22. Graph the equation $y^2 = x + 2$. (The easiest way to do this is to give y some simple values and determine the corresponding x values from the equation.)

23. Graph the equation $y^2 = -x$.

24. a) Graph the equations $y = x^2$ and $y = x + 2$ on the same axes. Estimate their points of intersection.
    b) Compute the points of intersection algebraically by solving the equations simultaneously.

25. The graph of the equation $x^2 + 4y^2 = 4$ is a smooth oval-shaped curve called an *ellipse*. It will be longest in the direction of the x-axis and shortest in the direction of the y-axis. Use this information to sketch the graph, starting with only its intercepts.

26. The graph of $9x^2 + 4y^2 = 36$ is also an ellipse, but this time it is longest in the y direction. Plot its intercepts and sketch and graph.

27. Graph $y = x^2/2$, using integer values of x from $x = -3$ to $x = 3$.

28. Graph $y = x^2/4$, using integer values of x from $x = -5$ to $x = 5$.

3. THE STRAIGHT LINE

Horizontal and vertical lines have the simplest equations. Consider the horizontal line 3 units above the x-axis. A point (x, y) lies on this line if and only if its y-

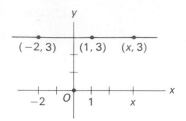

coordinate is 3, so this line is the graph of the equation $y = 3$. In general, the graph of

$$y = b$$

is the horizontal line with y-intercept b. If b is negative, say $b = -4$, then the line lies 4 units *below* the x-axis.
    Similarly,

$$x = a$$

is the equation of the vertical line with x-intercept a.

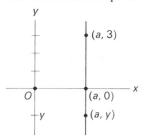

Note that although x is missing from an equation like $y = 3$, we can think of it as being there with a zero coefficient:

$$0x + y = 3.$$

    Consider next a nonvertical straight line l through the origin. Suppose, for the sake of definiteness, that l contains the point $(1,2)$. If a point (x, y) lies on l in the first quadrant, then the similar right triangles shown at the left in the next figure have proportional corresponding sides, so

$$\frac{y}{x} = \frac{2}{1}.$$

If (x, y) lies on l in the third quadrant, then the triangle side lengths are $-x$ and $-y$, so then $(-y)/(-x) = 2/1$. Each of these equations is equivalent to

$$y = 2x,$$

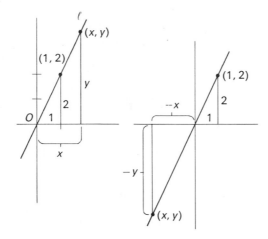

which thus holds for the coordinates of any point (x, y) on l. On the other hand, the reverse of the above line of reasoning shows that if $y = 2x$, then (x, y) lies on l. Therefore, l is the graph of the equation $y = 2x$.

Similarly, we can check that the graph of

$$y = -2x$$

is the straight line through the origin and $(1, -2)$. Here the two ways of rewriting the equation as statements about similar triangles are:

$$\frac{-y}{x} = \frac{2}{1} \quad \text{and} \quad \frac{y}{-x} = \frac{2}{1}.$$

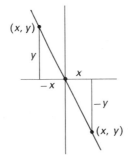

If we use an arbitrary constant m instead of the constants 2 and -2, the above reasoning shows that:

The graph of the equation

$$y = mx$$

is the straight line through the origin and the point $(1, m)$.

A different value of m gives a different line, and every nonvertical line through the origin is obtained in this way, because each such line intersects the vertical line $x = 1$ in a point $(1, m)$. (Strictly speaking, this geometric reasoning applies only when $m \neq 0$, so that the figure involves genuine triangles. But when $m = 0$, the equation $y = mx$ reduces to $y = 0$, which we already know to be the equation of the x-axis. Since the x-axis is the line through the origin and $(1,0)$, our general conclusion remains correct.)

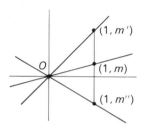

The constant m is called the *slope* of the line. Notice that $m > 0$ for a line *rising* to the right and that $m < 0$ for a line *falling* to the right. A line with large positive slope rises steeply and a line with small positive slope rises gently.

Now consider how the graph of

$$y = 2x - 3$$

is related to the straight line $y = 2x$. In order to consider these two graphs simultaneously, we use a different y variable for the second equation, say

$$\bar{y} = 2x.$$

Then for each value of x we can compare the points (x, y) and $(x, \bar{y})$ on the two graphs, as we did at the end of Section 2. Subtracting the second equation from the first gives

$$y - \bar{y} = -3.$$

That is, for each x the point (x, y) on the graph of $y = 2x - 3$ lies three vertical units below the point $(x, \bar{y})$ on the line $\bar{y} = 2x$. The points (x, y) satisfying

$$y = 2x - 3.$$

thus make up the parallel line three vertical units below the line $\bar{y} = 2x$, i.e., the parallel line that intersects the y-axis in the point $(0, -3)$.

If we carry through this argument using b and m instead of -3 and 2, we have the general conclusion in Theorem 2.

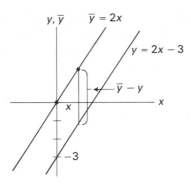

Theorem 2. *The graph of the equation*

$$y = mx + b$$

is the straight line that is parallel to the line $y = mx$ *and has y-intercept b.*

If l is any nonvertical line, then l is parallel to a unique nonvertical line through the origin, and l has a unique y-intercept b. Thus each nonvertical line l has a unique equation of the above form. As before, m is called the *slope* of l.

Also, since two lines are parallel to each other if and only if they are parallel to the same line through the origin, we have:

Corollary. *Two nonvertical lines are parallel if and only if they have the same slope.*

In summary: the slope m gives the direction of the line l, and the y-intercept b gives the point where l crosses the y-axis. The two numbers m and b together determine l and are determined by l.

EXAMPLE. Identify the graph of the equation $2x + 3y - 6 = 0$.

. . . .

Solution. Solving for y we obtain the equivalent equation

$$y = -\frac{2}{3}x + 2.$$

This is of the form

$$y = mx + b,$$

with $m = -2/3$ and $b = 2$. The graph is therefore a straight line, with slope $-2/3$ and y-intercept 2.

The equation $y = mx + b$ is called the *slope–intercept* form of the equation for l.

A line is determined if we know its slope m and one point (x_1, y_1) on it, and we should therefore be able to calculate its equation from this data. We can. For example, a line with slope 2 has the equation

$$y = 2x + b,$$

with b still to be determined. If we know that $(2,1)$ is on the line then

$$1 = 2 \cdot 2 + b,$$

giving $b = -3$, and

$$y = 2x - 3.$$

Thus, given the above data we can always solve numerically for b, and so obtain the slope–intercept equation. However, a more efficient procedure is to eliminate b in the general algebraic situation and obtain a new standard form for the equation of a line. Thus, if (x_1, y_1) is on the line

$$y = mx + b,$$

then it is true that

$$y_1 = mx_1 + b,$$

and if we subtract the second equation from the first then the b's cancel and we get the *point–slope* form of the equation:

$$y - y_1 = m(x - x_1).$$

We can write this form down directly whenever we know the slope of a line and the coordinates of a point on it. For example, the line through $(2,1)$ with slope 2 has the point–slope equation

$$y - 1 = 2(x - 2).$$

Solving this for y, we get back to the slope–intercept form

$$y = 2x - 4 + 1 = 2x - 3.$$

The earlier qualitative remarks about the significance of the slope m can be sharpened to an important quantitative fact:

The slope m is exactly the change in y divided by the change in x between any two points on the line.

That is,

Theorem 3. *If (x_1, y_1) and (x_2, y_2) are any two distinct points on a line with slope m, then*

$$m = \frac{y_2 - y_1}{x_2 - x_1}.$$

Proof. Since each pair of numbers satisfies the equation $y = mx + b$, we have

$$y_2 = mx_2 + b,$$

$$y_1 = mx_1 + b;$$

and so, subtracting,

$$y_2 - y_1 = m(x_2 - x_1).$$

Since the line is not vertical, the two x-coordinates must be different and we can divide by the nonzero number $x_2 - x_1$, getting $m = (y_2 - y_1)/(x_2 - x_1)$. ∎

Another procedure is to write the equation of the line in point–slope form as $y - y_1 = m(x - x_1)$, and then use the fact that (x_2, y_2) is on the line to get $(y_2 - y_1) = m(x_2 - x_1)$.

We can interpret this slope formula as follows: the slope m is the constant rate of change of y with respect to x along the line. If $m = 3$, then y increases 3 units per unit increase in x, and if $m = -\frac{1}{2}$, then y *decreases* one-half unit per unit *increase* in x.

Two distinct points (x_1, y_1) and (x_2, y_2) determine a line, and if the line is not vertical then the above formula gives its slope m. Thus the line through $(1,2)$ and $(3, -1)$ has slope

$$m = \frac{y_2 - y_1}{x_2 - x_1} = \frac{(-1) - 2}{3 - 1} = -\frac{3}{2}.$$

For the purposes of geometry it is useful to have a single form of equation that covers all lines, vertical or nonvertical. The equation

$$Ax + By + C = 0$$

does this (it being assumed that at least one of the coefficients A and B is not zero.)

In order to see that the graph of the above equation is always a line, we consider two cases. If $B \neq 0$ we can solve the equation for y and get an equivalent equation of the form $y = mx + b$. In this case the graph is a nonvertical line. For example, the equation

$$x + 2y - 6 = 0$$

can be solved for y, turning into the equivalent equation

$$y = \left(-\frac{1}{2}\right)x + 3.$$

Its graph is thus the nonvertical line with slope $-1/2$ and y-intercept 3.

If B is 0, then A cannot be 0 and we can solve for x, getting an equation of the form $x = a$. In this case the graph is the vertical line with x-intercept a. Thus $3x + 4 = 0$ becomes $x = -\frac{4}{3}$, the equation of the vertical line $\frac{4}{3}$ units to the left of the y-axis.

Conversely, the standard forms

$$y = mx + b \qquad \text{and} \qquad x = a$$

can be rewritten

$$mx - y + b = 0 \qquad \text{and} \qquad x - a = 0$$

respectively, both of which are of the form $Ax + By + C = 0$.

PROBLEMS FOR SECTION 3

1. Write the equation of the line with y-intercept -6 and slope 3. Find its x-intercept. Find the point where it intersects the vertical line $x = 1$. Same for the horizontal line $y = 1$.

Find the slope m and the y-intercept b of each of the following lines:

2. $x + y = 0$ 　　　　　　　　　　3. $x + 2y = 4$

4. $3x - 4y = 4$ 　　　　　　　　　5. $2x + 3y - 6 = 0$

6. $2y - 3x - 6 = 0$ 　　　　　　　7. $x - y + 1 = 0$

8. Show that the lines $4x + 2y = 1$ and $y = -2x + 3$ are parallel:
   a) by drawing both graphs;
   b) by proving algebraically that there is no point of intersection.

Draw the graphs of the following equations:

9. $x + y + 1 = 0$ 　　　　　　　　10. $y + 2x = 0$

11. $3y = x$ 　　　　　　　　　　　12. $y = x - 1$

13. $2x - 3y = 6$ 　　　　　　　　　14. $x + y = 1$

15. $x + 2y = 3$

16. Determine the slope and y-intercept for the line whose equation is $Ax + By + C = 0$ where A, B, and C are arbitrary constants, and $B \neq 0$.

Determine the slope of the line through the given pair of points, and then write the equation of the line:

17. $(0,0)$, $(2, 1)$ 　　　　　　　　18. $(2, -1)$, $(-4, 0)$

19. $(-2, 1)$, $(-2, 4)$ 　　　　　　20. $(3, 1)$, $(-2, 6)$

21. $(2, 1), (-3, 1)$ 22. $(1, 2), (-2, -1)$

Find the (equation of the) line:

23. through $(1,2)$, with slope $-(1/2)$;

24. through the points $(1,2)$ and $(2, -1)$;

25. through the points $(-2, -1)$ and $(1,1)$;

26. through the points $(1,2)$ and $(1,4)$;

27. through $(2,3)$, with x-intercept -1.

Find the point of intersection of the two lines:

28. $x + y = 1$, and $y = x$ 29. $x + y = 1$, and $x + 2y = 4$

30. $3x + y = 2$, and $x = -2$

31. Find the line through the point $(-1,2)$ parallel to the line through $(1,1)$ and $(0,2)$.

32. Which of the following pairs of lines are parallel?
     a) $2x + y - 2 = 0$, $4x + 2y + 18 = 0$
     b) $x + 2y - 3 = 0$, $2x - y + 3 = 0$
     c) $x + y - 1 = 0$, $y - x + 3 = 0$
     d) $x + 3y + 2 = 0$, $x + 3y - 2 = 0$
     e) $y + x = 1$, $2x + 2y + 5 = 0$

33. Determine the equation of the line passing through the point $(2,1)$ and parallel to the line through points $(4,2)$ and $(-2,3)$.

34. Determine the equation of the line passing through the point $(-2,5)$ and parallel to the line $2x + y + 6 = 0$.

35. Prove algebraically that two nonvertical lines are parallel if and only if their slopes are equal. (The lines l_1 and l_2 are parallel if they never intersect, or if $l_1 = l_2$.)

36. A point P moves along a line with slope $\frac{1}{3}$. What is the change in the y-coordinate of P when its x-coordinate increases by 10? Answer the same question for a line with slope -2.

In Problems 37–44, determine algebraically whether or not the three given points lie on a straight line.

37. $(0,0), (5,4) (-10, -8)$.

38. $(0, -1)$, $(1,0)$, $(3,2)$.

39. $(6,5)$, $(3,3)$, $(1,2)$.

40. $(7,5)$, $(3,3)$, $(-1,1)$.

41. Find the fourth vertex of the parallelogram with vertices at $(-1,0)$, $(0, -1)$, $(3,0)$ in that order.

42. A parallelogram has two sides lying on the lines $y = 2x$ and $3y = x$, and a vertex at $(3,3)$. Find the equations of the lines containing the other two sides, and find the remaining vertices.

43. If a line has nonzero intercepts a and b on the x and y axes, respectively, show that its equation can be written in the form

$$\frac{x}{a} + \frac{y}{b} = 1.$$

44. Show that the points $(1,1)$ and $(2,4)$ are on opposite sides of the line $x - y + 1 = 0$, by reasoning as follows: First, check the sign of $x_0 - y_0 + 1$ for each of the above two points (x_0, y_0). What, then, must happen to the sign of $x - y - 1$ as (x, y) moves along the line segment joining $(1,1)$ to $(2,4)$?

45. Show, by reasoning in a manner suggested by the above problem, that the points $(-1,1)$ and $(2,4)$ are on the same side of the line $x - y + 1 = 0$.

46. Suppose that the line $Ax + By + C$ does not go through the origin. Show that a point (x_0, y_0) and the origin are on the same side (opposite sides) of this line if $Ax_0 + By_0 + C$ and C have the same sign (opposite signs).

4. DISTANCE IN THE COORDINATE PLANE

We regard the numbers x and $-x$ as having the same *magnitude*. They are at the same distance from 0 on the number line, and this distance is x or $-x$, whichever one is positive. We call this distance the *absolute value* of x, and designate it $|x|$. Thus,

$$|x| = |-x|,$$

$$|x| = \begin{cases} x & \text{if } x \text{ is positive or zero} \\ -x & \text{if } -x \text{ is positive, i.e., if } x \text{ is negative.} \end{cases}$$

It follows that

$$|x| = \text{the larger of } x \text{ and } -x.$$

Also,

$$|x| = \sqrt{x^2},$$

since $\sqrt{a}$ is defined to be the positive square root of a. Theorem 1 says, in particular, that:

Theorem 4. *The distance between any two numbers x and t, considered as points on the number line, is*

$$|x - t|.$$

EXAMPLE. Graph the equation $y = |x|$.

. . . .

Solution. By the definition of absolute value, the equation $y = |x|$ says that

$$y = x \qquad \text{if } x \geq 0,$$
$$y = -x \qquad \text{if } x \leq 0.$$

Thus when $x \geq 0$ the graph runs along the straight line $y = x$, and when $x \leq 0$ it runs along the line $y = -x$, as shown below.

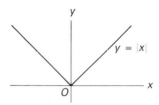

The laws relating absolute value to addition and multiplication are:

$$|x + y| \leq |x| + |y|,$$
$$|xy| = |x| \cdot |y|.$$

These properties of $|x|$ are important for numerical computations and for the rigorous treatment of the foundations of calculus. They will be used extensively in Chapter 20.

We now turn to the coordinate plane. Up until now we have not needed the same units of distance on the two axes, and it is frequently convenient to use different units, especially when we are graphing the relationship between a pair of unlike quantities. For example, if we wish to show graphically how the average daily temperature varies through the months of the year we would probably use scales somewhat as in the figure below.

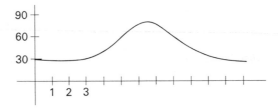

But when we wish to obtain a simple formula for distance in the plane, we have to use the same unit of distance on the two axes; and from now on a common unit will generally be assumed. Such a coordinate system is said to be *Cartesian*.

The Pythagorean theorem says that a triangle is a right triangle if and only if the square (of the length) of one side is equal to the sum of the squares (of the lengths) of the other two sides. Thus the triangle in the figure below is a right triangle with hypotenuse length h if and only if

$$h^2 = a^2 + b^2.$$

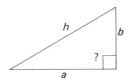

This numerical statement of the theorem depends on a unit of length common to the whole plane, and is the basic reason for using a Cartesian coordinate system. When we combine the Pythagorean theorem with Theorem 4, as shown in the figure below, we obtain a formula for the distance d between any two points (x_1, y_1) and (x_2, y_2) in the plane:

$$a = |x_2 - x_1|$$
$$b = |y_2 - y_1|$$
$$d^2 = a^2 + b^2$$
$$= (x_2 - x_1)^2 + (y_2 - y_1)^2.$$

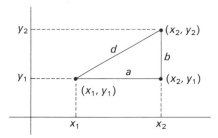

Theorem 5. *The distance between any two points* (x_1,y_1) *and* (x_2,y_2) *in the Cartesian plane is given by the formula*

$$d = \sqrt{(x_1 - x_2)^2 + (y_1 - y_2)^2} \, .$$

Strictly speaking, the Pythagorean theorem applies only when there is a genuine right triangle. However, the distance formula is correct in all cases. For example, if the two points lie on a horizontal line, then $y_1 - y_2 = 0$ and the formula reduces to

$$d = \sqrt{(x_1 - x_2)^2} = |x_1 - x_2|,$$

which is the correct one-dimensional distance formula.

The slope condition for perpendicular lines follows from these Pythagorean considerations.

Theorem 6. *Two nonvertical lines are perpendicular if and only if their slopes m and n satisfy*

$$mn = -1.$$

Proof. Suppose first that we are given two nonvertical perpendicular lines through the origin, with equations $y = mx$ and $y = nx$. These lines intersect the vertical line $x = 1$ in the points $(1,m)$ and $(1,n)$, respectively, and will be perpendicular if and only if the triangle thus formed is a right triangle. By the Pythagorean theorem, this will be the case if and only if

$$c^2 = a^2 + b^2,$$

or

$$(m - n)^2 = (1 + m^2) + (1 + n^2).$$

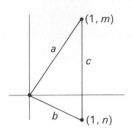

If we expand the square on the left, cancel the terms m^2 and n^2 and finally divide by -2, we end up with

$$mn = -1.$$

Conversely, if this equation holds, then we can work backward through the above steps and conclude that $c^2 = a^2 + b^2$ so that the a, b, c triangle is a right triangle. This proves the theorem for two lines through the origin. In the general case we replace two given lines by the two parallel lines through the origin. Since this replacement changes neither the slopes nor the angle between the lines, the general result follows from the above special case.

PROBLEMS FOR SECTION 4

Determine the distance between each of the following pairs of points:

1. $(2, 1)$, $(3, 3)$ 2. $(-1, 2)$, $(2, -3)$

3. $(0, 0)$, $(0, 4)$ 4. $(2, -5)$, $(2, 1)$

5. $(8, 3)$, $(-2, 1)$ 6. $(3, 4)$, $(-6, -1)$

7. $(0, 1)$, $(-2, 0)$ 8. $(5, 2)$, $(3, -1)$

9. $(1, 1)$, $(-1, -1)$ 10. $(2, -1)$, $(-1, 3)$

Find the lengths of the sides of the triangles with the given points as vertices.

11. $A(4, 1)$, $B(2, -1)$, $C(-1, 5)$ 12. $A(1, 2)$, $B(3, 1)$, $C(4, 2)$

13. $A(3, -4)$, $B(2, 1)$, $C(6, -2)$ 14. $A(0, 0)$, $B(2, 1)$, $C(1, 2)$

15. Show that the triangle with vertices at $P_1(1, -2)$, $P_2(-4, 2)$, and $P_3(1, 6)$ is isosceles.

16. Prove that the triangle with vertices at $(2,1)$, $(1,3)$, and $(8,4)$ is a right triangle, by checking the Pythagorean theorem.

17. Prove that the above triangle is a right triangle by computing slopes.

18. Find (the equation of) the line through the point $(1,1)$ perpendicular to the line $x + 2y = 0$.

19. Find the line through $(2,0)$ perpendicular to the line through $(2,0)$ and $(-1,1)$.

20. Find the equation of the line passing through point $(-2,3)$ and perpendicular to the line $2x - 3y + 6 = 0$.

21. The midpoint of the line segment joining the points (a_1, b_1) and (a_2, b_2) is the point

$$\left(\frac{a_1 + a_2}{2}, \frac{b_1 + b_2}{2} \right).$$

Verify this analytically in two steps:
a) Show that the three points are collinear.
b) Show that the segment endpoints are equidistant from the claimed midpoint.

22. Using the formula in Problem 21, write the equation of the perpendicular bisector of the segment between $(7,4)$ and $(-1, -2)$.

Given the vertices of a triangle $P_1(7,9)$, $P_2(-5,-7)$, and $P_3(12,-3)$, find:

23. The equation of side $P_1 P_2$.

24. The equation of the median through P_1.

25. The equation of the altitude through P_2.

26. The equation of the perpendicular bisector of the side $P_1 P_2$.

27. The points $P_1(3,-2)$, $P_2(4, 1)$, and $P_3(-3,5)$ are the vertices of a triangle. Show that the line through the midpoints of the sides $P_1 P_2$ and $P_1 P_3$ is parallel to the base $P_2 P_3$ of the triangle.

28. Prove the same result for the triangle with vertices (x_1, y_1), (x_2, y_2), (x_3, y_3).

29. Show that the points equidistant from two given points form a straight line (by equating two distances and simplifying algebraically).

30. Show that if two medians of a triangle are equal, then the triangle is isoceles. (Let the vertices be $(a,0)$, $(b,0)$, and $(0,c)$, and equate the lengths of the medians through the vertices $(a,0)$ and $(b,0)$.)

31. Graph the equation $y = x + |x|$.   32. Graph the equation $y = x - |x|$.

33. Graph the equation $|x| + |y| = 1$.

34. Prove algebraically that the three medians of a triangle are concurrent. (We can simplify the algebra by taking the y-axis through one vertex and the x-axis along the opposite side. The vertices are then $(a,0)$, $(b,0)$, $(0,c)$.)

35. Prove that the three perpendicular bisectors of the sides of a triangle are concurrent.

36. Find the distance from the point (2,3) to the line $x + 2y - 1 = 0$. (Find the point of intersection of the given line and the line perpendicular to it through the given point.)

37. Find the distance from $(1, -1)$ to the line $2x - y + 1 = 0$. (As above, first find the foot of the perpendicular dropped from the point to the line.)

38. Prove that the distance d from the point (x_0, y_0) to the line $Ax + By + C = 0$ is given by

$$d = \frac{|Ax_0 + By_0 + C|}{\sqrt{A^2 + B^2}}.$$

5. THE CIRCLE

A point (x, y) lies on the circle about (a, b) of radius r if and only if the distance between (x, y) and (a, b) is r;

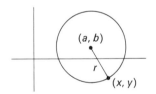

that is, if and only if

$$\sqrt{(x - a)^2 + (y - b)^2} = r.$$

This, then, is an equation for the circle. We usually eliminate the radical by squaring, to get the equivalent equation

$$(x - a)^2 + (y - b)^2 = r^2. \tag{*}$$

In particular, the equation for the circle of radius r about the origin is

$$x^2 + y^2 = r^2.$$

Remember that in each case we are saying that the graph of the equation is the circle.

EXAMPLES. 1. The circle about the origin with radius $r = 3$ has the equation

$$x^2 + y^2 = 9.$$

2. The equation

$$x^2 + y^2 = 3$$

has the form $x^2 + y^2 = r^2$ with $r = \sqrt{3}$. Therefore its graph is the circle about the origin of radius $r = \sqrt{3}$.

3. The circle about $(1, -2)$ with radius 5 has the equation

$$(x - 1)^2 + (y + 2)^2 = 25.$$

The above equation and the more general equation $(x - a)^2 + (y - b)^2 = r^2$ each can be directly interpreted as saying that a certain distance is r. But this form is artificial from the point of view of a general polynomial equation. We would more likely find the squares expanded and the constants collected. These steps transform the equation in Example 3 into

$$x^2 + y^2 - 2x + 4y - 20 = 0.$$

This is of the general form

$$x^2 + y^2 + Ax + By + C = 0, \tag{**}$$

and leads us to wonder if every equation of this form has a circle as its graph. Simple examples show us that this is too much to expect. Thus

$$x^2 + y^2 = 0$$

is satisfied only for $(x, y) = (0, 0)$, so that the graph of this equation consists of just one point, while

$$x^2 + y^2 + 1 = 0$$

is *never* satisfied and so has an empty graph.

How then can we tell whether an equation having the general form (**) has a circle for its graph or not? And what circle, if so? The procedure is to try to work back from the general form (**) to the special form (*). We gather together the x terms and make them into a perfect square by adding a suitable constant, and do the same with the y terms. Then the only question is whether the resulting constant on the right side of the equation is positive, zero, or negative.

EXAMPLE 4. If we try to recapture Example 3 from its expanded form by following this prescription, we start with

$$x^2 + y^2 - 2x + 4y - 20 = 0,$$

and first rewrite it as

$$(x^2 - 2x + K) + (y^2 + 4y + L) = 20 + K + L,$$

where we have to discover what values for the constants K and L will make the parentheses into perfect squares. Notice that we have to add these constants to *both sides* of the equation in order for the new equation to be equivalent to the old. In order to find K, we note that $(x^2 - 2x + K)$ must turn out to be $(x - a)^2 = x^2 - 2ax + a^2$. Thus the two expressions

$$x^2 - 2x + K \quad \text{and} \quad x^2 - 2ax + a^2$$

must be the same, so that $2 = 2a$ and $K = a^2$. This gives $a = 1$ and $K = 1^2 = 1$.
   Similarly, the two expressions

$$y^2 + 4y + L \quad \text{and} \quad (y - b)^2 = y^2 - 2by + b^2$$

must be the same, so that $4 = -2b$ and $L = b^2$, giving $b = -2$, and $L = (-2)^2 = 4$.
   This process is known as *completing the square* (a procedure first encountered in high-school algebra.) After a little practice you probably can do it in your head. Some people prefer to learn the rule: *to get the constant which must be added, we divide the coefficient of x by 2 and then square.*
   At this point we have

$$(x - 1)^2 + (y + 2)^2 = 20 + 1 + 4 = 25,$$

which is of the form (*) with $(a, b) = (1, -2)$ and $r = 5$. Now we can identify the graph as the circle with center at $(1, -2)$ and radius 5.

EXAMPLE 5. Find the graph of $x^2 + y^2 - 6x + 2y + 10 = 0$.

. . . .

Solution. Following the above rules, we write the equation in the form

$$(x^2 - 6x + K) + (y^2 + 2y + L) = -10 + K + L$$

and see that $K = 9$ and $L = 1$, so that the equation becomes

$$(x - 3)^2 + (y + 1)^2 = 0.$$

Since a sum of squares is zero only if each square is zero, the only solution of this equation is the point $(x, y) = (3, -1)$.

If 10 were replaced by 15 in this example, we would end up with -5 on the right, and since a sum of squares can never be negative there is no graph.

PROBLEMS FOR SECTION 5

Write the equation of each of the following circles in *standard* form. Then expand and collect coefficients to get the general form. Sketch each circle.

1. $C(2, 1)$ and $r = 3$
2. $C(-1, 0)$ and $r = \sqrt{2}$

3. $C(2, -3)$ and $r = 1$
4. $C(-2, -3)$ and $r = 4$

5. $C(-2, 4)$ and $r = 5$
6. $C(2, -1)$ and $r = 0$

7. $C(0, 0)$ and $r = |-4|$
8. $C(2, 5)$ and $r = \sqrt{3}$

By completing the square, convert each of the following equations into the standard form for the equation of a circle. Specify the coordinates of the *center* and the *radius* of the circle.

9. $x^2 + y^2 + 6x - 8y = 0$
10. $x^2 + y^2 - 4x + 2y + 5 = 0$

11. $x^2 + y^2 + 3x - 5y - \frac{1}{2} = 0$
12. $2x^2 + 2y^2 + 3x + 5y + 2 = 0$

Identify the graph of each of the following equations. That is, determine whether the graph is a circle, and if so, what circle.

13. $x^2 + y^2 + 2x = 0$
14. $x^2 + y^2 + 4x - 2y = 4$

15. $x^2 + y^2 + y = 1$
16. $x^2 + y^2 + 2y + 1 = 0$

17. $x^2 + y^2 + 2x + 2y + 4 = 0$

18. Determine the value of k such that $x^2 + y^2 - 8x + 10y + k = 0$ is the equation of a circle with radius 5.

19. Derive the equation of the circle whose center is at $(5, -2)$ and which passes through the point $(-1, 5)$.

20. Find the equation of the circle having as a diameter the line segment joining $P_1(-4, -3)$ and $P_2(2, 5)$.

21. Find the equation of the circle passing through the origin and the point $(4, 2)$ and having its center on the x-axis.

22. Find the equation of and identify the circle passing through the three points $(1, 0)$, $(-1, 0)$, and $(0, 2)$.

23. Same question for the points (0,0), (1,0), and (2,1).

24. Identify the locus of a point $P = (x, y)$ moving in such a way that the sum of the squares of its distances from $(-2, 0)$ and $(2, 0)$ is a constant k greater than 8.

25. Find the locus of a point $P = (x, y)$ whose distance from the origin is twice its distance from (3,0).

26. Let Q_1 and Q_2 be fixed points and let k be a positive constant. Prove that the locus of a point P, which is moving in such a way that its distance from Q_1 is k times its distance from Q_2, is either a circle or a straight line.

27. Prove algebraically that the line through the origin and the point (a, b) intersects the unit circle $x^2 + y^2 = 1$ in the point $(a/r, b/r)$, where $r = \sqrt{a^2 + b^2}$.

28. Show that the point $(1, -3)$ lies inside the circle $x^2 + y^2 + 6x + 8y = 0$.

29. Show that $x_0^2 + y_0^2 + Ax_0 + By_0 + C < 0$ if the point (x_0, y_0) lies inside the circle

$$x^2 + y^2 + Ax + By + C = 0.$$

30. Show that if $x_0^2 + y_0^2 + Ax_0 + By_0 + C < 0$, then the point (x_0, y_0) lies inside the circle

$$x^2 + y^2 + Ax + By + C = 0.$$

(Reason as follows: If x_1 and y_1 are very large, then

$$x_1^2 + y_1^2 + Ax_1 + By_1 + C$$

is positive. What, therefore, must happen to the sign of

$$x^2 + y^2 + Ax + By + C$$

as the point (x, y) moves along the line segment from (x_0, y_0) to (x_1, y_1)?)

6. SYMMETRY

Frequently we want to graph an equation in a situation where we don't know ahead of time what kind of curve to expect. The crudest approach is to calculate a few solution points, draw as smooth a curve as we can through them, and hope that we have a reasonable approximation to the graph. Unfortunately we can make gross errors this way, because the few points we choose may fail to indicate some essential feature which we therefore miss completely. We shall see later on how calculus gives us a systematic way to look for the general qualitative features of a graph with a minimum of effort. However, there are also some algebraic aspects to the art of graphing that are useful. Instead of plowing ahead and computing a table of values in a routine way, we can try first to draw some conclusions about the shape of the graph from the form of the equation. One of the most illuminating features to notice, if it is there, is *symmetry* with respect to the coordinate axes.

If a point *P* is not on a line *l*, then the *symmetric image* of *P* in *l* is the "mirror image" of *P* in *l*. It is the unique point *P'* such that *l* is the perpendicular bisector of the segment *PP'*. We obtain *P'* by dropping the perpendicular from *P* to *l* and then continuing an equal distance across *l*.

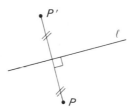

A geometric figure *F* has *l* as a *line of symmetry* if *F* contains the symmetric image in *l* of each of its points *P*. Then the 180° rotation of the plane over the axis *l* just interchanges each symmetric pair of points and carries the figure *F* exactly into itself.

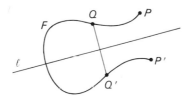

An equilateral triangle has three lines of symmetry; an isosceles triangle that is not equilateral has one line of symmetry; and a nonisosceles (scalene) triangle has no line of symmetry. A square has four lines of symmetry, and a rectangle that is not square has two lines of symmetry. A circle has an *infinite number* of lines of symmetry.

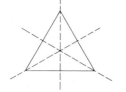
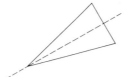

When we introduce coordinates we see that symmetry about the y-axis can be stated as follows: If a point (x, y) lies on the graph, then so does $(-x, y)$. The algebraic expression of this condition is that replacing x by $-x$ in the equation of the graph yields an equivalent equation.

EXAMPLE 1. Since $y = x^2$ is equivalent to $y = (-x)^2$, the graph of $y = x^2$ is symmetric about the y-axis.

Similarly, a graph is symmetric about the x-axis if replacing y by $-y$ in its equation yields an equivalent equation.

EXAMPLE 2. The graph of

$$\frac{x^2}{a^2} + \frac{y^2}{b^2} = 1$$

is symmetric about each axis since replacing either x or y by its negative leaves the equation unchanged.

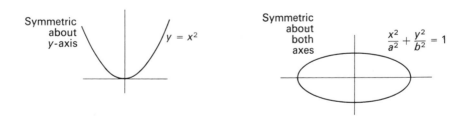

Symmetric about y-axis $y = x^2$

Symmetric about both axes $\dfrac{x^2}{a^2} + \dfrac{y^2}{b^2} = 1$

Another type of symmetry centers about a point Q. Two points P and P' are symmetric *in* Q, or *with respect to* Q, if Q is the midpoint of the segment PP'. Then P' is the symmetric image of P in Q, and vice versa. A figure F has Q as a *point of symmetry*, if F contains the symmetric image in Q of each of its points P.

A parallelogram has a point of symmetry. A triangle does not.

If a figure F has a pair of perpendicular lines of symmetry, then their intersection is necessarily a point of symmetry for F.

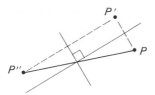

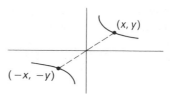

When we introduce coordinates, we see that symmetry with respect to the origin can be stated as follows: if (x, y) is on the graph then so is $(-x, -y)$. That is, replacing (x, y) by $(-x, -y)$ yields an equivalent equation.

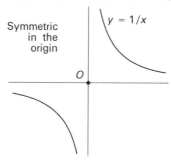

Symmetric
in the
origin

$y = 1/x$

O

EXAMPLE 3. Since $y = 1/x$ is equivalent to $-y - 1/(-x)$, its graph is symmetric in the origin.

As the above example shows, a curve symmetric in the *origin* need not be symmetric in *either axis*. However, a curve that is symmetric about *both* axes must of necessity also be symmetric in the origin.

PROBLEMS FOR SECTION 6

State what symmetry, if any, the following equations have with respect to the axes and/or the origin.

1. $9x^2 + 16y^2 = 144$.

2. $x^2 + 2xy + y^3 = 0$.

3. $2x^3 + 3y + 2y^3 = 0$.

4. $2y = x$.

5. $2x^2 + y = 0$.

6. $y^2 + 2x + 2x^2 + 5 = 0$.

7. $x^2 - y^2 = 0$.

8. $xy = 1$.

9. $x^2 + xy + y^2 = 1$.

10. a) Draw all the lines of symmetry of a square.
    b) Draw all the lines of symmetry of a rectangle that is not a square.
    c) Draw all the lines of symmetry of a rhombus that is not a square. (A rhombus is a quadrilateral with all sides equal.)

11. Prove that a triangle has a line of symmetry if and only if it is isosceles.

12. Show that a parallelogram has a point of symmetry.

13. Show that if a figure has a pair of perpendicular lines of symmetry, then it has their point of intersection as a point of symmetry.

14. Discuss the symmetry of a regular hexagon.

15. Discuss the symmetry of a regular pentagon.

16. If a figure has exactly n lines of symmetry, all passing through a common point 0, show that the angle between adjacent lines of symmetry is $180°/n$.

17. Prove that if a quadrilateral has a line of symmetry through a vertex, then its diagonals must be perpendicular.

18. Prove that if a quadrilateral has a point of symmetry then it is a parallelogram.

19. Show that a quadrilateral with two lines of symmetry is a rhombus or a rectangle.

20. Deduce another symmetry characterization of a rhombus from Problems 17 and 18.

*21. Show that if a configuration is bounded (i.e., lies entirely inside some circle), then it cannot have more than one point of symmetry.

22. Assuming the result in Problem 21, show that if a configuration has both a line of symmetry l and a point of symmetry P, and if the configuration is bounded (i.e., lies wholly inside some circle), then P lies on l.

7. SECOND-DEGREE EQUATIONS.

From time to time we shall want to work with one or another of the following standard equations, and you should be familiar with the general shapes of their graphs.

$$y = kx^2 \qquad\qquad\qquad \text{Parabola}$$

$$\frac{x^2}{a^2} + \frac{y^2}{b^2} = 1 \quad (a > b > 0) \qquad \text{Ellipse}$$

$$\frac{x^2}{a^2} - \frac{y^2}{b^2} = 1 \qquad\qquad \text{Hyperbola}$$

The ellipse is symmetric in both axes, and has a center of symmetry at the origin. It has a diameter of maximum length $2a$ along the x-axis and a diameter of minimum length $2b$ along the y-axis. The maximum diameter, and the axis along which it lies, are each referred to as the *major axis* of the ellipse. The

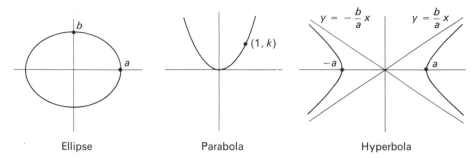

Ellipse   Parabola   Hyperbola

minimum diameter and its axis are each called the *minor axis*. The intercepts *a* and *b* are thus the lengths of the semimajor and semiminor axes respectively.

To graph an ellipse, we can calculate a few points in the first quadrant and then use symmetry to obtain the corresponding points in the other quadrants. But for a quick sketch we just plot the intercepts and draw a smooth, symmetric oval through them.

EXAMPLE 1. Quickly sketch the graph of

$$9x^2 + 16y^2 = 144.$$

. . . .

Solution. Dividing by 144 yields the standard form

$$\frac{x^2}{16} + \frac{y^2}{9} = 1,$$

so we have an ellipse with $a^2 = 16$, $b^2 = 9$. The intercepts are thus

$$\pm a = \pm 4 \quad \text{and} \quad \pm b = \pm 3,$$

and the graph looks roughly like this:

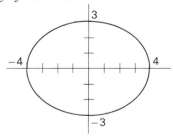

The parabola has one line of symmetry, here the y-axis, and one intersection with its line of symmetry, called its *vertex*, here the origin. Starting at the vertex it opens up along its line of symmetry in one direction, here the positive y-axis.

The hyperbola has two axes of symmetry, but it intersects only one of them and this is called the axis of the hyperbola. To sketch the hyperbola

$$\frac{x^2}{a^2} - \frac{y^2}{b^2} = 1$$

quickly, we use the x-intercepts $\pm a$ and the straight lines

$$y = \pm \frac{b}{a}x.$$

The hyperbola is *asymptotic* to these lines, in the sense that it hugs them more and more closely as $|x|$ gets large. (This is a limit statement that we will take up in the next chapter.)

EXAMPLE 2. Sketch the hyperbola

$$\frac{x^2}{9} - \frac{y^2}{4} = 1.$$

. . . .

Solution. Here $a^2 = 9$ and the x-intercepts are ± 3. The asymptotes are

$$y = \pm \frac{2}{3}x.$$

We plot the intercepts and symptotes and then sketch the hyperbola.

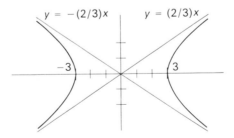

If we interchange x and y in the standard equations we just interchange the roles of the coordinate axes. For example, $x = y^2$ is a parabola opening up along the positive x-axis; $x^2 + y^2/4 = 1$ is an ellipse with major axis along the y-axis; and $y^2 - x^2 = 1$ is a hyperbola with its intercepts on the y-axis. These interchanged forms are equally simple equations, and can also be considered to be standard equations.

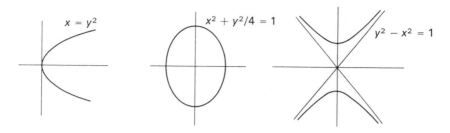

Conic Sections

The rest of this section is purely descriptive and is to be read only for "general education." Like the circle, each of the above curves can be given various direct geometric definitions, independent of coordinate systems and equations. For one thing, they are the curves that we get when we intersect a cone with a plane in various ways, and for this reason they are called the *conic sections*.

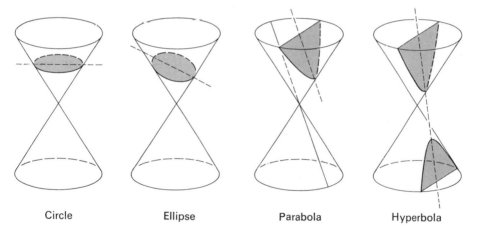

| Circle | Ellipse | Parabola | Hyperbola |

We should be more explicit. A curve in a geometric plane doesn't have an equation. We have to choose a coordinate system for the plane in order to describe a curve by an equation. Moreover, if we use a different coordinate system, then the same curve will acquire a different equation. Suppose, then, that we are given a plane intersecting a cone. We suppose also that we are given a fixed unit of length, so that we know, for example, whether a circle is large or small. However, we are allowed to place our axes wherever we wish in the plane. Then what we are saying is that the axis system can be so placed that the conic section will have one of the above standard equations.

The proof that these standard equations actually represent conic sections involves a lot of work, and we shall not go into it. However, the equation graphs sketched earlier ought to look the way you visualize the conic sections from the above figure.

If we move each of the intersection planes in the figure parallel to itself until it passes through the vertex of the cone then we get three corresponding "degenerate" conic sections, namely, a single point, a full straight line, and a pair of intersecting straight lines.

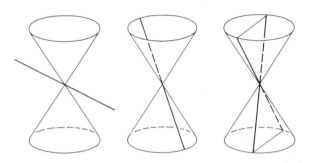

Their corresponding degenerate second-degree equations, in standard form, are

$$x^2 + y^2 = 0,$$

$$(x - y)^2 = 0,$$

$$a^2 x^2 - y^2 = 0.$$

It can be proved that *every* conic section, in standard position or not, has a second-degree equation of the general form

$$Ax^2 + Bxy + Cy^2 + Dx + Ey + F = 0.$$

Conversely, it can be proved that the graph of *any* such second-degree equation is a conic section or a degenerate configuration. This means in particular that a new and better placed axis system can be found with respect to which the graph acquires one of the standard equations. However, if the *xy* term is present in the equation, then the new axes will be *tilted* with respect to the original system.

EXAMPLE. The graph of

$$x^2 + 2xy + y^2 + x - y = 0$$

is the parabola shown in the next figure.

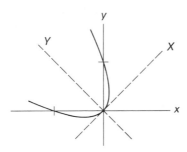

It can be proved that if the axes are rotated 45° counterclockwise, this parabola acquires the standard equation

$$Y = \sqrt{2}\, X^2.$$

PROBLEMS FOR SECTION 7

Use intercepts, symmetry and (possibly) asymptotes to make a quick sketch of the graph of each of the following equations.

1. $x^2 + 4y^2 = 1$.
2. $x^2 - 4y^2 = 1$.
3. $x^2 - 4y^2 = -1$.
4. $x^2 - 4y = 0$.
5. $4x^2 + 9y^2 = 36$.
6. $2x^2 - y^2 = 4$.
7. $x + 4y^2 = 0$.
8. $x^2 + 4y^2 = 0$.
9. $y + x^2 = 0$.
10. $y^2 - x^2 = 4$.
11. $x^2 + 9y^2 = 9$.
12. $9x^2 + 4y^2 = 36$.
13. $3x^2 + 2y^2 = 6$.
14. $3x^2 - 2y^2 = 6$.
15. $4x^2 + y^2 = 4$.
16. $x^2 - 4y^2 = 4$.
17. $4x^2 - y^2 = 4$.
18. $x^2 + 4y = 0$.
19. $y^2 + x^2 = 0$.
20. $9x - y^2 = 0$.
21. $4x^2 - y^2 = 0$.

8. TRANSLATION OF AXES

(This section can be postponed; it will not be mentioned until the third section of Chapter 4, at which point there is a reference to Theorem 7 below.)

In the last section it was stated that every conic section in the coordinate plane has an equation of the second degree, and that the equation of a tilted conic section contains an xy term. None of this was proved. However, it is relatively easy to see that we will always get a conic section that is not tilted as the graph of a second-degree equation without xy term. We shall do this here, and shall find that the graph of such an equation can be determined by completing squares and shifting to a new parallel-axis system.

We consider a second set of axes having the same unit of distance and the same directions as the original axes, but a new origin O' at the point having old coordinates (h, k). These new XY-axes can be viewed as obtained by *translating* the old axes, i.e., by sliding them parallel to themselves to a new position.

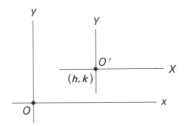

Each point in the plane will now have two sets of coordinates. For example, the new origin O' has old coordinates (h, k) and new coordinates $(0,0)$. Each curve in the plane will now have two equations. For example, the circle of radius r about the new origin has old and new equations

$$(x - h)^2 + (y - k)^2 = r^2,$$

$$X^2 + Y^2 = r^2.$$

This was shown in Section 1.5. As this example suggests, if the new coordinate system is in some way more favorably located with respect to the curve than the old, then the new equation for the curve may be simpler and more revealing of its nature than the old. The question is: How do we obtain the coordinates of a point and the equation of a curve in one coordinate system from those in another? The theorem below gives the coordinate change, and the change of equation will then follow.

Theorem 7. *If a translated XY-axis system is chosen with its origin O' at the point having old coordinates (h,k), then the new coordinates (X,Y) of a point P are obtained from its old coordinates (x,y) by the change-of-coordinate equations*

$$X = x - h,$$

$$Y = y - k.$$

Proof. We look at the first coordinates for both coordinate systems on the horizontal line through $O' = (h, k)$. Considering only the one-dimensional coordinate systems on this line, we know that when the new origin is put at h, then the point whose old coordinate is x acquires the new coordinate $X = x - h$ (by the corollary of Theorem 1). But then the identity $X = x - h$ holds for the whole plane, since the first coordinates x and X remain constant along every vertical line in the plane. Similarly, measuring second coordinates along the vertical line through $O' = (h, k)$, we see that a new origin at k replaces the old coordinate y of any point by the new coordinate $Y = y - k$, and this also holds for all points in the plane. We have thus proved the theorem. ∎

Now consider the equation

$$y = 2x^2 - 4x.$$

We wonder whether it can be simplified by using a translated coordinate system. Our experience with circles suggests that completing the square may lead to a simplifying new origin, and following this lead we get the equivalent equation

$$y + 2 = 2(x^2 - 2x + 1)$$

$$= 2(x - 1)^2.$$

Under the change of coordinates

$$X = x - 1,$$

$$Y = y + 2,$$

this equation becomes

$$Y = 2X^2.$$

That is, a point has new coordinates (X, Y) which satisfy the equation

$$Y = 2X^2$$

if and only if its old coordinates (x, y) satisfy

$$y = 2x^2 - 4x.$$

The graphs of the two equations are the same curve in the plane, and the common graph is now seen to be a parabola, located in the standard way with respect to the *new axes*, with new origin O' at $(1, -2)$.

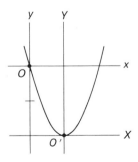

Instead of simplifying by completing the square, we can try a translation change of coordinates and see what happens to the equations. Starting from

$$y = 2x^2 - 4x,$$

the equation in the new coordinates X and Y is obtained by substituting from the change-of-coordinate equations

$$X = x - h,$$
$$Y = y - k.$$

Thus we set

$$x = X + h,$$
$$y = Y + k$$

in the above equation, and get

$$Y + k = 2(X + h)^2 - 4(X + h),$$

or

$$Y = 2X^2 + (4h - 4)X + (2h^2 - 4h - k).$$

This simplifies to

$$Y = 2X^2$$

if

$$4h - 4 = 0,$$
$$2h^2 - 4h - k = 0,$$

and this occurs when $h = 1$ and $k = 2h^2 - 4h = 2 - 4 = -2$, as before.

EXAMPLE. Identify the graph of the equation

$$x^2 + 4y^2 + 4x = 0.$$

. . . .

Solution. We gather together the x terms and complete the square, just as we did for the parabola above and for circles in Section 5. Thus,

$$(x^2 + 4x + 4) + 4y^2 = 4,$$
$$(x + 2)^2 + 4y^2 = 4,$$
$$\frac{(x + 2)^2}{4} + y^2 = 1.$$

This equation is of the form

$$\frac{X^2}{4} + \frac{Y^2}{1} = 1,$$

and it is thus the standard equation of an ellipse in the (X, Y) coordinate system, with semimajor axis 2 and semiminor axis 1. Since

$$X = x + 2 = x - h,$$
$$Y = y = y - k,$$

we see that

$$(h, k) = (-2, 0).$$

These are the (x, y) coordinates of the new origin, and we now have enough information to draw the graph.

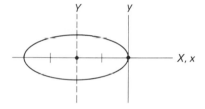

In general, if a curve C in the coordinate plane has a certain equation in x and y, and if we choose a translated XY-axis system with new origin O' at (h, k), then the equation of C in the new coordinates X and Y is obtained from the old equation in x and y by setting

$$x = X + h,$$
$$y = Y + k.$$

Proceeding just as we did for the equation $y = 2x^2 - 4x$, we can reduce any second-degree equation without an xy term to one of the standard forms (possibly with variables interchanged), or to a degenerate form. In particular we can identify the graph of any equation

$$Ax^2 + By^2 + Cx + Dy + E = 0$$

as a conic section or a degenerate conic section.

PROBLEMS FOR SECTION 8

1. a) Draw an xy-coordinate system and then draw new XY axes with a new
        origin at the point (1,4). From your sketch determine the new coordi-
        nates of the point P which has old coordinates (3,3).
    b) Write down the change-of-coordinate equations and compute the new
        coordinates of (3,3) from them.

Follow the above instructions in each of the following situations.

2. New origin at $(-2,3)$; $P = (-3,3)$.

3. New origin at $(2,-4)$; $P = (5,5)$.

4. New origin at $(-5,-2)$; $P = (-6,-3)$.

5. New origin at $(3,2)$; $P = (1,1)$.

6. New origin at $(6,-1)$; $P = (6,0)$.

7. New origin at $(-3,-2)$; $P = (-1,-1)$.

8. New origin at $(-3,2)$; $P = (2,1)$.

Reduce each of the following equations to a standard form by a translation of
axes. Identify and sketch the graph.

9. $x^2 + y^2 + 4x - 6y = 5$.   10. $25x^2 - 16y^2 = 400$.

11. $x^2 + 9y^2 = 9$.   12. $y^2 + 2y - 8x - 3 = 0$.

13. $4x^2 + 9y^2 + 16x - 18y - 11 = 0$.

14. $x^2 + y^2 - 2x + 3y + 3 = 0$.   15. $4x^2 - y^2 - 8x + 2y + 7 = 0$.

16. $x^2 + 2x + 4y - 7 = 0$.   17. $x^2 + 4y^2 - 2x - 16y + 13 = 0$.

18. $9x^2 - 16y^2 + 36x + 96y - 144 = 0$.

Find the new equation for the graph of each of the following equations when a
new axis system is chosen with new origin O' at the given point.

19. $x^2 + y^2 = 4$ $O' = (2,-1)$

20. $xy = 1$ $O' = (1,3)$

21. $xy - 3x + 2y = 6$ $O' = (-2,3)$

22. $x^2y - y^3 = 0$ $O' = (1,-1)$

9. INEQUALITIES

There are several important applications of calculus to computations, and
whenever such a topic comes up one must be able to use the elementary
properties of inequalities in a routine and easy manner. There is a summary of
the basic inequality laws in the Appendix, and this material could be reviewed
here if desired.

functions and limits

Calculus centers around a wholly new kind of calculation, having enormous potential for applications. This chapter introduces, in an intuitive way, the basic process that underlies these new calculations, but first we have to discuss the objects that we are going to calculate with. The operations of calculus apply not to numbers, the way arithmetic operations do, but to the more complicated things called *functions*. This, in fact, is the source of the great power of calculus in its applications. A function is a record of the way one varying quantity depends on another. The processes of calculus let us analyze such dependency relationships and make quantitative computations and predictions about them.

So the goal of this chapter is twofold: first, to review the notion of function; second, to introduce the basic new calculation that we make on functions, called the calculation of a limit. We shall be led to a list of properties of the limit operation called the *limit laws*. Ultimately, in Chapter 20, we shall see how to prove these laws. Until then they can be used as axioms to give logical structure to much of our early informal reasoning.

1. FUNCTIONS

A function can be thought of as being active or passive. On the one hand it acts on "input" numbers and determines "output" numbers. On the other hand it is just a kind of dependency relationship between variables. More often than not in calculus we view a function as active, and we shall therefore make our approach to the notion from this side.

Definition. *A function is an* operation *that determines a unique "output" number y when applied to a given "input" number x. The collection of numbers to which the operation can be applied make up its* domain.

EXAMPLE 1. The operation of taking the reciprocal of a number, i.e., the operation of going from x to $1/x$, is a function defined for all numbers except 0. Its domain is thus the collection of all nonzero numbers.

We use letters such as f and g to represent functions. If f is a function and x is a number in its domain, then we designate by $f(x)$ the unique number obtained by applying the operation f to x. The symbol $f(x)$ is read "f of x", although "f applied to x" would really be more appropriate.

EXAMPLE 2. If f is the reciprocal function considered above, then $f(x) = 1/x$. Therefore

$$f(3) = \frac{1}{3}, \quad f(-1) = -1, \quad f(1 + t) = \frac{1}{(1 + t)}, \quad \text{and} \quad f(\frac{1}{a}) = a.$$

Also, $1/(f(x)) = x$ and

$$f(x) + \frac{1}{f(x)} = \frac{1}{x} + x = \frac{1 + x^2}{x}.$$

EXAMPLE 3. Let g be the square-root operation, $g(x) = \sqrt{x}$ (where $\sqrt{x}$ is always the positive square root of x, as in

$$\sqrt{(-3)^2} = \sqrt{9} = +3.$$

Thus

$$g(4) = 2, \qquad g(x^8) = x^4, \qquad g(y^2) = |y|.$$

The domain of g is the collection of all numbers having square roots, i.e., the collection of all nonnegative numbers.

EXAMPLE 4. Functions frequently arise from relations among variables. Consider the equation

$$y^3 = x.$$

It has the property that *for each value of x there is a unique value of y making the equation true.* We then have the operation of going from x to the unique y that it determines. And we say that *the equation determines y as a function of x.*
    We can identify the above function. The unique y such that $y^3 = x$ is the cube root of x, by definition. Thus the equation determines y as a function of x, and the function f is the cube-root function,

$$y = f(x) = x^{1/3}.$$

EXAMPLE 5. Consider next the equation

$$xy = 1.$$

If $x = 0$ there is no value of y satisfying the equation, but for every other value of x there is a unique value of y which makes the equation true. (We could say that for every value of x there is *at most* one corresponding value of y.) The equation thus determines y as a function of x, the domain of the function being the set of all nonzero numbers.
    Here, again, we can identify the function. The unique y such that $xy = 1$ is the reciprocal of x, by definition. The equation determines y as a function of x, and the function f is the reciprocal function $y = f(x) = 1/x$, with domain the set of all nonzero numbers.

EXAMPLE 6. The equation

$$y = x^2$$

is already in the form $y = f(x)$, with

$$f(x) = x^2.$$

Here f is the squaring operation.

EXAMPLE 7. On the other hand, the equation

$$y^2 = x,$$

does *not* determine y as a function of x. For each positive value of x there are *two* values of y making the equation true: $y = \pm\sqrt{x}$; so y is not *uniquely* determined by x.

   In general, whenever two variable quantities x and y are so related that the value of y depends on and is uniquely determined by the value of x, we say that *y is a function of x*. The function f in question is the operation of going from x to the unique y it determines. We call x the *independent variable*, because its value is chosen arbitrarily, and we call y the *dependent variable*, because its value depends on the value of x.
   The phrase "*y* is a function of *x*" is very old, and antedates the notion of a function as a separate and distinct object. Thus for many years mathematicians were talking in terms of functionally related variables before they understood functions. This spawned some murky reasoning, and eventually led to such a backlash against the "*y* is a function of *x*" language that many authors avoid it entirely. When used *along with* the function concept, however, the old language is perfectly respectable. More than that, it is needed. It is the idiom that fits reality, for in practice functions generally arise as relationships among variables.

EXAMPLE 8. In a manufacturing process the cost C of producing x items is a function of x. In a simple situation there might be a fixed initial cost I to purchase a machine, and then an additional cost per item of amount k, for material, labor, etc. The total cost of producing x items is then

$$C = I + kx.$$

Thus, $C = f(x)$, where $f(x) = I + kx$. The average cost, or cost per item, is given by

$$c = \frac{C}{x},$$

and this also is a function of x:

$$c = g(x) = \frac{I}{x} + k.$$

Note that since x is the number of items manufactured, x is a positive integer. The domain of the cost function $f(x) = I + kx$ is thus a set of positive integers. However, if we were to graph the cost equation $C = I + kx$ in order to visualize the relationship between C and x, we would generally draw the smooth curve for all positive x.

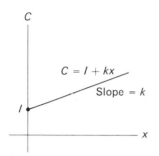
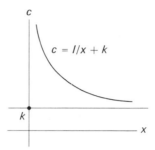

EXAMPLE 9. If a flexible container of gas is kept at a constant temperature, then it is observed that the volume V occupied by the gas is inversely proportional to the pressure p applied to it. That is,

$$pV = \text{constant.}$$

This relationship determines V as a function of p, and also determines p as a function of V. Here, again, the function domains are the *positive* real numbers, because of the physical interpretations of the variables.

The world is full of varying quantities, and the fact that two or more of them satisfy a dependency relationship of this sort amounts to a "law of nature" that we can study mathematically. In particular, we can view it as a cause-and-effect relationship that can be interpreted as a function, and we can analyze the behavior of this function by calculus.

EXAMPLE 10. An object allowed to fall from rest is observed to fall a distance

$$s = 16t^2$$

feet in t seconds. This even applies to a feather, provided it is allowed to fall in a vacuum. In other words, this equation is the universal law for all falling bodies, "neglecting air resistance." It is an observed law of nature expressing s as a function of t, and when this function is examined by calculus it reveals information about how the force of gravity acts.

The notion of a function f is not tied to any particular scheme for obtaining $f(x)$ from x.

EXAMPLE 11. Consider the area of the variable trapezoid drawn below at the left. The area A depends on the position $x = a$ of the righthand base, so A is a function of a, $A = f(a)$. In fact,

$$A = \frac{1}{2}a(1 + (a + 1)),$$

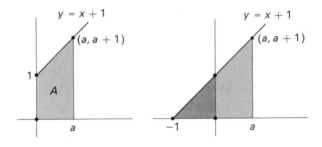

by the formula for the area of a trapezoid. But also

$$A = \frac{1}{2}[a - (-1)](a + 1) - \frac{1}{2}(1 \cdot 1),$$

where the right side is the difference of the two triangle areas in the right figure. We thus have two different explicit ways for expressing the operation f of *going from a to A*.

EXAMPLE 12. If f and g are the functions defined by the equations

$$f(x) = x^2 - 1 \quad \text{and} \quad g(x) = (x - 1)(x + 1),$$

then $f(x) = g(x)$ for all x, so f and g are the same operation. The two formulas present two different ways for carrying out this operation. The inner workings of a particular scheme for obtaining $f(x)$ from x can be thought of as indicating a method for computing the function value $f(x)$, and a given function f can generally be computed in many different ways.

PROBLEMS FOR SECTION 1

Determine the domain of each of the following functions:

1. $f(x) = \sqrt{x}$

2. $f(x) = \sqrt[3]{x}$

3. $f(x) = \dfrac{1}{x - 2}$

4. $f(x) = x^2 + 2x + 1$

5. $f(x) = \dfrac{x + 2}{x^2 - 4}$

6. $f(x) = \sqrt{1 - x^2}$

For each of the following functions find $f(z)$ when z is the given value:

7. $f(t) = t^2 + 2t - 9$; $z = 2$

8. $f(x) = x^3 + 2x^2 - \sqrt{x} + 3$; $z = \pi$

9. $f(u) = u^3 + 3u^2 + 2u$; $z = a + 1$

10. $f(x) = \sqrt{2x + 11}$; $z = 2.5$

11. $f(x) = \dfrac{x^2 - 1}{x + 1}$; $z = u - 1$

If $H(t) = (1 - t)/(1 + t)$, find

12. $H(-x)$

13. $H(1/t)$

14. $H(1/(1 + x))$

15. $H(H(t))$

Show that $x - y$ is a factor of $g(x) - g(y)$ for each of the following functions.

16. $g(x) = x^2$

17. $g(x) = 1/x$

18. $g(x) = x^3$

19. $g(x) = 1/(x^2 + 1)$

20. Let f be the reciprocal function, $f(x) = 1/x$. Show that $f(f(x)) = x$ for all x different from 0.

21. Let g be the function $g(x) = \sqrt{1 - x^2}$. Determine the domain D of g and show that $g(g(x)) = x$ for every x in "one-half" of D.

22. If f is the square-root function and g is the squaring function, show that

$$g(f(x)) = x \qquad \text{for every } x \text{ in domain } f;$$

$$f(g(x)) = |x| \qquad \text{for every } x.$$

23. If f is the cube-root function, $f(x) = x^{1/3}$, what is the function g such that $g(f(x)) = x$ for all x?

24. Find a function f such that

$$\frac{f(x) + 1}{f(x) - 1} = x.$$

(Solve the equation for $f(x)$.) What is the domain of f?

In each of the following cases, decide whether or not the equation determines y as a function f of x. If not, show why not. If so, find a formula for f and give its domain.

25. $x^2 + 4y^2 = 1$ 26. $x^2 + 4y = 1$ 27. $x + 4y^2 = 1$

28. $xy + x - y = 2$ 29. $\sqrt{x - y} = 1$ 30. $x^2 + y^3 = 0$

31. $x^3 + y^2 = 0$ 32. $x^3 + y^3 = 0$ 33. $\dfrac{x + y}{x - y} = x$

34. Refer to Example 10 in the text. If an object is dropped from a height h, then the time t it takes to fall to the ground is a function of h. Find a formula for this function $f(h)$.

35. The area A of an equilateral triangle is a function of the side length s, $A = f(s)$. Find this function f.

2. THE GRAPH OF A FUNCTION

As x runs across the domain of a function f, the point $(x, f(x))$ traces a curve in the coordinate plane called the *graph* of f. The graph is thus made up of the points (x, y) satisfying the equation $y = f(x)$, so:

The graph of the function f is the graph of the equation $y = f(x)$.

EXAMPLE 1. If $f(x) = x^3$, then the graph of f is the set of points (x, x^3), i.e., the set of points (x, y) such that $y = x^3$, i.e., the graph of the equation $y = x^3$.

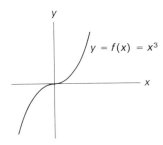

Here we can run afoul of standard axis labeling. As we have seen, functions frequently arise as relations between variables other than x and y. A function may also arise as a relation between x and y with their natural roles reversed: an equation may define x as a function of y. *When we graph a function, we require that the independent variable be graphed horizontally, even though this may mean labeling the axes in a nonstandard way.*

EXAMPLE 2. The equation

$$x = y^2 - 2y$$

expresses x as a function of y, and the graph of this function f is shown below.

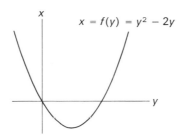

The *standard* graph of the above equation, shown next, is not the graph of the *function.*

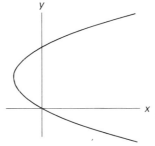

Of course, these two graphs are closely related. Each is obtained from the other by rotating, or flipping, the coordinate plane over the $45°$ line $y = x$. This brings the vertical positive y-axis down to the righthand horizontal axis, and takes the horizontal positive x-axis up to the vertical position. And it interchanges the two graphs above.

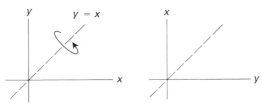

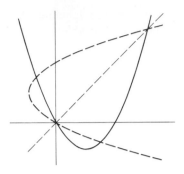

It may happen that a graph and its "flipped over" graph are *both* function graphs, as in our $x = y^3, y = x^{1/3}$ example. This will occur whenever an equation in x and y determines *each* variable as a function of the other, so that we get a function graph no matter which variable is plotted horizontally. Of course, the functions are generally different. In the above example one function is the cubing function $f(y) = y^3$, and the other is the cube-root function $g(x) = x^{1/3}$. Two functions related this way are said to be *inverse* to each other. This is an important notion and we shall devote a chapter to it later on (Chapter 7).

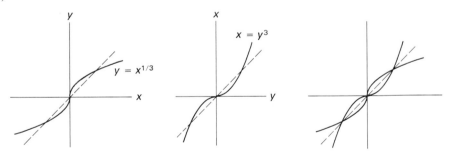

Our definition of a function as an operation leads us to view its action one number at a time. A function graph, on the other hand, spreads the whole input–output relationship out before us as a geometric curve. This gives us a simultaneous view of the whole function as a global, but passive, entity, and is quite different in nature from the operator-waiting-to-act notion. The two points of view reinforce each other, and are part of our total notion of a function.

It is possible to take the passive function as the basic notion. We note that a curve in the coordinate plane determines y as a function of x if each vertical line cuts it in only one point (or not at all for x not in the domain). For this just says geometrically that only one y value goes with a given x value. We can then define a function by this kind of configuration in the coordinate plane.

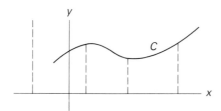

Definition. *A function is defined by a configuration C in the coordinate plane with the property that each vertical line intersects C in at most one point. That is, a function is given by a collection C of ordered pairs of numbers having the property that each first number x occurring in the points of C is paired with only one second number y.*

This set-of-ordered-pairs definition of a function has become popular recently as part of the emphasis on sets, but it doesn't fully convey the notion of a function.

PROBLEMS FOR SECTION 2

1. Graph the squaring function $f(x) = x^2$ by plotting the points over $x = 0$, $\pm\frac{1}{2}, \pm1, \pm\frac{3}{2}, \pm2$. Name this graph from the list of standard equations in Section 7 of Chapter 1.

2. The graph of $f(x) = \sqrt{x}$ is half of a parabola. How do we know this? Sketch the graph.

3. Show that the graph of $f(x) = \sqrt{1 - x^2}$ is a semicircle about the origin. Then plot one point on the graph and sketch it.

4. Show that the graph of $f(x) = \frac{1}{2}\sqrt{x^2 + 4}$ is half of a hyperbola (by comparing with a nearly equivalent equation in Section 7 of Chapter 1). Plot the intercept and asymptotes and quickly sketch the graph.

5. The equation $y = \sqrt{x}$ determines x as a function f of y, $x = f(y)$. What are $f(0)$ and $f(1)$? Using these two values only, sketch the graph of f, being sure that it has the right general shape. (The equation $y = \sqrt{x}$ is equivalent to $x = y^2$ and $y \geq 0$.)

6. The equation

$$2x - y + 1 = 0$$

determines y as a function f of x, and also determines x as a function g of y. Draw the graphs of these two functions f and g.

7. For positive x and y the equation

$$yx^2 = 1$$

determines each of the variables x and y as a function of the other. Draw the graphs of these two functions.

8. A function f is said to be *even* if

$$f(-x) = f(x)$$

for every number x in its domain. Note that if x is in the domain of f, then $-x$ must also be there if this condition is to make sense. Show that f is even if and only if its graph is symmetric about the y-axis.

9. A function f is *odd* if

$$f(-x) = -f(x)$$

for every x in its domain. Again it is understood that $-x$ must be in the domain for each x in the domain. Show that f is odd if and only if its graph is symmetric with respect to the origin.

10. Suppose that

$$f = g + h$$

where g is even and h is odd. Prove that

$$g(x) = \frac{f(x) + f(-x)}{2}, \quad \text{and} \quad h(x) = \frac{f(x) - f(-x)}{2}.$$

Now show that an *arbitrary* function f, defined on an interval of the form $[-a, a]$, can be written in a *unique* way as a sum of an even function g and an odd function h. We shall call g the *even component* of f, and h the *odd component* of f.

11. Find the even and odd components of the polynomial $1 + x + x^2$.

12. Find the even and odd components of the polynomial $p(x) = a_0 + a_1 x + \cdots + a_n x^n$. When is $p(x)$ itself even? odd? Your answer here should suggest where the words *even* and *odd* come from.

13. Show that the sum of two odd functions is odd, and that the product of two odd functions is even. Write down the rest of the similar statements about even and odd functions.

14. Show how the even and odd components of the product of two functions can be expressed in terms of the even and odd components of the factors. Use the notation

$$f = g + h, \qquad f_1 = g_1 + h_1, \qquad f_2 = g_2 + h_2, \qquad f = f_1 f_2.$$

15. Graph $f(x) = x/|x|$. What symmetry does the graph have?

16. Graph $f(x) = |x| + x$. Find and graph the even and odd components of f.

17. Sketch the graphs of $f(x) = x^2$ and $g(x) = x^4$ on the same axis system, using the function values at $x = 0, 1/2, 1, 3/2$, and symmetry. What configuration in the coordinate plane do the graphs of the even powers x^{2n} approach as n gets very large?

18. Answer the same question for $f(x) = x^3, g(x) = x^5$, and the family of all odd powers x^{2n+1}.

19. A polynomial of degree three,

$$p(x) = x^3 + a_2 x^2 + a_1 x + a_0,$$

behaves like x^3 when $|x|$ is very large. Morcover, for any polynomial p, $x - a$ is a factor of $p(x)$ if and only if $p(a) = 0$. (This is a theorem of polynomial algebra.) Classify the possible graphs of cubic polynomials p by the ways in which they can cross or touch the x-axis, and draw a sketch of each possibility. (Show that there are four possibilities, as follows:
a) cross 3 times;
b) cross once and touch once from above;
c) cross once and touch once from below;
d) cross once.)

The following polynomials exhibit the above four possibilities. Determine which is which and sketch very roughly. [*Hint.* Factor.]

20. $x^3 + x$ 21. $x^3 - x$

22. $x^3 + 2x^2 + x$ 23. $x^3 - 2x^2 + x$

24. Make the same analysis of the possible graphs of a fourth-degree polynomial

$$p(x) = x^4 + a_3 x^3 + a_2 x^2 + a_1 x + a_0.$$

25. A rubber strand occupying the segment $[a,b]$ is stretched to the new position $[c,d]$. If y is the new position in $[c,d]$ of the point originally at the position x in $[a,b]$, then y is a function of x. Derive a formula for this function. Draw its graph.

26. Suppose a rubber line segment is held under tension between endpoints a and b. We attach clips to the segment at points $x_1, x_2, \ldots, x_n$, and then move these clips to the new positions (along the same line) $y_1, y_2, \ldots, y_n$. If y is the new position of the point originally at x, then y is a function of x. Draw the graph of this function.

3. THE BEHAVIOR OF A FUNCTION

When we speak of the *behavior* of a function we are referring to the manner in which the value of $f(x)$ changes when the value of x is changed. Algebraically this will generally mean examining a change from a first value to a second value, but geometrically we can visualize x as a point sliding along the x-axis and $(x, f(x))$ as a point sliding along the graph. In fact, Newton's calculus was founded on viewing a curve as the path of a moving point. Behavior related to the notions of continuity and limits will be discussed in later sections. Here we shall look at some types of behavior that have simple geometric meanings in terms of function graphs. First we need to introduce intervals.

An *interval* is simply a segment on the number line. If its endpoints are the numbers a and b, then the interval consists of the numbers lying between a and b. Supposing that a is to the left of b, these are the points which are simultaneously to the right of a and to the left of b. However, we may or may not want to include the endpoints themselves in the interval. The *closed interval* from a to b, designated $[a,b]$, includes the endpoints, and so contains exactly those numbers x such that $a \leq x \leq b$. Thus $[1,2]$ consists of the real numbers from 1 to 2, *inclusive*.

We use a parenthesis to indicate an *excluded* endpoint. Therefore the interval $[a,b)$ contains a but not b, and so contains exactly those numbers x such that $a \leq x < b$. The interval (a,b), with both endpoints excluded, is called the *open* interval from a to b.

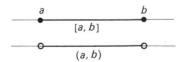

A half-line is considered to be an interval extending to infinity. We use the infinity symbol, ∞, to indicate the unterminated direction of such an interval. Thus $[a, \infty)$ consists of those points x such that $x \geq a$. And $(-\infty, \infty)$ is the whole real-number system. There is, of course, no question of having $[a, \infty]$: ∞ is not a number, and we can't include a number that doesn't exist.

The domains of the functions that we meet in calculus are almost always made up out of one or more intervals. For example:

the domain of $f(x) = \sqrt{x}$ is $[0, \infty)$;

the domain of $f(x) = \sqrt{1 - x^2}$ is $[-1, 1]$;

the domain of $f(x) = 1/x$ is the union of $(-\infty, 0)$ and $(0, \infty)$.

We frequently refer to intervals by letters such as I and J. We do this when we want to talk about an interval but don't need to refer specifically to its endpoints; and we may not care particularly whether the interval is open or closed.

Any point x in an interval I that is not an endpoint of I is called an *interior point* of I. We also say that x is *in the interior of I*. An interval extends *across* an interior point but lies entirely *on one side of* an endpoint. An open interval consists only of interior points.

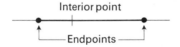

Intervals are very useful when we want to describe function behavior. A function will behave one way over one interval and another way over another interval. For example, the graph of $f(x) = x^2$ looks like this:

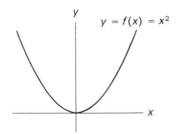

As x increases from $-\infty$ to 0 the function value $f(x)$ steadily *decreases*. We say that f *is decreasing on* $(-\infty, 0]$. The convention is that x is understood to be increasing unless explicitly stated otherwise. Similarly, we see that f is *increasing*

on the interval $[0, \infty)$. For a more complicated example consider the graph of $y = f(x) = 3x - x^3$, pictured below.

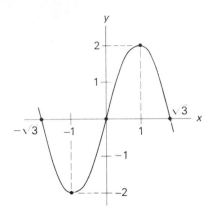

First, f is positive over the intervals $(-\infty, -\sqrt{3})$ and $(0, \sqrt{3})$, and negative over the intervals $(-\sqrt{3}, 0)$ and $(\sqrt{3}, \infty)$. Next, f is increasing over $[-1, 1]$ and decreasing over $(-\infty, -1]$ and $[1, \infty)$. Finally, the graph of f is everywhere *turning*, or *curving, in the upward direction* over $(-\infty, 0]$ and is everywhere *turning in the downward direction* over $[0, \infty)$. Note that over $(-\infty, -1)$, the graph is turning upward while it is decreasing, and that over $(0, 1)$ it is turning downward while it is rising.

Another property that a function can have over an interval is *boundedness*. In the present example we can say that f is *bounded by* 2 over the interval $[-\sqrt{3}, \sqrt{3}]$. This means that $|f(x)| \leq 2$ for every x in the interval. We also say that 2 is a *bound* for f on $[-\sqrt{3}, \sqrt{3}]$. Of course, f is also bounded on $[-\sqrt{3}, \sqrt{3}]$ by any number larger than 2. By the time x reaches 3, $f(x) = 3x - x^3$ has decreased to -18. Similarly, $f(-3) = 18$. Thus f is bounded by 18 over the interval $[-3, 3]$, but not by any smaller number.

Sometimes we want to divide up the way a function is bounded and consider separately the questions of *upper* and *lower* bounds. For instance, f is bounded *above* by 2 over the interval $[0, 3]$, meaning that $f(x) \leq 2$ for every x in the interval; but the largest *lower* bound for f on this interval is -18. Over $(-\infty, 0]$, f has *no upper* bound and we therefore say that f is *unbounded from above* over $(-\infty, 0]$. However, f has the lower bound -2 over this interval.

It is easy enough to look at smooth graphs like the above and see that they are curving downward here and curving upward there, but the big question is: How do we *know* that the graph should be drawn this way? How, for example, do we know that the graph of $f(x) = 1/(x^2 + 1)$, shown below, is turning downward exactly between $-\sqrt{3}/3$ and $+\sqrt{3}/3$, and turning upward everywhere else? The answer will be found in Chapter 5.

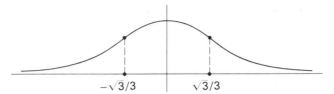

$$-\sqrt{3}/3 \qquad \sqrt{3}/3$$

When we restrict x to some particular set A of its values, the corresponding set of values of $f(x)$ is called the *image of A under f*, or the *f-image of A*. For the kind of function we meet in calculus, the image of an interval I will generally turn out to be some interval J. For example, it seems clear (and is true) that the image of $[1,2]$ under $f(x) = x^3$ is $[1,8]$. We then say that $f(x)$ runs across the interval J as x runs across the interval I. This imagery probably suggests that $f(x)$ runs straight across J, with no turning back, as below on the left. If $f(x)$ goes back and forth in the process of crossing J, as in the righthand figure, we can say that $f(x)$ *sweeps out J*. The f image of the whole domain of f is called the *range* of f.

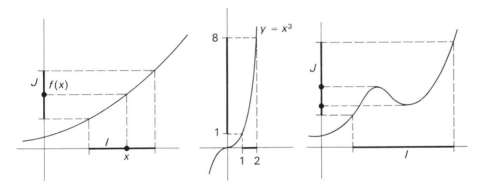

This terminology is related to our discussion of bounds for f. To say that the f-image of a set A lies entirely in an interval $[c,d]$ is to say that f is bounded above by d on the set A and that f is bounded below by c on the set A.

PROBLEMS FOR SECTION 3

Each of the following equations is of the form $y = f(x)$. Determine the domain of f in each case, and express it as an interval or a union of intervals.

1. $y = 8x$

2. $y = \dfrac{1}{x + 3}$

3. $y = \dfrac{x + 4}{(x - 2)(x + 3)}$

4. $y = \dfrac{1}{2x}$

5. $y = \sqrt{36 - x^2}$

6. $y = \dfrac{1}{9 - x^2}$

7. $y = \sqrt{x^2 - 16}$

Determine the range of each of the following functions (as intervals or unions of intervals):

8. $y = 8x$

9. $y = \dfrac{1}{x + 3}$

10. $y = \dfrac{1}{2x}$

11. $y = \sqrt{36 - x^2}$

12. $y = \dfrac{1}{9 - x^2}$

13. $y = \sqrt{x^2 - 16}$

14. Here is the graph of $f(x) = x^4 - 2x^2$.

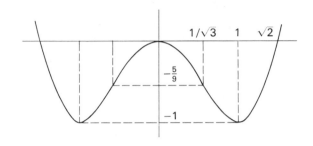

a) On what intervals is f positive? Negative?
b) Over what interval is f increasing? Decreasing?
c) On what intervals is f turning upward? Turning downward?

15. Here is the graph of $x/(1 + x^2)$. Analyze it as in the above example.

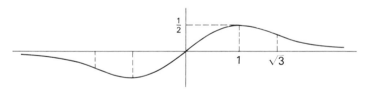

16. Returning to the graph in Problem 14, what is the greatest lower bound for f? What is the largest interval on which f is bounded above by 0? Find the largest interval on which f is bounded above by 3 (by solving the equation $f(x) = 3$).

17. The graph of $f(x) = 3x - x^3$ drawn in the text shows that the maximum value of f on the interval $[0, \infty]$ is 2. In the next chapter we shall see how calculus lets us determine such salient features of a graph. However, this particular fact can be established by algebra. Show that

$$2 - f(x) = x^3 - 3x + 2$$

is positive on the interval $[-2, \infty)$. [*Hint.* Try to factor the cubic polynomial.]

Investigate the behavior of the following functions over their domain. (In each case make a sketch on the basis of what information you have, and use the sketch to guess the endpoints of some of the "behavior intervals.")

18.　$y = 6 - 3x$ 　　　　　　　　　　　　19.　$y = x^2 - 1$

20.　$y = (x + 1)(x - 1)(x - 2)$ 　　　21.　$y = (x + 2)(x - 3)^2$

22.　$y = x^2 + 2x + 5$ 　　　　　　　　23.　$y = 2x - x^2$

24.　$y = (x + 2)^3$

4. GRAPHICAL DIFFERENTIATION

This section contains a brief geometric discussion of a process for obtaining a new function g from a given function f, that will turn out in our later work to be the basic process of calculus. Consider the graph of a fixed function f and suppose that the graph is *smooth*, in the sense of having a unique tangent line at each of its points. The tangent line at the point over $x = a$ is the line having the *direction of the curve at that point*. This is only an intuitive idea at the moment. We suppose that your visualization of a smooth curve includes the notion that a point moving along the curve is moving in a definite direction at each point. Therefore, the tangent line at the point over $x = a$ has a uniquely determined slope m. When we change a, it is clear from the picture that m will change. Thus m is a function of a.

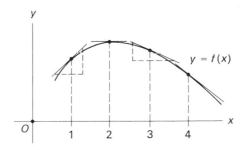

This new function g is called the *derivative* of the given function f. Our sole occupation for the next few chapters will be learning how to calculate a formula for g from a formula for f, and how to apply this process. What we can do here is to get a rough idea of the nature of g from the given graph of f by observing the general way in which m is varying with a and by estimating a few of its values.

In the above figure we note that the slope m_1 at the point $a = 1$ is positive, m_2 (at $a = 2$) is zero, m_3 is slightly negative and m_4 is more negative. It is evident that *m* is *decreasing* as *a* increases, and you should be able to visualize that this will happen whenever a graph is turning in the downward direction. We now make a rough estimate of the numerical value of *m* at two or three points, say at $a = 1$ and $a = 3$. It is helpful to sketch a small right triangle with the tangent line as hypotenuse, as in the figure above, and then estimate the slope *m* as the *ratio* of the triangle legs. Doing this, it appears that m_1 is about 1 and that m_3 is about $-\frac{1}{2}$. The graph of *m* as a function of *a* is therefore something like this:

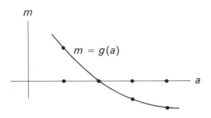

This process is called *graphical differentiation*.

EXAMPLE. Sketch the derivative *g* of the function *f* graphed below.

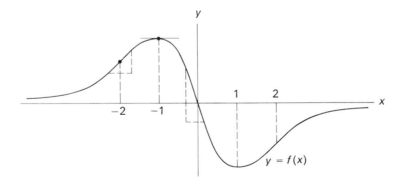

. . . .

Solution. The curve is turning up below $- 2$, so on the interval $(-\infty, -2)$ *m* is increasing, from values near zero to a value that we estimate to be approximately 1 at $a = -2$. From $- 2$ to 0 the curve is turning downward, so *m* decreases there, passing through the value 0 at $a = -1$ and reaching approximately the value -3 at the origin. We thus have an approximation of the left half of the graph of *m*, and since *m* is apparently symmetric with respect to the vertical axis we have the approximate graph in the next figure.

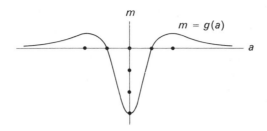

$m = g(a)$

PROBLEMS FOR SECTION 4

Sketch the graph of the derivative $m = g(a)$ for each of the functions shown below:

1.

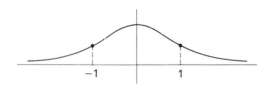

2.

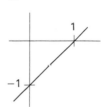

3.

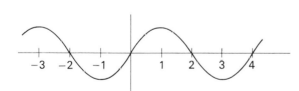

4.

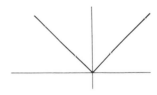

5.

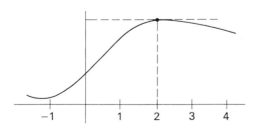

6.

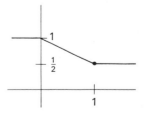

5. CONTINUITY

This section and the next will discuss certain ways in which a function can behave near a point, and introduce the language we use to describe this behavior. We shall not give precise definitions, nor is anything going to be proved; if something seems plausible we shall assume it. The idea is to convey a *qualitative* feeling for what is happening without going into the quantitative calculations that ultimately back up these intuitions.

The number x^3 varies "continuously" with the number x, in the sense that a small change in x produces only a small change in x^3. Moreover, we can keep the change in x^3 as small as we wish by holding the change in x to within a sufficiently small range.

You probably would agree on intuitive grounds that these statements are valid descriptions of the behavior of x^3. Your intuition might stem from looking at the very smooth graph of $y = x^3$, or you might have experimented numerically or algebraically to see what specific small changes in x do to the values of x^3. Since such experimentation gives a very concrete feel for the situation, we include an example.

EXAMPLE 1. We wish to compute how close x^3 is to $2^3 = 8$ when $x = 2.1$. However, instead of just cubing 2.1, we shall find it more illuminating to write $(2.1)^3$ as $(2 + .1)^3$ and use the expansion

$$(a + h)^3 = a^3 + 3a^2h + 3ah^2 + h^3.$$

The difference $(2.1)^3 - 2^3$ that we want to calculate is thus of the form

$$(a + h)^3 - a^3 = h(3a^2 + 3ah + h^2),$$

with $a = 2$ and $h = .1$, so

$$(2.1)^3 - 2^3 = .1(12 + .6 + .01)$$

$$= .1(12.61)$$

$$= 1.261.$$

For $x = 2.01$ we have

$$(2.01)^3 - 2^3 = .01(12 + .06 + .0001)$$

$$= .01(12.0601)$$

$$= .120601.$$

Such numerical examples make it especially clear that the reason for $(a + h)^3 - a^3$ becoming small when h becomes small is that it contains h as a factor.

In Chapter 20 we shall explicitly calculate *how* small we have to hold the change in x in order to keep the change in x^3 within prescribed limits. Such computations form the core of the technical study of continuity. For now, though, we shall rely on intuitions about continuous variation. Here, then, is our provisional and intuitive definition of continuity.

The function f is continuous *if a small change in x produces only a small change in f(x), and if we can keep the change in f(x) as small as we wish by holding the change in x sufficiently small.*

Practically every function that we shall want to study in the differential calculus will be continuous in the above sense. The reason is that in building up functions we customarily use operations, like addition and multiplication, that themselves react to small input changes with only small changes in output. Some basic continuity principles are given below. They will probably seem correct on intuitive grounds, and we shall assume them for now.

   I. $x^{1/n}$ *varies continuously with x, for any fixed positive integer n. Any constant function is trivially continuous (since it doesn't vary at all).*

  II.   *If f(x) and g(x) vary continuously with x, then so do f(x) + g(x) and f(x) g(x). So does 1/f(x), provided x is kept away from any value for which f(x) = 0.*

 III.   *If g(y) varies continuously with y and if y = f(x) varies continuously with x, then g(f(x)) varies continuously with x.*

For example, in order to convince ourselves that $\sqrt{1 - x^2}$ varies continuously with x, we visualize a small change in x, producing only a small change in x^2 and hence in $1 - x^2$, and this small change in $1 - x^2$ then produces only a small change in $\sqrt{1 - x^2}$. This intuitive analysis uses all three of the above principles. Starting off with the continuity of 1, x, and $\sqrt{y}$ (Principle I), we get the continuity of x^2 and then $1 - x^2$ from (II), and the continuity of $\sqrt{y} = \sqrt{1 - x^2}$ from (III).

In the same way, our intuitive feeling that a polynomial like

$$p(x) = x^4 + 2x^3 - x^2 + 5x + 1$$

varies continuously with x can be traced back to (I) and repeated applications of (II). A rational function $f(x)$ is a quotient of two polynomials,

$$r(x) = \frac{p(x)}{q(x)};$$

and, assuming that a polynomial is continuous, we can conclude that $r(x)$ varies continuously so long as we stay away from any value of x for which $q(x) = 0$ (by (II) again). However, not all functions that we meet in everyday life and in mathematics are continuous.

EXAMPLE 2. When we mail a letter we have to use one stamp of a certain denomination for anything up to and including one ounce in weight, then two stamps up to and including two ounces, etc. Thus, if $p(x)$ is the number of stamps required to mail a letter weighing x ounces, then the graph of p looks like this:

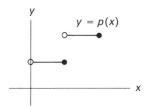

The empty dot ∘ shows a point *not* on the graph, whereas solid dots show points on the graph. The function $p(x)$ is called the postage-stamp function. It is defined for $x > 0$ and has a *jump discontinuity* at each of the points $x = 1, 2, 3,$ ... For example, the change in $p(x)$ *cannot* be made as small as we wish when x changes from slightly below 1 to slightly above 1, so $p(x)$ is *not* continuous as x moves through the value 1.

EXAMPLE 3. The greatest integer function $[x]$ is an analogue of the postage-stamp function in pure mathematics: $[x]$ is defined as the largest integer that is $\leq x$. Thus $[2.5] = 2$, $[2] = 2$, and $[1.9] = 1$. Although the domain of $[x]$ might naturally be taken to be $[0, \infty)$, we shall define $[x]$ for negative x in the same way. The integers less than -2.5 are all negative but among them there is a greatest one, namely -3. Thus $[-2.5] = -3$, $[-3] = -3$, etc. The graph of $[x]$ looks like this:

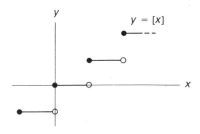

EXAMPLE 4. Here is a glimpse of the graph of $x - [x]$.

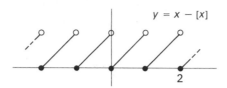

$$y = x - [x]$$

2

These examples show that even quite simple functions may have isolated points of discontinuity. This focuses our attention on the notion of continuity *at a point* $x = a$. The idea is the same, except that we restrict the variation of x to small changes away from $x = a$.

Definition. *The function f is continuous at the point $x = a$ in its domain if a small change in x away from $x = a$ produces only a small change in $f(x)$. More precisely, we require that we can keep the change $f(x) - f(a)$ as small as we wish in magnitude by restricting x to a suitably small interval about $x = a$.*

Then:

Definition *A function is continuous on an interval I in its domain if it is continuous at each point a in I.*

PROBLEMS FOR SECTION 5

1. a) Compute how close $f(x) = 2x - x^2$ is to $f(3) = -3$ when x is near 3. Do this first algebraically by writing $x - 3 = h$ and factoring h from $f(x) - f(3) = f(3 + h) - f(3)$. Then give $x - 3 = h$ the values .1, .01, .001, and compute the difference $f(x) - f(3)$.
    b) Now compute how close $f(x)$ is to $f(2) = 0$, following the same program as above.

2. Compute how close $f(x) = 1/x$ is to $f(2) = 1/2$ when x is near 2. Follow the same steps as in the above problem.

Verify the continuity of each of the following functions by applying the continuity principles I through III.

3. $3x^2 + 5x$ 4. $x/(1 + x^2)$

5. $1/\sqrt{x}$ 6. $\sqrt{x^2 + 4}$

7. $\sqrt{x^2}$ 8. $|x|$

9. $|x^{1/3}|$ (Sketch this graph.)

10. Draw the graph of $f(x) = x - [x/2]$. What are its points of discontinuity?

11. Define a function (by drawing its graph) that is constant over each of the intervals $(-\infty, -1)$, $(-1, 1)$, $(1, \infty)$, and has discontinuities at $x = -1$ and $x = 1$.

12. Show that when x is within a distance d of 2, then x^2 is within a distance $d(d + 4)$ of 4. [*Hint*: Write $x = 2 + e$, where $|e| < d$, and then expand x^2.]

13. Suppose that x is within a distance d of the point 2, where $d < 1$. Show that then x^2 is within a distance $5d$ of 4. [*Hint*: Show first that $1 < x < 3$. Then factor $x^2 - 4$.)

14. Suppose that x is within a distance d of 2, where $d < 1$. Show that then x^3 is within a distance $19d$ of 8.

15. Strictly speaking, the function $f(x) = (x^2 - x)/(x - 1)$ is not defined at $x = 1$. Draw its graph, indicating the missing domain point. Show that f becomes continuous if $f(1)$ is suitably defined.

16. Suppose that $f(x) = (x^2 + x - 6)/(x - 2)$ for $x \neq 2$. Define $f(2)$ so that f will be everywhere continuous.

17. The same problem for

$$f(x) = \frac{x^2 - 1}{x^3 - 1},$$

and the point $x = 1$.

18. The same problem for

$$f(x) = \frac{\sqrt{x} - 2}{x - 4}$$

at the point $x = 4$.

6. LIMITS

The new operation on which calculus rests is the calculation of a limit. In the beginning this calculation is very easy to make in terms of continuity, and the rules of algebra for limits are obvious. Later on, we shall have to make more difficult limit calculations, and this will force us to examine the nature of the limit operation more critically.

We start by describing the continuity of x^3 at the point $x = a$ in different language. It is reasonable to say that x^3 *approaches* a^3 as x *approaches* a. The arrow $\rightarrow$ is the symbol for the word *approaches*, so in symbols this statement is:

$$x^3 \rightarrow a^3 \qquad \text{as} \quad x \rightarrow a.$$

The slightly different connotation of "approaches" is that here we imagine x moving *toward* a, whereas before we imagined x moving slightly *away* from a. But we mean the same thing—we can see to it that x^3 is as close to a^3 as we wish merely by restricting x to a sufficiently small interval about a. So we are just restating the continuity of x^3 at the point $x = a$.

More generally, the continuity of a function f at the point $x = a$ can be described by saying that $f(x)$ *approaches* $f(a)$ *as* x *approaches* a:

$$f(x) \rightarrow f(a) \qquad \text{as} \quad x \rightarrow a.$$

The value that $f(x)$ approaches is called its *limit*, and another form of the above statement is:

$$f(a) \quad \text{is the limit of } f(x) \qquad \text{as } x \text{ approaches } a.$$

In symbols this form is written

$$f(a) = \lim_{x \to a} f(x).$$

We thus have a new way of viewing the continuity of f at a. The value $f(a)$ is like a target that $f(x)$ "zeroes in on" as x approaches a. It turns out that this new point of view expresses continuity in terms of a more general phenomenon, for we shall see that a function can exhibit such target-seeking behavior as x approaches a without the target having to be $f(a)$. That is, a function can have a limit as x approaches a without being continuous at a.

EXAMPLE 1. The function

$$f(x) = \frac{1 - \sqrt{x}}{1 - x}$$

is not defined at $x = 1$, so it cannot be continuous there. Yet $f(x)$ approaches the limit $\frac{1}{2}$ as x approaches 1, as we shall see a little later. That is, the difference $f(x) - \frac{1}{2}$ can be made as small as we wish by restricting x to a sufficiently small interval about 1, *but excluding the value $x = 1$*, since $f(1)$ is not defined.

This example suggests how the general description of a limit should be worded.

Definition. *We say that $f(x)$ has the limit l as x approaches a, and write*

$$\lim_{x \to a} f(x) = l,$$

if we can make $f(x)$ as close to l as we wish by restricting x to a sufficiently small interval about a, excluding the value $x = a$.

Continuity at a is then a special case:

A function f is continuous at $x = a$ if and only if a is in the domain of f and

$$\lim_{x \to a} f(x) = f(a).$$

Although the limit property is more general than continuity, it is not *much* more general: a function that has a limit at a is very close to being continuous at a. Comparison of the above two limit equations shows that there are only two possible ways for a function to have a limit at $x = a$ without being continuous there. First $f(a)$ may not be defined. We noted above that the function $f(x) = (1 - \sqrt{x})/(1 - x)$ is like this at $x = 1$, although we haven't established the limit yet.

The second possibility is that $f(a)$ exists but is different from the limit $\lim_{x \to a} f(x)$.

EXAMPLE 2. An example of a function exhibiting this bizarre behavior is

$$f(x) = p(x) - [x],$$

the difference between the postage-stamp function and the greatest-integer function. If you will reexamine the definitions of these functions carefully, especially around the integer values of x, you will see that their difference has the constant value 1, *except at the integers*, where its value is 0. In particular,

$$\lim_{x \to 2} f(x) = 1,$$

but

$$f(2) = 0,$$

so

$$\lim_{x \to 2} f(x) \neq f(2).$$

The same thing happens at every integer value of x.

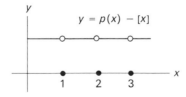

If f has a limit at a without being continuous there, then we can modify f so that it becomes continuous at a simply by changing what happens at the one point a. We just define (or redefine) $f(a)$ to be the limit l, and then have

$$\lim_{x \to a} f(x) = f(a).$$

Thus, if $f(x)$ has a limit as x approaches a, then the limit l is the number that f *ought* to have as its value *at* a, whether or not it does.

Functions don't *have* to have limits, and this anomolous behavior can occur in various ways. We have already met one example: The greatest-integer function $[x]$ has a limit *from the left* at $x = 2$ and also a limit *from the right* at 2, but the two "one-sided" limits are different. However, our principal concern here is evaluating limits that do exist.

Limit evaluations are trivial for continuous functions. (We assume the properties of continuity from the last section.)

EXAMPLE 3. Any polynomial $p(x)$ is everywhere continuous, so $\lim_{x \to a} p(x) = p(a)$ for any a. For example,

$$\lim_{x \to 4} (x^3 - 8x) = 4^3 - 8 \cdot 4 = 64 - 32 = 32.$$

EXAMPLE 4. A rational function is a quotient

$$r(x) = \frac{p(x)}{q(x)}$$

of two polynomials. It follows from our general principles for continuous functions that $r(x)$ is continuous everywhere except at points where the denominator is zero. For example, if

$$r(x) = \frac{x - 1}{x + 1}.$$

then

$$\lim_{x \to 3} r(x) = r(3) = \frac{3 - 1}{3 + 1} = \frac{1}{2}.$$

EXAMPLE 5

$$\lim_{x \to 2} \sqrt{x} = \sqrt{2}.$$

EXAMPLE 6. We return now to Example 1, which is a nontrivial limit typifying our early calculus computations. We wish to show that if

$$f(x) = \frac{1 - \sqrt{x}}{1 - x},$$

then

$$\lim_{x \to 1} f(x) = \frac{1}{2}.$$

. . . .

Solution. If we try substituting $x = 1$, we get $0/0$, which is meaningless; so f is not defined at $x = 1$ and we do not have a trivial limit. We can nevertheless calculate the limit as follows. Factoring $1 - x$ as a difference of squares, $1 - x = (1 - \sqrt{x})(1 + \sqrt{x})$, we see that

$$\frac{1 - \sqrt{x}}{1 - x} = \frac{1 - \sqrt{x}}{(1 - \sqrt{x})(1 + \sqrt{x})} = \frac{1}{1 + \sqrt{x}}$$

when $x \neq 1$. Now

$$g(x) = \frac{1}{1 + \sqrt{x}}$$

is continuous on $[0, \infty)$, and its limit at 1 is its value there:

$$\lim_{x \to 1} g(x) = g(1) = \frac{1}{1 + \sqrt{1}} = \frac{1}{2}.$$

Since $f(x) = g(x)$ for $x \neq 1$, we have

$$\lim_{x \to 1} f(x) = \lim_{x \to 1} g(x) = g(1) = \frac{1}{2}.$$

That is,

$$\lim_{x \to 1} \frac{1 - \sqrt{x}}{1 - x} = \lim_{x \to 1} \frac{1}{1 + \sqrt{x}} = \frac{1}{1 + \sqrt{1}} = \frac{1}{2}.$$

Thus, although f is not defined at $x = 1$, it is equal *everywhere else* to a function g that is continuous at 1, and we use this fact to evaluate the limit of f at 1.

_____

Although we can frequently use this trick to evaluate a limit, it isn't always available. Sometimes, instead of finding $f(x)$ to be *equal* to a known continuous function, we find it *squeezed between two* known continuous functions that come together at the limit point.

EXAMPLE 7. We shall need to evaluate the limit

$$\lim_{t \to 0} \frac{\sin t}{t}.$$

You may not know what the functions sin t and cos t are, but read on anyway.
Setting $t = 0$ again gives 0/0 so the function $f(t) = \sin t/t$ is not defined at
$t = 0$. This time it can be shown that

$$\cos t < \frac{\sin t}{t} < 1$$

on an interval about 0. Now cos t is continuous everywhere, and in particular

$$\lim_{t \to 0} \cos t = \cos 0 = 1.$$

Thus (sin $t)/t$ is squeezed between two continuous functions each having the
value 1 at $t = 0$. We conclude that therefore (sin $t)/t$ has the limit 1 at $t = 0$.

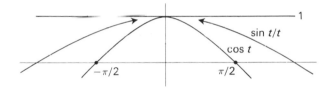

The above examples exhibit the methods by which we shall make most of our
explicit limit calculations. In each case, we exploited the connection between
limits and continuity, as follows:

a) $\lim_{x \to a} f(x) = f(a)$ if f is continuous at a;

b) $\lim_{x \to a} f(x) = g(a)$ if g is continuous at a and equal to f for all $x \neq a$ (in
some interval about a).

c) $\lim_{x \to a} f(x) = g(a) = h(a)$ if g and h are continuous at a, $g(a) = h(a)$, and

$$g(x) \leq f(x) \leq h(x)$$

for all $x \neq a$ (on some interval about a).

These devices are listed in order of increasing generality, and from the point of
view of strict logic, the squeeze device (c) includes both of the others.

PROBLEMS FOR SECTION 6

Evaluate each of the following limits, citing the relevant reasons in each case.

1. $\lim_{x \to 2} \sqrt{x}$

2. $\lim_{x \to 2} \dfrac{x + 1}{x^2 - 1}$

3. $\lim_{x \to 8} x^{1/3}$

4. $\lim_{x \to -1} 5x^3 + 4x^2 + 1$

5. $\lim_{x \to 0} \dfrac{1 - \sqrt{x}}{1 + \sqrt{x}}$

Use the trick of Example 6 to evaluate each of the following limits.

6. $\lim_{x \to 1} \dfrac{x - 1}{x^2 - 1}$

7. $\lim_{x \to 2} \dfrac{x^3 - 8}{x - 2}$

8. $\lim_{x \to 4} \dfrac{\sqrt{x} - 2}{x - 4}$

9. $\lim_{x \to 4} \dfrac{x - \sqrt{x} - 2}{x - 4}$

10. $\lim_{x \to 1} \dfrac{x^3 - 1}{x + \sqrt{x} - 2}$

11. $\lim_{x \to 0} \dfrac{\sqrt{1 + x} - 1}{x}$

Evaluate the following limits (if possible).

12. $\lim_{x \to -1} \dfrac{2x^2 - x - 3}{x + 1}$

13. $\lim_{x \to 0} \dfrac{2x^2 - x - 3}{x + 1}$

14. $\lim_{x \to 4} \dfrac{x - 4}{3(\sqrt{x} - 2)}$

15. $\lim_{x \to 4} \dfrac{x - 4}{x^2 - x - 12}$

16. $\lim_{x \to -3} \dfrac{x - 4}{x^2 - x - 12}$

17. $\lim_{x \to 4} \dfrac{\sqrt{x^2 + 9} - 5}{x^2 - 4x}$

18. $\lim_{x \to 2} \dfrac{\sqrt{x^2 + 9} - 5}{x^2 - 4x}$

19. $\lim_{x \to 0} \dfrac{\sqrt{x^2 + 9} - 5}{x^2 - 4x}$

20. $\lim_{x \to 2} \left(\dfrac{x^4 - 16}{x^3 - 8} \right)^{1/2}$

21. $\lim_{x \to -4} \dfrac{x^2 + x - 6}{x^2 + 7x + 12}$

22. $\lim_{x \to -3} \dfrac{x^2 + x - 6}{x^2 + 7x + 12}$

23. $\lim_{x \to -2} \dfrac{x^2 + x - 6}{x^2 + 7x + 12}$

24. Is it possible for a function f to satisfy the inequality $2x \le f(x) \le x^2 + 1$ on an interval containing $x = 1$? Supposing that this inequality does hold near $x = 1$, prove that $f(x) \to 2$ as $x \to 1$.

25. Evaluate $\lim_{x \to 0} f(x)$ if $\sqrt{1 + x^2} \le f(x) \le 1 + |x|$.

26. It was remarked in the text that although the greatest integer function is discontinuous at $x = 2$, nevertheless it does have the limit 1 as x approaches 2 *from the left*, and also the limit 2 as x approaches 2 *from the right*. Suppose that we designate these one-sided limits by using a minus superscript to indicate approach from below, and a plus superscript to indicate approach from above. Thus

$$\lim_{x \to 2^-} [x] = 1, \qquad \lim_{x \to 2^+} [x] = 2.$$

In this terminology, what are the values of the following one-sided limits?

a) $\lim_{x \to 4^-} [x]$ b) $\lim_{x \to 4^+} [x]$ c) $\lim_{x \to 4^-} p(x)$

d) $\lim_{x \to 4^+} p(x)$ e) $\lim_{x \to 4^-} [x] + p(x)$ f) $\lim_{x \to 4^+} [x] + p(x)$

27. Graph the function $f(x) = |x|/x$. What are its one-sided limits at $x = 0$?

Determine each of the one-sided limits below.

28. $\lim_{x \to 1^+} \dfrac{|x - 1|}{x - 1}$ 29. $\lim_{x \to 1^-} \dfrac{|x - 1|}{x - 1}$

30. $\lim_{x \to 2^-} \dfrac{|x - 2|}{x - 2}$ 31. $\lim_{x \to 3^+} \dfrac{x - 3}{|x - 3|}$

32. $\lim_{x \to 4^-} \dfrac{x^2 - 16}{|x - 4|}$ 33. $\lim_{x \to 1^+} \dfrac{|1 - x^2|}{1 - x}$

34. If $\lim_{x \to a^-} f(x) = f(a)$, we say that f is continuous from the left at $x = a$. Write down the definition for continuity at the right at $x = a$.

35. Examine the greatest integer function for one-sided continuity and state your conclusions.

36. The same for the postage stamp function $p(x)$.

37. Show that $[x] + p(x)$ is not one-sidedly continuous at $x = 4$.

38. Suppose that

$$\lim_{x \to a^-} f(x), \qquad f(a), \qquad \lim_{x \to a^+} f(x)$$

all exist. List the possibilities for equality and/or inequality among these three values.

39. Show that the following function is continuous everywhere. The crucial point is $t = -1$, where you have to show that the left- and righthand limits are both equal to $h(-1)$. Graph the function h.

$$h(t) = \begin{cases} 3t + 5 & \text{if } t < -1, \\ t^2 + 1 & \text{if } t \geq -1. \end{cases}$$

40. Determine the constant c so as to make the function

$$f(x) = \begin{cases} x^2 & \text{if } x < 1/2, \\ c - x^2 & \text{if } x \geq 1/2 \end{cases}$$

continuous at $1/2$. Sketch its graph.

7. THE LIMIT LAWS

When we reduce a limit calculation to the trivial limit of a continuous function, as in the last section, we don't need to make any explicit use of laws of algebra for limits. Algebraic combinations of continuous functions are continuous, according to the principles in Section 5, and this fact automatically accounts for the limiting behavior of such algebraic combinations. However, there are situations in which the reduction of a limit calculation to the evaluation of a continuous function doesn't occur quite so naturally and easily, and it is useful and customary to replace the continuity laws by their more general limit formulations. We do this below. At present we shall regard these improved versions as *axioms* about the continuous and limiting behavior of functions. They can play much the same role as the axioms of geometry. That is, we can start with them as an explicit list of properties that seem obviously true, and then we can deduce by logical reasoning other conclusions called *theorems*. In particular, we could use these axioms to justify formally all the steps in the examples in the last section.

In these laws there are implicit assumptions about domains. Generally it is understood that all the functions in question are defined throughout some interval about the limit point a except possibly at a itself. Occasionally we want to consider a function defined only on *one side* of a. Then the other functions in question will be defined on the *same* side of a, so that the combinations $f(x) + g(x)$, $f(x)g(x)$, etc., can still be formed.

Continuity and limit laws.

L1. *A function f is continuous at the point $x = a$ if and only if $f(a)$ is defined and $f(x) \rightarrow f(a)$ as $x \rightarrow a$.*

L2. *For each positive integer n, the function $f(x) = x^{1/n}$ is continuous at each point of its domain, as is the constant function $f(x) = c$.*

L3. *If $f(x) \rightarrow l$ and $g(x) \rightarrow m$ as $x \rightarrow a$, then*

a) $\qquad\qquad\qquad f(x) + g(x) \rightarrow l + m,$

b) $\qquad\qquad\qquad f(x)\, g(x) \rightarrow lm,$

c) $\qquad\qquad\qquad \dfrac{1}{f(x)} \rightarrow \dfrac{1}{l} \qquad (\text{provided } l \neq 0)$

as $x \rightarrow a$.

L4. *If $f(x) \rightarrow l$ as $x \rightarrow a$, and if $g(y)$ is continuous at $y = l$, then*

$$g(f(x)) \rightarrow g(l) \qquad \text{as } x \rightarrow a.$$

So far this is simply our list from Section 5 partially restated in terms of limits. We now add the squeeze principle.

L5. *If f(x) and h(x) have the same limit l as x → a, and if f(x) ≤ g(x) ≤ h(x) on some interval about a (except, possibly, at x=a), then g(x) → l as x → a. In particular, if f(x) = g(x) except possibly at x = a, and if f(x) → l as x → a, then g(x) → l as x → a.*

Our informal computation of the limit of $(1 - \sqrt{x})/(1 - x)$ as x approaches 1 could now be recast into a formal proof in which each step is justified by one of the limit laws. The same will be true of the many limit calculations to be found throughout the book; they will generally be presented informally, but will have formal versions that can be filled in by anyone interested.

Note that **L1** and **L3** imply:

L3′ *If f and g are continuous at a, then so are f + g, fg, and 1/f (provided f(a) ≠ 0).*

By repeatedly applying this principle, starting from the continuity of x and the constant functions, we establish the continuity of any function that can be built up from x and constants by using the two operations of addition and multiplication. Such functions are called *polynomials*. Examples are

$$p(x) = x^2 + 2x + 5,$$

$$q(x) = 4x^{13} - x^9 + 2x^5 + 3x.$$

Knowing that every polynomial is continuous, we can then conclude, from **L3′** again, that the quotient

$$r(x) = \frac{p(x)}{q(x)}$$

of any two polynomials is continuous, except where it is undefined (i.e., where $q(x) = 0$). A quotient of two polynomials is called a *rational function*. We thus have the theorem:

L6. Theorem. *A polynomial function p(x) is everywhere continuous. A rational function*

$$r(x) = \frac{p(x)}{q(x)},$$

where p and q are polynomials, is continuous except where undefined (i.e., where q(x) = 0).

More generally, by using also the continuity of $x^{1/n}$ and **L4**, it can be shown that any function that can be built up from x and the constants by using the operations of addition, multiplication, division, and root extraction, is continuous wherever it is defined.

Our discussion of continuity and limits has been intuitive and qualitative. We have not considered *how* close to a we have to take x in order to ensure that $f(x)$

is closer to l than some preassigned tolerance. In Chapter 20 we shall consider these quantitative calculations, and shall see that carrying them out for various combinations of functions amounts to quantitative proofs for the above laws.
    We conclude with two remarks.

A. To be completely formal, the proof of **L6** has to use a principle called *mathematical induction* that you may or may not know about. We shall not give any formal proofs requiring induction in this book.

B. A word about polynomial notation. A particular polynomial, like the ones displayed above, can always be written out explicitly. This becomes a chore when the degree is large—a polynomial of degree 18 would (normally) have 19 terms, for example—and in many situations there is no particular need to see each individual term. Then we can present the polynomial "in outline" by giving its first and last terms, and indicating the middle terms by three dots, . . . , that we read "and so on." Thus,

$$p(x) = a_{18} x^{18} + a_{17} x^{17} + \cdots + a_0$$

is the polynomial $a_{18} x^{18} + a_{17} x^{17} +$ and so on, down to a_0. Of course this notation is not needed for low-degree polynomials.

PROBLEMS FOR SECTION 7

1. List all the steps required to show that the polynomial $p(x) = x^3 - 4x + 2$ is continuous, justifying each step by referring to part of **L3'**.

2. List all the steps required to show that the rational function $r(x) = (x^2 - 1)/(x^2 + 1)$ is continuous, justifying each step by referring to part of **L3'**.

Justify the continuity of the following functions from **L2** and **L3'**, in the manner of the two problems above.

3. $\sqrt{x} + \dfrac{1}{\sqrt{x}}$

4. $\dfrac{1}{x^{1/3} + 1}$

5. $\dfrac{\sqrt{x}}{1 + x^2}$

6. $x^{1/2} + x^{1/3} + x^{1/4}$

7. Write out the limit law **L3** for limits from the left, using the notation:

$$\lim_{x \to a^-} f(x).$$

(See the remarks in Problem 26 in the last section.)

8. Show that $y^2 - 4y$ is continuous at $x = 2$ if $y = f(x)$ and

$$\lim_{x \to 2} f(x) = 1, \qquad f(2) = 3.$$

9. Show that $y^3 - 4y + 1$ is continuous at $x = a$ if $y = f(x)$ and

$$\lim_{x \to a^-} f(x) = -2, \qquad f(a) = 0, \qquad \lim_{x \to a^+} f(x) = +2.$$

10. Show by an example that the following "one-sided" version of **L4** is incorrect:

If $f(x) \to l$ as $x \to a^-$ and if $g(y)$ is continuous from the left at $y = l$, then

$$g(f(x)) \to g(l) \qquad \text{as } x \to a^-.$$

11. State a correct version of **L4** involving $x \to a^-$.

12. Prove that

$$\lim_{x \to 0} [1 - x^2] = 0$$

from the limit laws and the known behavior of $[x]$. Graph the function $f(x) = [1 - x^2]$.

8. INFINITE LIMITS AND LIMITS AT INFINITY

Although the symbol $+\infty$ does not represent a number, it is nevertheless used in limit statements with a special convention about what is being said.

EXAMPLE 1. The equation

$$\lim_{x \to 0} \frac{1}{x^2} = +\infty,$$

read literally, says that $1/x^2$ approaches and comes arbitrarily close to $+\infty$ as x approaches 0. We interpret this as meaning that $1/x^2$ *grows without bound* as x aproaches 0, in the sense that it becomes and remains larger than any given number when x is taken suitably small. This is true. For example, if we want

$$\frac{1}{x^2} > 10{,}000,$$

then we only have to take $|x| < 1/100$.
    Similarly,

$$\lim_{x \to 0} -\frac{1}{x^2} = -\infty,$$

meaning that $-1/x^2$ becomes and remains *smaller* than any given negative

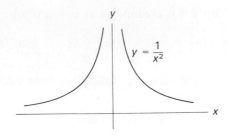

number when $|x|$ is taken suitably small. For example,

$$-\frac{1}{x^2} < -10,000$$

when $|x| < 1/100$.

The "approaching" symbolism seems especially appropriate for infinite limits. Thus,

$$\frac{1}{x^2} \to +\infty \qquad \text{as } x \to 0.$$

EXAMPLE 2. The function $f(x) = 1/x$ is more complicated in that what it does as x approaches 0 depends on which side of zero x is on. We use 0^- and 0^+ to indicate approach from below and from above respectively. Thus

$$\lim_{x \to 0^-} \frac{1}{x} = -\infty, \qquad \lim_{x \to 0^+} \frac{1}{x} = +\infty.$$

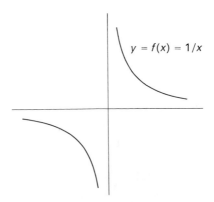

In the same way, the postage-stamp function $p(x)$ has unequal one-sided limits at each of its points of discontinuity. For example,

$$\lim_{x \to 1^-} p(x) = 1, \qquad \lim_{x \to 1^+} p(x) = 2,$$

as you will see upon reviewing the definition and graph of $p(x)$ on page 68.

EXAMPLE 3. Consider next the statement

$$\lim_{x \to +\infty} \frac{1}{x} = 0.$$

Here the limit is a number but the independent variable x is tending to ∞, i.e., growing without bound. The statement means that $1/x$ can be made as close to zero as desired by taking x sufficiently *large*. For example, to get

$$\frac{1}{x} < \frac{1}{1,000}$$

we take $x > 1,000$.

EXAMPLE 4. Similarly

$$\lim_{x \to +\infty} \frac{1}{\sqrt{x}} = 0.$$

But $1/\sqrt{x}$ gets small much more slowly than $1/x$. In order to get

$$\frac{1}{\sqrt{x}} < \frac{1}{1,000},$$

we have to take $x > (1,000)^2 = 1,000,000$.
  It is customary to combine the two statements

$$\lim_{x \to +\infty} \frac{1}{x} = 0 \qquad \text{and} \qquad \lim_{x \to -\infty} \frac{1}{x} = 0$$

into the single statement

$$\lim_{x \to \infty} \frac{1}{x} = 0$$

or

$$\frac{1}{x} \to 0 \qquad \text{as } x \to \infty.$$

Limits at infinity obey the same limit laws (**L3** through **L5**) as for limits at a, but sometimes a preliminary revision is needed before a computation can be made.

EXAMPLE 5. In order to verify that

$$\lim_{x \to \infty} \frac{3 + x^2}{3x^2 + 1} = \frac{1}{3},$$

we first rewrite the expression by dividing top and bottom by x^2, getting

$$\frac{\dfrac{3}{x^2} + 1}{3 + \dfrac{1}{x^2}}$$

(for $x \neq 0$). Now $3/x^2$ and $1/x^2$ both approach 0 as x tends to ∞. Thus

$$\lim_{x \to \infty} \frac{3 + x^2}{3x^2 + 1} = \lim_{x \to \infty} \frac{\dfrac{3}{x^2} + 1}{3 + \dfrac{1}{x^2}}$$

$$= \frac{0 + 1}{3 + 0}$$

$$= \frac{1}{3}.$$

EXAMPLE 6. Show that the hyperbola

$$y = \frac{b}{a}\sqrt{x^2 - a^2}$$

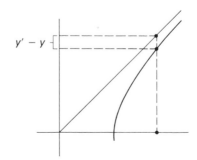

$y' - y$

is asymptotic to the line

$$y' = \frac{b}{a}x$$

in the strong sense that

$$y' - y \to 0 \qquad \text{as } x \to +\infty.$$

· ·

Solution. We rationalize a numerator:

$$y' - y = \frac{b}{a}\left[x - \sqrt{x^2 - a^2}\right]$$

$$= \frac{b}{a}\left[x - \sqrt{x^2 - a^2}\right]\frac{(x + \sqrt{x^2 - a^2})}{(x + \sqrt{x^2 - a^2})}$$

$$= \frac{ab}{x + \sqrt{x^2 - a^2}}.$$

In particular,

$$0 < y' - y < \frac{ab}{x},$$

and since $1/x \to 0$ as $x \to +\infty$, it follows that

$$y' - y \to 0$$

by the squeeze limit law.

If

$$\lim_{x \to \infty} \frac{f(x)}{g(x)} = 1,$$

then f and g must behave in essentially the same way "at ∞". For example, if g has a limit at infinity, then f must have the same limit:

If $g(x) \to l$, then

$$f(x) = \frac{f(x)}{g(x)} \cdot g(x) \to 1 \cdot l = l$$

as $x \to \infty$. This holds for the limits $+\infty$ and $-\infty$ as well.

A polynomial

$$p(x) = a_n x^n + a_{n-1} x^{n-1} + \cdots + a_0$$

behaves like its highest-degree term $a_n x^n$ at ∞. For (supposing $a_n \neq 0$, of course)

$$\frac{p(x)}{a_n x^n} = 1 + \frac{a_{n-1}}{a_n} \cdot \frac{1}{x} + \cdots + \frac{a_0}{a_n x^n}$$
$$\to 1 + 0 + \cdots + 0$$
$$= 1$$

as $x \to \infty$.

In technical terms, we say that f and g are *asymptotically equal* at ∞ if $f(x)/g(x) \to 1$ as $x \to \infty$. Thus we have just shown that any polynomial is asymptotically equal to its highest-degree term at infinity. In the same way you can show that any rational function

$$r(x) = \frac{p(x)}{q(x)}$$
$$= \frac{a_n x^n + a_{n-1} x^{n-1} + \cdots + a_0}{b_m x^m + b_{m-1} x^{m-1} + \cdots + b_0}$$

is asymptotically equal to $(a_n/b_m)x^{n-m}$ at ∞.

PROBLEMS FOR SECTION 8

Evaluate the following limits.

1. $\displaystyle \lim_{x \to +\infty} \frac{7x + 2}{3 - x}$

2. $\displaystyle \lim_{x \to +\infty} (\sqrt{x^2 + 4} - x)$

3. $\displaystyle \lim_{x \to -\infty} \left(1 + \frac{2}{x}\right)$

4. $\displaystyle \lim_{x \to +\infty} \sqrt{x^2 + x + 1} - x$

5. $\displaystyle \lim_{x \to \infty} \frac{2x^2 + 1}{6 + x - 3x^2}$

6. $\displaystyle \lim_{x \to \infty} \frac{x}{x^2 + 5}$

7. $\displaystyle \lim_{x \to +\infty} \frac{3^x - 3^{-x}}{3^x + 3^{-x}}$

8. $\displaystyle \lim_{x \to -\infty} \frac{3^x - 3^{-x}}{3^x + 3^{-x}}$

9. $\displaystyle \lim_{x \to \infty} \frac{\sqrt{x + 1}}{\sqrt{4x - 1}}$

10. $\displaystyle \lim_{x \to \infty} \left(\frac{1}{x} + 1\right)\left(\frac{5x^2 - 1}{x^2}\right)$

11. $\displaystyle\lim_{x\to\infty} \frac{1 + 99x + 2x^2}{3x^2 + 4}$

Evaluate the following limits.

12. $\displaystyle\lim_{x\to-1^-} \frac{3x^2 - 4x - 7}{x^2 + 2x + 1}$

13. $\displaystyle\lim_{x\to 5^+} \frac{5x^2 + x^3}{25 - x^2}$

14. $\displaystyle\lim_{x\to 1^+} \frac{x^2}{1 - x^2}$

15. $\displaystyle\lim_{x\to 2^+} \frac{\sqrt{x^2 - 4}}{x - 2}$

16. $\displaystyle\lim_{x\to 0} \frac{x + 1}{|x|}$

17. Evaluate

$$\lim_{x\to 3} \frac{2x^2 - 5x - 3}{x^2 - x - 6}$$

and

$$\lim_{x\to\infty} \frac{2x^2 - 5x - 3}{x^2 - x \quad 6}.$$

Justify both calculations.

18. Prove the assertion made at the end of Section 8   about $r(x)$ being asymptotically equal to

$$(a_n/b_m)x^{n-m} \qquad \text{at } \infty.$$

19. How large must a positive number x be taken in order to ensure that $1/x^3$ is less than .001?

20. How large must x be taken to ensure that $x^{-1/3}$ is less than .1?   .01?   .001?

21. Let d be a small positive number. Determine how large a positive number x must be taken in order to ensure that $1/x^3 < d$.

22. Determine how large x must be taken to ensure that $x^{-1/3} < d$.

23. Show that

$$\frac{1}{\sqrt{x}} > 1,000$$

for all x smaller than a certain number d. (Find an explicit value of d that will work.)

24. If M is any positive number, show that

$$\frac{1}{\sqrt{x}} > M$$

for all x smaller than a certain number d. (Find d in terms of M.)

25. Given a positive number d, determine a positive number M such that

$$\frac{1}{x^2} < d$$

for all x larger than M.

26. Given a positive number d, determine a positive number M such that

$$\frac{1}{x^{1/3}} < d$$

for all x larger than M.

chapter 3
the derivative

The derivative evolved historically as the common solution to two quite different problems: one, the problem of determining the tangent lines to curves other than circles; the other, the problem of computing the velocity and acceleration of a moving particle. The original success of the calculus in solving such problems was important, but it hardly suggested the impact calculus was to have on quantitative disciplines of all kinds. Most natural laws seem to involve relationships among the rates of change of varying quantities. Such relationships are stated as equations containing derivatives, i.e., as differential equations. The study of differential equations then enables us to understand and predict nature, and the extent to which this has been possible is hard to believe until it has been experienced. A first course cannot get deeply into the subject, but it can cover important applications in sufficient quantity to show the power of calculus.

We shall rely heavily on geometric reasoning and intuition in our development of the calculus. This does not mean we have a particular interest in geometry, although it is one of the subjects to which calculus can be applied. (For example, we shall be able to compute areas and volumes that are otherwise inaccessible to us.) Rather, the reason for the special role of geometry is that geometric intuition helps us understand calculus itself, in much the way that the graph of a function f lets us see in a very vivid manner how $f(x)$ varies with x. Furthermore, geometric reasoning will help us to apply calculus to *other* disciplines. Here, for example, is a problem of a type that we shall take up in Chapter 6.

The efficiency E of a certain machine varies with the viscosity x of the lubricating oil that is used. The machine runs badly if the oil is either too thin or too thick, i.e., if x is either too small or too large. Our theoretical calculations (or perhaps our empirical observations) show us that the relationship between E and x is given by

$$E = x - x^3.$$

The problem is to determine the viscosity x that gives the maximum efficiency E. When we plot E against x we find that the graph looks like this in the first quadrant.

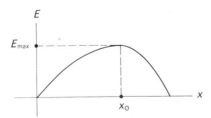

Now we see that our problem can be rephrased geometrically as that of finding the *highest point* on this graph.

This example shows how a nongeometric problem can be cast into geometric terms, but it doesn't suggest the role of calculus. Recall, however, that calculus will let us compute the slope m of the tangent line as a function of x. At the highest point on the above graph the tangent line appears to be horizontal, so we want the value of x for which $m=0$. Therefore, in order to solve our maximum problem, we first compute the slope function $m(x)$ by calculus, and then solve the equation $m(x) = 0$ to find the value $x = x_0$ giving maximum efficiency.

Thus, when we approach the derivative through the tangent-line problem, we shall not be solving *only* this geometric problem. We will also be creating a geometric interpretation of the derivative that will help us in calculus itself and in the applications of calculus to other disciplines.

Later in this chapter we shall take up the other major interpretation of the derivative, as a rate of change. Otherwise, the chapter treats the computation of derivatives from scratch, the beginning algorithms of the systematic theory, and a few elementary applications.

1. THE SLOPE OF A TANGENT LINE

The tangent-line problem was frustrating for classical geometers who knew only synthetic geometry. There just is no universal geometric construction for a tangent line that generalizes the classical circle constructon. We can visualize the tangent line as having the same *direction* that the curve has at the point of tangency, and we can find the tangent if we can find this common direction, but there is no purely geometric way to obtain the directions of curves other than circles.

The solution to the problem had to wait for the development of analytic geometry. In the coordinate plane a direction is specified by its *slope*, and the tangent problem is solved by the surprising discovery that we can in fact compute the slope of a smooth graph *at any point*.

By way of illustration we shall compute the slope of the graph of $f(x) = x^3$ at the point $(x_0, y_0) = (1, 1)$. We start by choosing a second nearby point on the graph of $y = x^3$, say $(x_1, y_1) = (r, r^3)$, and we compute the slope of the *secant* line through these two points on the graph:

$$m_{\text{sec}} = \frac{r^3 - 1}{r - 1}.$$

This is just the slope formula $(y_1 - y_0)/(x_1 - x_0)$ for the two points (x_0, y_0) $= (1, 1)$ and $(x_1, y_1) = (r, r^3)$. We are going to let the second point vary, and we use r instead of x_1 to suggest this "running point," and also to avoid subscripts. We don't replace x_1 by x because we want to use x for another purpose.

Now let r approach 1 and imagine the point (r, r^3) sliding along the graph toward the fixed point $(1,1)$. As this happens the secant line rotates about $(1,1)$, and we visualize it approaching the tangent line as its limiting position. Our intuition thus tells us that the slope m_{sec} of the secant line approaches the slope m of the tangent line as its limiting value.

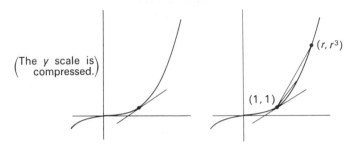

We can't compute this limit by simply setting $r=1$ in the formula m_{sec} $= (r^3 - 1)/(r - 1)$ because this would just give the meaningless 0/0. In fact, it isn't at all obvious what (if anything) m_{sec} approaches as r approaches 1. We have the quotient of two quantities each of which becomes arbitrarily small, and we can't guess how the quotient behaves. However, the reason that the numerator becomes small is that it contains the denominator $r - 1$ as a factor,

$$r^3 - 1 = (r - 1)(r^2 + r + 1).$$

After this common factor is cancelled from top and bottom we *can* see what goes on:

$$m_{sec} = \frac{r^3 - 1}{r - 1} = r^2 + r + 1 \qquad \text{for } r \neq 1,$$

and $r^2 + r + 1$ approaches $1 + 1 + 1 = 3$ as its limit as r approaches 1. We have thus found that the slope of the secant line approaches 3 as r approaches 1, so 3 is the slope m of the tangent line at the point $(1,1)$. The equation for the tangent line is then

$$y - 1 = m(x - 1),$$

with $m = 3$, and we end up with

$$y = 3x - 2.$$

As we noted above, the slope of the tangent line is the slope that should be assigned to the graph itself at the point of tangency. Thus, the graph of $y = x^3$ has the slope 3 at the point $(1,1)$.

In limit notation, the above computation of m is written

$$m = \lim_{r \to 1} m_{sec} = \lim_{r \to 1} \frac{r^3 - 1}{r - 1} = \lim_{r \to 1} (r^2 + r + 1) = 1 + 1 + 1 = 3.$$

The limit evaluation

$$\lim_{r \to 1} (r^2 + r + 1) = 1 + 1 + 1$$

is the equation

$$\lim_{r \to 1} q(r) = q(1)$$

for the function $q(r) = r^2 + r + 1$; it simply expresses the continuity of q at $r = 1$. We can "see" the value of the limit because the continuity of the polynomial $r^2 + r + 1$ is intuitively obvious.

Now let us rerun the above tangent–slope computation, but this time find the slope m of the tangent line at the general point $(x_0, y_0) = (x_0, x_0^3)$ on the graph of $y = x^3$. The secant slope is now

$$m_{sec} = \frac{r^3 - x_0^3}{r - x_0},$$

and we have agreed on intuitive grounds that the tangent slope m is the limit of m_{sec} as r approaches x_0. If we can keep in mind that r is the variable in this limit calculation, we can drop the subscript on x_0 and write the secant slope more simply as

$$m_{sec} = \frac{r^3 - x^3}{r - x}.$$

As before, the numerator has $r - x$ as a factor

$$r^3 - x^3 = (r - x)(r^2 + rx + x^2),$$

so

$$m = \lim_{r \to x} \frac{r^3 - x^3}{r - x} = \lim_{r \to x} (r^2 + rx + x^2) = x^2 + x^2 + x^2 = 3x^2.$$

Again, this limit is obvious because it's obvious that $r^2 + rx + x^2$ is a continuous function of r (x being held fixed). We have now obtained the formula

$$m = 3x^2$$

for the slope m of the graph and its tangent line at the point (x, x^3). This tells us, for example, that at $x = 2$ the curve has slope

$$m = 3 \cdot 2^2 = 12,$$

and that at $x = -1/2$ its slope is

$$m = 3(-\frac{1}{2})^2 = \frac{3}{4}.$$

The varying slope $m = 3x^2$ is a new function of x which is called the *derivative* of the original function x^3.

The derivative calculation that we have just made for the function $f(x) = x^3$ can be described for an arbitrary function f. We first compute the slope of the secant line through the two points $(x, f(x))$ and $(r, f(r))$ on the graph of f,

$$m_{sec} = \frac{f(r) - f(x)}{r - x}.$$

Then we calculate the limit of m_{sec} as r approaches x, obtaining a number that we interpret geometrically as the slope of the graph of f and its tangent line at the point $(x, f(x))$. This limit depends on x and is therefore a new function of x. The new function is called the *derivative* of the function f, and is designated f'. In this notation we calculated above that $f'(x) = 3x^2$ when $f(x) = x^3$.

PROBLEMS FOR SECTION 1

1. Compute the slope of the tangent line to the graph of $f(x) = x^2$ at the point $(1,1)$. (Imitate the text discussion of $f(x) = x^3$.)

2. Compute the slope of the tangent line to the graph of $y = x^2$ at the point over $x_0 = 2$. Write down the equation of the tangent line.

In each of Problems 3 through 9, compute the slope of the tangent line to the given graph at the point having the given x-coordinate. Then write down the equation of the tangent line.

3. $y = x^2, \quad x_0 = -1$.

4. $y = x^2, \quad x_0 = a$.   (For what value of a will the tangent line have slope 1?)

5. $y = 1/x, \quad x_0 = 1$.

6. $y = 1/x, \quad x_0 = 2$.

7. $y = x - x^2, \quad x_0 = 0$.

8. $y = x - x^2, \quad x_0 = 2$.

9. $y = x - x^2, \quad x_0 = a$. (For what value of a will the tangent be horizontal?)

10. Presumably a straight line $y = mx + b$ is its own tangent line at any point. Verify this by the slope-determining procedure, applied to the function $f(x) = mx + b$ at $x_0 = a$.

2. THE DERIVATIVE

Omitting the geometric motivation, we have the following formal definition.

Definition. *Given any function f, its derivative f' is the function whose value at x is defined by:*

$$f'(x) = \lim_{r \to x} \frac{f(r) - f(x)}{r - x}.$$

In making the limit calculation, x stays fixed while r varies. The domain of f' then consists of those numbers x for which this r limit exists.

The formal definition makes no reference to geometry. The fact that $(f(r) - f(x))/(r - x)$ and $f'(x)$ can be viewed as slopes constitutes a *geometric interpretation* of what we are doing. There are also, however, other fundamental interpretations related to the notion of velocity (see Sections 7 and 12). It is therefore important to understand that the above definition is a purely *analytic* definition, independent of interpretations. Given a function f, we derive a new function f' from f by making a certain limit calculation that we set up according to the instructions given in the above formula. In this neutral frame of mind we call

$$\frac{f(r) - f(x)}{r - x}$$

the *difference quotient* of the function f.

In applying this definition we always have the problem that the difference quotient is not in a form allowing the limit to be evaluated, because of the 0/0 difficulty that we experienced above. Thus we always have to manipulate the difference quotient into a new form that we can handle. (This may be easy or very hard. It is usually the heart of the problem; the subsequent limit evaluation can be anticlimactic.) So here is our scheme:

a) Write down the difference quotient $(f(r) - f(x))/(r - x)$ for the particular function in question;

b) Manipulate the difference quotient into a form allowing the limit to be evaluated (in these first examples this will always mean cancelling out $(r - x)$ from top and bottom);

c) Evaluate the limit as $r \to x$.

EXAMPLE 1. We compute $f'(x)$ by this procedure when $f(x) = 1/x$.

a) $\dfrac{f(r) - f(x)}{r - x} = \dfrac{\frac{1}{r} - \frac{1}{x}}{r - x}$

b) $= \dfrac{\frac{x - r}{rx}}{r - x} = -\dfrac{(r - x)}{rx} \cdot \dfrac{1}{r - x} = -\dfrac{1}{rx}.$

c) $f'(x) = \lim\limits_{r \to x} \dfrac{f(r) - f(x)}{r - x} = \lim\limits_{r \to x} \left(-\dfrac{1}{rx}\right) = -\dfrac{1}{xx} = -\dfrac{1}{x^2}.$

Thus $f'(x) = -1/x^2$ when $f(x) = 1/x$.

EXAMPLE 2. Find $f'(x)$ for $f(x) = \sqrt{x}$.

. . . .

Solution

a) $\dfrac{f(r) - f(x)}{r - x} = \dfrac{\sqrt{r} - \sqrt{x}}{r - x}.$

b) Here we shall "rationalize the numerator" by multiplying top and bottom by $\sqrt{r} + \sqrt{x}$, giving

$$\frac{\sqrt{r} - \sqrt{x}}{r - x} \cdot \frac{\sqrt{r} + \sqrt{x}}{\sqrt{r} + \sqrt{x}} = \frac{r - x}{(r - x)(\sqrt{r} + \sqrt{x})} = \frac{1}{\sqrt{r} + \sqrt{x}}.$$

c) Since

$$\frac{1}{\sqrt{r} + \sqrt{x}} \to \frac{1}{\sqrt{x} + \sqrt{x}} = \frac{1}{2\sqrt{x}}$$

as $r \to x$, we see that $f'(x) = 1/(2\sqrt{x})$ when $f(x) = \sqrt{x}$.

EXAMPLE 3. Find $f'(x)$ when $f(x) = x + x^2$.

. . . .

Solution

$$\frac{f(r) - f(x)}{r - x} = \frac{(r + r^2) - (x + x^2)}{r - x} = \frac{(r - x) + (r^2 - x^2)}{r - x}$$

$$= \frac{(r - x) + (r - x)(r + x)}{(r - x)}$$

$$= 1 + r + x,$$

which approaches $1 + 2x$ as its limit as r approaches x. Thus $f'(x) = 1 + 2x$.

EXAMPLE 4. Find $f'(x)$. when $f(x) = x^{1/3}$.

. . . .

Solution. Here we break our rule about factoring out $r - x$, but simplify in a similar way. In the difference quotient

$$\frac{f(r) - f(x)}{r - x} = \frac{r^{1/3} - x^{1/3}}{r - x},$$

we may treat $(r - x)$ as a *difference of cubes* and factor it:

$$\frac{r^{1/3} - x^{1/3}}{r - x} = \frac{r^{1/3} - x^{1/3}}{(r^{1/3} - x^{1/3})(r^{2/3} + r^{1/3}x^{1/3} + x^{2/3})}$$

$$= \frac{1}{r^{2/3} + r^{1/3}x^{1/3} + x^{2/3}}.$$

Now $r^{1/3} \to x^{1/3}$ as $r \to x$, since $r^{1/3}$ is a continuous function of r, and therefore

$$f'(x) = \lim_{r \to x} \frac{f(r) - f(x)}{r - x} = \lim_{r \to x} \frac{1}{r^{2/3} + r^{1/3}x^{1/3} + x^{2/3}} = \frac{1}{3x^{2/3}},$$

provided $x \neq 0$. Thus $f'(x) = \frac{1}{3}x^{-2/3}$ when $f(x) = x^{1/3}$.

EXAMPLE 5. The algebra in Example 4 can be simplified by setting $s = r^{1/3}$ and $y = x^{1/3}$. Then $r = s^3$, $x = y^3$ and

$$\frac{f(r) - f(x)}{r - x} = \frac{r^{1/3} - x^{1/3}}{r - x} = \frac{s - y}{s^3 - y^3}.$$

We can then proceed exactly as in the x^3 example, except that we factor out $(s - y)$ and the formulas are upside down.

In each of these examples the limit computation becomes completely formal if we justify each step by one of the limit laws (pp. 78,79). For example,

$$\lim_{r \to x} \frac{\frac{1}{r} - \frac{1}{x}}{r - x} = \lim_{r \to x} -\frac{1}{rx} \qquad \text{(L5)}$$

$$= -\frac{1}{x^2}. \qquad \text{(L1 and L6)}$$

Instead of relying on the general theorem L6, we could work out the particular case occurring here step by step from the earlier principles:

$$\lim_{r \to x} \left(-\frac{1}{rx}\right) = -\frac{1}{\lim_{r \to x}(rx)} = -\frac{1}{x(\lim_{r \to x} r)}$$

$$= -\frac{1}{x^2}.$$

Such formal arguments do not really make the conclusions any more convincing, because the limits in question are just as obvious as the limit laws from which they are deduced. However, all of the limit calculations occurring in the early chapters *can* be formalized this way if and when it becomes desirable to emphasize logical structure.

PROBLEMS FOR SECTION 2

Establish the following derivative formulas, using the above scheme in each case to evaluate the limit of the difference quotient.

1. If $f(x) = x^2$, then $f'(x) = 2x$.

2. If $f(x) = x^4$, then $f'(x) = 4x^3$.

3. If $f(x) = 1/x^2$, then $f'(x) = -2/x^3$.

4. If $f(x) = 1/x^{1/2}$, then $f'(x) = -1/2x^{3/2}$.

5. If $f(t) = t^{3/2}$, then $f'(t) = \dfrac{3}{2}t^{1/2}$.

6. If $f(u) = u^{1/4}$, then $f'(u) = \dfrac{1}{4}u^{-3/4}$.

7. If $f(s) = -3s$, then $f'(s) = -3$.

8. If $f(x) = 2x - x^2$, then $f'(x) = 2(1 - x)$.

9. If $f(x) = mx + b$, then $f'(x) = m$.

10. If $f(x) = ax^2 + bx + c$, then $f'(x) = 2ax + b$.

11. Each of the first six functions above is of the form $f(x) = x^a$. Examine the form of $f'(x)$ in each of these cases, and then conjecture what the general rule is for the derivative of a power of x: If $f(x) = x^a$, then $f'(x) = $ ___.

Calculate the derivatives of the following functions:

12. $f(x) = x + 1$.

13. $f(x) = x^2 + 4$.

14. $f(x) = x^3/3$.

15. $f(x) = \sqrt{2x}$.

16. $g(t) = \sqrt{3t - 2}$.

17. $f(y) = 1/(y - 1)$.

18. $f(x) = x + (1/x)$.

19. $g(x) = (x^2 + 1)/x$.

20. $f(s) = 1/(s^2 - 1)$.

21. $f(x) = \sqrt{x^2 + 1}$.

22. $f(x) = (x - 4)/(x + 1)$.

23. $h(x) = 1/\sqrt{x + 1}$.

24. $f(x) = (x + 2)^{1/3}$.

25. $f(x) = x/(x^2 + 1)$.

26. $f(x) = 1/(1 + \sqrt{x})$.

27. Calculate the derivative of $f(x) = x^{1/4}$ by the method of Example 5 in the text.

28. Justify the limit calculation in Example 2, appealing to appropriate laws about continuous functions and/or limits from Sections 5 through 7 in Chapter 2.

29. Calculate the derivative of $f(x) = x^{2/3}$ by setting $s = r^{1/3}$, $y = x^{1/3}$ in the difference quotient and then simplifying.

30. Find the derivative of $f(x) = x^{3/4}$, using substitutions similar to those suggested in the above problem.

31. Use the formula in Problem 10 to show that if $g(x)$ and $h(x)$ are quadratic polynomials and if $f(x) = g(x) + h(x)$, then $f'(x) = g'(x) + h'(x)$.

32. Using the same formula, show that if g and h are linear functions,

$$g(x) = mx + b, \qquad h(x) = nx + c,$$

and if f is the quadratic polynomial $f(x) = g(x)h(x)$, then

$$f'(x) = g(x)h'(x) + h(x)g'(x).$$

3. LEIBNIZ'S NOTATION

In situations where y is a function of x, Leibniz designated the derivative

$$\frac{dy}{dx},$$

which is read "the derivative of y with respect to x". This notation allows us to write down the derivatives of particular functions of x without using the function symbols f and f'. For example,

$$\text{if } y = x^3, \qquad \text{then } \frac{dy}{dx} = 3x^2.$$

Even more simply, we can write

$$\frac{d}{dx}x^3 = 3x^2.$$

In using this notation, however, we must remember that only a *function* can have a derivative. The symbol dy/dx implies that y is a certain function of x, and dy/dx is the derivative of that function. Thus:

$$\text{if } y = f(x), \qquad \text{then } \frac{dy}{dx} = f'(x).$$

Although the form of the Leibniz notation suggests a quotient, such an interpretation is not available until suitable separate meanings are given to dx and dy. This will come about eventually, but for the time being dy/dx is just a rather complicated (and indivisible) single symbol for the derivative. The letters x and y in dy/dx are not to be thought of as the variables x and y, but are there to suggest that dy/dx is related to these variables.

Because dy/dx doesn't display the variable x in the way that $f'(x)$ does, substitutions are harder to indicate, and we have to use some such awkward notation as

$$\left.\frac{dy}{dx}\right|_{x=5}$$

for the value of dy/dx when $x = 5$. Thus, if $dy/dx = 3x^2$, then $dy/dx|_{x=5} = 75$. In general,

$$\text{if } y = f(x), \qquad \text{then } \left.\frac{dy}{dx}\right|_{x=a} = f'(a).$$

This awkwardness in showing substitutions is the particular weakness of the Leibniz notation.

From now on we shall use the notations dy/dx and $f'(x)$ freely and interchangeably (as the situation warrants).

4. SOME GENERAL RESULTS ABOUT DERIVATIVES

From the several special cases that have been worked out above (and in the problems) you may have conjectured the general power rule

$$\frac{d}{dx}x^a = ax^{a-1}$$

for any constant a. We now prove this for *any positive integer n*. The proof is just like the case x^3 that we began with.

Theorem 1. *If $f(x) = x^n$, where n is a positive integer, then f is differentiable and $f'(x) = nx^{n-1}$.*

Proof. We use the factorization

$$r^n - x^n = (r - x)(r^{n-1} + r^{n-2}x + \cdots + x^{n-1}).$$

Then

$$\frac{f(r) - f(x)}{r - x} = \frac{r^n - x^n}{r - x} = r^{n-1} + r^{n-2}x + \cdots + x^{n-1}.$$

As r approaches x, this polynomial in r has the limit

$$\underbrace{x^{n-1} + x^{n-1} + \cdots + x^{n-1}}_{n \text{ terms}} = nx^{n-1}.$$

That is,

$$\lim_{r \to x} \frac{r^n - x^n}{r - x} = nx^{n-1};$$

as claimed. ∎

Another thing you may have noticed in the examples and problems is that the derivative of the *sum* of two functions seems always to be the *sum* of the two derivatives. We now check that this always happens.

Theorem 2. *If the functions g and h are differentiable at x, then so is their sum f = g + h, and*

$$f'(x) = g'(x) + h'(x).$$

Proof

$$\frac{f(r) - f(x)}{r - x} = \frac{[g(r) + h(r)] - [g(x) + h(x)]}{r - x}$$

$$= \frac{g(r) - g(x)}{r - x} + \frac{h(r) - h(x)}{r - x},$$

and this sum approaches $g'(x) + h'(x)$ as $r \to x$. Thus

$$f'(x) = \lim_{r \to x} \frac{f(r) - f(x)}{r - x} = g'(x) + h'(x). \quad \blacksquare$$

In this limit computation, the difference quotient of f rearranges itself into the sum of the difference quotients of g and h, and this turns into $f'(x) = g'(x) + h'(x)$ upon taking limits. In a similar way, we can show that the derivative of a constant times f is that constant times the derivative of f:

$$(cf)' = cf'.$$

Theorem 3. *If the function f is differentiable at x, then so is g = cf, and*

$$g'(x) = cf'(x).$$

For example,

$$\frac{d}{dx} 4x^5 = 4\frac{d}{dx}x^5 = 4 \cdot 5x^4 = 20x^4.$$

From the above theorems we can write down the derivative of any polynomial. For example,

$$\frac{d}{dx}(5x^3 + 3x^2) = \frac{d}{dx}(5x^3) + \frac{d}{dx}(3x^2)$$

$$= 5\frac{d}{dx}x^3 + 3\frac{d}{dx}x^2$$

$$= 5 \cdot 3x^2 + 3 \cdot 2x$$

$$= 15x^2 + 6x.$$

After a little practice one can omit the middle steps in this type of computation and just write the answer down.

EXAMPLE. Find the equation of the line tangent to the graph of $y = x^3 - 7x$ at the point given by $x_0 = 2$.

....

Solution. When $x_0 = 2$,

$$y_0 = x_0^3 - 7x_0 = 8 - 14 = -6.$$

Thus the point of tangency is $(x_0, y_0) = (2, -6)$. The slope of the graph at that point is

$$m = \left. \frac{dy}{dx} \right|_{x=2} = 3x^2 - 7 \Big|_{x=2} = 12 - 7 = 5.$$

The tangent line $y - y_0 = m(x - x_0)$ is thus

$$y - (-6) = 5(x - 2), \quad \text{or} \quad y = 5x - 16.$$

The use of Theorems 1 through 3 to differentiate polynomials is an example of the process of *systematic differentiation*, which frees us from always having to go back to the basic difference–quotient computation. We work out once and for all the derivatives of various special functions, like x^n, and then we develop formulas for computing the derivatives of *combinations* of functions in terms of the derivatives of their constituents, such as the formula $(f + g)' = f' + g'$. We can then write down the derivative of practically any function, using only these basic formulas and combining rules. This program is carried out systematically in Chapters 4 and 7.

We continue with the power rule for the case $a = 1/n$. This nth root situation generalizes the $x^{1/3}$ example.

Theorem 4. If $f(x) = x^a$, where $a = 1/n$ and n is a positive integer, then f is differentiable and

$$f'(x) = ax^{a-1}.$$

Proof. We proceed exactly as in Example 4 in Section 2. If we write $r - x$ as a difference of nth powers,

$$r - x = (r^{1/n})^n - (x^{1/n})^n,$$

then $r^{1/n} - x^{1/n}$ factors out and cancels against the numerator in the difference quotient, so

$$\frac{f(r) - f(x)}{r - x} = \frac{r^{1/n} - x^{1/n}}{r - x}$$

$$= \frac{1}{(r^{1/n})^{n-1} + (r^{1/n})^{n-2} x^{1/n} + \cdots + (x^{1/n})^{n-1}} .$$

The resulting expression is a continuous function of r, so its limit as r approaches x is its value at $r = x$ (supposing $x \neq 0$). Thus

$$f'(x) = \lim_{r \to x} \frac{r^{1/n} - x^{1/n}}{r - x} = \frac{1}{nx^{1-(1/n)}}$$

$$= ax^{a-1},$$

where $a = 1/n$. ∎

We conclude our discussion with a general fact that we shall have to use constantly, namely, that a *differentiable function is necessarily continuous*. This is intuitively clear: If a graph has a tangent line at a point, then it *cannot* be discontinuous there. The analytic proof is just a limit evaluation.

Theorem 5. *If f is differentiable at x, then f is continuous at x.*

Proof

$$\lim_{r \to x} [f(r) - f(x)] = \lim_{r \to x} \left[\frac{f(r) - f(x)}{r - x} \right] (r - x)$$

$$= \lim_{r \to x} \left[\frac{f(r) - f(x)}{r - x} \right] \cdot \lim_{r \to x} (r - x)$$

$$= f'(x) \cdot 0 = 0.$$

Therefore

$$\lim_{r \to x} f(r) = f(x),$$

so f is continuous at x. ∎

PROBLEMS FOR SECTION 4

Find dy/dx for each of the following functions:
1. $y = x^7$ 2. $y = 5x^{120}$ 3. $y = x^3 + 3x^2$

4. $y = x^3 + x - 1$ 5. $y = mx + b$ 6. $y = ax^2 + bx + c$

7. $y = 4x^4 - x^2 + 2$

8. $y = \dfrac{x^5}{5} + \dfrac{x^4}{4} + \dfrac{x^3}{3} + \dfrac{x^2}{2} + x + 1$

9. $y = x(x - 1)$

10. $y = (x - 2)(x + 3)$

11. $y = 4x^{1/2} + 2x^2 - 3x + 6$

12. Prove that if m is a positive integer then

$$\frac{d}{dx}x^{-m} = (-m)x^{(-m)-1}$$

(After a first step simplifying the difference quotient, the proof more or less copies that of Theorem 1.)

Using this new formula, differentiate the following functions:

13. $f(x) = x + \dfrac{1}{x}$ 14. $f(x) = x^3 + x^{-3}$

15. $f(x) = \dfrac{x^3 + 1}{x^2}$ 16. $f(x) = \dfrac{x + 1}{\sqrt{x}}$

17. If $f(x) = x - (1/x)$, prove that

$$f(x^2) = xf(x)f'(x).$$

18. Find the points on the graph of $y = x^3 + 3x^2 - 9x$ at which the tangent line is horizontal.

19. Sketch the graph of $y = x + (1/x)$ in the first quadrant. Find its lowest point. (Your sketch should suggest that the tangent line must be horizontal at the lowest point.)

20. The parabola $y = ax^2 + bx - 2$ is tangent to the line $y = 4x + 7$ at the point $(-1, 3)$. Find a and b.

21. Find the coordinates of the vertex of the parabola $y = x^2 - 6x + 1$. (Make use of the fact that the slope of the tangent to the curve is zero at the vertex.)

22. Find the points on the curve

$$y = x^3 + x^2$$

where the tangent has slope 1.

23. Find the equation of the line *normal* to the curve $y = x^2 + 3x + 2$ at the point where $x = 3$. (The normal line is perpendicular to the tangent.)

24. Prove that the tangents to the parabola $y = x^2$ at the ends of any chord through $(0, \frac{1}{4})$ are perpendicular to each other.

25. Find the vertex of the parabola

$$y = ax^2 + bx + c.$$

(See Problem 21.)

26. Show that the area of the triangle cut off in the first quadrant by a tangent line to the graph of $f(x) = 1/x$ is always 2, no matter what the point of tangency is. (Find the tangent line at the point $(a, 1/a)$ on the curve, then find the area of the triangle cut off, etc.)

27. Prove that the tangent line to the parabola $y = x^2$ at the point (x_0, y_0) has y-intercept $-y_0 (= -x_0^2)$.

28. Show that for any fixed point $(x_0, y_0) = (x_0, x_0^2)$ on the parabola $y = x^2$, the triangle with vertices (x_0, y_0), $(0, -y_0)$, and $(0, 1/4)$ is isosceles. (This is just a distance computation.)

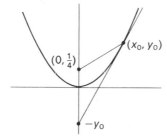

29. Assume that a ray of light reflects off a curve in such a way that the two rays make equal angles with the tangent line at the point of reflection. This is the law of reflection. Also assume the results of Problems 27 and 28. Prove that a point source of light placed at $(0, \frac{1}{4})$ will be reflected off the parabola $y = x^2$ to form a parallel beam of light shining vertically upward. (This is the principle of parabolic reflectors. The point $(0, 1/4)$ is called the focus of the parabola.)

30. Find the equation of the tangent line to the graph of $y = x^3$ at the point $(x_0, y_0) = (x_0, x_0^3)$. Then show that the tangent line intersects the graph again at the point where $x = -2x_0$.

31. Find the lines tangent to the parabola $y = x^2$ through the external point $(2,3)$. [*Hint:* Equate two expressions for the slope of the tangent line.]

32. Show that a line can be drawn through the point (a,b) tangent to the parabola $y = x^2$ if and only if $b \le a^2$. Interpret this requirement on (a,b) gometrically.

33. State the limit law (from Section 7 in Chapter 2) needed to make the proof of Theorem 2 logically complete.

34. Prove Theorem 3.

35. The proof of Theorem 4 used the continuity of a certain rather complicated function of r. How do we know this function is continuous?

36. Prove Theorem 4 again, in the manner of Example 5 in Section 2.

37. If $f(x) = x^{1/3}$, show that $f'(x) \to \infty$ as $x \to 0$. Sketch the graph of f.

38. Verify the product law

$$\frac{d}{dx} uv = u\frac{dv}{dx} + v\frac{du}{dx}$$

for the case $u = x^m$, $v = x^n$, where m and n are positive integers.

39. Suppose that the product law stated in the above problem holds for each pair of the four functions f, g, h, and k. Show that it holds for the product

$$(f + g)(h + k).$$

40. Determine the derivative of $f(x) = |x|$, and show that it can be expressed

$$\frac{d}{dx}|x| = \frac{|x|}{x}, \qquad x \neq 0.$$

41. Suppose that $f(x) = xg(x)$ on an interval I about the origin, and that $g(x)$ is continuous at $x = 0$. Prove that $f'(0)$ exists and equals $g(0)$.

42. Suppose that $-x^2 \leq f(x) \leq x^2$ on an interval I about the origin. Show that $f'(0)$ exists and has the value 0.

43. Suppose that $g(x) \leq f(x) \leq h(x)$ on an interval about x_0, and that

$$g(x_0) = h(x_0), \qquad g'(x_0) = h'(x_0).$$

Show that then $f'(x_0)$ exists and has the same value.

5. POWERS OF DIFFERENTIABLE FUNCTIONS

We have proved several special cases of the power rule

$$\frac{d}{dx}x^a = ax^{a-1},$$

in Sections 2 and 4. If we replace x by $g(x)$, each of these calculations proves the more general formula

$$\frac{d}{dx}[g(x)]^a = a[g(x)]^{a-1} \cdot g'(x).$$

Note the extra factor $g'(x)$. For example,

$$\frac{d}{dx}(1 + x^3)^2 = 2(1 + x^3) \cdot 3x^2$$

and

$$\frac{d}{dx}(1 + x^3)^{1/2} = \frac{1}{2}(1 + x^3)^{-1/2} \cdot 3x^2.$$

These more general calculations will all be subsumed under the chain rule in the next chapter, so in a sense this section is superfluous. However, one may feel more comfortable with the chain rule if independent proofs of some of its special cases have already been seen. Moreover, the general power rule is immediately useful.

First, a matter of notation. It is conventional to let g^a designate the ath power of the function g; its value at x is of course

$$g^a(x) = [g(x)]^a.$$

For example, g^2 is the square of g, so $g^2(x) = [g(x)]^2$. In this notation the general power formula is:

$$\frac{d}{dx}g^a(x) = ag^{a-1}(x) \cdot g'(x).$$

Setting $u = g(x)$ converts this to the simpler form:

$$\frac{d}{dx}u^a = au^{a-1}\frac{du}{dx}.$$

EXAMPLE 1. Prove the general power formula for the case $a = 3$:

$$\frac{d}{dx}u^3 = 3u^2\frac{du}{dx},$$

or

$$\frac{d}{dx}g^3(x) = 3g^2(x) \cdot g'(x).$$

. . . .

Solution. We proceed just as in Section 2, factoring a difference of cubes in order to simplify. If $f(x) = g^3(x)$, then

$$\frac{f(r) - f(x)}{r - x} = \frac{g^3(r) - g^3(x)}{r - x}$$

$$= \left[\frac{g(r) - g(x)}{r - x}\right][g^2(r) + g(r)g(x) + g^2(x)].$$

We know that a differentiable function is always continuous, so $g(r) \to g(x)$ as $r \to x$; and of course the difference quotient $(g(r) - g(x))/(r - x)$ approaches the derivative $g'(x)$. The right side above, therefore, has the limit

$$g'(x) \cdot 3g^2(x),$$

and this is $f'(x)$ by definition of the derivative.

Note that if $u = x$, then $du/dx = 1$ and we recapture the earlier formula for x^3.

The other special cases treated earlier generalize in exactly the same way. Thus, Examples 1 and 2 in Section 2 turn into the general power rule for

$$a = -1 \qquad \text{and} \qquad a = \frac{1}{2},$$

while Theorems 1 and 4 in Section 4 are the cases

$$a = \text{an arbitrary positive integer } m,$$

and

$$a = \text{the reciprocal of a positive integer,} \quad a = \frac{1}{n}.$$

EXAMPLE 2. For the case $a = -1$ we set

$$f(x) = g^{-1}(x) = \frac{1}{g(x)}$$

and see that

$$\frac{f(r) - f(x)}{r - x} = \frac{\dfrac{1}{g(r)} - \dfrac{1}{g(x)}}{r - x}$$

$$= -\frac{g(r) - g(x)}{r - x} \cdot \frac{1}{g(r)g(x)},$$

which approaches

$$-g'(x) \cdot \frac{1}{g^2(x)}$$

as $r \to x$. Therefore $f(x) = g^{-1}(x)$ is differentiable and its derivative is

$$-g^{-2}(x) \cdot g'(x). \quad \blacksquare$$

EXAMPLE 3. Find the line perpendicular to the curve

$$y = \sqrt{25 - x^2}$$

at the point $(x_0, y_0) = (3, 4)$.

. . . .

Solution. The general power formula for $a = 1/2$ tells us that

$$\frac{dy}{dx} = \frac{1}{2}(25 - x^2)^{-\frac{1}{2}}(-2x) = -\frac{x}{\sqrt{25 - x^2}}.$$

The tangent line at (3,4) therefore has the slope

$$m = -\frac{3}{\sqrt{25 - 9}} = -\frac{3}{4},$$

and the perpendicular line has slope

$$-\frac{1}{m} = \frac{4}{3}.$$

The perpendicular line thus has the point–slope equation

$$y - 4 = \frac{4}{3}(x - 3) \qquad \text{or} \qquad y = \frac{4}{3}x.$$

We can now do something new that will turn out in the next chapter to be a typical chain-rule procedure. We can apply the various powers already in hand *to each other*, and end up with the general power rule for any rational exponent a, i.e., any exponent a of the form $a = m/n$, where m and n are integers.

Theorem 6. *If u is a differentiable function of x and a is any rational number, then u^a is differentiable (wherever it is defined) and*

$$\frac{d}{dx}u^a = au^{a-1}\frac{du}{dx}.$$

Proof. Suppose first that $r = m/n$, where m and n are positive integers. Then

$$\frac{d}{dx}u^r = \frac{d}{dx}(u^m)^{1/n}$$

$$= \frac{1}{n}(u^m)^{(1/n)-1}\frac{d}{dx}u^m \qquad \text{(By the special case } a = 1/n\text{)}$$

$$= \frac{1}{n}u^{(m/n)-m}\cdot mu^{m-1}$$

$$= \frac{m}{n}u^{(m/n)-1} = ru^{r-1}.$$

This takes care of all *positive rational* exponents. If a is a *negative* rational number, then $a = -r$, where r is positive, and

$$\frac{d}{dx} u^a = \frac{d}{dx}\left(\frac{1}{u^r}\right) = -\frac{1}{(u^r)^2}\frac{d}{dx}(u^r) \qquad \text{(By Example 2)}$$

$$= -\frac{1}{u^{2r}} \cdot ru^{r-1} = -ru^{-r-1}$$

$$= au^{a-1}.$$

This finishes the proof of the theorem. ∎

Since we did not fully carry out the program of reproving the various special cases in more general form, the general power rule has not really been completely proved. You ought to be convinced of its validity, however, and the next chapter will develop a new rule that simultaneously covers all these generalizing steps.

PROBLEMS FOR SECTION 5

Compute dy/dx when:

1. $y = \sqrt{x^2 + 4x}$.

2. $y = \sqrt{x^2 + 4x + 4}$.

3. $y = 1/(x + 2)$.

4. $y = (1 - x)^{2/3}$.

5. $y = 1/(x^2 + 1)$.

6. $y = 1/\sqrt{x^2 + 1}$.

7. $y = \sqrt{1 - x^2}$.

8. $y = (x^2 + 1)^{1/3}$.

9. $y = (2x + 1)^{4/5}$.

10. $y = (3x^2 + 1)^{4/5}$.

11. $y = 1/(1 + x^3)^{1/3}$.

12. $y = \sqrt{x + (1/x)}$.

13. Prove from scratch that if $f(x) = \sqrt{g(x)}$, then

$$f'(x) = \frac{g'(x)}{2\sqrt{g(x)}}.$$

14. The half-hyperbola $y = \sqrt{4x^2 + 1}$ is asymptotic to the line $y = 2x$ in the first quadrant. Prove that its tangent line approaches the asymptote $y = 2x$ as its "limiting position," in the sense that its slope approaches 2 and its intercepts approach 0.

15. Prove that if n is a positive integer, then

$$\frac{d}{dx}g^n(x) = ng^{n-1}(x) \cdot g'(x)$$

by imitating the proof of Theorem 1 in the last section.

16. Prove that if $a = 1/m$, where m is a positive integer, then

$$\frac{d}{dx}g^a(x) = ag^{a-1}(x)g'(x),$$

by imitating the proof of Theorem 4 in the last section.

17. Find the smallest value of the function

$$f(x) = 16x^{2/3} + x^{-2/3}.$$

[Hint: The lowest point on the graph of f should have a horizontal tangent.]

18. Prove that a tangent to a circle is perpendicular to the radius drawn to the point of tangency.

19. Prove that the tangent line to the ellipse

$$\frac{x^2}{a^2} + \frac{y^2}{b^2} = 1$$

at the point (x_0, y_0) has the equation

$$\frac{xx_0}{a^2} + \frac{yy_0}{b^2} = 1.$$

20. Assuming the result in Problem 19, show that the normal line to the ellipse at (x_0, y_0) has the x-intercept $c^2 x_0/a^2$, where $c^2 = a^2 - b^2$.

21. Find the formula for the tangent line to the hyperbola

$$\frac{x^2}{a^2} - \frac{y^2}{b^2} = 1$$

analogous to that given in Problem 19 for the ellipse.

Compute $f'(x)$ when:

22. $f(x) = \sqrt{x + \sqrt{1 - x^2}}$.

23. $f(x) = (x^{2/3} + a^{2/3})^{3/2}$.

24. $f(x) = \dfrac{1}{x + \sqrt{x^2 - 1}}$.

25. $f(x) = 1/(x^2 + 2x + 1)$.

26. $f(x) = (x^a + 1)^{1/a}$

27. $f(x) = (1 + g(x))^{1/3}$.

28. $f(x) = 1/(x + g(x))$.

29. $f(x) = \sqrt{1 - (g(x))^2}$.

30. $f(x) = 1/(x^2 + g^2(x))$.

31. If $f(x) = (x^{1/3} + a^{1/3})^3$, show that

$$f'(x) = (1 + (\frac{a}{x})^{1/3})^2.$$

32. Verify by differentiating that $y = (c - x^2)^{1/3}$ satisfies the differential equation

$$3y^2\frac{dy}{dx} + 2x = 0,$$

for any constant c.

6. INCREMENTS

When we consider the difference quotient $(f(r) - f(x))/(r - x)$, it is often useful to think of r as a second value of x and the difference $(r - x)$ as a *change in x*. The classical notation for such a change is Δx, read "delta x". (Δ is capital δ.) Here Δx is not the product of a number Δ times a number x, but a wholly new numerical variable, called the *increment* in x. Its value is always to be thought of as the *change* in the value of x, from a *first* value to a *second* value, or from an *old* value to a *new* value. The most consistent notation in this situation would be to use subscripts on x, say x_0 and x_1, for the old and new values respectively. Then

$$\Delta x = x_1 - x_0$$

is the change in x in going from the old to the new value. We can also consider the new value as being obtained from the old value by adding the change:

$$x_1 = x_0 + \Delta x.$$

An increment Δx can be of either sign. Thus if $x_0 = 2$ and $\Delta x = -3$, then $x_1 = 2 + (-3) = -1$.

As often as not we discard the zero subscript, so that the old and new values are x and $x + \Delta x (= x_1)$. That is, we use x itself to represent the old value of x. This may not seem quite cricket, but it works out all right.

Now suppose that y is a function of x, $y = f(x)$. When x changes from x to $x + \Delta x$, then y changes from $f(x)$ to $f(x + \Delta x)$. The increment in y thus has the basic formula

$$\Delta y = f(x + \Delta x) - f(x).$$

EXAMPLE 1. If $y = x^2$, then

$$\Delta y = (x + \Delta x)^2 - x^2 = 2x\Delta x + (\Delta x)^2 = \Delta x(2x + \Delta x).$$

EXAMPLE 2. When $y = x^3$, we have

$$\Delta y = f(x + \Delta x) - f(x) = (x + \Delta x)^3 - x^3$$
$$= x^3 + 3x^2 \Delta x + 3x(\Delta x)^2 + (\Delta x)^3 - x^3$$
$$= \Delta x (3x^2 + 3x\Delta x + (\Delta x)^2).$$

These examples show that increment notation may change the way in which a difference factors. Often it makes the simplification more routine in nature. In Example 2 the simplification follows after *expanding the cube* $(x + \Delta x)^3$, whereas before it depended on *factoring* $r^3 - x^3$. Generally speaking, factoring is more elegant, but multiplication is easier and more automatic. And in many other situations the use of increments makes the steps to be taken more or less obvious, whereas the form $f(r) - f(x)$ may require ingenuity in its treatment.

In terms of increments, the difference quotient $(f(r) - f(x))/(r - x)$ becomes

$$\frac{f(x + \Delta x) - f(x)}{\Delta x} = \frac{\Delta y}{\Delta x},$$

which tells us, literally, that the difference quotient is the *change in y divided by the change in x*. The derivative can now be expressed

$$\frac{dy}{dx} = \lim_{\Delta x \to 0} \frac{\Delta y}{\Delta x}.$$

(Although of later origin, the Δ symbolism obviously fits well with Leibniz's symbolism.)

From the above expression for dy/dx we can obtain

$$\lim_{\Delta x \to 0} \Delta y = \lim_{\Delta x \to 0} \left(\frac{\Delta y}{\Delta x} \right) \Delta x$$
$$= \left(\lim_{\Delta x \to 0} \frac{\Delta y}{\Delta x} \right) \cdot \left(\lim_{\Delta x \to 0} \Delta x \right) = \frac{dy}{dx} \cdot 0 = 0.$$

This is just a reformulation in terms of increments of our earlier proof that a differentiable function is continuous.

EXAMPLE 3. We recompute the derivative formula for $y = x^3$ in increment notation. From Example 2,

$$\frac{\Delta y}{\Delta x} = 3x^2 + 3x\Delta x + (\Delta x)^2.$$

Therefore

$$\frac{dy}{dx} = \lim_{\Delta x \to 0} \frac{\Delta y}{\Delta x} = 3x^2 + 3x \cdot 0 + 0^2 = 3x^2.$$

PROBLEMS FOR SECTION 6

Recompute the derivatives of the following functions by the method of increments:

1. $f(x) = x^2$.
2. $f(x) = 1/x$.
3. $f(x) = \sqrt{x}$.
4. $f(x) = x^4$.
5. $f(x) = x + x^2$.
6. $f(x) = 1/(1 + x)$.
7. $f(x) = 1/(1 + x^2)$.

8. If $y = x^2$ and x is positive, then y can be interpreted as the area of a square of side x. Then Δy is the *increase in area* when the side is increased by Δx. Draw a figure and interpret the two terms of Δy as areas.

9. If $y = x^3$ and x is positive, then y can be interpreted as the volume of a cube of side x. Then Δy is the *increase in volume* when the side is increased by Δx. Draw a figure and interpret the three terms of Δy as volumes.

10. Write out the proof of Theorem 2 in Section 4 in increment notation. (Note first that if $u = g(x)$, $v = h(x)$, and $y = u + v$, then $\Delta y = \Delta u + \Delta v$.)

7. VELOCITY

Suppose that a particle is moving along a coordinate line. At each instant of time t the particle has a unique position coordinate s, so its position s is a function of t,

$$s = f(t).$$

If the particle is you in your car, and if it takes you half an hour to drive ten miles through a city, then you say that your average velocity is 20 miles per hour through the city, because

$$\frac{\text{Distance travelled}}{\text{Elapsed time}} = \frac{10}{1/2} = 20.$$

In general,

$$\text{Average velocity} = \frac{\text{Distance travelled}}{\text{Elapsed time}} = \frac{\Delta s}{\Delta t},$$

a formula that ought to be in agreement with your intuitive notion of average velocity. Since

$$\frac{\Delta s}{\Delta t} = \frac{f(t + \Delta t) - f(t)}{\Delta t},$$

average velocity is the difference quotient of position as a function of time.

EXAMPLE 1. An object falling from rest falls

$$s = 16t^2$$

feet in t seconds. Here the position function is $f(t) = 16t^2$, measured downward from the point of release. The average velocity of the object during the first second of fall is

$$\frac{\Delta s}{\Delta t} = \frac{f(1) - f(0)}{1} = \frac{16 \cdot 1^2 - 16 \cdot 0^2}{1} = 16 \text{ feet per second.}$$

During the *second* second it is

$$\frac{\Delta s}{\Delta t} = \frac{f(2) - f(1)}{1} = \frac{16 \cdot 2^2 - 16 \cdot 1^2}{1} = \frac{64 - 16}{1} = 48 \text{ ft/sec.}$$

Its average velocity during the *first two seconds* is

$$\frac{\Delta s}{\Delta t} = \frac{16 \cdot 2^2 - 16 \cdot 0^2}{2} = \frac{64}{2} = 32 \text{ ft/sec.}$$

There is also a direct intuitive notion of *instantaneous velocity* v. It is the number that measures how fast something is moving at a given instant. It is the varying reading on a perfect speedometer. In order to identify the instantaneous velocity v, let v_0 be its value at time t_0 and let Δt be a very small change in t. Then over the time interval between t_0 and $t_0 + \Delta t$, the velocity v varies only slightly from v_0, and its average value must therefore be very close to v_0. That is,

$$\frac{\Delta s}{\Delta t} \approx v_0,$$

where $\approx$ is read "is approximately equal to." For smaller Δt, the difference between initial value and average value is smaller still. Therefore,

$$v_0 = \lim_{\Delta t \to 0} \frac{\Delta s}{\Delta t} = \frac{ds}{dt}\bigg|_{t_0}.$$

That is, we see on intuitive grounds that the instantaneous velocity v must be the limit of the average velocity $\Delta s/\Delta t$ as the length of the time interval over which the average is computed tends to zero. So v is given by the derivative

$$v = \frac{ds}{dt}.$$

We thus have a second interpretation of the derivative. For Newton, who was struggling to understand the motions of the planets and other heavenly bodies, this was undoubtedly the major interpretation of the derivative.

EXAMPLE 2. If $s = 3t - 10$, where s is measured in feet and t in seconds, then $v = ds/dt = 3$ feet per second. The particle is moving in *uniform motion* with constant velocity $v = 3$ ft/sec.

EXAMPLE 3. If the motion of the particle is described by $s = t^3 - t$, in the same units as above, then the velocity v is

$$v = \frac{ds}{dt} = (3t^2 - 1) \text{ feet per second.}$$

Thus, at time $t = 2$, its velocity is $v = 3 \cdot 2^2 - 1 = 11$ ft/sec, and at time $t = 0$ it is $v = 3 \cdot 0^2 - 1 = -1$ ft/sec. Note that the velocity is negative at $t = 0$. Thus $\Delta s/\Delta t < 0$ for small Δt and the particle is moving *backward* at time $t = 0$. If the particle was moving and the clock running before the zero setting on the clock, we see that at time $t = -1$ the velocity was $v = 3(-1)^2 - 1 = 2$.

Although the particle in the above example is moving backward and forward on the line, the *graph* of the motion is the graph of the equation $s = t^3 - t$ in the ts -plane.

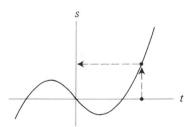

The particle is *not* moving along this curve, but along the s-coordinate line, pictured in this figure as the vertical axis. Its position at time t is obtained from the graph by the "up-and-over" procedure indicated in the figure. (The graph is called the "world line" of the particle in space–time, space being one-dimensional here.)

We say that a particle is *accelerating* if its velocity is increasing. Its *average acceleration* over an interval of time is given by the formula

$$\text{Average acceleration} = \frac{\text{Change in velocity}}{\text{Elapsed time}} = \frac{\Delta v}{\Delta t}.$$

Its *instantaneous acceleration a* is defined to be the limit of its average acceleration as the length of the time interval over which the average is computed tends to zero. The acceleration of the particle is thus the derivative of its velocity:

$$a = \lim_{\Delta t \to 0} \frac{\Delta v}{\Delta t} = \frac{dv}{dt}.$$

In the above example we had $v = 3t^2 - 1$ and so the acceleration is

$$a = \frac{dv}{dt} = 6t.$$

The particle is moving with increasing acceleration.

Since the velocity is so many feet per second, the acceleration, which is the *change in velocity per unit of time*, must be so many (feet per second) per second. We designate this ft/sec². Thus in the problem above the acceleration was $6t$ ft/sec².

EXAMPLE 4. An object is thrown vertically upward, and its position above the ground (after t seconds have elapsed) is given by $s = 50t - 16t^2$ feet. Find its initial velocity, its highest point, and its velocity when it strikes the ground, as well as its acceleration.

. . . .

Solution. The velocity is

$$v = \frac{ds}{dt} = 50 - 32t \text{ ft/sec .}$$

The initial velocity v_0 is the velocity at time $t_0 = 0$. Thus

$$v_0 = 50 \text{ ft/sec .}$$

The highest point will be reached just when the upward velocity has fallen away to zero. Setting $v = 0$, we have

$$0 = 50 - 32t,$$

$$t = \frac{50}{32} = \frac{25}{16} \text{ sec .}$$

The particle returns to ground at the moment when s becomes 0 again, i.e., when

$$0 = s = 50t - 16t^2.$$

This gives $t = 50/16$ seconds, and then

$$v = 50 - 32t = 50 - 32 \cdot \frac{50}{16}$$

$$= -50 \text{ ft/sec .}$$

The acceleration of the object is

$$a = \frac{dv}{dt} = \frac{d}{dt}(50 - 32t) = -32 \text{ ft/sec}^2 .$$

The acceleration is constant and it is *negative*. The velocity is *decreasing* at a constant rate.

PROBLEMS FOR SECTION 7

1. A particle moves along the x-axis and its position as a function of time is given by $x = t^3 - 2t^2 + t + 1$. Find its velocity. At what moments is it standing still? When is it moving forward and when backward?

2. A particle is moving along a horizontal line on which the positive direction is to the right. The equation of motion of the particle is $s(t) = t^3 - 9t^2 + 24t + 1$.

   a) What is the velocity of the particle at $t = 1$?

   b) Is the particle moving to the right or left at $t = 1$?

   c) What is the acceleration of the particle at $t = 1$?

   d) Is the speed of the particle increasing or decreasing at $t = 1$? Explain your reasoning.

   e) At what times, if any, is the velocity of the particle zero?

3. A ball thrown directly upward with a speed of 96 ft/sec moves according to the law

$$y = 96t - 16t^2,$$

   where y is the height in feet above the ground, and t is the time in seconds after it is thrown. How high does the ball go?

4. The movement of an airplane from the time it releases its brakes until takeoff is governed by the equation $s = 2t^2$, where s is the distance from the starting point in feet and t is the time in seconds since brake release.

   a) If takeoff velocity is 120 mph (176 fps), determine the time elapsed from brake release until takeoff.

   b) What is the distance covered during the time?

5. A coin is thrown straight up from the top of a building which is 300 feet tall. After t seconds, its position is described by $s = -16t^2 + 24t + 300$, where s is its height above the ground in feet. When does the coin begin to descend? What is its velocity when it is 305 feet above the ground?

6. In the first t seconds after an Apollo space shot is launched, the rocket reaches a height of $40\,t^2$ feet above the earth.

a) What is the rocket's velocity after 5 seconds?

b) If the speed of sound is 1050 ft/sec, calculate the height at which the space craft attains supersonic speed.

7. Given that the motion of a particle is expressed by the equation

$$s(t) = t^3 - 6t^2 + 2,$$

a) determine the velocity of the particle at $t = 2.3$;

b) determine the acceleration of the particle at 2.3;

c) determine the *average* velocity of the particle in the time interval $t = 1.3$ to $t = 3.3$.

8. The height above the ground of a bullet shot vertically upward with an initial velocity of 320 ft/sec is given by $s(t) = 320t - 16t^2$, where $s(t)$ is the height in ft and t is time in sec.

a) Determine the velocity of the bullet 4 sec after it is fired.

b) Determine the time required for the bullet to reach its maximum height, and the maximum height attained.

9. A ball is thrown down from the top of a 200 ft tower with an equation of motion

$$y = -16t^2 - 40t + 200.$$

What is the *average* velocity of the ball in its trip to the ground? At what time t does the velocity of the ball equal its average velocity?

10. A particle whose velocity is zero during an interval of time is standing still during that time interval. Restate this obvious fact as a theorem about differentiable functions.

11. The distance, rate, and time problems of elementary algebra refer to motions of constant velocity. The formula $d = rt$ would then be restated in our language as

$$\Delta s = r\,\Delta t,$$

where r is the constant velocity. Show that the most general description of such a motion is given by

$$s = rt + c,$$

where c is any constant.

8. HIGHER DERIVATIVES

The derivative f' may again be a differentiable function and we naturally use the notation f'' for *its* derivative $(f')'$. For example, if $f(x) = x^5$, then

$$f'(x) = 5x^4, \quad \text{and} \quad f''(x) = \frac{d}{dx}f'(x) = \frac{d}{dx}5x^4 = 20x^3.$$

Then $f'''(x) = 60x^2$, $f''''(x) = 120x$, $f'''''(x) = 120$, and $f''''''(x) = 0$.

When the number of primes begins to be unwieldy, we use a numeral in parentheses instead. Thus,

$$f^{(4)}(x) = f''''(x) = 120x, f^{(5)}(x) = 120, \text{ and } f^{(6)}(x) = 0.$$

in the above example.

The Leibniz notation is

$$\frac{d^2y}{dx^2} = \frac{d}{dx}\left(\frac{dy}{dx}\right).$$

Note that the superscript 2 is on d in the upper part of the term, and on dx in the lower, and that this is consistent with the right side of the equation. In Leibniz notation, the above example is

$$\frac{d}{dx}x^5 = 5x^4, \quad \frac{d^2}{dx^2}x^5 = 20x^3, \quad \frac{d^3}{dx^3}x^5 = 60x^2,$$

$$\frac{d^4}{dx^4}x^5 = 120x, \quad \frac{d^5}{dx^5}x^5 = 120, \quad \frac{d^6}{dx^6}x^6 = 0.$$

The higher-order derivatives of f play an important role in calculus and its applications. The second derivative d^2y/dx^2 is particularly important because it can be directly interpreted. We shall see in Chapter 5 that the sign of d^2y/dx^2 determines which way the graph of $y = f(x)$ is turning. Later on, in Chapter 11, this qualitative meaning of the second derivative will be sharpened into a formula for the *curvature* of a graph.

On the other hand, if $s = f(t)$ is the position coordinate of a particle at time t, then $d^2s/dt^2 = dv/dt$ is the *acceleration* of the particle, and this interpretation is the cornerstone of the mathematical study of motion. Newton's second law of motion says that the acceleration of a body is proportional to the force acting on it (the constant of proportionality being the mass), and this relationship is a *second-order differential equation* that determines the motion from the known forces. This application will be touched on in Chapters 6, 14 and 19. (There are whole books devoted to Newtonian dynamics.)

PROBLEMS FOR SECTION 8

Compute d^2y/dx^2 when:

1. $y = 2x$ 2. $y = 2x^2$ 3. $y = 1/x$

4. $y = x^{1/3}$ 5. $y = x^3 + 3/x$ 6. $y = x^n$

7. $y = x^{-3}$ 8. $y = x^{-n}$ 9. $y = ax + b$

10. $y = x^4 - x^2$ 11. $y = \sqrt{x}$ 12. $y = 1/(1 - x)$

13. What is d^4y/dx^4 if $y = x^3$?

14. The symbol $n!$ (n factorial) represents the product
$$n(n - 1)(n - 2) \cdots 2 \cdot 1.$$
Thus $4! = 4 \cdot 3 \cdot 2 \cdot 1 \cdot = 24$. Show that
$$\frac{d^n}{dx^n} x^n = n!$$
for $n = 1, 2, 3, 4, 5$.

15. Show that if the above formula holds for $n = m$, it holds for $n = m + 1$. It must therefore hold for every value of n. Why?

16. Find the formula for $d^n y/dx^n$ when $y = 1/x$.

17. If $f(x) = x^{3/2}$, show that $f''(x)f(x) = \frac{3}{4}x$.

18. A body dropped from rest falls $s = 16t^2$ feet in t seconds. What is its acceleration?

19. A particle moves along a line, its position coordinate at time t being
$$s = t^3 - t^2 + 2t.$$
If $t = 0$ is the initial time, what are the initial velocity and initial acceleration of the particle? At what moment does the particle stop decelerating and start accelerating?

20. Suppose that $f(x) = g(x) + h(x)$. Show that if the formula
$$f^{(n)}(x) = g^{(n)}(x) + h^{(n)}(x)$$
holds for some value of n, say $n = m$, then it holds for the next value $n = m + 1$. Then use this result to prove that the formula holds for every value of n.

9. ELEMENTARY GRAPH SKETCHING

If we draw some random smooth graphs and a few tangent lines to them, we are led to certain conclusions about the relationship between the behavior of f' and the behavior of f.

First, it seems clear that a graph with positive slope is rising and that a graph with negative slope is falling.

Second, it seems clear that a graph can change from sloping up to sloping down only by going over a summit point where the slope is zero. Similarly, it can change from sloping down to sloping up only by going through a bottom point where the slope is zero.

Combining these two observations we conclude that if the slope is never zero then the graph cannot change directions, and is either everywhere rising or everywhere falling. A function whose graph behaves this way is said to be *monotone*. Again, a function is monotone if its graph moves in only one direction, either always up or always down. Our combined conclusion is thus:

If f is a differentiable function on an interval I and if f' is never zero on I, then f is monotone on I.

This "shape principle" tells us that the values of x for which $f'(x) = 0$ are the critical values in determining the overall behavior of f, since $f(x)$ moves in a constant direction on each of the intervals marked off by such values. This idea enables us to find the general overall shape of a graph with a minimum of calculation.

EXAMPLE 1. Consider $f(x) = (x^3 - 3x + 2)/3$. The critical points of f are the values of x for which $f'(x) = 0$. Since

$$f'(x) = \frac{3x^2 - 3}{3} = x^2 - 1,$$

the critical points of f are the roots of $x^2 - 1 = 0$, i.e., the two points $x = -1, +1$.

We make a little table showing the values of f at the critical points and we add the behavior of f at $\pm\infty$. We have

$$f(-1) = \frac{-1 + 3 + 2}{3} = \frac{4}{3} \quad \text{and} \quad f(1) = \frac{1 - 3 + 2}{3} = 0.$$

Also, since $f(x)$ behaves like its highest power $x^3/3$ for large x, we see that $f(x) \to -\infty$ as $x \to -\infty$, and $f(x) \to +\infty$ as $x \to +\infty$. Our table is thus

x	$-\infty$	-1	1	$+\infty$
$f(x)$	$-\infty$	$4/3$	0	$+\infty$

We now apply the shape principle. Since $f'(x)$ is never zero in the interval $(-1, 1)$, the graph of f must be either always rising over $(-1, 1)$ or always falling over $(-1, 1)$. We see from our table that it must be falling since $f(x)$ goes from $4/3$ to 0 as x runs from -1 to $+1$. In the same way, we see from the shape principle and the table that the graph is rising over the whole interval $(-\infty, -1)$, and also over $(1, \infty)$. Schematically, the graph is therefore like this:

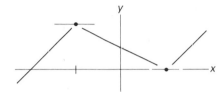

Now consider how we would actually sketch the graph from this information. There aren't going to be any sharp corners; the graph is going to be as smooth and simple as possible subject to the above conditions. It changes from rising to falling as it passes over $x = -1$, and so is smoothly turning downward there. (In Chapter 5 we shall consider what this really means.) Similarly, it is smoothly turning upward as it passes over $x = +1$. It cannot contain any flat stretches; only a linear graph $y = mx + b$ can do that. Therefore the graph is always turning, one way or the other, except at isolated points where it changes its direction of turning. We need only one such point and we guess $x = 0$. This gives us the smoothest, simplest graph.

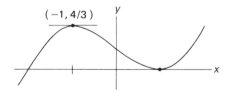

Note that the lefthand x-intercept is shown at $x = -2$. We didn't discuss how to see this; it amounts to being able to factor the original cubic. The roots of $f(x) = 0$ are always valuable graphing aids when they can be found, but the location of roots can be very difficult. As a matter of fact, we sometimes graph a function to help us guess approximately where its roots are, as a preliminary step to a calculation procedure for the roots.

EXAMPLE 2. Sketch the graph of $f(x) = x^4 - 2x^2$ in the above way.

. . . .

Solution. We have $f'(x) = 4x^3 - 4x = 4x(x^2 - 1)$, giving the critical points $x = -1, 0, +1$. When $|x|$ is very large, $f(x)$ behaves like x^4 and so approaches $+\infty$ as x approaches $-\infty$ or $+\infty$. We thus have the following determining table.

x	$-\infty$	-1	0	$+1$	∞
$f(x)$	∞	-1	0	-1	∞

      decr.   incr.   decr.   incr.

In this case also we can find the x-intercepts, i.e., the solution of $x^4 - 2x^2 = 0$, which are $-\sqrt{2}, 0, \sqrt{2}$. Plotting the critical points and intercepts, we get the graph below.

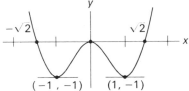

$(-1, -1)$ $(1, -1)$

Another method for discovering which way $f(x)$ is going on one of its intervals of monotonicity is to compute the value of $f'(x)$ *at one point* in the interval, because that will tell us the constant sign of f' over the whole interval and hence which way f is going there.

Consider the above example from this point of view.

We have to calculate the critical values of f in order to draw the graph, and they already determine which way f is moving on $(-1, 0)$ and on $(0, +1)$. But for $(-\infty, -1)$ we could replace the evaluation of the behavior of f at $-\infty$ by the evaluation

$$f'(-2) = 4x(x^2 - 1)|_{x=-2} = -8 \cdot 3 = -24.$$

Since this is negative, $f'(x)$ is negative on the whole interval $(-\infty, -1)$ and the graph of f is falling there. Similarly, the evaluation $f'(2) = 24$ shows that f must be increasing over $(1, \infty)$.

PROBLEMS FOR SECTION 9

Use calculus to sketch the graphs of the following equations:

1. $y = x^2 - x$.   2. $y = 2 - x - x^2$.   3. $y = x^2 - 4x + 4$.

4. $y = x^2 - 2x + 2.$ 5. $s = t^3 - t.$ 6. $y = x^4 - 3x^2 + 2.$

7. $u = 3v^4 + 4v^3.$

8. $y = x^3 - 4x^2 - 3x.$

9. $y = x^3 - 3x^2 + 3x.$

10. $y = x^3 - 4x^2 + 3x.$

11. $y = x + \dfrac{1}{x}.$ (Be careful around the origin.)

12. $y = x^2 - 3x^{2/3}.$ (Same warning.)

13. $3y = 3x^5 - 10x^3 + 15x + 3.$

14. $y = 3x^4 - 8x^3 + 6x^2 + 1.$

15. Discuss the graph of

$$y = ax^2 + bx + c$$

in light of the shape principle.

16. Sketch the graph of a function f defined for positive x and having the properties

$$f(1) = 0,$$
$$f'(x) = 1/x, \qquad \text{all } x > 0.$$

10. THE MEAN-VALUE PROPERTY

Consider again a particle moving along a coordinate line with position coordinate given by $s = f(t)$. We agreed that its instantaneous velocity is given by $ds/dt = f'(t)$, and that the average velocity over the time interval $[t_0, t_1]$ is

$$\frac{\Delta s}{\Delta t} = \frac{f(t_1) - f(t_0)}{t_1 - t_0}.$$

We assume that the velocity varies continuously with t, so as it varies from above its average value to below (or vice versa), it will cross the average value at some time $t = T$. That is, there must be some number T between t_0 and t_1 such that

$$f'(T) = \frac{f(t_1) - f(t_0)}{t_1 - t_0}.$$

We thus have an analytic conclusion arising from the velocity interpretation of the derivative. It is called the *mean-value property* of a differentiable function. (Average values used to be called mean values.)

The above plausibility argument is not a real proof, and it may not even seem very convincing. But the following *geometric* principle *is* intuitively obvious:

If we draw a secant line through two points on a smooth graph, then there is at least one point in between where the tangent line is parallel to the secant.

We visualize moving the secant line parallel to itself, and we see that it turns into the tangent line at its last point of contact with the graph.

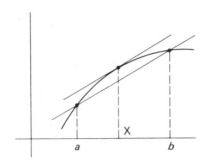
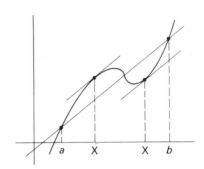

When we state this analytically as an equality of two slopes, it is again the mean-value property.

The mean-value property thus appears to be true in the contexts of the two principal interpretations of the derivative. On this basis we shall assume it for the time being, i.e., treat it as a temporary axiom.

Mean-value property. *If f is differentiable on the closed interval [a,b], then there is at least one point X strictly between a and b at which*

$$f'(X) = \frac{f(b) - f(a)}{b - a}.$$

Replacing a and b by x and $x + \Delta x$, we have the increment form

$$f'(X) = \frac{f(x + \Delta x) - f(x)}{\Delta x} = \frac{\Delta y}{\Delta x},$$

where X lies strictly between x and $x + \Delta x$. Also useful are the cross-multiplied forms:

$$\Delta y = f'(X)\Delta x;$$

$$f(x + \Delta x) = f(x) + f'(X)\Delta x.$$

It is clear from the above figure that there may be more than one such point X. Moreover, the mean-value theorem says nothing about where X is, except that it lies somewhere between a and b. One might feel that anything so vague as this couldn't be very helpful, but we shall see that it is extremely useful. In the usual context, $f'(x)$ is known to have some property throughout the interval (a,b), and it therefore has this property at the point X, no matter where it is. So the mere fact that there must *be* such a point X is often enough to get us off the ground. We shall give two important examples, here and in the next section, to show how this works, but most of the applications of the mean-value property will come later.

Theorem 7. *If f' exists and is everywhere zero on an interval I, then f is a constant function on I. That is, there is a constant c such that $f(x) = c$ for every x in I.*

This is geometrically obvious. If every tangent line to a graph is horizontal, then the graph can never rise or fall, and so must be a horizontal straight line.

Proof. Analytically, the theorem is an immediate corollary of the mean-value property. Choose any point x_0 in I and set $c = f(x_0)$. For any other x the mean-value property says that

$$f(x) - f(x_0) = f'(X)(x - x_0),$$

where X is some number between x_0 and x. But since f' is everywhere 0 this shows that $f(x) - f(x_0) = 0$, and so

$$f(x) = f(x_0) = c$$

for all x. ∎

This fact is crucially important when it comes to reversing the differentiation procedure. We shall see a simple example of its use in Section 12.

PROBLEMS FOR SECTION 10

For each of the following functions and the given values of a and b, find a value for X such that $f'(X) = (f(b) - f(a))/(b - a)$.

1. $f(x) = x^2 - 6x + 5$, $a = 1$, $b = 4$.

2. $f(x) = 3x^2 + 4x - 3$, $a = 1$, $b = 3$.

3. $f(x) = x^3$, $a = 0$, $b = 1$.

4. $f(x) = -\dfrac{2}{x}$, $a = 1$, $b = 3$.

5. $f(x) = x^{-2}$, $a = 1$, $b = 2$.

6. Show that if you average 40 miles per hour on a trip from Chicago to St. Louis, then for at least one instant during your trip your speedometer reads 40 miles per hour.

7.   Given that $f(x) = (x + 2)/(x + 1)$ and $a = 1$, $b = 2$, find all values X in the interval $(1,2)$ such that

$$f'(X) = \frac{f(b) - f(a)}{b - a}.$$

8.   Prove that a particle whose velocity is zero during an interval of time is standing still during that time interval.

9.   Prove that if $f'(x) = 3x^2$, then $f(x) = x^3 + c$ for some constant c. [*Hint.* What is $(f - g)'$ if $g(x) = x^3$?]

10. Prove that if f and g have the same derivative on an interval I, then $f(x) = g(x) + c$ for some constant c.

11. Prove that if f' is constant, say $f' = m$, then $f(x) = mx + b$.

12. What is the most general function f such that $f''(x) = x$? [*Hint.* Use Problem 10 twice, first finding what f' must be, and then f.]

13. A body moving vertically under the influence of gravity has constant acceleration $d^2s/dt^2 = -32$ ft/sec^2 (if the positive s direction is upward).

a) Show that its motion must be of the form

$$s = -16t^2 + bt + c$$

for some constants b and c.

b) Determine the motion explicitly if $s = 0$ and

$$v = \frac{ds}{dt} = 50 \text{ ft/sec}$$

when $t = 0$.

14. If $f(x) = x^{1/3}$, find the unique number X (depending on x) for which

$$f(x) - f(0) = f'(X)(x - 0).$$

15. If $f(x) = x^a (a \neq 1)$, find the unique number X (depending on x) for which

$$f(x) - f(0) = f'(X)(x - 0).$$

16. Show that there is at most one function f defined on the interval $(0, \infty)$ such that

$$f(1) = 0 \quad \text{and} \quad f'(x) = 1/x$$

for all x. (Show that if f and g both have these properties, then $f(x) = g(x)$ for all x.)

17. Show that a function f is uniquely determined on an interval I if we know its derivative f' and its value at one point. That is, if $f' = g'$ on I and if $f(x_0) = g(x_0)$, then $f = g$ on I.

11. SOME ELEMENTARY ESTIMATION

We shall now apply the mean-value property directly to some elementary calculations. In practical computations we almost always replace each number by a finite decimal that approximates it. For example, we know that

$$3.14 < \pi < 3.15,$$

and on the basis of this inequality we know that each of 3.14 and 3.15 approximates π wth an error less than 0.01, i.e., with an error less than 1 in the second decimal place.

Let us be clear about what we are saying here. When we regard a number x as an *approximation of a number* a, we call the difference $(a - x)$ the *error in the approximation*; it is *how much we miss the mark by*. The choice of sign is arbitrary, and some people would say that $(x - a)$ is the error; but since we are usually concerned with the *magnitude* of the error, $|a - x|$, this ambiguity in sign is seldom important.

Now if we subtract 3.14 from all terms of the inequality

$$3.14 < \pi < 3.15,$$

we obtain the inequality

$$0 < \pi - 3.14 < 0.01,$$

and this says *exactly* that the error in the approximation of π by 3.14 is positive and less than .01. If we subtract 3.15 instead, we get

$$-.01 < \pi - 3.15 < 0,$$

So this time the error is *negative* but less than .01 in magnitude. These conclusions depend on a basic (and familiar) inequality law:

If $x < y$, *then* $x + c < y + c$, *for any number c.*

Other inequality laws also will be needed in these examples. They are mostly in accord with common sense, and will be applied without comment, but you can check Appendix 1 later in case of doubt.

Actually, 3.14 is the two-place decimal *closest* to π, i.e., the *best* approximation of π by a two-place decimal. This means that

$$|\pi - 3.14| < 0.005,$$

and it is a consequence of the inequality

$$3.141 < \pi < 3.142.$$

(This is what we mean when we write $\pi = 3.141 \cdots$) This inequality shows, in particular, that

$$3.14 < \pi < 3.142,$$

so 3.14 approximates π with an error less than 0.002.

We don't want to discuss "rounding off" here, so we shall adopt the following temporary definition:

The number x approximates the number a correctly to two decimal places if $|x - a| < 0.005$. More generally, the approximation is correct to n decimal places if

$$|x - a| < \frac{10^{-n}}{2} \qquad (= 5 \cdot 10^{-(n+1)}).$$

Most of the following examples can also be worked out by algebra, but this will not be possible for our later and more sophisticated applications of the mean-value property to computations.

EXAMPLE 1. Show that $(3.14)^3$ approximates π^3 with an error less than 1 in the first decimal place, i.e., with an error less than 0.1. Use the inequality $\pi - 3.14 < 0.002$ that we noted above.

. . . .

Solution. We use the mean-value property with $f(x) = x^3$ and the x-values 3.14 and π. It guarantees a number X lying between these two x values such that

$$\pi^3 - (3.14)^3 = f(\pi) - f(3.14) = f'(X)(\pi - 3.14) = 3X^2(\pi - 3.14).$$

Since $X < 4$ and $\pi - 3.14 < 0.002$, we can conclude that $\pi^3 - (3.14)^3 < 48(0.002) = 0.096 < .1$, as we claimed.

EXAMPLE 2. A table gives $a = 1.414$ as the value of $\sqrt{2}$ to the nearest three decimal places. Justify this, using the value $(1.414)^2 = 1.999396$.

. . . .

Solution A. We use the mean-value property with $f(x) = x^2$ and the x values $\sqrt{2}$ and 1.414. It guarantees a number X lying between these X values such that

$$2 - (1.414)^2 = f'(X)(\sqrt{2} - 1.414)$$
$$= 2X(\sqrt{2} - 1.414).$$

Since $X > 1.4$, and $2 - (1.414)^2 = 2 - 1.999396 = 0.000604$, we conclude that

$$\sqrt{2} - 1.414 = \frac{0.000604}{2X} < \frac{0.000604}{2.8} < 0.0003,$$

which meets our criterion for three-decimal-place accuracy.

. . . .

Solution B. We use $g(x) = \sqrt{x}$ and the x-values 2 and 1.999396. Then there is a number X lying between these two x values such that

$$\sqrt{2} - 1.414 = g(2) - g(1.999396) = g'(X)[2 - 1.999396]$$

$$= \frac{1}{2\sqrt{X}}(0.000604) < \frac{0.000604}{2.8} < 0.0003,$$

which is the same conclusion as before.

EXAMPLE 3. Show that $1/3.14$ approximates $1/\pi$ accurately to three decimal places. Use the fact that $\pi - 3.14 < 0.002$.

. . . .

Solution A. This can be done by algebra as easily as by calculus:

$$\frac{1}{3.14} - \frac{1}{\pi} = \frac{\pi - 3.14}{3.14\pi} < \frac{0.002}{3.14\pi} < \frac{0.002}{9} < 0.0003.$$

Since this is less than 0.0005, the approximation is accurate to three decimal places.

. . . .

Solution B. We use the mean-value property for $f(x) = 1/x$ and $f'(x) = -1/x^2$. Thus,

$$\frac{1}{3.14} - \frac{1}{\pi} = f(3.14) - f(\pi) = f'(X)(3.14 - \pi)$$

$$= \frac{1}{X^2}(\pi - 3.14) < \frac{0.002}{X^2} < \frac{0.002}{9} < 0.0003,$$

as before.

EXAMPLE 4. How closely should we approximate π by a number r in order to ensure that r^3 approximates π^3 accurately to two decimal places?

. . . .

Solution. We want to know how small to take $|\pi - r|$ in order to ensure that

$$|\pi^3 - r^3| < 0.005.$$

But if $f(x) = x^3$, then

$$\pi^3 - r^3 = f(\pi) - f(r) = f'(X)(\pi - r) = 3X^2(\pi - r)$$

for some X between π and r, by the mean-value property. If we restrict r to the interval $[3,4]$, then $3X^2 < 3(4)^2 = 48$, so

$$|\pi^3 - r^3| < 48|\pi - r|.$$

Therefore, in order to ensure that $|\pi^3 - r^3| < 0.005$, it is sufficient to require that

$$48|\pi - r| < 0.005 \qquad \text{or} \qquad |\pi - r| < \frac{0.005}{48}.$$

Since

$$\frac{0.005}{48} > \frac{0.005}{50} = 0.0001,$$

it is sufficient to take r so that

$$|\pi - r| < 0.0001,$$

i.e., to take r as an approximation to π that is accurate to within 1 in the fourth decimal place. You may know that $\pi = 3.1415\cdots$, which means that

$$3.1415 < \pi < 3.1416.$$

We can thus take r as either 3.1415 or 3.1416, and be certain that r^3 approximates π^3 accurately to two decimal places.

Example 4 is harder than the preceding examples, where we were asked only to verify that $f(x) - f(r)$ is as small as we claimed for two given numbers x and r. Here r is *not* given, and we have to *find* it: we have to find how close r should be taken to x in order to ensure that $f(x) - f(r)$ is as small as required. This kind of calculation answers the question of *how* continuous f is at x. We noted in Theorem 5 that any differentiable function is automatically continuous, but here we use the mean-value property to compute *how* small $(x - r)$ has to be in order to ensure that the error $f(x) - f(r)$ lies within preassigned bounds. This whole question of the quantitative aspect of continuity is examined in Chapter 20.

PROBLEMS FOR SECTION 11

1. We know that 3.1416 approximates π with an error less than 10^{-5} in magnitude. Show, then, that $(3.1416)^5$ approximates π^5 to the nearest two decimal places. (Assume $3.15 < \sqrt{10}$).

2. Assume again that 3.1416 approximates π with an error less than 10^{-5} in magnitude. We find by long division that $1/3.1416 = 0.31831$ with an error less than 10^{-6}. Show, therefore, that 0.31831 approximates $1/\pi$ to the nearest five decimal places.

3. Show that 10 approximates $(100,001)^{1/5}$ to the nearest four decimal places.

4. Show that, if r is the closest eight-place decimal to π, then $r^{1/3}$ approximates $\pi^{1/3}$ with an error less than 1 in the ninth decimal place.

5. Show that, if r is the closest n-place decimal to π, then $r^{1/3}$ approximates $\pi^{1/3}$ with an error less than 1 in the $(n+1)$st decimal place.

6. a) How closely should we approximate a positive number a by a finite decimal r $(r < a)$ in order to ensure that r^3 approximates a^3 with an error less than 0.01? The answer will be in terms of a.
   b) Same question if the tolerable error in the approximation of a^3 by r^3 is e.

7. How closely should the finite decimal r approximate π in order to ensure that $1/r$ approximates $1/\pi$ with an error at most e?

8. Prove from the mean-value property that

$$x^n - 1 > n(x - 1)$$

   if $x > 1$ and $n > 1$.

9. Prove that if $b > 1$ then $b^n \to \infty$ as $n \to \infty$. [*Hint.* Use the preceding problem, with x fixed and n variable, to show that b^n is larger than any preassigned number M when n is chosen large enough.]

10. Prove that if $0 < a < 1$, then $a^n \to 0$ as $n \to \infty$. (Use the preceding problem.)

11. Prove that if $x > 0$, then

$$\sqrt{1 + x} < 1 + \frac{x}{2}.$$

12. Prove that if $x > 1$, then

$$x^{1/5} < \frac{4}{5} + \frac{x}{5}.$$

12. RATE OF CHANGE

So far we have interpreted the derivative as the *slope* of the tangent line to a graph, and as the *velocity* of a moving particle. Now we shall see that the increment point of view leads to a generalization of the velocity interpretation: Given a differentiable function $y = f(x)$, we can always interpret $f'(x_0)$ as the *rate of change of y with respect to x as x passes through the value* x_0, or *the rate of change of the function f at the point* x_0.

Consider first a linear function, say

$$y = f(x) = 3x - 5.$$

Here

$$\Delta y = f(x_0 + \Delta x) - f(x_0) = [3(x_0 + \Delta x) - 5] - [3x_0 - 5]$$

$$= 3\Delta x,$$

so that the change in y is always exactly 3 times the change in x, no matter what values we use for x_0 and Δx. We can say that y changes exactly three times *as fast as x* changes, but it must be understood that we are not necessarily talking about *motion*. All we mean is that if x is changed, then y will change; and the change in y will always be three units per unit change in x. In this situation we say that 3 is the constant *rate* at which y changes as compared to x, or the *constant rate of change of y with respect to x.*

For the general linear function

$$y = ax + b$$

we have, similarly,

$$\Delta y = a\Delta x$$

and a is the constant rate of change of y with respect to x. Knowing the rate a we can obtain Δy from Δx by the above equation. Knowing Δx and Δy we obtain the rate a from

$$\frac{\Delta y}{\Delta x} = a.$$

Note that these two equations in Δx, Δy, and a are not quite equivalent, since Δx cannot be zero in the second.

When $f(x)$ is not a linear function, then the increment ratio

$$\frac{\Delta y}{\Delta x} = \frac{f(x_0 + \Delta x) - f(x_0)}{\Delta x}$$

depends on x_0 and Δx. If its value is r, then the equation

$$\Delta y = r \Delta x$$

says that over the x interval between x_0 and $x_0 + \Delta x$ the total change in y is exactly r times the total change in x. We therefore say that, over this interval, y has been changing *on the average* r times as fast as x, and we call r the average rate of change of y with respect to x.

Definition. *The averate rate of change of y with respect to x, over the interval between x_0 and $x_1 = x_0 + \Delta x$, is*

$$\frac{\Delta y}{\Delta x} = \frac{f(x_0 + \Delta x) - f(x_0)}{\Delta x}.$$

The derivative

$$\frac{dy}{dx} = \lim_{\Delta x \to 0} \frac{\Delta y}{\Delta x}$$

is now viewed as the limit of the average rate of change as the width of the interval over which the average is computed tends to zero. We therefore interpret

$$\frac{dy}{dx}\bigg|_{x=x_0} = f'(x_0)$$

as the **true** *rate of change of y with respect to x at the point x_0, or the true rate of change as x passes through the value $x = x_0$.*

Average velocity and true (or instantaneous) velocity are the special case when y is distance and x is time. We saw in Section 7 that our intuitions about velocity lead us pretty directly to the formula $v = ds/dt$ for the true velocity; and this should add conviction to the general interpretation of $dy/dx = f'(x)$ as the true rate of change.

The rate of change of any quantity with respect to time is called its *time rate of change*. In general, a time rate of change answers the question of how *fast* something is changing.

EXAMPLE 1. A stone is dropped into a pond and it is observed that the spreading circular ripple has the radius $r = 2t$ ft after t seconds. How fast is the area increasing at the end of 5 seconds?

Solution. We have the formula

$$A = \pi r^2 = \pi(2t)^2 = 4\pi t^2$$

for the area at the end of t seconds. Its rate of change at time t is therefore

$$\frac{dA}{dt} = 8\pi t,$$

and at $t = 5$ the area is increasing at the rate of

$$\left.\frac{dA}{dt}\right|_{t=5} = 40\pi,$$

or about 120 sq. ft/sec.

We now give some further examples of the general rate-of-change interpretation, starting with some words on rate of change as a *coefficient*.

If f is a differentiable function whose rate of change is constant over an interval I, then f must be linear over I. That is, f must be of the form

$$f(x) = mx + b.$$

For, if f' has the constant value m, then $f(x) - mx$ has derivative zero and so must be a constant function b, by Theorem 7.

Because of the above fact, when a rate of change is constant or nearly constant in the applications, it is often called a *coefficient*.

EXAMPLE 2. We consider a metal rod that would be one unit long at zero degrees (centigrade). When it is heated it expands. The rate of change of its length l with respect to the temperature T is called the *coefficient of thermal expansion* for the given metal. This rate of change is the derivative dl/dT, and calling it a *coefficient* implies that it is very nearly a constant c, so that l is given approximately by

$$l = cT + k$$

(where $k = 1$ since $l = 1$ when $T = 0$). Actually, the coefficient of thermal expansion is variable to the extent that its variability is mentioned in tables and handbooks. The notion of a variable coefficient seems paradoxical, but when we keep in mind that what we really are talking about is the derivative dl/dT, the fact that it turns out to be a nonconstant function of T ceases to bother us.

EXAMPLE 3. We experiment with an electrical circuit and plot the voltage E required to produce a current I. For simple circuits we find that E is proportional to I, so that

$$E = RI,$$

where the constant of proportionality R is called the *resistance* of the circuit. Of course the rate of change dE/dI then has the constant value R, and R is a true *coefficient* in the above sense.

However, if E and I are the plate voltage and plate current of a vacuum tube, then the graph of E as a function of I looks like this. It is called a characteristic of the tube.

Now we must define the resistance R as the rate of change dE/dI, and we note that it increases as the current I increases.

EXAMPLE 4. In economics the word *marginal* signifies a rate of change. For example, if it costs C dollars to produce x tons of coal, then the *marginal* cost is the increase in total cost C *per extra ton produced*, i.e., the rate of change of C with respect to x, dC/dx. If the relationship between C and x has the simple form

$$C = I + kx,$$

then the marginal cost is the coefficient k. Here I is a fixed initial cost, for machinery etc., and the marginal cost k is the constant running cost per ton, for labor, expended materials like fuel, etc.

Normally the relationship between C and x is more complicated than this. We impose one more condition on our problem, namely, that C is the cost to produce x tons of coal *in a given fixed time interval*, say one week. There will generally be a fixed cost of I dollars per week. But the *extra* running cost per ton, instead of being constant, will probably decrease as production increases, because it is possible to achieve greater internal efficiency with greater volume. The trend will continue only up to a certain point, after which the cost per ton will begin to increase again, because the larger demand will make it necessary to utilize older machinery, pay overtime wages, etc. Thus a typical cost curve might look like this:

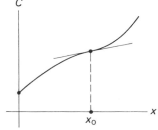

For example, $C = x^3 - 3x^2 + 4x + 1$ has a graph like this. (We shall not try to invent a precise cost situation that would lead to this formula.) The marginal cost is

$$c_M = \frac{dC}{dx} = 3x^2 - 6x + 4.$$

It is variable, and has a minimum value at the point labelled x_0. We shall see later that the minimum value of the marginal cost occurs where *its* derivative dc_M/dx is zero. Since

$$\frac{dc_M}{dx} = 6x - 6 = 6(x - 1),$$

it follows that c_M has its minimum value at $x_0 = 1$, and that its minimum value is

$$\text{Min } c_M = 3(1)^2 - 6 \cdot 1 + 4 = 1.$$

This is the point at which the producer is operating most efficiently, but, because of other factors, it will *not* be the point at which he will choose to operate. The marginal cost rises to very high values if the producer strains his capacity in an effort to put out a very large weekly tonnage.

EXAMPLE 5. We continue in the coal vein. Our producer must also consider how many tons of coal he can sell at a given price. Presumably, the lower his price the more coal he can sell (per week), so his *demand curve* is the graph of a decreasing function.

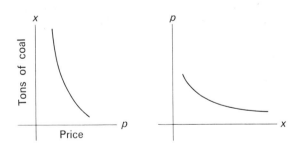

In order to compare this with his cost function, he would plot this relationship with axes interchanged, and so consider p as a function of x.

This function is the producer's *demand function*. Its value $p = d(x)$ is the price the producer must charge in order to sell exactly x tons (per week).

Then $R = px = d(x)x$ is his *total weekly revenue* from selling x tons, and the rate of change of r with respect to x, dR/dx, is called his *marginal revenue*.

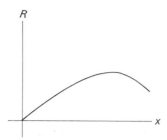

Our scales are all wrong. In order to plot a curve that will react to moderate changes in x, the unit of quantity would probably be a thousand tons, and the unit of money might be $10,000. The revenue R would then be the number of $10,000 units obtained from selling x thousands of tons. (Continued in Chapter 6.)

PROBLEMS FOR SECTION 12

1. a) What is the rate of change of the volume of a sphere with respect to its radius?
    b) What is the rate of change of the radius of a sphere with respect to its volume?

2. Find the rate of change of the volume of a sphere with respect to its radius when the radius is 5 inches. (The answer should be expressed in cubic inches per inch.) Find the rate of change when $r = 10$ inches.

3. A ladder 20 ft long leans against a vertical wall. Its base is the distance from the foot of the ladder to the wall. Find the rate of change of the height of the ladder with respect to its base when the base is 5 feet. (The answer should be expressed as so many feet per foot.)

4. a) What is the rate of change of the area of a circle with respect to its radius?
    b) What is the value of this rate of change when the radius is 5 inches? (The answer should be expressed in square inches per inch.)
    c) What is the rate when $r = 10$ inches?

5. What is the rate of change of the volume of a cube with respect to its edge length, the unit of length being the centimeter? By how much must the volume be increased in order to double this rate of change?

6. A balloon is being filled with air at the rate of 10 cubic feet per minute. How fast is its diameter expanding when its volume is 8 cubic feet? 125 cubic feet?

7. A growing tree increases its diameter at the rate of $\frac{1}{4}$ inch per year, and its height at the rate of 1 foot per year. Assuming that the shape of the tree is approximately conical, at what rate is new wood being added when the tree is 10 years old? 50 years old?

8. A growing cubical crystal increases its edge length at the rate of one millimeter per day. How fast is its surface area increasing at the end of the first week? How fast is its volume increasing then?

9. If l_0 is the length of a piece of platinum wire at $0°$ centigrade, then its length l_t at $t°C$ is given by

$$l_t = l_0(1 + \alpha t + \beta t^2)$$

where $\alpha = 0.0868 \times 10^{-4}$ and $\beta = 0.013 \times 10^{-7}$. Discuss the sense in which platinum has a coefficient of thermal expansion.

10. A manufacturer finds that it costs him

$$10^{-4}(x^3 - 1500x^2) + 150x + 5000$$

dollars per month to produce x items per month.

a) What is his marginal cost c_M if he is producing 250 items per month? (The answer should be in dollars per item.) What is his marginal cost if he produces 500 items per month? 1000 items per month?

b) Now compute his *average* cost per item when his monthly production is 250 items; 500 items; 1000 items.

11. The marginal cost of production will generally have a graph like the one at the left below. That is, the marginal cost will normally decrease with increasing production until it reaches a minimum value, and then will increase. Show by an intuitive argument that the total cost function $C(x)$ must necessarily look like the graph at the right.

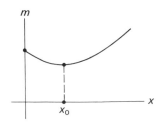 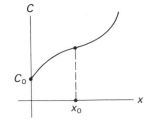

12. A marginal cost graph like the one shown in the preceding problem is approximately a quadratic graph, i.e., approximately the graph of

$$m(x) = ax^2 + bx + c$$

for suitable values of the coefficients a, b, and c.

a) Find this quadratic function if $m(0) = 10$ and if m has the minimum value $m(1000) = 8$ at $x_0 = 1000$. [*Hint*. Cast the general expression for $m(x)$ above into a new form by completing the square.]

b) Find the corresponding cost function $C(x)$, assuming that there is a fixed initial cost of 500 dollars even if no items are produced. (The problem is to find a cubic function $C(x)$ whose derivative is a given quadratic function $m(x)$, and such that $C(0) = 500$.)

13. A manufacturer's average cost per item is $f(x)$ dollars. Show that his marginal cost is

$$f(x) + xf'(x)$$

dollar per item. (This is really a problem for Section 4. You are asked to prove that if $C(x) = xf(x)$, then $C'(x) = f(x) + xf'(x)$.)

13. THE ANTIDERIVATIVE

At this point, it is possible and appropriate to start on Chapter 8, especially if there is some reason why an early introduction to integration is desirable. A few of the examples toward the end of the chapter involve functions we haven't yet discussed and these would have to be omitted; but otherwise the whole chapter can be studied now.

chapter 4
systematic
differentiation

One of the contributions of Newton and Leibniz was their realization that derivatives could be computed systematically. Starting with the derivative formulas for a few basic functions, and with the derivative rules for four or five ways of combining functions, we can compute in a systematic, and almost automatic, way the derivative of practically any function that is likely to come up in everyday mathematical affairs.

What we know so far can be summarized in the formulas

$$\frac{d}{dx}c = 0, \qquad \frac{d}{dx}x = 1,$$

and the rules

$$(f + g)' = f' + g',$$

$$(cf)' = cf',$$

$$(f^a)' = af^{a-1} \cdot f'.$$

In this chapter we shall complete the mechanism. The remaining rules are the product and quotient rules

$$(fg)' = f \cdot g' + g \cdot f',$$

$$\left(\frac{f}{g}\right)' = \frac{g \cdot f' - f \cdot g'}{g^2},$$

and the chain rule, which we shall state later. One corollary of the chain rule, the inverse-function rule, will be taken up in Chapter 7.

The list of basic functions will be completed with the functions e^x, $\sin x$, and $\cos x$, and their derivative formulas

$$\frac{d}{dx}e^x = e^x, \qquad \frac{d}{dx}\sin x = \cos x, \qquad \frac{d}{dx}\cos x = -\sin x.$$

These three functions are very intimately related. A comparison of the graph of e^x with the graph of $\sin x$ or $\cos x$ would not suggest any connection, and it requires complex numbers and complex variables to show what this relationship really is, but we shall find traces of it, especially in the chapter on power series. Here we shall see certain "structural" similarities: Each function is characterized by a fundamental *addition formula* that it satisfies, and these formulas lead in almost identical ways to the derivative formulas.

The chain rule is put off until after these functions have been studied because it is of little use until then. However, it can be taken up earlier with the other rules if desired.

Finally, there is the problem of differentials. Although modern mathematics develops differentials conceptually, we are better off in a first course with the simpler approach in which differentials are treated in a purely formal, or symbolic, manner. Section 7 takes this symbolic point of view. Then Section 8 presents a brief description of the way in which such formal differentials can be conceptualized.

1. THE PRODUCT AND QUOTIENT RULES

You might guess offhand that the derivative of the product of two functions ought to be the product of their derivatives. But if you try this out on practically any product for which you already know the answer, you will see that it doesn't work. For instance,

$$x^2 = x \cdot x, \qquad \frac{d(x^2)}{dx} = 2x, \qquad \text{and} \qquad \frac{dx}{dx} = 1,$$

but $2x$ is not $1 \cdot 1$. The quotient rule can't be guessed either.

These rules were stated above in function notation. Here are their Leibniz forms.

Theorem 1. *If u and v are differentiable functions of x (say, $u = f(x)$ and $v = g(x)$), then so are uv and u/v, and*

$$\frac{d}{dx}(uv) = u\frac{dv}{dx} + v\frac{du}{dx};$$

$$\frac{d}{dx}\left(\frac{u}{v}\right) = \frac{v\dfrac{du}{dx} - u\dfrac{dv}{dx}}{v^2}.$$

It is understood that these formulas hold for those values of x at which both f and g are differentiable and, in the case of f/g, where also $g(x) \neq 0$.

It may be easier to remember "word" versions, such as the following:

*The derivative of a product is the **first** times the derivative of the **second** plus the* **second** *times the derivative of the* **first**.

The derivative of a quotient is the **denominator** *times the derivative of the* **numerator minus** *the* **numerator** *times the derivative of the* **denominator**, *all divided by the* **denominator squared**.

The product and quotient rules enable us to compute derivatives of more complicated functions than we could handle before. We shall practice them now and prove them later.

EXAMPLE 1. Find dy/dx when $y = (x - 3)^9(x + 2)^{15}$. If we insisted on doing this with our "bare hands," so to speak, we could expand the binomials and so express y as a polynomial of degree 24. Then we could write the answer down, using the more elementary rules from the last chapter; but the product rule lets us do it directly:

$$\frac{dy}{dx} = (x - 3)^9 \frac{d}{dx}(x + 2)^{15} + (x + 2)^{15} \frac{d}{dx}(x - 3)^9$$

$$= (x - 3)^9 \cdot 15(x + 2)^{14} + (x + 2)^{15} \cdot 9(x - 3)^8$$

$$= (x - 3)^8(x + 2)^{14}[15(x - 3) + 9(x + 2)]$$

$$= (x - 3)^8(x + 2)^{14}[24x - 27]$$

$$= 3(x - 3)^8(x + 2)^{14}(8x - 9).$$

If we know how to write down the derivative of the two factors we would not generally bother with the first line above, but would start by writing out the second line.

EXAMPLE 2. Find the derivative of $x^2\sqrt{x - 1}$.

. . . .

Solution. Write $\sqrt{x - 1} = (x - 1)^{1/2}$ and apply the product rule:

$$\frac{d}{dx}x^2(x - 1)^{1/2} = x^2\left(\frac{1}{2}(x - 1)^{-1/2}\right) + (x - 1)^{1/2} \cdot 2x \ \dagger$$

$$= \frac{\dfrac{x^2}{2} + (x - 1) \cdot 2x}{(x - 1)^{1/2}} = \frac{x^2 + 4x^2 - 4x}{2(x - 1)^{1/2}}$$

$$= \frac{5x^2 - 4x}{2\sqrt{x - 1}}.$$

EXAMPLE 3

$$\frac{d}{dx}\frac{(x^2 - 2x)}{x - 1} = \frac{(x - 1)\dfrac{d}{dx}(x^2 - 2x) - (x^2 - 2x)}{(x - 1)^2}$$

$$= \frac{(x - 1)(2x - 2) - (x^2 - 2x) \cdot 1}{(x - 1)^2}$$

$$= \frac{x^2 - 2x + 2}{(x - 1)^2}.$$

---

† This is already a correct answer, but we generally try to express a function in as simple a form as possible, so we go on.

EXAMPLE 4. Calculate

$$\frac{d}{dx}\left(\frac{x}{\sqrt{x-1}}\right).$$

. . . .

Solution

$$\frac{d}{dx}\frac{x}{(x-1)^{1/2}} = \frac{(x-1)^{1/2}\cdot 1 - x\cdot\frac{1}{2}(x-1)^{-1/2}}{x-1}$$

$$= \frac{(x-1)^{-1/2}[(x-1)-\frac{x}{2}]}{x-1}$$

$$= \frac{2(x-1)-x}{2(x-1)^{3/2}} = \frac{x-2}{2(x-1)^{3/2}}.$$

Some students have trouble factoring out a *negative* power because it goes against their feeling that when you "factor something out" there should be *less* of it left behind. For example, when a^2 is factored out of a^3, what is left is a^1:

$$a^3 = a^2\cdot a^1;$$

the remaining power is *less* by the amount factored out. But when $a^{-2/3}$ is factored out of $a^{1/3}$ what is left is a^1,

$$a^{1/3} = a^{-2/3}\cdot a^1.$$

This is correct by the law of exponents, but it also fits the above pattern because *when something is decreased by a negative amount it is* **increased.** Here the remaining power is less than 1/3 by the amount $-2/3$; it is

$$\frac{1}{3}-\left(-\frac{2}{3}\right) = 1.$$

EXAMPLE 5. Work Example 4 by the product rule.

$$\frac{d}{dx}\left(\frac{x}{(x-1)^{1/2}}\right) = \frac{d}{dx}(x(x-1)^{-1/2})$$

$$= x\cdot -\frac{1}{2}(x-1)^{-3/2} + (x-1)^{-1/2}\cdot 1$$

$$= (x-1)^{-3/2}[-\frac{x}{2} + (x-1)]$$

$$= \frac{x-2}{2(x-1)^{3/2}}.$$

EXAMPLE 6. For

$$f(x) = \frac{x^2(x-1)}{x+2}$$

we have a choice as to whether we first take $f(x)$ to be the quotient $[x^2(x-1)]/(x+2)$, or a product, say $x^2[(x-1)/(x+2)]$. The problem can be worked either way. If we start with the quotient rule, we get

$$f'(x) = \frac{(x+2)\frac{d}{dx}[x^2(x-1)] - x^2(x-1)\cdot 1}{(x+2)^2},$$

where we still have to compute the product derivative,

$$\frac{d}{dx}x^2(x-1) = x^2\cdot 1 + (x-1)(2x) = 3x^2 - 2x,$$

and substitute it. We end up with

$$f'(x) = \frac{(x+2)(3x-2)x - x^2(x-1)}{(x+2)^2}$$

$$= \frac{x[3x^2 + 4x - 4 - x^2 + x]}{(x+2)^2} = \frac{x(2x^2 + 5x - 4)}{(x+2)^2}.$$

After a little experience you would probably do this all at once. In applying the quotient rule, the term $v\,du/dx$ involves the above product computation, and this can be written out before going on to complete the quotient rule. This would give

$$f'(x) = \frac{(x+2)[x^2\cdot 1 + (x-1)2x] - x^2(x-1)\cdot 1}{(x+2)^2},$$

which needs only algebraic simplification.

We turn now to the proofs of the product and quotient rules. These proofs can be given in the r,x-notation of Chapter 3, but the special forms of the product and quotient rules are better understood in terms of increments. Recall from Chapter 3 that if $y = f(x)$, then Δy depends on Δx (and x) according to the formula

$$\Delta y = f(x + \Delta x) - f(x).$$

If we add $y = f(x)$, we obtain the equivalent equation

$$y + \Delta y = f(x + \Delta x),$$

which just says that $y + \Delta y$ is the new value of y when $x + \Delta x$ is the new value of x.

Similarly, we can express the new value of any variable u as its old value plus its increment, $u + \Delta u$. This notation follows the philosophy that computations become most transparent when all changes are expressed in terms of increments.

EXAMPLE 7. If the variables u and v are given increments Δu and Δv, what is the increment in the product $y = uv$?

. . . .

Solution. The new values of u and v are $u + \Delta u$ and $v + \Delta v$, so the new value of y is

$$y + \Delta y = (u + \Delta u)(v + \Delta v).$$

Then

$$\Delta y = (y + \Delta y) - y = (u + \Delta u)(v + \Delta v) - uv,$$

$$= u \cdot \Delta v + v \cdot \Delta u + \Delta u \cdot \Delta v.$$

This expression for $\Delta(uv)$ can be pictured when everything is positive by interpreting uv as the area of a rectangle with sides u and v. Note that the added area is shown in three pieces, corresponding to the three terms in the above sum.

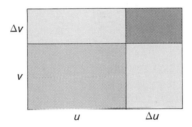

Proof of Theorem 1. We are assuming that the functions $u = g(x)$ and $v = h(x)$ are differentiable. That is, if we give x an increment Δx, getting the new values

$$u + \Delta u = g(x + \Delta x) \quad \text{and} \quad v + \Delta v = h(x + \Delta x),$$

then our hypothesis is that the limits

$$\lim_{\Delta x \to 0} \frac{\Delta u}{\Delta x} = \frac{du}{dx}, \qquad \lim_{\Delta x \to 0} \frac{\Delta v}{\Delta x} = \frac{dv}{dx}$$

both exist. We computed above that the product $y = uv$ has the increment

$$\Delta y = u\Delta v + v\Delta u + \Delta u \cdot \Delta v.$$

Thus,

$$\frac{\Delta y}{\Delta x} = u\frac{\Delta v}{\Delta x} + v\frac{\Delta u}{\Delta x} + \Delta u \cdot \frac{\Delta v}{\Delta x}.$$

As $\Delta x \to 0$, the right side of this equation has the limit

$$u\frac{dv}{dx} + v\frac{du}{dx} + 0 \cdot \frac{dv}{dx}.$$

That is, the limit $dy/dx = \lim\limits_{\Delta x \to 0} \Delta y/\Delta x$ exists, and

$$\frac{dy}{dx} = u\frac{dv}{dx} + v\frac{du}{dx}.$$

Similarly, the quotient $y = u/v$ turns out to have the increment

$$\Delta y = \frac{v\Delta u - u\Delta v}{v(v + \Delta v)}.$$

Therefore,

$$\frac{\Delta y}{\Delta x} = \frac{v\dfrac{\Delta u}{\Delta x} - u\dfrac{\Delta v}{\Delta x}}{v(v + \Delta v)},$$

and taking the limit as $\Delta x \to 0$ we get, similarly,

$$\frac{dy}{dx} = \frac{v\dfrac{du}{dx} - u\dfrac{dv}{dx}}{v^2}. \quad \blacksquare$$

Note in each case how the formula for dy/dx is dictated by the form of the expression for $\Delta y/\Delta x$.

There is an alternative derivation of the quotient rule from the product rule and the power rule with exponent $a = -1$:

$$\frac{d}{dx}\left(\frac{u}{v}\right) = \frac{d}{dx}u \cdot v^{-1} = \cdots$$

(computation to be finished by the reader). The quotient rule is thus not really a basic rule, but it is so useful that it should be learned anyway.

PROBLEMS FOR SECTION 1

Differentiate the following functions.

1. $y = (x - 1)^3(x + 2)^4$

2. $y = x^m \cdot x^n$ (Do it two ways and compare the answers)

3. $s = 3t^5(t^2 + 2t)$ (Two ways and compare)

4. $y = x^{21}(1 + x)^{21}$ (Two ways, using $x(1 + x) = x + x^2$)

5. $y = (x^3 + 6x^2 - 2x + 1)(x^2 + 3x - 5)$

6. $h(x) = (x + 1)^2(x^2 + 1)^{-3}$

7. $u = v^{-4}(1 + 2v)^4$ (Two ways)

8. $y = x\sqrt{1 + x^2}$ (Two ways)

9. $u = (x^2 + 2x + 1)^2(3x^2 - 1)^{-1/2}$

10. $s = (t + 1)^2(t + 1)^{-2}$ (Two ways)

11. $y = (x^2 - 1)(x + 1)^{-1}$ (Two ways)

12. $y = x^{1/3}(1 + x)^{1/3}$ (Two ways)

13. $w = (z + 1)^{1/2}(z - 1)^{-1/2}$

14. $y = (ax^2 + bx + c)\sqrt{2ax + b}$

15. $y = (1 - x)^{2/3}(1 + x)^{1/3}$

16. $y = x^3(x - 1)^4(x + 2)^2$

17. Given that $f(x) = u(x) \cdot v(x) \cdot w(x)$, show that $f' = uvw' + uv'w + u'vw$.

Use the formula derived in Problem 17 to find the derivative of the following functions:

18. $s = (2t^2 + t^{-2})(t^2 - 3)(4t + 1)$

19. $f(x) = (x - 3)(x + 1)(x + 2)$

Differentiate the following functions using the quotient rule (although not necessarily as the first step).

20. $y = \dfrac{x}{1 + x}$

21. $s = \dfrac{t}{1 + t^2}$

22. $y = \dfrac{x^2 - 1}{x^2 + 1}$

23. $y = \dfrac{x^3}{1 + 3x^2}$

24. $s = \dfrac{2t}{(1 + t^2)^2}$

25. $y = \dfrac{x + 1}{(x^2 + 2x + 2)^{3/2}}$

26. $u = \dfrac{1 + v^2}{1 + v}$

27. $y = \sqrt{\dfrac{1 + x}{1 - x}}$

28. $y = \dfrac{\sqrt{1 - x^2}}{1 - x}$

29. $w = \dfrac{u}{\sqrt{1 + u^2}}$

30. $y = \dfrac{f(x)}{x}$

31. a) $y = \dfrac{1}{1 - x^2}$ b) $y = \dfrac{x^2}{1 - x^2}$ (Compare the answers and explain.)

32. $y = \dfrac{6x^2 - 2}{(x^2 + 1)^3}$

33. $y = \left(\dfrac{1 - x}{1 + x}\right)^2$

34. $y = \left(\dfrac{1 - x}{1 + x}\right)^n$

35. $y = \dfrac{1 - f(x)}{1 + f(x)}$

36. $y = \left(\dfrac{f(x)}{1 + f(x)}\right)^a$

37. $y = \dfrac{f(x) - g(x)}{f(x) + g(x)}$

38. $y = \dfrac{1 + x}{1 - x}(3x + 2)$

39. $y = \dfrac{x\sqrt{1 + x}}{1 + x^2}$

40. $y = \dfrac{x}{(1 + 3x)\sqrt{1 + x^2}}$

41. Prove from the product rule that if it is true that

$$(f^m)' = mf^{m-1}f',$$

then it is true that

$$(f^{m+1})' = (m + 1)f^m f'.$$

42. Verify that the above formula is true for $m = 1$. Then show how its truth for an arbitrary positive integer p can be established by repeatedly applying the result in the above problem, starting from $m = 1$. This gives a new way of approaching the general power rule.

43. Assuming that f and $\sqrt{f}$ are both differentiable, prove the formula

$$(\sqrt{f})' = \dfrac{1}{2\sqrt{f}} \cdot f'$$

from the product rule.

44. Assuming that f and $1/f$ are both differentiable, prove the formula

$$\left(\dfrac{1}{f}\right)' = -\dfrac{1}{f^2} \cdot f'$$

from the product rule.

45. If $f(x) = g(x)h(x)$, show that

$$f'' = g''h + 2g'h' + gh''; \qquad f''' = g'''h + 3g''h' + 3g'h'' + gh'''.$$

Find d^2y/dx^2 when:

46. $y = \sqrt{1 - x^2}$,

47. $y = 1/(1 + x^2)$.

48. Call $(m - n)$ the degree of the rational function

$$\frac{a_m x^m + a_{m-1} x^{m-1} + \cdots + a_0}{b_n x^n + b_{n-1} x^{n-1} + \cdots + b_0},$$

where a_m and b_n are both nonzero. Show that if $r(x)$ is a rational function with *nonzero* degree p, then $r'(x)$ is of degree $(p - 1)$.

49. Show by a simple example that if $r(x)$ is a rational function of degree zero, then the degree of $r'(x)$ can be any negative integer except -1.

50. Prove the quotient rule from the product rule and the general power rule $(n = -1)$.

51. Prove the formula for Δy when $y = u/v$.

52. If $y = uv$, what is the formula for $\Delta y/y$ (when u and v are given increments Δu and Δv, respectively)?

53. If $y = uvw$, find the formula for $\Delta y/y$, when u, v, and w are given the increments Δu, Δv, and Δw, respectively.

54. Formalize the proof of the product law given in the text by quoting the limit law from Section 7 of Chapter 2 that justifies each step in the limit computation. (You will also have to quote Theorem 5 of Chapter 3.)

55. Formalize the proof of the quotient law in the same way.

56. Compute $f'(a)$ if $f(x) = (x - a)g(x)$. Can this result be true if g is not differentiable at $x = a$? Discuss.

57. Show by using the product law (and not the mean-value property) that the derivative of $(x - a)^m (x - b)^n$ is zero at a point between a and b.

58. Suppose that f is a differentiable function such that

$$tf'(t) = nf(t).$$

Show that $f(t) = ct^n$. (Compute the derivative of $f(t)/t^n$.)

59. Write out the proof of the product rule using function notation.

2. THE EXPONENTIAL FUNCTION

The exponential function $f(x) = e^x$ is one of the most important functions in mathematics and in the applications of mathematics. For example, we shall see in Chapter 6 that a population grows exponentially, whether it be a population of bacteria growing in a culture, or a bank account growing at continuously compounded interest.

The General Base a

In elementary algebra one learns how to manipulate powers a^m and roots $a^{1/n}$ of a positive "base" number a. These manipulations are governed by the following *laws of exponents*:

$$a^{x+y} = a^x a^y,$$

$$(a^x)^y = a^{xy},$$

$$a^1 = a.$$

There is also a law for mixed bases, namely

$$a^x b^x = (ab)^x.$$

In calculus we need a^x to be defined for every real number x, and in this chapter we shall simply assume that this is all right. Specifically, we assume:

If a is any fixed positive number, then $f(x) = a^x$ is a positive, continuous function that is defined for all real numbers x and satisfies the above laws of exponents.

In principle, anyway, any particular value a^x can be computed by using the above properties. If x is a positive rational number, i.e., a quotient of positive integers $x = m/n$, then $a^x = a^{m/n}$ is obtained from a by raising to powers and extracting roots: The number

$$a^{m/n} = (a^m)^{1/n} = (a^{1/n})^m$$

is the nth root of the mth power of a, and also the mth power of the nth root of a. These are the familiar exponents of elementary algebra.

When x is a positive irrational number, then a^x is approximated by the values of a^r for rational numbers r near x: a^x is the limit of a^r as r approaches x through rational values.

Finally,

$$a^0 = 1, \quad \text{and} \quad a^{-x} = \frac{1}{a^x}.$$

Actually, we could *define* a^x to be the number computed in the above way. If we were to try this, we would have to show that everything works out all right, and, in particular, that the function $f(x) = a^x$ defined this way satisfies the laws of exponents. We don't do this here for two reasons. First, it would bog us down in an excessive amount of detailed calculations. Second, it will turn out ultimately to be unnecessary, for if we take the exponential function on faith for a while and see what calculus properties it has to have, we are led ultimately to a much more efficient way of defining and calculating it. (This will be by infinite series, in Chapters 13 and 20.)

The Derivative of 2^x

The function $f(x) = a^x$ is increasing if the base a is greater than 1. To get an idea of the shape of such a graph, we plot 2^x for integer values of x. In the right figure we have filled in the values 2^x for $x = 1/2, 3/2$, etc., using

$$2^{1/2} = \sqrt{2} \approx 1.414, \qquad 2^{3/2} = 2 \cdot 2^{1/2} \approx 2.818, \qquad \text{etc.}$$

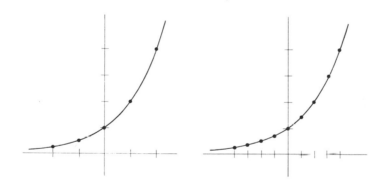

The graph of 2^x appears to be very smooth, and we shall accept this for now. That is, we now add the assumption that 2^x is differentiable, so that our only task is to find the formula for its derivative. It turns out that it is enough to assume that $f(x) = 2^x$ is differentiable at $x = 0$. Then the laws of exponents can be used to prove that $f'(x)$ exists for *all* x and to find the formula for this derivative.

When we are using the increments of only one or two variables, we sometimes forsake the Δ notation in favor of simpler symbols, and it is traditional to set $\Delta x = h$. In this notation we are assuming the existence of the limit

$$f'(0) = \lim_{h \to 0} \frac{f(0 + h) - f(0)}{h} = \lim_{h \to 0} \frac{2^h - 1}{h}.$$

Theorem 2. *Suppose that $f(x) = 2^x$ is differentiable at $x = 0$ and let α be the value of its derivative there. Then the derivative $d2^x/dx$ exists for all x, and*

$$\frac{d}{dx} 2^x = \alpha 2^x.$$

Proof

$$f'(x) = \lim_{h \to 0} \frac{f(x + h) - f(x)}{h} = \lim_{h \to 0} \frac{2^{x+h} - 2^x}{h}$$

$$= \lim_{h \to 0} 2^x \cdot \frac{2^h - 1}{h} = 2^x \lim_{h \to 0} \frac{2^h - 1}{h}$$

$$= 2^x f'(0) = \alpha 2^x. \quad \blacksquare$$

We don't know the value of $\alpha = f'(0)$. However, the two chords in the figure below have slopes $1/2$ and 1 respectively, and the tangent line slope α appears to lie between:

$$\frac{1}{2} < \alpha < 1.$$

Actually, you are in a position to prove this, and hints are given in Problems 24 through 26.

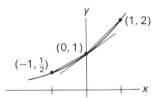

(0, 1)

(1, 2)

$(-1, \frac{1}{2})$

The Base *e*

For the exponential function $g(x) = 4^x$ we can show similarly that $1 < g'(0) < 2$. (See Problem 27.) Thus, as we increase the base a from $a = 2$ to $a = 4$, the graphs of the various exponential functions a^x have slopes over $x = 0$ that increase from below 1 to above 1. We therefore expect that there is exactly one base between 2 and 4 for which this slope is 1, that is, exactly one base a such that $da^x/dx\big|_{x=0} = 1$. We now prove this.

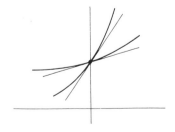

Theorem 3. *There is a unique base, which we shall designate e, such that*

$$\frac{d}{dx}e^x = e^x.$$

Proof. The unknown base e can be expressed in the form $e = 2^b$, where b has to be determined. Then $e^x = (2^b)^x = 2^{bx}$. If we had the general power rule for all exponents a, we could now see immediately that

$$\frac{d}{dx}e^x = \frac{d}{dx}2^{bx} = \frac{d}{dx}(2^x)^b = b(2^x)^{b-1} \cdot \alpha2^x = \alpha b2^{bx} = \alpha be^x,$$

so that b must have the value $1/\alpha$ to make the theorem true.

Instead, we start from scratch and rerun the proof of Theorem 1, obtaining

$$\frac{d}{dx}2^{bx} = 2^{bx}\lim_{h \to 0}\frac{2^{bh} - 1}{h}.$$

But

$$\lim_{h \to 0}\frac{2^{bh} - 1}{h} = b\lim_{h \to 0}\frac{2^{bh} - 1}{bh}$$

$$= b\lim_{t \to 0}\frac{2^t - 1}{t} \qquad \text{(where we have set } t = bh)$$

$$= b\alpha.$$

Therefore

$$\frac{d}{dx}2^{bx} = 2^{bx}$$

if and only if

$$b\alpha = 1, \quad \text{or} \quad b = \frac{1}{\alpha}.$$

That is, $e = 2^{1/\alpha}$ is the unique base such that $de^x/dx = e^x$. ∎

It is because of the simplicity of this law, as compared with $d2^x/dx = \alpha2^x$, that the base e is always used in calculus.

All we know about the number e at the moment is that it lies between 2 and 4 (because $e = 2^{1/\alpha}$ and $1 < 1/\alpha < 2$). Later on we shall discover how to compute e. Its decimal expansion starts off

$$e = 2.71828 \cdots.$$

EXAMPLE 1.

$$\frac{d}{dx}e^{2x} = \frac{d}{dx}e^x \cdot e^x = e^x \cdot e^x + e^x \cdot e^x$$

$$= 2e^{2x}.$$

EXAMPLE 2. Show that

$$\frac{d}{dx}e^{ax} = ae^{ax}$$

if a is a rational number.

. . . .

Solution. By the general power rule,

$$\frac{d}{dx}e^{ax} = \frac{d}{dx}(e^x)^a = a(e^x)^{a-1} \cdot e^x$$

$$= ae^{ax}.$$

EXAMPLE 3. If $y = xe^x$, show that

$$\frac{d^2y}{dx^2} - 2\frac{dy}{dx} + y = 0.$$

. . . .

Solution. We have

$$\frac{dy}{dx} = xe^x + e^x,$$

$$\frac{d^2y}{dx^2} = xe^x + e^x + e^x$$

$$= xe^x + 2e^x,$$

$$\frac{d^2y}{dx^2} - 2\frac{dy}{dx} + y = (xe^x + 2e^x) - 2(xe^x + e^x) + xe^x = 0.$$

EXAMPLE 4. Show that a differentiable function $y = f(x)$ satisfies the equation

$$\frac{dy}{dx} = y \tag{1}$$

over an interval I if and only if $y = ce^x$ on I, for a suitable constant c.

. . . .

Solution. That the function ce^x satisfies the equation (1) is exactly what the derivative rule $de^x/dx = e^x$ says. Now suppose that f is some other function satisfying (1) on an interval I. Then

$$\frac{d}{dx}(f(x)e^{-x}) = f(x)(-e^{-x}) + f'(x)e^{-x}$$

$$= -f(x)e^{-x} + f(x)e^{-x} = 0.$$

Thus,

$$f(x)e^{-x} = c$$

by Theorem 7 of Chapter 3, so $f(x) = ce^x$.

If we ask for a solution of (1) that has a given value at a given point, then we get an uniquely determined answer. Thus $f(x) = e^x$ is the unique solution f of (1) satisfying

$$f(0) = 1.$$

(The equations $f(x) = ce^x$ and $f(0) = 1$ imply that

$$1 = f(0) = ce^0 = c \cdot 1 = c$$

and $c = 1$.) The extra requirement $f(0) = 1$ is called an *initial condition*.

EXAMPLE 5. Let us find the solution f of (1) that satisfies the initial condition $f(3) = 5$.

. . . .

Solution. Since $f(x) = ce^x$, the initial condition becomes $5 = f(3) = ce^3$. Then

$$c = 5e^{-3},$$

and

$$f(x) = ce^x = 5e^{-3}e^x = 5e^{x-3}$$

is the unique solution.

PROBLEMS FOR SECTION 2

Differentiate (assuming the rule from Example 2 in the text).

1. $xe^x - e^x$.

2. e^x/x.

3. $x^2 e^x - 2xe^x + 2e^x$.

4. $e^{2x} - e^{-x}$.

5. $\sqrt{e^x - 1}$.

6. $(e^x - e^{-x})/(e^x + e^{-x})$.

7. $e^{x/2}\sqrt{x - 1}$.

8. $e^x + e^{2x} + e^{3x}$.

9. $(e^x)^a$.

10. x/e^x.

11. $(e^{3x} + x)^{1/3}$.

12. $e^x/(1 - e^x)$.

13. $(e^{2x} + 2 + e^{-2x})^{1/2}$.

14. $\left(\dfrac{1 - e^x}{1 + e^x}\right)^{1/2}$.

15. $\dfrac{xe^{-x}}{1 + x^2}$.

16. Show that $(d/dx)e^{x+k} = e^{x+k}$.

In Problems 17 through 20 determine the value or values of k for which $y = e^{kx}$ satisfies the given equation. Use Example 2 in the text.

17. $\dfrac{d^2y}{dx^2} + \dfrac{dy}{dx} - 2y = 0$

18. $\dfrac{d^2y}{dx^2} + 2\dfrac{dy}{dx} + y = 0$

19. $\dfrac{d^2y}{dx^2} + 2\dfrac{dy}{dx} - 3y = 0$

20. $\dfrac{d^2y}{dx^2} - y = 0$

21. If $y = xe^{-2x}$, show that

$$\frac{d^2y}{dx^2} + 4\frac{dy}{dx} + 4y = 0.$$

22. Suppose that the tangent line to the graph of $y = e^x$ at the point x_0 intersects the x-axis at $x = x_1$. Show that $x_0 - x_1 = 1$.

23. Using Example 2, show that if k is rational, then every solution of the equation

$$\frac{dy}{dx} = ky$$

is of the form $y = ce^{kx}$. (Look at Example 4.)

24. If $f(x) = a^x$, where a is any positive base, prove that $f'(x) = f'(0)f(x)$. (Repeat the proof of Theorem 2.)

25. Use the above result to show that if a is any positive base different from 1, then $f'(x) = d(a^x)/dx$ is an increasing function of x.

26. We claimed in the text that $1/2 < \alpha < 1$, where α is the value at the origin of $d(2^x)/dx$. Prove this, by first applying the mean-value property to the chords of the graph $y = 2^x$ over the pairs of points $x = -1, 0,$ and $x = 0, 1,$ and then using Problem 25.

27. Prove that if $g(x) = 4^x$, then $1 < g'(0) < 2$, using the steps outlined in the above problem.

28. Assuming the result in Example 4 in the text, find the solution of the differential equation $dy/dx = y$ that satisfies the initial condition $f(1) = 1$.

29. a) Compute

$$e^{-kx}\left(\frac{d}{dx}e^{kx}f(x)\right).$$

   b) Assume that if $g'(x) = e^{rx}$, then $g(x) = (e^{rx}/r) + c$. Keeping this in mind, together with the result of (a), solve the equation

$$f'(x) + 3f(x) = e^{5x}.$$

30. Prove from scratch that

$$\frac{d}{dx}e^{x^2} = 2xe^{x^2}.$$

31. Prove that

$$\frac{d}{dx}e^{f(x)} = e^{f(x)} \cdot f'(x).$$

(Follow the proof of Theorem 2, along the line used in Section 5 of Chapter 3.)

32. A particle moves along the x-axis from the origin toward $x = 1$, its position at time t being given by $x = 1 - e^{-t}$. Show that its acceleration is the negative of its distance from 1.

33. A particle moves along the x-axis, its position at time t being given by $x = te^{-t}$.

   a) How far to the right does it get? (This should be its position when its velocity has fallen to zero.)
   b) Show that its position x, velocity v, and acceleration a are related by

$$a + 2v + x = 0.$$

3. SOME TRIGONOMETRY

In approaching the exponential function it was necessary to assume quite a lot before making any analytic progress. The functions $\sin t$ and $\cos t$ are somewhat more accessible; because of their origins in geometry, their basic properties can be established by geometric reasoning. The crux of the matter is that we can put a coordinate system on the circumference of the unit circle that behaves much like the basic coordinate system on a line.

Recall that the coordinate correspondence on a line assigns to each real number x the uniquely determined point P_x that we get by measuring off the signed distance x from the origin. That is, P_x is at the distance $|x|$ from the origin, and is to the right or left of the origin depending on whether x is positive or negative.

Now we do the same thing on the unit circle $x^2 + y^2 = 1$. The conventional origin here is the point $Q = (1,0)$ where the circle intersects the positive x-axis,

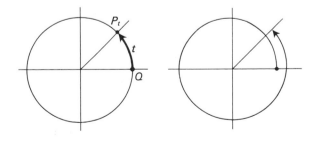

and the positive direction is the *counterclockwise* direction along the circle. We assign to each real number t the uniquely determined point P_t that we get by measuring off the signed distance t along the circumference from Q. That is, we go in the counterclockwise or clockwise direction, depending on whether t is positive or negative, and we measure off the distance $|t|$.

We shall call t *an angular coordinate* of the point P_t. We can't call it *the* angular coordinate because this coordinate correspondence is not one-to-one. For example, if $t = 2\pi$, we go counterclockwise around the circle and get back exactly to the origin point Q, since the length of the full unit circumference is 2π. If $t = 4\pi$ we go *twice* around and get back exactly to Q. If $t = -2\pi$, we go clockwise around and come back to Q. Thus $P_{2\pi} = P_{4\pi} = P_{-2\pi} = P_0 = Q$, and $0, 2\pi, 4\pi$, and -2π are all angular coordinates of Q.

On the other hand, each point P on the unit circle $x^2 + y^2 = 1$ has exactly one angular coordinate t in the interval $[0, 2\pi)$; t is the length of the arc from Q around to P in the counterclockwise direction. Then $t + 2n\pi$ is also an angular coordinate of P, for any integer n, positive or negative, and every coordinate of P is of this form.

We define cos t (cosine t) and sin t (sine t) as the x and y coordinates of P_t:

$$\cos t = x\text{-coordinate of } P_t,$$

$$\sin t = y\text{-coordinate of } P_t.$$

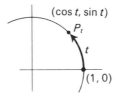

We shall see shortly that these definitions are substantially the same as others with which you may be more familiar. Note that cos t and sin t are functions of t, defined for all values of t, and related by the equation of the unit circle.

$$\cos^2 t + \sin^2 t = 1.$$

The graphs of these functions have the general features shown below. Remember that if t is large we may "wrap around" the circle several times before coming to the terminal point $P_t = (x, y)$; and if t is negative then we go in the *clockwise* direction a distance $|t|$. You ought to be able to visualize the x-coordinate of this terminal point P_t oscillating between $+1$ and -1 in the general manner shown, as P_t moves around the circle. In particular, x first becomes 0 when t is the length

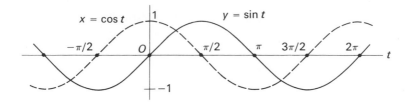

of one quarter of the full circumference,

$$t = \frac{(2\pi)}{4} = \frac{\pi}{2}.$$

The graphs of $\sin t$ and $\cos t$ sketched above suggest that

$$\cos(-t) = \cos t,$$

$$\sin(-t) = -\sin t,$$

for all t; and these conjectures can be verified from the geometric definitions, as indicated below.

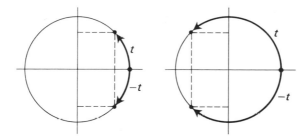

The graphs also show other properties that can be directly checked. Thus, both sine and cosine are *periodic* with *period* 2π, meaning that

$$\sin(t + 2\pi) = \sin t,$$

$$\cos(t + 2\pi) = \cos t$$

for all t. This follows from our earlier observations that

$$P_{t+2\pi} = P_t.$$

A stronger property of each function suggested by the graphs is that

$$\sin(t + \pi) = -\sin t,$$

$$\cos(t + \pi) = -\cos t.$$

These equations express the fact that $P_{t+\pi}$ is diametrically opposite P_t on the unit circle. Finally, it appears from the graphs that if the cosine graph were slid $\pi/2$ units to the right it would become the sine graph, i.e., that

$$\cos t = \sin(t + \pi/2)$$

for all t. This is true but not quite so easy to prove.

It is very helpful to think geometrically about $\sin t$ and $\cos t$ as we did above, and to regard the general shapes of their graphs as forming a storehouse of various symmetry properties and other relationships that can be retrieved by inspection. But the facts suggested and recalled by such geometric reasoning all have analytic origins in the *addition formulas* for sine and cosine, which we shall come to shortly.

Each point P_t determines a geometric angle which we shall designate θ_t, namely, the angle from the positive x-axis around to the ray OP_t. When θ_t is acute, then the older right-triangle definitions of sine and cosine for θ_t agree with the present ones. For, referring to the right triangle in the next figure, we have

$$\cos t = x = \frac{x}{1} = \frac{\text{Adjacent}}{\text{Hypotenuse}},$$

$$\sin t = y = \frac{y}{1} = \frac{\text{Opposite}}{\text{Hypotenuse}}.$$

The number t is called the *radian measure* of the angle θ_t. In order to convert from degrees to radians, note that θ_t is a right angle when P_t is one quarter of the way around the circle, i.e., when $t = (1/4)(2\pi) = \pi/2$. This gives the first line in the table below, and the rest of the table follows from it.

θ_t in degrees	t	$\sin t$	$\cos t$
90°	$\pi/2$		
1°	$\pi/180$		
$d°$	$(\pi/180)d$		
30°	$\pi/6$	$1/2$	$\sqrt{3}/2$
45°	$\pi/4$	$\sqrt{2}/2$	$\sqrt{2}/2$
60°	$\pi/3$	$\sqrt{3}/2$	$1/2$

The last three lines record the standard facts for a 30°-60°-90° triangle and for an isosceles right triangle (45°-45°-90°).

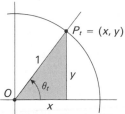

Now consider an arbitrary point $P = (x, y)$ different from the origin $O = (0,0)$, and let t be the radian measure of an angle from the positive x-axis around to the ray OP.

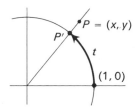

What now is the relationship between x, y, and t? The connecting link is the point P' where the ray OP intersects the unit circle $x^2 + y^2 = 1$. On the one hand,

$$P' = \left(\frac{x}{r}, \frac{y}{r}\right),$$

where $r = \sqrt{x^2 + y^2}$, while on the other hand P' has angular coordinate t. Therefore

$$\cos t = \frac{x}{r}, \qquad \sin t = \frac{y}{r}$$

by the definition of cosine and sine. (For an acute angle, these formulas can also be derived from the right triangle definitions of $\cos t$ and $\sin t$.) Note that the coordinates (x,y) can be recovered from r and t by the equations

$$x = r \cos t,$$

$$y = r \sin t.$$

The numbers r and t are called *polar coordinates* of P, and the above equations express the change of coordinates for P from polar coordinates to rectangular coordinates. Polar coordinates will be discussed further in Section 6 of Chapter 5.

Finally, if P is on the unit circle $(x - a)^2 + (y - b)^2 = 1$ with center $O' = (a, b)$, then an angle t from the positive x direction around to the ray $O'P$ satisfies

$$\cos t = x - a, \qquad \sin t = y - b,$$

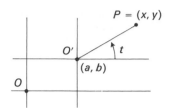

because $x - a$ and $y - b$ are the *new coordinates* of P when we put a new origin at O' (Theorem 7 in Chapter 1). If P is not on the unit circle about O', then, as above,

$$\cos t = \frac{x - a}{r}, \qquad \sin t = \frac{y - b}{r},$$

where $O' = (a, b)$, $P = (x, y)$, and $r = \sqrt{(x - a)^2 + (y - b)^2}$.

Underlying all the properties of the trigonometric functions (and implicit already in some of our remarks above) is the following basic fact:

If we measure off the signed distance u along the unit circumference, starting from the point P_t, then we end up at the point P_{t+u}.

This should appear geometrically obvious from looking at a few special cases, and we shall simply assume it. It is the analog for angular coordinates of the Addition Law in Section 1 of Chapter 1, since it can be interpreted as the statement that $u = s - t$ is a signed distance from P_t to P_s along the unit circle.

According to this basic principle, if we measure off the same signed arc length $(s - t)$ from each of the two initial points P_t and P_0, we obtain the terminal points P_s and P_{s-t}, respectively. And since equal arcs have equal chords, we then have the identity

$$\overline{P_t P_s} = \overline{P_0 P_{s-t}},$$

where we have used the notation $\overline{PQ}$ for the length of the line segment from P to Q.

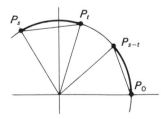

The above identity is equivalent to the trigonometric addition laws! When the segment lengths are expressed by the distance formula, and both sides are squared, the identity becomes

$$(\cos s - \cos t)^2 + (\sin s - \sin t)^2 = (\cos(s - t) - 1)^2 + (\sin(s - t) - 0)^2.$$

Now just simplify algebraically: expand the squares, use the identity $\sin^2 x + \cos^2 x = 1$ three times, and cancel common terms and factors. What is left is the basic law

$$\cos s \cos t + \sin s \sin t = \cos(s - t).$$

In conjunction with the table

	0	$\pi/2$	π
sin	0	1	0
cos	1	0	-1

this law yields *all* the remaining trigonometric identities. A scheme for working these out is suggested in the problems. Meanwhile we shall make use of the addition forms of the above law, namely:

$$\sin(x + y) = \sin x \cos y + \cos x \sin y,$$

$$\cos(x + y) = \cos x \cos y - \sin x \sin y.$$

The functions $\sin x$ and $\cos x$ are the basic trigonometric functions, but there are four others, the tangent, cotangent, secant and cosecant functions, defined as follows:

$$\tan x = \frac{\sin x}{\cos x} \qquad \cot x = \frac{\cos x}{\sin x} \left(= \frac{1}{\tan x} \right)$$

$$\sec x = \frac{1}{\cos x} \qquad \csc x = \frac{1}{\sin x}.$$

It follows from these definitions that the graphs of $\tan x$ and $\sec x$ have the general appearance shown below. Some other properties will be derived in the problems.

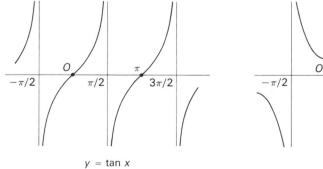

$y = \tan x$

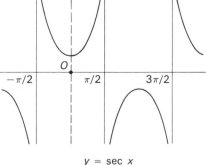

$y = \sec x$

PROBLEMS FOR SECTION 3

In Problems 1 through 11 below, draw a unit circle, locate the point p_t with the given angular coordinate t, and determine $\cos t$ and $\sin t$ (exactly if t involves π, and approximately otherwise).

1. $t = 3\pi/2$ 2. $t = 5\pi/4$ 3. $t = -\pi/4$

4. $t = 1$ 5. $t = 2$ 6. $t = 7\pi$

7. $t = -3\pi$ 8. $t = -3\pi/2$ 9. $t = 4$

10. $t = 7$ 11. $t = -7\pi/4$

In Problems 12 through 20 plot the given point (x,y) on the unit circle, and estimate (or determine exactly if you can) the unique number t in $[0,2\pi)$ for which $(\cos t, \sin t) = (x,y)$.

12. $(0,-1)$ 13. $(-\sqrt{2}/2, -\sqrt{2}/2)$ 14. $(1/2, -\sqrt{3}/2)$

15. $(-1/2, \sqrt{3}/2)$ 16. $(\sqrt{3}/2, -1/2)$ 17. $(4/5, 3/5)$

18. $(3/5, 4/5)$ 19. $(-1/\sqrt{5}, 2/\sqrt{5})$ 20. $(2/\sqrt{5}, -1/\sqrt{5})$

21. What trigonometric identities are suggested by the following figure?

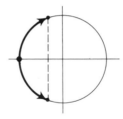

22. What trigonometric identities are suggested by the following figure?

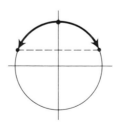

Give the geometric proof (from right triangles) for the identities:

23. $\sin \pi/6 = 1/2$, $\cos \pi/6 = \sqrt{3}/2$.

24. $\sin \pi/4 = \sqrt{2}/2$.

25. Let θ be the angle that the segment from $(1,2)$ to $(2,4)$ makes with the direction of the positive x-axis. Compute $\sin \theta$ and $\cos \theta$.

26. Same question for the angle θ that the segment from $(1, -1)$ to $(0,4)$ makes with the direction of the positive x-axis.

In Problems 27 through 33 estimate the smallest positive angle from the positive x-axis to the ray OP, for the given point P.

27. $P = (1, 1)$ 28. $P = (1, 2)$

29. $P = (-3, 3)$ 30. $P = (2, -4)$

31. $P = (-4, 2)$ 32. $P = (-1/2, -1/2)$

33. $P = (1, -1/2)$

Use the $\sin(x + y)$ and $\cos(x + y)$ laws stated in the text, together with the special values for $\sin x$ and $\cos x$ at $x = \pm\pi/2, \pi$ (these can be read off from a diagram), to prove the following laws:

34. $\sin(x + \pi/2) = \cos x$.   35. $\sin(x - \pi/2) = -\cos x$.

36. $\cos(x + \pi/2) = -\sin x$.   37. $\cos(x - \pi/2) = \sin x$.

38. $\sin(x + \pi) = -\sin x$.   39. $\cos(x + \pi) = -\cos x$.

40. Prove the identities:

$$\sin(x + \pi/4) = \frac{\sqrt{2}}{2}(\sin x + \cos x),$$

$$\cos(x + \pi/4) = \frac{\sqrt{2}}{2}(\cos x - \sin x).$$

41. Prove the identities:

$$\sin(x + \pi/3) = \frac{1}{2}(\sin x + \sqrt{3}\cos x),$$

$$\cos(x + \pi/3) = \frac{1}{2}(\cos x - \sqrt{3}\sin x).$$

42. Show that $\tan x$ is periodic with period π. That is, prove that

$$\tan(x + \pi) = \tan x$$

for all x.

43. Show that if t is (the measure of) a positive acute angle, then $\tan t$, $\cot t$, $\sec t$, and $\csc t$ have the right-triangle characterizations

$$\tan t = \frac{\text{opposite}}{\text{adjacent}}, \qquad \cot t = \frac{\text{adjacent}}{\text{opposite}},$$

$$\sec t = \frac{\text{hypotenuse}}{\text{adjacent}}, \qquad \csc t = \frac{\text{hypotenuse}}{\text{opposite}}.$$

44. Prove the identities:

$$\tan^2 x + 1 = \sec^2 x, \qquad 1 + \cot^2 x = \csc^2 x.$$

It was claimed in the text that the $\cos(s - t)$ law, together with the values of $\cos t$ and $\sin t$ at $t = 0$, $\pi/2$, and π, determine all the trigonometric identities. The order in which things are done is important. Here is one sequence, with some hints:

45. $\cos(-t) = \cos t$ (Take $s = 0$ in the formula for $\cos(s - t)$)

46. $\cos(s - \pi/2) = \sin s$ (Take $t = \pi/2$)

47. $\cos(s - \pi) = -\cos s$ (Take $t = \pi$)

48. $\sin(s - \pi/2) = -\cos s$ (Write $s - \pi = (s - \pi/2) - \pi/2$ in (47) and apply (46).)

49. $\sin(x - y) = \sin x \cos y - \cos x \sin y$ (Use 46, the $\cos(s - t)$ law, 46, and 48, in that order.)

50. $\sin(-y) = -\sin y$ (Take $x = 0$)

51. $\sin(x + y) = \sin x \cos y + \cos x \sin y$

52. $\cos(x + y) = \cos x \cos y - \sin x \sin y$ (From the $\cos(s - t)$ law, using (45) and (50).)

53. $\tan(x + y) = \dfrac{\tan x + \tan y}{1 - \tan x \tan y}$ (Divide (51) by (52).)

54. $\tan(x - y) = \dfrac{\tan x - \tan y}{1 + \tan x \tan y}$

Double-angle Formulas

55. $\sin 2x = 2 \sin x \cos x$

56. $\begin{aligned}\cos 2x &= \cos^2 x - \sin^2 x \\ &= 1 - 2 \sin^2 x \\ &= 2 \cos^2 x - 1\end{aligned}$

57. $\tan 2x = \dfrac{2 \tan x}{1 - \tan^2 x}$

Half-angle Formulas

58. $\cos^2\left(\dfrac{y}{2}\right) = \dfrac{1 + \cos y}{2}$

59. $\sin^2\left(\dfrac{y}{2}\right) = \dfrac{1 - \cos y}{2}$

60. $\tan\left(\dfrac{y}{2}\right) = \dfrac{\sin y}{1 + \cos y} = \dfrac{1 - \cos y}{\sin y}$ (Verify by setting $y = 2x$ and using double-angle formulas.)

Finally, if we set $x + y = A$ and $x - y = B$, so that

$$A + B = 2x \quad \text{and} \quad A - B = 2y,$$

the addition formulas give us:

61. $\sin A - \sin B = 2 \cos\left(\dfrac{A + B}{2}\right)\sin\left(\dfrac{A - B}{2}\right)$

62. $\cos A - \cos B = -2 \sin\left(\dfrac{A + B}{2}\right)\sin\left(\dfrac{A - B}{2}\right)$

· 63.   Show that $a \sin x + b \cos x$ can be written in the form

$$A \sin(x + \theta).$$

Here a and b are arbitrary constants, and A and θ are to be determined from them.

64.   What trigonometric identities are suggested by the following figures?

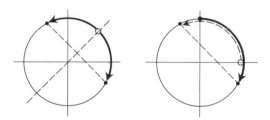

4. THE DERIVATIVES OF sin x AND cos x,

The functions $\sin x$ and $\cos x$ are closely related to e^x. The relationship appears here in the similar way in which their derivative formulas can be established; under the assumption that they are differentiable at the origin, the addition laws can be used to show that they are everywhere differentiable and to calculate their derivative formulas. The limits in question are:

$$\sin'(0) = \lim_{h \to 0} \frac{\sin(0 + h) - \sin(0)}{h} = \lim_{h \to 0} \frac{\sin h}{h},$$

$$\cos'(0) = \lim_{h \to 0} \frac{\cos(0 + h) - \cos(0)}{h} = \lim_{h \to 0} \frac{\cos h - 1}{h}.$$

Here also we can do more than we could for the exponential function. The geometric definitions of $\sin t$ and $\cos t$ allow us to establish and evaluate these limits geometrically. The $\sin h/h$ limit is the basic one.

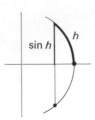

For positive h, the figure above shows that

$$\frac{\sin h}{h} = \frac{2 \sin h}{2h}$$

can be interpreted as the ratio of a chord length to its arc length. It should seem geometrically clear that the chord becomes a very good approximation to the arc when h is small. That is,

Lemma

$$\frac{\sin h}{h} \to 1 \qquad \text{as } h \to 0.$$

Proof. The geometric proof of this limit is easier in terms of area than it is in terms of arc length. We assume that the area A of a circular sector is proportional to its central angle. Since the area of the full unit circle is π, this proportion is

$$\frac{A}{\pi} = \frac{t}{2\pi},$$

from which we see that $A = t/2$. We note that A lies between the two triangular areas shown below, so

$$\frac{\sin t}{2} < \frac{t}{2} < \frac{\tan t}{2}.$$

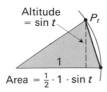

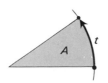

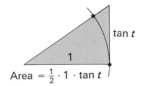

Dividing throughout by sin $t/2$, we obtain the inequality

$$1 < \frac{t}{\sin t} < \frac{1}{\cos t},$$

and this becomes

$$\cos t < \frac{\sin t}{t} < 1$$

upon taking reciprocals. Assuming for the moment that cos t is continuous at $t = 0$, so that $\lim_{t \to 0} \cos t = \cos 0 = 1$, we conclude that

$$\frac{\sin t}{t} \to 1 \qquad \text{as} \quad t \to 0$$

by the squeeze limit law. The proof that cos t is continuous at $t = 0$ will be left as an exercise.

Since

$$\frac{\sin(-t)}{-t} = \frac{\sin t}{t},$$

we do not need a separate argument for negative h. ∎

The other limit is 0:

$$\cos'(0) = \lim_{h \to 0} \frac{\cos h - 1}{h} = 0.$$

This can be derived from the sin h/h limit by using the addition law for cos $h = \cos(h/2 + h/2)$; but it should also seem obvious from the graph of the cosine function that it has a horizontal tangent at $t = 0$.

Now the addition law for $\sin(x + h)$ lets us rewrite the difference quotient for sin x in a form that permits its limit to be evaluated.

Theorem 4. *The functions* sin x *and* cos x *are everywhere differentiable, and*

$$\frac{d}{dx} \sin x = \cos x,$$

$$\frac{d}{dx} \cos x = -\sin x.$$

Proof. We have

$$\frac{\sin(x + h) - \sin x}{h} = \frac{\sin x \cos h + \cos x \sin h - \sin x}{h}$$

$$= \sin x \left(\frac{\cos h - 1}{h} \right) + \cos x \left(\frac{\sin h}{h} \right).$$

We have just seen that the quantities in parentheses have limits 0 and 1 respectively. Thus

$$\frac{d}{dx} \sin x = \lim_{h \to 0} \frac{\sin(x + h) - \sin x}{h}$$

$$= (\sin x) \cdot 0 + (\cos x) \cdot 1$$

$$= \cos x.$$

The proof for $\cos x$ goes in exactly the same way, starting from the addition law for $\cos(x + h)$. ∎

The remaining trigonometric functions are defined from $\sin x$ and $\cos x$, and their derivatives are computed from these definitions. Here again are the definitions:

$$\tan x = \frac{\sin x}{\cos x}, \qquad \cot x = \frac{\cos x}{\sin x},$$

$$\sec x = \frac{1}{\cos x}, \qquad \csc x = \frac{1}{\sin x}.$$

These are the tangent, secant, cotangent, and cosecant functions.
    Dividing the identity $\sin^2 x + \cos^2 x = 1$ first by $\cos^2 x$ and then by $\sin^2 x$ gives the useful identities

$$\tan^2 x + 1 = \sec^2 x,$$

$$1 + \cot^2 x = \csc^2 x.$$

The derivative formulas for these four functions are calculated from the sine and cosine formulas by the quotient rule and/or the general power rule. The remaining formulas are:

$$\frac{d}{dx} \tan x = \sec^2 x,$$

$$\frac{d}{dx} \sec x = \tan x \sec x,$$

$$\frac{d}{dx} \cot x = -\csc^2 x,$$

$$\frac{d}{dx} \csc x = -\cot x \csc x.$$

The tangent and secant derivative formulas come up often enough to be worth memorizing, but they can always be worked out if forgotten.

EXAMPLE 1. Prove that

$$\frac{d}{dx} \tan x = \sec^2 x.$$

Solution. By definition

$$\tan x = \frac{\sin x}{\cos x}.$$

We can therefore compute its derivative by the quotient law:

$$\frac{d}{dx} \tan x = \frac{\cos x \cdot \cos x - \sin x(-\sin x)}{\cos^2 x}$$

$$= \frac{\cos^2 x + \sin^2 x}{\cos^2 x}$$

$$= \frac{1}{\cos^2 x} = \sec^2 x$$

(since $\sec x = 1/\cos x$).

EXAMPLE 2. If $y = e^x \sin x$, find constants b and c such that

$$y'' + by' + cy = 0$$

(where $y' = dy/dx$ and $y'' = d^2y/dx^2$).

Solution. Since

$$y' = e^x \cos x + e^x \sin x,$$

$$y'' = -e^x \sin x + e^x \cos x + e^x \cos x + e^x \sin x$$

$$= 2e^x \cos x,$$

we see that

$$y'' + by' + cy = (2 + b)e^x \cos x + (b + c)e^x \sin x.$$

This will be 0 if and only if

$$2 + b = 0,$$

$$b + c = 0,$$

i.e., if $b = -2$, $c = 2$. Thus

$$y'' - 2y' + 2y = 0$$

when $y = e^x \sin x$.

We saw at the end of Section 3 that $y = e^x$ is essentially the only function satisfying the equation

$$\frac{dy}{dx} = y,$$

in the sense that every solution of the equation is of the form $y = ce^x$ for some constant c. Since this equation relates an unknown function to one or more of its derivatives, it is called a *differential equation*, and it is of the *first order* because only the first-order derivative occurs.

The functions $\sin x$ and $\cos x$ have a similar property. They are both solutions of the *second-order* differential equation

$$\frac{d^2y}{dx^2} = -y.$$

Obviously neither sine nor cosine can be the unique solution in the above sense, but the two functions together are essentially the only solutions in a similar sense: Every other solution f is of the form

$$f(x) = A \sin x + B \cos x$$

for some constants A and B. The proof of this important fact is outlined below, but the details of the individual steps are left as exercises.

a) If g and h are any two solutions of the differential equation $y'' = -y$, then so is every linear combination of g and h, i.e., every function f of the form

$$f(x) = Ag(x) + Bh(x),$$

where A and B are constants.

b) If f is any solution then

$$f^2 + (f')^2 \equiv \text{constant}.$$

Thus every solution f satisfies an identity of the type

$$\sin^2 x + \cos^2 x = 1.$$

c) It follows from (b) that if ϕ is a solution such that $\phi(0) = \phi'(0) = 0$, then ϕ is the zero function: $\phi(x) = 0$ for all x.

d) If f is any solution, then

$$f(x) = f(0)\cos x + f'(0)\sin x$$

(set $\phi = f - f(0)\cos -f'(0)\sin$ and apply (c)). Thus the general solution of the differential equation $d^2y/dx^2 + y = 0$ is of the form $A \sin x + B \cos x$.

Essentially the same conclusions hold for the differential equation

$$\frac{d^2y}{dx^2} + by = 0,$$

where b is positive. We check that $\sin \sqrt{b}x$ and $\cos \sqrt{b}x$ are solutions, and can then show that

$$A \sin \sqrt{b}x + B \cos \sqrt{b}x$$

is the general solution by repeating the above development. An alternative procedure is to observe that a function f satisfies the new equation if and only if

$$g(x) = f(x/\sqrt{b})$$

satisfies the original equation. However, these calculations can't be made until we acquire the chain rule.

PROBLEMS FOR SECTION 4

Prove the formulas for the derivatives of the remaining trigonometric functions, as follows:

1. $\dfrac{d}{dx} \sec x = \sec x \tan x$

2. $\dfrac{d}{dx} \cot x = -\csc^2 x$

3. $\dfrac{d}{dx} \csc x = -\csc x \cot x$

Differentiate the following functions:

4. $x \sin x$

5. $e^x \cos x$

6. $\sin x/x$

7. $\sin^2 x$

8. $\sin x \cos x$

9. $\cos^2 x + \sin^2 x$

10. $\dfrac{\sin x}{1 + \cos x}$

11. $\dfrac{1 - \cos x}{\sin x}$

12. $\cos^2 x - \sin^2 x$

13. $\tan^2 x$

14. $\sec^2 x$

15. $\sqrt{1 + \cos x}$

16. $\sqrt{1 + \sin^2 x}$

17. $\sin^3 x \cos^3 x$

18. $\cos^4 x - \sin^4 x$

19. $\sec^2 x - \tan^2 x$

20. $x \sin x - \cos x$

21. $\dfrac{e^x(\sin x + \cos x)}{2}$

22. $x^2 \cos x - 2x \sin x - 2 \cos x$

23. $(1/2)(x + \sin x \cos x)$

With the aid of the addition formulas, prove that:

24. $\dfrac{d}{dx} \sin 2x = 2(\cos 2x)$

25. $\dfrac{d}{dx} \cos 2x = -2(\sin 2x)$

26. $\dfrac{d}{dx} \sin 3x = 3(\cos 3x)$ (using (24) and (25))

27. $\dfrac{d}{dx} \sin(x + a) = \cos(x + a)$ 28. $\dfrac{d}{dx} \cos(x + a) = -\sin(x + a)$

29. $\dfrac{d}{dx} \tan(x + a) = -\sec^2(x + a)$ (using (27) and (28))

30. Complete the proof of Theorem 4. That is, prove that $d \cos x/dx = -\sin x$ from scratch, using the addition law for $\cos(x + h)$.

31. Show that the formula $d \cos x/dx = -\sin x$ follows from Problem 27 above and the addition laws.

32. Show that if $g(x)$ and $h(x)$ are any two solutions of the differential equation

$$\frac{d^2y}{dx^2} = -y, \tag{I}$$

then so is any linear combination

$$f(x) = Ag(x) + Bh(x),$$

where A and B are constants.

33. Show that $f^2 + (f')^2$ is constant for any solution f of (I). Conclude that if ϕ is a solution such that $\phi(0) = \phi'(0) = 0$, then ϕ is identically zero.

34. Show that every solution f of (I) is of the form

$$f(x) = A \cos x + B \sin x.$$

[*Hint.* Set $\phi(x) = f(x) - f(0)\cos x - f'(0)\sin x$ and apply the above problems.]

35. Show that $e^x \sin x$ satisfies the second-order differential equation

$$\frac{d^2y}{dx^2} - 2\frac{dy}{dx} + 2y = 0.$$

36. If $y = f(x)$ is any solution of the above differential equation, show that $u = e^{-x}f(x) = e^{-x}y$ satisfies the differential equation

$$\frac{d^2u}{dx^2} + u = 0,$$

and conversely.

Using the most general solution of this equation, as determined at the end of the section, write down the most general solution of the equation in Problem 35.

37. Show that $y = x \sin x$ satisfies the fourth-order differential equation

$$\frac{d^4y}{dx^4} + 2\frac{d^2y}{dx^2} + y = 0.$$

38. a) If $f(x) = Ae^x \sin x + Be^x \cos x$, show that

$$f'(x) = (A - B)e^x \sin x + (A + B)e^x \cos x.$$

b) Without differentiating any further, write down what $f''(x), f'''(x)$, $f^{(4)}(x)$, and $f^{(5)}(x)$ are.

Some more derivatives:

39. $\dfrac{\sin x + \cos x}{\sin x - \cos x}$

40. $\sqrt{\dfrac{1 - \cos x}{1 + \cos x}}$

41. $xe^x \sin x$

42. $\sqrt{1 - \cos^2 x}$

43. $\dfrac{\tan x}{1 + \tan^2 x}$

44. $\dfrac{1 - \sin 2x}{1 + \sin 2x}$

45. $(1/5)\sin^5 x - (2/3)\sin^3 x + \sin x$ (Simplify the answer)

46. $(1/3)\tan^3 x - \tan x + x$ (Simplify the answer.)

47. $(1/5)\tan^5 x + (2/3)\tan^3 x + \tan x$ (Simplify.)

48. A particle moves back and forth along the x-axis, its position at time t being given by $x = 3 \sin 2t$. Show that its acceleration a is proportional to its position x but oppositely directed.

49. A particle moves back and forth along the x-axis in a "decaying vibration," its position at time t being given by $x = e^{-t}\sin t$. Express its acceleration a in terms of its velocity v and position x.

50. We saw in the text that $0 < \sin t < t$ when t is positive. Use this inequality to prove that $\cos t \to 1$ as $t \to 0$. (There are a couple of ways to start this. For one thing, $\sin t = \sqrt{1 - \cos^2 t}$.)

51. Show that the unit circle about the origin lies entirely below the graph of $\cos x$ except for the point of tangency of the two graphs at $x = 0$.

52. Use the identity for $\sin A - \sin B$ (Problem 61 in the last section) to give another proof that the derivative of $\sin x$ is $\cos x$.

53. The 6 trigonometric functions can be grouped into three pairs: $(\sin x, \cos x)$, $(\tan x, \cot x)$, $(\sec x, \csc x)$. In a given pair, each function is called the cofunction of the other. Thus, $\cot x$ is the cofunction of $\tan x$, but also $\tan x$

is the cofunction of cot x: tan $x = $ co(cot x). This relationship is carried out to combinations of functions. For example, sin $x + $ cot x is the cofunction of cos $x + $ tan x, and csc x cot x is the cofunction of sec x tan x: csc x cot $x = $ co (sec x tan x).

With the above remarks in mind, show that the derivative formulas of the six trigonometric functions satisfy the following rule:

$$\frac{d}{dx} \text{co}(f(x)) = -\text{co}\left(\frac{d}{dx}f(x)\right).$$

The hyperbolic sine of x (sinh x) and the hyperbolic cosine of x (cosh x) are defined as follows:

$$\sinh x = \frac{e^x - e^{-x}}{2}, \quad \cosh x = \frac{e^x + e^{-x}}{2}.$$

The remaining hyperbolic functions are defined from sinh x and cosh x in the standard way. Show that the hyperbolic functions have the following properties.

54. $\dfrac{d}{dx} \sinh x = \cosh x$ 55. $\dfrac{d}{dx} \cosh x = \sinh x$

56. $\dfrac{d}{dx} \tanh x = \text{sech}^2 x$ 57. $\dfrac{d}{dx} \text{sech } x = -\text{sech } x \tanh x$

58. $\cosh(-x) = \cosh x$ 59. $\sinh(-x) = -\sinh x$

60. $\sinh(x + y) = \sinh x \cosh y + \cosh x \sinh y$.

61. $\cosh(x + y) = \cosh x \cosh y + \sinh x \sinh y$.

5. THE CHAIN RULE

When we start combining the exponential and trigonometric functions with other functions we discover that there are very simple combinations that we can't yet differentiate, such as cos πx and $e^{(x^2)}$. These are examples of functions *inside of* functions, and in order to handle this general method of combination we need one final rule, called the *chain rule*.

If y is a function of u and if u is a function of x, then y is a function of x. For example, if $y = u^{3/2}$ and $u = 3 + x^2$, then $y = (3 + x^2)^{3/2}$. The question is: How does the derivative dy/dx of such a final "composite" function relate to the derivatives dy/du and du/dx of its two "constituent" functions?

It is easy to see from the rate-of-change interpretation of the derivative what the answer ought to be. Consider, first, what happens around $x = 1$ in the above example. If we compute the derivatives at $x = 1$ and at $u = 3 + x^2 = 3 + (1)^2 = 4$, we get

$$\frac{dy}{du} = 3, \quad \frac{du}{dx} = 2.$$

Thus, as x passes through 1, y is increasing three times as fast as u and u is increasing twice as fast as x. Therefore, altogether y is increasing $3 \cdot 2 = 6$ times as fast as x, and so $dy/dx = 6$.

Now consider the general situation. Suppose that at $x = a$ the derivative dy/du has the value r and du/dx has the value s. Thus, as x passes through a, y is increasing r times as fast as u and u is increasing s times as fast as x. Therefore, altogether y is increasing rs times as fast as x. That is, $dy/dx = rs = (dy/dy)(du/dx)$.

Theorem 5. *If y is a differentiable function of u and if u is a differentiable function of x, then y is a differentiable function of x and*

$$\frac{dy}{dx} = \frac{dy}{du} \cdot \frac{du}{dx}.$$

This is the Leibniz version of the chain rule. Our heuristic derivation needs to be bolstered by a real proof but, as we have done with the other rules, we shall state the chain rule in various ways and practice using it before taking up its proof.

EXAMPLE 1. If

$$y = \sin u \qquad \text{and} \qquad u = 3 + x^2,$$

we have

$$\frac{dy}{du} = \cos u, \qquad \frac{du}{dx} = 2x,$$

and the chain rule gives

$$\frac{dy}{dx} = \frac{dy}{du} \cdot \frac{du}{dx} = (\cos u)2x = 2x \cos(3 + x^2).$$

The intermediate variable u may not be part of the original problem. For instance, the above example would typically arise as the problem of computing dy/dx when $y = \sin(3 + x^2)$. Here we have a function that is too complicated for our earlier rules to handle. But setting $u = 3 + x^2$ breaks the complicated function down into two simpler stages that we do know how to differentiate,

$$y = \sin u \qquad \text{and} \qquad u = 3 + x^2,$$

and we can now compute dy/dx by the chain rule, as above.

Here is the same example without commentary.

EXAMPLE 2. Compute dy/dx when $y = \sin(3 + x^2)$.

· ·

Solution. Setting $u = 3 + x^2$, we have $y = \sin u$. Then by the chain rule,

$$\frac{dy}{dx} = \frac{dy}{du} \cdot \frac{du}{dx} = (\cos u) \cdot 2x = 2x \cos(3 + x^2).$$

Here the derivatives dy/du and du/dx are so simple that there is no need to compute them separately before multiplying them in the chain rule. But if one of them has to be worked out, then computing it separately may help to keep track of what is going on.

EXAMPLE 3. If $y = [x/(1 + x)]^{1/3}$, we have a power rule situation, but here we want to apply the chain rule. Setting $u = x/(1 + x)$ breaks it up suitably, into

$$y = u^{1/3} \quad \text{and} \quad u = \frac{x}{1 + x}.$$

Then

$$\frac{dy}{du} = \frac{1}{3} u^{-2/3}, \quad \frac{du}{dx} = \frac{(1 + x) \cdot 1 - x \cdot 1}{(1 + x)^2} = \frac{1}{(1 + x)^2},$$

and so

$$\frac{dy}{dx} = \frac{dy}{du} \cdot \frac{du}{dx} = \frac{1}{3} u^{-2/3} \cdot \frac{1}{(1 + x)^2}$$

$$= \frac{1}{3} \left(\frac{x}{1 + x} \right)^{-2/3} \frac{1}{(1 + x)^2} = \frac{1}{3 x^{2/3} (1 + x)^{4/3}}.$$

EXAMPLE 4. We now use the chain rule to prove the power rule for a positive rational exponent $a = m/n$, where m and n are positive integers. The problem is to compute dy/dx when

$$y = x^{m/n} = (x^{1/n})^m.$$

Setting $u = x^{1/n}$, we have

$$y = u^m, \quad u = x^{1/n}.$$

From Theorems 1 and 4 in Chapter 3,

$$\frac{dy}{du} = mu^{m-1}, \quad \frac{du}{dx} = \frac{1}{n} x^{(1/n)-1}.$$

Therefore,

$$\frac{dy}{dx} = \frac{dy}{du} \cdot \frac{du}{dx} = mu^{m-1} \cdot \frac{1}{n} x^{(1/n)-1}$$

$$= \frac{m}{n} (x^{1/n})^{m-1} x^{1/n-1}$$

$$= \frac{m}{n} x^{(m/n)-(1/n)+(1/n)-1} = \frac{m}{n} x^{(m/n)-1}$$

$$= ax^{a-1}.$$

As you get more practice, and in situations where you know the two derivatives dy/du and du/dx, you will find that you can write down the chain-rule derivative in one or two steps.

EXAMPLE 5. As you look at

$$y = (\underbrace{3 + x^2}_{u})^{3/2},$$

you can mentally take $3 + x^2$ as u, and see that you then have $y = u^{3/2}$, without writing it down separately. Nor will you need to write down the Leibniz formula. You would probably just write

$$\frac{dy}{dx} = \frac{3}{2} u^{1/2} \cdot 2x$$

or

$$\frac{dy}{dx} = \frac{3}{2} (3 + x^2)^{1/2} \cdot 2x.$$

As you put down the righthand side, you first think dy/du, see that it is $(3/2)u^{1/2}$ and write this down, and then think du/dx, see that it is $2x$ and write that down.

EXAMPLE 6. Of course, a chain-rule calculation may involve other variables, or the same variables used in different ways. Thus, if $s = (3 + t^2)^{3/2}$ and if we set $x = 3 + t^2$, then

$$\frac{ds}{dt} = \frac{ds}{dx} \cdot \frac{dx}{dt} = \frac{3}{2} x^{1/2} \cdot 2t = 3t(3 + t^2)^{1/2}.$$

In the example $y = (3 + x^2)^{3/2} = u^{3/2}$, we can replace y by what it is equal to and write this case of the chain rule as

$$\frac{d}{dx} (\underbrace{3 + x^2}_{u})^{3/2} = \frac{d}{du} u^{3/2} \cdot \frac{du}{dx} = \frac{3}{2} u^{1/2} \frac{du}{dx}.$$

In general, if $y = g(u)$ and u is a function of x then the chain rule can be rewritten

$$\frac{d}{dx}g(u) = \frac{d}{du}g(u) \cdot \frac{du}{dx} = g'(u)\frac{du}{dx}.$$

This form is useful enough to be given special status.

Theorem 5'. *If $y = g(u)$ and u is a function of x, and if both functions are differentiable, then y is a differentiable function of x and*

$$\frac{dy}{dx} = g'(u) \cdot \frac{du}{dx},$$

or

$$\frac{d}{dx}g(u) = g'(u)\frac{du}{dx}.$$

The "Built in" Chain Rule.

When we take g in the chain rule

$$\frac{d}{dx}g(u) = g'(u)\frac{du}{dx}$$

to be one of the functions u^a, e^u, $\sin u$, and $\cos u$, we obtain the *general* differentiation formulas

$$\frac{d}{dx}u^a = au^{a-1}\frac{du}{dx},$$

$$\frac{d}{dx}e^u = e^u\frac{du}{dx},$$

$$\frac{d}{dx}\sin u = \cos u\frac{du}{dx},$$

$$\frac{d}{dx}\cos u = -\sin u\frac{du}{dx}.$$

The first of these formulas is the general power rule from Chapter 3.

Some people like to learn all of the basic derivative formulas in this general form, incorporating the chain rule. Then most routine applications of the chain rule occur automatically. For example

$$\frac{d}{dx}\sin(x^{1/3}) = [\cos(x^{1/3})]\frac{d}{dx}x^{1/3}$$

is just an application of the general sine formula, and needs no special thought of the chain rule. This procedure is very efficient, but it can also be costly if it leads one to lose sight of the fundamental chain rule itself and so possibly to flounder when a less routine application of it is needed.

The other possible approach, and the one we have been following in this section, is to make every application of the chain rule a conscious act. Then we start with the formulas

$$\frac{d}{dx}x^a = ax^{a-1}, \qquad \frac{d}{dx}e^x = e^x, \qquad \frac{d}{dx}\sin x = \cos x, \qquad \frac{d}{dx}\cos x = -\sin x,$$

etc., and combine any one of them with the chain rule whenever it is necessary.

Composition of Functions

In Theorem 5′, u is still some unspecified function of x. If $u = h(x)$, so that $du/dx = h'(x)$, then we can replace all mention of u and obtain the pure function form

$$\frac{d}{dx}g(h(x)) = g'(h(x))h'(x).$$

This allows the most precise statement of the chain rule:

Theorem 5″. *If the function h is differentiable at the point $x = x_0$ and if g is differentiable at the point $u_0 = h(x_0)$, then the function $f(x) = g(h(x))$ is differentiable at x_0 and*

$$f'(x_0) = g'(u_0) \cdot h'(x_0)$$
$$= g'(h(x_0))h'(x_0).$$

The function

$$f(x) = g(h(x))$$

is called the *composition product* of g and h. As a function of x it is formed by applying the outside function g to the inside function of x, $h(x)$.

EXAMPLE 7. The function $f(x) = (3 + x^2)^{3/2}$ is the composition product of the outside function $g(u) = u^{3/2}$ and the inside function $h(x) = 3 + x^2$.

The composition product of g with h is sometimes designated by the special notation $g \circ h$. This means that $g \circ h$ is the function defined by

$$(g \circ h)(x) = g(h(x)).$$

This notation allows us to write the function form of the chain rule without any variables at all:

$$(g \circ h)' = (g' \circ h)h'.$$

Proof of Theorem 5'. We are considering

$$y = g(u),$$

where u is a differentiable function of x, and we shall assume that g has a continuous derivative. An increment Δx in x produces an increment Δu in u, and this in turn produces the increment

$$\Delta y = g(u + \Delta u) - g(u)$$

in y. We rewrite the right side of this equation by the mean-value theorem, and have

$$\Delta y = g'(U)\Delta u,$$

where U is some point between u and $u + \Delta u$. Now divide by Δx,

$$\frac{\Delta y}{\Delta x} = g'(U)\frac{\Delta u}{\Delta x},$$

and let $\Delta x \to 0$. Then $\Delta u/\Delta x \to du/dx$ by hypothesis. Also, $\Delta u \to 0$ because a differentiable function is continuous, $U \to u$ because U lies between u and $u + \Delta u$, and $g'(U) \to g'(u)$ because g' is continuous. The right side above thus approaches $g'(u)(du/dx)$ as Δx approaches 0. Thus the limit

$$\frac{dy}{dx} = \lim_{\Delta x \to 0} \frac{\Delta y}{\Delta x}$$

exists, and

$$\frac{dy}{dx} = g'(u)\frac{du}{dx}. \quad \blacksquare$$

We have thus proved Theorem 5', but under the extra hypothesis that g' is continuous in an interval about u_0. This will cover practically any situation that comes up. For the theorem as stated we can still proceed in essentially the above way because we can concoct a function $G(\Delta u)$ that plays the same role that $g'(U)$ played above. We define G artificially as follows:

$$G(\Delta u) = \frac{g(u_0 + \Delta u) - g(u_0)}{\Delta u} \qquad \text{when } \Delta u \neq 0,$$

$$G(0) = g'(u_0).$$

It follows from this definition that

$$\Delta y = g(u_0 + \Delta u) - g(u_0) = G(\Delta u) \cdot \Delta u$$

for all values of Δu, and we can then proceed exactly as before, using the fact that $G(\Delta u) \to g'(u_0)$ as $\Delta u \to 0$.

PROBLEMS FOR SECTION 5

Prove the following differentiation rules:

1. $\dfrac{d}{dx} \sin(ax + b) = a \cos(ax + b).$

2. $\dfrac{d}{dx} \cos(ax + b) = -a \sin(ax + b).$

3. $\dfrac{d}{dx} e^{ax+b} = ae^{ax+b}.$

Compute the derivatives of the following functions.

4. e^{x^2}　　　　　　　　5. $\sin \sqrt{x}$　　　　　　　6. $\sqrt{\sin x}$

7. $\cos(e^x)$　　　　　　　8. $e^{(\cos x)}$　　　　　　9. $e^{(e^x)}$

10. $e^{ax} \sin bx$　　　　　11. $\sin(x^3 - x)$　　　　12. $\sin(e^{ax})$

13. Suppose that there is a differentiable function f such that $e^{f(x)} = x$ for all positive x. Show that $f'(x) = 1/x$.

14. Suppose that there is a differentiable function f such that $\tan f(x) = x$ for all x. Show that $f'(x) = 1/(1 + x^2)$.

15. Suppose that f and g are differentiable functions such that $f(g(x)) = x$ for all x. Show that $g'(x) = 1/f'(u)$, where $u = g(x)$.

16. Show that

$$\frac{d}{dx} \sin(f(x)) = \cos(f(x))$$

only if $f(x)$ is of the form

$$f(x) = x + b.$$

17. From the sketches of graphs of even and odd functions decide how such functions behave under differentiation, and then prove your conjecture from the chain rule.

18. Show that $y = e^{kx}$ satisfies the differential equation

$$y'' + ay' + by = 0$$

if and only if

$$k^2 + ak + b = 0.$$

(Here $y' = dy/dx$ and $y'' = d^2y/dx^2$.)

19. Show that $y = xe^{kx}$ satisfies the above equation if and only if $a^2 = 4b$ and $k = -a/2$.

20. Show that $y = e^{mx}\sin rx$ satisfies the second-order differential equation

$$\frac{d^2y}{dx^2} - 2m\frac{dy}{dx} + (m^2 + r^2)y = 0.$$

Differentiate:

21. $y = e^{\sqrt{1-x^2}}$

22. $s = \sqrt{1 - e^{2t}}$

23. $y = \sin(1/x)$

24. $y = \cos(\tan\theta)$

25. $y = x^2\sin(1/x)$

26. $w = \dfrac{e^{(u^2)}}{u}$

27. $y = e^{-x^2}(x + 1)$

28. $y = \sin(\sin x)$

29. $y = \sin(\cos(\sin x))$

30. $y = e^{-1/x^2}$

31. $y = xe^{1/x}$

32. $x = a\sin b\theta + b\sin a\theta$

33. $y = \sqrt{a + \dfrac{b}{x}}$

34. $y = \sqrt{\sin x} - \sqrt{\cos x}$

35. $s = \sqrt{1 + \sin^2(\sqrt{t})}$

36. $y = e^{a\sqrt{x}}$

37. $y = \dfrac{1 + \sin(x^2)}{1 + \cos(x^2)}$

38. $y = e^{|x|}$

39. $y = f(x^2)$

40. $y = f(e^x)$

41. $y = e^{2f(x)}$

42. $y = f(x + g(x))$

43. $y = f(e^{g(x)})$

44. $y = f([g(x)]^n)$

45. If $f(x) = x^2\sin(1/x)$ when $x \neq 0$ and $f(0) = 0$, then f is everywhere differentiable. (See Problem 25 above and Problem 42 in Section 4 of Chapter 3.) Prove that $f'(x)$ is not continuous at $x = 0$.

46. Prove that a function $y = f(x)$ satisfies the differential equation

$$\frac{d^2y}{dx^2} + b^2y = 0,$$

where b is a constant, if and only if $f(x)$ is of the form

$$y = A \sin bx + B \cos bx$$

for some constants A and B.

6. FUNCTIONS DEFINED IMPLICITLY

If n is an odd integer, or if n is even and y is positive, then the equation

$$y^n = x$$

defines y as a function of x. In this case we can solve for y,

$$y = f(x) = x^{1/n},$$

and we computed the derivative of this function in Theorem 4 of Chapter 3. However, the chain rule lets us find a formula for the derivative *without* solving for y. We just differentiate the original equation *with respect to x, treating y as a differentiable function of x.* Here this involves computing dy^n/dx, and hence involves the general power rule. Thus, differentiating

$$y^n = x,$$

we get

$$ny^{n-1}\frac{dy}{dx} = 1;$$

and so

$$\frac{dy}{dx} = \frac{1}{ny^{n-1}} = \frac{1}{n}y^{1-n}.$$

The answer involves the *dependent* variable y. Here we know what y is in terms of x, so we can recapture our earlier formula:

$$\frac{dy}{dx} = \frac{1}{n}y^{1-n} = \frac{1}{n}(x^{1/n})^{1-n} = \frac{1}{n}x^{(1/n)-1} = ax^{a-1}.$$

If an equation which determines y as a function of x cannot be solved for y, then the above procedure is the only one available to give us a formula for the derivative dy/dx.

EXAMPLE 1. Suppose that the equation

$$e^y = x$$

determines y as a differentiable function of x. Find a formula for dy/dx.

. . . .

Solution. We differentiate the equation with respect to x, treating y as a function of x. This involves the chain rule (or the exponential formula with "built in" chain rule). We get

$$e^y \frac{dy}{dx} = 1,$$

$$\frac{dy}{dx} = \frac{1}{e^y} = \frac{1}{x}.$$

Here again (and this will always happen), the answer first appears involving the dependent variable y. When we can't solve the original equation for y in terms of x, we won't generally be able to express the derivative dy/dx entirely in terms of x, Here we were lucky.

_____

It is possible for a function $y = f(x)$ to satisfy an equation in x and y without being uniquely determined by the equation. For example, the function

$$y = f(x) = \sqrt{1 - x^2}$$

satisfies the equation

$$x^2 + y^2 = 1.$$

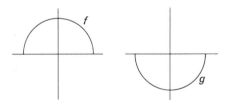

However, the function graph is only the upper half of the equation graph, and there is a second function satisfying the same equation whose graph is the lower half of the equation graph. This is, of course, the negative of the above function,

$$y = g(x) = -\sqrt{1 - x^2} \, .$$

Since both functions satisfy the same equation, we can't say that either one of them is uniquely determined by the equation. We can call them *function solutions* of the equation.

If we allow discontinuity, then we can find still more function solutions of $x^2 + y^2 = 1$. For example, the function h whose graph is drawn below is one. This function h is made up artificially by choosing part of the function f above and part of g. Such extraneous solution functions h can be eliminated by requiring that a solution function be continuous.

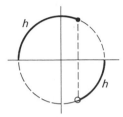

The functions defined *implicitly* by an equation are simply its solution functions. We shall consider the domain of an implicitly defined function to be an interval, although this slightly oversimplifies a rather fussy question.

The point of our discussion is that we can obtain derivative formulas for implicitly defined functions by the same procedure that we used above for an equation defining y as a function of x.

EXAMPLE 2. Suppose that y is a differentiable function of x defined implicitly by the equation

$$x^2 + y^2 = 1.$$

Differentiating with respect to x, we get

$$2x + 2y\frac{dy}{dx} = 0,$$

and so

$$\frac{dy}{dx} = -\frac{x}{y}.$$

Note that this is *always* correct. If $y = f(x) = \sqrt{1 - x^2}$, it gives

$$\frac{dy}{dx} = \frac{-x}{\sqrt{1 - x^2}};$$

while if $y = g(x) = -\sqrt{1 - x^2}$, it gives

$$\frac{dy}{dx} = \frac{x}{\sqrt{1 - x^2}}.$$

EXAMPLE 3. Find a formula for du/dx, supposing that u is a differentiable function of x defined implicitly by the equation

$$u^3 + ux^2 + x^3 = 1.$$

. . . .

Solution. Differentiating the equation with respect to x, we have

$$3u^2 \frac{du}{dx} + [u \cdot 2x + x^2 \frac{du}{dx}] + 3x^2 = 0,$$

$$\frac{du}{dx} = -\frac{2xu + 3x^2}{3u^2 + x^2}.$$

If $u = h(x)$ is the function we are talking about, then this equation becomes

$$h'(x) = -\frac{2xh(x) + 3x^2}{3(h(x))^2 + x^2}.$$

Higher derivatives of implicitly defined functions can also be computed. The following simple example illustrates the process.

EXAMPLE 4. Compute d^2y/dx^2 when y is a differentiable function of x implicitly defined by the equation

$$\frac{x^2}{a^2} - \frac{y^2}{b^2} = 1.$$

. . . .

Solution. We first compute dy/dx implicitly as in our earlier examples. We have

$$\frac{2x}{a^2} - \frac{2y \frac{dy}{dx}}{b^2} = 0,$$

$$\frac{dy}{dx} = \frac{b^2 x}{a^2 y}.$$

Differentiating this equation with respect to x gives

$$\frac{d^2y}{dx^2} = \frac{b^2}{a^2} \left[\frac{y - x \frac{dy}{dx}}{y^2} \right].$$

Now all we have to do is substitute the formula already obtained for dy/dx in the right side above. Then,

$$\frac{d^2y}{dx^2} = \frac{b^2}{a^2}\left[\frac{y - x\left(\frac{b^2 x}{a^2 y}\right)}{y^2}\right]$$

$$= \frac{b^2}{a^2}\left[\frac{a^2 y^2 - b^2 x^2}{a^2 y^3}\right]$$

$$= \frac{b^2}{a^2}\left[-\frac{a^2 b^2}{a^2 y^3}\right] = \frac{-b^4}{a^2 y^3}.$$

In order to reach the last line above we used the equation

$$a^2 y^2 - b^2 x^2 = -a^2 b^2,$$

which is another form of the equation of the hyperbola

$$\frac{x^2}{a^2} - \frac{y^2}{b^2} = 1.$$

We saw above how to calculate the derivative of a function h if we assume that h is a differentiable function defined implicitly by an equation in x and y. Such an equation can always be written in the form

$$f(x, y) = 0,$$

where f is a function of *two* variables. Then $y = h(x)$ is a function satisfying the identity

$$f(x, h(x)) \equiv 0.$$

We shall take an abbreviated look at some calculus of functions of more than one variable in Chapters 16 and 17, and we will see there what kind of property f must have in order to guarantee that the equation $f(x, y) = 0$ has such a smooth solution function $y = h(x)$. The theorem in question is naturally called the *implicit function theorem*.

PROBLEMS FOR SECTION 6

For Problems 1 through 9 find (dy/dx) by:

a) Solving explicitly for y and then differentiating;

b) Using implicit differentiation.

Show that your two answers are equivalent.

1. $2xy + 5 = 0$

2. $y^3 = 4x^2 + 2x + 1$

3. $x^2 + y^2 = 16$

4. $y^2 = 2x + 1$

5. $3x^2 + 2x + y^2 = 10$

6. $y^5 = x^3$

7. $y^2 = \dfrac{x^2 - 1}{x^2 + 1}$

8. $x^2 + 2xy = 3y^2$

9. $x^{3/2} + y^{3/2} = 2$

In Problems 10 through 17 find (dy/dx) by implicit differentiation.

10. $2x^3 + 3y^3 + 6 = 0$

11. $x^2y - xy^2 + x^2 + y^2 = 0$

12. $x^4y^3 - 5xy = 100$

13. $\cos y = x$

14. $xy + x - 2y - 1 = 0$

15. $\sin y = xy$

16. $e^y = x^2$

17. $y^3 + y = \sin x$

18. Write the equation of the line tangent to the curve $xy + y^2 - 2x = 0$ at the point $(1,1)$.

19. Find y'' in terms of x and y when $x^2 + 4y^2 = 25$.

20. Find the equation of the tangent line to the curve $x^3 - 3xy^3 + xy^2 = xy + 6$ at the point $(1,-1)$.

21. Find the equation of the normal line to $2x^2 - y^2 = 1$ at the point $(1,1)$.

22. Find y' and y'' when $2xy - x^2 + 3y^2 + 6 = 0$.

23. Prove by implicit differentiation that the tangent line to the ellipse

$$\frac{x^2}{a^2} + \frac{y^2}{b^2} = 1$$

at the point (x_0, y_0) has the equation

$$\frac{xx_0}{a^2} + \frac{yy_0}{b^2} = 1.$$

24. Find the equation for the tangent line to the hyperbola

$$\frac{x^2}{a^2} - \frac{y^2}{b^2} = 1$$

at the point (x_0, y_0) in a form analogous to the one given above for the ellipse.

In the following problems compute dy/dx and the second derivative d^2y/dx^2 by implicit differentiation.

25. $y = e^{(x+y)}$ 26. $y^3 + y = x$ 27. $\cos x = \sin y$

7. THE DIFFERENTIAL FORMALISM

You are likely to run into differentials such as dx and dy in your reading, and we should say something about them. They will be treated here in a purely formal way, as a useful symbolism. In fact, they are just going to be new variables that are related to each other by certain equations.

For anyone interested, the next section suggests a conceptual basis for differentials. Something like this is the "right way" to approach differentials but, unfortunately, it seems unfeasible in a first course. Many mathematicians would prefer omitting the subject entirely until it can be developed conceptually, making do meanwhile with the following coincidences in the Leibniz symbolism:

I. Although we have insisted that the Leibniz symbol dy/dx is not a quotient, nevertheless the chain rule

$$\frac{dy}{dt} = \frac{dy}{dx} \cdot \frac{dx}{dt}$$

looks as though it were obtained by cancelling the two occurrences of dx on the right, and this formal cancellation is a good way of remembering the rule.

II. A similar situation will be discovered in Chapter 9, when an integration is performed by changing variables. If $x = g(t)$ and if an integral has the form

$$\int \cdots g'(t)\, dt,$$

then we are tempted to make the formal cancellation

$$g'(t)dt = \frac{dx}{dt}\, dt = dx,$$

and so to write the integral as

$$\int \cdots dx.$$

This formal cancellation is found to be correct, again because of the chain rule.

Thus the chain rule permits us to treat dy/dx as though it were a quotient in several important situations. We shall now proceed actually to make dy/dx into a quotient.

Suppose that we had been using only function notation up to this point. In

particular, we would then have the chain rule only in the following form:
If $y = f(x)$ and $x = g(t)$, and if we set $h(t) = f(g(t))$, then

$$h'(t) = f'(g(t)) \cdot g'(t)$$
$$= f'(x) \cdot g'(t).$$

We now reintroduce the Leibniz symbols, but in a different way. Suppose there is some underlying independent variable, say t. We associate with t a new independent variable dt, called the differential of t. Like Δt, this is not a product of something d times something t, but a wholly new variable, independent of t, but to be used along with t. We can give dt any value we wish, and in practice we often set $dt = \Delta t$.

Now suppose that x is a differentiable function of t, say

$$x = g(t).$$

Then the differential dx is the function of the two variables t and dt defined by

$$dx = g'(t) \, dt.$$

The differential of any other function of t is defined in the same way. Thus if

$$y = h(t),$$

then

$$dy = h'(t) \, dt.$$

Finally, let us suppose that y also happens to be a differentiable function of x, say

$$y = f(x).$$

Then

$$h(t) = y = f(x) = f(g(t)),$$

so

$$h'(t) = f'(x)g'(t)$$

by the chain rule. Then

$$dy = h'(t) \, dt = f'(x)g'(t) \, dt = f'(x) \, dx.$$

We thus see that the equation

$$dy = f'(x) \, dx$$

holds if y = f(x), regardless of whether x is independent or both x and y depend on some other variable t. So long as y = f(x), then also dy = f'(x) dx, no matter what variable ultimately turns out to be the underlying independent variable.

Moreover, if $dx \neq 0$, then we can divide by it and have

$$\frac{dy}{dx} = f'(x),$$

where now the Leibniz symbol dy/dx is a quotient, a quotient of two related differentials.

The Differential Rules

The derivative rules now acquire very useful differential formulations. For example, if we multiply the product rule

$$\frac{d}{dx}(uv) = u\frac{dv}{dx} + v\frac{du}{dx}$$

by dx, it becomes the differential product rule

$$d(uv) = u\,dv + v\,du.$$

This form has the advantage of being "independent of the independent variable." Moreover, it turns out to be just as valid for functions of several variables as it is for functions of one variable.

Here is our basic list in differential form:

$$d(u^a) = au^{a-1}\,du \qquad \text{(for any rational exponent } a\text{)},$$

$$d(e^u) = e^u\,du,$$

$$d(\sin u) = \cos u\,du,$$

$$d(\cos u) = -\sin u\,du,$$

$$d(u + v) = du + dv,$$

$$d(cu) = c\,du,$$

$$d(uv) = u\,dv + v\,du,$$

$$d\left(\frac{u}{v}\right) = \frac{v\,du - u\,dv}{v^2}.$$

We recover the derivative forms of these rules by dividing through by the differential of the independent variable.

PROBLEMS FOR SECTION 7

Find the following differentials.

1. $d(x^3 + 2x + 3)$ 2. $d(3x + 2)$

3. $d(x^2 + 2)(2x + 1)^3$ 4. $d\left(\dfrac{x - 3}{x + 2}\right)$

5. de^{x^2} 6. $d\sqrt{1 + \sqrt{1 + x}}$

7. $d\left(\dfrac{x^3 + 2x + 1}{x^2 + 3}\right)$ 8. $d(\cos^2 2x + \sin 3x)$

9. $d(x^{-1} + 2x + 5)$ 10. $d\left(\dfrac{\sin x}{\cos x}\right)$

Using the method of differentials, find the derivatives of the following functions.

11. $y = 2 + 2x$

12. $y = \dfrac{1}{(2 + 3x)^3}$

13. $y = (x + 2)^{2/3}(x - 1)^{1/3}$

14. $y = \dfrac{x^{2/3}}{(x + 1)^{2/3}}$

Use differentials to find the derivative dy/dx in each of the following situations, assuming in each case that the equation defines y as a differentiable function of x. Also find the derivative dx/dy, assuming that the equation defines x as a differentiable function of y.

15. $e^y = x^2$ 16. $y^3 + x = \sin xy$

17. $y^3 + xy - x^5 = 5$

18. $2x^2 + xy - y^2 + 2x - 3y + 5 = 0$

8. DIFFERENTIALS AS FUNCTIONS

Modern advanced mathematics takes the point of view that only functions can have differentials. The differential of the function ϕ is a new function $d\phi$ of *two* independent variables x and h, defined by

$$d\phi(x, h) = \phi'(x) \cdot h.$$

(The symbol ϕ is the lower-case Greek letter phi, and is pronounced "fee" or "fie".) Does this do violence to our symbolic treatment of differentials? The answer is "no," and we can see this as follows:

Whenever

$$y = f(x),$$

it is possible to view the variables x and y as functions. For example, we could imagine the graph of f being traced by a moving point that has the x-coordinate $x(t)$ at time t. Then the equation $y = f(x)$ must be viewed as the composite function equation

$$y(t) = f(x(t)).$$

(In terms of the composition operation $\circ$, this is the equation $y = f \circ x$.) According to our new definition, the functions x and y have differentials dx and dy defined by

$$dx(t, h) = x'(t) \cdot h,$$

$$dy(t, h) = y'(t) \cdot h.$$

But

$$y'(t) = f'(x(t))x'(t),$$

by the chain rule, and if we substitute this into the equation for $dy(t, h)$ we see that

$$dy(t, h) = f'(x(t)) \cdot dx(t, h).$$

We thus end up with the same equation

$$dy = f'(x) \, dx,$$

but now interpreted as an identity between the *functions dx* and *dy*.

The two points of view toward differentials are therefore notationally consistent. This is one aspect of the general principle that it doesn't matter how we view variables like x and y in a first course. At one extreme we can view the variable x as just a symbol, i.e., as the letter x used in a certain way. At the other extreme we can view x as a function, as above, or in other ways. For example, if A is the area of a triangle, then we can view A as a numerically-valued function whose domain is the collection of all triangles in the plane.

What does matter about variables is that they can be given values. We must be able to say "Let x have the value 5" and "If $x = 5$, then $x^2 = 25$". In other words, what matters about variables is that they can be *used* in the familiar way.

the shape

of a graph

In Chapter 3 we saw how we could determine much about the overall behavior of a function f just by finding the points on its graph where the derivative f' is zero. In a similar way, we can use the zeros of the *second* derivative f'' to give our sketch of the graph of f the right *shape*. This is the first topic in the present chapter. Then we shall go in the other direction and examine the impact of an *isolated singlular point* at which *no* derivative exists. Finally, we shall apply these new ideas to study the "local" behavior of f. First we look at the question of how closely a tangent line fits the graph of f near the point of tangency. Then we analyze the graph in the neighborhood of a critical point.

The chapter concludes with a section on polar coordinates. This material does not particularly belong here, but it has to go somewhere.

1. CONCAVITY

One of the most distinctive features of a graph is the direction in which it is turning or curving. The first two patches of graphs shown below are *concave up* (turning upward) and the second two are *concave down* (turning downward). Note that the graph of a decreasing function can be turning in *either* direction (first and third patches), as can the graph of an increasing function (second and fourth patches).

Knowing the concavity of a graph contributes more to getting a correct overall picture than anything else except locating the critical points. Knowing how the graph is curving is really what tells us its shape. For example, if we know that the three points shown below lie on a graph that is concave down then we know how the graph must look.

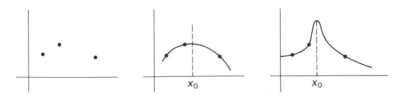

But if we don't know that the graph is concave down then it may look entirely different, as at the right. Roughly speaking, *constant* concavity rules out oscillatory, or wiggly, behavior.

Most graphs will be concave up over some intervals and concave down over other intervals. Points across which the direction of concavity changes are called *inflection points*. In the following figure the graph on page 123 is relabelled for concavity.

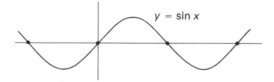

The only point of inflection is at $x = 0$, the curve being concave down over $(-\infty, 0]$ and concave up over $[0, \infty)$.

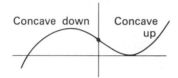

The graph of sin x has a point of inflection at every multiple of π. It is concave down whenever it lies above the x-axis and concave up whenever it lies below.

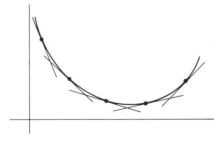

Geometrically, it appears that a graph is *concave up* if its tangent line rotates *counterclockwise* and its slope *increases* as the point of tangency runs from left to right along the curve. You should be able to "see" the slope increasing. If you visualize a succession of tangent lines as in the figure above, then you see that each slope is greater than the one before. Or you can visualize holding a straight edge against the bottom of the curve and "rolling" it counterclockwise along the curve, in which case the tangent segments above are successive positions of the rolling line. We make this property our definition.

Definition. A function f is concave up over the interval I if f′ exists and is an increasing function on I. Similarly, f is concave down on I if f′ is decreasing on I.

In view of the above definition, the shape principles of Section 9, Chapter 3, relate the *concavity* of f to its *second* derivative f'', as follows.

1. If f'' is never zero on an interval I, then f is of constant concavity on I (since then f' is monotone on I). The intervals of constant concavity of f (the intervals of monotonicity of f') are thus marked off by the values of x where $f''(x) = 0$, i.e., by the critical points of f'.

Of course, if we happen to know the sign of f'', then we know which concavity we have:

2. If $f''(x) > 0$ on an interval I, then f is concave up on I (since then f' is increasing on I), and if $f''(x) < 0$, then f is concave down on I.

EXAMPLE 1. Since $d^2 \sin x / dx^2 = -\sin x$, it follows that the graph of $\sin x$ is concave down whenever it is above the x-axis and concave up whenever it is below. We have been drawing the graph this way, but until now we couldn't be absolutely certain that there wasn't some extra wiggle we were overlooking.

EXAMPLE 2. Let us sketch the graph of

$$f(x) = x^4 - 2x^3 = x^3(x - 2),$$

using the concavity shape principle. We have

$$f'(x) = 4x^3 - 6x^2 = 2x^2(2x - 3),$$

$$f''(x) = 12x^2 - 12x = 12x(x - 1).$$

The intervals of constant concavity for f are marked off by the values of x for which $f''(x) = 0$ (the critical points of f') as we noted above, and we see that these values are $x = 0$ and $x = 1$. The intervals of constant concavity are thus $(-\infty, 0]$, $[0, 1]$, and $[1, \infty)$. In order to determine which way the graph of f is concave over these intervals (i.e., whether f' is increasing or decreasing), we compute the values of f' at the ends of the intervals and the behavior of f' at infinity, and get the table

x	$-\infty$	0	1	∞
$f'(x)$	$-\infty$	0	-2	∞

Thus f is concave up over $(-\infty, 0]$. The reason, again, is that we already know that f' is monotone on this interval and now we see that it is increasing, since its values run from $-\infty$ to 0. Similarly, f is concave down on $[0,1]$, because f' runs from 0 to -2 and hence is decreasing there. Finally f is concave up over $[1,\infty)$.

We also need the critical points of f, and find that $f'(x) = 0$ at $x = 0$ and $x = 3/2$. In Chapter 3 we then went on to determine whether f is increasing or decreasing on each of the intervals these two points determine. However, this is unnecessary when we know the concavity of f. For example, a curve that is concave up must be increasing just after a horizontal tangent and decreasing just after a vertical tangent.

We compute the values of f at all these special points and end up with the following combined table.

Concavity		Up		Down		Up	
x	$-\infty$		0	1	3/2		$+\infty$
$f(x)$	$+\infty$		0	-1	$-27/16$		$+\infty$
$f'(x)$	$-\infty$		0	-2	0		$+\infty$
$f''(x)$			0	0			

These data determine the shape and position of the graph.

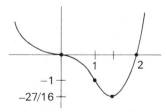

EXAMPLE 3. We shall sketch the graph of

$$f(x) = \frac{1}{1 + x^2}.$$

The first two derivatives are

$$f'(x) = \frac{-2x}{(1 + x^2)^2},$$

$$f''(x) = \frac{(1 + x^2)^2(-2) - (-2x)2(1 + x^2)(2x)}{(1 + x^2)^4}$$

$$= \frac{-2(1 + x^2) + 8x^2}{(1 + x^2)^3}$$

$$= \frac{2(3x^2 - 1)}{(1 + x^2)^3}.$$

This time we go directly to the combined table. Remember, we want the critical points of both f and f' (the zero of both f' and f'') and their values at infinity. You may be able to do the arithmetic mentally, or you may want to check it with pencil and paper.

Concavity	Up			Down			Up
x	$-\infty$		$-1/\sqrt{3}$	0	$1/\sqrt{3}$		$+\infty$
$f(x)$	0		$3/4$	1	$3/4$		0
$f'(x)$	0		$9/8\sqrt{3}$	0	$-9/8\sqrt{3}$		
$f''(x)$			0		0		

The direction of concavity depends on whether f' is increasing or decreasing, and this is determined by comparing the values of f' at the endpoints of its critical intervals.

In plotting the above values we use the approximation $1/\sqrt{3} \approx 3/5$. As x tends to infinity, the graph approaches, but never touches, the x-axis. Two curves related this way appear to be "tangent at infinity." We say they are *asymptotic*.

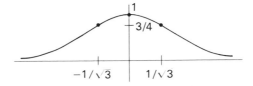

EXAMPLE 4. The function

$$f(x) = \frac{1}{x^2 - 1}$$

has the added complication of "blowing up," along with its derivatives, at ± 1. Now we have to determine whether $f(x)$ approaches $+\infty$ or $-\infty$ as x approaches one of these points from the left and from the right. But again concavity considerations save us from calculating these facts, as we shall see in a minute. We have

$$f'(x) = \frac{-2x}{(x^2 - 1)^2},$$

$$f''(x) = \frac{(x^2 - 1)^2(-2) - (-2x)2(x^2 - 1)2x}{(x^2 - 1)^4}$$

$$= \frac{-2(x^2 - 1) + 8x^2}{(x^2 - 1)^3}$$

$$= \frac{2(3x^2 + 1)}{(x^2 - 1)^3}.$$

Notice that f' has *no* critical points unless we include the points where it blows up. The intervals of constant concavity are thus $(-\infty, -1)$, $(-1, 1)$, $(1, \infty)$. This time we shall determine the direction of concavity from the sign of f'' at one point in each of these intervals. Then we have the following table, including points where the function is not defined (nd):

Concavity		Up		Down		Up	
x	$-\infty$	-2	-1	0	$+1$	$+2$	$+\infty$
$f(x)$	0		nd	-1	nd		0
$f'(x)$			nd	0	nd		
$f''(x)$		$+$	nd	$-$	nd	$+$	

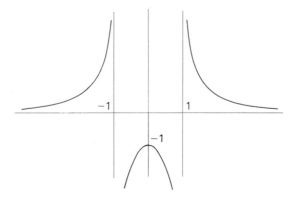

We now know that the graph is concave down over $(-1, 1)$ and that it blows up at each endpoint. It therefore must approach $-\infty$ at both endpoints. Similarly it is concave up over $(-\infty, -1)$ and blows up at -1, and so must approach $+\infty$ there. We thus have the graph shown above. The graph is asymptotic to the x-axis and to the two vertical lines at $x = 1$ and $x = -1$. Note how these three asymptotes combine with the known concavity to govern the overall shape of the graph.

The Notion of Concavity

Probably you would agree that a concave upward graph has all four of the properties suggested by the following figures,

These properties are:

1. *The slope $f'(x)$ of the tangent line is an increasing function of x ($f'(x_1) < f'(x_2)$ if $x_1 < x_2$).*

2. *The graph of f lies strictly above each tangent (except for the point of tangency).*

3. *Each arc of the graph of f lies strictly below its chord (except for the two endpoints).*

4. *The second of two nonoverlapping chords always has the larger slope.*

But then a graph can't be considered to be concave up unless it has all four of these properties. That is, the intuitive notion of upward concavity is equivalent to the sum total of these properties. Fortunately, at least for those who like simple definitions, there is a theorem to the effect that these four properties are all logically equivalent to each other. So if the graph of a differentiable function has any one of these properties, then of necessity it has them all. It is by virtue of this theorem that the definition of concavity given earlier is adequate.

PROBLEMS FOR SECTION 1

1. Show in the manner of Example 1 that we have been drawing the graph of $y = e^x$ with the right general shape.

Determine the intervals of constant concavity and the critical points for the graph of each of the following equations. Then sketch the graph.

2. $x^2 - 4 = y$

3. $3y = 8x^3 - 6x + 1$

4. $y = 8x^3 - 6x^2 + 1$

5. $y = x^3 + x$

6. $y(x - 2) = 1$

7. $x^2y - 2x^2 - 16y = 0$

8. $xy - y - x - 2 = 0$ 9. $y = \sin 3x$

10. $x^2 y - x^2 + 4y = 0$ 11. $y = x^3 - 3x + 2$

12. $x^2 y - x - 4y = 0$ 13. $y = x^2 - 2x + 3$

14. $x^2 y + y - 4x = 0$ 15. $y = x^3 - (21/4)x^2 + 9x - 4$

16. $y = x^4 - 2x^2$ 17. $9y = (x + 2)(x - 2)^3$

18. $y = (x - 1)(x + 1)^3$ 19. $y = (1/3)x^3 - 4x^2 + 12x - 8$

20. $y = xe^{1-x}$, over the interval $[0,\infty)$. Use the approximate value $e = 2.7$.

21. $y = x - x^3$, over the interval $[0,1]$. Be sure to have the right slopes at the endpoints.

22. $y = x/2 - \sin x$, over the interval $[0,2\pi]$.

23. $y = 2 \sin x + \sin 2x$, over $[0,2\pi]$.
     a) Do this first by sketching the graphs of $\sin 2x$ and $2 \sin x$ and visually adding them.
     b) Then apply the methods of this section.

24. $y = x/\sqrt{1 + x^2}$. (This function has finite limits as $x \to \pm\infty$, which determine horizontal asymptotes toward which the graph tends as $x \to \infty$ and $x \to -\infty$, respectively. Be sure to get these values right.)

25. $y = e^{-x^2/2}$ 26. $y = x^2/(1 + x^2)$

27. $y = x^3/(1 + x^2)$ 28. $y = xe^x$

29. $y = x^2 e^x$.

30. $y = \tan x$, over $(-\pi/2, \pi/2)$. Your results should justify the general features shown in Section 3 of the last chapter.

31. $y = \sec x$, over $(-\pi/2, 3\pi/2)$.

32. $y = \dfrac{1}{x} + x$. Note that this graph is asymptotic to the line $y = x$, since the difference between the two y-coordinates $(1/x)$ approaches 0 as $x \to \infty$.

33. $y = \dfrac{1}{x} - x$ 34. $y = \dfrac{1}{x^2} + 2x$

35. $y = \dfrac{1}{x} + \sqrt{x}$. (This graph is asymptotic to the half parabola $y = \sqrt{x}$.)

36. A function graph turns in the same direction whether it is traced forward or backward, so $f(x)$ and $f(-x)$ should show the same concavity behavior. State this fact more carefully and verify it analytically.

37. Show that the cubic graph

$$y = x^3 + ax^2 + bx + c$$

has three possible shapes, depending on whether $a^2 < 3b$, $a^2 = 3b$, or $a^2 > 3b$. Sketch the three possible shapes.

In the following four problems we consider a differentiable function f over an interval I and show that the various possible definitions for f being *concave up* are all equivalent.

38. Show that if $f'(x)$ is increasing, then the graph of f lies strictly above each tangent line (except for the point of tangency). [*Hint.* Try to use the fact that if $g'(x) > 0$, then g is increasing, for a suitably chosen function g.]

39. Show that if the graph of f lies strictly above each of its tangent lines, then the second of any pair of nonoverlapping chords always has the larger slope.

40. Show that if the graph of f has the property that the second of any pair of nonoverlapping chords always has the larger slope, then each arc of the graph of f lies strictly below its chord (except for the endpoints).

41. Show, finally, that if each arc of the graph of f lies strictly below its chord, then $f'(x)$ is increasing.

42. Show that a rational function of degree at most 1 has an asymptote as x approaches infinity.

2. SINGULAR POINTS AND THE GEOMETRIC PRINCIPLES

According to Theorem 5 of Chapter 3, a differentiable function is necessarily continuous. This theorem does not have a converse, however. A continuous function can fail to be differentiable, and this can happen at isolated points even for functions that are otherwise very smooth. For example, if $f(x) = x^{1/3}$, then

$$f'(x) = \frac{1}{3}x^{-2/3} = \frac{1}{3x^{2/3}}$$

except at $x = 0$. Since $f'(x) \to \infty$ as $x \to 0$, we visualize the tangent line rotating into a vertical positiom as the point of tangency approaches the origin. We therefore expect that at the origin the graph has a vertical tangent. In fact, at $x = 0$, we have the secant slope

$$\frac{f(r) - f(0)}{r - 0} = \frac{r^{1/3} - 0}{r - 0} = r^{-2/3},$$

and therefore

$$\lim_{r \to 0} \frac{f(r) - f(0)}{r - 0} = \lim_{r \to 0} r^{-2/3} = +\infty,$$

as predicted. However, $f(x) = x^{1/3}$ is continuous *everywhere*. Here are the graphs:

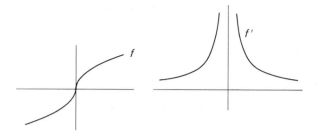

The function $f(x) = x^{2/3}$ is somewhat similar, except that here the derivative

$$f'(x) = \frac{2}{3}x^{-1/3} = \frac{2}{3x^{1/3}}$$

approaches $-\infty$ as x approaches 0 from the left $(x \to 0^-)$, while $f'(x) \to +\infty$ as $x \to 0^+$. The graph of $f(x)$ has a *cusp* at the origin.

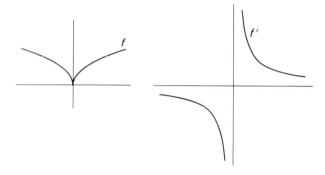

Another way in which a continuous function can fail to be differentiable is shown by $f(x) = |x|$. Again we have an everywhere continuous function, with graph as shown below on the left. When $x < 0$, we have $f(x) = -x$ and $f'(x) = -1$. If $x > 0$, then $f(x) = x$ and $f'(x) = 1$. The graph of f' is thus as shown on the right.

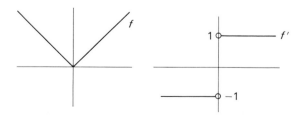

At $x = 0$ we can obviously expect trouble. The secant slope (difference quotient) there is

$$\frac{f(r) - f(0)}{r - 0} = \frac{|r|}{r} \begin{cases} = \dfrac{-r}{r} = -1, & \text{if } r < 0, \\[2mm] = \dfrac{r}{r} = +1, & \text{if } r > 0. \end{cases}$$

The difference quotient thus has the *one-sided* limits -1 on the left and $+1$ on the right, and we could say that f has *one-sided derivatives*

$$f'(0^-) = -1, \qquad f'(0^+) = +1.$$

But $f'(0)$, the two-sided limit, does not exist.

A more interesting function with the same behavior at $x = 0$ is $f(x) = |x| + x^2$. Its graph looks like this:

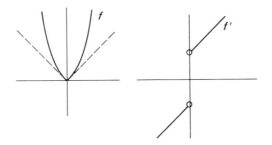

Each of the above functions has an *isolated singularity* at the origin. The function

$$f(x) = x^{2/3} + (x - 1)^{2/3} - 1$$

has isolated singularities at $x = 0$ and $x = 1$.

Definition. *A function f has an isolated singularity at the point x_0 if its derivative f' exists throughout some interval I about x_0 except at the point x_0 itself.*

The mean-value property can be spoiled by a single interior singularity. For example, consider $f(x) = x^{2/3}$ over the interval $[a, b] = [-1, 1]$. Then the secant slope $[f(1) - f(-1)]/(1 - (-1))$ is 0, but there is no point where the derivative

$f'(x) = 2/3x^{-1/3}$ is 0. This failure of the mean-value property occurs because f has a singularity at the interior point where we would expect f' to be zero, namely, at the origin. On the other hand, the figure on the right shows that a singularity at an endpoint of $[a,b]$ does no damage, provided f is continuous there.

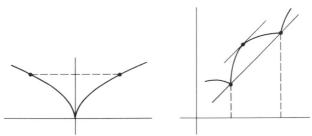

This observation is an important strengthening of the mean value property. However, before stating the stronger form we interpose one more remark. We have not yet proved the mean-value property because it seems obviously true and there is no compelling need to prove the obvious. Later on, however, in Chapters 11 and 20, we shall decide that such principles, when arrived at only intuitively, do require analytic justification. So we shall revise the status of the mean-value property to that of a *theorem* (whose proof is yet to be considered).

Theorem 1 (***Mean-Value Theorem***). *If f is continuous on the closed interval $[a, b]$ and differentiable everywhere except possibly at one or both endpoints, then there is an interior point X at which*

$$f'(X) = \frac{f(b) - f(a)}{b - a}.$$

Next we must make explicit a distinction between a graph that is *sloping up* and a graph that is *rising*. Of course, sloping upward means positive slope, i.e., $f' > 0$. On the other hand, to say that a graph is *rising* over an interval I means, in terms of function values, that $f(x)$ increases when x increases, i.e., that

$$f(x_1) < f(x_2) \qquad \text{whenever } x_1 < x_2 \text{ in } I.$$

A function having this property is said to be *increasing* on I. These two notions are not quite equivalent. For example, the quarter-circle drawn below is rising over the interval $I = [-1,0]$, but its slope is not positive at either endpoint. However, there is a close connection.

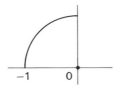

Theorem 2. *If if is differentiable and $f' > 0$ on the open interval (a, b), and if f is continuous on the closed interval $[a, b]$, then f is increasing on $[a, b]$.*

Proof. This follows analytically from the mean-value theorem in the same way as did Theorem 7 in Chapter 3. If x_1 and x_2 are any two points of $[a,b]$, then, by the mean-value theorem,

$$f(x_2) - f(x_1) = f'(X)(x_2 - x_1),$$

where X is some point lying strictly between x_1 and x_2. Since $f'(X)$ is positive, $x_2 - x_1$ and $f(x_2) - f(x_1)$ have the same sign. In particular, if $x_1 < x_2$, then $f(x_1) < f(x_2)$. That is, f is increasing on $[a,b]$. ∎

In exactly the same way we see that if $f' \geq 0$ throughout an interval I, then $f(x) \leq f(y)$ whenever $x < y$ in I. This leads to a result that we shall find useful in the next section. Suppose that $f'(x) \leq g'(x)$ in an interval I. Then $(g - f)' \geq 0$ in I and so

$$g(x) - f(x) \leq g(y) - f(y)$$

whenever $x < y$ in I. If we rearrange this inequality we obtain the following result.

Theorem 3. *If $f'(x) \leq g'(x)$ throughout an interval I, then*

$$f(y) - f(x) \leq g(y) - g(x)$$

whenever $x \leq y$ in I. In loose terms, a function with a smaller derivative changes less.

The following useful "local" result also seems to be related to Theorem 2, but it does not involve the mean-value theorem. The question is:
What can we conclude from the inequality $f'(x) > 0$ at the single point $x = x_0$?
Since $f'(x_0)$ is the limit of the difference quotient $(f(r) - f(x_0))/(r - x_0)$ as $r \to x_0$, it follows that if $f'(x_0)$ is positive, then so is the difference quotient for all r sufficiently close to x_0. And since a quotient is positive exactly when its numerator and denominator have the same sign, we conclude that

$$f(r) - f(x_0) \qquad \text{and} \qquad r - x_0$$

have the same sign when r is close to x_0. Thus:

Theorem 4. *If $f'(x_0) > 0$, then $f(x) - f(x_0)$ changes sign from $-$ to $+$ as x crosses x_0. That is, there is a small interval about x_0 on which we have*

$$f(x) < f(x_0) \qquad \text{if } x < x_0,$$

and

$$f(x) > f(x_0) \qquad \text{if } x > x_0.$$

(If $f'(x_0) < 0$, the f inequalities are interchanged.)

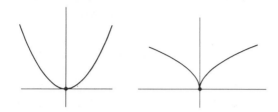

Finally, consider the notion of a critical point. When we compare functions like $f(x) = x^2$ and $g(x) = x^{2/3}$, graphed above, we see that a singularity can be just as critical for the behavior of a graph as a point where $f'(x) = 0$. We therefore redefine the notion of critical point.

Definition. *A critical point of a function f is a point x interior to its domain at which either $f'(x) = 0$ or $f'(x)$ fails to exist.*

For example, each of the functions $f(x) = x^2$ and $g(x) = x^{2/3}$ has $x = 0$ as its only critical point, but for different reasons.

In these terms, the shape principle of Chapter 3 has the following more general form.

Theorem 5. *If f is continuous on an interval I and if f has no critical point interior to I, then f is monotone on I.*

PROBLEMS FOR SECTION 2

Sketch the graphs of the following functions, using the procedures of Section 1. In each case the new feature will be one or more points where the graph has a vertical tangent, or two different one-sided tangents.

1. $f(x) = x^{1/3}$ 2. $f(x) = x^{2/3}$

3. $f(x) = x^{2/3}\left(\dfrac{5}{2} - x\right)$ 4. $f(x) = x^{1/3} + x^{2/3}$

5. $f(x) = \sin(2x^{1/3})$ on $[-4, 4]$

6. $f(x) = \sin\left(\dfrac{\pi}{2}x^{2/3}\right)$ on $[-3, 3]$

7. $f(x) = \sin|x|$

8. $f(x) = (x - 1)^{1/3} + (x + 1)^{1/3}$ 9. $f(x) = (x - 1)^{2/3} - (x + 1)^{2/3}$

10. Define a function f such that $f'(x) = |x|$, and draw its graph. This is a continuous function with a continuous derivative, for which f'' is discontinuous.

11. The mean-value theorem may fail if f is discontinuous at one of the endpoints a and b. Show this by drawing the graph of a function that has a jump discontinuity at $x = a$ but is otherwise smooth, and for which the mean-value theorem fails.

12. The function $f(x) = x^3$ is increasing on $(-\infty, \infty)$ despite the fact that $f'(0) = 0$.
    a) Prove this algebraically from the inequality laws.
    b) Prove it by applying Theorem 2 separately on the intervals $(-\infty, 0]$ and $[0, \infty)$ and then combining these two results.

13. The function $f(x) = x^{1/3}$ is increasing on $(-\infty, \infty)$ despite the fact that $f'(0)$ is not defined. Prove this in two different ways, as in the above problem.

14. Prove the following theorem.

    Theorem. *If $f(x)$ is continuous on the interval I, and if $f'(x)$ exists and is positive everywhere except possibly at a finite number of points, then f is increasing on I.*

15. Rephrase the definition of an increasing function $y = f(x)$ in terms of Δx and Δy.

A polygonal function is a continuous function whose graph is made up of a number of straight line segments.

16. Show that $f(x) = |x - 2| + (x - 2)$ is a polygonal function.

17. Show that if f is a polygonal function whose graph has just one vertex, at $x = 2$, then $f(x)$ can be expressed in the form

$$f(x) = a|x - 2| + bx + c.$$

18. Show that a polygonal function with just one vertex, at $x = x_0$, is of the form

$$f(x) = a|x - x_0| + bx + c.$$

19. Show that

$$f(x) = |x + 1| + |x - 1| - 1$$

is a polygonal function. (Show, for example, that if x lies in the interval $(-\infty, -1)$ then $f(x) = -(x + 1) - (x - 1) - 1 = -2x - 1)$. A polygonal function can be graphed exactly from its vertices and its two slopes before the first vertex and after the last vertex. Graph the above function in this manner.

20. Show that any function of the form

$$f(x) = a_1|x - x_1| + a_2|x - x_2| + \cdots + a_n|x - x_n| + bx + c$$

is a polygonal function.

Graph the following polygonal functions

21. $f(x) = \frac{1}{2}(x + |x|)$.   22. $f(x) = |x + 1| - |x - 1|$.

23. $f(x) = |x + 1| + |x - 1| - 2x$.

Find a formula for each of the polygonal functions graphed below.

24.

25.

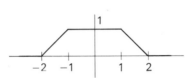

26.

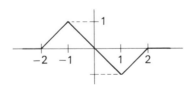

27. a) Prove that $\sin x \le x$ for all nonnegative x (use Theorem 3).
    b) Use part (a) and Theorem 3 again to conclude that

$$1 - \cos x \le \frac{x^2}{2}.$$

28. Prove the analogue of Theorem 2 for a strictly decreasing function.

29. Prove that if $f'(x) \ge 0$ for every x in an interval I, then $f(x) \le f(y)$ whenever $x < y$ in I.

30. Suppose that $g(x)$ is defined on an interval containing the origin, with

$$g(0) = 0 \quad \text{and} \quad g'(0) = 1.$$

Show that there is an interval I about 0 on which

$$g(x) > \frac{x}{2} \text{ when } x > 0; \qquad g(x) < \frac{x}{2} \text{ when } x < 0.$$

[*Hint.* Try to apply Theorem 4.]

31. Suppose we know that f is continuous everywhere and differentiable except possibly at the point $x = a$. Suppose also that the limit

$$\lim_{x \to a} f'(x)$$

exists; call it l. Prove that then f is differentiable at a and that $f'(a) = l$. [*Hint.* Try to apply the mean-value theorem.]

32. Let f be any function at all satisfying the inequality

$$x \le f(x) \le x + x^2.$$

Prove that $f'(0)$ exists and has the value 1.

33. Carefully sketch the graphs of $y = x$ and $y = x + x^2$ over the interval $[-1, 1]$. Then draw part of the graph of a function f that will lie between these two curves but will not be monotone on any interval containing $x = 0$. You won't be able to draw the whole graph of f, but describe what feature it will have that will prevent it from being monotone on any interval about 0.

34. What conclusion can you draw from the results in the above two problems?

35. Rewrite the inequality proved in Problem 27(b) as

$$1 - \frac{x^2}{2} \le \cos x,$$

and continue applying Theorem 3, showing first that

$$x - \frac{x^3}{3!} \le \sin x$$

for all nonnegative x, and then that

$$\frac{x^2}{2} - \frac{x^4}{4!} \le 1 - \cos x.$$

36. Carry this process on for one more step, proving that

$$\sin x \leq x - \frac{x^3}{3!} + \frac{x^5}{5!}$$

for all nonnegative x.

37. a) Use inequalities from the above two problems to show that

$$\sin x \approx x - \frac{x^3}{3!}$$

with a positive error that is at most $x^5/5!$.

b) Show that $\sin 0.3 \approx 0.2955$ to the nearest four decimal places.

38. A function may have a derivative from the right at a point x_0 and also a derivative from the left at x_0, and yet $f'(x_0)$ may fail to exist because the one-sided derivatives are not equal. The function $|x|$ is an example. Show, however, that any such function is continuous at x_0. (See the proof of Theorem 5 in Chapter 3.)

3. THE TANGENT LINE APPROXIMATION

The line tangent to the graph of

$$y = f(x)$$

at the point $(a, f(a))$ has the equation

$$y_t = f(a) + f'(a)(x - a).$$

This is the point–slope equation of the tangent line, slightly modified by throwing the y_0 term to the right side. We have used the different dependent variable y_t so that we can compare the two graphs for all values of x.

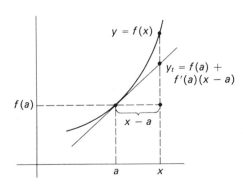

The tangent line has the direction of the graph at the point of tangency, and therefore the tangent line is a good approximation to the graph near the point of tangency. In analytic terms, the linear function $f(a) + f'(a)(x - a)$ is a good approximation to $f(x)$ near $x = a$:

$$f(x) \approx f(a) + f'(a)(x - a),$$

where the symbol $\approx$ is read "is approximately equal to," or just "is approximately."

This tangent-line approximation is related to the calculations in Section 11 of Chapter 3, but here the approximation occurs at the next stage. There we approximated $f(x)$ by $f(a)$. The error was $f(x) - f(a)$, and we used the mean-value property to find a bound for it. Here we are using the better approximation $f(a) + f'(a)(x - a)$ for $f(x)$. This is equivalent to approximating $f(x) - f(a)$ by $f'(a)(x - a)$, so *here we are, in effect, approximating the former error term.* The error in *this* approximation is

$$[f(x) - f(a)] - f'(a)(x - a),$$

and when we use the mean-value property to find a bound for this "second generation" error, it will be the *second* derivative f'' that will be involved.

For the moment, let us put aside the fundamental problem of estimating the error and simply use the above approximate equality as a rough estimate of $f(x)$ when x is near a.

EXAMPLE 1. The approximation

$$f(x) \approx f(a) + f'(a)(x - a)$$

becomes

$$\sqrt{x} \approx \sqrt{25} + \frac{1}{2\sqrt{25}}(x - 25)$$

when $f(x) = \sqrt{x}$ and $a = 25$. If $x = 27$, this gives us the rough estimate

$$\sqrt{27} \approx 5 + \frac{2}{10} = 5.2.$$

This example shows the circumstances under which the above approximation is useful. *If we know or can easily find the values of f and f' at a point $x = a$, then the formula estimates the value of f at a nearby point x.*

EXAMPLE 2. Estimate $63^{1/3}$.
. . . .

Solution. Since $64^{1/3} = 4$, we set $f(x) = x^{1/3}$ and evaluate the formula at $a = 64$

and $x = 63$. Here $f'(x) = \frac{1}{3}x^{-2/3}$, and the approximation becomes

$$63^{1/3} \approx 64^{1/3} + \frac{1}{3 \cdot 64^{2/3}} \cdot (-1) = 4 - \frac{1}{48} \approx 4 - \frac{1}{50} = 3.98.$$

(When we replace $1/48$ by $1/50$ we increase the error by

$$\frac{1}{48} - \frac{1}{50} = \frac{2}{48 \cdot 50} = \frac{1}{1200} < 0.001.)$$

The estimate above is correct to two decimal places, although we have no way of showing this at the moment.

If we set $\Delta x = x - a$, then $f(x) - f(a) = \Delta y$, and the approximate equality

$$f(x) \approx f(a) + f'(a)(x - a)$$

turns into

$$\Delta y \approx f'(a)\Delta x$$

upon subtracting $f(a)$ from both sides. This form also can be used to make rough estimates.

EXAMPLE 3. By approximately how much does the volume of a sphere increase when its radius is increased from 10 to 11 feet?

. . . .

Solution. Here $V = (4/3)\pi r^3$ and $dV/dr = 4\pi r^2$, so the formula is

$$\Delta V \approx 4\pi a^2 \Delta r.$$

When $a = 10$ and $\Delta r = 11 - 10 = 1$, we have the estimate

$$\Delta V \approx 4\pi \cdot 100 \cdot 1 = 400\pi \approx 1200 \text{ cu. ft.}$$

This is just an order-of-magnitude estimate, but such rough approximations can be very useful in giving a "feel" for what is happening. The actual value of ΔV here is

$$\Delta V = \frac{4}{3}\pi(11^3 - 10^3)$$

$$= \frac{4}{3}\pi \cdot 1 \cdot [11^2 + 11 \cdot 10 + 10^2]$$

$$= \frac{4}{3}\pi[331] = 441\frac{1}{3}\pi \text{ cu. ft.}$$

Now let us consider the error E in these approximations. It is the difference $E = y - y_t$, for which we have the two expressions

$$E = f(x) - [f(a) + f'(a)(x - a)]$$
$$= \Delta y - f'(a)\Delta x.$$

Note that the second form of E is obtained from the first by rebracketing. The error E has a very simple geometric interpretation. Of course, $E \to 0$ as $\Delta x \to 0$, but it appears from the figure that $E = \Delta y - f'(a)\Delta x$ becomes small *even in comparison to* Δx, and thus approaches 0 *faster than* Δx.

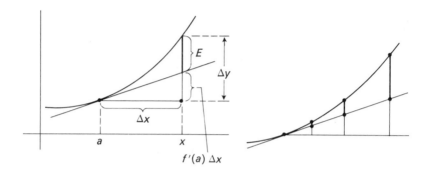

Let us calculate E explicitly for $y = f(x) = x^3$. Then

$$\Delta y = (a + \Delta x)^3 - a^3 = 3a^2\Delta x + 3a(\Delta x)^2 + (\Delta x)^3,$$
$$f'(a)\Delta x = 3a^2\Delta x,$$
$$E = \Delta y - f'(a)\Delta x = (\Delta x)^2[3a + \Delta x].$$

Such calculations make it appear that $E = \Delta y - f'(a)\Delta x$ approaches 0 faster than Δx by virtue of having $(\Delta x)^2$ as a factor. But something else is involved too.

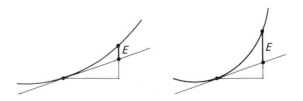

It is apparent that E will be comparatively small for a graph that bends away from its tangent line slowly, and our experience with concavity suggests that this will occur if the *second* derivative $f''(x)$ is small near $x = a$. We can verify these reasonable conjectures by working backward using Theorem 3.

Theorem 6. *If* f'' *exists and is bounded by* B *on an interval* I, *i.e., if* $|f''(x)| \leq B$ *for every* x *in* I, *then*

$$|E| \leq \frac{B}{2}(\Delta x)^2$$

for any two points a *and* $x = a + \Delta x$ *in* I.

Proof. We shall prove a somewhat more general result. Suppose that $f''(x) \leq K$ throughout I. Since $f''(x)$ is the derivative of $f'(x)$, and K is the derivative of Kx, we can apply Theorem 3 and conclude that

$$f'(x) - f'(a) \leq K(x - a).$$

But $f'(x) - f'(a)$ is the derivative of $f(x) - f'(a)x$, and $K(x - a)$ is the derivative of $K(x - a)^2/2$. So, applying Theorem 3 a second time, we conclude that

$$[f(x) - f'(a)x] - [f(a) - f'(a)a] \leq K\frac{(x - a)^2}{2}.$$

The left side can be rearranged to $f(x) - f(a) - f'(a)(x - a)$, which is the error E, and we have $E \leq K(x - a)^2/2$.

In the above, we have been working under the hypothesis that $a < x$. If $x < a$, then the first application of Theorem 3 yields the reversal of the first inequality, but it reverses back again after the second application of Theorem 3.

Arguing in exactly the same way, we find that if $f''(x) \geq L$ throughout I, then $E \geq L(x - a)^2/2$. Thus, if $L \leq f'' \leq K$ on I, then

$$\frac{L}{2}(\Delta x)^2 \leq E \leq \frac{K}{2}(\Delta x)^2.$$

The theorem is now just the special case where $K = B$ and $L = -B$. ∎

EXAMPLE 4. We reconsider Example 1. Taking $f(x) = \sqrt{x}$, $a = 25$ and $\Delta x = 2$, we had

$$\sqrt{27} \approx 5 + \frac{1}{2}\frac{1}{\sqrt{25}} \cdot 2 = 5.2,$$

but now we can go on. When $x \geq 25$, the second derivative $f''(x) = -1/(4x^{3/2})$ is bounded by $1/(4(25)^{3/2}) = 1/500 = 0.002$. Thus

$$|E| \leq \frac{B}{2}(\Delta x)^2 = \frac{0.002}{2}4 = 0.004,$$

so

$$\sqrt{27} \approx 5.20,$$

correct to the nearest two decimal places.

The reasoning used in the proof of Theorem 6 is straightforward, but it was written out for only one of several similar cases, so the proof was not given in full detail. A shorter but more sophisticated proof will be found in Section 6 of Chapter 11.

PROBLEMS FOR SECTION 3

Use the method of differentials (the tangent-line approximation) to estimate each of the following values.

1. $\sqrt{65}$ 2. $\sqrt[3]{124}$ 3. $\sqrt[4]{80}$

4. $\sqrt[5]{31}$ 5. $(.98)^{-1}$ 6. $(17)^{1/4}$

7. $26^{1/3}$ 8. $\sqrt{24}$ 9. $1/\sqrt{15}$

10. $1/80$

11. Find the approximate change in the volume of a sphere of radius r caused by increasing the radius by 2%.

12. A metal cylinder is found by measurement to be 2 ft in diameter and 5 ft long. What, approximately, is the error in the computed volume of this cylinder if the error in measuring the diameter was $1/2$ in?

13. Let V and A be the volume and surface area of a sphere of radius r. If the radius is changed by a small amount Δr, show that

$$\Delta V \approx A \cdot \Delta r.$$

14. Let V and A be the volume and surface area of a cube of edge length x. If x is given a small increment Δx, show that

$$\Delta V \approx A \frac{\Delta x}{2}.$$

15. Solve Problem 1 again, along the lines of Example 4 in the text. That is, obtain a bound for the error E by applying Theorem 6, and then state the tangent-line estimate in decimal form, with some statement about E.

16. Same for Problem 2.

17. Same for Problem 3.

18. Same for Problem 4.

19. Same for Problem 5.

20. Same for Problem 6.

21. Using the fact that $(9/2)^2 = 20\frac{1}{4}$, show that

$$\sqrt{20} = 4.4722\ldots$$

with an error less than 1 in the fourth decimal place.

22. Using the fact that $(3/2)^4 = 5\frac{1}{16}$, and assuming that $5^{3/4} > 3$, show that

$$5^{1/4} \approx \frac{3}{2} - \frac{1}{216}$$

with an error less than $1/10 \cdot 2^{12}$, and hence less than $3(10)^{-5}$.

23. Show that

$$\sin t \approx t$$

with an error less than $t^3/2$. (Use the fact that $|\sin t| \leq |t|$.)

24. Show that

$$1 - \cos x \leq \frac{x^2}{2}$$

for all x.

25. Prove for any positive number x that

$$\sqrt{1 + x} \approx 1 + \frac{x}{2}$$

with an error less than $x^2/8$.

26. Show that if $h > 0$ and $a < 2$, then

$$(1 + h)^a \approx 1 + ah$$

with an error at most $|a(a - 1)h^2/2|$ in magnitude.

27. Write out the proof of Theorem 6 for the case $f''(x) \leq K$ and $x < a$.

28. Write out the proof of Theorem 6 for the case $f''(x) \geq L$ and $x > a$.

29. Draw the figure showing E as the length of a vertical line segment for the case when Δx is negative.

30. The proof of Theorem 6 shows that the error E has the same sign as f'' (supposing that f'' has constant sign). The figure in the text and Problem 29

show this geometrically for the concave up case. Draw the corresponding figures for the case that E and f'' are both negative.

31. By comparing $\sqrt{10}$ to $\sqrt{9} = 3$, and referring to the above problem, show that $\sqrt{10} \simeq 19/6$, with an error that is negative and less than $1/200$ in magnitude. Conclude that

$$3.16 < \sqrt{10} < 3.17.$$

32. By comparing $\sqrt{3}$ to $\sqrt{25/9} = 5/3$, show that $\sqrt{3} \simeq 26/15$, with an error that is negative and less than $1/750$ in magnitude. Conclude that

$$1.732 < \sqrt{3} < 1.734.$$

33. In Chapter 4 we defined the differential dy in terms of a new independent variable dx (supposing that $y = f(x)$). Look up that definition, and show:

If we set $dx = \Delta x$, then $\Delta y - dy$ is the error E in the tangent-line approximation.

Draw the figure for the error E and relabel it so as to exhibit the identity $\cdot E = \Delta y - dy$.

4. NEWTON'S METHOD

Assuming that we know $f(a)$ and $f'(a)$, the approximate equality

$$f(x) \approx f(a) + f'(a)(x - a)$$

was used in the last section to estimate $f(x)$ in terms of x. But it can also be used in the opposite direction to estimate x from $f(x)$. That is, we can estimate where f takes on a given value, supposing that we know a reasonably close initial estimate a.

EXAMPLE 1. Estimate the positive solution of the equation $x^2 = 27$.

Solution. Here $f(x) = x^2$, and we want to estimate the value of x for which $f(x) = 27$ from the approximation

$$27 = f(x) \approx f(a) + f'(a)(x - a) = a^2 + 2a(x - a),$$

which is good when x and a are close together. Since 27 lies between $25 = 5^2$ and $36 = 6^2$ it is reasonable to take $a = 5$, this being probably the closest integer to the solution. So we solve the approximate equation

$$27 \approx 25 + 10(x - 5),$$

obtaining

$$x - 5 \approx \frac{2}{10} = 0.2,$$

and

$$x \approx 5.2.$$

Thus $x_1 = 5.2$ is our new estimate of the solution of $x^2 = 27$.

Note that we are again approximating $\sqrt{27}$, obtaining the same estimate as in the last section, but by a different method. This time we are approximating the solution of the equation

$$f(x) = 27$$

by the solution $x = x_1$ of the equation

$$f(5) + f'(5)(x - 5) = 27.$$

In general, we use the solution $x = x_1$ of the equation

$$f(a) + f'(a)(x - a) = k$$

as an estimate of a root r of the equation

$$f(x) = k.$$

Again, there is a very simple geometric interpretation. The root r and its approximation x_1 are simply the x-values where the graph of f and its tangent line at $(a, f(a))$ intersect the horizontal line $y = k$, as pictured below.

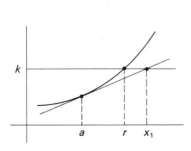 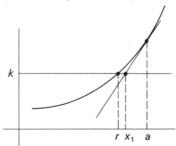

In practice, a will be a preliminary, very rough estimate of the root r, and x_1 will be an improved estimate. The figures above show that *if the graph of f is*

*concave up, then the estimate x_1 will always be larger than the root r, regardless of
whether the initial estimate a is larger or smaller than r.*

We can now *iterate* the approximation, replacing the initial estimate a by x_1,
and thus obtaining a "second generation" estimate $x = x_2$ by solving

$$f(x_1) + f'(x_1)(x - x_1) = k.$$

It appears from the figure that x_2 is an exceedingly close estimate of r.

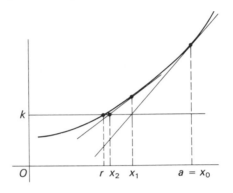

If we go on in this way, we obtain a sequence $x_0, x_1, x_2, x_3, \ldots,$ of rapidly
improving approximations, i.e., an infinite sequence that converges very fast on
the true root $x = r$. This iterative procedure is called *Newton's method*. We shall
see that it can be extraordinarily accurate at even the second step.

___

EXAMPLE 2. Compute a "second-generation" approximation to the solution of
$x^2 = 27$ by starting from the estimate x_1 obtained in Example 1.

. . . .

Solution. Now we want to solve

$$27 = f(x) \approx f(x_1) + f'(x_1)(x - x_1)$$

for x, where $x_1 = 5.2, f(x_1) = (5.2)^2 = 27.04,$ and $f'(x_1) = 2x_1 = 2(5.2)$
$= 10.4$. Thus

$$27 \approx 27.04 + (10.4)(x - 5.2),$$

giving

$$x - 5.2 \approx \frac{-.04}{10.4} = -.003846 \cdots,$$

$$x \approx 5.196154.$$

Thus

$$x_2 = 5.196154$$

is our improved estimate. We carried the long division .04/10.4 out to six decimal places because it can be shown that *this second estimate is correct to within two digits in the sixth decimal place*. That is, the error $x_2 - r$ is less than $2(10)^{-6}$.

Estimating the Error

We shall now find a bound for the error $E = r - x_1$, where

$$f(r) = k$$

$$f(a) + f'(a)(x_1 - a) = k.$$

We have

$$f(x_1) - f(r) = f(x_1) - k = f(x_1) - f(a) - f'(a)(x_1 - a).$$

We now apply the mean-value theorem on the left and Theorem 6 on the right, and conclude that

$$|f'(X)(x_1 - r)| \le \frac{B}{2}(x_1 - a)^2,$$

where B is any bound for f'' over the interval between a and x_1, and X is some number between x_1 and r. Now suppose that we also know that $|f'(x)| > L > 0$ for all the x-values in question. Then the left member above is larger than $L|x_1 - r|$, and dividing by L gives us

$$|E| \le \frac{B}{2L}(x_1 - a)^2$$

as a bound for the error $E = r - x_1$ in the new estimate of r. In applying this formula, remember:

a) *a* is a preliminary, rough estimate of the root r of the equation $f(x) = k$;
b) x_1 is the new estimate obtained from a by one application of Newton's method;
c) B is an upper bound for $|f''(x)|$ over the values of x in question;
d) L is a *lower* bound for $|f'(x)|$ over these values of x.

EXAMPLE 3. Use this formula to estimate the error in Example 2.

. . . .

Solution. Here $f(x) = x^2$, and we are operating with values of x larger than 5. Therefore $f'(x) = 2x \ge 10$, while $f''(x) \equiv 2$. Using these constants in the error formula above, we get

$$|E| \le \frac{2}{2 \cdot 10}(x_2 - x_1)^2 = \frac{(x_2 - x_1)^2}{10}.$$

(We have replaced a and x_1 by x_1 and x_2.) Since $x_1 - x_2 = 0.0038 \cdots < 0.004$, we see that

$$|E| < \frac{(0.004)^2}{10} = 0.0000016 = 1.6(10)^{-6}.$$

Therefore

$$\sqrt{27} \approx 5.196154,$$

with an error less than 2 in the sixth decimal place.

PROBLEMS FOR SECTION 4

1. Estimating where $g(x)$ has the value k is the same as estimating where $f(x) = g(x) - k$ has the value 0, and this is the context in which we generally apply the tangent-line approximation to this problem. Show that

$$x = a - \frac{f(a)}{f'(a)} \qquad (*)$$

is the estimate of the solution of $f(x) = 0$ provided by the tangent-line approximation.

In each of the following problems, locate the nearest integer a to the solution of $f(x) = 0$ by sketching the graph of f, and then estimate the solution by the equation $(*)$ from the problem above.

2. $f(x) = x^2 - 5, \quad x > 0$ 3. $f(x) = x^3 + x - 1$

4. $f(x) = x^3 + 2x - 5$

5. a) In Problem 4 above the answers are $a = 1$, $x_0 = 1\frac{2}{5} = 1.4$. Now use this estimate for x_0 as a new value of a, and obtain the second-generation estimate

$$x_1 = 1.331.$$

   b) Use the error formula, with $f''(x) \le 6(1.4) = 8.4$ and $f'(x) \ge 5$, to conclude that $x = 1.33$ is the solution of the equation $x^3 + 2x = 5$ to the nearest two decimal places.

6. The answers in Problem 3 above are $a = 1$, $x_0 = 3/4$. Now show that

$$x = 0.686$$

is the solution of the equation $x^3 + x = 1$ with an error at most 5 in the third decimal place. (Use x_0 as a new starting estimate a, to obtain a second-generation estimate x_1, and then apply the error formula.)

7. Show that the positive solution of the equation

$$\sin x = \frac{2}{3}x$$

is given by $x = 3/2$, with an error at most 0.01 in magnitude. (Take $a = \pi/2$ and use the approximation once.)

8. Show by two figures that if the graph of f is concave down, then Newton's estimate of the solution of $f(x) = 0$ is always too small, regardless of whether the initial estimate is too small or too large.

9. Starting from the nearest integer solution of the equation $x^2 - 10 = 0$, and applying the reverse tangent line approximation twice (Newton's method) show that

$$\sqrt{10} = 3.16228,$$

correct to five decimal places. (The error is less than 5 in the sixth decimal place.)

10. Since $(2.2)^2 = 4.84$ and $(2.3)^2 = 5.29$, we know that $2.2 < \sqrt{5} < 2.3$. Using 2.2 as the initial estimate, apply Newton's method once. Show that your answer is accurate to 3 decimal places.

11. Compute $\sqrt{101}$ to three decimal places by a single application of Newton's method. (Start with $a = 10$.)

In the following problems apply Newton's method twice, starting from the given rough approximation. Estimate the error in the final approximation.

12. $\sqrt{3} \approx 2$.   13. $\sqrt{2} \approx \frac{3}{2}$.   14. $\sqrt{5} \approx \frac{5}{2}$.   15. $\sqrt{3} \approx \frac{7}{4}$.

5. THE CLASSIFICATION OF ISOLATED CRITICAL POINTS

If x_0 is an isolated critical point of f, and if the interval $I = [u, v]$ has been chosen small enough about x_0 so that it contains no other critical point of f, then f is strictly monotone on each half interval $[u, x_0]$ and $[x_0, v]$, by Theorem 5. We therefore have only the four possibilities shown schematically below.

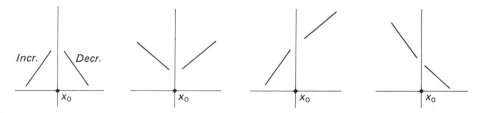

If f is increasing on $[u, x_0]$ and decreasing on $[x_0, v]$ then $f(x_0)$ is the maximum value of f on the interval $[u,v]$. This may not be the absolute maximum value of f because f may very well have larger values outside of $[u,v]$. We therefore say that f has a *local maximum*, or a *relative maximum* at x_0.

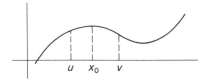

Similarly, if f is decreasing on $[u, x_0]$ and increasing on $[x_0, v]$, then x_0 is a *relative minimum* point for f.

If f has neither a local maximum nor a local minimum at the isolated critical point x_0, then f is monotone in the interval $[u,v]$ about x_0. That is, f is then monotone "in the neighborhood of x_0". This can happen whether $f'(x_0)$ is zero or is undefined, as in the following figures.

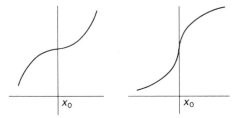

The derivative f' presumably has only isolated critical points also, and therefore we can further restrict the size of the interval $[u,v]$ about x_0 so that f' has no critical point in $[u,v]$ except possibly x_0. Then f' is monotone in each half interval $[u, x_0]$ and $[x_0, v]$, so now these are also intervals of *constant concavity* for f.

If $f'(x_0) = 0$, then the four possibilities for concavity before and after x_0 imply the same fourfold classification for the behavior of f in the neighborhood of x_0 that we found earlier, except that the ordering of the cases is different. For example, if the concavity changes from up to down at x_0, then x_0 is a horizontal point of inflection, and this is the earlier case where f is decreasing in the neighborhood of the isolated critical point x_0. Or, if $f'(x_0) = 0$ and f is concave down in an interval about x_0, then f has a relative maximum at x_0.

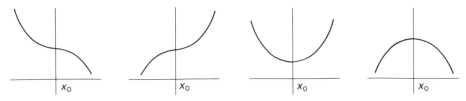

If x_0 is a singular point ($f'(x_0)$ is not defined), then there are many more possibilities. For example, each of the four concavity possibilities can now occur around a relative maximum point.

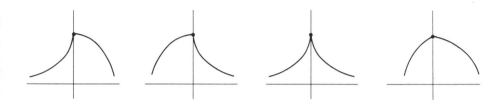

PROBLEMS FOR SECTION 5

1. By applying Theorem 4 to $f'(x)$, prove the following *second derivative test for a relative minimum*:

   If $f'(x_0) = 0$ and $f''(x_0) > 0$, then f has a relative minimum at x_0.

2. Similarly, prove that if $f'(x_0) = 0$ and $f''(x_0) < 0$, then f has a relative maximum at x_0.

3. Show that if $f'(x_0) = f''(x_0) = 0$ but $f'''(x_0) > 0$, then f is increasing on an interval about x_0. (This time apply Theorem 4 to $f''(x)$, or apply Problem 1 to $f'(x)$.)

4. Show that if $f'(x_0) = f''(x_0) = f'''(x_0) = 0$ but $f''''(x_0) > 0$, then f has a relative minimum at x_0. (Apply Theorem 4 to $f'''(x)$ and work backward, or apply Problem 3 to $f'(x)$.)

5. Reread Problems 1, 3 and 4. Now state and prove the next in this chain of results.

6. Draw figures to show that if x_0 is a singular point, then the four concavity possibilities on the two sides of x_0 are all consistent with f having a relative minimum at x_0.

Find and classify all critical points of the following functions, using the higher derivative tests from Problems 1 through 4.

7. $y = x^3 - 3x^2$.
8. $y = x^3 - x^2 - x + 2$
9. $y = (x - 1)^2(x + 1)$
10. $y = x^4 - 2x^3 + 3$
11. $y = 2x^3 - 24x$
12. $y = 3x^4 - 4x^3 - 6x^2 + 4$
13. $y = x^3(x + 3)^2$
14. $y = x^4 - 4x^3 + 16x$
15. $y = x^6 - 2x^3 + 1$
16. $y = (x^3 - 1)^2$

6. POLAR COORDINATES

Sometimes a graph seems to have a special affinity for the origin. It loops about and seems to be constantly pulled inward toward the origin.

Such a curve will not be a function graph in a Cartesian coordinate system, and its equation may be complicated and unrevealing of what is going on. When a point traces such a curve, the most natural description may be in terms of its distance from and orientation to the origin, i.e., in terms of its *polar coordinates*.

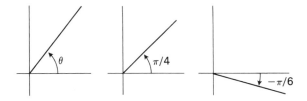

Recall that any number θ, considered as an angular coordinate, determines a unique ray, or half-line, drawn from the origin. Then a nonnegative number r determines the unique point P on the ray at the distance r from the origin. The numbers r and θ together determine the point P uniquely, and are called *polar coordinates* of P.

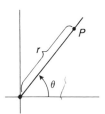

The rectangular coordinates of P are obtained from the polar coordinates r,θ by the change-of-coordinates equations

$$x = r \cos \theta,$$

$$y = r \sin \theta.$$

We considered all of this in Chapter 4, but did not pursue the matter there.
    The polar-coordinate correspondence

$$(r, \theta) \rightarrow P$$

is not one-to-one. That is, it is not true, conversely, that P determines a unique
pair of polar coordinates. An angle θ can always be modified by adding a
multiple of 2π without changing the ray it determines, so P determines its polar
coordinates only "up to" the addition of a multiple of 2π to its angular
coordinate θ.
    So far it has been implicitly assumed that r is positive or zero. But the equations
$x = r \cos \theta, y = r \sin \theta$, define a point P for *any* numbers r and θ. If r is
negative, then P is obtained geometrically by first measuring off the angle θ to
obtain a ray and then proceeding along this ray *backward across the origin* a
distance $|r|$ on the other side. Such negative values of r are not normally used,
but we shall see that some polar graphs seem incomplete unless negative r is
allowed.

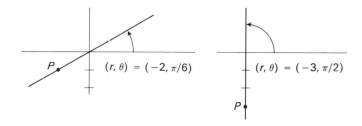

$(r, \theta) = (-2, \pi/6)$ $(r, \theta) = (-3, \pi/2)$

    Any equation in r and θ has a *polar graph*, consisting of all points in the plane
whose polar coordinates satisfy the equation. The polar graph of a function f is
the polar graph of the equation

$$r = f(\theta).$$

EXAMPLE 1. The polar graph of $r = 1$ is the unit circle about the origin:

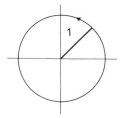

EXAMPLE 2. The polar graph of $r = \theta$ is a spiral, as shown here:

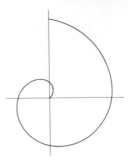

EXAMPLE 3. The graph of $r = \cos \theta$ is another circle, but this is not so obvious. We can try to get an idea of the shape of an unknown polar graph by computing and plotting a few selected points. We do this below for half the graph, and see that the graph is probably some kind of *oval*.

θ	$r = \cos \theta$
0	1
$\pi/6$	$\sqrt{3}/2 \approx 0.86$
$\pi/4$	$\sqrt{2}/2 \approx 0.70$
$\pi/3$	$1/2$
$\pi/2$	0

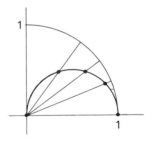

A better procedure, when it works, is to find and recognize the Cartesian equation of the graph. Here we use $x = r \cos \theta$ to get, in order,

$$r = \cos \theta = x/r,$$

$$r^2 = x,$$

$$x^2 + y^2 = x,$$

$$x^2 - x + y^2 = 0,$$

$$(x - \tfrac{1}{2})^2 + y^2 = \tfrac{1}{4},$$

and finally we recognize that the graph is the circle of radius $1/2$ about the center $(1/2, 0)$.

EXAMPLE 4. If e is a positive number less than 1, show in a similar way that the polar graph of

$$r = \frac{1}{1 - e \cos \theta}$$

is an ellipse.

. . . .

Solution. Setting $\cos \theta = x/r$ and cross-multiplying, the equation becomes, in turn,

$$r\left(1 - e\frac{x}{r}\right) = 1,$$

$$r = 1 + ex,$$

$$x^2 + y^2 = r^2 = (1 + ex)^2 = 1 + 2ex + e^2x^2,$$

$$x^2(1 - e^2) - 2ex + y^2 = 1,$$

$$(1 - e^2)\left(x - \frac{e}{1 - e^2}\right)^2 + y^2 = 1 + \frac{e^2}{1 - e^2} = \frac{1}{1 - e^2},$$

or

$$\frac{(x - \alpha)^2}{a^2} + \frac{y^2}{b^2} = 1,$$

where $a = 1/(1 - e^2)$, $b = 1/\sqrt{1 - e^2}$, $\alpha = e/(1 - e^2)$. If $e = 4/5$, then $a = 25/9$, $b = 5/3$, $\alpha = 20/9$, and the ellipse is shown below.

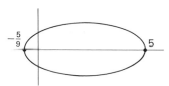

EXAMPLE 5. If negative r is not allowed, then the graph of $r = \cos 2\theta$ is as shown at the left below. At the right is the "full" graph, using negative r.

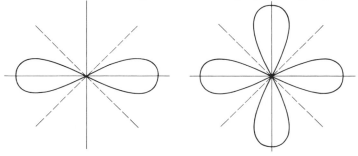

PROBLEMS FOR SECTION 6

Sketch the polar graphs of each of the following equations.

1. $r = \cos 3\theta$ 2. $r = \sin 2\theta$

3. $r = 2 \cos 5\theta$ 4. $r = 2 \sin 4\theta$

5. $r^2 = \cos 2\theta$ 6. $r^2 = \cos \theta$

7. $r = 2 \sin \theta$ (Find the x,y equation first)

8. $r = 1 + \cos \theta$

9. $r = 1 - \cos \theta$ 10. $r = 1 - 2 \cos \theta$

11. Show that $r \sin \theta = 2$ is the equation of the horizontal straight line two units above the x-axis.

12. In view of the above problem, consider the equation

$$r \sin(\theta - \theta_0) = 2.$$

It has the above form if we set $\phi = \theta - \theta_0$. On this basis, show by a geometric argument that it must be the equation of the line which is at a distance 2 from the origin and which makes the angle θ_0 with the direction of the x-axis.

13. Show that the polar graph of the equation

$$r = \frac{1}{1 - \cos \theta}$$

is a parabola. (Follow Example 4.)

14. Show that the polar graph of the equation

$$r = \frac{1}{1 - \epsilon \cos \theta}$$

is a hyperbola if $\epsilon > 1$.

15. Show more generally that for any positive constants e and d the polar graph of

$$r = \frac{d}{1 - e \cos \theta}$$

is an ellipse if $e < 1$, a parabola if $e = 1$, and a hyperbola if $e > 1$.

16. Consider the locus traced by a point whose distance from the origin is a constant e times its distance from the line $x = -k$. Assuming the above problem, show that this locus is an ellipse if $e < 1$, a parabola if $e=1$, and a hyperbola if $e > 1$.

chapter 6
optimization problems and

rate problems

In an application of mathematics, the step that may be the hardest of all is formulating the problem mathematically, i.e., translating the nonmathematical situation into suitable mathematical terms.

In this chapter we shall consider two general classes of such problems, maximum–minimum problems and rate problems. In each case, the appropriate mathematical technique is a relatively simple derivative computation, and the hardest part of a problem is apt to be "setting it up."

We shall also develop the theory appropriate for the max–min problems.

1. FINDING A MAXIMUM OR MINIMUM VALUE

We start with a simple but important application of the derivative to the problem of finding the maximum (or minimum) value that a varying quantity can have, and "where" it occurs. Historically, this was one of the very first successes of calculus.

Suppose, for example, that a rancher has one mile of fence that he can use to fence off a rectangular grazing area along the bank of a straight river. He doesn't need to fence the bank of the river itself, so his fence will run along only three sides of the rectangle. His problem is to choose the dimensions of the rectangle so as to maximize its area. Note that if he chose a *square*, then each of the 3 fenced sides would be 1/3 of a mile long and the area would be 1/9 square mile; but he can do better than that.

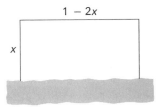

If x is the fence length *perpendicular* to the river, then the length available to run parallel to the river is $1 - 2x$ and the area enclosed is

$$A = x(1 - 2x) = x - 2x^2.$$

The physical limitation on x is that $0 \leq x \leq 1/2$. The graph of A as a function of x over this interval must look someting like this figure,

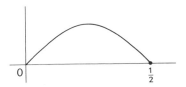

because $A = 0$ when $x = 0$ or when $x = 1/2$, and $A > 0$ in between. It seems clear that A will have its maximum value at a point where the tangent line is horizontal, i.e., where $dA/dx = 0$. Assuming that this is correct, we have

$$\frac{dA}{dx} = 1 - 4x,$$

and $dA/dx = 0$ when $x = 1/4$. The maximum area is thus

$$A = x - 2x^2 \Big|_{x=\frac{1}{4}} = \frac{1}{4} - \frac{1}{8} = \frac{1}{8} \text{ sq. mi.}$$

Before considering more examples, we shall show that the simple procedure used above is correct.

Suppose that a quantity to be maximized is called y and that we have managed to express y as a continuous function of some other variable x, where the conditions of the problem restrict x to a closed interval $[a,b]$. It may or may not seem obvious to you that such a continuous function *must necessarily* have a maximum value somewhere in the interval $[a,b]$. Supposing that it does, however, we can classify the ways in which it can happen.

The following diagrams illustrate the possible ways in which a function $y = f(x)$ can achieve a maximum value over an interval I:

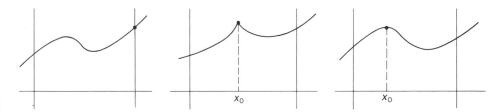

1. At an endpoint of the interval; or

2. At an interior point where the derivative is not defined; or

3. At an interior point x_0 where the derivative exists.

In the third case it probably seems geometrically clear that $f'(x_0)$ must be 0, and this is easily verified analytically.

Theorem 1. *If a function f assumes a maximum (or minimum) value at an interior point x_0 in its domain, and if $f'(x_0)$ exists, then $f'(x_0) = 0$.*

Proof. Theorem 4 of Chapter 5 shows that if $f'(x_0)$ is *not* zero, then $f(x_0)$ is *not* a maximum value. So if $f(x_0)$ is the maximum value, then $f'(x_0)$ must be zero. ∎

The latter two possibilities for a maximum point occur at what are called critical points of the function in question. The definition below was given in Chapter 5.

Definition *A* **critical point** *of a function f is an interior point x in its domain at which either* $f'(x) = 0$ *or* $f'(x)$ *is not defined.*

We saw above that if *f* has a maximum value on an interval *I*, then it can occur only at a critical point or at an endpoint of *I*. To find the maximum value for *f*, we should therefore calculate the values of *f* at all such points and choose the largest of them. This might be impossible if there were infinitely many critical points, and there also might be trouble if *f* were not defined at an endpoint. However, supposing that the number of critical points is finite, and that the domain of *f* is a closed interval, then we can prove directly that the above procedure does give the maximum value of *f*. The following figure shows how this goes.

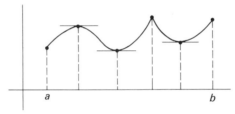

Here the maximum value of *f* on the interval [*a,b*] is the largest of the six key values shown. Any other value $f(x)$ lies *between* two adjacent values in this list, because *f* is monotone between any two adjacent key points.

Theorem 2. *Let f be a continuous function on a closed interval* [*a, b*] *and suppose that f has only a finite number of critical points, say n. Let M be the maximum of the* $(n + 2)$ *values of f at the critical points and at the two endpoints. Then M is the maximum value of f on the whole interval* [*a, b*]. *Similarly, the minimum of these* $(n + 2)$ *values is the minimum value of f on* [*a, b*].

Proof. The *n* critical points divide [*a,b*] up into $(n + 1)$ nonoverlapping closed subintervals. If *I* is one of these subintervals, then *f* has no critical point interior to *I*, and hence *f* is monotone on *I*, by Theorem 5 of Chapter 5. This means that the values of *f* at the two endpoints of *I* are the maximum and minimum values of *f* on *I*. Therefore, the largest and smallest of the values of *f* at all $(n + 2)$ of these endpoints are the maximum and minimum values of *f* on [*a,b*]. ∎

If there is only *one* critical point x_0 in *I*, then it may not be necessary to check the endpoint values at all. Suppose we know that *f'* changes sign from plus to minus as *x* crosses the single critical point x_0. Then we know that *f* is increasing just before x_0 and therefore on the whole of *I* up to x_0, and that *f* is decreasing just after x_0 and therefore on the whole of *I* after x_0. Therefore we know that $f(x_0)$ is the maximum value of *f* on *I*. Similarly, $f(x_0)$ must be the minimum value of f if f' changes sign from $-$ to $+$ as x crosses x_0. This reasoning doesn't require that the endpoints of I even belong to the domain of f, and I might perfectly well extend to infinity. The situation is particularly simple if $f''(x_0)$ exists and is not zero.

Theorem 3. *Suppose that f is a differentiable function with exactly one critical point x_0 on an interval I. Then $f(x_0)$ is the maximum value of f on I if $f''(x_0)$ is negative, and $f(x_0)$ is the minimum value of f if $f''(x_0)$ is positive.*

Proof. Suppose first that $f''(x_0) > 0$. Thus the function f' is zero at x_0 and its derivative is positive there. Then Theorem 4 of Chapter 5 says exactly that $f'(x)$ changes sign from $-$ to $+$ as x crosses x_0. Therefore $f(x_0)$ must be the minimum value of f on I, as we saw above. Similarly, if $f''(x_0) < 0$, then f' changes from $+$ to $-$ as x crosses x_0 and $f(x_0)$ is the maximum value. ∎

EXAMPLE 1. Find the maximum and minimum values of $f(x) = x^4 - 4x$ on the interval $[0,2]$.

. . . .

Solution. The derivative $f'(x) = 4x^3 - 4 = 4(x^3 - 1)$ is zero when $x^3 - 1 = 0$, i.e., when

$$x^3 = 1, \quad \text{or} \quad x = 1.$$

Thus $x = 1$ is the only critical point of f. Since the second derivative $f''(x) = 12x^2$ is positive at $x = 1$, Theorem 3 says that $f(1) = -3$ is the absolute minimum value of f over its whole domain. In particular, this is the minimum value of f on the interval $[0,2]$.

Theorem 2 says that the maximum value of f on $[0,2]$ is either this critical point value or one of the two endpoint values, $f(0) = 0$ and $f(2) = 16 - 8 = 8$. So the maximum value is the right hand endpoint value 8.

EXAMPLE 2. Find the maximum and minimum values of $f(x) = x^3 - x^2$ on the interval $[1,4]$.

. . . .

Solution. Since

$$f'(x) = 3x^2 - 2x = x(3x - 2),$$

the critical points of f are $x = 0, \frac{2}{3}$. Neither of these critical points belongs to the interval $[1,4]$, so the maximum and minimum values of f on this interval are the two endpoint values,

$$f(1) = 0 \quad \text{and} \quad f(4) = 48,$$

by Theorem 2.

EXAMPLE 3. Apply Theorem 3 to $f(x) = x + (1/x)$ on the interval $(0, \infty)$.

. . . .

Solution. Since $f'(x) = 1 - (1/x^2)$, the only critical point of f in the interval $(0, \infty)$ is at $x = 1$. At this point $f''(x) = 2/x^3$ is positive. Therefore $f(1) = 2$ is the minimum value of f, by Theorem 3.

PROBLEMS FOR SECTION 1

Use Theorem 2 to find the maximum and minimum values of each of the following functions over the specified interval. (There may be none.)

1. $f(x) = 3 + x - x^2$; $[0, 2]$
2. $f(x) = x^3 - x^2 - x + 2$; $[0, 2]$
3. $F(x) = x^3 - 5x^2 - 8x + 20$;
   $[-1, 5]$
4. $g(x) = x\sqrt{x + 3}$; $[-3, 3]$
5. $h(x) = x^{2/3}(x - 5)$; $[-1, 1]$
6. $G(x) = \sqrt[3]{x^2 - 2x}$; $(0, 2)$
7. $y = x\sqrt{2x - x^2}$; $[0, 2]$
8. $y = x^{3/2}(x - 8)^{-1/2}$; $[10, 16]$
9. $f(x) = 2x^4 + x$; $[-1, 1]$
10. $f(x) = \sin x + \cos x$; $[0, \pi]$
11. $f(x) = x^2 e^{-x}$; $[0, 4]$
12. $f(x) = xe^x$; $[-2, 0]$
13. $f(x) = \sin x + \cos^2 x$; $[0, 2\pi]$

Use Theorem 3 to find the maximum or minimum value of $f(x)$ in each of the following situations.

14. $f(x) = 2x^2 + x$
15. $f(x) = 3 + x - x^2$
16. $f(x) = x^3 - 3x$, over $[-1, 2]$.
17. $f(x) = x^3 - x^2 - x + 2$,
    over $[-1, 0]$.
18. $f(x) = x\sqrt{x + 3}$, over $[-3, 3]$.
19. $f(x) = 2x^4 + x$
20. $f(x) = \sin x + \cos x$, over $[0, \pi]$.
21. $f(x) = x^2 e^{-x}$, over $[0, 4]$.
22. $f(x) = 2 \sec x + \tan x$, over $(-\pi/2, \pi/2)$.

2. MAXIMUM–MINIMUM PROBLEMS

EXAMPLE 1. We continue with our coal producer from Examples 4 and 5 in Section 12 of Chapter 3. Suppose that his revenue function is

$$R = 9x - 2x^2,$$

and that his cost function is

$$C = x^3 - 3x^2 + 4x + 1.$$

Then

$$P = R - C = -x^3 + x^2 + 5x - 1$$

is his profit, and, being selfish, he proposes to operate so as to maximize his profit. Since

$$\frac{dP}{dx} = -3x^2 + 2x + 5 = -(3x - 5)(x + 1),$$

and since his production x is always positive, we see that the only critical point for his profit occurs at $x = 5/3$. The correspoinding value of his profit works out to be $P = 148/27 \approx 5\frac{1}{2}$. It seems clear that this must be his maximum profit, and this is easily checked. For one thing, $d^2 P/dx^2 = -6x + 2$ is negative at $x = 5/3$, so Theorem 3 affirms that this is a maximum point. In the units we chose before, his maximum profit will be \$5,500 per week (he is probably a corporation) and this will be achieved by producing 1,667 tons of coal per week.

EXAMPLE 2. In general, any entrepreneur will experience losses when his production is too low, becase of fixed costs, and also when his production is too high, because of very high marginal costs. Unless he can operate profitably at some in-between production he won't be in business at all, so we can suppose his profit curve looks like this:

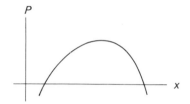

Then his maximum profit will always occur at a critical point, i.e., supposing P to be a differentiable function of x, it will occur at a point where $dP/dx = 0$.
Now $P = R - C$, so

$$0 = \frac{dP}{dx} = \frac{dR}{dx} - \frac{dC}{dx}$$

if and only if $dR/dx = dC/dx$. That is,

Profit will be maximum when the marginal revenue equals the marginal cost.

EXAMPLE 3. It will generally happen, in setting up a maximum problem, that the variable to be maximized (or minimized) is most naturally expressed in terms of more than one independent variable; but then the conditions of the problem will allow all but one of these variables to be eliminated, as the present example will show.

Consider the problem of finding the most efficient shape for a one-cubic-foot cylindrical container, the most efficient cylinder being the one using the least material, i.e., having the *smallest total surface area*. Now the volume of a cylinder is $V = \pi r^2 h$, and its total surface area is $A = 2\pi r^2 + 2\pi rh$. Our problem, therefore, is so to determine r and h that A has the smallest possible value subject to the requirement that $V = 1$. The restriction

$$\pi r^2 h = 1$$

allows us to solve for either h or r as a function of the other, say

$$h = \frac{1}{\pi r^2};$$

and when this is substituted in the area formula, the area becomes a function of r alone,

$$A = 2\pi r^2 + \frac{2}{r}.$$

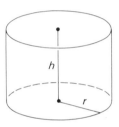

The physical limitation on r is that it be positive. To find the critical points we compute dA/dr and set it equal to 0,

$$\frac{dA}{dr} = 4\pi r - \frac{2}{r^2} = 0$$

which gives $4\pi r^3 = 2$. Thus $r = (1/2\pi)^{1/3}$ is the only critical point. Since

$$\frac{d^2 A}{dr^2} = 4\pi + \frac{4}{r^3} > 0,$$

we know from Theorem 3 that A has its minimum value at the critical point.

Since we asked for the *shape* of the most efficient container, we want h also, and we compute

$$h = \frac{1}{\pi r^2} = \frac{(2\pi)^{2/3}}{\pi} = \frac{2}{(2\pi)^{1/3}} = 2r.$$

The most efficient shape is therefore that for which the altitude h equals the diameter $2r$ of the base. The can has a square cross section along its vertical axis.

Here are some general suggestions for tackling problems of maxima and minima. Unfortunately, they do not guarantee success, but it is difficult to imagine getting anywhere without them. (That is, they are *necessary* but not *sufficient* conditions for success.)

1. In a problem that involves relationships among geometric magnitudes, draw a reasonably accurate figure. Be sure to show the *general* configuration. For instance, if the problem involves a general triangle, don't draw one that looks isosceles, because that might lead you astray in seeing what is going on.

2. Label your figure with appropriate constants and variables, being sure that you know what is being held fixed and what is varying.

3. Derive from known geometric relationships and formulas the equations connecting the variables involved in your problem.

4. If Q is the quantity to be maximized or minimized, then these equations should determine Q as a function of *one* of the variables. Then apply Theorem 2 or Theorem 3.

Sometimes an angle can be taken as the independent variable, as in the following example.

EXAMPLE 4. A corridor of width a meets a corridor of width b at right angles. Workmen wish to push a heavy beam of length c on dollies around the corner, but they want to be sure it will be able to make the turn before starting. How long a beam will go around the corner (neglecting the width of the beam)?

. . . .

Solution. This is essentially the problem of minimizing the length l of the segment cut off by the corridor on a varying line through the corner point P. The beam will go around the corner if its length c is less than the minimum of l.

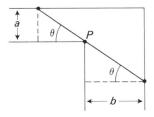

In terms of the angle θ shown in the figure, the length l is given by

$$l = \frac{a}{\sin \theta} + \frac{b}{\cos \theta}.$$

The domain of θ is $(0, \pi/2)$, and l approaches $+\infty$ as θ approachs 0 or $\pi/2$. The minimum of l will therefore occur at a point where $dl/d\theta = 0$. We see that

$$\frac{dl}{d\theta} = -\frac{a \cos \theta}{\sin^2 \theta} + \frac{b \sin \theta}{\cos^2 \theta}$$

$$= \frac{b \sin^3 \theta - a \cos^3 \theta}{\sin^2 \theta \cos^2 \theta},$$

so $dl/d\theta = 0$ if and only if

$$b \sin^3 \theta - a \cos^3 \theta = 0 \qquad \text{or} \qquad \tan \theta = \left(\frac{a}{b}\right)^{1/3}.$$

The corresponding value of l then works out to be

$$l = (a^{2/3} + b^{2/3})^{3/2}.$$

This is a good exercise in algebra. Anyway, the beam will go around the corner if its length is not greater than this value of l.

REMARK. Strictly speaking, this example requires that Theorem 2 be modified so as to apply to a function f defined on an interval that is missing one or both endpoints. We replace any such missing endpoint value of f by the limit of $f(x)$ as x approaches that end of the interval, and take the maximum of this modified collection of numbers. If this maximum is an actual value of f, then we have found the maximum value of f. But if the maximum of the modified collection occurs only as the limit of $f(x)$ at a missing endpoint, then f has no maximum value on the interval. Three such situations are pictured below.

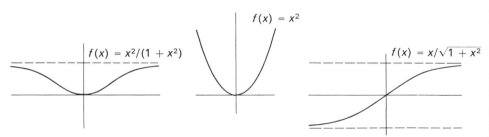

$f(x) = x^2/(1 + x^2)$ $\qquad$ $f(x) = x^2$ $\qquad$ $f(x) = x/\sqrt{1 + x^2}$

In the first example, $\lim_{x \to \infty} f(x) = 1 = \lim_{x \to -\infty} f(x)$, and this is larger than the value 0 at the only critical point, $x = 0$; f has no maximum value. In the second case, $\lim_{x \to \infty} f(x) = \infty = \lim_{x \to -\infty} f(x)$, which is larger than the value of f at the only critical point 0. Again, no maximum. In the third case, there are *no* critical points *or* endpoint values, and hence no maximum or minimum value.

Going back to Example 3, we had

$$A = 2\pi r^2 + \frac{2}{r}$$

on the domain $(0,\infty)$, and we see that $A \to \infty$ as $r \to 0$ and as $r \to \infty$. Thus, the minimum of A must be at a critical point, by the modified Theorem 2; the same is true for Example 4.

The method of auxiliary variables

It was noted earlier that the variable to be maximized or minimized may be most naturally expressed as a function of two or more other variables, but that the conditions of the problem would always allow the elimination of all these variables but one. We now treat this situation in another way.

EXAMPLE 5. In Example 3 we wanted to minimize

$$A = 2\pi r^2 + 2\pi rh = 2\pi[r^2 + rh]$$

subject to the condition that

$$V = \pi r^2 h = 1.$$

This restriction determines either h or r as a function of the other, and what we did before was to solve for h, substitute in the first equation, and go on.

The new procedure will be to treat h as a function of r that is determined by the second equation, but leave both h and r in both equations. We differentiate the two equations with respect to r, keeping in mind that h is a function of r:

$$\frac{dA}{dr} = 2\pi[2r + r\frac{dh}{dr} + h]$$

$$\frac{dV}{dr} = \pi[r^2\frac{dh}{dr} + 2rh] = 0.$$

Now solve the second equation for dh/dr and substitute in the first. We get $dh/dr = -2h/r$ and

$$\frac{dA}{dr} = 2\pi\left[2r + r\left(\frac{-2h}{r}\right) + h\right]$$

$$= 2\pi[2r - h].$$

The critical-point equation $dA/dr = 0$ now appears as $2r - h = 0$, or

$$h = 2r.$$

This example typifies what one can expect from the new approach. There are two things to notice. First, the answer appears as a relationship among the supporting variables, and for some problems this is really all we want. The above problem, for instance, asked for the most efficient shape and this is exactly what the answer $h = 2r$ gives us. If we want to know what the minimum area is, and for what value of r it occurs, then we have to substitute $h = 2r$ in the equation $\pi r^2 h = 1$, and solve for r. The second point about this method is that it really only gives us the critical point configuration, and does not (at least not without further work) let us conclude with assurance that this configuration is where the maximum (or minimum) value occurs.

EXAMPLE 6. Find the rectangle of largest area (with sides parallel to the axes) that can be inscribed in the ellipse

$$\frac{x^2}{a^2} + \frac{y^2}{b^2} = 1.$$

. . . .

Solution. If (x,y) is the vertex of the rectangle lying in the first quadrant, then the area of the rectangle is

$$A = 4xy,$$

where

$$\frac{x^2}{a^2} + \frac{y^2}{b^2} = 1.$$

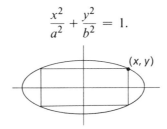

The second equation determines y as a function of x, and we can therefore differentiate both equations with respect to x:

$$\frac{dA}{dx} = 4x\frac{dy}{dx} + 4y,$$

$$\frac{2x}{a^2} + \frac{2y\,dy/dx}{b^2} = 0.$$

We can then solve for dy/dx from the second equation and substitute in the first, obtaining

$$\frac{dy}{dx} = -\frac{b^2 x}{a^2 y},$$

$$\frac{dA}{dx} = 4\left(-\frac{b^2 x^2}{a^2 y} + y\right).$$

The critical-point equation $dA/dx = 0$ is thus

$$0 = \frac{-b^2 x^2 + a^2 y^2}{a^2 y},$$

from which we concude that

$$\frac{y}{x} = \frac{b}{a}$$

at the critical point. That is, the critical rectangle has sides proportional to the axes of the ellipse. In order to find the rectangular area, we have to substitute $y = bx/a$ in the equation of the ellipse and solve for x and y. We find that

$$x = \frac{a}{\sqrt{2}}, \qquad y = \frac{b}{\sqrt{2}},$$

so the critical (maximum) area is

$$A = 4xy = 2ab.$$

PROBLEMS FOR SECTION 2

1. In a certain physical situation, it is found that a variable quantity Q can be expressed in the form

    $$Q = x + y,$$

    where x and y are related by

    $$x^2 + y^2 = 1.$$

    Find the maximum value of Q.

2. In a certain situation it is found that a variable quantity Q that is to be maximized is expressed in terms of variable quantities x and y by

    $$Q = xy^2,$$

    where x and y are related by

    $$x^2 + y^2 = 1.$$

    Find the maximum value of Q.

3. Find the maximum value of $Q = xy$, if x and y are related by the condition

    $$2x^2 + y^2 = 1.$$

4. A quantity Q is given by

$$Q = x^3 + 2y^3,$$

where x and y are positive variables related by the equation

$$x + y = 1.$$

Show that the minimum value of Q is $6-4\sqrt{2}$, and this is greater than $1/3$.

5. Find the point on the graph $y = x^3 - 6x^2 - 3x$ at which the tangent line has minimum slope.

6. Find the maximum value of $x - 2y$ on the unit circle $x^2 + y^2 = 1$.

7. Find the minimum value of $3x + y^3$ on the circle $x^2 + y^2 = 2$.

8. Find the rectangle of maximum area that can be inscribed in the circle

$$x^2 + y^2 = 1.$$

9. Find the circular cylinder of maximum volume that can be inscribed in the cone of altitude H and base radius R.

10. Find the circular cylinder of maximum volume that can be inscribed in a sphere of radius r.

11. A rectangular garden is to be laid out with one side adjoining a neighbor's lot, and is to contain 48 square yards. If the neighbor pays for half the dividing fence, what dimensions of the garden will minimize the cost of the fence?

12. The post office places a limit of 120 in. on the combined length and girth of a package. What are the dimensions of a rectangular box with square cross section, that will contain the largest mailable volume?

13. Find the proportions for a rectangle of given area A that will minimize the distance from one corner to the midpoint of a nonadjacent side.

14. You are the owner of an 80-unit motel. When the daily charge for a unit is $20, all units are occupied. If the daily charge is increased by d dollars then $3d$ of the units become vacant. Each occupied unit requires $3 daily for service and repairs. What should be your daily charge to realize the most profit?

15. A wire 24 in. long is cut in two, and then one part is bent into the shape of a circle and the other into the shape of a square. How should it be cut if the sum of the areas of the circle and the square is to be a *minimum?*

16. Find the positive number for which the sum of its reciprocal and four times its square is the smallest possible.

17. Find the shortest distance from the point $(0,2)$ to the hyperbola $x^2 - y^2 = 1$.

18. Find the rectangle of maximum perimeter than can be inscribed in the ellipse $4x^2 + 9y^2 = 36$. The sides of the rectangle are parallel to the axes of the ellipse.

19. The strength of a wooden beam of rectangular cross section is proportional to the width of the beam and the square of its depth. Determine the proportions of the strongest beam that can be cut from a circular log.

20. A small loan company is limited by law to an 18% interest charge on any loan. The amount of money available for loans is proportional to the interest rate the company will pay investors. If the company can loan out all the money that is invested with it, what interest rate should it pay its investors in order to maximize profits?

21. A rectangular box with a square base and a cover is to be built to contain 640 cu. ft. If the cost per square foot for the bottom is 15¢ and for the top and sides 10¢, what are the dimensions for a minimum cost?

22. Find the minimum distance from the point $(1,4)$ to the parabola $x^2 = 3y$.

23. In connecting a water line to a building at A (see figure below), the contractor finds that he must connect to a certain point C on the water main which lies under the paved parking lot of a shopping center. It will cost him $20 a foot to dig, lay pipe, fill, and resurface the parking lot, but only $12 a foot to lay pipe along the edge. Find the distance from the store water inlet to the point B, at which he should turn the water line and go directly to point C in order to minimize his cost.

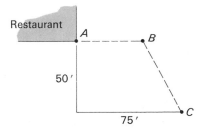

24. A frame for a cylindrically shaped lamp shade is made from a piece of wire 16 ft. long. The frame consists of two equal circles, two diametral wires in the upper circle, and four equal wires from the upper to the lower circle. For what radius will the volume of the cylinder be a maximum?

25. A rectangular box is to be made from a sheet of tin 20 in. by 20 in. by cutting a square from each corner and turning up the sides. Find the edge of this square which makes the volume a maximum.

26. Two towers 40 ft apart are 30 and 20 ft high respectively. A wire fastened to the top of each tower is guyed to the ground at a point between the towers, and is tightened so that there is no sag. How far from the tallest tower will the wire touch the ground if the length of the wire is a minimum?

27. A steel mill is capable of producing x tons per day of a low-grade steel and y tons per day of a high-grade steel, where

$$y = \frac{40 - 5x}{10 - x}.$$

If the fixed-market price of low-grade steel is half that of the high-grade steel, find the amount of low-grade steel to be produced each day that will yield maximum receipts.

28. A window in the shape of a rectangle surmounted by a semicircle has a perimeter of 24 ft. If the amount of light entering through the window is to be as large as possible, what should the dimensions of the window be?

29. Two roads intersect at right angles, and a spring is located in an adjoining field 10 yds from one road and 5 yds from the other. How should a straight path just passing the spring be laid out so as to cut off the least amount of land? How much land is cut off?

30. Find the shortest path in the above situation.

31. Which is the more efficient container, a cube, or a circular cylinder? (Find the most efficient cylinder and then compare with the cube.)

32. Solve the Norman window problem (Problem 28) if colored glass is used for the semicircle that admits only half as much light per unit area as does the clear glass in the rectangle.

33. A silo consists of a circular cylinder with a hemispherical top. Find its most efficient shape. That is, find the relative dimensions that maximize the volume for a given total area (base area plus lateral area of the cylinder, plus the area of the hemisphere).

34. Find the most economical shape for a silo like that in the above problem if the construction of the hemisphere costs twice as much per unit area as the cylinder.

35. A smooth graph not passing through the origin always has a point (x_0, y_0) closest to the origin. Show that the segment from the origin to (x_0, y_0) is perpendicular to the graph.

36. A tank has hemispherical ends and a cylindrical center. Find its most efficient shape. That is, find the proportions of the cylinder that will maximize the volume for a given surface area.

37. If the hemispherical ends to the tank in the above problem cost twice as much (per unit area) as the cylindrical center, find its most economical shape.

38. It costs a manufacturer $x^3 - 3x^2 + 4x + 1$ hundreds of dollars to produce x thousands of an item per week, and he can sell them for $10x$ hundreds of dollars. How many should he sell per week in order to maximize his profit, and what is this maximum profit?

39. A manufacturer's cost function is $4\sqrt{x} + 1$, and his revenue function is $9x - x^2$, both in thousands of dollars per thousand items. Find his maximum profit.

40. The illumination from a light source is inversely proportional to the square of the distance from the light and directly proportional to the sine of the angle of incidence. How high should a light be placed on a pole in order to maximize the illumination on the ground along the circumference of a circle of radius 25 feet?

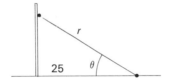

3. RELATED RATES

Suppose that y is a function of x, say $y = f(x)$, and that x varies with time t. Because y depends on x, it will also vary with time. That is, if y is a function of x and x is a function of t, then y is a function of t. The chain rule says that

$$\frac{dy}{dt} = f'(x)\frac{dx}{dt}.$$

Thus the rate of change of y is related to the rate of change of x because of the relationship between y and x.

EXAMPLE 1. Suppose that a circular ripple is spreading out over a pond, and that the radius r is increasing at the rate of $\frac{1}{2}$ foot per second at the moment when r goes through the value $r = 5$ feet. How fast is the disturbed area increasing at that moment?

. . . .

Solution. The disturbed area A and the radius r are varying with time, but at all times they are related by the equation

$$A = \pi r^2.$$

We take the time t to be the underlying independent variable, and differentiate this identity with respect to t:

$$\frac{dA}{dt} = (2\pi r)\frac{dr}{dt}.$$

At the moment in question we are given $dr/dt = \frac{1}{2}$ and $r = 5$. Therefore, at that moment

$$\frac{dA}{dt} = 2\pi 5 \cdot \frac{1}{2} = 5\pi,$$

or approximately 15 square feet per second.

EXAMPLE 2. A tank in the shape of an inverted cone with equilateral cross section is being filled with water at the constant rate of 10 cubic feet per minute. How fast is the water rising when its depth is 5 feet?

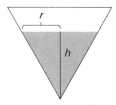

. . . .

Solution. The amount of water in the cone is related to its depth h and its surface radius r by the formula for the volume of a cone:

$$V = \frac{1}{3}\pi r^2 h.$$

Here V, r, and h are all increasing with time as water pours in, but are always related by the above equation. We are given $dV/dt = 10$, and we want to know the value of dh/dt when $h = 5$. Our general principle is to *relate, by an equation that holds throughout the variation, the quantity whose rate is wanted to the quantity whose rate is given, and then differentiate with respect to t.* The above volume equation is such an equation, but it isn't quite right because it contains the extra variable r. However, we can eliminate r by using the fact that the tank cross section is equilateral: We have

$$h^2 + r^2 = 4r^2$$

by the Pythagorean theorem, and so

$$h^2 = 3r^2.$$

Therefore

$$V = \frac{1}{3}\pi\frac{h^2}{3} \cdot h = \frac{\pi}{9}h^3,$$

$$\frac{dV}{dt} = \frac{\pi}{9} \cdot 3h^2 \cdot \frac{dh}{dt}.$$

We are given $dV/dt = 10$, and we want dh/dt when $h = 5$. At that moment

$$10 = \frac{\pi 25}{3}\frac{dh}{dt},$$

and

$$\frac{dh}{dt} = \frac{6}{5\pi},$$

or approximately 2/5 ft/min.

PROBLEMS FOR SECTION 3

1. Two variable quantities Q and R are found to be related by the equation

   $$Q^3 + R^3 = 9.$$

   What is the rate of change dQ/dt at the moment when $Q = 2$, if $dR/dt = 3$ at that moment?

2. Two variable quantities x and y are found to be related by the equation

   $$x^2 + y^2 = 25.$$

   If x increases at the constant rate of 3 units per unit of time, what is the rate at which y is increasing at the moment when $x = 0$? When $x = 3$? When $x = 5$?

3. A ladder 26 ft long leans against a vertical wall. If the lower end is being moved away from the wall at the rate of 5 ft/sec, how fast is the top descending when the lower end is 10 ft from the wall?

4. A conical funnel is 8 in. across the top and 12 in. deep. A liquid is flowing in at a rate of 60 cu. in./sec and flowing out at a rate of 40 cu. in./sec. Find how fast the surface is rising when the liquid is 6 in. deep.

5. A rope 32 ft long is attached to a weight and passed over a pulley 16 feet above the ground. The other end of the rope is pulled away along the ground at the rate of 3 ft/sec. At what rate does the weight rise at the instant when the other end of the rope is 12 ft from its initial point?

6. A man 6 ft tall walks away from a street light 15 ft high at the rate of 3 mi/hr.

   a) How fast is the far end of his shadow moving when he is 30 ft from the light pole?
   b) How fast is his shadow lengthening?

7. Two cars start from the same point at the same time. One travels north at 25 mi/hr, and the other travels east at 60 mi/hr. How fast is the distance between them increasing as they start out? At the end of an hour?

8. A revolving beacon located 3 miles from a straight shore line makes 2 revolutions/min. Find the speed of the spot of light along the shore when it is two miles away from the point on the shore nearest the light.

9. A drawbridge with two 10′ spans is being raised at a rate of 2 radians per minute. How fast is the distance increasing between the ends of the spans when they both are at an elevation of $\pi/4$ radians?

10. A bridge is 30 ft above a canal. A motor boat traveling at 10 ft/sec passes under the center of the bridge at the same instant that a man walking 5 ft/sec reaches that point. How rapidly are they separating 3 sec later?

11. If, in Problem 10, the man reaches the center of the bridge 5 sec before the boat passes under it, find the rate at which the distance between them is changing 4 sec later.

12. A conical reservoir 8 ft across and 4 ft deep is filled at a rate of 2 cu. ft/sec. A leak allows fluid to run out at a rate depending on the depth of the water, the loss being at the rate of $h^4/8$ cubic feet per minute when its depth is h feet. At what depth will the surface level stabilize?

13. A hemispherical bowl of diameter 18 inches is being filled with water. If the depth of the water is increasing at the rate of 1/8 in./sec when it is 8 in. deep, how fast is the water flowing in? (The volume of a segment of a sphere of radius r is

$$\pi h^2 \left(r - \frac{h}{3} \right),$$

where h is the height of the segment.)

14. A light is at the top of a tower 80 ft high. A ball is dropped from the same height from a point 20 ft from the light. Assuming that the ball falls according to the law $s = 16t^2$, how fast is the shadow of the ball moving along the ground two seconds after release?

15. A balloon is rising vertically over a point A on the ground at the rate of 15 ft/sec. A point B on the ground is level with and 30 ft from A. The angle of elevation is being measured at B. At what rate is the elevation changing when the balloon is $30\sqrt{3}$ ft above A?

16. Water is being pumped into a trough 8 ft long with a triangular cross section 2 ft × 2 ft × 2 ft at the rate of 16 cu. ft/min. How fast is the surface level rising when the water is 1 ft deep?

17. Gas is escaping from a spherical balloon at the rate of 1/2 cu. in./sec. At what rate is the surface area decreasing when the radius is 6 in.?

18. If $uv = 1$ and if v is increasing at the constant rate of 2 units per unit time, what is the rate of change of u when $v = 4$?

19. A particle moves along the parabola $y = 3x^2$ in such a way that its x-coordinate is increasing at the constant rate of 5 inches per second (the inch being the unit of distance in the coordinate plane). What is the rate at which its y-coordinate is changing when $x = -1$? When $x = 0$? When $x = 1/3$?

20. A particle moves along the parabola $y = x^2$, and dy/dt is found to have the value 3 when $x = 2$. How fast is x changing at that moment?

21. A circular wheel of radius 5 feet lies over the xy-plane, with its center at the origin, and revolves steadily in the counterclockwise direction at the rate of 25 revolutions per minute. How fast is the y-coordinate of a particle on the wheel increasing as the particle goes through the point (3,4)? The same question for its x-coordinate.

22. Referring to the above problem, suppose that the wheel is rotating at the constant rate of one radian per minute. Show that then $dy/dt = x$, for all positions of the particle.

23. Suppose a particle traces a circle about the origin in such a way that $dy/dt = x$. Show that the particle is moving along the circle at the constant rate of 1 radian per unit of time.

24. A rolling snowball picks up new snow at a rate proportional to its surface area. Assuming that the snowball always remains spherical, show that its radius is increasing at a constant rate.

25. A ladder 20 feet long leans against a wall 10 feet high. If the lower end of the ladder is pulled away from the wall at the rate of 3 feet per second, how fast is the angle between the ladder and the top of the wall changing at the moment when the angle is $45° = \pi/4$ radians?

4. GROWTH AND DECAY

We saw in Chapter 4, Section 3, Example 4, that "up to a multiplicative constant," $y = e^x$ is the unique function satisfying the differential equation

$$\frac{dy}{dx} = y.$$

That is, if f is a function such that

$$f' = f,$$

then $f(x) = ce^x$ for some constant c.

If we carry through the same argument, using e^{kx} instead of e^x, we find that:

Theorem 4. *A function* $y = f(x)$ *satisfies the equation*

$$\frac{dy}{dx} = ky$$

if and only if

$$y = ce^{kx}$$

for some constant c.

The equation

$$\frac{dy}{dt} = ky$$

is the basic law of population growth. Neglecting special inhibiting and stimulating factors, a population normally reproduces itself at a rate proportional to its size, and this is exactly what the above equation says. Bacteria colonies grow this way as long as they have normal environment, and so does a bank balance with a fixed rate of continuously compounded interest. Theorem 4 shows that a population normally grows exponentially, and if you will look back at the shape of an exponential graph (page 155) you will see what "population explosion" is all about.

The constant k is sometimes called the *rate of exponential growth*. This is not the rate of change of the population size, which is

$$\frac{dy}{dt} = ky,$$

but the constant that y must be multiplied by to get its rate of change. It is thus a different use of the word *rate*. It is like the *interest rate* paid by a bank. If the interest rate is 0.05, we do not mean that your bank balance y is growing at the rate of 0.05 dollars per year but at the rate of $0.05y$ dollar per year. We therefore express the rate as 5% per year, rather than 0.05 dollars per year. We could say that the rate is 0.05 dollars *per dollar* per year. (Where the interest is compounded continuously, the interest rate is a true exponential growth rate.)

When a quantity Q is growing exponentially,

$$Q = ce^{kt},$$

the time T it takes to double depends only on the growth rate k. Conversely, the growth rate k can be found from the doubling time T.

EXAMPLE. Suppose that a culture of bacteria doubles its size from 0.15 grams to 0.3 grams during the seven-hour period from $t = 15$ hrs to $t = 22$ hrs. What is its growth rate k?

. . . .

Solution. Assuming normal exponential growth

$$y = ce^{kt},$$

our data are

$$0.15 = ce^{k \cdot 15},$$

$$0.30 = cc^{k \cdot 22}.$$

Dividing the second equation by the first gives us

$$2 = e^{k(22-15)} = e^{k \cdot 7}.$$

Solving this equation for k requires logarithms, which haven't come up yet but may be familiar to you from high school. In any event, if $y = e^x$, then x is called the natural logarithm of y, and is written

$$x = \log y.$$

Thus, from $2 = e^{k \cdot 7}$ we get

$$k \cdot 7 = \log 2,$$

$$k = \frac{\log 2}{7}.$$

The table on page 891 gives $\log 2 = 0.6931$ to four decimal places. Thus the growth rate k is approximately 1

In general the exponential growth rate k and the doubling time T are related by

$$kT = \log 2 \approx 0.6931,$$

so that either determines the other. You should be able to show this.

If the exponential growth rate is *negative*, say $-k$ where k is positive, then the equation $dy/dt = -ky$ shows y to be *decreasing* as a function of time, and the solution $y = ce^{-kt}$ shows it to be decreasing exponentially. This is called exponential *decay*. A radioactive element behaves this way; it disintegrates at a rate proportional to the amount present. In this context, we have the *half-life* of the element instead of the doubling time. This is the time it takes for $y = ce^{-kt}$ to decrease to half of an original value. That is, if

$$y_0 = ce^{-kt_0},$$

$$\frac{y_0}{2} = ce^{-kt_1},$$

then $T = t_1 - t_0$ is the half-life. Again we find that

$$kT = \log 2,$$

so that the half-life T depends only on the decay rate k. In particular, it is independent of the initial population size y_0 and the initial time t_0.

If we know the growth (or decay) rate k, then the remaining constant c is determined by a single initial condition (t_0, y_0), as before.

PROBLEMS FOR SECTION 4

1. A colony of bacteria grows at the rate of 1% per hour. What is its doubling time?

2. A culture of bacteria is found to increase by 41% in 12 hours. What is its growth rate per hour? (Note that $1.41 \simeq \sqrt{2}$.)

3. A growth rate of 100% per day corresponds to what growth rate per hour?

4. A bank advertises that it compounds interest continuously and that it will double your money in 10 years. What is its annual interest rate?

5. A certain radioactive material has a half-life of 1 year. When will it be 99% gone? ($\log 10 \simeq 2.3$)

6. A quantity Q_1 grows exponentially with a doubling time of one week. Q_2 grows exponentially with a doubling time of three weeks. If the initial amounts of Q_1 and Q_2 are the same, when will Q_1 be twice the size of Q_2?

7. There is nothing sacred about doubling time. Any two measurements of an exponentially growing population $y = ce^{kt}$ will determine both parameters k and c. Show that if y has the values y_0 and y_1 at times t_0 and t_1, then

$$k = \frac{\log(y_1/y_0)}{t_1 - t_0}.$$

8. In a chemical reaction a substance A decomposes at a rate proportional to the amount of A present. It is found that 8 pounds of A will reduce to 4 pounds in 3 hours. At what time will there be only 1 pound left?

9. In a chemical reaction a substance B decomposes at a rate proportional to the amount of B present It is found that 8 pounds of B will reduce to 7 pounds in one hour. Make a rough calculation to show that more than four hours are needed for 8 pounds of B to reduce to 4 pounds.

10. The temperature T of a cooling body drops at a rate that is proportional to the difference $T - C$, where C is the constant temperature of the surrounding medium. Use Theorem 3 to show that

$$T = ae^{-kt} + C.$$

11. The temperature of a cup of freshly poured coffee is 200° and the room temperature is 70°. The coffee cools 10° in 5 minutes. How much longer will it take to cool 20° more? (Use the above formula.)

12. If a cold object is placed in a warm medium whose temperature is held constant at $C°$, the object warms up according to the same general principle. State a general law of temperature change that includes both the warming and cooling situations.

5. SIMPLE HARMONIC MOTION

When a weight suspended at the end of a coiled spring is disturbed, it bobs up and down with a definite frequency. The motion dies out as time passes, due to air resistance and internal friction in the spring. Presumably, if the motion were to occur in a perfect vacuum, and if the spring were perfectly elastic, then the oscillatory motion would continue undiminished forever.

The motion described above is a simple example of *elastic vibration*, and we can learn a lot about the nature of this universal phenomenon by studying such a simple mass–spring prototype. At the moment we do not have the technique to handle friction, but the ideal frictionless model is still very instructive and useful.

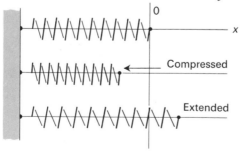

Consider, then, a coiled spring lying along a frictionless x-axis, with one end held fixed and the free end at the origin. When we compress the spring it resists and pushes back. When we extend the spring it resists and pulls back. That is, the spring always exerts a "restoring force" when it is deformed. Moreover, we find after careful measurement that the restoring force F is exactly proportional to the amount of deformation x (within the so-called elastic limits). Since F is directed against the deformation, it is given by

$$F = -sx,$$

where s is the constant of proportionality. This is Hooke's Law. If we use a stronger spring, s will be larger, and we call s the "*stiffness* of the spring."

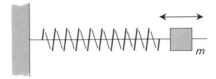

If we now attach a mass m to the end of the spring and disturb the system somehow, then our ideal oscillatory motion will occur. In order to study this motion mathematically, we need to know how a mass m moves when it is acted on by a force F. We find that m is *accelerated* by F, and that its acceleration $d^2 x/dt^2$ is proportional to F:

$$F = k \frac{d^2 x}{dt^2}.$$

Moreover, the constant of proportionality k is a measure of the mass m: if we double the mass, then it requires double the force to produce the same acceleration. This means that we can write the force law as

$$F = ma = m \frac{d^2 x}{dt^2}$$

(when suitable units are used for all these quantities). This is Newton's second law of motion.

The laws of Newton and Hooke combine to show that the motion of our model mass–spring system satisfies the differential equation

$$m \frac{d^2 x}{dt^2} = F = -sx, \qquad \text{or} \qquad \frac{d^2 x}{dt^2} + \frac{s}{m} x = 0.$$

We already know the general solution of this equation from Section 4 in Chapter 4. It is

$$x = A \sin \omega t + B \cos \omega t,$$

where $\omega = \sqrt{s/m}$, and A and B are constants of integration. Without knowing A and B, we don't know the motion exactly, but we do know its frequency: since sine and cosine are periodic with period 2π, it follows that x is a periodic function of t with period

$$T = \frac{2\pi}{\omega} = 2\pi\sqrt{\frac{m}{s}}.$$

The frequency of the motion is the number of cycles (periods) per unit time:

$$f = \frac{1}{T} = \frac{1}{2\pi}\sqrt{\frac{s}{m}}.$$

These equations show exactly how the period and frequency depend on the physical constants of the elastic system. The frequency varies as the square root of the stiffness of the spring, and is *inversely* proportional to the square root of the mass.

A particular motion is generally determined by known "initial conditions." We shall neglect mentioning units in the examples below.

EXAMPLE 1. A particle of mass $m = 4$ is attached to a spring of stiffness $s = 2$ in the configuration discussed above. Starting at the equilibrium position, the particle is struck and given an initial velocity $v_0 = -3$. What is the motion?

. . . .

Solution. Here $\omega = \sqrt{s/m} = 1/\sqrt{2}$, and the initial conditions are $x = 0$ and $dx/dt = -3$ when $t = 0$. Substituting these values in the equations

$$x = A \sin \omega t + b \cos \omega t,$$

$$\frac{dx}{dt} = \omega A \cos \omega t - \omega B \sin \omega t,$$

we see that

$$0 = A \cdot 0 + B \cdot 1,$$

$$-3 = \frac{1}{\sqrt{2}}A \cdot 1 - \frac{1}{\sqrt{2}}B \cdot 0,$$

so that $B = 0$ and $A = -3\sqrt{2}$. Thus

$$x = -3\sqrt{2} \sin(t/\sqrt{2})$$

is the equation of motion. The particule oscillates with frequency

$$f = \frac{\omega}{2\pi} = \frac{1}{2\sqrt{2}\,\pi} \approx \frac{1}{9} \quad \text{cycles per unit time,}$$

back and forth across the interval $[-3\sqrt{2}, 3\sqrt{2}]$. Its maximum displacement $3\sqrt{2}$ is called the *amplitude* of its motion.

EXAMPLE 2. Show that any particular solution

$$x = A \sin \omega t + B \cos \omega t$$

can be written uniquely in the form

$$x = a \sin(\omega t + \alpha,)$$

where a is positive and $0 \le \alpha < 2\pi$.

. . . .

Solution. Expanding the required form by the sine addition law,

$$x = a[\cos \alpha \sin \omega t + \sin \alpha \cos \omega t],$$

and comparing with the given equation

$$x = A \sin \omega t + B \cos \omega t,$$

we see that the requirements on α and a reduce to

$$a \cos \alpha = A,$$

$$a \sin \alpha = B.$$

Then $a^2 = A^2 + B^2$, so a and α are determined by:

$$a = \sqrt{A^2 + B^2}$$

$$(\cos \alpha, \sin \alpha) = (A/a, B/a).$$

If we consider $x = \sin \omega t$ to be the normal mode of oscillation, then the motion $x = \sin(\omega t + \alpha)$ *leads* the normal oscillation by the fixed angle α. We call α the *leading phase angle*. Recapitulating:

The motion of our ideal frictionless elastic system is always sinusoidal. If it is given by

$$x = A \sin \omega t + B \cos \omega t,$$

then its *frequency* is

$$f = \frac{\omega}{2\pi},$$

its *amplitude* is

$$a = \sqrt{A^2 + B^2},$$

and its *leading phase angle* α is determined by

$$(\cos\alpha, \sin\alpha) = \left(\frac{A}{a}, \frac{B}{b}\right).$$

The motion

$$x = a\sin(\omega t + \alpha)$$

is called *simple harmonic motion*. The development above shows it to be the basic type of periodic motion occurring when elastic objects vibrate.

PROBLEMS FOR SECTION 5

In each of the following motions calculate the period, amplitude, and phase angle. (Rewrite each in the form $x = a\sin(\omega t + \alpha)$, computing a exactly, but possibly only estimating α.)

1. $x = \sin(3t) - \sqrt{3}\cos(3t)$ 2. $x = \cos t + \sin t$

3. $x = 5\cos t - 5\sin t$ 4. $x = \sin 2t + 2\cos 2t$

5. A particle is moving in simple harmonic motion with amplitude a and frequency f (in cycles per second). Show that its velocity at $x = 0$ is $\pm 2\pi a f$.

6. A particle of mass 10 is attached to a spring of stiffness 4 and oscillates freely in simple harmonic motion. Suppose that at time $t = 0$, the position and velocity of the particle are

$$x_0 = 2, \qquad v_0 = -3.$$

Find the motion.

7. The same system is given a new motion with initial conditions $x_0 = 2$, $v_0 = 3$. Find the new motion.

8. Show that a mass m hanging at the bottom of a vertical spring of stiffness k executes simple harmonic motion about its rest position. (Gravity acts with the constant downward force $mg = 32.2m$. Calculate the total force, and then put a new origin at the point where the total force is zero.)

A pendulum consists of a pendulum bob (a weight) suspended at the end of a light rigid rod, and allowed to swing back and forth under the action of gravity. We consider it *ideally* to be a point mass m at the end of a weightless rod of length l. The acceleration caused by gravity in a freely falling body is $g \simeq 32.2$ ft/sec^2, so by Newton's law $F = ma$, the force exerted by gravity on a body of mass m is $mg \simeq 32.2m$. In the case of the pendulum, most of the force of gravity is counteracted by the support of the rod, but a small amount of sidewise force is left over to make the pendulum oscillate. Assuming that forces combine according to the parallelogram law, this residual force is $mg \sin \theta$. Newton's second law $F = ma$ thus becomes

$$-mg \sin \theta = m\frac{d^2s}{dt^2} = m\frac{d^2(l\theta)}{dt^2} \quad \text{or} \quad \frac{d^2\theta}{dt^2} = -\left(\frac{g}{l}\right)\sin \theta.$$

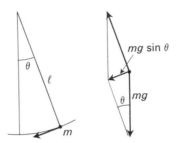

Because of the appearance of $\sin \theta$ rather than θ on the right, we cannot solve this equation. However, when θ is small,

$$\sin \theta \approx \theta$$

(since $\sin \theta/\theta \approx 1$), and small-amplitude motions are therefore governed by the "linearized" equation

$$\frac{d^2\theta}{dt^2} = -\left(\frac{g}{l}\right)\theta.$$

9. The above discussion is important to clockmakers. A one-second pendulum takes one second for each swing, so its period is 2 sec. Show that the length l of such a *second-pendulum* is 39.2 inches.

10. Find the length of a one-half-second pendulum.

11. Find the length of a two-second pendulum.

12. Show that the period of a pendulum is proportional to the square root of its length, and calculate the constant of proportionality.

13. On Planet X the force of gravity is one-third what it is on earth. Find the length of a one-second pendulum on X.

chapter 7

inverse functions

1. THE NOTION OF AN INVERSE FUNCTION

The functions $g(x) = x^{1/3}$ and $f(y) = y^3$ have the property that each one undoes what the other one does. Each one cancels the effect of the other. Thus, if we apply g to the number x, getting $x^{1/3}$, and then apply f to *that* number, it takes us back to the number x that we began with:

$$f(g(x)) = f(x^{1/3}) = (x^{1/3})^3 = x.$$

Similarly,

$$g(f(y)) = g(y^3) = (y^3)^{1/3} = y.$$

Two functions related in this way are said to be *mutually inverse* to each other, and each is called the *inverse* of the other. Thus, f and g are mutually inverse if they satisfy the cancellation equations:

$$f(g(x)) = x \qquad \textit{for every x in domain g,}$$

and

$$g(f(y)) = y \qquad \textit{for every y in domain f.}$$

Note also that the equations

$$y = x^{1/3} \qquad \text{and} \qquad x = y^3$$

are equivalent; that is, they are satisfied by exactly the same pairs of values (x,y). This is another way of saying that two functions are mutually inverse.

Theorem 1. *If the equations*

$$y = g(x) \qquad \text{and} \qquad x = f(y)$$

are equivalent, then f and g are mutually inverse functions.

Proof. Since the two equations hold for the same pairs (x,y), we can substitute from each equation into the other. For example, we can set $x = f(y)$ in the first equation, and this gives the cacellation equation

$$y = g(f(y)).$$

The other cancellation equation is obtained in the same way. ∎

It is usually in this context that we first meet the inverse of a function.

Theorem 2. *If the equation $y = g(x)$ determines x as a function of y, then this new function f is the inverse of g.*

Proof. By hypothesis, for each number y in the range of g there is a unique number x making the equation $y = g(x)$ true, and f is the operation of going from y to the uniquely determined x. Thus, $x = f(y)$ holds for a pair of numbers x and y if and only if $y = g(x)$, so the two equations

$$x = f(y) \qquad \text{and} \qquad y = g(x)$$

are equivalent. ∎

In particular, if we are given a function g and if we can *solve* the equation $y = g(x)$ uniquely for x in terms of y, getting the equivalent equation $x = f(y)$, then f is automatically the inverse of g.

EXAMPLE. We show that $g(x) = x/(1 - x)$ has an inverse by this method. Solving for x in

$$y = \frac{x}{1 - x},$$

we have, first cross-multiplying,

$$y - yx = x,$$

$$y = x(y + 1),$$

$$x = \frac{y}{1 + y}.$$

Thus $f(y) = y/(1 + y)$ is the inverse of $g(x) = x/(1 - x)$.

We know that these functions have to cancel each other, but it is interesting to check it directly. We have

$$f(g(x)) = f\left(\frac{x}{1 - x}\right) = \frac{\dfrac{x}{1 - x}}{1 + \dfrac{x}{1 - x}} = \frac{x}{(1 - x) + x} = x.$$

The other identity $g(f(y)) = y$ works out similarly.

Since the equations $y = g(x)$ and $x = f(y)$ are equivalent, they have the same graph in a standard x,y coordinate system. However, this is not the standard graph of the function f because the independent variable y in the equation for f is not plotted horizontally. To get the graph of f we have to replot with relabelled axes, and geometrically this amounts to rotating, or *flipping*, the original configuration of graph and axes over the 45° line $y = x$. We plot the same pairs of values (x_0, y_0) but with order interchanged, as (y_0, x_0). We have already considered this situation in Section 2 of the second chapter.

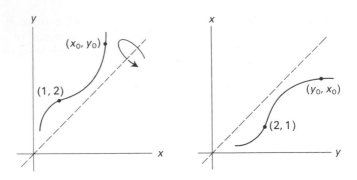

A comment about our use of x and y is in order here. When we are looking at a single function f, it is natural to write its equation in the form

$$y = f(x),$$

with x independent and y dependent, this being normal usage for these two variables. But if this equation determines x as a function of y, then the equivalent equation for the inverse function g is

$$x = g(y),$$

with y independent and x dependent. We are then simultaneously faced with a function g that may be entirely new to us and with abnormal usage of the standard variables. Therefore, when we are looking toward the unfamiliar inverse g of a familiar function f, we tend to present f by the equation

$$x = f(y),$$

so that the variables will be used normally in the equivalent equation for g,

$$y = g(x).$$

This is a very minor point, but it will determine our use of variables when we define the functions $\log x$, arc sin x, and arc tan x as inverse functions, a little later on.

Finally, we note that the inverse of a function f is its inverse under the operation of *composition*. (See the end of Section 5 in Chapter 4.) For the cancellation identities

$$f(g(x)) = x \qquad \text{and} \qquad g(f(y)) = y,$$

just say that the composition product of f and g, in either order, is the identity function. Thus, each is the *composition inverse* of the other.

PROBLEMS FOR SECTION 1

1. Show that $f(x) = (x + 4)/3$ and $g(x) = 3x - 4$ are mutually inverse functions by computing $f(g(x))$ and $g(f(x))$.

2. Show that $f(x) = (4 - x)/(1 + x)$ is its own inverse by computing $f(f(x))$.

3. Compute $f(g(x))$ and $g(f(x))$, where

$$f(x) = \frac{x}{\sqrt{1 + x^2}} \quad \text{and} \quad g(x) = \frac{x}{\sqrt{1 - x^2}}.$$

   What is your conclusion?

4. Find the values of a and b such that the linear function $f(x) = ax + b$ is self-inverse. There is more than one solution. Graph each type.

Show that each of the following functions f has an inverse function g by solving the equation for x in terms of y. Give $g(x)$ in each case, and sketch the two graphs.

5. $y = f(x) = 3x - 5$ 6. $y = f(x) = 1/x$

7. $y = f(x) = x^{3/5}$ 8. $y = f(x) = 1 + x^{1/7}$

9. $y = f(x) = \dfrac{1 - x}{1 + x}$ 10. $y = f(x) = x^3 + 3x^2 + 3x + 1$

11. $y = f(x) = \dfrac{x}{\sqrt{1 + x^2}}$ 12. $y = f(x) = \sqrt{x}$
    (Be careful here.)

13. Compute $f(g(x))$ and $g(f(x))$, where

$$f(x) = x^2 - 2x + 1, \qquad g(x) = 1 + \sqrt{x}.$$

   Show that f and g are mutually inverse if the domain of f is suitably restricted.

14. Show that

$$f(x) = (x^2 + 1)^{1/4} \quad \text{and} \quad g(x) = \sqrt{(x^2 + 1)(x^2 - 1)}$$

   are mutually inverse if their domains are suitably restricted.

15. Show that $f(x) = \sqrt{1 - x^2}$ is its own inverse, provided that its domain is suitably restricted.

16. a) Show that if $g(f(x)) = x$ for every x in the domain of f, then the domain of f is included in the range of g and the range of f is included in the domain of g.

b) Show therefore that if f and g are mutually inverse, then

$$\text{domain } f = \text{range } g,$$

$$\text{range } f = \text{domain } g.$$

17. The symmetric form of Theorem 2 is as follows:

If an equation in x and y determines each of x and y as a function of the other, then these two functions are mutually inverse.

Show that this is so. (Argue in any manner, using function symbols only if you want to.)

18. Show that if f and g are both invertible, then so is their composition product $f \circ g$, and

$$\text{inv } (f \circ g) = (\text{inv}) \, g \circ (\text{inv}) \, f \, .$$

19. Show that

$$f(x) = \frac{ax + b}{cx + d}$$

is self-inverse if and only if:

a) $d = -a,$ or

b) $d = a, c = b = 0.$

20. Show that if f and g are mutually inverse, then so are $f(ax)$ and $g(x)/a$, for any nonzero constant a.

2. THE EXISTENCE OF INVERSE FUNCTIONS

What kind of function f has an inverse? If the domain of f is an interval and if we restrict ourselves to inverse functions f and g that are *differentiable*, then there is a simple condition.

Theorem 3. *If f is a differentiable function with domain an interval I, and if f has a differentiable inverse g, then f is monotone.*

Proof. We differentiate the identity

$$g(f(x)) = x$$

by the chain rule, and conclude that

$$g'(y) \cdot f'(x) = 1$$

on I, where $y = f(x)$. In particular f' can never be zero. Therefore f must be monotone, by Theorem 5 in Chapter 5. ∎

Now let us go in the opposite direction and use geometric reasoning. Suppose that the domain of f is an interval I and that f is continuous and monotone on I, say increasing. It then seems clear on geometric grounds that the range of f is an interval J and that for each y in J there is exactly one x in I such that $y = f(x)$ That is, the equation $y = f(x)$ determines x as a function of y over the interval J.

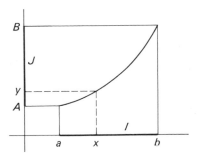

Then this new function g is automatically the inverse of f, as we saw in Theorem 2. Diagrammatically, we can exhibit the operations f and g as following the arrows in the figure below. Clearly these operations undo each other and are mutually inverse.

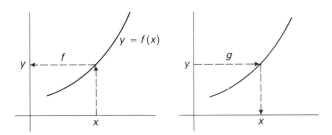

These geometric considerations make it more or less clear that the inverse function g is continuous. But we can check this directly, and will do so a little later.

As we noted earlier, we get the graph of the inverse function g by flipping the graph of f over the 45° line $y = x$, In doing this, a line tangent to the graph of f over a point $x = x_0$ rotates to a line tangent to the graph of g over the point $y = y_0 = f(x_0)$. If the first line is not horizontal, then the second line will not be vertical, and its slope will then be $g'(y_0)$. That is, if $f'(x_0)$ exists and is not zero, then $g'(y_0)$ exists, where $y_0 = f(x_0)$. It follows from the chain-rule calculation in Theorem 3 that $g'(y_0) = 1/f'(x_0)$. This also could be shown geometrically.

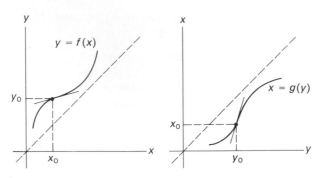

Altogether, the above geometric reasoning adds up to the following theorem.

Theorem 4. *Let f be continuous and monotone on an interval I. Then the range of f is an interval J and f has a continuous, monotone inverse function g with domain J. If f is differentiable at a point x_0 in I and if $f'(x_0) \neq 0$, then g is differentiable at $y_0 = f(x_0)$ and*

$$g'(y_0) = \frac{1}{f'(x_0)}.$$

In applying Theorem 4 to get an inverse function we have to know that f is increasing on an interval I (or decreasing on I). This may follow directly from the definition of f. For example, the exponential addition law guarantees that $f(x) = 2^x$ is increasing, since if h is positive, then $2^{x+h} = 2^h \cdot 2^x > 2^x$. Similarly, the geometric definition of $f(t) = \sin t$ shows it to be increasing over the interval $[-\pi/2, \pi/2]$. But once we know how to differentiate f, we have an alternative method for showing that f is increasing: we can simply check that f' is positive over I. There can even be a finite number of exceptional points. (See Problem 14, Section 2, Chapter 5.) That is:

Corollary. *If f is continuous on the interval I, and if $f'(x)$ is everywhere positive (with the possible exception of a finite number of points), then f is invertible over I, and its inverse function g is of the same type. That is, g is continuous, and has an everywhere positive derivative (except possibly at a finite number of points).*

At some point Theorem 4 has to be given an analytic proof. However, we shall first apply it, and then return to the proof in Section 8.

So far we have started with the functions x, e^x, $\sin x$, and $\cos x$, and have considered the class of function built up from the above list by using the algebraic operations and composition. The derivative of any such function involves the same operations and belongs to the same class. The above theorem gives the first method for obtaining new functions outside this class. In the next four sections we shall investigate three such new functions, obtained as inverses of e^x, $\sin x$, and $\tan x$.

PROBLEMS FOR SECTION 2

Some of the functions in Problems 1 through 15 are already known to be invertible. But give a (new) proof in each case by applying the Corollary of Theorem 4.

1. $f(x) = x^3$

2. $f(x) = x^2$, on the interval $[0, \infty)$

3. $f(x) = x^5$

4. $f(x) = x^3 + x$

5. $f(x) = 2x - x^2$, on $[-1, 1]$

6. $f(x) = x^3 + x^2$, on $[0, \infty)$

7. $f(x) = x^3 - 3x^2 + 3x$

8. $f(x) = \dfrac{x^5}{5} - \dfrac{x^4}{2} + \dfrac{x^3}{3}$

9. $f(x) = x^3 - x^2 + x$

10. $f(x) = x\sqrt{1 + x^2}$

11. $f(x) = \dfrac{x}{\sqrt{1 + x^2}}$

12. $f(x) = \sin x$, on $[-\pi/2, \pi/2]$

13. $f(x) = \tan x$, on $(-\pi/2, \pi/2)$

14. $f(x) = x \cos x$, on $[-\pi/4, \pi/4]$

15. $f(x) = \sin^3 x$, on $[-\pi/2, \pi/2]$

16. Prove that $f(x) = x^3/(1 + x^2)$ is invertible, and show that its inverse function is everywhere defined, i.e., its domain is $(-\infty, \infty)$. (Apply Theorem 4.)

17. Prove that $f(x) = e^x$ is invertible and show that its inverse function has domain $(0, \infty)$.

18. Show that if f is increasing and equal to its own inverse, then $f(x) = x$.

19. Suppose that an equation E in x and y determines each of x and y as a function of the other. Show that these mutually inverse functions are the same function if and only if interchanging x and y in the equation E yields an equivalent equation.

20. Keeping in mind the principles stated in Problems 18 and 19, draw the graph of a self-inverse function f that is decreasing and concave up. Find an equation for such a function.

21. Draw a graph and find an equation for a decreasing self-inverse function f that is concave down.

22. What can you say about the function solution of the equation

$$x^{1/3} + y^{1/3} = 2 ?$$

Sketch its graph.

23. Suppose that f is defined for all positive numbers x and that $f'(x) = 1/x$. Show that f is invertible, that its inverse function g necessarily satisfies the equation

$$g'(y) = g(y),$$

and hence that $g(y) = ce^y$ for some constant c.

24. a) Suppose f is a differentiable function defined on the interval $(-1,1)$ such that $f'(x) = 1/\sqrt{1-x^2}$. Show that f is invertible, and that its inverse function g necessarily satisfies the equation

$$g^2 + (g')^2 = 1.$$

   b) Prove from this equation that $g'' = -g$. What can you then say about g?

3. THE NATURAL LOGARITHM FUNCTION

Logarithms to the base 10 are often taught in secondary school, starting from the definition of $\log_{10} x$ as that number y such that $x = 10^y$. Similarly, for any positive number x, $\log_e x$ is defined to be *that number y such that $x = e^y$*. The graph of $x = e^y$ makes it seem clear that there is such a number y uniquely determined by x, no matter what positive number x may be. So $\log_e x = y$ if and only if $x = e^y$, which means, in the language of the present chapter, that we have defined the function $g(x) = \log_e x$ as the inverse of $f(y) = e^y$.

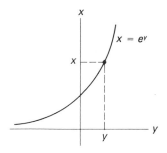

It is Theorem 4 that makes all of this legitimate. Since e^x is increasing and has an everywhere positive derivative, Theorem 4 says that it has a differentiable inverse.

The number $\log_e x$ is called the *natural logarithm* of x, and is frequently designated $\ln x$. We also write simply $\log x$, the base e being understood unless a different base is explicitly indicated. Thus,

$$\ln x = \log x = \log_e x.$$

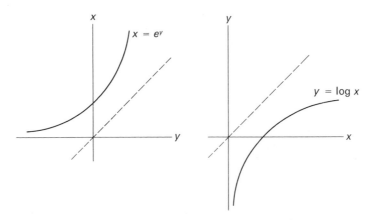

The graph of log x is obtained geometrically from the graph of e^y by reflecting in, or rotating about, the 45° line $y = x$. The domain of log x is $(0, +\infty)$, as shown in the graph. This follows from Theorem 4 and our knowledge of e^y. The theorem says that the domain is an interval, and is the range of e^y. Since $e^y \to +\infty$ as $y \to +\infty$, and $e^y \to 0$ as $y \to -\infty$, this interval is $(0, +\infty.)$ The cancelling identities are

$$e^{\log x} = x \qquad \text{and} \qquad \log(e^y) = y.$$

The cancellation laws in effect define the logarithm function from the exponential function, so we should be able to use them to deduce the logarithm law from the law of exponents. This is short but tricky. We have

$$\log uv = \log(e^{\log u} e^{\log v}) = \log(e^{(\log u + \log v)})$$
$$= \log u + \log v,$$

where we have used both cancelling identities and the law of exponents. The law

$$\log uv = \log u + \log v$$

is the characteristic property of a logarithm, and the above proof shows why the inverse of any exponential function is a logarithm.

Also useful are the properties:

$$\log 1 = 0,$$

$$\log \frac{1}{u} = -\log u.$$

These follow from the logarithm law and the identities $1 \cdot 1 = 1$, $u(1/u) = 1$. The details are left to the reader.

We can compute the derivative of log x from Theorem 4, but it is probably better to recompute the derivative of any inverse function directly from the chain rule. Differentiating the cancellation identity

$$x = e^{\log x},$$

we get

$$1 = \frac{d}{dx}(e^{\log x}) = e^{\log x}\frac{d}{dx}\log x = x\frac{d}{dx}\log x,$$

so the derivative formula for log x is

$$\frac{d}{dx}\log x = \frac{1}{x}.$$

In order to perform the above computation, we need to know that log x is differentiable so we have used part of Theorem 4. With the built-in chain rule, this formula becomes

$$\frac{d}{dx}\log u = \frac{1}{u}\frac{du}{dx},$$

$$d(\log u) = \frac{du}{u}.$$

EXAMPLES

1. $\dfrac{d}{dx}\log(1 + x^2) = \dfrac{1}{1 + x^2}\dfrac{d}{dx}x^2 = \dfrac{2x}{1 + x^2}.$

2. $\dfrac{d}{dx}\log kx = \dfrac{k}{kx} = \dfrac{1}{x}.$

3. $\dfrac{d}{dx}\log(\cos x) = \dfrac{1}{\cos x}\cdot(-\sin x) = -\tan x.$

4. $\dfrac{d}{dx}\log(\log x) = \dfrac{1}{\log x}\left(\dfrac{1}{x}\right) = \dfrac{1}{x\log x}.$

5. $\dfrac{d}{dx}(x\log x - x) = x\cdot\dfrac{1}{x} + \log x\cdot 1 - 1 = \log x.$

6. $\dfrac{d}{dx}\log\left(\dfrac{1-x}{1+x}\right) = \dfrac{1}{\left(\dfrac{1-x}{1+x}\right)}\cdot\dfrac{(1+x)(-1) - (1-x)\cdot 1}{(1+x)^2}$

$$= \dfrac{-2}{(1-x)(1+x)} = -\dfrac{2}{1-x^2}.$$

7. Since $\log[(1-x)/(1+x)] = \log(1-x) - \log(1+x),$

$$\dfrac{d}{dx}\log\left(\dfrac{1-x}{1+x}\right) = \dfrac{1}{(1-x)}(-1) - \dfrac{1}{(1+x)} = \dfrac{-2}{(1-x)(1+x)}.$$

REMARK. The calculation of the derivative of $\log x$ is really an implicit differentiation. To see this, remember that the function $\log x$ exists only because the equation

$$x = e^y$$

determines y as a function of x. If we leave y in this equation, instead of setting $y = \log x$, then the calculation given earlier amounts to differentiating this equation implicitly:

$$1 = \frac{d}{dx} e^y = e^y \frac{dy}{dx} = x \frac{dy}{dx},$$

and

$$\frac{dy}{dx} = \frac{1}{x}.$$

REMARK. A function f is uniquely determined if we know its derivative and its value at one point. In particular, $\log x$ is the unique function $f(x)$ satisfying the equations

$$f'(x) = \frac{1}{x} \quad \text{and} \quad f(1) = 0.$$

Therefore, the addition law

$$\log ax = \log a + \log x$$

must somehow be implied by these equations. There are hints on how this goes in Problem 27.

PROBLEMS FOR SECTION 3

Reformulate the following expressions by using the logarithm law.

1. $\log(x - 1) + \log(x + 1)$ 2. $2 \log x + 3 \log(x - 2)$
3. $\log(x^2 - 2x + 1)$
4. $\log(x - 1) - \log(x + 1) + \log(x^2 + x + 1) - \log(x^2 - x + 1)$

5. Prove the more general formula

$$\frac{d}{dx} \log|x| = \frac{1}{x} \quad \text{(if } x \neq 0\text{)}.$$

Differentiate the following functions.

6. $\log(x + 1)$ 7. $\log(x^2 + 1)$

8. $\log \sqrt{x^2 + 4}$ 9. $x \log x$

10. $e^x \log x$ 11. $(\log x)/x$

12. $\log(e^x + 1)$ 13. $\log(e^x)^2$

14. $x^2 \log x - x^2/2$ 15. $\dfrac{x^{n+1}}{n + 1}\left(\log x - \dfrac{1}{n + 1}\right)$

16. $(\log x)^2$

Prove the following formulas.

17. $\dfrac{d}{dx}\log(\sin x) = \cot x$ 18. $\dfrac{d}{dx}\log(\sec x) = \tan x$

19. $\dfrac{d}{dx}\log(\sec x + \tan x) = \sec x$ 20. $\dfrac{d}{dx}\log\left(\dfrac{1 + \sin x}{1 - \sin x}\right) = 2\sec x$

21. $\dfrac{d}{dx}\log(x + \sqrt{1 + x^2}) = \dfrac{1}{\sqrt{1 + x^2}}$

22. $\dfrac{d}{dx}\log(x + \sqrt{x^2 - 1}) = \dfrac{1}{\sqrt{x^2 - 1}}$

23. $\dfrac{d}{dx}\log\left(\dfrac{1 + \sqrt{1 + x^2}}{x}\right) = -\dfrac{1}{x\sqrt{1 + x^2}}$

24. Prove that $a \log x = \log x^a$ for any rational number a. [*Hint.* Show that both sides have the same derivative.]

25. Prove the identities

$$\log 1 = 0,$$

$$\log(1/u) = -\log u$$

from the addition law $\log(ab) = \log a + \log b$.

26. Use the cancellation identities to show that the law of exponents follows from the logarithm law.

27. Prove that if f is a function defined for positive x such that $f'(x) = 1/x$ and $f(1) = 0$, then

$$f(ab) = f(a) + f(b)$$

for all positive a and b. [*Hint.* What is $f'(ax)$?]

28. Using the notation $u' = du/dx$, etc., show that

$$\frac{d}{dx}\log(uv) = \frac{u'}{u} + \frac{v'}{v},$$

$$\frac{d}{dx}\log(uvw) = \frac{u'}{u} + \frac{v'}{v} + \frac{w'}{w}.$$

29. Prove that $a \log x < x^a - 1$ when $x > 1$ for any positive rational number a. (Use the mean-value theorem.)

30. Using the above inequality, show that for any positive rational number b,

$$\log x < x^b$$

when x is large enough. (Take $a = b/2$ and manipulate a little.)

31. Show then that

$$x^n < e^x$$

when x is large enough.

32. Show from the preceding problem that

$$x^m e^{-x} \to 0$$

as $x \to +\infty$, for any positive integer m.

33. Find the minimum value of $f(x) = x \log x$.

34. Find the minimum value of $f(x) = x^2 \log x$.

4. THE GENERAL EXPONENTIAL FUNCTION

In Chapter 4 we briefly discussed the exponetial function a^x with a general positive base a, but since then we have concentrated on the base e. We can now reduce other bases to the base e by using the cancelling law $a = e^{\log a}$, which gives us the identity

$$a^x = e^{x \log a}.$$

We always use this identity to reduce problems involving other bases to the known rules for the base e.

EXAMPLE 1. We start by proving the "mixed" law

$$\log a^x = x \log a.$$

Proof. $\log(a^x) = \log(e^{x \log a}) = x \log a$, by the cancellation law.

EXAMPLE 2.

$$\frac{d}{dx}(2^x) = \frac{d}{dx}e^{(x\log 2)}$$

$$= e^{x\log 2} \cdot \frac{d}{dx}(x \log 2)$$

$$= 2^x \cdot \log 2.$$

Thus the constant α that we met in Chapter 4 when we first considered the derivative of 2^x is $\alpha = \log 2$.

EXAMPLE 3. At last we can prove the power rule for *arbitrary* exponent a. We have

$$\frac{d}{dx}x^a = \frac{d}{dx}e^{a\log x}$$

$$= e^{a\log x}\frac{d}{dx}(a \log x)$$

$$= x^a \cdot \frac{a}{x}$$

$$= ax^{a-1}.$$

EXAMPLE 4

$$\frac{d}{dx}x^x = \frac{d}{dx}e^{x\log x}$$

$$= e^{x\log x}\frac{d}{dx}(x \log x)$$

$$= x^x[1 + \log x].$$

The function

$$[f(x)]^{g(x)} = e^{g(x)\log f(x)}$$

is defined only when $f(x)$ is positive; it is then differentiable whenever f and g are differentiable. One way of computing its derivative is illustrated in the examples above. There is another procedure, called *logarithmic differentiation*, where we compute the derivative of the logarithm of the function in question.

EXAMPLE 5. Find the derivative of $y = x^x$.

Solution. Since

$$\log y = \log x^x = x \log x,$$

differentiating with respect to x gives

$$\frac{1}{y}\frac{dy}{dx} = x \cdot \frac{1}{x} + \log x \cdot 1 = 1 + \log x,$$

so

$$\frac{dy}{dx} = y[1 + \log x] = x^x[1 + \log x],$$

as before.

EXAMPLE 6. Products and quotients can be attacked by logarithmic differentiation. We shall illustrate this by finding the rule for the differential of a product
with three factors. If

$$y = uvw,$$

then

$$\log y = \log u + \log v + \log w.$$

Applying the differential rules to this equation, we see that

$$\frac{dy}{y} = \frac{du}{u} + \frac{dv}{v} + \frac{dw}{w},$$

or

$$dy = uvw\left(\frac{du}{u} + \frac{dv}{v} + \frac{dw}{w}\right).$$

PROBLEMS FOR SECTION 4

Differentiate the following.

1. $x^{\sqrt{x}}$

2. 3^{2x}

3. 9^x

4. $x^{\log x}$

5. $(x^a)^x$

6. $x^{(a^x)}$

7. $(\sin x)^x$

8. $(\log x)^{\log x}$

9. $10^{(ax^2+bx+c)}$

10. $(e^e)^x$

11. $x^{\tan x}$

12. $(x^2)^{x^2}$

13. We define $\log_a x$ as the function inverse to a^x. Show that

$$\log_a x = \frac{\log x}{\log a}.$$

14. Strictly speaking, the equation

$$a^x = e^{x \log a}$$

should be taken as the *definition* of a^x. Prove, then, that:

$$a^{x+y} = a^x a^y, \qquad (a^x)^y = a^{xy}, \qquad \text{and} \qquad a^1 = a.$$

In the following problems, compute dy/dx by logarithmic differentiation

15. $y = \sqrt{1 + x^2}$

16. $y^3 = x(x - 1)$

17. $y^n = x^m$

18. $y = x^{(e^x)}$

19. $(xy)^x = 4$

20. $y = (\sin x)^x$

21. $y = x^{\tan x}$

22. $y^3 + y = x^3$

23. $y = \sqrt{\left(\dfrac{1 - x}{1 + x}\right)}$

24. $y = \dfrac{(1 + x)^{1/3}(1 - x)^{2/3}}{(1 + x^2)^{1/6}}$

5. THE ARCSINE FUNCTION

We think of the functions $f(x) = \sqrt{x}$ and $g(y) = y^2$ as being inverse to each other, but this is only partly correct. The function $\sqrt{x}$ is strictly increasing and has a differentiable inverse, but its inverse is only the "righthand half" of $g(y) = y^2$. This is clear geometrically from the flip-over principle. It must also show up in the cancelling identities. We find that one is all right,

$$g(f(x)) = (\sqrt{x})^2 = x,$$

but the other is not, for

$$f(g(y)) = \sqrt{y^2} = |y|.$$

Thus we have $f(g(y)) = y$ only when $y \geq 0$.

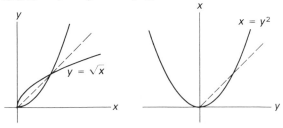

According to Theorems 3 and 4 in Section 2, if g is a differentiable function whose domain is an interval I, then g is invertible if and only if it is monotone.

From this viewpoint, the function $g(y) = y^2$ in its entirety doesn't have an inverse because it isn't monotone, and what we have done above, in order to get an inverse for y^2, is to restrict its domain to an interval on which it *is* monotone, choosing as large an interval as possible.

Now consider $x = g(y) = \sin y$. Again we write the equation with variables reversed so that the eventual inverse will be expressed in standard form. In order to get an inverse function we have to restrict $\sin y$ to an interval on which it is monotone. There are many possible choices, but the standard one is $[-\pi/2, \pi/2]$. The inverse of this restricted sine function is called the *arcsine* function, and its value at x is designated arcsin x. Thus, for each number x in the closed interval $[-1, 1]$, arcsin x is defined to be *that number y in $[-\pi/2, \pi/2]$ such that $x = \sin y$*. Put another way, when y is restricted to this interval, the equation $x = \sin y$ determines y as a function of x, the function in question being called the arcsin function. The domain of arcsin x is the closed interval $[-1, 1]$.

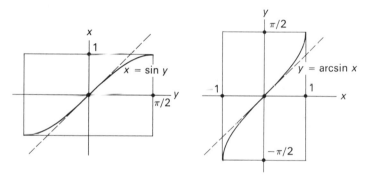

The cancellation identities are

$$\sin(\arcsin x) = x \qquad \text{(for } x \text{ in } [-1, 1]),$$

$$\arcsin(\sin y) = y \qquad \text{(for } y \text{ in } [-\pi/2, \pi/2]),$$

and the mutually equivalent equations are

$$x = \sin y \qquad \text{and} \qquad y = \arcsin x.$$

According to Theorem 4, $y = \arcsin x$ is everywhere differentiable in $[-1, 1]$ except at the two endpoints. To compute dy/dx, we differentiate the equivalent equation

$$x = \sin y$$

with respect to x, and get

$$1 = \frac{d}{dx} \sin y = \cos y \cdot \frac{dy}{dx}.$$

Therefore

$$\frac{dy}{dx} = \frac{1}{\cos y} = \frac{1}{\pm\sqrt{1 - \sin^2 y}} = \frac{1}{\pm\sqrt{1 - x^2}}.$$

Moreover, $\cos y > 0$ when $-\pi/2 < y < \pi/2$ and so the plus sign before the radical is correct. Thus

$$\frac{d}{dx}\arcsin x = \frac{1}{\sqrt{1 - x^2}},$$

$$\frac{d}{dx}\arcsin u = \frac{1}{\sqrt{1 - u^2}} \cdot \frac{du}{dx},$$

$$d(\arcsin u) = \frac{du}{\sqrt{1 - u^2}}.$$

EXAMPLES

1. If a is positive, then

$$\frac{d}{dx}\arcsin \frac{x}{a} = \frac{1}{\sqrt{1 - \left(\frac{x}{a}\right)^2}} \cdot \frac{1}{a} = \frac{1}{\sqrt{a^2 - x^2}}.$$

2.

$$\frac{d}{dx}(\sqrt{1 - x^2}\,\arcsin x - x) = \frac{\sqrt{1 - x^2}}{\sqrt{1 - x^2}} + \arcsin x \cdot \frac{-x}{\sqrt{1 - x^2}} - 1$$

$$= -\frac{x\,\arcsin x}{\sqrt{1 - x^2}}.$$

3.

$$\frac{d}{dx}\arcsin \frac{1}{x} = \frac{1}{\sqrt{1 - (1/x)^2}} \cdot \frac{-1}{x^2} = \frac{-\sqrt{x^2}}{x^2\sqrt{x^2 - 1}} = \frac{-1}{|x|\sqrt{x^2 - 1}}.$$

The domain of this function consists of two separate intervals: $(-\infty, -1]$ and $[1, \infty)$.

The inverse of $\cos x$, called $\arccos x$, is similar to $\arcsin x$. We have to choose an interval on which $\cos x$ is monotone, and the usual choice is $[0, \pi]$, so that this time we are dealing with *decreasing* functions. The graph follows.

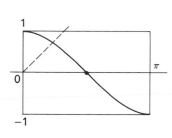 

Proceeding as before, we would find that

$$\frac{d}{dx} \arccos x = -\frac{1}{\sqrt{1 - x^2}}.$$

Therefore the derivative of arccsin x + arccos x is zero, and the sum is a constant! Once our attention has been directed to this fact, we find that we can easily establish it directly from the definition of the two functions. It depends on the identity

$$\cos(\pi/2 - y) = \sin y$$

for all y, and the fact that $\pi/2 - y$ runs from π to 0 as y runs from $-\pi/2$ to $\pi/2$. Thus, if x is the common value in the above identity, then

$$\arcsin x = y,$$

$$\arccos x = \frac{\pi}{2} - y,$$

and

$$\arcsin x + \arccos x = \frac{\pi}{2}.$$

The derivative formula for arccos x can now be obtained by differentiating $\pi/2 - \arcsin x$.

PROBLEMS FOR SECTION 5

Find the derivatives of the following functions.

1. $\arcsin \sqrt{x}$
2. $\arcsin kx$
3. $\arcsin x^2$
4. $\arcsin(\sin^2 x)$
5. $\arcsin(\cos x)$
6. $\arcsin(1 - 2x)$
7. $x \arcsin x + \sqrt{1 - x^2}$
8. $\arcsin x + \sqrt{1 - x^2}$

9. $\arcsin x + x\sqrt{1 - x^2}$ 10. $\arcsin\left(\dfrac{1 - x}{1 + x}\right)$

11. $\arcsin\left(\dfrac{\cos x}{1 + \sin x}\right)$ 12. $\cos^2(\arcsin x)$

13. $\cos(\arcsin x) - \sqrt{1 - x^2}$ 14. $\arcsin \dfrac{x}{\sqrt{1 + x^2}}$

15. If we don't restrict the domain of $g(y) = y^2$, then g fails to be the inverse of $f(x) = \sqrt{x}$, as was noted in the text. But g is a *left inverse* of f, in the sense that $g(f(x)) = x$ for all x in the domain of f. What breaks down is that g fails to be a right inverse of f. Show, similarly, that $\sin y$ is a left inverse of $\arcsin x$ but not a right inverse.

16. Show that $\pi - \arcsin x$ is another right inverse of $\sin x$.

17. Describe the most general differentiable right inverse of $\sin x$ with domain $[-1, 1]$.

18. Draw the graph of $\arcsin(\sin x)$ for $-2\pi \le x \le 2\pi$.

19. Do Problems 1 and 6, and compare their answers. Try to figure out what the explanation is.

20. Show that

$$\arccos x = \arcsin \sqrt{1 - x^2}$$

for all x in the interval $[0,1]$.

21. Derive the formula for the derative of $\arccos x$ by differentiating the right side of the above identity.

6. THE ARCTANGENT FUNCTION

We use the same restricted domain to invert the tangent function as for the sine function, except that now the endpoints are missing. Thus, if y is restricted to the

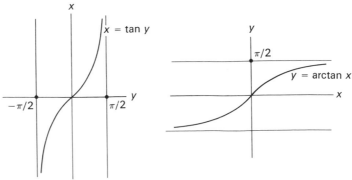

open interval $(-\pi/2)$, $\pi/2)$, then $x = \tan y$ is an increasing function with an everywhere positive derivative, and so has a differentiable inverse, by Theorem 4. This new inverse function is called the *arctangent* function, and its value at x is designated arctan x.

The domain of arctan x *is the whole real-number system* $(-\infty, \infty)$. This fact makes arctan x a more generally useful function than arcsin x.

The cancellation identities are:

$$\tan(\arctan x) = x \qquad \text{for every } x,$$

$$\arctan(\tan y) = y \qquad \text{for } y \text{ in } (-\pi/2, \pi/2),$$

and the equivalent equations are

$$x = \tan y \qquad \text{and} \qquad y = \arctan x.$$

Differentiating the first of these equivalent equations with respect to x gives

$$1 = \frac{d}{dx}(\tan y) = \sec^2 y \frac{dy}{dx},$$

so

$$\frac{dy}{dx} = \frac{1}{\sec^2 y} = \frac{1}{1 + \tan^2 y} = \frac{1}{1 + x^2}.$$

Thus,

$$\frac{d}{dx} \arctan x = \frac{1}{1 + x^2},$$

$$\frac{d}{dx} \arctan u = \frac{1}{1 + u^2} \frac{du}{dx}$$

$$d \arctan u = \frac{du}{1 + u^2}.$$

EXAMPLES

1.
$$\frac{d}{dx} \arctan \frac{x}{a} = \frac{1}{1 + \left(\dfrac{x}{a}\right)^2} \cdot \frac{1}{a} = \frac{a}{a^2 + x^2}.$$

2.
$$\frac{d}{dx}(1 + x^2)\arctan x = \frac{1 + x^2}{1 + x^2} + (\arctan x)2x$$
$$= 2x \arctan x + 1.$$

3.
$$\frac{d}{dx} \arctan \frac{x-1}{x+1} = \frac{1}{1 + \frac{(x-1)^2}{(x+1)^2}} \cdot \frac{(x+1) - (x-1)}{(x+1)^2}$$

$$= \frac{1}{(x+1)^2 + (x-1)^2} \cdot 2 = \frac{1}{x^2 + 1}.$$

You might like to consider what is behind this interesting result.

PROBLEMS FOR SECTION 6

Differentiate the following functions.

1. $\arctan x^2$

2. $\arctan 3x$

3. $\arctan ax$

4. $\arctan \sqrt{x}$

5. $\arctan \sqrt{x^2 - 1}$

6. $\arctan\left(\dfrac{\sin x}{1 + \cos x}\right)$

7. $\arctan x + \log \sqrt{1 + x^2}$

Prove the following identities.

8. $\dfrac{d}{dx}\left[x \arctan x - \log \sqrt{1 + x^2}\right] = \arctan x.$

9. $\dfrac{d}{dx}\left[\sqrt{x} - \arctan \sqrt{x}\right] = \dfrac{\sqrt{x}}{2(1 + x)}.$

10. $\dfrac{d}{dx} \arctan\left(\dfrac{e^x - e^{-x}}{2}\right) = \dfrac{2}{e^x + e^{-x}}.$

11. If we invert $\cot x$ on the interval $(0,\pi)$, show that

$$\text{arccot } x = \frac{\pi}{2} - \arctan x,$$

and hence that

$$\frac{d}{dx} \text{arccot } x = -\frac{1}{x^2 + 1}.$$

12. Show that for positive values of x,

$$\text{arccot } x = \arctan \frac{1}{x}.$$

Then compute the derivative of arccot x from this identity, and compare with the above problem.

13. If we invert $x = \sec y$ on the interval $[0,\pi]$, excluding $\pi/2$, show that

$$\frac{d}{dx} \text{arcsec } x = \frac{1}{|x|\sqrt{x^2 - 1}}.$$

14. Prove similarly that

$$\frac{d}{dx} \operatorname{arccsc} x = -\frac{1}{|x|\sqrt{x^2 - 1}}.$$

Differentiate (using the formulas in Problems 11 through 14 when necessary).

15. $\operatorname{arccot}(\tan x)$ 16. $\operatorname{arcsec} \sqrt{x^2 + 1}$

17. $\operatorname{arcsin} \dfrac{x}{\sqrt{1 + x^2}}$ 18. $\operatorname{arcsec} \dfrac{1}{x}$

19. $\operatorname{arccot}(\log x)$

20. It appeared from Example 3 in the text that

$$\arctan \frac{x - 1}{x + 1} = \arctan x + C.$$

Show that this is so and that $C = -\pi/4$.

21. Draw the graph of $\arctan(\tan x)$ over $[-2\pi, 2\pi]$.

22. Show that

$$\arctan x = \arcsin \frac{x}{\sqrt{1 + x^2}}.$$

23. Show that

$$\arcsin x = \arctan \frac{x}{\sqrt{1 - x^2}}.$$

24. Find the maximum value of

$$f(x) = \arctan(x + 1) - \arctan(x - 1).$$

25. Find the value of x where

$$f(x) = \arctan 2x - \arctan x$$

has its maximum value on the interval $[0, \infty)$.

26. Sketch the graph of

$$y = 2 \arctan \frac{x}{2} - \arctan x$$

(critical points, concavity, behavior at infinity). You will have to guess at a couple of y values.

27. Sketch the graph of

$$y = \arctan 2x - \arctan x.$$

You will have to guess at a couple of y values.

28. A statue a feet tall stands on a pedestal whose top is b feet above an observer's eye level. If the observer stands x feet away from the pedestal, show that the angle θ subtended by the statue at his eyes is given by

$$\theta = \arctan \frac{a + b}{x} - \arctan \frac{b}{x}.$$

29. Find where the observer must stand in order to maximize the angle of vision θ given by the above formula.

7. CONCAVITY AND INVERSE FUNCTIONS

The graphs below suggest what happens to concavity when a graph is reflected in, or rotated over, the 45° line $y = x$. It appears that *an increasing graph reverses its concavity when reflected*, and that *a decreasing graph preserves concavity*.

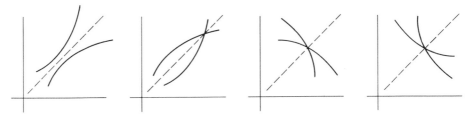

These conjectures are easy to prove. We saw earlier that two functions related this way are mutually inverse and, in particular, satisfy

$$f(g(x)) = x.$$

Then $f'(g(x))g'(x) = 1$ and

$$g'(x) = \frac{1}{f'(g(x))},$$

by the chain rule. We have been through all of this before. Differentiating once more, and then substituting from the above equation, we have

$$g''(x) = -\frac{1}{[f'(g(x))]^2} \cdot f''(g(x)) \cdot g'(x)$$

$$= -\frac{f''(g(x))}{[f'(g(x))]^3} = -\frac{f''(y)}{[f'(y)]^3},$$

where $y = g(x)$. We are supposing that f' is never 0. Thus if f is increasing, then f' is everywhere positive and the formula above shows that g'' and f'' have opposite signs, so that concavity reverses. Similarly, if f is decreasing, then f' and $(f')^3$ are everywhere negative, and the formula shows that concavity is preserved.

8. THE INVERSE FUNCTION THEOREM

Now let us look more critically at the process of forming a new function as the composition inverse of a suitable known function. Theorem 4 seemed plausible on purely geometric grounds. In fact, it made implicit use of a new geometric principle that should be singled out and labelled, for later analysis. This is the fact that if f is continuous on an interval I, then its graph cannot "jump over" a horizontal line $y = k$. If the graph crosses from below the line to above it, then in the process the graph must intersect the line.

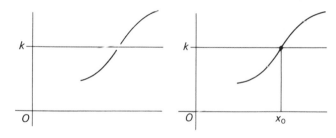

If we view the graph of a continuous function as the path of a moving particle (a view that was predominant during the early history of calculus), then the above property is part of our intuitive feeling about continuous motion.

Consider, for example, $f(x) = x^3$. As the point $p = (x, x^3)$ moves continuously along the graph from $(1,1)$ to $(2,8)$, then $y = x^3$ moves continuously along the y-axis from 1 to 8, and *of necessity passes through every intermediate value k in the process.*

This is why we feel sure that cube roots exit. For example, as $y = x^3$ moves from below 2 to above 2, it has to go through 2. That is, there has to be a number X such that $X^3 = 2$, and this X is the cube root of 2.

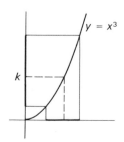

Another example occurred in Chapter 4, when we were looking for a certain number e lying between 2 and 4. Since 2^x runs from 2 to 4 as x runs from 1 to 2, it was obvious to us that we could express e in the form $e = 2^b$ for some number b between 1 and 2.

And in Chapter 3 we reasoned that if the continuously varying velocity $V = f'(t)$ assumes values both above and below its average value, then there must be a time T at which $f'(T)$ is exactly equal to the average value.

Here is the general property.

Intermediate-value property   *If f is continuous on the closed interval $[a, b]$ and if k is a number between $f(a)$ and $f(b)$, then there is a number X in $[a, b]$ such that $f(X) = k$.*

This is an *existence* principle. It says that a number having a certain property exists, but doesn't specify where it is, or how to calculate it. We shall take up the problem of calculation in Chapter 20, and there is a general discussion of geometric properties in Chapter 11.

Note also that the hypothesis is *continuity*, not differentiability. The property appears to hold for any *continuous* function that we can draw, as in the figure below.

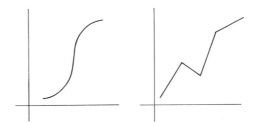

The intermediate-value property assures us that if f is a continuous increasing function on a closed interval $[a,b]$, then its range is exactly the closed interval $[A, B] = [f(a), f(b)]$, for its range contains every value k between $f(a)$ and $f(b)$.

Similarly, if f is a continuous increasing function on an *open* interval (a,b), then its range is exactly an open interval (A,B). However, this is a little trickier to show. For one thing, we have to obtain the endpoint B as the limit of $f(x)$ as x approaches b. Proving that this limit exists and calculating its value is generally more difficult than computing a function value $B = f(b)$. Similar remarks apply to $A = \lim_{x \to a} f(x)$. Fortunately, in the particular situations that we studied above, there was no problem in evaluating these limits.

Next, are we sure that the range of f fills out the open interval (A,B)? That is, if k is in (A,B), must there be an X in (a,b) for which $f(X) = k$? We can see that there is, as follows.

Since $k < B$, and since $f(x) \to B$ as $x \to b$, $f(x)$ must eventually become greater than k. We can therefore choose some particular b_0 at which $f(b_0) > k$. Similarly, we can choose a_0 such that $f(a_0) < k$. Then k is a value of f on the *closed* interval $[a_0, b_0]$, by the intermediate-value property. Thus we do know in this situation that the open interval (A,B) is the range of f.

Theorem 4 further asserted that the inverse g of a continuous monotone function f is continuous, and that if $f'(x)$ exists and is not zero, then $g'(y)$ exists and

$$g'(y) = \frac{1}{f'(x)},$$

where $y = f(x)$.

Let us assume the continuity of g for the moment and consider the derivative formula. If we give x a nonzero increment Δx, then the increment

$$\Delta y = f(x + \Delta x) - f(x)$$

is also different from zero, because f is monotone. We can therefore consider both quotients $\Delta y / \Delta x$ and $\Delta x / \Delta y$, and the derivative calculation looks simple:

$$\frac{dx}{dy} = \lim_{\Delta y \to 0} \frac{\Delta x}{\Delta y} = \frac{1}{\lim_{\Delta y \to 0} \frac{\Delta y}{\Delta x}} = \frac{1}{\lim_{\Delta x \to 0} \frac{\Delta y}{\Delta x}} = \frac{1}{\frac{dy}{dx}}.$$

There are, however, two gaps to be filled:

1. In order for $\Delta x / \Delta y$ to be the difference quotient for the function g, Δx must be the increment in the value of g given by the increment Δy in y:

$$\Delta x = g(y + \Delta y) - g(y).$$

This is all right, because we know that $y + \Delta y = f(x + \Delta x)$, so

$$x + \Delta x = g(y + \Delta y).$$

We then subtract $x = g(y)$ to get the above Δx formula.

2. We have to know that $\Delta x \to 0$ as $\Delta y \to 0$, so that the limit can be rewritten as we did above. But this is just the continuity of g, which we have been putting off. This is at once the easiest and the hardest step. There is almost nothing to the reasoning, but what there is involves the *concept* of continuity more directly than most of our continuity arguments. In the argument below we assume that f and g are increasing functions.

In order to show that g is continuous at $y_0 = f(x_0)$ we have to show that $x = g(y)$ can be kept as close as we please to $x_0 = g(y_0)$ by holding y sufficiently close to y_0. So suppose we wish to keep x within the interval (x_1, x_2) containing x_0. Then it is sufficient to hold y within the interval

$$(y_1, y_2) = (f(x_1), f(x_2)),$$

which is an interval containing y_0. (Remember that when f is restricted to the closed interval $[x_1, x_2]$, then its range is exactly the closed interval $[y_1, y_2]$ $= [f(x_1), f(x_2)]$, and that this is the domain of the restricted inverse function.) Thus, g is continuous at y_0.

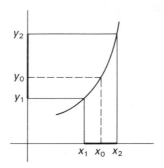

In a sense we have been "dotting i's" in this section. Mathematicians frequently don't bother to dot their i's, but they must be ready to, on demand.

chapter 8
the antiderivative

A second major discovery exploited by Newton and Leibniz was that many problems in geometry and physics can be solved by "backwards differentiation," or *anti-differentiation*.

This chapter corresponds roughly to Chapter 3. It is concerned with the same derivative rules—but now run backwards—and leads in particular to the "integration" of polynomials. Even these beginning results have interesting applications, and some of them are developed here.

1. THE ANTIDERIVATIVE

What do we know about the motion of a particle travelling along a coordinate line with constant velocity 5? In Chapter 3 we decided that if s is the position coordinate of the particle, then its velocity v is given by the derivative ds/dt. So here we are assuming that s is a function of t which has the constant derivative 5. The question is: What does this tell us about the function itself?

You undoubtedly can visualize *one* function of t having the constant derivative 5, namely $5t$. Moreover, adding a constant term doesn't change the derivative, so if s is any one of

$$5t + 3, \quad 5t - 10, \quad 5t + \sqrt{2}, \quad 5t + c,$$

then s is a function of t for which $ds/dt = 5$. Are there any others? The answer is *no*:

Theorem 1. *If h_1 and h_2 are differentiable functions having the same derivative on an interval I, then there is a constant C such that*

$$h_2(x) = h_1(x) + C$$

for all x in I.

Proof. If $h_1' = h_2'$ on I, then $(h_1 - h_2)' = 0$ on I, and any function whose derivative is identically zero must be a constant. (This is Theorem 7 of Chapter 3. In any case, it should seem plausible that if every tangent line to a graph is horizontal, then the graph can neither rise nor fall and so must be a horizontal straight line.) ∎

Using this principle we see that if $ds/dt = 5$, then s must differ from $5t$ by a constant, so

$$s = 5t + c$$

for some constant c. (This is all we can conclude without further information.)

The problem above involved finding a function whose derivative is known. A differentiable function F such that $F' = f$ is called an *antiderivative* of f, and the process of finding F from f is *antidifferentiation* We saw above that:

A given function f does not have a uniquely determined antiderivative, but if we can find one antiderivative F, then every other antiderivative is of the form $F +$ constant.

For example, $x^3/3$ is one antiderivative of x^2, and every other antiderivative is then necessarily of the form

$$\frac{x^3}{3} + c.$$

An antiderivative of f is also called an *integral* of f, and antidifferentiation is also called *integration*. The most general integral of x^2 shown above is called its *indefinite integral*, because of the arbitrary constant c. This constant is called the *constant of integration*. The Leibniz notation for the indefinite integral of a function f is

$$\int f(x)\,dx,$$

a notation that will prove to be especially useful in Chapter 9. For example,

$$\int x^2\,dx = \frac{x^3}{3} + c.$$

In this context, x^2 is called the *integrand*. The origin of the Leibniz notation will become apparent in Chapter 10.

The polynomial rules of Chapter 3, read backwards, now become the following integration rules:

$$\int x^r\,dx = \frac{x^{r+1}}{r+1} \qquad (\text{if } r \neq -1),$$

$$\int \left(af(x) + bg(x) \right) dx = a \int f(x)\,dx + b \int g(x)\,dx$$

We'll check them shortly, but consider first an example of how they are used.

EXAMPLE

$$\int (3x + 2x^3)\,dx = 3 \int x\,dx + 2 \int x^3\,dx \qquad \text{(By the second rule)}$$

$$= 3\left(\frac{x^2}{2} + c_1\right) + 2\left(\frac{x^4}{4} + c_2\right) \qquad \text{(By the first rule)}$$

$$= \frac{3x^2 + x^4}{2} + (3c_1 + 2c_2).$$

But we can write this as

$$\frac{3x^2 + x^4}{2} + c,$$

because $3c_1 + 2c_2$ is itself just an arbitrary constant.

In practice, when we have to compute the indefinite integral of a sum of functions, we just add up the particular integrals (antiderivatives) that we find, and then add an arbitrary constant at the end. For example, the above calculation would be done like this:

$$\int (3x + 2x^3)\,dx = 3 \cdot \frac{x^2}{2} + 2 \cdot \frac{x^4}{4} + c$$

$$= \frac{3x^2 + x^4}{2} + c.$$

The integration rules are verified by differentiation. Since

$$\frac{d}{dx}\left(\frac{x^{r+1}}{r+1}\right) = \frac{(r+1)x^r}{r+1} = x^r,$$

we see that $x^{r+1}/(r+1)$ is a particular antiderivative of x^r, and the first rule follows.

The second rule is just the integral form of the corresponding derivative identity. For if F and G are antiderivatives of f and g, respectively, then

$$a \int f(x)\,dx + b \int g(x)\,dx = a(F(x) + C_1) + b(G(x) + C_2)$$

$$= aF(x) + bG(x) + C,$$

and this is equal to

$$\int \Big(af(x) + bg(x)\Big)\,dx,$$

because

$$(aF + bG)' = aF' + bG' = af + bg.$$

PROBLEMS FOR SECTION 1

Compute the following integrals. Be sure to include the constant of integration in each answer.

1. $\int x^6\,dx$

2. $\int \sqrt{x}\,dx$

3. $\int x\,dx$

4. $\int dx$

5. $\int x^{1/3}\,dx$

6. $\int (x^2 + x + 1)\,dx$

7. $\int \left(3x^3 - \dfrac{1}{3x^3}\right)dx$

8. $\int (t^2 - 2t + 3)\,dt$

9. $\int (2x + 3)(x - 2)\,dx$

10. $\int x^{2/3}\,dx$

11. $\int \dfrac{dx}{\sqrt{x}}$

12. $\int \dfrac{dx}{\sqrt[3]{x}}$

13. $\int 3ay^2\,dy$

14. $\int (x^{3/2} - 2x^{2/3} + 5\sqrt{x} - 3)\,dx$

15. $\int \dfrac{4x^2 - 2\sqrt{x}}{x}\,dx$

16. $\int \sqrt{x}\,(3x - 2)\,dx$

17. $\int (9t^2 + 25t + 14)\,dt$

18. $\int \dfrac{dx}{x^2}$

19. $\int \dfrac{dx}{\sqrt[3]{x^2}}$

20. $\int (x^3 + 2)^2 3x^2\,dx$

21. Prove that $\displaystyle\int (ax + b)^r\,dx = \dfrac{(ax + b)^{r+1}}{a(r + 1)} + c \quad (r \neq -1)$.

22. $\int \sqrt{x + 1}\,dx$

23. $\int \dfrac{1}{\sqrt{x + 1}}\,dx$

24. $\int \sqrt{2x + 1}\,dx$

25. $\int (3x + 2)^{-2/3}\,dx$

26. Find an antiderivative of x^2 that has the value 1 at $x = 2$. [*Hint.* Every antiderivative of x^2 is of the form $y = (x^3/3) + C$. Now see what value C must have to make $y = 1$ when $x = 2$.]

27. Find an antiderivative $F(x)$ of $\sqrt{x}$ such that $F(4) = 4$.

28. Find a function F such that $F'(x) = x$ and $F(1) = 0$. Show that F is uniquely determined by these conditions.

2. AREAS

One of the unsolved problems of ancient mathematics was the *area* problem. The Greeks knew much about the areas of triangles and circles, and configurations derivable from them, but any other figure represented a new, and generally insoluble, problem. Archimedes was able to apply a method that eventually became known as *exhaustion* to compute the area of a parabolic segment, and to compute a few other particular geometric magnitudes. But for nearly two thousand years this cluster of computations by Archimedes stood as an isolated achievement. By the end of the seventeenth century, mathematicians understood better that the method of exhaustion represented a systematic approach, and eventually it evolved into the calculational procedure that today is associated with the definite integral. But meanwhile Newton and Leibniz showed that if a quantity could be calculated exactly by exhaustion, then it could be computed *much more easily* from antiderivatives. This remarkable fact, when stated in more precise terms, is called the Fundamental Theorem of Calculus. Our temporary and incomplete version will be that no matter how we consider the area of a figure to be *defined*, we can frequently *calculate* it by antiderivatives.

Suppose, for example, that we want to find the exact area of the region bounded by the half-parabola $y = \sqrt{x}$, the x-axis, and the vertical line $x = 1$, as sketched at the left below.

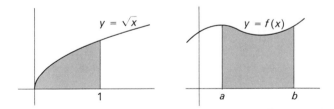

More generally, we will be interested in the area of the region bounded by a graph $y = f(x)$, the x-axis, and two vertical lines

$$x = a \qquad \text{and} \qquad x = b,$$

as shown at the right. Intuition tells us that any such region *has* an exact area, and our problem is to *find* it.

The key to the Newton–Leibniz approach seems paradoxical. In order to solve this problem, we replace it by an apparently harder problem. Instead of asking what the *fixed* area is, we ask how the area *varies* when we *change* the righthand boundary line. It is convenient to denote the position of this variable edge by x, and the whole figure will be clearer if we use a different independent variable, say t, for the graph, as below. The varying area A is then a function of the varying righthand edge coordinate x. Let us see how this helps in our parabolic problem.

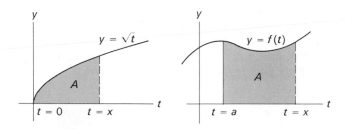

When we give x an increment Δx, the area changes by an increment ΔA, as shown below, and ΔA is squeezed between two rectangular areas,

$$\sqrt{x} \cdot \Delta x < \Delta A < \sqrt{x + \Delta x} \cdot \Delta x.$$

But then

$$\sqrt{x} < \frac{\Delta A}{\Delta x} < \sqrt{x + \Delta x}.$$

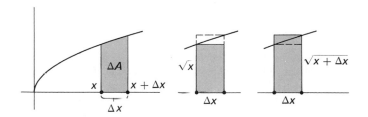

(This assumes Δx positive. If $\Delta x < 0$, then the inequality is reversed.) Since $\sqrt{x + \Delta x} \to \sqrt{x}$ as $\Delta x \to 0$, the difference quotient $\Delta A / \Delta x$ is squeezed between $\sqrt{x}$ and something approaching $\sqrt{x}$ and so must have the same limit $\sqrt{x}$. This shows that *the variable area A is a differentiable function of x and that*

$$\frac{dA}{dx} = \lim_{\Delta x \to 0} \frac{\Delta A}{\Delta x} = \sqrt{x}.$$

Therefore,

$$A = \int x^{1/2} dx = \frac{2}{3} x^{3/2} + c,$$

where the constant c has a particular value that still has to be determined. To do this we use the fact that $A = 0$ when $x = 0$, which gives

$$0 = \frac{2}{3} \cdot 0 + c$$

and $c = 0$. Thus,

$$A = \frac{2}{3} x^{3/2}.$$

Now we can solve the original problem. Setting $x = 1$, we have $A = 2/3$ for the exact value of the area of the region we began with.

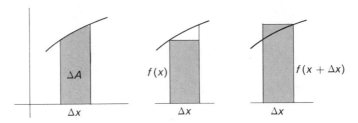

Exactly the same reasoning applies to *any* increasing, positive, continuous function f. When we give x an increment Δx, the area changes by an increment ΔA, as shown above, and ΔA is squeezed between two rectangular areas:

$$f(x) \cdot \Delta x < \Delta A < f(x + \Delta x)\Delta x.$$

But then

$$f(x) < \frac{\Delta A}{\Delta x} < f(x + \Delta x)$$

(if Δx is positive). Since $f(x + \Delta x) \to f(x)$ as $\Delta x \to 0$, the difference quotient $\Delta A/\Delta x$ is squeezed between $f(x)$ and something approaching $f(x)$, and must have the same limit $f(x)$. This shows that *the variable area A is a differentiable function of x and that*

$$\frac{dA}{dx} = \lim_{\Delta x \to 0} \frac{\Delta A}{\Delta x} = f(x).$$

If f is a *decreasing positive* function, the squeeze inequality is reversed:

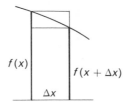

but similar reasoning applies, with the same conclusion.

We will not always be dealing with a monotone function. However, if there is a sequence of intervals on which a positive continuous function f is alternately increasing and decreasing, then the conclusion $dA/dx = f(x)$ follows by one or the other of the above arguments, depending on which interval contains x. But it is better to describe this situation in a more general way.

Assuming $\Delta x > 0$, let m and M be the minimum and maximum values of $f(x)$ on the interval $[x, x + \Delta x]$. Then in every case

$$m\Delta x < \Delta A < M\Delta x,$$

and

$$m < \frac{\Delta A}{\Delta x} < M.$$

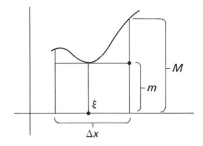

Now m is a function value $f(\xi)$ for some ξ between x and $x + \Delta x$. Since $\xi \to x$ as $\Delta x \to 0$ and since f is continuous, it follows that $m \to f(x)$ as $\Delta x \to 0$. Similarly for M. Thus $\Delta A/\Delta x$ is squeezed to the same limit, and

$$\frac{dA}{dx} = \lim \frac{\Delta A}{\Delta x} = f(x)$$

as before. If Δx is negative, the first inequality is reversed, but it reverses again on dividing by Δx, so

$$m < \frac{\Delta A}{\Delta x} < M$$

in this case, too.

It is customary to state this result for an arbitrary continuous function. The trouble is that we don't know enough about what an arbitrary continuous function can be like. If "continuous" brings to your mind a graph that smoothly rises and falls, increasing over one interval, then decreasing over a succeeding interval, etc., then our discussion above is adequate. But there are continuous functions that don't behave so nicely. Try to imagine a continuous graph that is everywhere "infinitely krinkly." This may be practically impossible to conceive of, but it can happen. Then the simple pictures that have been supporting our arguments don't apply, and we lose confidence in our intuition. So it is best—and really necessary for logical completeness—to be very explicit about what assumptions are needed to make the above argument work.

First, we assumed that the values of a continuous function on a closed interval $[a,b]$ run from a minimum value m to a maximum value M. This is called the *extreme value property* of a continuous function; it will be discussed in Chapter 11.

Of course, we also assumed that each region that we describe has a uniquely determined numerical area, and that this area varies with the region in the natural way. For example, we implicitly used the fact that the area of a union of two nonoverlapping regions is exactly the sum of their two areas.

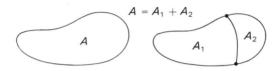

At any rate, subject to these cautionary qualifications, we have established the following theorem.

Theorem 2. *Let f be a positive continuous function and let A be the area of the region bounded by the graph of $y = f(t)$, the t-axis, and the vertical lines $t = a$ and $t = x$, where $x > a$. Then A is a differentiable function of x, and*

$$\frac{dA}{dx} = f(x).$$

EXAMPLE 1. Find the area of the region below the graph of $f(x) = 2 - x - x^2$ and above the x-axis.

. . . .

Solution. The graph is shown below. The x-intercepts are the solutions of $2 - x - x^2 = 0$, and are $x = -2$ and $x = 1$, as shown. These are the limits between which we want the area. According to the theorem, the area A from -2 to x is a function of x such that

$$\frac{dA}{dx} = 2 - x - x^2.$$

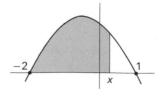

Therefore,

$$A = \int (2 - x - x^2)\, dx = 2x - \frac{x^2}{2} - \frac{x^3}{3} + C,$$

where C has to be determined. Since $A = 0$ when $x = -2$,

$$0 = 2(-2) - \frac{(-2)^2}{2} - \frac{(-2)^3}{3} + C = -\frac{10}{3} + C,$$

and $C = 10/3$. So

$$A = 2x - \frac{x^2}{2} - \frac{x^3}{3} + \frac{10}{3}.$$

The particular area required was from $x = -2$ to $x = 1$, and is

$$A_1 = 2 - \frac{1}{2} - \frac{1}{3} + \frac{10}{3} = 4\frac{1}{2}.$$

In addition to showing us how to compute an exact area when we can find an antiderivative, Theorem 2 also has an important theoretical consequence.

Corollary *Every positive continuous function (on an interval I) has an antiderivative.*

For A as a function of x is one. This is an *existence* statement. It says that even though we may be unsuccessful in looking for an explicit formula for the antiderivative, nevertheless an antiderivative does exist.

The corollary extends easily to an arbitrary (not necessarily positive) continuous function on a closed interval. (See Problem 22.) This is so important that we give it theorem status.

Theorem 3. *Every continuous function on a closed interval $[a, b]$ has an antiderivative on $[a, b]$.*

Remember, however, that we have this result only by virtue of assuming certain other properties of a continuous function.

Remember, also, that we have to be able to *find* an antiderivative of f in order to use Theorem 2 to evaluate the area under the graph of f. We know how to do this for simple functions by means of our polynomial rules, but it can be very difficult for more complicated functions f, and we shall spend the whole of the next chapter developing the procedures that are available.

PROBLEMS FOR SECTION 2

A sketch is an almost necessary part of the solution of a problem involving a geometric magnitude, and you should get in the habit of drawing one as a matter of course, always trying to be reasonably accurate. In these area problems, your sketch should show both the area to be computed and the variable area which lets us use calculus in the computation.

1. Find, by integration, the area of the triangle bounded by the line $y = 2x$, the x-axis, and the line $x = 4$. Verify your answer by using the formula $A = \frac{1}{2}bh$.

2. Find, by integration, the area of the trapezoid bounded by the line $x + y = 15$, the x-axis, and the lines $x = 3$ and $x = 10$. Verify your answer by use of the formula $A = \frac{1}{2}(a + b)h$.

Find the area bounded by the given curve, the x-axis, and the given vertical lines.

3. $y = x^3$; $x = 0, x = 4$.   4. $y = x^2 + x + 1$; $x = 2, x = 3$.

5. $y = x^2 + 4x$; $x = -4, x = -2$.   6. $y^2 + 4x = 0$; $x = -1, x = 0$.

7. $y = 2x + \dfrac{1}{x^2}$; $x = 1, x = 4$.   8. $y = x^2$; $x = 2, x = 5$.

9. $y = \dfrac{1}{\sqrt{x + 4}}$; $x = 0, x = 5$.   10. $x = 3y^2 - 9$; $x = 0$.

11. Find the area of the region bounded by the coordinate axes and the line $x + y = 2$.

12. Verify the formula for the area of a triangle in the case of the right triangle with vertices at the origin, the point (b,h) and the point $(b,0)$.

13. Verify the formula for the area of a trapezoid in the case of the vertices $(0, b_1)$, (h, b_2), and the points on the x-axis under these two points.

14. The variable area A between the graph of $x = f(y)$ and the y-axis, from a to y, is a differentiable function of y satisfying

$$\frac{dA}{dy} = f(y).$$

This can be proved simply by repeating the proof of Theorem 2 in this slightly different context. Instead, show by a geometric argument that the variant above is a corollary of Theorem 2.

Assuming the formula in the preceding problem, find the area of the region bounded by each of the following graphs and the given lines.

15. $x = 9y - y^3$; $y = 0, y = 3$.   16. $y^2 = 4x$; $y = 0, y = 4$.

17. $x = -(y^2 + 4y)$; $x = 0$.   18. $y = 4 - x^2$; $y = 0, y = 3$.

19. Find the area under one arch of the curve $y = \cos x$.

20. Find the area of the region under the graph of $y = \sec^2 x$, from $x = 0$ to $x = \pi/4$.

21. Find the area of the region under the graph of $y = e^x$ from $x = -N$ to $x = 1$. What is its limit as $N \to \infty$? What would you say is the area of the region under the graph of e^x extending from $x = 1$ all the way to $-\infty$?

22. Prove Theorem 3 from the preceding corollary. [*Hint.* Let f be continuous on $[a,b]$ and let m be its minimum value. It may happen that m is negative. However, what can we say about the function $g(x) = f(x) - m$?]

3. THE VOLUME OF A SOLID OF REVOLUTION

If the upper half-plane is rotated about the x-axis, then each point on the graph $y = \sqrt{x}$ has a circular path, and the whole graph sweeps out a certain surface, called a *surface of revolution*. The plane region between the graph, the x-axis, and the vertical line $x = 1$ sweeps out a *solid of revolution*. The boundary of the solid lies partly on the surface of revolution and partly on the vertical plane through $x = 1$. We propose to calculate the volume of this solid. More generally, we shall be interested in the volume swept out by rotating about the x-axis the region bounded by a graph $y = f(x)$, the x-axis, and the vertical lines $x = a$ and $x = b$, as shown at the right in the accompanying figure.

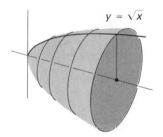
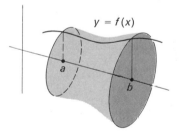

Proceeding just as before, we solve this problem by first solving the apparently harder problem of how the volume varies when we vary its righthand edge. We relabel the horizontal axis with t, in order to leave x free to mark the variable right edge, and we give x an increment Δx. The volume then changes by an increment ΔV, as shown below, and ΔV is squeezed between two cylindrical volumes. A circular cylinder with base radius r and altitude h has volume $\pi r^2 h$, so this squeeze on ΔV is the inequality

$$\pi(\sqrt{x})^2 \Delta x < \Delta V < \pi(\sqrt{x + \Delta x})^2 \Delta x,$$

or

$$\pi x \Delta x < \Delta V < \pi(x + \Delta x)\Delta x.$$

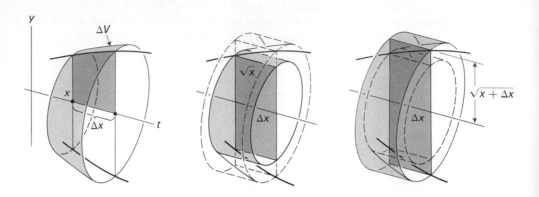

If Δx is positive then

$$\pi x < \frac{\Delta V}{\Delta x} < \pi(x + \Delta x).$$

If Δx is negative, the inequality is reversed. In either case $\Delta V/\Delta x$ is squeezed between πx and something approaching πx, and so it must approach πx, too. That is,

$$\frac{dV}{dx} = \lim_{\Delta x \to 0} \frac{\Delta V}{\Delta x} = \pi x,$$

by the squeeze limit law. Thus

$$V = \int \pi x\, dx = \frac{\pi x^2}{2} + c,$$

where the constant c is to be determined by the fact that $V = 0$ when $x = 0$. This gives $c = 0$ and

$$V = \frac{\pi x^2}{2}.$$

Setting $x = 1$, we get

$$V = \frac{\pi}{2}$$

as the answer to the original problem.

If $\sqrt{x}$ is replaced by $f(x)$, then the squeeze inequality boxing ΔV between two cylindrical volumes has to be set up in more general terms, but it gives the same conclusion:

$$\frac{dV}{dx} = \lim_{\Delta x \to 0} \frac{\Delta V}{\Delta x} = \pi(f(x))^2.$$

Theorem 4. *If f is a positive continuous function over the interval* $[a, b]$ *and if V is the volume of the solid of revolution generated by the graph of* $y = f(t)$ *between* $t = a$ *and* $t = x$, *then V is a differentiable function of x and*

$$\frac{dV}{dx} = \pi(f(x))^2.$$

EXAMPLE. Find the volume of the solid of revolution generated by rotating the region under the graph of $y = x^2$, from $x = 1$ to $x = 3$.

. . . .

Solution. We first find the volume V from 1 to x. The theorem says that

$$\frac{dV}{dx} = \pi(x^2)^2 = \pi x^4,$$

so

$$V = \int \pi x^4 \, dx = \frac{\pi x^5}{5} + c.$$

Since $V = 0$ when $x = 1$, we have

$$0 = \pi\left(\frac{1}{5}\right) + c$$

and

$$c = -\frac{\pi}{5}.$$

Thus,

$$V = \frac{\pi(x^5 - 1)}{5},$$

and substituting $x = 3$ gives

$$V = \pi\frac{(243 - 1)}{5} = (48\tfrac{2}{5})\pi$$

as the answer to our original problem.

PROBLEMS FOR SECTION 3

1. Prove, by integration, the formula

$$V = \frac{4}{3}\pi r^3$$

for the volume of a sphere. (The sphere is generated by revolving the circle $x^2 + y^2 = r^2$ about a diameter.)

2. Find, by integration, the volume of the truncated cone generated by revolving the area bounded by $y = 6 - x$, $y = 0$, $x = 0$, and $x = 4$ about the x-axis.

Find the volume generated by revolving about the x-axis the regions bounded by the following graphs.

3. $y = x^3$; $y = 0$, $x = 1$.

4. $9x^2 + 16y^2 = 144$.

5. $y = x^2 - 6x$; $y = 0$.

6. $y^2 = (2 - x)^3$; $y = 0$, $x = 0$, $x = 1$.

7. $(x - 1)y = 2$; $y = 0$, $x = 2$, $x = 5$.

8. $y = x^{1/3}$; $y = 0$, $x = 0$, $x = 8$.

9. $y = \sqrt{1 - x^4}$; $y = 0$.

10. $y = \sin x$; $y = 0$, $x = 0$, $x = \pi$.   [*Hint.* $\sin^2 x = (1 - \cos 2x)/2$.]

11. $y = \sec x$; $y = 0$, $x = 0$, $x = \pi/4$.

12. $y = e^x$; $y = 0$, $x = 0$, $x = 1$.

13. Show that the volume generated by rotating the ellipse

$$\frac{x^2}{a^2} + \frac{y^2}{b^2} = 1$$

about the x-axis is

$$V = \frac{4}{3}\pi b^2 a.$$

(Note that this generalizes the formula for the volume of a sphere.)

14. Find the limit as $n \to \infty$ of the volume obtained by rotating about the x-axis the region between the x-axis and the graph of $y = 1 + x^n$, from $x = 0$ to $x = 1$. What is the geometric interpretation of this limit?

15. Prove the formula

$$V = \frac{1}{3}\pi r^2 h$$

for the volume of a cone. (Rotate the region bounded by the straight line $y = rx/h$, the x-axis, and the vertical line $x = h$, about the x-axis.)

16. Prove Theorem 4 by imitating the earlier proof of Theorem 2. That is, let m and M be the minimum and maximum of $f(x)$ on the increment interval $[x, x + \Delta x]$, and build the proof around the inequality

$$\pi m^2 \Delta x < \Delta V < \pi M^2 \Delta x.$$

17. Find the formula for the volume of a spherical cap.

4. THE INITIAL-VALUE PROBLEM

The graphs of the antiderivatives of x^2 are the graphs of the functions

$$y = \int x^2\, dx = \frac{x^3}{3} + c$$

for the various values of the constant c, as shown below. They form a family of curves filling up the plane, exactly one curve going through any given point (x_0, y_0).

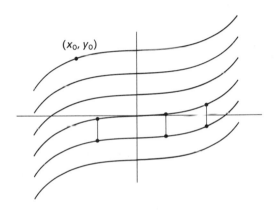

Analytically, this means that if we ask for an antiderivative $F(x)$ of x^2 having a given value at a given point, say $F(-1) = 2$, then there is a unique solution. We find it as follows. The function F must be of the form

$$F(x) = \int x^2\, dx = \frac{x^3}{3} + c,$$

for some value of the constant c, and c is determined by the extra requirement that $F(-1) = 2$. Thus

$$2 = F(-1) = -\frac{1}{3} + c,$$

so that $c = 2 + 1/3 = 7/3$, giving the unique solution

$$F(x) = \frac{x^3 + 7}{3}.$$

The requirement that $F(-1) = 2$ is called an *initial condition*, and the total problem is called an *initial-value problem*. The above initial-value problem, then, is to determine the unique function F such that

$$F'(x) = x^2, \qquad F(-1) = 2.$$

In pure Leibniz notation it is the problem of determining what function y must be of x if

$$\frac{dy}{dx} = x^2,$$

$$y = 2 \qquad \text{when } x = -1.$$

In this notation the solution would be written out as follows. First,

$$y = \int x^2\, dx = \frac{x^3}{3} + c,$$

where the constant c is to be determined by the initial condition. Substituting the initial values $(x,y) = (-1,2)$ gives

$$2 = \frac{(-1)^3}{3} + c,$$

so $c = 7/3$, as before.

In the same way, Theorem 3 leads to the general conclusion:

Theorem 5. *If f is a function continuous over an interval I and if (x_0, y_0) is any fixed point in the plane lying over I (i.e., x_0 is in I), then the initial-value problem*

$$F' = f,$$

$$F(x_0) = y_0,$$

has a unique solution F over I.

The variable areas and volumes of the last two sections were solutions of initial-value problems:

1. If A is the area of the region bounded by the graph of f, the t-axis, and the vertical lines $t = a$ and $t = x$, then A is the unique solution of the initial-value problem

$$\frac{dA}{dx} = f(x),$$

$$A = 0 \qquad \text{when } x = a.$$

It is assumed here that f is positive and that $x > a$.

2. The corresponding solid-of-revolution volume is the unique solution of the initial-value problem

$$\frac{dV}{dx} = \pi[f(x)]^2,$$

$$V = 0 \qquad \text{when } x = a.$$

We now turn to initial-value problems in which we determine the motion of an object from its known velocity or acceleration.

EXAMPLE 1. A particle travels along the number line with constant velocity 5. Find its position s as a function of t if $s = -10$ when $t = 1$.

This is an initial-value problem. From $ds/dt = 5$ we obtain $s = 5t + c$, as before, although we might now write this

$$s = \int 5 \, dt = 5t + c.$$

Substituting the initial values $(t, s) = (1, -10)$ gives

$$-10 = 5 \cdot 1 + c,$$

so that $c = -15$ and

$$s = 5t - 15.$$

It is said that in the seventeenth century Galileo experimented by dropping objects from the tower in Pisa, and concluded that, neglecting air resistance, which would slow down light objects more, all falling objects drop with a constant acceleration of 32 ft per sec^2 .

If the position of the falling body is measured along a vertical coordinate system with the positive direction upward, then its constant acceleration is $-$ 32, because its velocity is becoming increasingly negative. The acceleration due to gravity is usually designated g, so we have $g = -32$ in the present axis system.

EXAMPLE 2. A stone is thrown vertically upward with an initial velocity of 50 ft/sec. What is its velocity two seconds later? How high will it rise?

. . . .

Solution. Since $dv/dt = g = -32$, we have

$$v = \int g \, dt = \int (-32) \, dt = -32t + c.$$

If we start measuring time at the instant the stone is thrown, then the initial condition is that $v = 50$ when $t = 0$. Thus $50 = -32 \cdot 0 + c$ and $c = 50$, giving the velocity equation

$$v = 50 - 32t.$$

At $t = 2$ we have $v = 50 - 64 = -14$ ft/sec. The stone is already falling.

The highest point in its trajectory will be reached when the stone just stops rising and starts to fall, i.e., when $v = 0$. This occurs when

$$0 = 50 - 32t$$

or $t = 25/16$ seconds. But in order to find how far the stone has risen we have to integrate the equation

$$v = \frac{ds}{dt} = 50 - 32t.$$

We get

$$s = 50t - 16t^2 + c.$$

If we measure s from the point at which the stone is thrown, then the initial condition is $s = 0$ at $t = 0$. This gives $c = 0$, and at $t = 25/16$ we have

$$s = \frac{50 \cdot 25}{16} - 16\left(\frac{25}{16}\right)^2 = \frac{625}{16} \text{ ft,}$$

or approximately 39 ft.

The highest point is the maximum value of s, and can be determined by finding where $dx/dt = 0$ (see Chapter 6). But this is just the condition $v = 0$ that we arrived at intuitively above.

In general, an initial-value problem is the combination of a *differential equation* and one or more *side conditions* that single out a unique solution from among the many solutions to the differential equation. A differential equation is just an equation involving the derivatives of an unknown function. In the first example above, the differential equation is

$$\frac{dy}{dx} = x^2$$

for the unknown function $y = f(x)$. The differential equation for the exponential function is

$$\frac{dy}{dx} = y$$

for the unknown function $y = f(x)$. It is shown in Chapter 4 that the exponential function is the unique solution of the initial-value problem

$$\frac{dy}{dx} = y,$$

$$y = 1 \qquad \text{when } x = 0.$$

PROBLEMS FOR SECTION 4

Solve the following initial-value problems.

1. $f'(x) = x - 3, \quad f(2) = 9.$

2. $g'(x) = x + 3 - 5x^2, \quad g(6) = -20.$

3. $f'(y) = y^3 - b^2 y, \quad f(2) = 0.$

4. $f'(x) = bx^3 + ax + 4, \quad f(b) = 10.$

5. $f'(t) = \sqrt{t} + \dfrac{1}{\sqrt{t}}, \quad f(4) = 0.$

6. Given $dy/dx = (2x + 1)$, $y = 7$ when $x = 1$. Find the value of y when $x = 3$.

7. Given $dA/dx = \sqrt{2px}$, $A = p^2/3$ when $x = p/2$. Find the value of A when $x = 2p$.

8. Given the expressions below for acceleration, find the relation between s (displacement) and t if $s = 0$, $v = 20$, when $t = 0$.
   a) $a = 32$ b) $a = 4 - t$

9. With what velocity will a stone strike the ground if dropped from the top of a building 100 ft high? ($g = 32$ if distance is measured positively in the downward direction.)

10. If the stone in Problem 9 is thrown from the top of the building with an initial velocity of 100 ft/sec downward, what will be its velocity upon hitting the ground? What if it has an *upward* initial velocity of 100 ft/sec?

11. A train leaving a railroad station has an acceleration of

$$\frac{1}{2} + \frac{2}{100} t \text{ ft/sec}^2 .$$

How far will the train move in the first 20 sec of motion?

12. What constant acceleration is required to
   a) Move a particle 50 ft in 5 sec?
   b) Slow a particle from a velocity of 45 ft/sec to a stop in 15 ft?

13. Prove Theorem 5.

5. INTEGRATION BETWEEN LIMITS

It is convenient to use the definite-integral notation $\int_a^b g(x)\,dx$ even though we are not yet ready for the basic definition of the integral. In order to do this we shall adopt a provisional definition that will eventually be replaced. (Of course, our temporary definition must give the right answers or it would be useless.)

Let g be a continuous function on an interval I. We know from Theorem 3 that g has an antiderivative on I. If G and H are any two antiderivatives of g and if a and b are any two points of I, then

$$G(b) - G(a) = H(b) - H(a).$$

The reason is that there must be a constant c such that $G(x) = H(x) + c$ for all x in I, and then c cancels when the above differences are formed:

$$G(b) - G(a) = (H(b) + c) - (H(a) + c) = H(b) - H(a).$$

Thus the difference $G(b) - G(a)$ has the same value for all antiderivatives of g.

Definition. *We designate by $\int_a^b g$, or $\int_a^b g(x)\,dx$, the number that is the common value of all the differences $G(b) - G(a)$, where G is any antiderivative of g.*

Because $\int_a^b g$ is a definite number we call it the *definite integral* of g from a to b, and it is natural to call a and b the *limits of integration*.

EXAMPLE 1. Using the antiderivative $G(x) = x^3/3$ of $g(x) = x^2$, the defining equation

$$\int_a^b g(x)\,dx = G(b) - G(a)$$

becomes

$$\int_a^b x^2\,dx = \frac{b^3}{3} - \frac{a^3}{3}.$$

There is an intermediate step that is helpful in more complicated problems. Instead of the above, we would write

$$\int_a^b x^2\,dx = \frac{x^3}{3}\bigg]_a^b = \frac{b^3}{3} - \frac{a^3}{3}.$$

That is, we first write down an antiderivative, together with a bracket labelled as above to show how it is to be evaluated. The convention is that $f(x)]_a^b = f(b) - f(a)$. Then we carry out the evaluation as a second step.

EXAMPLE 2.

$$\int_1^2 (x^2 - x)\, dx = \frac{x^3}{3} - \frac{x^2}{2} \Big]_1^2$$

$$= \left(\frac{8}{3} - \frac{4}{2} \right) - \left(\frac{1}{3} - \frac{1}{2} \right)$$

$$= \frac{7}{3} - \frac{3}{2} = \frac{5}{6}.$$

These examples of $\int_a^b g(x)\, dx$ show how the variable x disappears in the evaluation. It is a *dummy* variable, and any other variable can be used just as well. For example

$$\int_a^b t^2\, dt = \frac{t^3}{3} \Big]_a^b = \frac{b^3}{3} - \frac{a^3}{3} = \int_a^b x^2\, dx.$$

When we are using the function symbol f, we can dispense with the dummy variable entirely, and write $\int_a^b f$. Thus

$$\int_a^b f = \int_a^b f(x)\, dx = \int_a^b f(t)\, dt.$$

The areas and volumes that we computed earlier have natural expressions as definite integrals.

Theorem 6. *If f is a positive continuous function over the closed interval $[a, b]$, then the area of the region between the graph of f and the x-axis, from $x = a$ to $x = b$, is $\int_a^b f$. The volume of the solid of revolution generated by rotating this region about the x-axis is $\int_a^b \pi f^2$.*

Proof. Suppose, for the sake of definiteness, that $0 < a < b$, and let $A = q(x)$ be the area of the region over the interval $[0, x]$. We proved earlier that $dA/dx = f(x)$, i.e., that

$$q'(x) = f(x).$$

Thus $q(x)$ is an antiderivative of $f(x)$, and

$$\int_a^b f(x)\,dx = q(b) - q(a).$$

But $q(b) - q(a)$ is the area over $[0,b]$ minus the area over $[0,a]$, and hence is the area over $[a,b]$, as claimed in the theorem. The volume proof goes the same way. ∎

This gives a streamlined way to calculate areas and volumes of revolution. For if we can find *any* antiderivative F of f, then

$$A = \int_a^b f = F(b) - F(a).$$

EXAMPLE 3. The area under (the graph of) $y = x^2$ from 1 to 4 is

$$\int_1^4 x^2\,dx = \frac{x^3}{3}\Bigg]_1^4 = \frac{64}{3} - \frac{1}{3} = 21.$$

EXAMPLE 4. The area under $y = \sqrt{x}$ from 0 to 1 is

$$\int_0^1 x^{1/2}\,dx = \frac{2x^{3/2}}{3}\Bigg]_0^1 = \frac{2}{3}.$$

EXAMPLE 5. The volume of the solid of revolution generated by the region in Example 4 is

$$\int_a^b \pi f^2 = \int_0^1 \pi(x^{1/2})^2\,dx = \int_0^1 \pi x\,dx = \pi\frac{x^2}{2}\Bigg]_0^1 = \frac{\pi}{2}.$$

Of course, any derivative formula becomes an integral formula when reversed. Thus

$$\int \frac{dx}{x} = \log x + c,$$

$$\int e^x\,dx = e^x + c,$$

$$\int \cos x\,dx = \sin x + c,$$

etc.

EXAMPLE 6. What is the area under the graph of $f(x) = 1/x$ from $x = 1$ to $x = 2$?

Solution. This is

$$\int_1^2 \frac{dx}{x} = \log x \Big]_1^2 = \log 2 - \log 1 = \log 2.$$

EXAMPLE 7. The area under $y = \cos t$ from $t = 0$ to $t = x$ is

$$\int_0^x \cos t \, dt = \sin t \Big]_0^x = \sin x.$$

EXAMPLE 8. The area under one complete arch of $\cos x$ is

$$\int_{-\pi/2}^{\pi/2} \cos x \, dx = \sin x \Big]_{-\pi/2}^{\pi/2} = 1 - (-1) = 2.$$

EXAMPLE 9. Find the total area under the graph of $y = 1/x^2$ from $x = 1$ to $x = \infty$.

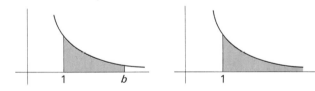

Solution. As worded, the problem doesn't make sense, since the only area formula we know is for a finite region. However, the area from $x = 1$ to $x = b$ is

$$\int_1^b \frac{dx}{x^2} = -\frac{1}{x}\Big]_1^b = 1 - \frac{1}{b}.$$

Then

$$\lim_{b \to \infty} [\text{area from 1 to } b] = \lim_{b \to \infty} \left(1 - \frac{1}{b}\right) = 1.$$

This limit is what we *mean* by the area from 1 to ∞. So here we have an example of an infinitely long region with a finite area.

EXAMPLE 10. Show that the region under the graph of $y = 1/\sqrt{x}$ from $x = 1$ to ∞ has infinite area.

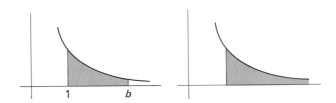

Solution. By definition the area A from 1 to ∞ is the limit as $b \to \infty$ of the area from 1 to b. So

$$A = \lim_{b \to \infty} \int_1^b \frac{dx}{\sqrt{x}} = \lim_{b \to \infty} 2\sqrt{x} \Big]_1^b$$

$$= \lim_{b \to \infty} (2\sqrt{b} - 2) = \infty.$$

An integral of the form $\int_a^\infty f(x)\,dx$ is an example of an *improper* integral. It is defined as the limit

$$\int_a^\infty f(x)\,dx = \lim_{b \to \infty} \int_a^b f(x)\,dx.$$

If the limit exists (as a finite number l) we say that the improper integral *converges*, and that its value is l. If the limit does not exist, we say that the improper integral *diverges*.

EXAMPLE 11. Find the area under the graph of $y = 1/\sqrt{x}$ from $x = 0$ to $x = 1$.

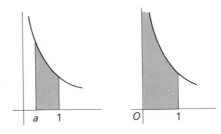

. . . .

Solution. This region is infinitely long along the *y*-axis, rather than along the *x*-axis. In other words, the integral that should give its area, $\int_0^1 dx/\sqrt{x}$, is improper because the integrand tends to infinity as *x* approaches 0, rather than because it is integrated over an infinitely long interval. However, we can follow essentially the same scheme. We first compute the area over the interval $[a,1]$, where *a* is a small positive number. Over this interval the integrand $1/\sqrt{x}$ is continuous and there is no problem. Then we can let *a* tend to zero and see what happens "in the limit." We have:

$$A = \int_0^1 \frac{dx}{\sqrt{x}} = \lim_{a\to0} \int_a^1 \frac{dx}{\sqrt{x}} = \lim_{a\to0} 2\sqrt{x}\,\Big]_a^1$$

$$= \lim_{a\to0} (2 - 2\sqrt{a}) = 2.$$

So the improper integral is convergent, and its value is 2. The area we were looking for is thus 2.

EXAMPLE 12. The improper integral $\int_0^1 dx/x^2$ is divergent:

$$\int_0^1 \frac{dx}{x^2} = \lim_{a\to0} \int_a^1 \frac{dx}{x^2} = \lim_{a\to0} -\frac{1}{x}\,\Big]_a^1$$

$$= \lim_{a\to0} \left(\frac{1}{a} - 1\right) = \infty.$$

PROBLEMS FOR SECTION 5

Evaluate the following definite integrals.

1. $\displaystyle\int_0^1 (x - x^2)\,dx$ 2. $\displaystyle\int_1^2 (x - x^2)\,dx$

3. $\displaystyle\int_{-1}^1 (x^4 - x^2)\,dx$ 4. $\displaystyle\int_0^a (ax - x^2)\,dx$

5. $\displaystyle\int_x^y 3t^2\,dt$ 6. $\displaystyle\int_0^\pi \cos t\,dt$

7. $\displaystyle\int_0^1 (x^3 - x^{1/3})\,dx$ 8. $\displaystyle\frac{1}{b-a}\int_a^b x\,dx$

Find the area bounded by the given curve, the *x*-axis, and the given vertical lines.

9. $y = x^3$; $x = 0, x = 4$. 10. $y = x^2 + x + 1$; $x = 2, x = 3$.

11. $y = x^2 + 4x$; $x = -4, x = -2$. 12. $y^2 + 4x = 0$; $x = -1, x = 0$.

13. $y = 2x + \dfrac{1}{x^2}$; $x = 1, x = 4$.   14. $x = 3y^2 - 9$; $x = 0$.

15. $y = \dfrac{1}{\sqrt{x + 4}}$; $x = 0, x = 5$.

Find the volume generated by revolving about the x-axis the regions bounded by the following graphs.

16. $y = x^3$; $y = 0, x = 1$.   17. $9x^2 + 16y^2 = 144$.

18. $y = x^2 - 6x$; $y = 0$.

19. $y^2 = (2 - x)^3$; $y = 0, x = 0, x = 1$.

20. $(x - 1)y = 2$; $y = 0, x = 2, x = 5$.

21. $y = x^{1/3}$; $y = 0, x = 0, x = 8$.

22. $y = \sqrt{1 - x^4}$; $y = 0$.   23. $y = \sec x, y = 0, x = 0, x = \pi/4$.

24. $y = e^x, y = 0, x = 0, x = 1$.

25. Prove the second half of Theorem 6.

26. Show that

$$\frac{d}{dx} \int_a^x f(t)\, dt = f(x).$$

27. Show that

$$\frac{d}{dx} \int_x^b f(t)\, dt = -f(x).$$

Compute the derivatives of the following functions.

28. $f(x) = \displaystyle\int_0^{x^2} t^2\, dt.$ 29. $f(x) = \displaystyle\int_{-x}^{x} \cos t\, dt.$

30. $f(x) = \displaystyle\int_1^{\sqrt{x}} \dfrac{dt}{t}.$ 31. $f(x) = \displaystyle\int_0^{\log x} e^t\, dt.$

Calculate the following integrals.

32. $\displaystyle\int_0^1 \left[\int_0^{\sqrt{x}} t^3\, dt \right] dx.$ 33. $\displaystyle\int_{-1}^{2} \left[\int_{x^2}^{2+x} t\, dt \right] dx.$

Determine whether each of the following improper integrals is convergent or divergent, and calculate its value if convergent.

34. $\displaystyle\int_1^{\infty} \dfrac{dx}{x^3}$ 35. $\displaystyle\int_1^{\infty} \dfrac{dx}{x^{3/2}}$

36. $\displaystyle\int_1^\infty \frac{dx}{x^{2/3}}$

37. $\displaystyle\int_0^\infty \frac{dx}{1 + x^2}$

38. $\displaystyle\int_0^\infty \frac{dx}{1 + x}$

39. $\displaystyle\int_0^1 \frac{dx}{x^{2/3}}$

40. $\displaystyle\int_0^1 \frac{dx}{x^3}$

41. $\displaystyle\int_0^1 \frac{dx}{\sqrt{1 - x^2}}$

42. $\displaystyle\int_0^{\pi/2} \sec x \tan x\, dx$

43. $\displaystyle\int_0^{\pi/2} \sec^2 x\, dx$

44. $\displaystyle\int_{-1}^1 \frac{dx}{x^2}$

45. $\displaystyle\int_{-1}^1 \frac{dx}{x^{2/3}}$

46. We know (Problem 36) that the region under the graph of $y = x^{-2/3}$ from $x = 1$ to ∞ has infinite area. Yet the volume generated by revolving this region about the x-axis is finite. Prove this, by calculating the volume.

6. DISTRIBUTIONS OVER THE AXIS; MORE AREAS

Many quantities in geometry, physics, and other applications of mathemetics have natural formulations as definite integrals. Area and volume are typical examples of such quantities, and they illustrate how such definite integral formulations come about.

The area of the region under a function graph is a quantity that is "spread out" over the x-axis. We mean by this statement that there is a definite amount of area over each interval I on the axis, and that if $a < b < c$, then the area over $[a,c]$ is the sum of the area over $[a,b]$ and the area over $[b,c]$.

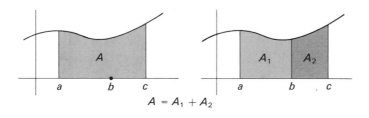

$$A = A_1 + A_2$$

The volume of a solid of revolution is distributed over the axis in the same way. The slab between $x = a$ and $x = b$ has a definite volume, and if $a < b < c$, then the volume from a to c is the sum of the volume from a to b and the volume from b to c.

Now consider any quantity that is distributed over the x-axis in this sense, and let Q be the amount of the quantity over the interval $[a,x]$. Then Q is a function of x,

$$Q = q(x),$$

called the *distribution function* of the quantity. Note that if $c < a$, and if the amount of the quantity over $[c,a]$ is k, then the amount over the interval $[c,x]$ is $k + q(x)$, the sum of the amount from c to a and the amount from a to x. Thus changing the point from which the quantity is measured only changes the distribution function by an additive constant.

We say that a quantity is *smoothly distributed* over an interval I if its distribution function $q(x)$ has a continuous derivative on I. By the remark above, this property does not depend on the point from which Q is measured.

A smooth distribution can always be computed as a definite integral, in exactly the way we computed areas and volumes in Theorem 6.

Theorem 7. *If a quantity is smoothly distributed over the x-axis, with distribution derivative $\rho(x)$, then the amount of the quantity over the interval $[a, b]$ is given by $\int_a^b \rho(x)\,dx$.*

Proof. Suppose for the sake of definiteness that $0 < a < b$, and let $Q = q(x)$ be the amount of the quantity over the interval $[0,x]$. By hypothesis $\rho(x) = q'(x)$, so

$$\int_a^b \rho(x)\,dx = q(b) - q(a).$$

But $q(b) - q(a)$ is the amount of Q on $[0,b]$ minus the amount of Q on $[0,a]$, and hence is the amount of Q on $[a,b]$.   ∎

We shall take up several more applications of Theorem 7 in the remaining sections of this chapter. The problem in every case will be to show that the distribution in question is smooth; i.e., that there is a continuous function $\rho(x)$ such that

$$\frac{\Delta Q}{\Delta x} \approx \rho(x),$$

where the approximation can be made as good as we wish by choosing Δx small enough. This estimate will always depend in an essential way on the fact that ΔQ is the amount of the quantity over the increment interval $[x, x + \Delta x]$, so that ΔQ can be directly compared with Δx. In the case of area and volume, this comparison led to an inequality of the form

$$l\Delta x \leq \Delta Q \leq u\Delta x,$$

where l and u both depend on Δx and both approach $\rho(x)$ as Δx approaches 0. Then the squeeze inequality

$$l \leq \frac{\Delta Q}{\Delta x} \leq u$$

forces $\Delta Q / \Delta x$ to approach $\rho(x)$ also, by the squeeze limit law. We shall see that the same scheme works in several other situations. One is treated below, and others will be taken up in later sections.

We shall now show that the area *between* two graphs is smoothly distributed over the x-axis. This slightly generalizes the earlier situation where we considered the area between the x-axis and a function graph.

Theorem 8. *Let f and g be continuous functions and suppose that $g(x) \leq f(x)$ over the interval $[a, b]$. Then the area of the region between the two graphs from $x = a$ to $x = b$ is $\int_a^b (f - g)$.*

Sketch of proof. The proof depends on squeezing ΔA between the areas of inner and outer rectangles, just as before. The figures here show that

$$(m_f - M_g)\Delta x \leq \Delta A \leq (M_f - m_g)\Delta x,$$

where m and M are the minimum and maximum values on the increment interval. This inequality is of the form

$$l\Delta x \leq \Delta A \leq u\Delta x,$$

where l and u each approach $f(x) - g(x)$ as Δx approaches 0. It follows, as before, that

$$\frac{dA}{dx} = \lim_{\Delta x \to 0} \frac{\Delta A}{\Delta x} = f(x) - g(x),$$

so $\int_a^b (f - g)$ is the area from a to b, by Theorem 7.

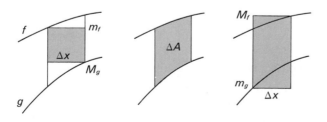

The missing details in the above proof will be left as an exercise.

EXAMPLE 1. Find the area of the finite region bounded by the graphs of $y = x^2$ and $y = \sqrt{x}$.

. . . .

Solution. You will generally want to draw a reasonably accurate sketch in a problem of this sort, to ensure that you have the right configuration in mind. Here, for example, we note that $y = \sqrt{x}$ is the *upper* graph between $x = 0$ and the point of intersection of the two curves. You may have noticed that the intersection point is (1,1),—since (1,1) clearly lies on both graphs,—but in any case it can be found algebraically by solving the two equations simultaneously. Here this reduces to solving the equation $x^2 = \sqrt{x}$, or $x^4 = x$, from which we find the solutions $x = 0$ and $x = 1$. These are the left and right edges of the region in question, so our area is

$$A = \int_0^1 (\sqrt{x} - x^2)\,dx = \left[\frac{2}{3} x^{3/2} - \frac{x^3}{3} \right]_0^1 = \frac{2}{3} - \frac{1}{3} = \frac{1}{3} \, .$$

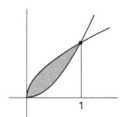

In Theorem 8 there is no requirement that the functions be positive.

EXAMPLE 2. Find the area between the parabola $x = y^2$ and the vertical line $x = 1$.

. . . .

Solution. Solving for y, we have

$$y = \pm\sqrt{x},$$

and we see that the area lies between the lower function graph $y = -\sqrt{x}$ and the upper function graph $y = \sqrt{x}$. So

$$A = \int_0^1 [\sqrt{x} - (-\sqrt{x})]\,dx = 2 \cdot \frac{2}{3} x^{3/2} \bigg]_0^1 = \frac{4}{3} \, .$$

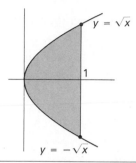

The arguments leading to Theorems 2 and 8 can be carried out just as well with the roles of the axes interchanged. Thus:

Theorem 8′. *If* $g(y) \leq f(y)$ *for all* y *in the interval* $[a, b]$, *then the area between the graphs* $x = g(y)$ *and* $x = f(y)$, *from* $y = a$ *to* $y = b$, *is*

$$\int_a^b [f(y) - g(y)]\, dy.$$

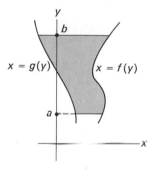

Instead of running through the same proof in this new configuration, we can reduce the new situation directly to the old as follows. The standard function graphs for the functions of Theorem 8′ would have the y-axis horizontal, and then the geometric configurations are exactly the same as for Theorem 8. So the area is

$$\int_a^b (f - g) = \int_a^b [f(y) - g(y)]\, dy,$$

by Theorem 8. Now rotate the whole configuration over the 45° line $y = x$, obtaining the configuration of Theorem 8′. The area remains unchanged, so it is still given by $\int_a^b (f - g)$. ∎

EXAMPLE 3. Solve Example 2 by integration with respect to y.

. . . .

Solution. Now the region is viewed as extending from $x = y^2$ up to $x = 1$, between the limits -1 and 1. Thus,

$$A = \int_{-1}^{1} (1 - y^2)\, dy = y - \frac{y^3}{3}\bigg]_{-1}^{1} = \left(1 - \frac{1}{3}\right) - \left(-1 + \frac{1}{3}\right) = \frac{4}{3}.$$

PROBLEMS FOR SECTION 6

In each of the following examples, find the area of the finite region bounded by the given graphs.

1. $y = 4x$; $y = 2x^2$.   2. $y^2 = x$; $y = 4$; $x = 0$.

3. $y^2 = 2x$; $x - y = 4$.   4. $y^2 = 6x$; $x^2 = 6y$.

5. $y^2 = 4x$; $x^2 = 6y$.   6. $y^2 = 4x$; $2x - y = 4$.

7. $y = 4 - x^2$; $y = 4 - 4x$.   8. $y = 6x - x^2$; $y = x$.

9. $y = x^3 - 3x$; $y = x$.   10. $y^2 = 4x$; $x = 12 + 2y - y^2$.

11. $x^2 y = x^2 - 1$; $y = 1$, $x = 1$, $x = 4$.

12. $y = x^2$; $y = x^3$, $x = 1$, $x = 2$.

13. $y = x^2$; $y = x^3$.   14. $w = 2 - x^2$; $w = x$.

15. $s = 1 - t^2$; $s = t^2$.   16. $t = y^4$; $t = 0, y = 1$.

17. $y = x^{1/3}$; $y = x/4$.   18. $y = \sin x$; $y = 2x/\pi$.

19. Find the area of the region between the graph of $y = x^3$ and its tangent line at $x = 1$.

20. Find the area of the region between $y = 1$ and $y = \cos x$, between two successive points of contact of these two graphs.

21. Write out the proof of Theorem 8 in detail.

22. Suppose that we have been studying a quantity Q that is distributed over the x-axis, and have found that the incremental amount ΔQ lying over the increment interval $[x, x + \Delta x]$ satisfies the inequalities

$$l \Delta x \leq \Delta Q \leq u \Delta x,$$

where $l \leq \rho(x) \leq u$.

Show that then

$$\Delta Q \approx \rho(x)\Delta x,$$

with an error at most $(u - l)\Delta x$ in magnitude.

7. AREA IN POLAR COORDINATES

Polar area is measured between given values of the *angular* coordinate, and the geometric picture is thus different, but the analysis is the same. A varying quantity is shown to have a derivative and is therefore computed as a definite integral.

Theorem 9. *Let f be a positive continuous function and let A be the area of the region bounded by the polar graph of the function f and the two rays $\theta = a$ and $\theta = b$. Then*

$$A = \frac{1}{2}\int_a^b f^2(\theta)\,d\theta.$$

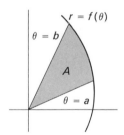

EXAMPLE 1. Find the area swept over in the first revolution of the spiral $r = \theta$.

Solution. Here $r = f(\theta) = \theta$, and θ runs from 0 to 2π. Therefore,

$$A = \frac{1}{2}\int_0^{2\pi} f(\theta)^2\,d\theta = \frac{1}{2}\int_0^{2\pi} \theta^2\,d\theta = \frac{1}{2}\left[\frac{\theta^3}{3}\right]_0^{2\pi} = \frac{4}{3}\pi^3,$$

by the theorem.

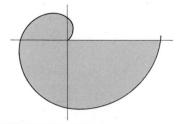

Proof of theorem. When θ is given an increment $\Delta\theta$, the area A acquires an increment ΔA as shown below. We have pictured the situation where f is increasing with θ and $\Delta\theta$ is positive. The area increment ΔA is squeezed between the areas of two circular sectors. In the situation shown below,

$$\frac{1}{2}f^2(\theta)\Delta\theta \leq \Delta A \leq \frac{1}{2}f^2(\theta + \Delta\theta)\Delta\theta,$$

so that

$$\frac{1}{2}f^2(\theta) \leq \frac{\Delta A}{\Delta\theta} \leq \frac{1}{2}f^2(\theta + \Delta\theta).$$

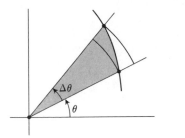
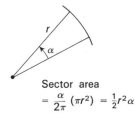

Sector area
$$= \frac{\alpha}{2\pi}(\pi r^2) = \frac{1}{2}r^2\alpha$$

In the general case, if m and M are the minimum and maximum values of f on the interval $[\theta, \theta + \Delta\theta]$, then

$$\frac{1}{2}m^2 \leq \frac{\Delta A}{\Delta\theta} \leq \frac{1}{2}M^2.$$

Since f is continuous, $f(\theta + \Delta\theta)$, m, and M all approach $f(\theta)$ as $\Delta\theta \to 0$. Therefore, $\Delta A/\Delta\theta$ approaches $f^2(\theta)/2$, by the squeeze limit law. That is, $dA/d\theta$ exists, and

$$\frac{dA}{d\theta} = \frac{1}{2}[f(\theta)]^2.$$

The integral formula then follows in the usual way (by Theorem 7.) ∎

EXAMPLE 2. Find the area enclosed by the cardioid $r = 1 + \cos\theta$.

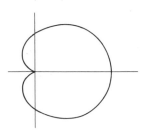

. . . .

Solution.

$$A = \frac{1}{2} \int_0^{2\pi} r^2 \, d\theta = \frac{1}{2} \int_0^{2\pi} (1 + \cos \theta)^2 \, d\theta$$

$$= \frac{1}{2} \int_0^{2\pi} [1 + 2 \cos \theta + \cos^2 \theta] \, d\theta.$$

We are now stuck if we don't know how to integrate $\cos^2 \theta$. The saving device is the addition law for $\cos(x + y)$, which gives us the identity

$$\cos^2 \theta = \frac{1 + \cos 2\theta}{2}.$$

Since we find that $\sin 2\theta / 2$ is an antiderivative of $\cos 2\theta$, we can now go back and finish integrating:

$$A = \frac{1}{2} \int_0^{2\pi} [1 + 2 \cos \theta + \cos^2 \theta] \, d\theta$$

$$= \frac{1}{2} \int_0^{2\pi} \left[1 + 2 \cos \theta + \frac{1}{2} + \frac{\cos 2\theta}{2} \right] d\theta$$

$$= \left[\frac{3}{4} \theta + \sin \theta + \frac{\sin 2\theta}{8} \right]_0^{2\pi} = \frac{3}{4} 2\pi = \frac{3\pi}{2}.$$

In the next chapter we shall practice integration devices like the one used above, and learn what to expect from a few basic tricks of the trade.

PROBLEMS FOR SECTION 7

1. Find the area swept out in the first revolution of the spiral $r = \theta^2$.

2. The area formula of Theorem 9 should be consistent with the formula for the area of a circle. Show that it is. (Use Theorem 9 to retrieve the value πr^2 as the area of a circle of radius r.)

3. Find the area of the region bounded by the polar graph $r = \cos \theta$ and the rays $\theta = 0$, $\theta = \pi/2$. (See Example 2 in the text for the integration trick.)

4. Find the area bounded by the polar graph $r = \sec \theta$ and the rays $\theta = 0$, $\theta = \pi/4$.

5. Show that the area cut off by the spiral $r = e^\theta$ in successive quadrants increases by the constant factor e^π.

6. Find the area of the region bounded on the outside by the cardioid $r = 1 + \cos\theta$ and on the inside by $r = 2\cos\theta$.

8. THE LENGTH OF A CURVE

In tackling the area problem we assumed that *area* exists and has certain natural properties, and then proved that the area under a graph could be computed by integration.

The same approach can be taken to the calculation of the *length of curves*. If we assume that the graph of a smooth function f over an interval $[a,x]$ has a length s, and that the variation of s with x meets a certain intuitively obvious criterion, then we can prove that s is a smoothly distributed quantity and that

$$\frac{ds}{dx} = \sqrt{1 + \left(\frac{dy}{dx}\right)^2}$$

The length of the graph over $[a,b]$ is therefore

$$s = \int_a^b \sqrt{1 + \left(\frac{dy}{dx}\right)^2}\, dx.$$

The intuitive criterion here is that the length of a very short arc is approximately equal to the length of its chord, in the sense that

$$\frac{\text{Arc length}}{\text{Chord length}} \to 1$$

as the length of the chord approaches 0. We apply this principle to the increment in graph length Δs arising from an increment Δx. The chord of the incremental arc has length $l = \sqrt{(\Delta x)^2 + (\Delta y)^2}$ by the Pythagorean theorem. Therefore,

$$\frac{\Delta s}{\Delta x} = \frac{\Delta s}{l} \cdot \frac{l}{\Delta x} = \frac{\Delta s}{l} \cdot \sqrt{1 + \left(\frac{\Delta y}{\Delta x}\right)^2}.$$

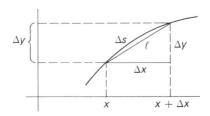

Since, by hypothesis, $\Delta s/l \to 1$ as $\Delta x \to 0$, we conclude that

$$\frac{ds}{dx} = \lim_{\Delta x \to 0} \frac{\Delta s}{\Delta x} = \lim_{\Delta x \to 0} \sqrt{1 + \left(\frac{\Delta y}{\Delta x}\right)^2} = \sqrt{1 + \left(\frac{dy}{dx}\right)^2},$$

as announced above.

The arc-length formula serves as an additional warning that we need more integration technique: At the moment there are hardly any functions for which we can work out the integral!

Moreover, it is customary to adopt a more fundamental approach to arc length, even in high-school geometry, and we shall reconsider the length question in Chapter 10.

PROBLEMS FOR SECTION 8

1. Find the length of the graph of $3y = 2x^{3/2}$ from $x = 0$ to $x = 2$.

2. a) Compute the derivative of $(x^{2/3} + 1)^{3/2}$.
   b) Find the length of the graph of $2y = 3x^{2/3}$ from $x = 0$ to $x = 8$. [*Hint.* Try to make use of your answer to part (a).]

3. Find the length of the graph of $2y = e^x + e^{-x}$ from $x = 0$ to $x = 1$.

4. If $f(x) = (e^x + e^{-x})/2$, show that the length of the graph of f from $x = 0$ to $x = a$ is $f'(a)$.

5. Find the length of the graph of $8y = x^4 + 2/x^2$ from $x = 1$ to $x = 2$.

9. VOLUMES BY SLICING

Suppose we put a coordinate line along beside a solid body S (or, possibly, skewering S). We shall call the coordinate line the x-axis. We want to consider how the solid intersects various planes perpendicular to the x-axis. If the plane perpendicular to the x-axis at $x = x_0$ slices the solid, then the cross section of the solid in this plane will be a plane region R_{x_0}. Nearby cross sections can be

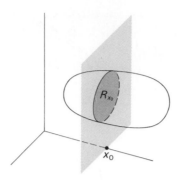

projected onto the same plane (by moving them parallel to the x-axis), so we can compare how the cross section R_x changes as we move the slicing plane. In general, the cross section R_x will vary with x. If it does not, then we have a *cylinder*.

Definition. *If the solid S runs from the plane at $x = a$ to the plane at $x = b$, and if all its cross sections R_x at points x between a and b are the same plane region R, then we say that S is a cylinder, with base R and altitude $h = b - a$.*

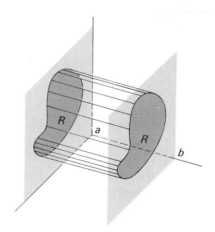

The volume of a cylinder is the product of its base area and its altitude:

$$V = Ah,$$

where A is the area of the base and h is the altitude. We shall assume this formula, and use it to derive a very general formula for the volume of a solid body.

To see how this goes, consider the incremental slab ΔS of width Δx that is sliced from the solid by the planes at $x = x_0$ and $x = x_0 + \Delta x$. We assume that the cross-sectional area $A(x)$ varies continuously with x. So if Δx is small, then $A(x)$ differs only slightly from $A(x_0)$ for all x between x_0 and $x_0 + \Delta x$, and the slab ΔS is approximately a cylinder of altitude Δx and base area $A(x_0)$.

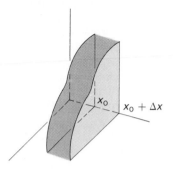

More precisely, we assume that the cross section R_x varies continuously with x, in the following sense. The incremental slab ΔS completely includes a cylinder C' whose base area A' is only slightly less than $A(x_0)$, and is completely included in a cylinder C'' whose base area A'' is only slightly greater than $A(x_0)$. That is, the incremental slab ΔS is "squeezed between" two cylinders C' and C'', whose base areas A' and A'' can be made as close to $A(x_0)$ as desired by taking Δx sufficiently small. (The inner and outer cylinders C' and C'' will not generally be circular cylinders. The figure below is drawn with circular cross sections for simplicity and clarity.)

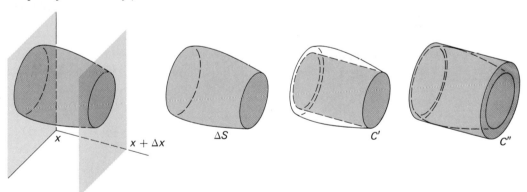

Consequently,

$$\text{Volume of } C' \le \text{Volume of } \Delta S \le \text{Volume of } C'',$$

or

$$A'\Delta x \le \Delta V \le A''\Delta x,$$

or

$$A' \le \frac{\Delta V}{\Delta x} \le A''.$$

Then, since A' and A'' both approach $A(x_0)$ as Δx approaches 0, $\Delta V/\Delta x$ is squeezed to the same limit.

The volume of the solid S is thus smoothly distributed over the x-axis, with distribution derivative $A(x)$, and we have proved (by virtue of Theorem 7):

Theorem 10. *Let S be a solid whose cross section R_x varies continuously with x as x runs from a to b. Let $A(x)$ be the area of the cross section R_x. Then the volume of the slab between $x = a$ and $x = b$ is given by*

$$V = \int_a^b A(x)\,dx.$$

EXAMPLE 1. The solid of revolution generated by rotating the graph of f about the x-axis has the cross-sectional area

$$A(x) = \pi[f(x)]^2.$$

Therefore,

$$V = \int_a^b A(x)\,dx = \int_a^b \pi[f(x)]^2\,dx,$$

and we recover the formula of Section 3 as a special case of the above "volumes-by-slicing" formula.

EXAMPLE 2. The figure below shows a solid whose cross sections perpendicular to the x-axis are all isosceles right triangles with one vertex on the half-parabola $y = \sqrt{1-x}$. The area of the triangle at x is

$$\frac{1}{2}bh = \frac{1}{2}\sqrt{1-x}\,\sqrt{1-x} = \frac{1-x}{2}.$$

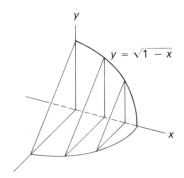

This is the cross-sectional area $A(x)$, and the volume of the solid between $x = 0$ and $x = 1$ is therefore

$$V = \int_0^1 A(x)\,dx = \int_0^1 \frac{1-x}{2}\,dx = \frac{x}{2} - \frac{x^2}{4}\bigg]_0^1 = \frac{1}{4}.$$

PROBLEMS FOR SECTION 9

Find the volume of each of the solids described below. You may assume that the cross section R_x varies continuously with x and that Theorem 10 can therefore be applied. Cross-section continuity is one of those annoying conditions that are obviously true but fussy to prove. We shall have more to say on this point in Chapter 18.

In the first four problems, the base of the solid is the unit circular disk $x^2 + y^2 \leq 1$, and its cross section perpendicular to the x-axis is the given figure:

1. A semicircle.

2. A square.

3. An equilateral triangle.

4. A rectangle of height $|y^3|$ (where $x^2 + y^2 = 1$).

5. The base of the solid is the square centered at the origin with one vertex at $(1,1)$, and its cross sections perpendicular to the x-axis are triangles of altitude $(1 - x^2)$.

6. The same base, with triangular cross sections of altitude $1 + x$.

7. The base of the solid is the square centered at the origin with one vertex at $(1,0)$, and its cross sections perpendicular to the x-axis are triangles with altitude $1 - x^4$.

8. The same base, with triangular cross sections of altitude 1.

9. The pyramid with square base of side a, and altitude h.

10. The pyramid whose base is an equilateral triangle of side a and whose altitude is h.

11. A cone with an irregular base region R is formed by drawing line segments from boundary points of R to a fixed vertex P. Prove that its volume is

$$V = \frac{1}{3}Ah,$$

where A is the area of R and h is the altitude of the cone. (Assume that similar plane figures have areas proportional to the squares of their diameters. The diameter of a plane figure is the length of the longest segment that can be drawn between two of its points.)

12.   The "wood chopper's wedge" pictured below, is cut from a cylinder of radius r by a plane through a diameter of the base circle, making an angle of $30° = \pi/6$ radians with the base plane.

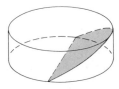

13.   A tent is made by stretching canvas from a circular base of radius r to a semicircular rib, erected at right angles to the base at the ends of a diameter. Compute the volume of the tent.

10. VOLUMES BY SHELLS

There is another way of viewing a volume of revolution as being spread out over an interval. It leads to a second formula for the volume of a solid of revolution, expressing as an x-integral the volume which is generated by revolving a region about the y-axis. (See left, below.)

Theorem 11. *If the region under the graph of f, from $x = a$ to $x = b$, is revolved around the y-axis, then the volume swept out is*

$$\int_a^b 2\pi x f(x)\, dx.$$

It is understood that $0 \le a < b$ and that $f \ge 0$ on $[a,b]$, so the back half of the solid can be pictured as in the figure at the right below.

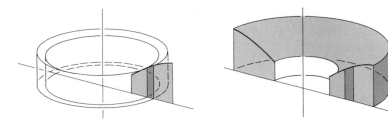

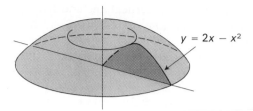

Suppose, for example, that $f(x) = 2x - x^2$. Then

$$V = \int_0^2 2\pi x(2x - x^2)\,dx$$

$$= 2\pi\left[\frac{2x^3}{3} - \frac{x^4}{4}\right]_0^2$$

$$= 2\pi\left[\frac{16}{3} - \frac{16}{4}\right] = \frac{8}{3}\pi.$$

When we use this formula we say that we are computing the volume by *cylindrical shells*, because the incremental volume ΔV associated with the Δx interval $[x, x + \Delta x]$ is approximately the volume of a thin cylindrical shell, obtained by rotating the thin vertical Δx strip about the y-axis, as shown in the figure on page 342.

We proceed now to estimate the increment ΔV. The *surface area* of a circular cylinder of altitude h and base radius r is

$$h \times \text{circumference} = h \cdot 2\pi r.$$

Here, $h = f(x)$ and $r = x$, so the cylindrical area A is

$$A = 2\pi x f(x).$$

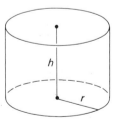

If we now thicken the surface slightly by an amount Δx, then we obtain a thin shell of volume approximately

$$A\Delta x = 2\pi x f(x)\Delta x.$$

(Think of cutting the shell along a vertical line and flattening it out. It becomes a rectangular solid whose dimensions are $2\pi x$, $f(x)$ and Δx, and whose volume is therefore $2\pi x \cdot f(x) \cdot \Delta x$.)

This is not exactly ΔV, for a couple of reasons; nevertheless, we can show that it is approximately ΔV,

$$\Delta V \approx 2\pi x f(x) \Delta x,$$

with an error that is small in comparison with Δx. That is, $\Delta V/\Delta x \to 2\pi x f(x)$ as $\Delta x \to 0$, so $V = \int_a^b 2\pi x f(x)\, dx$, by Theorem 7.

The details of this proof are worked out in Problems 8 through 10.

PROBLEMS FOR SECTION 10

In the following problems, use the method of shells to compute the volume generated by rotating about the y-axis the region bounded by the given graphs. Draw the plane region in each case.

1. $y = x$; $y = 0$, $x = a$.

2. $y = x^2$; $y = 1$, $x = 0$. (Compute this volume also by the earlier formula.)

3. $y^2 = x^3$; $x = 4$.

4. $y = x^2 - 2x$; $y = 2x$.

5. $y = x - x^3$; $y = 0$.

6. a) Differentiate $(r^2 - x^2)^{3/2}$.
   b) With the above formula in mind, find the volume generated by rotating about the y-axis the region to the right of the line $x = a$ and inside the circle $x^2 + y^2 = r^2$ (where $a < r$).

7. a) Differentiate $(4 - x^2)^{3/2}$.
   b) With this result in mind, compute the volume generated by rotating about the y-axis the region outside the hyperbola $x^2 - y^2 = 4$ and inside the line $x = 5$.

We turn now to the proof of Theorem 11.

8. Show that the exact volume of a cylindrical shell of altitude h, inner radius r, and outer radius $r + \Delta r$, is

$$\pi h(2r + \Delta r)\Delta r.$$

(Subtract the volume of a cylinder from the volume of a slightly larger cylinder.)

9. Suppose that f is an increasing function, and let ΔV be the incremental volume obtained by rotating the region under the graph of f over the increment interval $[x, x + \Delta x]$ about the y-axis. Use Problem 8 to show that

$$\pi(2x + \Delta x)f(x)\Delta x \leq \Delta V \leq \pi(2x + \Delta x)f(x + \Delta x)\Delta x.$$

Show, therefore, that

$$\lim \frac{\Delta V}{\Delta x} = 2\pi x f(x),$$

$$V = \int_a^b 2\pi x f(x)\,dx.$$

10. Modify the above argument to cover the case when f is *not* increasing, as in the area discussion in Section 2.

11. DENSITY

If Q is a quantity distributed over the x-axis, then the increment quotient $\Delta Q/\Delta x$ can be interpreted as the average amount of Q per unit length lying over the interval $[x, x + \Delta x]$, in which case it is called the *average density* of Q over the increment interval. From this point of view, the limit

$$\rho(x) = \lim \frac{\Delta Q}{\Delta x}$$

is called the *density* of the distribution at x. In these terms, we showed that the area A of the region under the graph of f is a smoothly distributed quantity with distribution density $f(x)$ at x, and that the volume V generated by rotating this region about the x-axis is a smoothly distributed quantity with distribution density $\pi[f(x)]^2$ at x. Generally, however, we don't use the density terminology unless the distribution actually lies *on* the axis. Here are some examples.

EXAMPLE 1. Let a piece of wire of varying constitution lie along the x-axis from $x = a$ to $x = b$, and let m be the mass of the segment $[a,x]$. If we give x a positive increment Δx, then the corresponding increment Δm is the mass of the segment $[x, x + \Delta x]$, and the difference quotient

$$\frac{\Delta m}{\Delta x}$$

is thus the average density of the wire throughout the incremental segment. The density at x is defined to be the limit of the average density as Δx tends to zero, and it is thus the derivative dm/dx. It could be measured in ounces per inch, or grams per centimeter, or, for a very fine wire, in milligrams per centimeter.

EXAMPLE 2. A mass distribution on $[0,1]$ has density $\rho(x) = x - x^2$. By Theorem 7 the amount of mass on $[1/4, 1/2,]$ is then

$$\int_{1/4}^{1/2} \rho(x)\,dx = \int_{1/4}^{1/2} (x - x^2)\,dx = \frac{x^2}{2} - \frac{x^3}{3}\Big]_{1/4}^{1/2}$$

$$= \left(\frac{1}{8} - \frac{1}{24}\right) - \left(\frac{1}{32} - \frac{1}{192}\right)$$

$$= \frac{11}{192} .$$

EXAMPLE 3. A distribution of electric charge lies along the interval $[0,1]$, there being $C(x) = x^2 - x^3$ units of charge in the interval $[0,x]$. What is the charge density at $x = 1/4$? At $x = 1/2$? At $x = 3/4$?

. . . .

Solution. The charge density $\rho(x)$ is given by

$$\rho(x) = \frac{dC(x)}{dx} = 2x - 3x^2,$$

expressed as units of charge per unit length. Thus,

$$\rho\left(\frac{1}{4}\right) = \frac{5}{16}, \qquad \rho\left(\frac{1}{2}\right) = \frac{1}{4}, \qquad \text{and} \qquad \rho\left(\frac{3}{4}\right) = -\frac{3}{16}.$$

The amount of charge in an incremental interval $[1/4, 1/4 + \Delta x]$ is approximately $\rho(1/4)\Delta x = 5\Delta x/16$. The interval $[3/4, 3/4 + \Delta x]$ contains negative charge, of approximately $-3\Delta x/16$ units.

EXAMPLE 4. Another important quantity Q that is "spread out" along the axis is a probability distribution. Suppose we randomly choose points from an interval $[a,b]$ by some mechanism. We assume that for each x there is a number $P = p(x)$ representing the *probability* of a point landing in the subinterval $[a,x]$. Furthermore, we assume that $p(x)$ is an increasing function of x such that $p(a) = 0$ and $p(b) = 1$, and that $p(y) - p(x)$ is the probability of landing in the subinterval $[x,y]$. So if we increase x by Δx, then $\Delta P = p(x + \Delta x) - p(x)$ is the probability of landing in the increment interval $[x, x + \Delta x]$. Thus, probability is a distribution like mass or charge. If $p(x)$ is a differentiable, then $\rho(x) = dP/dx$ is the *probability density*. It is interpreted as the probability per unit length at x, in the sense that

$$\frac{\Delta P}{\Delta x} \approx \rho(x)$$

and

$$\Delta P \approx \rho(x)\Delta x.$$

EXAMPLE 5. A probability distribution on the interval $[0,1]$ is defined by $p(x) = x^2$. Thus, the probability of landing in the interval $[0, 1/2]$ is $p(1/2) = 1/4$. The probability density is

$$\rho(x) = 2x,$$

and the probability of landing in the incremental interval $[(3/4), (3/4) + \Delta x]$ is approximately $\rho(x)\Delta x = 3\Delta x/2$.

EXAMPLE 6. A probability distribution on $[-1,1]$ has density $\rho(x) = 3x^2/2$. What is the probability of landing in the interval $[-1/2, 0]$?

. . . .

Solution. We should first check that we really do have a probability distribution. The basic requirement is that the probability of $[-1, x]$,

$$p(x) = \int_{-1}^{x} \rho(t)\, dt,$$

should increase from 0 to 1 in value as x crosses the interval $[-1, 1]$. Since $p'(x) = \rho(x) = 3x^2/2 \geq 0$, p does increase. Also,

$$p(1) = \int_{-1}^{1} (3x^2/2)\, dx = \frac{x^3}{2}\bigg]_{-1}^{1} = \frac{1}{2} - \left(-\frac{1}{2}\right) = 1,$$

as required. Finally, the probability of $[-1/2, 0]$ is

$$\int_{-1/2}^{0} \rho(x)\, dx = \frac{x^3}{2}\bigg]_{-1/2}^{0} = 0 - \left(-\frac{1}{16}\right) = \frac{1}{16}.$$

We conclude with some remarks about the general notion of density. Consider a quantity Q, such as a mass distribution or a distribution of charge, that is "spread out" in space. We say that the distribution has the constant density ρ if the amount of Q contained in any region R is exactly ρ times the volume V of R. Thus,

$$\rho = \frac{Q(R)}{V(R)},$$

no matter what space region R we consider. If Q is a mass distribution, and if we measure mass in ounces and volume in cubic inches, then the density ρ is measured in ounces per cubic inch. In the metric system density is measured in grams per cubic centimeter.

EXAMPLE 7. We find from a handbook that the density of aluminum is 2.7 g/cm^3. This means that the mass of any aluminum object, measured in grams, is 2.7 times the volume it occupies, measured in cubic centimeters. The density of iron is 7.9 and the density of gold is 19.3.

Unfortunately, a distribution Q will not generally have a constant density, and we have to consider density as something that varies from point to point. In order to define the density of Q at the point p_0, we first choose some small region ΔR around p_0, such as a small ball or a small cube centered at p_0. Let ΔQ be the amount of Q contained in ΔR and let ΔV be the volume of ΔR. The quotient

$$\frac{\Delta Q}{\Delta V}$$

is then called the *average density* of Q throughout ΔR. For a mass distribution in the above units, $\Delta Q/\Delta V$ is the average number of ounces per cubic inch throughout ΔR. *The space density of Q at p_0* is then defined to be the limit

$$\rho(p_0) = \lim_{\Delta V \to 0} \frac{\Delta Q}{\Delta V},$$

supposing that the limit exists. This is not a derivative in the ordinary sense, and we are in no position to discuss how we might evaluate such a limit. Nevertheless, we can conceive of the limit existing, and if it happened to have the value 5, we would say that the Q density at p_0 is 5 ounces per cubic inch.

Similar remarks apply to a two-dimensional distribution of some quantity Q. We could consider a charge distribution spread out over a two-dimensional sheet, or we could think of a flat sheet of metal with varying constitution as giving us a two-dimensional mass distribution. Now the average plane density in a small region ΔR is the quotient

$$\frac{\Delta Q}{\Delta A},$$

where ΔA is the *area* of ΔR. For a mass distribution with the same units as before, the average density is measured in ounces per *square* inch (or in grams per *square* centimeter). So is the plane point density

$$\rho(p_0) = \lim_{\Delta A \to 0} \frac{\Delta Q}{\Delta A}.$$

Here again we fail to have an ordinary derivative, although we are taking the limit of something like a difference quotient.

The computation of the total amount of a quantity in a plane region R from its density function ρ requires double integration, and is taken up in Chapter 18.

PROBLEMS FOR SECTION 11

1. a) Show that $\rho(x) = x + x^3 - 2x^2$ is a possible density for a mass distribution along the interval $[0,4]$. (Show that $\rho(x) \geq 0$ in the interval.)
    b) Find the mass of the segment $[1,2]$.

2. a) A triangular sheet of metal with vertices $(0,0)$, $(2,0)$, $(2,1)$ has constant density 1. What is its mass?
    b) What is the density $\rho(x)$ if the metal triangle is viewed as a mass distribution over the x-axis (or along the x-axis)? Compute its mass by integrating the density $\rho(x)$.

3. A distribution of electric charge along a rod has density $\rho(x) = x - x^2$.
    a) How much charge is on the segment $[0,1]$?
    b) What is the total charge on the segment $[-1/2, 1]$?

4. a) Show that $\rho(x) = 3(1 - x^2)/2$ is a probability density on the interval $[0,1]$.
    b) Find the probability that a number will be in the subinterval $[1/2,1]$.

5. The function $\rho(x) = x^3$ is a probability density on an interval $[0,b]$. What is b?

6. a) The function $\rho(x) = 12x^2$ is a probability density on an interval $[-a, a]$.
        Find a.
    b) Find the probability of the subinterval $[0, a/2]$.

7. Show that $\rho(x) = e^x$ is a probability density on the interval $(-\infty, 0]$.

chapter 9
finding antiderivatives

If we start with the five functions x, e^x, $\log x$, $\sin x$, and $\arcsin x$, and build new functions by using the algebraic operations and composition, possibly repeatedly and in combination with each other, we generate the class of *elementary functions*. Such functions are said to have *closed form* because they can be explicitly written down.

We were able to omit $\cos x$ and $\arctan x$ from the basic list because they arise as simple compositions:

$$\cos x = \sin(x + \pi/2),$$

$$\arctan x = \arcsin \frac{x}{\sqrt{1 + x^2}}.$$

The derivative of an elementary function is always an elementary function, and it can be found in a systematic way by applying the differentiation rules. When we first turn to antidifferentiation, we probably expect the same pattern; that is, we expect every elementary function to have an antiderivative in "closed form" that we can find in an explicit and systematic manner. We saw in Chapter 8 that polynomials behave this way; in fact, polynomials can be integrated as easily as they can be differentiated.

This is not so, however, in general. There is, in general, no systematic integration procedure that can be followed step by step to a guaranteed answer. There may not even *be* an answer, at least until we invent a new function for the purpose. For example, such a simple looking function as $f(x) = e^{-x^2}$ has no antiderivative at all within the class of elementary functions. It does have an antiderivative: If we define $F(x)$ to be the area under the graph of $y = e^{-t^2}$ from $t = 0$ to $t = x$, then we know from Chapter 8 that $F(x)$ is an antiderivative of e^{-x^2}, and we shall learn how to calculate $F(x)$ in Chapter 13. But it can be proved that there is no way of representing F in terms of functions we already know about.

Despite such uncertainties, it is very useful to find antiderivatives explicitly in terms of known functions when it is possible and feasible. That this process must be viewed as an art rather than a systematic routine should not be discouraging. Many students find, after they get the hang of the three or four basic tricks of the trade, that integrating is more interesting than differentiating because it is more like solving puzzles.

We recall that systematic differentiation involves:

a) The derivative formulas for the functions x^a, e^x, $\sin x$, $\cos x$, $\log x$, $\arcsin x$, and $\arctan x$;

b) The basic differentiation rules:
   1. linearity, $(af + bg)' = af' + bg'$;
   2. the product rule, $(fg)' = fg' + gf'$;
   3. the chain rule, $(f \circ g)' = (f' \circ g) \cdot g'$.

The quotient rule is omitted because it really is a secondary rule derived from the above: f/g is the product $f(1/g)$, and $1/g$ is the composition g^{-1}, a special case of the general power function g^a.

Integration involves these same formulas and rules, but run backwards. We start with the following formulas:

$$\int x^a dx = \frac{x^{a+1}}{a+1} + C, \qquad \text{provided } a \neq -1;$$

$$\int \frac{dx}{x} = \log x + C \qquad \text{(the case } a = -1);$$

$$\int \frac{dx}{1+x^2} = \text{arc tan } x + C;$$

$$\int \frac{dx}{\sqrt{1-x^2}} = \text{arc sin } x + C;$$

$$\int e^x dx = e^x + C;$$

$$\int \sin x\, dx = -\cos x + C;$$

$$\int \cos x\, dx = \sin x + C.$$

Then more complicated functions are attacked by the backwards rules. We have already considered linearity in Chapter 8, and this leaves only the chain rule and the product rule. The backwards chain rule is called *integration by substitution* or *change of variable*. The backwards product rule is called *integration by parts*.

In addition, there is one integration procedure that does not correspond to a differentiation rule: Occasionally we can subdue a balky integrand by algebraically manipulating it into a more docile form. The systematic integration of rational functions is accomplished this way (in conjunction with the rules), and there are a couple of other situations in which we can benefit by such tactics.

However, no matter how clever we are, there will always be more functions that we cannot integrate than ones we can, and some technique of numerical integration is a necessary companion to the art of integrating. The next chapter is devoted to this *estimation* approach to integration.

The logarithm integral has a complication that we must discuss. Although the integrand $1/x$ is defined for all x except $x = 0$, the formula $\int dx/x = \log x + c$ is valid only for $x > 0$, since negative numbers don't have logarithms. However, if $x < 0$, then $\log(-x)$ exists, and

$$\frac{d}{dx}\log(-x) = \frac{1}{-x} \cdot (-1) = \frac{1}{x}.$$

Therefore,

$$\int \frac{dx}{x} = \log(-x) + c \qquad \text{over the interval } (-\infty, 0).$$

It is customary to combine these two formulas and write

$$\int \frac{dx}{x} = \log|x| + c, \qquad x \neq 0.$$

But this is a dangerous formula, and it must be used very carefully. It seems to say that $\log|x| + c$ is the most general antiderivative of $1/x$ over the full domain of $1/x$. This is false. The function

$$f(x) = \begin{cases} \log|x|, & \text{when } x < 0, \\ \log|x| + 2, & \text{when } x > 0, \end{cases}$$

is an antiderivative which is not in the above form. The theorem guaranteeing that the most general antiderivative is of the form $F(x) + c$ was only established over an *interval* lying in the domains of both F and $f = F'$. It fails for $\log|x|$ and $d \log|x|/dx = 1/x$, because their domain is not an interval. Therefore the standard formula

$$\int \frac{dx}{x} = \log|x| + c$$

must be understood *subject to this correction*, and must be applied very carefully. The same considerations apply to

$$\int dx/x^2 = -1/x + c,$$

and to other negative powers of x.

1. CHANGE OF VARIABLE; DIRECT SUBSTITUTION

So far the dx in Leibniz's notation $\int f(x)\,dx$ has played no role, but now we shall find it useful to take dx at its face value as the differential of x.
    Consider the integral

$$\int (\sin x)^2 \cos x\,dx.$$

If we set $u = \sin x$, then $du = \cos x\,dx$, and the integral takes on the new form

$$\int u^2\,du.$$

Since

$$\int u^2\,du = \frac{u^3}{3} + c,$$

and since $u = \sin x$, the answer we are looking for ought to be

$$\int (\sin x)^2 \cos x\,dx = \frac{(\sin x)^3}{3} + c.$$

We see that it is, by differentiating. This is a chain-rule differentiation, and our change of variables from x to u can thus be considered as a backward application of the chain rule.

The general pattern of the above calculation is this. We have a complicated function $f(x)$ that can be seen to be in the form $f(x) = g(h(x))h'(x)$, and we want to compute the integral

$$\int f(x)\,dx = \int g(h(x))h'(x)\,dx.$$

If we set $u = h(x)$, then $du = h'(x)dx$, and the integral takes on the new form

$$\int g(u)\,du.$$

We suppose that we can find an antiderivative G of g, so that

$$\int g(u)\,du = G(u) + c.$$

Then, since $u = h(x)$, the answer we are looking for ought to be

$$\int f(x)\,dx = G(h(x)) + c.$$

We see that it is, by the chain rule:

$$\frac{d}{dx}G(h(x)) = G'(h(x))h'(x) = g(h(x))h'(x) = f(x).$$

Thus, if we take Leibniz's notation at its face value, as involving a *differential*, then it automatically gives us the correct integration procedure that corresponds to the chain rule.

One might think that this device requires an integrand of such special form that it would hardly ever be applicable, but it is surprising how often it can be used. Here are some further examples:

EXAMPLE 1

$$\int \frac{2x\,dx}{1 + x^2} = \int \frac{du}{u} \qquad \left(\begin{array}{l}\text{Set } u = 1 + x^2;\\ \text{then } du = 2x\,dx\end{array}\right)$$

$$= \log u + c = \log(1 + x^2) + c.$$

EXAMPLE 2

$$\int \frac{2x\,dx}{(1 + x^2)^2} = \int \frac{du}{u^2} = -\frac{1}{u} + c = -\frac{1}{1 + x^2} + c.$$

EXAMPLE 3

$$\int \frac{\log x}{x}\,dx = \int u\,du \qquad\qquad \left(\begin{array}{l} u = \log x, \\ du = \dfrac{dx}{x} \end{array}\right)$$

$$= \frac{u^2}{2} + c = \frac{(\log x)^2}{2} + c.$$

EXAMPLE 4. Frequently we will have to multiply by a constant to get $h'(x)dx$. Thus, in

$$\int xe^{x^2}dx,$$

we try setting $u = x^2$. Then $du = 2x\,dx$, which we don't quite have. However, we do have

$$x\,dx = \frac{du}{2},$$

and this is good enough because a constant can be factored out:

$$\int xe^{x^2}dx = \int e^{x^2}(x\,dx) = \int \frac{e^u\,du}{2} = \frac{1}{2}\int e^u\,du$$

$$= \frac{1}{2}e^u + c = \frac{1}{2}e^{x^2} + c.$$

The simplest substitution of all occurs when we have the derivative of a known function except for an added constant inside.

EXAMPLE 5

$$\int \cos(x + 3)\,dx = \int \cos u\,du \qquad\qquad \left(\begin{array}{l} u = x + 3, \\ du = dx \end{array}\right)$$

$$= \sin u + c = \sin(x + 3) + c.$$

EXAMPLE 6

$$\int \frac{dx}{x + a} = \int \frac{du}{u} \qquad\qquad \left(\begin{array}{l} u = x + a, \\ du = dx \end{array}\right)$$

$$= \log|u| + c = \log|x + a| + c.$$

EXAMPLE 7. A multiplicative constant inside is about as simple,

$$\int \cos 2x \, dx = \frac{1}{2} \int \cos u \, du \qquad \left(\begin{array}{l} u = 2x, \\ du = 2dx, \\ dx = \dfrac{du}{2} \end{array} \right)$$

$$= \frac{1}{2} \sin 2x + c.$$

Note that we could have viewed this substitution as setting $x = u/2$, $dx = du/2$. In the next example this is the *natural* procedure. (See Section 4.)

EXAMPLE 8

$$\int \frac{dx}{\sqrt{4 - x^2}} = \int \frac{du}{\sqrt{1 - u^2}} \qquad \left(\begin{array}{l} x = 2u, \\ dx = 2du \end{array} \right)$$

$$= \arcsin u + c = \arcsin \frac{x}{2} + c.$$

EXAMPLE 9

$$\int \frac{dx}{(x + 1)^2 + 2} = \frac{1}{\sqrt{2}} \int \frac{du}{u^2 + 1} \qquad \left(\begin{array}{l} x + 1 = \sqrt{2}\,u, \\ dx = \sqrt{2}\,du \end{array} \right)$$

$$= \frac{1}{\sqrt{2}} \arctan u = \frac{1}{\sqrt{2}} \arctan \frac{x + 1}{\sqrt{2}} + c.$$

After a little practice it is perfectly feasible to make simple substitutions mentally and just write the answer down, as in

$$\int \cos(x + 3) \, dx = \sin(x + 3) + c.$$

PROBLEMS FOR SECTION 1

Integrate the following.

1. $\int \sin 2x \, dx$

2. $\int e^{x/2} \, dx$

3. $\int \sin^2 x \cos x \, dx$

4. $\int x \cos x^2 \, dx$

5. $\int t^2 e^{-t^3} \, dt$

6. $\int \sqrt{ax + b} \, dx$

7. $\int \cos(2 - x) \, dx$

8. $\int e^x \cos(e^x) \, dx$

9. $\int \dfrac{\sin x}{1 + \cos x}\,dx$

10. $\int \tan x\,dx$

11. $\int \dfrac{\log x\,dx}{x}$

12. $\int \dfrac{dx}{x \log x}$

13. $\int \dfrac{2x + 1}{x^2 + x - 1}\,dx$

14. $\int \dfrac{\sin \theta}{\cos^2 \theta}\,d\theta$

15. $\int y\sqrt[3]{a + y}\,dy$

16. $\int \dfrac{x^{1/2}\,dx}{1 + x^{3/4}}$

17. $\int \dfrac{(x + 2)\,dx}{x\sqrt{x - 3}}$

18. $\int (x^3 + 2)^{1/3} x^2\,dx$

19. $\int \dfrac{8x^2\,dx}{(x^3 + 2)^3}$

20. $\int \dfrac{dx}{4 - 2x}$

21. $\int \dfrac{e^{\sqrt{x}}}{\sqrt{x}}\,dx$

22. $\int \dfrac{dx}{(1 + x)^{3/2} + (1 + x)^{1/2}}$

23. $\int (e^x + 1)^3 e^x\,dx$

24. $\int \dfrac{x\,dx}{x^2 - 1}$

25. $\int \dfrac{(\log x)^2\,dx}{x}$

26. $\int \dfrac{x\,dx}{1 + x^4}$

27. $\int \dfrac{\arctan t}{1 + t^2}\,dt$

28. $\int \dfrac{dx}{x^{1/2}(1 + x)}$

29. $\int \dfrac{\sin \sqrt{x}}{\sqrt{x}}\,dx$

30. $\int \dfrac{\cos(\log x)\,dx}{x}$

31. $\int \dfrac{\cos t}{\sqrt{1 - \sin t}}\,dt$

32. $\int \dfrac{x^2}{\sqrt{1 - x}}\,dx$

33. $\int \sqrt{x^4 + x^2}\,dx$

34. $\int \dfrac{e^{2t}\,dt}{1 + e^{2t}}$

35. $\int \dfrac{e^t}{1 + e^{2t}}\,dt$

36. $\int \tan \theta \sec^2 \theta\,d\theta$

37. $\int \tan^n \theta \sec^2 \theta\,d\theta$

38. $\int \sec^n \theta \tan \theta\,d\theta$

39. $\int \sin^n \theta \cos \theta\,d\theta$

40. $\int \dfrac{\sin t\,dt}{(a + b \cos t)^n}$

2. RATIONAL FUNCTIONS

In this section and the next, we shall look at two general situations in which an integrand can be algebraically manipulated into a form that can be integrated.

First, we look at some simple rational functions. The basic device employed here—called a *partial fraction decomposition*—can be used to integrate any rational function whatsoever. We shall discuss this systematic process further in Section 6.

EXAMPLE 1

$$\int \frac{x\,dx}{(x-1)(x+3)}.$$

The crucial fact is that there are constants a and b such that

$$\frac{x}{(x-1)(x+3)} = \frac{a}{x-1} + \frac{b}{x+3}.$$

In order to determine these constants, first cross-multiply on the right:

$$\frac{x}{(x-1)(x+3)} = \frac{a(x+3) + b(x-1)}{(x-1)(x+3)}.$$

Then the two numerators have to be equal:

$$x = (a+b)x + (3a-b).$$

But two polynomials are equal only if their corresponding coefficients are equal, so

$$1 = a+b, \qquad 0 = 3a - b.$$

Solving these equations simultaneously, we find that $a = 1/4$ and $b = 3/4$. Therefore,

$$\int \frac{x\,dx}{(x-1)(x+3)} = \frac{1}{4}\int \frac{dx}{x-1} + \frac{3}{4}\int \frac{dx}{x+3}$$

$$= \frac{1}{4}\log|x-1| + \frac{3}{4}\log|x+3| + c$$

$$= \log|x-1|^{1/4}|x+3|^{3/4} + c.$$

EXAMPLE 2. In order to integrate

$$\int \frac{x\,dx}{(x-1)^2}$$

we have to modify the above scheme slightly. This time we look for constants a and b such that

$$\frac{x}{(x-1)^2} = \frac{a}{(x-1)} + \frac{b}{(x-1)^2}.$$

Proceeding as in the first example we find that $a = b = 1$, so

$$\int \frac{x\,dx}{(x-1)^2} = \int \frac{dx}{(x-1)} + \int \frac{dx}{(x-1)^2}$$

$$= \log|x-1| - \frac{1}{(x-1)} + c.$$

An alternative procedure here is to write $x = (x-1) + 1$ in the numerator:

$$\frac{x}{(x-1)^2} = \frac{(x-1)+1}{(x-1)^2} = \frac{1}{(x-1)} + \frac{1}{(x-1)^2}.$$

EXAMPLE 3. In order for the above device to work, the integrand must be a *proper* rational function; that is, the degree of its numerator must be less than the degree of its denominator. If it isn't proper, we use long division to reduce to the proper case. For example, we find by long division that

$$\frac{x^3 - x + 2}{x^2 - x} = x + 1 + \frac{2}{x^2 - x}.$$

Therefore,

$$\int \frac{x^3 - x + 2}{x^2 - x}\,dx = \frac{x^2}{2} + x + 2 \int \frac{dx}{x(x-1)},$$

and the latter integrand is then computed as above. Thus

$$\frac{1}{x(x-1)} = -\frac{1}{x} + \frac{1}{(x-1)},$$

and the final answer is

$$\frac{x^2}{2} + x + 2 \log\left|\frac{x-1}{x}\right| + c = \frac{x^2}{2} + x + \log\left(1 - \frac{1}{x}\right)^2 + c.$$

EXAMPLE 4

$$\int \frac{x\,dx}{x^2 + 2x - 3} = \int \frac{x\,dx}{(x-1)(x+3)},$$

which is the same as the first example.

EXAMPLE 5

$$\int \frac{dx}{x^2 + 2x + 3}.$$

Here we can't factor the denominator, and in fact this is an arctangent integral rather than a logarithm. In order to see what is going on, we complete the square in the denominator, and have

$$\int \frac{dx}{x^2 + 2x + 3} = \int \frac{dx}{(x + 1)^2 + 2} = \frac{1}{\sqrt{2}} \int \frac{du}{u^2 + 1}$$

(where we have set $x + 1 = \sqrt{2}u$, $dx = \sqrt{2}\,du$). This is the arctan integral

$$\frac{1}{\sqrt{2}} \arctan u + c = \frac{1}{\sqrt{2}} \arctan\left(\frac{x + 1}{\sqrt{2}}\right) + c.$$

Note that if the numerator in the above example were $(2x + 2)$, then the answer would be $\log|x^2 + 2x + 3|$, by simple substitution. If the numerator were $2x$, we would rewrite it as $(2x + 2) - 2$, and so have a combination of a logarithm and arctangent.

The device of completing the square could be used even for a quadratic that can be factored, such as in Example 4:

$$x^2 + 2x - 3 = (x + 1)^2 - 4$$
$$= [(x + 1) - 2][(x + 1) + 2]$$
$$= (x - 1)(x + 3).$$

EXAMPLE 6. We have just seen that there are three distinct possibilities for a proper rational function with a quadratic denominator, depending on how the denominator factors. If the degree of the denominator is three, then there will be three unknown coefficients to determine in the partial-fraction decomposition, and we consider three different possible setups for them.

1. *Three different linear factors:*

$$\frac{1}{(x - 1)(x + 2)(x + 3)} = \frac{a}{x - 1} + \frac{b}{x + 2} + \frac{c}{x + 3}.$$

2. *Two different linear factors,* but *one repeated:*

$$\frac{1}{(x - 1)^2(x + 3)} = \frac{a}{(x - 1)} + \frac{b}{(x - 1)^2} + \frac{c}{(x + 3)}.$$

3. *One linear factor* and *one irreducible quadratic factor:*

$$\frac{1}{(x - 1)(x^2 + 1)} = \frac{a}{(x - 1)} + \frac{bx + c}{(x^2 + 1)}.$$

The coefficients are computed in the same way as before. Thus (3) goes on:

$$\frac{1}{(x-1)(x^2+1)} = \frac{a(x^2+1) + (bx+c)(x-1)}{(x-1)(x^2+1)};$$

$$1 = (a+b)x^2 + (c-b)x + (a-c);$$

$$
\left.\begin{aligned}
a+b &= 0 \\
c-b &= 0 \\
a-c &= 1
\end{aligned}\right\}
\qquad
\begin{aligned}
a &= \frac{1}{2}, \\
b &= -\frac{1}{2}, \\
c &= -\frac{1}{2}.
\end{aligned}
$$

$$\int \frac{dx}{(x-1)(x^2+1)} = \frac{1}{2}\int \frac{dx}{(x-1)} - \frac{1}{2}\int \frac{x\,dx}{x^2+1} - \frac{1}{2}\int \frac{dx}{x^2+1}$$

$$= \frac{1}{2}\log|x-1| - \frac{1}{4}\log(x^2+1) - \frac{1}{2}\arctan x + c$$

$$= \frac{1}{2}\left[\log \frac{|x-1|}{\sqrt{x^2+1}} - \arctan x\right] + c.$$

In each of the above three cases, the numerator 1 can be replaced by any polynomial of degree less than 3 (the quotient must be proper) without changing the procedure. For example, if we set

$$\frac{x^2+2}{(x-1)(x^2+1)} = \frac{a}{x-1} + \frac{bx+c}{x^2+1},$$

we just change the final set of three equations in the three unknowns a, b, and c. When we equate numerators this time, we have

$$x^2 + 2 = (a+b)x^2 + (c-b)x + (a-c),$$

so

$$
\left.\begin{aligned}
a+b &= 1 \\
c-b &= 0 \\
a-c &= 2
\end{aligned}\right\}
\qquad
\begin{aligned}
a &= \frac{3}{2}. \\
b &= -\frac{1}{2}, \\
c &= -\frac{1}{2}.
\end{aligned}
$$

The general rational function is treated in essentially the same way, but the complications can be greater. Further discussion will be postponed to Section 6.

PROBLEMS FOR SECTION 2

Integrate

1. $\displaystyle\int \frac{dx}{x^2 - 9}$

2. $\displaystyle\int \frac{dx}{x^2 + 9}$

3. $\displaystyle\int \frac{dx}{x^2 - 9x}$

4. $\displaystyle\int \frac{dx}{x^2 + x - 2}$

5. $\displaystyle\int \frac{dx}{x^2 + 2x - 2}$

6. $\displaystyle\int \frac{dx}{x^2 + 2x + 2}$

7. $\displaystyle\int \frac{dx}{x^2 + 3x + 2}$

8. $\displaystyle\int \frac{dx}{x^2 + 2x + 1}$

9. $\displaystyle\int \frac{dx}{4x^2 + 9}$

10. $\displaystyle\int \frac{dx}{2x^2 + x - 3}$

11. $\displaystyle\int \frac{x\,dx}{x^2 + 4x - 5}$

12. $\displaystyle\int \frac{4\,dx}{x^3 + 4x}$

13. $\displaystyle\int \frac{x^2 dx}{x^2 + 2x + 1}$

14. $\displaystyle\int \frac{y^2 + 1}{y^2 + y + 1}\,dy$

15. $\displaystyle\int \frac{(x + 1)\,dx}{(x^2 + 4x - 5)}$

16. $\displaystyle\int \frac{2x^3 + x + 3}{(x^2 + 1)}\,dx$

17. $\displaystyle\int \frac{2x^2 + x + 2}{x(x - 1)^2}\,dx$

18. $\displaystyle\int \frac{dx}{x^3 + 8}$

19. $\displaystyle\int \frac{\cos\theta\,d\theta}{1 - \sin^2\theta}$

20. $\displaystyle\int \frac{\sin\theta\,d\theta}{\cos^2\theta + \cos\theta - 2}$

21. $\displaystyle\int \frac{e^t\,dt}{e^{2t} + 3e^t + 2}$

22. $\displaystyle\int \frac{dx}{x^3 + x^2 + 5x}$

23. $\displaystyle\int \frac{dx}{x^3 - 4x}$

24. $\displaystyle\int \frac{dx}{x^3 - 4x^2}$

25. $\displaystyle\int \frac{x^2 + 1}{x^3 - 4x^2}\,dx$

26. $\displaystyle\int \frac{dx}{x^{1/2}(x - 1)}$

27. $\displaystyle\int \frac{dx}{x^{1/2}(x^{1/3} + 1)}$

3. TRIGONOMETRIC POLYNOMIALS

Trigonometric polynomials can also be integrated in a systematic way. The new integrating device will again be algebraic manipulation of the integrand, this time by using the trigonometric identities:

$$\cos^2 x + \sin^2 x = 1$$

$$\cos^2 x - \sin^2 x = \cos 2x \qquad \text{(From the addition law for } \cos(x + x)\text{.)}$$

Adding and subtracting these two laws gives the "power-reducing" formulas:

$$\cos^2 x = \frac{1 + \cos 2x}{2} \qquad \text{and} \qquad \sin^2 x = \frac{1 - \cos 2x}{2}.$$

A trigonometric polynomial is a sum of products of trigonometric functions and, by using the definitions of these functions in terms of sine and cosine, any such product can be reduced to the form

$$\sin^m x \cos^n x,$$

where m and n are integers (positive, zero, or negative). The problem is therefore to integrate any such reduced product.

If m and n are *even* positive integers, we use the power-reducing formulas.

EXAMPLE 1

$$\int \cos^2 x \, dx = \int \frac{(1 + \cos 2x)}{2} dx = \frac{x}{2} + \frac{\sin 2x}{4} + c.$$

Since $\sin 2x = 2 \sin x \cos x$ (by the addition law for $\sin(x + x)$), this answer can be rewritten

$$\frac{x + \sin x \cos x}{2} + c.$$

EXAMPLE 2. Two applications of the power-reducing identity yield

$$\cos^4 x = \left(\frac{1 + \cos 2x}{2}\right)^2 = \frac{1}{4} + \frac{1}{2} \cos 2x + \frac{\cos^2 2x}{4}$$

$$= \frac{1}{4} + \frac{\cos 2x}{2} + \frac{1}{4}\left(\frac{1 + \cos 4x}{2}\right) = \frac{3}{8} + \frac{\cos 2x}{2} + \frac{\cos 4x}{8}.$$

Therefore

$$\int \cos^4 x \, dx = \frac{3x}{8} + \frac{\sin 2x}{4} + \frac{\sin 4x}{32} + c.$$

EXAMPLE 3. The above two examples combine to give

$$\int \sin^2 x \cos^2 x \, dx = \int (1 - \cos^2 x)\cos^2 x \, dx$$

$$= \int \cos^2 x \, dx - \int \cos^4 x \, dx$$

$$= \frac{x}{2} + \frac{\sin 2x}{4} - \frac{3x}{8} - \frac{\sin 2x}{4} - \frac{\sin 4x}{32} + c$$

$$= \frac{x}{8} - \frac{\sin 4x}{32} + c.$$

In this way we can integrate any product $\sin^m x \cos^n x$, where m and n are *even positive integers*.

If either exponent is *odd*, then the integration can be performed by simple substitution, by the following device:

$$\cos^{2p+1} x \, dx = \cos^{2p} x \cos x \, dx = (1 - \sin^2 x)^p \cos x \, dx$$

$$= (1 - u^2)^p \, du \qquad\qquad (u = \sin x).$$

EXAMPLE 4

$$\int \cos^3 x \, dx = \int (1 - \sin^2 x) \cos x \, dx = \int \cos x \, dx - \int \sin^2 x \cos x \, dx$$

$$= \sin x - \frac{\sin^3 x}{3} + c.$$

The same idea lets us integrate any product $\sin^m x \cos^n x$ where m and n are integers and at least one is odd. But when the odd integer is *negative*, the final integrand is a rational function, as in the following example.

EXAMPLE 5

$$\int \frac{dx}{\cos x} = \int \frac{\cos x \, dx}{\cos^2 x} \qquad \left(\begin{matrix} \text{Set } u = \sin x, \\ du = \cos x \, dx \end{matrix} \right)$$

$$= \int \frac{du}{1 - u^2} = \frac{1}{2} \int \left[\frac{1}{1 - u} + \frac{1}{1 + u} \right] du$$

$$= \frac{1}{2} \log \left[\frac{1 + u}{1 - u} \right] + c = \log \left[\frac{1 + \sin x}{1 - \sin x} \right]^{1/2} + c.$$

This answer can be simplified by multiplying top and bottom inside the brackets by $(1 + \sin x)$. It becomes

$$\log \left[\frac{1 + \sin x}{|\cos x|} \right] + c = \log |\sec x + \tan x| + c.$$

EXAMPLE 6

$$\int \frac{dx}{\cos^5 x} = \int \frac{\cos x \, dx}{\cos^6 x} = \int \frac{du}{(1 - u^2)^3}.$$

Here the partial-fraction decomposition is

$$\frac{1}{(1 - u)^3 (1 + u)^3} = \frac{a}{1 - u} + \frac{b}{(1 - u)^2} + \frac{c}{(1 - u)^3}$$

$$+ \frac{d}{1 + u} + \frac{e}{(1 + u)^2} + \frac{f}{(1 + u)^3},$$

and a lot of algebra is required.

EXAMPLE 7. If there is a *positive* odd power, then the other exponent does not have to be an integer.

$$\int \sqrt{\sin x}\ \cos^3 x\, dx = \int \sqrt{\sin x}\,(1 - \sin^2 x)\cos x\, dx$$

$$= \int (u^{1/2} - u^{5/2})\, du = \frac{2}{3}u^{3/2} - \frac{2}{7}u^{7/2} + c$$

$$= \frac{2}{3}(\sin x)^{3/2} - \frac{2}{7}(\sin x)^{7/2} + c.$$

EXAMPLE 8

$$\int \frac{\sin x\, dx}{\cos^2 x} = -\int \frac{du}{u^2} = \frac{1}{u} + c = \frac{1}{\cos u} + c.$$

The above example shows how we easily recapture the formula

$$\int \tan x\ \sec x\, dx = \sec x + c$$

in case we have forgotten that

$$\frac{d}{dx}\sec x = \sec x \tan x.$$

But the formula $d \tan x / dx = \sec^2 x$ doesn't reappear in this way. As an integral in $\sin x$ and $\cos x$, it is

$$\int \frac{dx}{\cos^2 x} = \frac{\sin x}{\cos x} + c,$$

and this represents the one case of $\sin^m x \cos^n x$ that we haven't covered so far: both m and n even and at least one of them negative. In order to handle such integrals we add the formulas

$$\int \sec^2 x\, dx = \tan x + c, \qquad \int \csc^2 x\, dx = -\cot x + c$$

to our basic list, and we convert the identity $\sin^2 x + \cos^2 x = 1$ to the equivalent forms

$$\tan^2 x + 1 = \sec^2 x,$$

$$1 + \cot^2 x = \csc^2 x,$$

in order to apply these additional integrals.

EXAMPLE 9

$$\int \frac{dx}{\cos^4 x} = \int \sec^4 x\, dx = \int (1 + \tan^2 x)\sec^2 x\, dx$$

$$= \int (1 + u^2)\, du \qquad (u = \tan x)$$

$$= u + \frac{u^3}{3} + c = \tan x + \frac{\tan^3 x}{3} + c.$$

EXAMPLE 10

$$\int \tan^2 x \, dx = \int (\sec^2 x - 1) \, dx = \tan x - x + c.$$

EXAMPLE 11

$$\int \frac{dx}{\sin^2 x \cos^2 x} = \int \csc^2 x \sec^2 x \, dx = \int \csc^2 x (\tan^2 x + 1) \, dx$$

$$= \int (\sec^2 x + \csc^2 x) \, dx = \tan x - \cot x + c.$$

PROBLEMS FOR SECTION 3

Integrate the following.

1. $\int \sin^2 x \, dx$
2. $\int \sin^3 x \, dx$
3. $\int \sin^4 x \, dx$
4. $\int \cos^2 3x \, dx$
5. $\int \sin^3 6x \cos 6x \, dx$
6. $\int \cos^3 2\theta \sin 2\theta \, d\theta$
7. $\int \cos^3 x \sin^{-4} x \, dx$
8. $\int \frac{\sin^5 y \, dy}{\sqrt{\cos y}}$
9. $\int \cos^6 \theta \, d\theta$
10. $\int \cos^3(\frac{x}{2}) \sin^2(\frac{x}{2}) \, dx$
11. $\int \frac{\sin^3 2x}{\sqrt[3]{\cos 2x}} \, dx$
12. $\int \sin^3 x \cos^3 x \, dx$
13. $\int \frac{dx}{\sin x \cos x}$
14. $\int \sin 3x \cot 3x \, dx$
15. $\int \csc^4 x \cot^2 x \, dx$
16. $\int \tan^4 x \, dx$
17. $\int \csc^4 \frac{x}{4} \, dx$
18. $\int \sec^3 x \, dx$
19. $\int \tan^3 x \sec^{5/2} x \, dx$
20. $\int \tan x \sqrt{\sec x} \, dx$
21. $\int \tan^3 x \, dx$. Do this in two ways:
    a) By substitution, since $\sin x$ (and $\cos x$ also) occurs to an odd power;
    b) By using $\tan^2 x = \sec^2 x - 1$.
22. $\int \sec x \, dx$. Do this in two ways:
    a) By substitution, since $\cos x$ occurs to an odd power;
    b) By tricky substitution, multiplying top and bottom by $(\sec x + \tan x)$.
23. $\int \tan^5 x \, dx$ (Do it the easier way.)
24. $\int \frac{\cos^5 x}{\sin^5 x} \, dx$ (Ditto.)
25. $\int \tan^6 x \, dx$
26. $\int \sec^6 x \, dx$

27. In Example 5 the absolute-value sign was not introduced inside the logarithm until the very end. Why was this permissible?

4. INVERSE SUBSTITUTIONS

In making the direct substitutions of Section 1, we set $u = h(x)$ where $h(x)$ was a part of the integrand. But then we also had to find $du = h'(x)dx$ as part of the integrand, and this meant that altogether the integrand had to be in a rather special form.

A much easier way to change variables in the integral $\int f(x)\,dx$ is simply to set $x = k(u)$ and $dx = k'(u)du$, where $k(u)$ is some function that is suggested by the form of the integrand.

EXAMPLE 1. In order to integrate

$$\int \sqrt{4 - x^2}\,dx,$$

we try setting $x = 2 \sin u$, which will at least eliminate the radical. Then

$$dx = 2 \cos u\,du,$$

$$\sqrt{4 - x^2} = 2\sqrt{1 - \sin^2 u} = 2 \cos u,$$

$$\int \sqrt{4 - x^2}\,dx = 4 \int \cos^2 u\,du = 4 \int \frac{1 + \cos 2u}{2}\,du. \qquad \text{(as in Section 3)}$$

$$= 2u + \sin 2u + c$$

$$= 2 \sin u \cos u + 2u + c.$$

Now we have to write the answer in terms of the original variable x, which means that we have to know the inverse of the substitution function. In this case $u = \arcsin x/2$, so

$$\int \sqrt{4 - x^2}\,dx = \frac{1}{2}x\sqrt{4 - x^2} + 2 \arcsin x/2 + C$$

is the final answer. Note that $x = 2 \sin u$ must be restricted to the domain $[-\pi/2, \pi/2]$ to make it invertible.

This process is called *inverse substitution*. Instead of setting $h(x) = u$ as we did in Section 1, we now set $x = k(u)$. This makes for a very easy transformation of the integral, but we have to know an inverse function $u = h(x)$ in order to express the u answer in terms of the original variable x.

There are two other useful substitutions similar to the one used in Example 1:

1. *to rationalize* $\sqrt{a^2 + x^2}$, *set* $x = a \tan u$, *with* $a > 0$;

2. *to rationalize* $\sqrt{x^2 - a^2}$, *set* $x = a \sec u$.

EXAMPLE 2

$$\int \frac{dx}{\sqrt{a^2 + x^2}}.$$

We set $x = a \tan u$, $dx = a \sec^2 u \, du$. Then

$$\int \frac{dx}{\sqrt{a^2 + x^2}} = \int \sec u \, du = \int \frac{\cos u \, du}{\cos^2 u}$$

$$= \int \frac{dv}{1 - v^2} = \frac{1}{2} \int \left[\frac{1}{1 - v} + \frac{1}{1 + v} \right] dv$$

$$= \log \left[\frac{1 + v}{1 - v} \right]^{1/2} + c = \log \left[\frac{1 + \sin u}{1 - \sin u} \right]^{1/2} + c$$

$$= \log |\sec u + \tan u| + c \qquad \text{(as in Section 3).}$$

There remains the question: What is $\sec u$ when $\tan u = x/a$? We could write

$$\sec u = \sec(\text{arc} \tan x/a)$$

and have an answer. But if we label a right triangle in the simplest way possible consistent with $\tan u = x/a$, we see that when $\tan u = x/a$, then

$$\sec u = \sqrt{a^2 + x^2}/a.$$

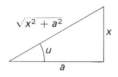

This could also be worked out directly,

$$\sec u = \sqrt{\sec^2 u} = \sqrt{1 + \tan^2 u} = \sqrt{1 + \left(\frac{x}{a} \right)^2},$$

but the right-triangle device is easier. In any case, the final answer is

$$\int \frac{dx}{\sqrt{a^2 + x^2}} = \log \left[\frac{\sqrt{a^2 + x^2} + x}{a} \right] + c = \log[\sqrt{a^2 + x^2} + x] + C.$$

EXAMPLE 3

$$\int \frac{dx}{\sqrt{x^2 - a^2}}.$$

Set $x = a \sec u$, $dx = a \sec u \tan u$. Then

$$\int \frac{dx}{\sqrt{x^2 - a^2}} = \int \frac{a \sec u \tan u \, du}{a \tan u} = \int \sec u \, du$$

$$= \log |\sec u + \tan u| + c$$

$$= \log \left| \frac{x + \sqrt{x^2 - a^2}}{a} \right| + c$$

$$= \log |x + \sqrt{x^2 - a^2}| + C.$$

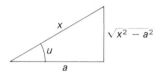

EXAMPLE 4. The same substitution and diagram is used here:

$$\int \frac{\sqrt{x^2 - a^2} \, dx}{x} = \int \frac{a^2 \sec u \tan^2 u}{a \sec u} \, du = a \int \tan^2 u \, du$$

$$= a \int (\sec^2 u - 1) \, du = a \tan u - au + c$$

$$= \sqrt{x^2 - a^2} - a \arctan \frac{\sqrt{x^2 - a^2}}{a} + c.$$

EXAMPLE 5

$$\int \sqrt{2x - x^2} \, dx.$$

We complete the square under the radical, and then proceed as in Example 1.

$$\int \sqrt{2x - x^2} \, dx = \int \sqrt{1 - (1 - x)^2} \, dx \qquad \left(\begin{array}{l} \text{set } 1 - x = \sin u \\ dx = -\cos u \, du \end{array} \right)$$

$$= -\int \cos^2 u \, du = -\frac{u + \sin u \cos u}{2} + c$$

$$= -\frac{\arcsin(1 - x) + (1 - x)\sqrt{2x - x^2}}{2} + c.$$

The computations above have an important feature that we have not remarked on. It was careless to write

$$\cos u = \sqrt{1 - \sin^2 u}$$

without checking signs, because cos u is negative as often as it is positive. However, $u = \arcsin x$ here, so u is restricted to the interval $[-\pi/2, \pi/2]$, and cos u is positive, as was assumed.

Similarly, sec u is positive over the interval $(-\pi/2, \pi/2)$. Since this is the range of the inverse function $u = \arctan x$, it is correct to write

$$\sqrt{1 + x^2} = \sqrt{1 + \tan^2 u} = \sqrt{\sec^2 u} = \sec u.$$

However,

$$\sqrt{x^2 - 1} = \sqrt{\sec^2 u - 1} = \sqrt{\tan^2 u} = \tan u$$

is correct only when u is in $[0, \pi/2)$, i.e., when $x = \sec u$ is ≥ 1. This restriction was therefore implicit in Examples 3 and 4. However, see Problems 38 and 39.

Here are a few more rationalizing substitutions.

EXAMPLE 6

$$\int \frac{x\,dx}{(1 + x)^{1/3}}.$$

We would like to set $u = (1 + x)^{1/3}$ in order to eliminate the radical. Fortunately, we can easily solve the substitution equation for x in terms of u,

$$x = u^3 - 1,$$

$$dx = 3u^2\,du,$$

so

$$\int \frac{x\,dx}{(1 + x)^{1/3}} = \int \frac{(u^3 - 1)3u^2\,du}{u} = 3\int (u^4 - u)\,du$$

$$= 3\left(\frac{u^5}{5} - \frac{u^2}{2}\right) + c = \frac{3u^2}{10}(2u^3 - 5) + c$$

$$= \frac{3}{10}(1 + x)^{2/3}(2x - 3) + c.$$

EXAMPLE 7

$$\int \frac{dx}{x^{1/2}(1 + x^{1/3})} = \int \frac{5u^5\,du}{u^3(1 + u^2)} \qquad \left(\begin{matrix} x = u^6, \\ dx = 5u^5\,du \end{matrix}\right)$$

$$= 5\int \frac{u^2\,du}{1 + u^2} = 5\int \left[1 - \frac{1}{1 + u^2}\right]du$$

$$= 5[u - \arctan u] + c$$

$$= 5[x^{1/6} - \arctan x^{1/6}] + c.$$

EXAMPLE 8. We can do Example 4 in this way, too.

$$\int \frac{\sqrt{x^2 - a^2}\,dx}{x} = \int \frac{\sqrt{x^2 - a^2}\,x\,dx}{x^2} \qquad \left(\begin{array}{l} \text{set } u^2 = x^2 - a^2, \\ u\,du = x\,dx \end{array}\right)$$

$$= \int \frac{u \cdot u\,du}{u^2 + a^2}$$

$$= \int \left[1 - \frac{a^2}{u^2 + a^2}\right]\,du$$

$$= u - a \arctan \frac{u}{a} + c$$

$$= \sqrt{x^2 - a^2} - a \arctan \frac{\sqrt{x^2 - a^2}}{a} + c.$$

This device can be used whenever one of the radicals

$$\sqrt{x^2 - a^2}, \qquad \sqrt{a^2 - x^2}, \qquad \sqrt{a^2 + x^2}$$

is combined with an *odd* power of x.

We have not yet justified the inverse substitution process. Here is a general description of what we have been doing. Setting $x = k(u)$ and $dx = k'(u)du$, we write

$$\int f(x)\,dx = \int f(k(u))k'(u)\,du = \int g(u)\,du,$$

where $g(u) = f(k(u))k'(u)$. We now find an antiderivative $G(u)$ of $g(u)$, and claim that then

$$\int f(x)\,dx = G(h(x)) + c,$$

where h is the inverse of the function k.

In order to see that this is correct, remember that in the direct substitution process of Section 1, we used the same integral transformation in the other direction to find an antiderivative of g. We showed that if F is any antiderivative of f then $F(k(u))$ is an antiderivative of $f(k(u))k'(u) = g(u)$. Thus, in the present context, where G is also an antiderivative of g, we must have

$$F(k(u)) = G(u) + c.$$

Setting $u = h(x)$, and remembering that $k(h(x)) = x$, we see that

$$F(x) = G(h(x)) + c.$$

That is, $G(h(x))$ is an antiderivative of f, as we claimed above.

PROBLEMS FOR SECTION 4

In the following integrations a labeled right triangle will often be helpful when substituting back, at the end.

1. $\displaystyle\int \frac{x^3\,dx}{\sqrt{4-x^2}}$

2. $\displaystyle\int \frac{dx}{x\sqrt{x^2-a^2}}$

3. $\displaystyle\int \frac{dx}{(x^2+a^2)^{3/2}}$

4. $\displaystyle\int \frac{dx}{(x^2-9)^{3/2}}$

5. $\displaystyle\int x\sqrt{16-x^2}\,dx$

6. $\displaystyle\int \frac{x\,dx}{\sqrt{(x^2+a^2)^3}}$

7. $\displaystyle\int \frac{dx}{x\sqrt{x^2+1}}$

8. $\displaystyle\int \frac{\sqrt{9-4x^2}}{x}\,dx$

9. $\displaystyle\int \frac{dx}{x\sqrt{25-x^2}}$

10. $\displaystyle\int \frac{dy}{y^2\sqrt{y^2-7}}$

11. $\displaystyle\int \frac{dx}{x\sqrt{9+4x^2}}$

12. $\displaystyle\int \frac{\sqrt{16-t^2}}{t^2}$

13. $\displaystyle\int \frac{x^2\,dx}{(a^2-x^2)^{3/2}}$

14. $\displaystyle\int \frac{x^2\,dx}{\sqrt{2x-x^2}}$

15. $\displaystyle\int \frac{dx}{(x^2-2x-3)^{3/2}}$

The following problems would normally be tackled in a different way, but this time do them using the trigonometric rationalizing substitutions.

16. $\displaystyle\int \frac{dy}{\sqrt{a^2-y^2}}$

17. $\displaystyle\int \frac{x\,dx}{\sqrt{1-x^2}}$

18. $\displaystyle\int \frac{dx}{x^2+4}$

19. $\displaystyle\int \frac{x\,dx}{\sqrt{1+x^2}}$

20. $\displaystyle\int \frac{dx}{x^2-a^2}$

21. $\displaystyle\int \frac{x\,dx}{\sqrt{x^2-a^2}}$

22. $\displaystyle\int \frac{dx}{(x^2+a^2)^2}$

Integrate by finding a suitable substitution.

23. $\displaystyle\int \frac{dx}{1+\sqrt{x}}$

24. $\displaystyle\int \frac{dx}{1+x^{1/3}}$

25. $\displaystyle\int \frac{1+x^{1/2}}{1+x^{1/3}}\,dx$

26. $\displaystyle\int \frac{1+x}{\sqrt{1-x}}\,dx$

27. $\displaystyle\int \frac{\sqrt{1-x}}{1+x}\,dx$

28. $\displaystyle\int \frac{\cos t}{2-\sin^2 t}\,dt$

29. $\displaystyle\int \frac{\sin t\,dt}{1+\cos^2 t}$

30. $\displaystyle\int \sqrt{\frac{1-x}{1+x}}\,dx$

31. $\displaystyle\int \frac{x^3\,dx}{a^2-x^2}$

32. $\displaystyle\int \frac{x^3\,dx}{\sqrt{a^2-x^2}}$

33. $\displaystyle\int \frac{\sqrt{x}}{1+x}\,dx$

34. $\displaystyle\int \frac{x^{1/3}}{1+x}\,dx$

35. Show that a right triangle as labeled in the text for reading off the values of the trigonometric functions when $x = a\tan u$ will give the right signs for these functions when x is negative $(-\pi/2 < u < 0)$.

36. Show that a right triangle labeled to read off the values of the trigonometric functions when $x = a\sin u$ will give the right signs of these functions when x is negative $(-\pi/2 \le u < 0)$.

37. The right triangle labeled to show the substitution $x = a\sec u$ does not give the correct signs for the other trigonometric functions when x is negative $(\pi/2 < u < \pi)$. Show that the signs come out right for x negative if one leg is labeled $-\sqrt{x^2 - a^2}$.

38. a) Example 4 in the text was worked under the assumption that x is positive. Show that the answer is also correct for x negative.
    b) Work Example 4 out again for the case of negative x. [*Hint.* The proper procedure is related to the conclusion in Problem 37.]

39. a) Example 3 in the text was worked out under the assumption that x is positive. Show by differentiation that the conclusion is correct for negative x.
    b) Work Example 3 out again for the case that x is negative.

5. INTEGRATION BY PARTS

We make the product rule

$$\frac{d}{dx}uv = u\frac{dv}{dx} + v\frac{du}{dx}$$

into an integration rule by solving for the term $u\,dv/dx$ and integrating:

$$\int u\frac{dv}{dx}\,dx = uv - \int v\frac{du}{dx}\,dx.$$

Although the indefinite integral of $d(uv)/dx$ is $uv + c$, we have suppressed the constant c since a constant of integration is contained implicitly in the remaining integrals.

The above formula states a very special integration procedure called *integration by parts*. We can try this device whenever an integrand is a product of two functions, *provided that we already know how to integrate one of them*. For example

$$\int x \cos x \, dx$$

can be considered to be in the form

$$\int u \frac{dv}{dx} dx$$

with $u = x$, $dv/dx = \cos x$. We can then use the formula because we know how to integrate $dv/dx = \cos x$ to find v. Any antiderivative is suitable, and we generally (but not always) choose the simplest. So here we set $v = \sin x$. Since $du/dx = 1$, the formula gives

$$\int x \cos x \, dx = x \sin x - \int \sin x \cdot 1 \cdot dx$$
$$= x \sin x + \cos x + c.$$

In terms of differentials, the integration-by-parts formula takes the neater form

$$\int u \, dv = uv - \int v \, du.$$

For instance, in the above example we have $x = u$ and $\cos x \, dx = dv$, from which we can conclude that $du = dx$. and $v = \sin x$

In addition to the requirement that we be able to integrate one factor, we also need to have the resulting new integral simpler than the original, or at least more amenable. This doesn't necessarily happen. For example, if we had reversed the roles of x and $\sin x$ in the above example, we would get

$$\int (\sin x) x \cdot dx = (\sin x) \frac{x^2}{2} - \int \frac{x^2}{2} \cos x \, dx,$$

which has made things worse rather than better. Thus, when we know how to integrate *both* factors of the integrand, and accordingly have a choice as to how to apply the integration-by-parts formula, we can expect that only one of the possibilities will be useful (and maybe neither will).

As we have said, integration by parts is a very special device, but sometimes it works when nothing else at all will.

EXAMPLE 1

$$\int \log x \, dx = x \log x - \int x \frac{dx}{x}$$

$$= x \log x - \int dx$$

$$= x \log x - x + c.$$

EXAMPLE 2. In Section 3 we had the most trouble with even powers of sin x and cos x. These can successfully be tackled by parts.

$$\int \cos^2 x \, dx = \int \cos\ x \cdot \cos x \, dx$$

$$= \cos x \sin x - \int \sin\ x(-\sin\ x \, dx)$$

$$= \cos x \sin x + \int \sin^2 x \, dx$$

$$= \cos x \sin x + \int (1 - \cos^2 x) \, dx$$

$$= \cos x \sin x + x + c - \int \cos^2 x \, dx.$$

Solving for $\int \cos^2 x \, dx$, we have

$$\int \cos^2 x \, dx = \frac{\cos x \sin x + x}{2} + c.$$

EXAMPLE 3

$$\int \tan^2 x \, dx = \int \frac{\sin^2 x}{\cos^2 x} \, dx = \int \sin\ x \cdot \frac{\sin x \, dx}{\cos^2 x}$$

$$= \sin\ x\left(\frac{1}{\cos x}\right) - \int \cos\ x\left(\frac{1}{\cos x}\right) dx$$

$$= \tan x - x + c.$$

EXAMPLE 4. In order to integrate $\int x^2 \cos x \, dx$ we have to apply the integration-by-parts process twice.

$$\int x^2 \cos x \, dx = x^2 \sin x - \int \sin x \, 2x \, dx$$

$$\int x \sin x \, dx = x(-\cos x) - \int (-\cos x) \, dx$$

$$= -x \cos x + \sin x + c;$$

$$\int x^2 \cos x \, dx = x^2 \sin x - 2[-x \cos x + \sin x + c]$$

$$= x^2 \sin x + 2x \cos x - 2 \sin x + c.$$

Note that we have replaced $2c$ by c. We should probably use something like c' or c_1, but we say to ourselves that $2c$ is just an arbitrary constant, and so relabel it and call it c.

EXAMPLE 5. Higher even powers of $\sin x$ and $\cos x$ can be approached as in Example 2, but more than one step will be needed.

$$\int \cos^4 x \, dx = \int \cos^3 x \cos x \, dx$$

$$= \cos^3 x \sin x - \int \sin x \cdot 3 \cos^2 x (-\sin x) \, dx$$

$$= \cos^3 x \sin x + \int 3 \cos^2 x \sin^2 x \, dx$$

$$= \cos^3 x \sin x + \int 3 \cos^2 x (1 - \cos^2 x) \, dx$$

$$= \cos^3 x \sin x - 3 \int \cos^4 x \, dx + 3 \int \cos^2 x \, dx.$$

Solving in this equation for $\int \cos^4 x \, dx$, we get

$$\int \cos^4 x \, dx = \frac{\cos^3 x \sin x}{4} + \frac{3}{4} \int \cos^2 x \, dx,$$

which reduces to the integral in Example 2. Similarly, $\int \cos^6 x \, dx$ would involve two such reducing steps.

EXAMPLE 6. In this trigonometric context, when the odd power occurred in the denominator we got involved with rational functions, possibly quite complicated ones. Here, also, integration by parts may lead to a reducing step. For example,

$$\int \frac{dx}{\cos^5 x} = \int \sec^5 x \, dx = \int \sec^3 x \cdot \sec^2 x \, dx$$

$$= \sec^3 x \tan x - \int \tan x \cdot 3 \sec^2 x \cdot \sec x \tan x \, dx$$

$$= \sec^3 x \tan x - 3 \int \sec^5 x \, dx + 3 \int \sec^3 x \, dx.$$

Therefore

$$\int \sec^5 x \, dx = \frac{1}{4} \sec^3 x \tan x + \frac{3}{4} \int \sec^3 x \, dx.$$

PROBLEMS FOR SECTION 5

Integrate.

1. $\int x \sin x \, dx$

2. $\int e^x \cos x \, dx$

3. $\int x\sqrt{x + 1} \, dx$

4. $\int x \log x \, dx$

5. $\int y^2 \sin y \, dy$

6. $\int x e^{ax} \, dx$

7. $\int x^2 e^{-x} \, dx$

8. $\int x \cos ax \, dx$

9. $\int \sin^2 x \, dx$

10. $\int \arctan x \, dx$

11. $\int \arcsin x \, dx$

12. $\int x^n \log x \, dx$

13. $\int \frac{\log x \, dx}{x}$

14. $\int \sec^3 x \, dx$

15. $\int \sin \sqrt{x} \, dx$

16. $\int \sin(\log x) \, dx$

17. $\int \log(1 + x^2) \, dx$

18. $\int \frac{x e^x \, dx}{(1 + x)^2}$

19. $\int x \arctan x \, dx$

20. $x \arcsin x \, dx$

21. $\int \frac{\log x}{(x + 1)^2} \, dx$

22. $\int x^3 \sin x \, dx$

23. $\int \arctan \sqrt{x} \, dx$

24. $\int x^2 \log(x + 1) \, dx$

25. $\int e^{ax} \cos bx \, dx$

26. $\int (\log x)^2 \, dx$

In the following examples find degree-reducing formulas analogous to the special cases worked out in Examples 5 and 6.

27. $\int \sec^n x \, dx$

28. $\int \cos^n x \, dx$

29. $\int (\log x)^n \, dx$

6. RATIONAL FUNCTIONS (CONCLUDED)

Partial-fraction decompositions depend on a lemma from polynomial algebra that we shall assume.

Lemma. *If $r(x) = p(x)/q(x)$ is a proper rational function, and if its denominator $q(x)$ can be written in the form*

$$q(x) = q_1(x)q_2(x),$$

where $q_1(x)$ and $q_2(x)$ are polynomials that have no factor in common (i.e., are relatively prime), then $r(x)$ can be written as the sum of two proper rational functions having denominators $q_1(x)$ and $q_2(x)$, respectively:

$$\frac{p(x)}{q_1(x)q_2(x)} = \frac{p_1(x)}{q_1(x)} + \frac{p_2(x)}{q_2(x)}.$$

The lemma doesn't tell us how to find the new numerators $p_1(x)$ and $p_2(x)$, but this is easy once we know that the decomposition is possible. Consider, for example, the rational function $r(x) = x/(x - 1)(x + 3)$, where $q_1(x)$ and $q_2(x)$ are each of the form $x + c$.

A rational function with such a denominator will be proper only if its numerator is a constant, i.e., a polynomial of degree zero. The lemma therefore guarantees that there are constants a and b such that

$$\frac{x}{(x - 1)(x + 3)} = \frac{a}{x - 1} + \frac{b}{x + 3},$$

and we proceed as in Section 2.

Consider next

$$r(x) = \frac{x^2 - 3}{(x - 1)(x^2 + 1)}.$$

A rational function with $x^2 + 1$ in the denominator is proper only if its numerator has the form $bx + c$, this being the most general polynomial of degree less than 2. The lemma therefore guarantees constants a, b, and c, such that

$$\frac{x^2 - 3}{(x - 1)(x^2 + 1)} = \frac{a}{x - 1} + \frac{bx + c}{x^2 + 1},$$

again a situation that we worked out in Section 2.

If $q_2(x)$ in turn could be factored into relatively prime factors, then the lemma could be applied again to the rational function $p_2(x)/q_2(x)$. Proceeding in this way, the lemma extends to denominators having many relatively prime factors $q_1(x), q_2(x), \ldots, q_n(x)$, and shows that we can always find polynomials $p_1(x), \ldots, p_n(x)$ such that

$$\frac{p(x)}{q(x)} = \frac{p_1(x)}{q_1(x)} + \frac{p_2(x)}{q_2(x)} + \cdots + \frac{p_n(x)}{q_n(x)},$$

each quotient being a proper rational function.

For example, we know that there are constants a, b, and c, such that

$$\frac{x-2}{x(x+1)(x+2)} = \frac{a}{x} + \frac{b}{x+1} + \frac{c}{x+2},$$

another type considered in Section 2.

Determining the constants in the above examples falls under the general heading of methods involving *undetermined coefficients*. In the first example, the answer we are looking for is a *linear combination* of $1/(x-1)$ and $1/(x+3)$, i.e., a sum of constants times these two functions. We start off with a linear combination having undetermined coefficients a and b, and then determine the coefficients by algebra.

The second example also involves a linear combination having undetermined coefficients, this time a linear combination

$$a\left(\frac{1}{x-1}\right) + b\left(\frac{x}{x^2+1}\right) + c\left(\frac{1}{x^2+1}\right)$$

of the three functions

$$\frac{1}{x-1}, \quad \frac{x}{x^2+1}, \quad \frac{1}{x^2+1}.$$

If a denominator factor is repeated there is more to do. For example, the lemma guarantees constants a, b, and c, such that

$$\frac{3x^2+2}{x(x-1)^2} = \frac{a}{x} + \frac{bx+c}{(x-1)^2},$$

because x and $(x-1)^2$ are relatively prime. We thus end up with a linear combination of the three functions

$$\frac{1}{x}, \quad \frac{x}{(x-1)^2}, \quad \frac{1}{(x-1)^2}.$$

We can write down the integrals of the first and third of these functions, but not the second. However,

$$\frac{x}{(x-1)^2} = \frac{(x-1)+1}{(x-1)^2} = \frac{1}{x-1} + \frac{1}{(x-1)^2},$$

and now we can integrate.

Instead of resorting to this second change of form, it is customary to incorporate it into the original linear combination decomposition. That is, we look for a, b, and c, such that

$$\frac{3x^2+2}{x(x-1)^2} = \frac{a}{x} + \frac{b}{x-1} + \frac{c}{(x-1)^2},$$

again as in Section 2. Similarly, if $(x^2+1)^2$ is a denominator factor, we look for two terms

$$\frac{ax + b}{x^2 + 1} + \frac{cx + d}{(x^2 + 1)^2},$$

instead of one term

$$\frac{ax^3 + bx^2 + cx + d}{(x^2 + 1)^2}.$$

We then have a linear combination of the following four functions:

$$\frac{x}{x^2 + 1}, \qquad \frac{1}{x^2 + 1}, \qquad \frac{x}{(x^2 + 1)^2}, \qquad \frac{1}{(x^2 + 1)^2}.$$

The first three can be directly integrated, but the last falls into the one category of elementary partial fractions that we don't yet know how to integrate:

$$\int \frac{dx}{(x^2 + 1)^n},$$

where n is an integer ≥ 2.

What is involved is a step-by-step reduction process, each step requiring an integration by parts. The procedure runs parallel to the treatment of the even powers of cos x that we discussed at the end of the last section. As a matter of fact, the substitution $x = \tan \theta$ reduces the present problem to the former one. The direct attack is illustrated below.

EXAMPLE. We consider the case $n = 3$. Since

$$\int \frac{x\, dx}{(x^2 + 1)^3} = -\frac{1}{4} \frac{1}{(x^2 + 1)^2}$$

by simple substitution, we try to get this integrand as dv in the integration-by-parts formula. This can be done by writing

$$\frac{1}{(x^2 + 1)^3} = \frac{-x^2 + x^2 + 1}{(x^2 + 1)^3} = \frac{-x^2}{(x^2 + 1)^3} + \frac{1}{(x^2 + 1)^2}.$$

Then

$$\int \frac{dx}{(x^2 + 1)^3} = -\int x \cdot \frac{x\, dx}{(x^2 + 1)^3} + \int \frac{dx}{(x^2 + 1)^2}$$

$$= \frac{1}{4}\left[\frac{x}{(x^2 + 1)^2} - \int \frac{dx}{(x^2 + 1)^2}\right] + \int \frac{dx}{(x^2 + 1)^2}$$

$$= \frac{1}{4}\frac{x}{(x^2 + 1)^2} + \frac{3}{4}\int \frac{dx}{(x^2 + 1)^2}.$$

The power of $(x^2 + 1)$ has been reduced by 1. The next reduction gives arctan x.

There is a theorem, called the Fundamental Theorem of Algebra, which asserts that every polynomial with real coefficients can be expressed as a product of linear polynomials such as $x - 1$ and quadratic polynomials such as $x^2 + 1$ and $x^2 + 2x + 2$ which cannot be factored into linear polynomials with real coefficients. Therefore, in theory, every rational function has a denominator like those in the examples we have been working, except that the number of factors might be large. This is why, in theory, every rational function can be integrated.

In practice, finding the linear and irreducible quadratic factors of a polynomial may be very difficult. For example,

$$x^4 + 4 = (x^2 + 2x + 2)(x^2 - 2x + 2),$$

but this factorization is by no means obvious, and a more complicated fourth-degree polynomial would be even more difficult to factor.

PROBLEMS FOR SECTION 6

Integrate.

1. $\int \dfrac{dx}{x^4 - x^3}$

2. $\int \dfrac{(x^2 + 1)\,dx}{x^4 - x^3}$

3. $\int \dfrac{dx}{x^4 - x^2}$

4. $\int \dfrac{x^2 + x + 1}{(x^4 - x^2)}\,dx$

5. $\int \dfrac{dx}{x^4 + x^2}$

6. $\int \dfrac{x^3 + 1}{x^4 + x^2}\,dx$

7. $\int \dfrac{dx}{(1 - x^2)^2}$

8. $\int \dfrac{x^3 + x}{(1 - x^2)^2}\,dx$

9. $\int \dfrac{dx}{(x^2 - 1)(x^2 - 4)}$

10. $\int \dfrac{x^2 + 2}{(x^2 - 1)(x^2 - 4)}\,dx$

11. $\int \dfrac{dx}{(x^2 + 1)(x^2 + 4)}$

12. $\int \dfrac{x^2 + 2}{(x^2 + 1)(x^2 + 4)}\,dx$

13. a) Show that

$$\frac{x^3}{(x^2 + 1)^2} = \frac{ax + b}{(x^2 + 1)} + \frac{cx + d}{(x^2 + 1)^2}$$

for suitable constants a, b, c, and d.

b) Using the result in (a), compute

$$\int \frac{x^3}{(x^2 + 1)^2} dx.$$

14. Show that

$$\frac{A x^3 + B x^2 + C x + D}{(x^2 + 1)^2}$$

can always be written in the form

$$\frac{ax + b}{x^2 + 1} + \frac{cx + d}{(x^2 + 1)^2}.$$

15. Find the degree-reducing formula for

$$\int \frac{dx}{(x^2 + a^2)^n},$$

using integration-by-parts as in the example in the text.

16. Find the degree-reducing formula for

$$\int \frac{dx}{(x^2 + a^2)^n},$$

by using the substitution $x = a \tan \theta$.

7. SYSTEMATIC INTEGRATION

We shall now describe several other classes of functions which can be integrated systematically in closed form. The main point here is to be able to recognize these functions. The integration procedures themselves can get involved enough to lead us to look for shortcuts (or to use integral tables). Then we are back to ingenuity, but with the important difference that now we *know* the function has a closed-form integral that can be systematically calculated, and we are merely challenged to find a more direct route to it.

A *monomial* in two variables u and v is a product of a power of u times a power of v, such as $u^2 v^3$ and $u^{19} v^5$. A *polynomial* in u and v, $p(u, v)$, is a sum of constants times such monomials. That is, a polynomial in two variables is a linear combination of monomials in these two variables. A *rational function* $R(u, v)$ in u and v is a quotient $p(u, v)/q(u, v)$ of two polynomials in u and v. Finally, a rational function in $\sin x$ and $\cos x$, $R(\sin x, \cos x)$, is obtained by setting $u = \sin x$ and $v = \cos x$ in a rational function $R(u, v)$. For example, if

$$R(u, v) = \frac{uv^2}{u^3 - 1},$$

then

$$R(\sin x, \cos x) = \frac{\sin x \cos^2 x}{\cos^3 x - 1}.$$

EXAMPLE. The function $\tan^3 x - \sec x$ is of the form $R(\sin x, \cos x)$:

$$\tan^3 x - \sec x = \frac{\sin^3 x}{\cos^3 x} - \frac{1}{\cos x}$$

$$= \frac{\sin^3 x - \cos^2 x}{\cos^3 x} = R(\sin x, \cos x),$$

where $R(u, v) = (u^3 - v^2)/v^3$.

We shall use $R(u)$ and $R(u, v)$ for arbitrary rational functions in one and two variables. In this terminology, here are some classes of functions that can be systematically integrated in closed form:

a) $R(e^x)$ 　　　　　　　　 b) $R(\sin x, \cos x)$

c′) $R(x, \sqrt{1 - x^2})$ 　　　 d) $R(x, \sqrt{ax^2 + bc + c})$

c″) $R(x, \sqrt{1 + x^2})$ 　　　 e) $R(x, (ax + b)^{1/n})$

c‴) $R(x, \sqrt{x^2 - 1})$

EXAMPLES

1. $\dfrac{x^3}{\sqrt{1 - x^2}}$ is of type (c′), with $R(u, v) = \dfrac{u^3}{v}$.

2. $\dfrac{e^x + e^{-x}}{e^x - e^{-x}}$ is of type (a), with $R(u) = \dfrac{u + 1/u}{u - 1/u} = \dfrac{u^2 + 1}{u^2 - 1}$.

3. $\dfrac{x^2 - 2x + 1}{x\sqrt{x^2 + x + 1}}$ is of type (d), with $R(u, v) = \dfrac{u^2 - 2u + 1}{uv}$.

Finally, a word about integral tables. Because finding an antiderivative is such an uncertain business, often requiring an unrealistic combination of ingenuity and stamina to push a calculation through, it has been found worthwhile to gather together the answers that are reasonable and/or important into tables, where they are classified by the form of the integrand and can simply be looked up. You ought to spend a little time examining such a table, to see how the classification goes and what kinds of functions are treated. A short table of integrals is given in Appendix 3.

8. PROPERTIES OF THE DEFINITE INTEGRAL

The final solution to a problem involving integration is frequently expressed as a definite integral that must be evaluated, or at least estimated. In this connection, and also in using the definite integral as a theoretical tool, it is important to know its principal properties. Some of these properties are merely restatements in definite integral form of the rules we have been using to find antiderivatives, but one or two properties belong uniquely to the definite integral. We start with linearity:

I)
$$\int_a^b [cf(x) + dg(x)]\,dx = c\int_a^b f(x)\,dx + d\int_a^b g(x)\,dx.$$

This merely reflects the linearity of antidifferentiation. We know that if F and G are antiderivatives of f and g, respectively, then $cF + dG$ is an antiderivative of $cf + dg$. The above equation is thus simply the evaluation identity

$$cF(x) + dG(x)\Big]_a^b = c\Big[F(x)\Big]_a^b + d\Big[G(x)\Big]_a^b.$$

Next, there is the additivity of the integral with respect to changing limits of integration:

II)
$$\int_a^b f + \int_b^c f = \int_a^c f.$$

The hypothesis here is that the limits a, b, and c all belong to an interval I on which f is defined and at least continuous. Then f has an antiderivative F on I, and the equation above reduces to the identity

$$[F(b) - F(a)] + [F(c) - F(b)] = F(c) - F(a).$$

As a corollary of this property we note that

$$\int_a^a f = 0,$$

$$\int_a^b f = -\int_b^a f.$$

(We didn't make a point of it, but the definition of $\int_a^b f$ did not require a to be less than b. Of course, in the area interpretation we needed both $a < b$ and $f > 0$.)

Now consider substitution. When we are looking for an antiderivative $\int f(x)\,dx$, we can substitute $x = k(u)$, $dx = k'(u)du$, but it is necessary that k be invertible (over the x interval in question), because we have to substitute back the inverse function $u = h(x)$ to get the final answer. This was the gist of Section 4. But, in the computation of a definite integral $\int_a^b f$, the final step is unnecessary, and we can substitute with almost no worries at all.

Theorem 1. *If the integral $\int f(x)\,dx$ transforms into the integral $\int g(u)\,du$ under the substitution $x = k(u)$, $dx = k'(u)du$, then*

III) $$\int_a^b f(x)\,dx = \int_A^B g(u)\,du,$$

where $a = k(A)$ and $b = k(B)$.

Of course, we have to be able to find the new limits A and B, but this is frequently easier than finding the whole inverse function. In fact, the transformation function k isn't required to be invertible here.

EXAMPLE 1. In Section 4 we found that

$$\int \frac{dx}{\sqrt{a^2 + x^2}} = \int \sec u\,du = \int \frac{dv}{1 - v^2} = \log\left[\frac{1 + v}{1 - v}\right]^{1/2},$$

where the substitutions were

$$x = a\tan u, \qquad v = \sin u.$$

We then had to substitute back for the answer. However,

$$\int_0^a \frac{dx}{\sqrt{a^2 + x^2}} = \int_0^{\pi/4} \sec u\,du = \int_0^{\sqrt{2}/2} \frac{dv}{1 - v^2} = \log\left[\frac{1 + \sqrt{2}/2}{1 - \sqrt{2}/2}\right]^{1/2},$$

and we are done, except possibly for algebraic simplification. We have used the fact that $a\tan(\pi/4) = a$ and that $\sin(\pi/4) = \sqrt{2}/2$. The limits of integration here are obviously rather special ones in relation to the integrand, but such special limits occur frequently in applications.

Incidentally, the answer given above can be simplified by "rationalizing the denominator" to give, finally

$$\int_0^a \frac{dx}{\sqrt{a^2 + x^2}} = \log(\sqrt{2} + 1).$$

Proof of theorem. By hypothesis,

$$g(u) = f(k(u))k'(u).$$

If F is an antiderivative of f, then $F(k(u))$ is an antiderivative of $g(u)$, by the chain rule. (This is just *direct* substitution.) Therefore,

$$\int_A^B g = F(k(B)) - F(k(A)).$$

But if $a = k(A)$ and $b = k(B)$, then

$$F(k(B)) - F(k(A)) = F(b) - F(a) = \int_a^b f.$$

Therefore, $\int_a^b f = \int_A^B g$, as claimed. ∎

There are some implicit hypotheses in this situation. We would assume that f and g are both at least continuous in order for the integrals to exist, and then, in order to be sure that $g(u) = f(k(u))k'(u)$ is continuous between A and B, we would require $k'(u)$ to be continuous there. In particular, A and B must not straddle any singularity of the substitution function k.

EXAMPLE 2. If we were careless about this condition, we might set $x = 1/y$, $dx = -dy/y^2$, and conclude that

$$2 = \int_{-1}^1 dx = -\int_{-1}^1 \frac{dy}{y^2}.$$

This is nonsense, since $f(y) = 1/y^2$ has no antiderivative at all on $[-1, 1]$. In fact, the integral on the right must be $-\infty$ under any reasonable definition of its value, for the area under the graph of $1/y^2$ from $y = -1$ to $y = 1$ can be shown to be infinite.

The integration-by-parts formula, in its definite-integral garb, offers no surprises. It is:

IV) $$\int_a^b f(x)g'(x)\,dx = f(x)g(x)\Big]_a^b - \int_a^b f'(x)g(x)\,dx,$$

or

$$\int_a^b fg' = f(b)g(b) - f(a)g(a) - \int_a^b f'g.$$

The power of this formula may be surprising, though. In many advanced areas of mathematics it is a crucial tool.

In one typical maneuver, the "boundary" term $f(x)g(x)]_a^b$ is zero and the formula transforms $\int_a^b fg'$ into $-\int_a^b gf'$. For example, if $f(a) = f(b) = 0$, then $\int_a^b f(x)\,dx = -\int_a^b (x + c)f'(x)\,dx$. Then a second integration by parts gives a new integrand consisting of $f''(x)$ multiplied by a quadratic polynomial with leading term $x^2/2$. If the quadratic is $(x - a)(x - b)/2$, then again the boundary term is zero. Thus:

Lemma 1. *If f has a continuous second derivative over $[a, b]$, and if $f(a) = f(b) = 0$, then*

$$\int_a^b f(x)\,dx = \frac{1}{2}\int_a^b (x - a)(x - b)f''(x)\,dx.$$

Proof. Start with the right side and integrate by parts twice. The boundary term will be zero at each step. The details are left as an exercise. ∎

In exactly the same way, four successive applications of integration by parts result in the following lemma.

Lemma 2. *If f has a continuous fourth derivative over the interval $[a, b]$, and if $f(a)$, $f(b)$, $f'(a)$, and $f'(b)$ are all zero, then*

$$\int_a^b f(x)\,dx = \frac{1}{24}\int_a^b (x - a)^2(x - b)^2 f^{(4)}(x)\,dx.$$

(This would be checked by starting with the right side and repeatedly integrating by parts.)

These apparently esoteric results have important consequences, as we shall see in the next chapter.

Finally, there are important inequality properties of the definite integral. They all stem from the basic lemma.

V) **Lemma 3.** *If $a < b$ and if $f > 0$ between a and b, then $\int_a^b f > 0$. If $a < b$ and $g(x) \leq h(x)$ on $[a, b]$, then*

$$\int_a^b g \leq \int_a^b h.$$

Proof. The first assertion is obvious on the basis of the area interpretation, since it just says that the region in question has positive area. Analytically, we note that an antiderivative F has the positive derivative f between a and b and hence is increasing on $[a, b]$. In particular, $F(a) < F(b)$. Thus

$$\int_a^b f = F(b) - F(a) > 0.$$

If we know only that $f \geq 0$ on $[a, b]$, then we conclude similarly that $\int_a^b f \geq 0$. So if $g \leq h$ then $f = h - g \geq 0$ and

$$\int_a^b h - \int_a^b g = \int_a^b f \geq 0. \blacksquare$$

The ordinary mean-value theorem can be given the integral form:

$$\int_a^b f = f(X)(b - a)$$

for some X strictly between a and b, for in terms of an antiderivative F of f, this just says that

$$F(b) - F(a) = F'(X)(b - a).$$

But the "weighted" mean-value theorem below requires an inequality argument depending on Lemma 3.

VI) **Theorem 2.** *Suppose that f and g are continuous on $[a, b]$, and that $g(x)$ is never zero on the open interval (a,b). Then there is a number X in $[a,b]$ such that*

$$\int_a^b fg = f(X) \int_a^b g.$$

Proof. By hypothesis, g cannot change sign on $[a,b]$. Suppose first that $g > 0$ on (a,b), and let m and M be the minimum and maximum values of f on $[a,b]$. The inequality

$$m \leq f(x) \leq M$$

can be multiplied through by the nonnegative member $g(x)$, and becomes

$$mg(x) \leq f(x)g(x) \leq Mg(x).$$

This holds for all x in $[a,b]$, so we can integrate the inequality, by Lemma 3:

$$m \int_a^b g \leq \int_a^b fg \leq M \int_a^b g.$$

We can now divide by $\int_a^b g$, which is positive, by the same lemma, so

$$m \leq \frac{\int_a^b fg}{\int_a^b g} \leq M.$$

And since each number between m and M is a value of f (by the intermediate-value property), we conclude that

$$\int_a^b fg \Big/ \int_a^b g = f(X),$$

for some X in $[a,b]$. Multiplying by $\int_a^b g$ then gives the weighted mean-value identity.

This was under the hypothesis that $g > 0$ on (a,b). If, instead, $g < 0$, then the above conclusion holds for the positive function $-g$, after which the minus signs cancel out, again giving the weighted mean-value identity.   ∎

Here is an application of Theorem 2.

Lemma 4. *If $f(a) = f(b) = 0$ and if $f''(x)$ is continuous on $[a,b]$, then*

$$\int_a^b f = \frac{-f''(X)}{12}(b - a)^3$$

for some number X in $[a,b]$.

Proof. We know from Lemma 1 that

$$\int_a^b f = \frac{1}{2}\int_a^b (x - a)(x - b)f'',$$

and from Problem 27 that

$$\frac{1}{2}\int_a^b (x - a)(b - x)\,dx = \frac{(b - a)^3}{12}.$$

We then just apply Theorem 2 above, with $g(x) = (x - a)(b - x)$, and have the lemma.   ∎

This lemma and the similar conclusion in Problem 33 have roles to play in the next chapter.

PROBLEMS FOR SECTION 8

Integrate.

1. $\int_0^{\pi/2} \cos^3 \theta \, d\theta$

2. $\int_0^{\pi} \cos^2 x \, dx$

3. $\int_0^{\pi/3} \tan^3 x \sec^2 x \, dx$

4. $\int_0^{\pi/4} \tan^2 x \sec^2 x \, dx$

5. $\int_0^{\pi/6} \tan^3 \theta \sec \theta \, d\theta$

6. $\int_0^{\pi} \frac{dx}{1 + \sin x}$

7. $\displaystyle\int_{-1}^{1}\frac{dx}{4-x^2}$

8. $\displaystyle\int_{-1}^{1}\frac{dx}{x^2+2x+5}$

9. $\displaystyle\int_{0}^{\log 2}\frac{dx}{1+e^x}$

10. $\int_{0}^{1}\arctan x\,dx$

11. $\int_{0}^{1/2}\arctan\sqrt{x}\,dx$

12. $\int_{0}^{\pi/3}x\sec^2 x\,dx$

13. $\displaystyle\int_{a}^{2a}x^3\sqrt{x^2-a^2}\,dx$

14. $\displaystyle\int_{0}^{1}\frac{dx}{(1+x^2)^{3/2}}$

15. $\displaystyle\int_{a}^{\sqrt{3}a}\frac{dx}{x^2\sqrt{a^2+x^2}}$

16. $\displaystyle\int_{0}^{1}\frac{dx}{(2-x^2)^{3/2}}$

17. $\displaystyle\int_{0}^{a}\frac{x\,dx}{\sqrt{a^2-x^2}}$

18. $\displaystyle\int_{0}^{r}\sqrt{r^2-x^2}\,dx$

19. $\displaystyle\int_{0}^{1}\sqrt{1+x^2}\,dx$

20. $\displaystyle\int_{1}^{2}\sqrt{x^2-1}\,dx$

21. $\displaystyle\int_{0}^{\infty}\frac{dx}{1+x^2}$ (Compute $\int_{0}^{r}$ and let $r\to\infty$)

22. $\displaystyle\int_{0}^{\infty}\frac{dx}{4+x^2}$

23. $\displaystyle\int_{0}^{\infty}\frac{dx}{(1+x^2)^{3/2}}$

24. a) Show that

$$\int_{0}^{a}f(-x)\,dx=\int_{-a}^{0}f(x)\,dx$$

for any function f.

b) Show that therefore

$$\int_{-a}^{a}f(x)\,dx=0$$

if f is an odd function on the interval $[-a,a]$, and that

$$\int_{-a}^{a}f(x)\,dx=2\int_{0}^{a}f(x)\,dx$$

if f is even on $[-a,a]$.

25. Show that

$$\int_{A}^{B}f(cx+d)\,dx=\frac{1}{c}\int_{cA+d}^{cB+d}f(x)\,dx.$$

26. Show that

$$\int_{0}^{1}x(1-x)\,dx=\frac{1}{6}.$$

27. Show that

$$\int_{a}^{b}(x-a)(b-x)\,dx=\frac{(b-a)^3}{6}.$$

28. Show that Problem 27 follows from Problem 26 by proving the identity

$$\int_a^b (x - a)(b - x)\,dx = (b - a)^3 \int_0^1 y(1 - y)\,dy.$$

(Set $x = y(b - a) + a$, and use Theorem 1 or problem 25.)

29. Show that

$$\int_0^1 x^2(1 - x)^2\,dx = \frac{1}{30}.$$

30. Show that

$$\int_a^b (x - a)^2(b - x)^2\,dx = \frac{(b - a)^5}{30},$$

a) by a direct computation;
b) by proving the identity analogous to the one in Problem 28.

31. Prove Lemma 1.

32. Prove Lemma 2.

33. Prove the following Theorem.

    Theorem. *If $f(a)$, $f'(a)$, $f(b)$, and $f'(b)$ are all 0, and if the fourth derivative $f^{(4)}(x)$ is continuous on $[a, b]$, then*

$$\int_a^b f = \frac{f^{(4)}(X)}{720}(b - a)^5$$

    for some number X in $[a, b]$.
    (Apply Theorem 2 to Lemma 2 and Problem 30.)

34. Prove from Lemma 3 that if $|f(x)| \le g(x)$ on the interval $[a,b]$, then

$$\left| \int_a^b f(x)\,dx \right| \le \int_a^b g(x)\,dx.$$

MISCELLANEOUS PROBLEMS

Area and volume problems

1. Prove that the area of the ellipse

$$\frac{x^2}{a^2} + \frac{y^2}{b^2} = 1$$

is πab.

2. One arch of the curve $y = \sin x$ is revolved about the x-axis. Compute the volume of the region thus generated.

3 Compute the volume of the solid generated by rotating about the x-axis the region under the graph of $y = \log x$, from $x = 1$ to $x = 2$.

4. Calculate the area under the graph of $f(x) = \arctan x$ from $x = 0$ to $x = 1$.

5. A torus is a doughnut-like solid formed by rotating a circular disk about an axis not touching it. If the radius of the circle is r and the distance from the center of the circle to the axis is R $(R > r)$, show that the volume of the torus is

$$V = 2\pi^2 r^2 R.$$

6. A cylindrical core of radius a is removed from a spherical apple of radius r. Find the volume of the cored apple.

In each of the problems below, find the volume generated by revolving about the y-axis the region bounded by the given graphs. In each case, draw the plane region to be rotated.

7. $y = \sin x,$ $y = 0,$ $x = 0,$ $x = \pi.$

8. $y = \log x,$ $y = 0,$ $x = 1,$ $x = 2.$

9. $y = e^{-x^2},$ $y = 0,$ $x = 0,$ $x = \infty.$

10. $y = e^x,$ $y = -e^x,$ $x = 0,$ $x = a.$

11. Calculate the area of the region between the graph of $y = \log x$, its tangent line at $x = 1$, and the line $x = 2$.

12. Find the volume swept out when the above area is rotated about the y-axis.

13. Same problem when the region is rotated about the x-axis.

14. Find the formula for the volume of a sphere by the shell method.

15. Show that the volume of the ellipsoid

$$\frac{x^2}{a^2} + \frac{y^2}{b^2} + \frac{z^2}{c^2} = 1$$

is $(4/3)\pi abc$. [*Hint.* The cross section of this solid perpendicular to the z-axis at the point $z = z_0$ is the ellipse

$$\frac{x^2}{a^2} + \frac{y^2}{b^2} = \left(1 - \frac{z_0^2}{c^2}\right).$$

Use Problem 1 to find its area $A(z_0)$, and then compute the volume by the volumes-by-slicing formula.]

chapter 10
the definite integral

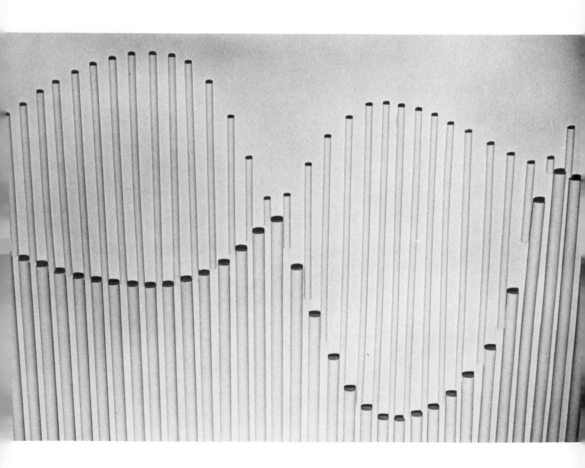

This chapter develops a method for estimating the number $\int_a^b f$ directly in terms of the values of f between a and b. We thus obtain a direct numerical calculation of $\int_a^b f$ that is independent of antiderivatives and that constitutes a more fundamental approach to the definite integral. (We will then be on the verge of redefining $\int_a^b f$ as the number furnished by the new calculational procedure. However, we shall put off actually taking this step until Chapter 20.) The direct method for estimating $\int_a^b f$ will lead to new applications of the definite integral, and in the process will improve our understanding of why the definite integral is the natural way to express so many of the magnitudes that we meet in the applications.

1. THE IDEA OF THE DEFINITE INTEGRAL

Our definition of $\int_a^b f$ as $F(b) - F(a)$, where F is any antiderivative of f, depended on the fact that the difference $F(b) - F(a)$ is the same for all antiderivatives of f. That is, the number $\int_a^b f$ depends on f, a, and b, but not on the particular antiderivative F. It ought to be possible, therefore, to give an analytic description of $\int_a^b f$ directly in terms of the values $f(x)$ between $x = a$ and $x = b$.

EXAMPLE 1. In order to see how this might go, let us consider $f(x) = 1/x$ and its antiderivative $F(x) = \log x$, and see if we can estimate the number

$$\int_1^{3/2} \frac{dx}{x} = \log(\frac{3}{2}) - \log 1 = \log \frac{3}{2}$$

directly in terms of the values of $f(x) = (1/x)$ between 1 and $(3/2)$.

The best estimating device we have available so far is the increment (tangent-line) approximation $\Delta y \approx (dy/dx)\Delta x$, or

$$F(x) - F(a) \approx F'(a)(x - a),$$

from Chapter 5. Thus,

$$\log(x) - \log(a) \approx \frac{1}{a}(x - a)$$

when $x - a$ is small. We wouldn't apply this estimate to the whole interval $[1, 3/2]$ and write

$$\log(\frac{3}{2}) - \log(1) \approx \frac{1}{1}\left(\frac{3}{2} - 1\right) = \frac{1}{2},$$

because we can't expect $\Delta x = (3/2) - 1 = 1/2$ to be small enough for the estimate to have any value. Suppose, however, that we subdivide the interval $[1, 3/2]$ into a number of small equal subintervals, say four subintervals of length $1/8$ each. We can then estimate the change in $\log x$ over each small subinterval

$$\begin{array}{ccc} 9/8 & 5/4 & 11/8 \end{array}$$

$$\begin{array}{cc} 1 & 3/2 \end{array}$$

by the increment approximation. From right to left the four approximations are:

$$\log\left(\frac{3}{2}\right) - \log\left(\frac{11}{8}\right) \approx \frac{8}{11} \cdot \frac{1}{8},$$

$$\log\left(\frac{11}{8}\right) - \log\left(\frac{5}{4}\right) \approx \frac{4}{5} \cdot \frac{1}{8},$$

$$\log\left(\frac{5}{4}\right) - \log\left(\frac{9}{8}\right) \approx \frac{8}{9} \cdot \frac{1}{8},$$

$$\log\left(\frac{9}{8}\right) - \log(1) \approx 1 \cdot \frac{1}{8}.$$

If we now add these approximate equalities, *all the intermediate values of* log *x drop out*, and we end up with

$$\log\left(\frac{3}{2}\right) - \log 1 \approx \frac{1}{8}\left[1 + \frac{8}{9} + \frac{4}{5} + \frac{8}{11}\right].$$

This is the kind of approximation we are aiming at.

EXAMPLE 2. Such an estimate has little value unless we can say something about how good the approximation is. In order to do this, we remember that the error in the increment approximation is at most

$$\frac{B}{2}(x - a)^2,$$

where B is a bound for the values of F'' between a and x. This was discussed in Chapter 5, Section 3. In each of our four increment estimates we have $x - a = \frac{1}{8}$, and $|F''(x)| = |-1/x^2| < 1$ (since $x > 1$). So each of the four increment approximations has an error less than

$$\frac{1}{2}\left(\frac{1}{8}\right)^2 = 2^{-7}$$

in magnitude. The accumulation of these four errors is then at most $4 \cdot 2^{-7} = (1/32)$ in magnitude. Thus

$$\frac{1}{8}\left[1 + \frac{8}{9} + \frac{4}{5} + \frac{8}{11}\right]$$

is an approximation of log $(3/2)$ that is in error by at most $(1/32)$.

This estimate isn't very accurate. Since $1/32 = 0.031 \cdots$, we can say that the estimate is correct in the first decimal place but may be off by as much as 3 in the second decimal place. The natural thing to try, in order to improve the estimate, is subdividing $[1, \frac{3}{2}]$ into a larger number of smaller intervals, say into 8 intervals of length $(1/16)$ each. Then each individual error would be at most

$$\frac{B}{2}(x - a)^2 = \frac{1}{2}\left(\frac{1}{16}\right)^2 = 2^{-9}$$

in magnitude, and the accumulated error would be at most

$$8 \cdot 2^{-9} = 2^{-6} = \frac{1}{64}.$$

So by doubling our effort we gain a factor of 2 in our accuracy.

In order to consider such approximating sums in general terms we shall need some new notation, and it may be as well to see how it looks in the $\log(3/2)$ example.

Suppose we label the subdividing points x_1, x_2, and x_3, and include the endpoints 1 and $\frac{3}{2}$ as x_0 and x_4. Then we can indicate the even spacing of the subdivision by writing that

$$x_k - x_{k-1} = \frac{1}{8}$$

for $k = 1, \ldots, 4$. That is, we use the subscript letter k as a variable that takes on integer values; and as k runs from 1 to 4 in value, the equation above represents the four equations $x_1 - x_0 = \frac{1}{8}, \ldots, x_4 - x_3 = \frac{1}{8}$.

x_0	x_1	x_2	x_3	x_4
1	9/8	5/4	11/8	3/2

Also, we can indicate the sum on the right in Example 1 by using the *summation notation*, involving the Greek letter capital sigma, Σ. Here we would indicate the sum

$$\frac{1}{x_0} + \frac{1}{x_1} + \frac{1}{x_2} + \frac{1}{x_3}$$

by writing

$$\sum_{k=0}^{3} \frac{1}{x_k},$$

which is read "the sum of the numbers $1/x_k$ from $k = 0$ to $k = 3$." The estimate in Example 1 is thus

$$\log \frac{3}{2} = \log\left(\frac{3}{2}\right) - \log 1 \approx \frac{1}{8}\left[\sum_{k=0}^{3} \frac{1}{x_k}\right].$$

Of course we have to know that $x_0 = 1$, $x_1 = 9/8$, etc., in order to evaluate the sum.

In general, $\sum_{k=1}^{n} a_k$ is the sum of the terms a_k from $k = 1$ to $k = n$:

$$\sum_{k=1}^{n} a_k = a_1 + a_2 + \cdots + a_n.$$

Now consider the above estimation procedure in general terms. The idea is to add up a number of successive approximations to small changes $F(x) - F(x')$ in order to estimate the large change $F(b) - F(a) = \int_a^b f$. We subdivide the closed interval $[a,b]$ into n equal subintervals of length $\Delta x = (b - a)/n$. If the subdividing points are

$$a = x_0, \quad x_1, \quad \cdots, \quad x_n = b,$$

then $x_k - x_{k-1} = \Delta x$ for $k = 1, \ldots, n$.

The increment approximations are

$$F(x_n) - F(x_{n-1}) \approx f(x_{n-1})\Delta x,$$
$$F(x_{n-1}) - F(x_{n-2}) \approx f(x_{n-2})\Delta x,$$
$$\vdots \qquad\qquad \vdots$$
$$F(x_2) - F(x_1) \approx f(x_1)\Delta x,$$
$$F(x_1) - F(x_0) \approx f(x_0)\Delta x.$$

When we add these approximate equalities, all the intermediate values of $F(x)$ cancel, and we end up with

$$F(x_n) - F(x_0) \approx \Delta x[f(x_0) + f(x_1) + \cdots + f(x_{n-1})].$$

Since $x_0 = a$ and $x_n = b$, the left side is $F(b) - F(a) = \int_a^b f$. Thus

$$\int_a^b f \approx \Delta x \sum_{k=0}^{n-1} f(x_k).$$

The sum on the right, including the common factor Δx, is called a *Riemann sum* for the function f over the interval $[a,b]$. For each integer n we have defined one such Riemann sum S_n, by subdividing $[a,b]$ into n equal subintervals and adding up the corresponding increment approximations.

Our experience with $\log x = \int_1^x dt/t$ suggests that the Riemann sum approximates the integral, and that the the approximation improves as n increases. We can verify this conjecture without too much trouble.

If $|f'(x)| \leq K$ over the whole parent interval $[a,b]$, then each of the n increment errors is at most

$$\frac{K}{2}(\Delta x)^2$$

in magnitude. This is simply Theorem 6 of Chapter 5 again, since $f'(x) = F''(x)$. The accumulated error in the final approximation is thus at most

$$n\frac{K}{2}(\Delta x)^2.$$

But $n\Delta x = b - a$. The final error is therefore at most

$$\frac{1}{2}K(b - a)\Delta x.$$

Altogether, we have established the following theorem.

Theorem 1. *If f is differentiable and f' is bounded by K on the interval $[a,b]$, then $\int_a^b f$ is approximated by the nth Riemann sum $\Delta x \sum_{k=0}^{n-1} f(x_k)$ with an error at most*

$$\frac{K(b - a)}{2}\Delta x.$$

It follows from the theorem that we can approximate $\int_a^b f$ by the Riemann sum as closely as we wish by taking Δx sufficiently small (i.e., by taking n sufficiently large). That is, $\int_a^b f$ is the *limit* of its Riemann sum S_n as n tends to infinity (or as $\Delta x \to 0$). This is the fundamental phenomenon underlying the definite integral, and in due course we shall *redefine* $\int_a^b f$ along these lines. At that time, the basic evaluation identity

$$\int_a^b f = F(b) - F(a)$$

will become a theorem to be proved.

Meanwhile, however, we know that $\int_a^b f$ can be characterized as the unique number that is approximated arbitrarily closely by Riemann sums in the manner discussed above, and we shall find this new description of $\int_a^b f$ very useful. The next two sections will illustrate how this goes.

EXAMPLE 3. If we want an estimate of $\log 2 = \int_1^2 dx/x$ that is accurate to the nearest two decimal places, then the error must be at most 0.005. Here, $b - a = 2 - 1 = 1$, and again $K = 1$, since 1 is the smallest bound for

$f'(x) = -1/x^2$ on the interval $[1,2]$. Therefore the subdivision spacing width Δx must satisfy the inequality $(1/2)K(b - a)\Delta x = (1/2)\cdot 1 \cdot \Delta x \leq 0.005$, or

$$\Delta x \leq 0.01 = \frac{1}{100}.$$

So one hundred terms are needed. That is, we know by virtue of the theorem that

$$\frac{1}{100}\left[\frac{1}{1.00} + \frac{1}{1.01} + \frac{1}{1.02} + \cdots + \frac{1}{1.99}\right]$$

$$= \frac{1}{100} + \frac{1}{101} + \frac{1}{102} + \cdots + \frac{1}{199}$$

$$= \sum_{k=0}^{99} \frac{1}{100 + k}$$

approximates log 2 with an error less than 0.005. Moreover, we cannot be sure, from the present argument, that a coarser subdivision of $[1,2]$ would be adequate.

It appears from such examples that estimating a definite integral by a Riemann sum is a very inefficient way of calculating. We would expect that there must be some way to calculate log 2 to two decimal places that doesn't involve summing 100 terms. However, except for the large number of terms, the arithmetic of a Riemann sum calculation is extremely simple and straightforward, and this is a kind of calculation, tedious but trivial, that a computing machine handles most easily.

For hand calculation it is worthwhile to look for ways of improving the Riemann-sum estimates, and a couple of methods will be discussed in Section 7 and in Chapter 13.

Another point: So far the validity of the Riemann-sum approximation depends on having a bound K for f' over the interval in question. Eventually we shall see that any continuous function having only a finite number of critical points in $[a,b]$ is also all right. This will take care of every function that will come up in practice. However, in theory anyway, the Riemann-sum approximation of $\int_a^b f$ is valid for *any continuous function whatever*. That is:

If f is continuous on the interval $[a,b]$, and if S_n is the nth Riemann sum for f over $[a,b]$, then $\int_a^b f$ is the limit of S_n as n tends to infinity.

We shall feel free to invoke this general result. It saves us from having to dig out and state the special conditions from which we can furnish actual error estimates, although in practice such special conditions are always present.

Finally, a word about inequalities. We argued above that if the error at each of n steps is at most e in magnitude, then the accumulated error at most is ne in magnitude. Presumably this all seemed reasonable and, most error estimates can be correctly established in the same common-sense way. But the formal justification for such conclusions lies in the inequality laws, which are reviewed in Appendix 1.

The evaluation of the definite integral $\int_a^b f$ as the limit of the Riemann sum

$$\sum_{k=1}^n f(x_k)\Delta x$$

as $\Delta x \to 0$ (i.e., as $n \to \infty$) suggests that

$$\int_a^b f(x)\,dx$$

be thought of as an infinite sum of infinitely small numbers. This viewpoint explains the Leibniz notation, in which the integral sign $\int$ is an elongated S, standing for a sum, with dx interpreted as the infinitesimal (infinitely small but not zero) limit of Δx. Unfortunately, such an interpretation of the definite integral is nonsensical, because there is no infinitely small number except zero, and any sum of zeros is zero. It is amazing that the early workers in calculus, constantly having to make their way through clouds of mysticism, found the right paths as often as they did.

PROBLEMS FOR SECTION 1

1. Show that the Riemann sum in Example 1 in the text can be written

$$\sum_{k=0}^3 \frac{1}{8 + k}.$$

2. Show that if $[0,\frac{3}{2}]$ is divided into eight equal subintervals, then the Riemann sum for $1/x$ is

$$\sum_{k=0}^7 \frac{1}{16 + k}.$$

3. In the manner of the above two problems, write down a formula for the Riemann sum for $1/x$ obtained by dividing $[1,2]$ into 1000 equal subintervals.

4. a) Write down the fourth Riemann sum S_4 for $f(x) = x$, over the interval $[1,\frac{3}{2}]$, leaving its terms as fractions.
    b) Compute $\int_1^{3/2} x\,dx$ and show that the error in using S_4 as an estimate of the integral is exactly $1/32$.
    c) Show therefore that the error estimate given in Theorem 1 cannot be improved.

5. a) Write down the fourth Riemann sum S_4 for $f(x) = x^2$ over the interval $[1,\frac{3}{2}]$, leaving its terms as fractions.

b) Compute $\int_1^{3/2} x^2 dx$ and show that the actual error is less than the error bound given in Theorem 1.

6. We wish to use a Riemann sum S_n for $f(x) = 1/(1 + x^2)$ over the interval $[0,1]$ to compute

$$\frac{\pi}{4} = \arctan 1 = \int_0^1 \frac{dx}{1 + x^2}.$$

Show that S_n will approximate $\pi/4$ accurately to the nearest two decimal places if $n = 67$. (Prove first that the maximum value of $|f'(x)|$ on $[0,1]$ is less than $2/3$, and then use the error estimate in Theorem 1.)

7. a) Compute the fourth Riemann sum S_4 for $f(x) = 1/(1 + x^2)$ over the interval $[0,1]$.
   b) Use facts from Problem 6 and the error estimate in Theorem 1 to show that

$$3.0 < \pi < 3.8.$$

8 Suppose we wish to compute $\pi = 4 \int_0^1 dx/(1 + x^2)$ to the nearest four decimal places, i.e., with an error less than $5(10)^{-5}$, by using a Riemann sum. How fine a subdivision would have to be used?

2. APPROXIMATING OTHER QUANTITIES BY RIEMANN SUMS

Suppose that $T = f(t)$ is the temperature at time t recorded at a weather station on a certain day. The station uses a 24-hour clock, so the domain of the temperature function f is the time interval $[0,24]$.

In order to find the average temperature for the given day, we might take six temperature readings at 4-hour intervals, starting at midnight: $T_0 = f(0)$, $T_1 = f(4)$, $\cdots$, $T_5 = f(20)$. The average reading would then be the sum of these six readings divided by 6:

$$T_{av} = \frac{T_0 + T_1 + T_2 + T_3 + T_4 + T_5}{6}$$

$$= \frac{f(0) + f(4) + f(8) + f(12) + f(16) + f(20)}{6}$$

$$= \frac{1}{6} \sum_{k=0}^5 f(4k).$$

This computation of the average temperature probably does not give the right answer. For example, suppose it is a hot summer day, and that at 2:00 in the afternoon (hour 14 on the 24-hour clock) there is a short thunderstorm that cools the air for an hour. This temporary dip in temperature wouldn't even show in the average computed above. And there might be other fluctuations that wouldn't be taken into account adequately by reading temperatures at 4-hour intervals.

Clearly we would do better by taking 24 readings on the hour, starting at midnight and using their average

$$T_{av} = \frac{f(0) + f(1) + \cdots + f(23)}{24} = \frac{1}{24} \sum_{k=0}^{23} f(k).$$

But this probably wouldn't be exactly right either, for the same reason that small temporary fluctuations may not be adequately represented in the average. Still better would be the average of 48 readings taken on the half-hour,

$$T_{av} = \frac{f(0) + \cdots + f(23\frac{1}{2})}{48} = \frac{1}{48} \sum_{k=0}^{47} f(\frac{k}{2}),$$

and so on.

We thus have various ways of computing an average daily temperature, none of them exactly right. Our intuition tells us that there is a *true average temperature* $\overline{T}$ that these approximating averages come close to, and that by taking a sufficiently large number of equally spaced temperature readings, we could, in principle, make the approximation as good as we wished.

On the other hand, the above finite sums look somewhat like Riemann sums for f. In fact, each of them is the corresponding Riemann sum divided by 24. For example, the first sum can be written

$$\frac{1}{6} \sum_{k=0}^{5} f(4k) = \frac{1}{24} \left[4 \sum_{k=0}^{5} f(4k) \right] = \frac{S_6}{24}.$$

where S_6 is the sixth Riemann sum for f over the interval $[0,24]$. Now we know that $\int_0^{24} f(t)\,dt$ is the unique number approximated arbitrarily closely by the Riemann sums S_n themselves. So each of the numbers $\overline{T}$ and $(\int_0^{24} f)/24$ is the unique number approximated arbitrarily closely by $S_n/24$ as n increases, and hence

$$\overline{T} = \frac{1}{24} \int_0^{24} f(t)\,dt.$$

We have thus established an integral formula for the average temperature $\overline{T}$ by the device of shared finite sum approximations.

We can reason this way about the average value of any continuous function $y = f(x)$ over a closed interval $[a,b]$, and conclude:

If f is continuous on the closed interval $[a,b]$, then its average value over the interval is

$$\frac{1}{b-a} \int_a^b f(x)\,dx.$$

EXAMPLE The average value of $f(x) = x^2$ over the interval $[0,2]$ is

$$\frac{1}{2}\int_0^2 x^2\,dx = \frac{1}{2}\frac{x^3}{3}\Big]_0^2 = \frac{4}{3}.$$

Note that although the value of $f(x)$ increases from 0 to 4, we wouldn't expect the average value to be 2 because we see from the graph that $f(x)$ is less than 2 more than half the time.

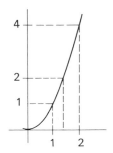

We now have a new procedure, complementing and reinforcing the one followed in Chapter 8, for showing that quantities Q in geometry, physics, and other disciplines are expressible as definite integrals. Here, again, is how it goes. We observe that Q can be estimated by a finite sum of a certain type, and we notice that this finite sum is also a Riemann sum for a certain function g over the interval $[a,b]$. Then Q and $\int_a^b g(x)\,dx$ are *both* approximated as closely as we wish by the *same* number, and this means that

$$Q = \int_a^b g(x)\,dx.$$

For example, the figure below shows how the volume V of a solid of revolution can be approximated by a finite sum of volumes of thin circular cylinders:

$$V \approx \sum \pi r_i^2 \Delta x = \sum \pi f(x_i)^2 \Delta x.$$

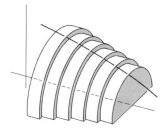

But this sum is also a Riemann sum $\sum g(x_i)\Delta x$ for the function $g(x) = \pi f(x)^2$. Therefore, V is given exactly by

$$V = \int_a^b \pi f(x)^2 \, dx.$$

We have glossed over the important step of proving that the Riemann sum actually does estimate the volume, with an error that approaches zero as n tends to infinity. The approximation seemed intuitively clear on the basis of a picture, and we let it go at that.

This will be our point of view throughout the present chapter. It will be intuitively clear that a quantity Q can be estimated arbitrarily closely by a finite sum that turns out to be a Riemann sum for a function g, and we accept the evaluation of Q by $\int_a^b g$ on this basis. In some situations this is the best we can do. But plausibility does not constitute a proof, and it can even lead one astray. So it is important, wherever possible, to justify the estimate by showing that the error in the incremental approximation $\Delta Q_i = g(x_i)\Delta x_i$ is small in comparison with Δx_i. In Chapter 18 we shall be more careful about these estimates, and we were more careful in Chapter 8.

PROBLEMS FOR SECTION 2

Find the average value of the following functions over the specified interval.

1. $y = 2x^3$; $[-1, 1]$.

2. $y = 4 - x^2$; $[-2, 2]$.

3. $y = x^2 - x + 1$; $[0, 2]$.

4. $y = (x/2) + 1$; $[2, 6]$.

5. $y = 2x + 1$; $[-1, 3]$.

6. $y = x^n$; $[0, 1]$. (Why is the answer reasonable?)

7. A man travels 20 miles an hour for $\frac{1}{2}$ hour, 30 miles per hour for 2 hours, and 40 miles per hour for $\frac{1}{2}$ hour. What is the average velocity with respect to time?

8. A typist's speed over a four-hour interval increases as she warms up and decreases as she tires. Her speed in words per minute can be approximated by $w(t) = 6[4^2 - (t - 1)^2]$. Find her speed at the beginning of the interval, the end of the interval, her maximum speed, and her average speed over the 4-hour period.

9. If a particle is moving along a coordinate line, and if its position at time t is given by $s = f(t)$, then we concluded in Chapter 3 that its average

velocity over the interval $[t_0, t_1]$ is given by

$$\frac{s_1 - s_0}{t_1 - t_0} = \frac{f(t_1) - f(t_0)}{t_1 - t_0},$$

and that its instantaneous velocity at time t is

$$V = \frac{ds}{dt} = f'(t).$$

Show now that the average velocity really is the average value of the instantaneous velocity, according to the notion of average value developed in this section.

10. If $y = f(x)$ and if $f'(x_0)$ is interpreted as the rate of change of y with respect to x at x_0, show that

$$\frac{\Delta y}{\Delta x} = \frac{f(x + \Delta x) - f(x)}{\Delta x}$$

is the average rate of change of y with respect to x over the interval $[x, x + \Delta x]$.

11. A particle moves along a straight line with varying velocity $v = g(t)$. Give a direct justification, in terms of finite-sum approximations, for the fact that the distance travelled during the time interval $[a,b]$ is given by

$$s = \int_a^b v\, dt = \int_a^b g(t)\, dt.$$

Start with the basic idea that over a very short time interval from t_0 to $t_0 + \Delta t$ the velocity is nearly constant, with value $v_0 = g(t_0)$, so the distance travelled is approximately $v_0 \Delta t = g(t_0)\Delta t$.

12. Give a direct justification, in terms of finite-sum approximations, for the fact that $\int_a^b f$ is the area under the graph of f from a to b. Start with the basic idea that, over a very short interval from x_0 to $x_0 + \Delta x$, the height $f(x)$ is nearly constant, so the area over this incremental interval is approximately a rectangular area.

3. MORE GENERAL RIEMANN SUMS

We return to the interpretation of the definite integral $\int_a^b f$ as the area of the region between the graph of f and the x-axis, from $x = a$ to $x = b$ (supposing that $a < b$ and $f > 0$). A Riemann sum for f over $[a,b]$ also has an area interpretation, as we shall check in a minute. We thus obtain a geometric picture of the approximation of $\int_a^b f$ by Riemann sums, and this area interpretation is a fruitful source of new ideas. Here it will provide visual evidence that we can approximate $\int_a^b f$ by a more general kind of sum than what we have called a Riemann sum.

We first check the area interpretation of the Riemann sum S_n. We subdivide the interval $[a,b]$ into n equal subintervals of width $\Delta x = (b - a)/n$ by the points $a = x_0, x_1, x_2, \ldots, x_n = b$, and consider the rectangles illustrated in the figure below (for the case $n = 5$). The areas of these rectangles, in order, are

$$f(x_0)\Delta x, \quad f(x_1)\Delta x, \quad \cdots, \quad f(x_{n-1})\Delta x.$$

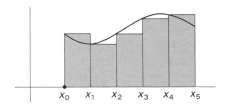

The sum of these areas,

$$f(x_0)\Delta x + f(x_1)\Delta x + \cdots + f(x_{n-1})\Delta x = \Delta x[f(x_0) + f(x_1) + \cdots + f(x_{n-1})]$$

$$= \Delta x \sum_{k=0}^{n-1} f(x_k),$$

is exactly the nth Riemann sum for f over $[a,b]$. It is clear geometrically that this sum of rectangular areas is a good approximation to the area under the graph of f from a to b. We thus have a direct geometrical visualization of the approximation of $\int_a^b f$ by its Riemann sum S_n.

The next figure illustrates a more general way of approximating the same integral, $\int_a^b f$, by a finite sum. First, we allow unequal spacing: We subdivide the interval $[a,b]$ into n subintervals of (possibly) varying lengths $\Delta x_1, \Delta x_2, \ldots, \Delta x_n$. Of course,

$$b - a = \Delta x_1 + \Delta x_2 + \cdots + \Delta x_n = \sum_{k=1}^{n} \Delta x_k.$$

Next, we choose an *arbitrary* evaluation point ξ_k in the kth interval, for

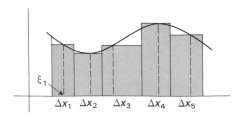

$k = 1, 2, \ldots, n$. Then we form the sum

$$S = f(\xi_1)\Delta x_1 + f(\xi_2)\Delta x_2 + \cdots + f(\xi_n)\Delta x_n$$

$$= \sum_{k=1}^{n} f(\xi_k)\Delta x_k.$$

This sum can be interpreted as the sum of the rectangular areas illustrated above, and it obviously can be just as good an approximation to the area $\int_a^b f$ as the Riemann sum. Thus, on the basis of area interpretations, we can write

$$\int_a^b f \approx \sum_{k=1}^{n} f(\xi_k)\Delta x_k.$$

The new sum is called a *general Riemann sum* for the function f over the interval $[a,b]$.

To be consistent, we should now back up the new approximation by an error estimate. What can be shown here, going back to the tangent-line approximation again, is this:

If δ is the largest of the interval widths $\Delta x_1, \ldots, \Delta x_n$, and if the derivative f' is bounded by K over the interval $[a,b]$, then the error in the above approximate equality is at most

$$\frac{1}{2}K(b - a)\delta$$

in magnitude.

Note that the earlier error estimate

$$\frac{1}{2}K(b - a)\Delta x$$

is now a special case. For if all the subintervals happen to have the same width Δx then $\delta = \Delta x$.

The proof of the new estimate is obviously going to be a little more complicated. We shall just sketch how it goes. First we have to consider the tangent line approximation itself in the more general form

$$g(x + \Delta x) - g(x) \approx g'(\xi)\Delta x,$$

as illustrated below. We have to show that if g'' is bounded by K between x and

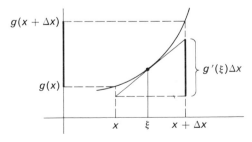

$x + \Delta x$ then the error in this more general estimate is still at most

$$\frac{K}{2}(\Delta x)^2.$$

To do this we just break the x-interval into two subintervals at the point ξ, and combine the backward and forward tangent line approximations based at ξ. The calculation is straightforward but tedious, and will be left to the interested reader.

Next we have to add up these new tangent line approximations corresponding to the pieces of a general subdivision of $[a,b]$. The accumulated error is then at most

$$\frac{K}{2}\sum_{k=1}^{n}(\Delta x_k)^2$$

in magnitude. But this, in turn, is at most

$$\frac{1}{2}K(b-a)\delta,$$

because of the inequality

$$\sum_{k=1}^{n}(\Delta x_k)^2 \leq (b-a)\delta.$$

This inequality is obvious geometrically when its terms are interpreted as rectangular areas:

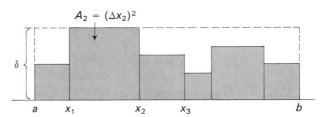

The picture even suggests how the algebraic proof goes, but the algebraic details will be left to the interested reader.

Because of the error estimate, we can say that $\int_a^b f$ is the limit of the general Riemann sum S as δ approaches 0. Here S is *not* a function of δ, and we are using the limit notion in a new way. But it is perfectly meaningful and appropriate.

In review, for any set of interval widths $\Delta x_1, \ldots, \Delta x_n$ adding to $b-a$, and any associated set of evaluation points $\xi_1, \ldots, \xi_n$, we form the two numbers

$$S = \sum_{k=1}^{n} f(\xi_k)\Delta x_k$$

$$\delta = \max\{\Delta x_1, \ldots, \Delta x_n\}.$$

The error estimate shows that we can cause the difference $\int_a^b f - S$ to be as small as we please in magnitude merely by requiring that δ be suitably small. This is what we *mean* by:

$$S \to \int_a^b f \quad \text{as } \delta \to 0.$$

PROBLEMS FOR SECTION 3

Let

$$\overline{S}_n = \Delta x \sum_{k=1}^{n} f(\overline{x}_k)$$

be the Riemann sum obtained by evaluating f at the *midpoints* of n equal subintervals of length $\Delta x = (b - a)/n$ each. This midpoint-evaluation Riemann sum is generally a much better estimate of $\int_a^b f$ than the simple Riemann sum S_n, as the following theorem shows.

Theorem. *If the second derivative of f is bounded by B on the interval $[a,b]$, then the nth midpoint-evaluation Riemann sum $\overline{S}_n$ approximates $\int_a^b f$ with an error at most*

$$\frac{B}{24}(b - a)(\Delta x)^2.$$

The following exercises center around this theorem. First we use it; then we prove it.

1. Show that

$$\log 2 \approx 2\left[\frac{1}{17} + \frac{1}{19} + \frac{1}{21} + \frac{1}{23} + \frac{1}{25} + \frac{1}{27} + \frac{1}{29} + \frac{1}{31}\right]$$

   with an error less than $(1/768)$ in magnitude.

2. If we wish to use the midpoint-evaluation Riemann sum $\overline{S}_n$ to approximate $\log 2 = \int_1^2 dx/x$ accurately to the nearest two decimal places, show that it is sufficient to take $n = 5$.

3. Show geometrically that $\overline{S}_n < \int_a^b f$ if f is decreasing and concave up. (Draw a typical subinterval and compare the errors in the area estimate on the two halves of the interval.)

4. Show that

$$\log \frac{3}{2} \approx \left[\frac{2}{17} + \frac{2}{19} + \frac{2}{21} + \frac{2}{23}\right]$$

   with an error less than $(1/1536)$. The estimate is too small, by Problem 3. Divide out the fractions to four decimal places and hence show that

$$\log \frac{3}{2} \approx 0.405$$

   correct to three decimal places.

5. Show that

$$\frac{\pi}{4} \approx \frac{1}{5}[\frac{1}{1.01} + \frac{1}{1.09} + \frac{1}{1.25} + \frac{1}{1.49} + \frac{1}{1.81}]$$

with an error less than $(1/300)$. (Assume that $|f''(x)| \le 2$ on $[0,1]$, where $f(x) = 1/(1 + x^2)$.)

6. Show that the rectangular area $f(\bar{x})\Delta x$ is the same as the trapezoidal area under the tangent line at $\bar{x}$, as suggested by the figure, when $\bar{x}$ is the midpoint of the increment integral.

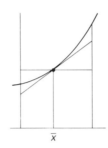

$$\bar{x}$$

Show therefore that

$$\left(\int_x^{x+\Delta x} f\right) - f(\bar{x})\Delta x$$

is equal to the integral

$$\int_x^{x+\Delta x} [f(t) - f(\bar{x}) - f'(\bar{x})(t - \bar{x})]\, dt.$$

7. Use the tangent-line error formula (Theorem 6, Chapter 5) to show that if f'' is bounded by B between x and $x + \Delta x$, then the above integral is at most

$$\frac{B}{24}(\Delta x)^3$$

in magnitude. This bounds the error over each subdivision interval. Now add these errors up and hence prove the theorem.

4. ARC LENGTH

In high-school geometry the length s of a circular arc is defined roughly like this: we inscribe a polygonal arc composed of n segments, compute its length s_n, and

then define s as the limit of s_n as n tends to infinity. We shall now show that this definition works for the graph of any smooth function.

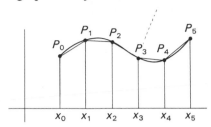

Consider, then, the graph of a smooth function f from $x = a$ to $x = b$, and inscribe a polygonal arc, with vertices $P_0, P_1, \ldots, P_n$. The x-coordinates of the vertices, $x_0, x_1, \ldots, x_n$ subdivide the interval $[a,b]$ into n subintervals of lengths

$$\Delta x_k = x_k - x_{k-1},$$

where $k = 1, 2, \ldots, n$.

Consider, to begin with, the kth polygon side. We isolate and magnify it in the figure below.

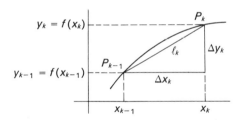

Its length l_k is the distance between P_{k-1} and P_k,

$$l_k = \sqrt{(\Delta x_k)^2 + (\Delta y_k)^2}.$$

By the mean-value theorem, there is a point ξ_k between x_{k-1} and x_k such that

$$\Delta y_k = f'(\xi_k)\Delta x_k,$$

and, when this is substituted in the radical above, we end up with

$$l_k = \sqrt{1 + [f'(\xi_k)]^2}\,\Delta x_k$$

as the length of the segment $P_{k-1}P_k$.

The length of the inscribed polygon is thus

$$l = \sum_{k=1}^{n} l_k = \sum_{k=1}^{n} \sqrt{1 + [f'(\xi_k)]^2}\,\Delta x_k.$$

But this is simply a Riemann sum for the function

$$g(x) = \sqrt{1 + [f'(x)]^2},$$

and therefore approaches the limit

$$\int_a^b \sqrt{1 + [f'(x)]^2}\, dx$$

as the subdivision of $[a,b]$ is taken finer and finer, and the maximum subinterval length δ approaches 0. We have thus not only shown that the arc length definition works, but also have derived the formula

$$\int_a^b \sqrt{1 + [f'(x)]^2}\, dx$$

for the length of the graph of f over the interval $[a,b]$. If $y = f(x)$, this can be written

$$\int_a^b \sqrt{1 + \left(\frac{dy}{dx}\right)^2}\, dx.$$

Having thus *defined* arc length in the appropriate way, we can now *prove* that it has the key property that we leaned on earlier. It is essentially a question of running the earlier argument backward. We first note that the length s from $t = a$ to $t = x$,

$$s = \int_a^x \sqrt{1 + [f'(t)]^2}\, dt,$$

is an antiderivative of the integrand:

$$\frac{ds}{dx} = \sqrt{1 + \left(\frac{dy}{dx}\right)^2}.$$

Then both $\Delta s/\Delta x$ and

$$\frac{l}{\Delta x} = \frac{\sqrt{(\Delta x)^2 + (\Delta y)^2}}{\Delta x} = \sqrt{1 + \left(\frac{\Delta y}{\Delta x}\right)^2}$$

have the same limit $\sqrt{1 + (dy/dx)^2}$ as $\Delta x \to 0$, so their quotient has the limit 1. Thus

$$\lim \frac{\Delta s}{l} = \lim \frac{\Delta s/\Delta x}{l/\Delta x} = 1.$$

That is,

$$\frac{\text{arc length}}{\text{chord length}} \to 1$$

as chord length $\rightarrow 0$, and we have proved our earlier geometric assumption.

The arc length formula leads to complicated integrands and there aren't very many functions for which we can work out the integral. Even simple changes of scale can affect the complexity of the calculation. For example, we can easily find the length of the graph of $f(x) = (e^x + e^{-x})/2$, but not $g(x) = e^x + e^{-x}$!

EXAMPLE 1. Find the length of the graph of

$$f(x) = \frac{e^x + e^{-x}}{2}$$

from $x = 0$ to $x = a$.

. . . .

Solution. Here $f'(x) = (e^x - e^{-x})/2$, and

$$1 + [f'(x)]^2 = 1 + \frac{e^{2x} - 2 + e^{-2x}}{4} = \frac{e^{2x} + 2 + e^{-2x}}{4}$$

$$= \left[\frac{e^x + e^{-x}}{2}\right]^2.$$

The arc length is therefore

$$\int_0^a \sqrt{1 + [f'(x)]^2}\, dx = \int_0^a \frac{e^x + e^{-x}}{2}\, dx = \frac{e^x - e^{-x}}{2}\bigg]_0^a$$

$$= \frac{e^a - e^{-a}}{2}.$$

EXAMPLE 2. Find the length of the graph of $\log x$ from $x = 1$ to $x = 2$.

. . . .

Solution. We have to integrate

$$\int \sqrt{1 + [f'(x)]^2}\, dx = \int \sqrt{1 + (\tfrac{1}{x})^2}\, dx = \int \frac{\sqrt{x^2 + 1}}{x}\, dx.$$

Setting $x = \tan\theta$, $dx = \sec^2\theta\, d\theta$, this becomes

$$\int \frac{\sec^3\theta\, d\theta}{\tan\theta} = \int \frac{\sin\theta\, d\theta}{\sin^2\theta\,\cos^2\theta} = -\int \frac{du}{(1 - u^2)u^2}$$

$$= -\int \left[\frac{1}{2(1 + u)} + \frac{1}{2(1 - u)} + \frac{1}{u^2}\right] du$$

$$= \frac{1}{2}\log\frac{1 - u}{1 + u} + \frac{1}{u}.$$

Retracing the substitutions $u = \cos\theta$, $x = \tan\theta$, we obtain $u = 1/\sqrt{x^2 + 1}$. The arc length is then

$$\left[\frac{1}{2}\log\left(\frac{\sqrt{x^2 + 1} - 1}{\sqrt{x^2 + 1} + 1}\right) + \sqrt{x^2 + 1}\right]_1^2 = \left[\log\left(\frac{\sqrt{x^2 + 1} - 1}{x}\right) + \sqrt{x^2 + 1}\right]_1^2$$

$$= \log\frac{\sqrt{5} - 1}{2(\sqrt{2} - 1)} + \sqrt{5} - \sqrt{2}.$$

PROBLEMS FOR SECTION 4

Find the lengths of the graphs of the following equations, over the given x intervals.

1. $y = \log\cos x$, $[0, \pi/4]$
2. $y = x^{1/2} - x^{3/2}/3$, $[1, 2]$
3. $y = ax^{3/2}$, $[0, 1]$
4. $3y = 2(x^2 + 1)^{3/2}$, $[0, a]$
5. $y^3 = x^2$, $[0, 8]$
6. $y = (1 - x^{2/3})^{3/2}$, $[0, 1]$
7. $y = e^x$, $[0, 1]$
8. $y = ax^2$, $[0, 1]$
9. Let f be a smooth decreasing function whose graph runs from $(0,1)$ to $(1,0)$. Prove that the length of the graph of f necessarily lies between $\sqrt{2}$ and 2. [Hint: Show that the length of any inscribed polygonal arc necessarily lies between these numbers.]
10. In the same way, show, for any two positive numbers a and b, that if the graph of f decreases from $(0,b)$ to $(a,0)$, then the length of the graph necessarily lies between $\sqrt{a^2 + b^2}$ and $a + b$. What would be the corresponding statement for an increasing graph?
11. Over the interval $[0,1]$, the graph of $f(x) = x^n$ increases smoothly from 0 to 1. Let a_n be the point where f has the value $\sqrt{1/n}$: $f(a_n) = \sqrt{1/n}$.
    a) Prove that $a_n > 1 - \sqrt{1/n}$. [Hint: Show that if $a_n \leq 1 - \sqrt{1/n}$, then the area under the graph must be larger than $1/n$. But compute the area.]
    b) Show that the length of the graph of f over $[0,1]$ is greater than $2 - 2\sqrt{1/n}$ (because there is a simple inscribed polygon with length greater than this).
12. Show, partly on the basis of the above result, that the bounds 2 and $\sqrt{2}$ in Problem 9 cannot be improved upon.

5. CENTER OF MASS.

Let us consider the x-axis as a rigid, weightless, horizontal rod pivoting at the point p, and suppose that we place masses m_i at the positions x_i. According to the principle of the lever, or the "seesaw" principle, the system will exactly balance if

$$\sum m_i(x_i - p) = 0.$$

More generally, the sum $\sum m_i(x_i - p)$ measures the tendency of the system to turn about p. It is called the *moment* of the system about p, and if the moment is zero, then the system is in equilibrium. If we are given the masses m_i sitting at positions x_i, and if we are free to move the pivot point p, then we can find a unique position $p = \bar{x}$ such that the moment about $\bar{x}$ is zero. For this, the requirement on $\bar{x}$ is

$$\sum m_i(x_i - \bar{x}) = 0,$$

or

$$\sum m_i x_i - \bar{x} \sum m_i = 0,$$

or

$$\bar{x} = \frac{\sum m_i x_i}{\sum m_i}.$$

This balancing point is called the *center of mass* of the system. The formula above says that the center of mass is obtained as the quotient

$$\bar{x} = \frac{\text{moment of system about the origin}}{\text{total mass of the system}}.$$

Now consider a rod with variable density $\rho(x)$ extending along the x-axis from $x = a$ to $x = b$. Our intuition tells us that there will be a unique pivot position $\bar{x}$ where the rod will exactly balance, and which we can therefore call the center of mass of the rod. The question is how to compute it.

What we do is approximate the rod by a finite collection of point masses. We subdivide the rod $[a,b]$ by the points

$$a = x_0 < x_1 < \cdots < x_n = b,$$

and we choose an arbitrary point x_i' in the ith subrod $[x_{i-1}, x_i]$ for each i. Then the mass of this subrod is approximately

$$m_i = \rho(x_i')\Delta x_i,$$

and we consider it to be concentrated at the point x_i' (We discussed the relationship between density and mass in Chapter 8.) We thus end up with the finite collection of masses m_i at the points x_i' as an approximation of the rod. The moment of the rod about p should then be approximately the moment of this finite system of point masses,

$$\Sigma\, m_i(x_i' - p) = \Sigma\, \rho(x_i')(x_i' - p)\Delta x_i.$$

This approximation should improve when we break the rod up into a larger number of smaller pieces and "in the limit" should give the moment exactly. On the other hand, the sum above is a Riemann sum for the function $\rho(x)(x - p)$ over the interval $[a,b]$ and hence approaches $\int_a^b \rho(x)(x - p)\,dx$ as its limit. We therefore conclude:

If a rod with density $\rho(x)$ lies along the interval $[a,b]$, then its moment about the point p is $\int_a^b (x - p)\rho(x)\,dx$.

Going on, we define the center of mass of the rod as the point $\bar{x}$ about which the moment is zero, and we can solve for $\bar{x}$ as before:

$$0 = \int_a^b (x - \bar{x})\rho(x)\,dx = \int_a^b x\rho(x)\,dx - \bar{x}\int_a^b \rho(x)\,dx,$$

$$\bar{x} = \frac{\displaystyle\int_a^b x\rho(x)\,dx}{\displaystyle\int_a^b \rho(x)\,dx} = \frac{\text{moment about }0}{\text{total mass}}.$$

EXAMPLE 1. A rod extends along the x-axis from $x = 0$ to $x = 1$ and has density $(x) = x^2$. Find its center of mass $\bar{x}$.

. . . .

Solution.

$$\bar{x} = \frac{\displaystyle\int_0^1 x\rho(x)\,dx}{\displaystyle\int_0^1 \rho(x)\,dx} = \frac{\displaystyle\int_0^1 x^3\,dx}{\displaystyle\int_0^1 x^2\,dx} = \frac{1/4}{1/3} = \frac{3}{4}.$$

We gave above an argument that is typical of the way scientists pass from laws and formulas governing finite discrete systems to their continuous analogues involving definite integrals. Such arguments are not mathematical proofs, and their conclusions are physical laws subject to verification.

We could make the mathematics completely sound by starting from a set of postulates for moments. This would be an improvement, but it does not guarantee a physically correct conclusion. It just shifts the onus to the postulates. That is, if the result of a completely logical argument is found to be in disagreement with experience, then some premise must already be out of line.

Here is such a postulate system. We fix the pivot point p and assume that every finite mass distribution along the line has a uniquely determined moment about p satisfying the following two properties:

1. The moment is additive. That is, if two mass distributions are superimposed to form a single "sum" distribution, then the moment of the sum is the sum of the moments.

2. If the mass distribution lies entirely between $x = x_1$ and $x = x_2$ (where $x_1 < x_2$), then

$$m(x_1 - p) \leq \text{moment} \leq m(x_2 - p),$$

where m is the total mass.

It follows from (2) that a mass m at a point x has the moment $m(x - p)$. Then (2) says that if we concentrate all the mass at a point to the left of the actual distribution, then we decrease the moment, and if we concentrate all the mass at a point to the right then we increase the moment.

Starting from these axioms it is possible to *prove* the formula that we obtained heuristically. However, the modifications that have to be made in our earlier argument are quite delicate, and we shall postpone this line of reasoning until Chapter 18. (But see Problem 17.)

PROBLEMS FOR SECTION 5

In each of the following problems, a rod with variable density $\rho(x)$ lies along the given interval. Find its center of mass.

1. $\rho(x) = 1 + x$; $[0, 2]$ 2. $\rho(x) = x^2$; $[1, 2]$

3. $\rho(x) = 2x - x^2$; $[0, 2]$ 4. $\rho(x) = x^3$; $[0, 1]$

5. $\rho(x) = x^4$; $[1, 2]$ 6. $\rho(x) = x^4$; $[0, 2]$

7. $\rho(x) = \sin x$; $[0, \pi]$ 8. $\rho(x) = e^x$; $[0, 1]$

9. $\rho(x) = \log x$; $[1, 2]$

10. A rod with density $\rho(x) = x$ lies over the interval $[0, 2]$. If the rod is broken in two at $x = 1$, find the mass and center of mass of the whole rod and of

each of its two pieces. If these are respectively m, $\bar{x}$, m_1, $\bar{x}_1$, m_2, $\bar{x}_2$, show that

$$\bar{x} = \frac{m_1 \bar{x}_1 + m_2 \bar{x}_2}{m_1 + m_2}.$$

That is, the center of mass of the whole rod can be computed by replacing each of its two pieces by a point mass located at the center of mass of the piece, and then computing the moment of this system of two point masses.

11. A rod of varying density $\rho(x)$ lies over the interval $[a,c]$. If it is broken into two pieces at the point $x = b$, and if these pieces have masses m_1 and m_2, and centers of mass $\bar{x}_1$ and $\bar{x}_2$, show that

$$\bar{x} = \frac{m_1 \bar{x}_1 + m_2 \bar{x}_2}{m_1 + m_2}$$

is the center of mass of the whole rod.

12. A finite number of point masses m_i sit at positions x_i on the x-axis. Let M be the moment of this system about the pivot point $x = p$, as given by the formula in the text. Let $\bar{x}$ be the center of mass of the system, and m its total mass. Show that in the moment computation the system behaves as though it consisted of a single point mass m at $x = \bar{x}$. That is, show that

$$M = m(\bar{x} - p).$$

13. A rod with variable density $\rho(x)$ has total mass m, center of mass $\bar{x}$, and moment M about the pivot point $x = p$, all given by the integral formulas in the text. Show that, as far as the computation of M is concerned, the rod behaves as though it were a point mass m at $x = \bar{x}$. That is, show that $M = m(\bar{x} - p)$.

14. A rod with variable density $\rho(x)$ lies over the interval $[a,c]$, and is broken into two pieces at $x = b$. If M, M_1 and M_2 are, respectively, the moments about $x = p$ of the whole rod and its two pieces, as given by the integral formula in the text, show that

$$M = M_1 + M_2.$$

15. Use the results of Problems 13 and 14 to give a new proof of Problem 11.

16. If the moment axiom (1) is applied repeatedly it acquires the following more general formulation:

1′. *If a mass distribution is broken up into a finite number of pieces, then the moment of the whole is equal to the sum of the moments of its pieces.*

Show that the formula $M = \sum m_i(x_i - p)$ for the moment about p of a system of point masses m_i at positions x_i is a consequence of the moment principles (1′) and (2).

17. Assuming the moment axioms (1) and (2), we can use the methods of Chapter 8 to prove the formula

$$M = \int_a^b \rho(x)(x - p)\,dx$$

for the moment about p of a continuous mass distribution along the interval $[a,b]$. In fact, Principle (1) tells us that the moment of a mass distribution is itself a distribution over the axis. So we can apply Theorem 7 (of Chapter 8) if we can show that this distribution has the derivative $\rho(x)(x - p)$. What we have to show, then, is that if ΔM is the moment of the piece lying over the increment interval $[x, x + \Delta x]$, then

$$\lim_{\Delta x \to 0} \frac{\Delta M}{\Delta x} = \rho(x)(x - p).$$

Do this, as follows. First show, using Principle (2), that if Δm is the mass on the increment interval, then

$$(x - p)\Delta m \leq \Delta M \leq (x + \Delta x - p)\Delta m.$$

Then finish off, as usual, by the squeeze limit law.

6. THE CENTROID OF A PLANE LAMINA

We can view a region G in the plane as representing a thin sheet of material of constant density k. That is, the mass of any part of G is k times the area of that part. When G is interpreted this way, it is often called a *homogeneous lamina*. We may as well suppose that $k = 1$, so that the mass is just the area.

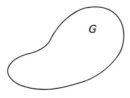

Suppose now that we view the coordinate plane as lying horizontally and that we interpret the region between the graph of a function f and the x-axis as a horizontal homogeneous lamina. We fix a line $x = p$ parallel to the y-axis and we consider whether or not the lamina G will "balance" on this new axis. That is, we consider the moment of the lamina G with respect to the axis $x = p$.

The seesaw principle for mass distributions in the plane has exactly the same formula as before: If masses m_i are placed at the points (x_i, y_i) on a rigid, weightless, horizontal sheet, then the system will exactly balance about the axis at $x = p$ if

$$\Sigma \, m_i(x_i - p) = 0.$$

That is, the moment of a mass m_i depends only on its distance from the axis at $x = p$, and hence is independent of its y-coordinate. We therefore treat the homogeneous lamina by dividing it into thin strips parallel to the balancing axis, each strip then having an approximately constant distance from the balancing axis. As usual, this slicing up of the lamina is discussed in terms of a subdivision

$$a = x_0 < x_1 < \cdots < x_n = b$$

of the interval $[a,b]$.

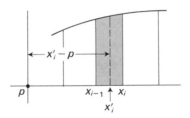

If Δm_i is the mass of the ith strip, lying between $x = x_{i-1}$ and $x = x_i$, and if an evaluation point x_i' is chosen from this interval, then the moment of this strip about the axis $x = p$ is approximately

$$\Delta m_i(x_i' - p).$$

We are assuming that the lamina has constant density equal to 1, so Δm_i is the area ΔA_i of the strip. Also, $\Delta A_i \approx f(x_i')\Delta x_i$ by the usual rectangular approximation to the area. We thus end up with

$$(x_i' - p)f(x_i')\Delta x_i$$

as an approximation to the moment of the ith strip, and the total moment is thus approximately

$$\Sigma \, (x_i' - p)f(x_i')\Delta x_i.$$

As the subdivision is taken finer we assume that this approximation will improve and have the exact moment as its limit. On the other hand, the above sum is a

Riemann sum for the function $(x - p)f(x)$, so its limit is known to be the definite integral of this function. Thus,

$$M = \int_a^b (x - p)f(x)\,dx$$

is the exact moment of the lamina about the axis $x = p$. Here, again, the moment axioms can be used to make the mathematics precise, but the physical conclusion is a law in the sense of physics, subject to constant verification.

This is the same formula that we obtained for the rod with variable density. The point is that since only distances in the x-direction count in determining this moment, the contribution of the whole strip over $[x_{i-1}, x_i]$ is the same as though it were a mass distribution on the x-axis of density approximately $f(x_i')$.

To get the axis $p = \bar{x}$ where the lamina will balance, we set the moment equal to zero and find

$$\bar{x} = \frac{\int_a^b xf(x)\,dx}{\int_a^b f(x)\,dx} = \frac{\text{moment about } y\text{-axis}}{\text{area of region}},$$

just as before. For a region bounded by two function graphs, say above by f and below by g, we just replace $f(x) = f(x) - 0$ by $f(x) - g(x)$ in the above computations. Thus,

$$\bar{x} = \frac{\int_a^b x(f(x) - g(x))\,dx}{\int_a^b (f(x) - g(x))\,dx}$$

is the more general form of the formula for $\bar{x}$. Actually, this more general form is a direct consequence of the earlier special form and the addition principle for moments (Axiom (1)), because of the identity

$$\int_a^b x(f(x) - g(x))\,dx = \int_a^b xf(x)\,dx - \int_a^b xg(x)\,dx.$$

This expresses the integral on the left as the difference of the moments about the y-axis of the regions R and R_2 below, and this is the moment of R_1, by Axiom (1).

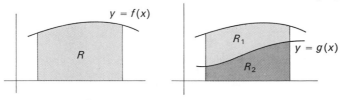

A plane lamina G will also have a balancing axis $y = \bar{y}$ parallel to the x-axis, and then $(\bar{x}, \bar{y})$ is the balancing *point* for G. This may not be obvious, but it can be proved. The balancing point $(\bar{x}, \bar{y})$ of a plane lamina G is called its *center of mass*. In the case of a homogeneous lamina, such as we are considering here, the point $(\bar{x}, \bar{y})$ is also called the *centroid* of the region G.

EXAMPLE 1. Find the centroid of a semicircular disk.

Solution. Let G be the right half-disk bounded by the circle $x^2 + y^2 = r^2$ and the y-axis.

Then

$$\bar{x} = \frac{\int_0^r x \cdot 2\sqrt{r^2 - x^2}\, dx}{\frac{1}{2}\pi r^2}$$

$$= \frac{-\frac{2}{3}(r^2 - x^2)^{3/2}\big|_0^r}{\frac{1}{2}\pi r^2}$$

$$= \frac{\frac{2}{3}r^3}{\frac{1}{2}\pi r^2} = \frac{4r}{3\pi}.$$

By symmetry $\bar{y}$ must be 0, so $(4r/3\pi, 0)$ is the centroid.

If a region G is not symmetric about the x-axis, then $\bar{y}$ must be computed. When G can be described as lying between the y-axis and the graph of $x = h(y)$, from $y = c$ to $y = d$, then exactly the same reasoning that gave us the formula for $\bar{x}$ will now lead to the corresponding formula for $\bar{y}$:

$$\bar{y} = \frac{\int_c^d y h(y)\, dy}{\text{area of } G} = \frac{\text{moment about } x\text{-axis}}{\text{area of } G}.$$

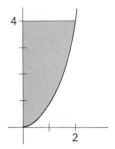

EXAMPLE 2. If G is the region between the y-axis, the parabola $y = x^2$, and the line $y = 4$, then we can compute $\bar{x}$ as before:

$$\bar{x} = \frac{\int_0^2 x(4 - x^2)\,dx}{\int_0^2 (4 - x^2)\,dx} = \frac{2x^2 - \dfrac{x^4}{4}\Big]_0^2}{4x - \dfrac{x^3}{3}\Big]_0^2} = \frac{4}{16/3} = \frac{3}{4}.$$

But G can also be described as lying between the y-axis and the graph of $x = \sqrt{y}$, from $y = 0$ to $y = 4$. So we can compute $\bar{y}$ by the corresponding y integral:

$$\bar{y} = \frac{\int_c^d yh(y)\,dy}{\text{area}} = \frac{3}{16} \int_0^4 y^{3/2}\,dy = \frac{3}{16} \cdot \frac{2}{5} y^{5/2}\Big]_0^4 = \frac{12}{5}.$$

Unfortunately, a region that can easily be described when it is viewed as lying over the x-axis may be hard to describe when viewed as lying over the y-axis. One wonders therefore whether it is possible to compute $\bar{y}$ by some kind of an x integral. It is, and in Problems 25 and 26 a proof is sketched for the following result:

If G is bounded above and below by the graphs of $y = f(x)$ and $y = g(x)$, and runs from $x = a$ to $x = b$, then its moment about the x-axis is given by

$$\frac{1}{2} \int_a^b (f^2(x) - g^2(x))\,dx.$$

Then $\bar{y}$ is found by dividing this moment by the area of G.

EXAMPLE 3. Find the y coordinate of the centroid of the region bounded by the graphs of $y = e^x$ and $y = 2e^x$, the y-axis and the line $x = 1$.

Solution. The area of G is

$$\int_0^1 (2e^x - e^x)\,dx = e^x\Big]_0^1 = e - 1.$$

According to the new formula above, the moment of G about the x-axis is

$$\frac{1}{2}\int_0^1 (4e^{2x} - e^{2x})\, dx = \frac{3}{4}e^{2x}\Big]_0^1 = \frac{3}{4}(e^2 - 1).$$

Then

$$\bar{y} = \frac{\text{moment about } x\text{-axis}}{\text{area}}$$

$$= \frac{\frac{3}{4}(e^2 - 1)}{e - 1} = \frac{3}{4}(e + 1).$$

PROBLEMS FOR SECTION 6

Find the center of mass of the homogeneous plane lamina G (centroid of the plane region G) where:

1. G is the rectangle with two sides along the axes and a vertex at (a,b).
2. G is an arbitrary rectangle with sides parallel to the axes.
3. G is bounded by the graph $y = |x|$ and the line $y = 2$.
4. G is bounded by the lines $y = x$, $y = 2$, and the y-axis.
5. G is bounded by the y-axis and the parabola $y^2 = 4 - x$.
6. G is the triangle formed by the coordinate axes and the line $x + 2y = 2$.
7. G is bounded by the parabola $y = x^2$ and the line $y = x$.
8. G is bounded by the two parabolas $y = x^2$, $x = y^2$.
9. G is the region in the first quadrant bounded by the cubic $y = x^3$ and the line $y = 2x$.
10. G is bounded by the curve $y^2 = x^3$ and the line $x = 2$.
11. G is bounded by the curves $y = e^x$, $y = -e^x$, $x = 0$, and $x = a$.
12. G is the part of the circular disk of radius r about the origin lying in the first quadrant.
13. G is the right half of the interior of the ellipse

$$\frac{x^2}{a^2} + \frac{y^2}{b^2} = 1.$$

14. G is the region $0 \le y \le \log x$, $x \le 2$.
15. G is the region $x \ge 0$, $-e^{-x} \le y \le e^{-x}$. (Here G extends to infinity and the integrals in the formula for $\bar{x}$ have to be evaluated as $\lim_{a\to\infty} \int_0^a \cdots$)

16. G is the triangle with vertices at $(a,0)$, $(b,0)$, and $(0,c)$, where $a < b$ and $c > 0$. [Hint: The numerator in the formula for $\bar{x}$ (the moment about the y-axis) can conveniently be evaluated as the sum of two moments.]

17. G is bounded by the parabola $y = x^2$ and the line $y = x + 2$.

18. Prove the following Theorem of Pappus.

Theorem. *If a plane region G is revolved about a line in the plane that does not cut into G, then the volume generated is equal to the area of G times the length of the circle traced by the centroid of G.*

Suggestion. Taking the axis of revolution as the y-axis, the center of mass $(\bar{x}, \bar{y})$ traces a circle of length $2\pi\bar{x}$ and the formula to be proved is

$$V = 2\pi\bar{x}A,$$

where A is the area of G and V is the volume of the solid of revolution generated by rotating G. In order to prove this formula, write down the formula for $\bar{x}$ and note that its numerator is very close to being the formula for computing V by shells.

19. We know the formulas for the area of a circle and the volume of a sphere. Use the Theorem of Pappus to compute the centroid of a semicircular disk.

20. Compute the formula for the volume of a torus by the Theorem of Pappus. Take the torus to be the solid generated by rotating the circle

$$(x - R)^2 + y^2 = r^2$$

about the y-axis, where $0 < r < R$.

21. The area of an ellipse is πab, where a and b are its semiaxes. Compute the volume of the "elliptical torus" generated by revolving the ellipse

$$\frac{(x - R)^2}{a^2} + \frac{y^2}{b^2} = 1$$

about the y-axis, where $a < R$.

22. Use the Theorem of Pappus to compute the volume of a cylindrical shell with rectangular cross section.

23. The sides of a square S lie along lines having slopes ± 1. Its left vertex is the point $(a,0)$ and its area is A. Show that the volume generated by rotating S about the y-axis is

$$V = 2\pi A(a + \sqrt{\frac{A}{2}}).$$

24. Prove from the integral formula for moments that the moment of a region G about the axis $x = p$ is $A(\bar{x} - p)$, where A is the area of G and $\bar{x}$ is the x-coordinate of its centroid.

25. Consider the region G lying between the graphs $y = f(x)$ and $y = g(x)$ and running from $x = a$ to $x = b$, where $a < b$, and $g(x) \le f(x)$ over the interval $[a,b]$.

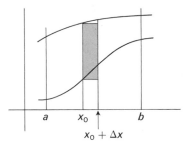

The incremental slice of G between x_0 and $x_0 + \Delta x$ is approximately a rectangle of altitude $f(x_0) - g(x_0)$. Show that the moment of this rectangle about the x-axis is

$$\frac{1}{2}[f^2(x_0) - g^2(x_0)]\Delta x.$$

(Assume the moment formula analogous to the one in Problem 24.)

26. Continuing the above problem, write down an approximation for the moment of G about the x-axis as a sum of moments of rectangles, and then conclude that the exact moment is

$$\frac{1}{2}\int_a^b [f^2(x) - g^2(x)]dx.$$

Find the centroid of each of the following regions G.

27. G is bounded by the x-axis and the first arch of the graph of $\cos x$.

28. G is bounded by the y-axis and the lines $y = 2x$, $y = x + 2$.

29. G is bounded by the parabolas $y = x^2$, $y = 2 - x^2$.

30. G is bounded by the parabolas $y = 2x^2$, $y = x^2 + 1$.

31. For a region G running from the x-axis up to the graph of $y = f(x)$ the moment M given by the formula in Problem 26 has the simpler expression

$$M = \frac{1}{2}\int_a^b f^2(x)dx.$$

Show that if $0 \le g(x) \le f(x)$, then the more general formula of Problem 26 follows from the above special case and the addition Axiom (1) for moments.

32. Assuming the Theorem of Pappus (Problem 18), show that the above moment formula (in Problem 31) follows from the formula for the volume of a solid of revolution.

7. THE TRAPEZOIDAL APPROXIMATION

By now you will have a good feeling for the difficulties one can meet in evaluating a definite integral $\int_a^b f$, and can better appreciate the need for numerical estimates. If f is a function, like $f(x) = e^{-x^2}$, for which no elementary antiderivative exists, then a direct numerical evaluation is our only recourse. But even if we can integrate f, the antiderivative F that we find may be very complicated, and the evaluation

$$\int_a^b f = F(b) - F(a)$$

may be more difficult than using some direct numerical estimate.

The only numerical method that we have considered so far is the approxima tion of $\int_a^b f$ by a Riemann sum. However, this approximation is inefficient, at least for hand computation, since it requires considerable labor to guarantee even a moderately good estimate. The trouble is that we are, in effect, approximating f locally by a constant function, and this takes no account at all of the shape of the graph of f.

When f is very smooth, in the sense of having several derivatives, then unexpected fluctuations are impossible and the values of f at successive subdivision points ought to determine the whole graph to within very narrow margins. Therefore, it should be possible to compute the integral $\int_a^b f$ from the subdivision values $f(x_i)$, where $i = 0, 1, \ldots, n$, very accurately.

The most obvious way to make some use of the shape of the graph is to replace each rectangle associated with a Riemann sum by the corresponding trapezoid, as shown below. The area of a trapezoid is $h(l_1 + l_2)/2$, where l_1 and l_2 are the

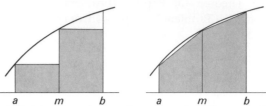

$$a \quad\quad m \quad\quad b \quad\quad\quad a \quad\quad m \quad\quad b$$

lengths of the parallel sides, i.e., the two base lengths. Thus, the figure on the right above suggests the estimates

$$\int_a^m f \approx \Delta x \frac{f(a) + f(m)}{2},$$

$$\int_m^b f \approx \Delta x \frac{f(m) + f(b)}{2},$$

and so

$$\int_a^b f \approx \Delta x \left[\frac{f(a)}{2} + f(m) + \frac{f(b)}{2} \right].$$

We expect from the look of the geometric figures that this estimate will be much better than our original Riemann-sum estimate

$$\int_a^b f \approx \Delta x [f(a) + f(m)].$$

In general, we subdivide the interval $[a,b]$ into n equal subintervals of common length $\Delta x = (b - a)/n$ the subdividing points being $a = x_0, x_1, \ldots, x_n = b$.

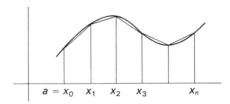

$$a = x_0 \quad x_1 \quad x_2 \quad x_3 \quad\quad\quad x_n$$

Then we add up all the individual trapezoidal estimates

$$\int_{x_{i-1}}^{x_i} f \approx \Delta x \frac{f(x_{i-1}) + f(x_i)}{2},$$

from $i = 1$ to $i = n$, and end up with the approximation

$$\int_a^b f \approx \Delta x \left[\frac{f(x_0)}{2} + f(x_1) + f(x_2) + \cdots + f(x_{n-1}) + \frac{f(x_n)}{2} \right]$$

$$= \Delta x \left[\frac{f(a) + f(b)}{2} + \sum_{k=1}^{n-1} f(x_k) \right].$$

Again, this is the analytic estimate corresponding to the approximation of the area under the graph of f by the sum of all the trapezoidal areas associated with the subdivision. This formula is called the *trapezoidal rule*. It is a much better estimate of $\int_a^b f$, as we shall now show.

Theorem 2. *If the second derivative of f exists and is bounded by K over the interval $[a,b]$, then the error in using the trapezoidal rule to estimate $\int_a^b f$ is at most*

$$\frac{K(b-a)}{12}(\Delta x)^2$$

in magnitude.

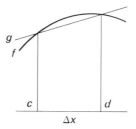

Proof. We first consider the error arising from a single trapezoid. Let g be the linear function that is equal to f at the two points c and d, as shown in the figure. Then $\int_c^d g$ is the area of the trapezoid in question. Therefore, if we set $h = f - g$, then

$$\int_c^d h = \int_c^d f - \int_c^d g = \int_c^d f - \Delta x \cdot \frac{f(c) + f(d)}{2}$$

$$= \text{error in trapezoidal estimate.}$$

On the other hand,

$$h(c) - h(d) = 0,$$

and

$$h'' = f'' - g'' = f'' - 0 = f'',$$

so that

$$|h''| \leq K$$

on (a,b). We saw in Section 8 of Chapter 9 that then

$$\left| \int_c^d h \right| \leq \frac{K(\Delta x)^3}{12}.$$

This, then, bounds the error in a trapezoidal estimate over a single interval of width Δx. If we subdivide $[a,b]$ into n equal subintervals of width $\Delta x = (b-a)/n$

each, and use the trapezoidal estimate over each one, then the total error resulting from summing the n trapezoidal estimates is at most

$$n\left[\frac{K(\Delta x)^3}{12}\right] = \frac{K(b-a)}{12}(\Delta x)^2,$$

since $n(\Delta x) = b - a.$ This proves the theorem. ∎

EXAMPLE 1. We use the same fourfold subdivisions of $[1, 3/2]$ as in Example 1 of Section 1 to estimate $\log(3/2)$ by the trapezoidal rule. We have

$$\log\frac{3}{2} = \int_1^{3/2}\frac{dt}{t} \approx \frac{1}{8}\left[\frac{x_0}{2} + x_1 + x_2 + x_3 + \frac{x_4}{2}\right]$$

$$= \frac{1}{8}\left[\frac{1}{2} + \frac{8}{9} + \frac{4}{5} + \frac{8}{11} + \frac{1}{2}\cdot\frac{2}{3}\right]$$

$$= \frac{1}{16} + \frac{1}{9} + \frac{1}{10} + \frac{1}{11} + \frac{1}{24}.$$

Here $f''(x) = 2/x^3$, which is bounded by $K = 2$ over the interval $[1,3/2]$. The error is thus at most

$$\frac{K}{12}(b-a)(\Delta x)^2 = \frac{2}{12}\cdot\frac{1}{2}\cdot\left(\frac{1}{8}\right)^2 = \frac{1}{768}.$$

In Section 1 we found that the simple Riemann-sum estimates $\log(3/2)$ with an error at most $(1/32)$. So here we are guaranteed a better estimate by a factor of 24.

You can check that $(1/768)$ is less then 0.002, so we can say that the error in our trapezoidal estimate of $\log(3/2)$ is less than 2 in the third decimal place.

In order to see how large the error actually is we can compare our estimates with the value from a log table. Dividing out the fractions in the trapezoidal estimate shows that its decimal expansion begins 0.4061. (We haven't discussed decimal expansions yet but you may know how to do this.) From a four-place table of natural logarithms we find that

$$\log\frac{3}{2} = 0.4055$$

to the nearest four decimal places. The actual error in our trapezoidal estimate of $\log 3/2$ is thus less than $0.4062 - 0.4054 = 0.0008$ and so is less than 1 in the third decimal place.

We shall see in Chapter 13 that the same fourfold subdivision of $[0, 3/2]$ leads to still closer estimates of $\log 3/2$ when we make use of bounds on higher derivatives. Thus, with essentially the same amount of arithmetic as required by a Riemann sum, but more sophistication, we can get very good approximations to a definite integral.

PROBLEMS FOR SECTION 7

1. Write down the trapezoidal approximation to log 2 arising from a tenfold subdivision of $[1,2]$. Leave the answer as a sum of rational numbers, as in Example 1 in the text. Show that this sum approximates log 2 with an error less than 0.002.

2. How fine a subdivision do we need to take in order to ensure that the trapezoidal sum will approximate log 2 to the nearest four decimal places (i.e., with an error less than $0.00005 = 5(10)^{-5}$)?

3. Write down the trapezoidal approximation to

$$\frac{\pi}{4} = \int_0^1 dx/(1 + x^2)$$

arising from a tenfold subdivision of $[0,1]$. Leave the answer as a sum of rational numbers, as in Example 1 in the text. Show that the error is less than 0.002.

4. Show that the trapezoidal approximation is exact (i.e., has zero error) for a linear integrand $f(x) = Ax + B$.

5. Compute the trapezoidal approximation to $\int_0^b x^2\,dx$ for the trivial subdivision of $[0,b]$ (no subdividing points). Compare with the value of the integral, and hence show that the error formula in Theorem 2 cannot be improved.

6. Compute the trapezoidal approximation to $\int_0^1 x^3\,dx$ for the fivefold subdivision of $[0,1]$. Show that the actual error is exactly one half the error bound provided by Theorem 2.

7. The proof of Theorem 2 didn't use the full amount of information available from Lemma 4 in the last chapter. Modify the proof to show that the error in the trapezoidal approximation can be exactly expressed as

$$\frac{-\sum_1^n f''(X_i)}{12}(\Delta x)^3.$$

8. The above expression contains a Riemann sum for f'' over the same subdivision, to which we can apply Theorem 1. Prove the following theorem.

Theorem. *If the third derivative f''' is bounded by K on $[a,b]$, and if the trapezoidal estimate is modified by adding the correction term*

$$-\frac{f'(b) - f'(a)}{12},$$

then the modified estimate of $\int_a^b f$ has an error at most

$$\frac{K}{24}(b - a)(\Delta x)^3.$$

(The modified estimate is actually much better than this, as we shall see in Chapter 13.)

9. Show that if the estimate of $\log 3/2$ in Example 1 is modified as in the above Theorem, then the error is less than $0.003 = 3(10)^{-4}$.

chapter 11

paths

This chapter further develops the themes of Chapter 5, with some variations. A curve in the plane is now viewed as a locus, or a path traced by a moving point. From this vantage point we can learn how to sketch the shapes of curves more general than function graphs, and in later chapters we can analyze the notions of velocity and acceleration. In this connection we add *curvature* as a new magnitude that can be computed for a curve.

The path idea lets us visualize geometrically a powerful generalization of the mean-value theorem, called the *parametric mean-value theorem.*

Finally, we look once more at the group of geometric principles upon which we have been basing so much of our work, and develop their interconnections as analytic theorems.

1. PATHS

EXAMPLE 1. It is often useful to view a graph as a path that is traced in some manner. Consider the two equations

$$x = t^3 - 3t,$$

$$y = t^2.$$

For any given value of t, the corresponding values of x and y, taken together, give a point in the xy-coordinate plane. For instance, when $t = -2$, we have $(x, y) = (-2, 4)$. As t varies, the point

$$(x, y) = (t^3 - 3t, t^2)$$

varies and traces out the curve in the xy-plane shown below. That is, the collection of all points (x, y) of the form $(x, y) = (t^3 - 3t, t^2)$ make up this curve. We shall see later how we know the graph looks like this.

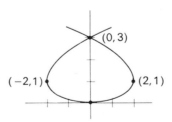

Note that we don't plot the variable t, but plot coordinates x and y that are determined by t. Thus, t is the independent variable, but it doesn't appear in the geometric picture. We call the equations $x = t^3 - 3t$, $y = t^2$ a *parametric representation* of the curve, and t is the *parameter.*

In general, any pair of functions f and g can be used together to define a curve parametrically in the above way. Using the same notation, the parametric equations are

$$x = g(t),$$
$$y = h(t),$$

and the curve is made up of the points $(x, y) = (g(t), h(t))$. We imagine the parameter t changing, and visualize the varying point $(x, y) = (g(t), h(t))$ tracing the curve. Probably the most vivid picture, and one of the most important interpretations, comes from thinking of t as time and $(x, y) = (g(t), h(t))$ as the position at time t of a moving point or particle. Then the path of this moving particle lies along the curve. Because of this interpretation, a parametric representation of a curve is often called a *path*. Sometimes we call the curve itself a path.

We shall consider only paths for which the parameter functions are continuously differentiable. The path of a moving particle is always like this.

The example above shows that a path does not necessarily lie along the graph of a function. In fact, one reason for introducing parametric representations is to obtain a new way of discussing curves more general than function graphs.

Sometimes we can eliminate the parameter t from the parametric equations

$$x = g(t),$$
$$y = h(t),$$

and obtain a single equation in x and y for the graph of the path. In the above example,

$$x = t^3 - 3t,$$
$$y = t^2,$$

if we square x, then the right side involves only powers of $t^2 = y$, and we obtain the equation

$$x^2 = y^3 - 6y^2 + 9y.$$

Thus, the point $(x, y) = (t^3 - 3t, t^2)$ moves along the graph of this equation.

Often we can recognize a path by eliminating the parameter in the above way. In each of the following examples, the parameter is easily eliminated to get an equation in x and y, and in each case we can easily sketch the graph of the equation. Then the path of the point along this known curve can be indicated by adding arrows.

EXAMPLE 2. $x = t^3$; $y = t^2$.

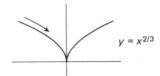

This example is closely related to the first example.

EXAMPLE 3. $x = \cos t$; $y = \sin t$.

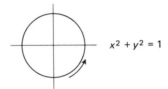

EXAMPLE 4. $x = \cos t^3$; $y = \sin t^3$.

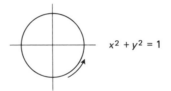

The point of this last example is the use of the word *path*. Some people would say that Examples 3 and 4 involve the same path, although it is traced somewhat differently in the two situations. Others feel that the idea of a path includes the way it is traced, and that these examples are, therefore, *different* paths. This ambiguity in the meaning of "path" causes no difficulty. We just have to be careful to be clear in cases where it matters.

In Example 2, even though the coordinate functions

$$x = g(t) = t^3 \qquad \text{and} \qquad y = h(t) = t^2$$

are everywhere continuously differentiable, the graph itself has a kink (called a *cusp*) at the origin, and the path direction along the graph takes a sudden 180° turn there. We shall see later that such a discontinuity in path direction can be ruled out by requiring path smoothness in the following sense.

Definition. *A path* $(x,y) = (g(t), h(t))$ *is smooth if* g *and* h *have continuous derivatives and if* g' *and* h' *are never simultaneously zero.*

The theorem that we are assuming for the moment is:

Theorem 1. *A smooth path has a continuously varying path direction.*

Note that $g'(0) = h'(0) = 0$ in Example 2, and according to the theorem it is this common critical point that allows the cusp.

We can sketch smooth paths much as we sketched graphs in Chapter 3. When we first considered function graphs, we decided that a graph could change from rising to falling (or vice versa) only by going through a point where the tangent is *horizontal*.

A path like the first one drawn above appears to have a complementary property: It can change from moving forward to moving backward (or vice versa) only by going through a point where the tangent is *vertical*.

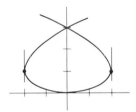

This suggests the following definition.

Definition. *The critical points of a smooth path are the points where the tangent line is either vertical or horizontal.*

Then we conjecture the theorem:

Theorem 2. *The critical points of a smooth path divide the path up into pieces, each one of which runs in one direction along a monotone function graph.*

This is evident in the drawing above, where the three critical points divide the path into four monotone function graphs. The theorem is true for all smooth paths, and is the principal tool for sketching paths. However, before it can be proved or even used, the path critical points have to be connected to derivative zeros of the coordinate functions g and h. What we need, therefore, is a formula for the slope of the path in terms of g' and h'.

In order to find such a formula, we consider a general path point (x, y) $= (g(t), h(t))$, and give the parameter t a (nonzero) increment Δt, obtaining a new path point

$$(x + \Delta x, y + \Delta y) = (g(t + \Delta t), h(t + \Delta t)).$$

The secant line between these two path points has the slope $\Delta y/\Delta x$, and the tangent slope m is the limit of the secant slope as the second point approaches the first point along the path, i.e., as Δt approaches 0. Thus

$$m = \lim_{\Delta t \to 0} \frac{\Delta y}{\Delta x} = \lim_{\Delta t \to 0} \frac{\frac{\Delta y}{\Delta t}}{\frac{\Delta x}{\Delta t}} = \frac{h'(t)}{g'(t)}.$$

This calculation is valid only if $g'(t) \neq 0$. Then $\Delta x \neq 0$ when Δt is sufficiently small (Theorem 4, Chapter 5), and there is no division by zero. However, the formula is still correct in the case $g'(t) = 0$, $h'(t) \neq 0$, provided it is interpreted as saying that the slope is infinite and the tangent line is vertical. For in this case the *reciprocal* slope is zero,

$$\lim_{\Delta t = 0} \frac{\Delta x}{\Delta y} = \frac{g'(t)}{h'(t)} = 0,$$

as above. And the fact that $\Delta x / \Delta y$ approaches zero means geometrically that the secant line approaches a vertical line as $\Delta t \to 0$. We have proved the following theorem.

Theorem 3. *At every point of a smooth path* $(x, y) = (g(t), h(t))$, *the path slope m is given by*

$$m = \frac{h'(t)}{g'(t)}.$$

If $g'(t) = 0$, *the formula is interpreted as saying that the slope is infinite and the tangent line is vertical.*

Corollary. *A smooth path has a tangent at every point, and its critical points correspond exactly to the values of the parameter t for which either*

$$g'(t) = 0 \quad \text{or} \quad h'(t) = 0.$$

REMARK. We have defined a path critical point as a point on the curve, whereas a function critical point was defined as the corresponding value of the independent variable. This ambiguity does no harm since it is always clear which way we are using the term.

Before proving Theorem 2 we shall look at some examples.

EXAMPLE 5. We know that the path

$$x = \cos t,$$
$$y = \sin t,$$

traces the unit circle $x^2 + y^2 = 1$ counterclockwise. Let us see, however, what Theorem 2 by itself would tell us about the path.

The path critical points are the points where either derivative is zero. We find

$$\frac{dx}{dt} = -\sin t = 0 \qquad \text{at } t = 0,\ \pi,\ 2\pi,\ \text{etc.}$$

$$\frac{dy}{dt} = \cos t = 0 \qquad \text{at } t = \pi/2,\ 3\pi/2,\ \text{etc.}$$

We compute the corresponding path points, and get the table

t	0	$\pi/2$	π	$3\pi/2$	2π	. . .
x	1	0	-1	0	1	. . .
y	0	1	0	-1	0	. . .

These points are in consecutive order on the path, which runs along a monotone graph between each pair. The general configuration must therefore be like this:

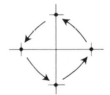

The slope formula $m = g'(t)/h'(t)$ shows whether the critical point tangent is vertical or horizontal. However, it can also be decided geometrically. For example, the tangent at the critical point $(1,0)$ must be vertical because the path doubles back from that point. The simplest curve that we can draw that will meet these requirements is some kind of oval, and we thus get the right general shape from these path considerations.

EXAMPLE 6. We now show that we have been using the correct figure for our first path

$$x = g(t) = t^3 - 3t,$$

$$y = h(t) = t^2.$$

First,

$$\frac{dx}{dt} = 3(t^2 - 1) = 0 \quad \text{at } t = \pm 1,$$

$$\frac{dy}{dt} = t^2 = 0 \quad \text{at } t = 0,$$

so that altogether these critical parameter values break the t-axis up into four intervals.

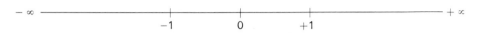

We then make up the corresponding critical point table:

t	$-\infty$	-1	0	1	∞
x	$-\infty$	2	0	-2	∞
y	$+\infty$	1	0	1	∞

Note also that $x = t^3 - 3t = 0$ when $t = 0, \pm\sqrt{3}$, so we can add the y-intercept at $y = (\sqrt{3})^2 = 3$. Plotting these points gives the framework shown at the left below, and we draw as smooth a path as we can, following these directions.

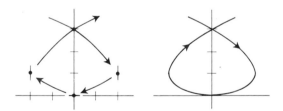

PROBLEMS FOR SECTION 1

In each of the following problems eliminate the parameter to obtain an equation in x and y for the graph along which the path runs. Sketch and identify (if possible) the graph, and indicate by one or more arrows how it is being traced. Compute the path slope dy/dx as a function of t and determine the path critical points (points where $dy/dx = 0$ or ∞, or is undefined).

1. $x = t, y = t^2$

2. $x = t - 1, y = 2t + 3$

3. $x = t - 1, y = t^3$

4. $x = 3 \sec t, y = 2 \tan t$
5. $x = 2t^{1/3}, y = 3t^{1/3}$
6. $x = t^2 - 1, y = t + 1$
7. $x = 2 \sin t, y = 3 \cos t$
8. $x = t + 1, y + t(t + 4)$
9. $x = 2 \cos t, y = 4 \sin t$
10. $x = a \cos^3 t, y = a \sin^3 t$
11. $x = 2 \cos t, y = 2 \sin t$
12. $x = a \cos t, y = b \sin t$

13. The function f has the graph shown below.

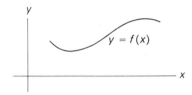

Sketch the following paths.
a) $x = t, y = f(t)$.
b) $x = -t, y = f(-t)$.
c) $x = f(t), y = t$.

14. Sketch the path $x = -t, y = f(t)$, where f is the function of Problem 13.

15. Sketch the path $x = t^2, y = t^4$, including some arrows to show how it is traced.

16. Do the same for the path $x = t^3 - 3t, y = (t^3 - 3t)^2$.

Parametric equations can be used to great advantage in a practical way when analyzing the motion of projectiles. In this case, the motion may be represented by:

$$x = v_0 t \cos \theta,$$

$$y = v_0 t \sin \theta - \frac{1}{2} g t^2,$$

where v_0 is the initial velocity, θ is the initial angle of inclination from the horizontal, and g is the acceleration of gravity (32 ft/sec^2).

17. A shell is fired with an initial velocity of 100 ft/sec and $\cos \theta = 3/5$ and $\sin \theta = 4/5$. Find the time t when the shell hits the ground, and the distance between the gun and the spot where the shell lands; also, find the maximum height reached by the shell.

18. A ball is thrown off the top of a building 100 ft high with a velocity of 60 ft/sec and at an angle of 45° above the horizontal. Find
    a) The maximum height attained by the ball.
    b) The distance from the building to the point where the ball hits the ground.
    c) The velocity of the ball when it strikes the ground.

Analyze each of the following paths by Theorem 2, following the scheme used in Example 6. Include the x and y intercepts in your table if you can determine them, and indicate any singular points that you find (points where the path is not smooth).

19. $x = t^2, y = t^2$.

20. $x = t^2, y = t^2 - 2t$.

21. $x = t^3 - 3t, y = t^3 - 3t^2$.

22. $x = t^2, y = 2t^3 - 9t^2 + 12t$.

23. $x = t^3 - 3t^2 + 3t, y = t^3 - 3t^2$.

24. $x = t^2 - t, y = t^4 - 2t^2$.

25. $x = 4t^3 - 3t^2, y = t^4 - 2t^2$.

26. Elsewhere we have called a function f smooth if it is continuously differentiable. Show that a function is smooth if and only if its graph is a smooth path when it is traced in the standard way (with the independent variable as parameter).

27. It is reasonable to call a *curve* smooth if there is a smooth path running along it. It is always possible to trace a smooth curve by a different path that fails to meet the smoothness requirement at one or more points. (Later we shall see that this amounts to tracing the curve by a particle that comes to a stop one or more times.) Show that Examples 3 and 4 in the text illustrate this possibility.

2. PROOF OF THEOREM 2

If a path $(x, y) = (g(t), h(t))$ runs for a while along the graph of a differentiable function F, then the coordinate-parameter functions

$$x = g(t) \quad \text{and} \quad y = h(t)$$

are related by the equation

$$y = F(x),$$

or

$$h(t) = F(g(t)),$$

along this part of the path. Therefore

$$\frac{dy}{dt} = F'(x)\frac{dx}{dt}$$

along this piece. This derivative relationship has two consequences.

First, if the path is smooth, then dx/dt can never be zero along this piece. For if $dx/dt = 0$, then the formula shows that $dy/dt = 0$ also, and this contradicts the definition of a smooth path. Second, if $dx/dt \neq 0$, then

$$F'(x) = \frac{dy/dt}{dx/dt} = \frac{h'(t)}{g'(t)},$$

along this piece. Since $F'(x)$ is the slope m of the graph of F, we thus recover this case of the formula

$$m = \frac{h'(t)}{g'(t)}$$

for the slope of a path in terms of the parameter functions.

We need a partial converse of the above remarks in order to prove Theorem 2.

Theorem 4. *If f and g are differentiable functions on an interval I and if $g'(t)$ is never zero in the interior of I, then as t runs across I the path $(x, y) = (g(t), h(t))$ runs along the graph of a continuous function F, forwards if $x = g(t)$ is increasing, and backwards if g is decreasing. Moreover, F is differentiable for t in the interior of I.*

Proof. Since g has no critical points interior to I, it is monotone on I. Since $x = g(t)$ is monotone and continuous on I, it has a monotone, continuous inverse function $t = k(x)$, by the inverse function theorem (Chapter 7, Theorem 4). Then

$$y = h(t) = h(k(x))$$

while t is in I. That is, if we set $F(x) = h(k(x))$, then the equation

$$y = F(x)$$

holds along the path while t is crossing I. If $x = g(t)$ is increasing, then the path runs along the graph of F in the normal left-to-right direction. If $x = g(t)$ is decreasing, then the path traces the graph of F backwards.

Since g' is never zero interior to I, its inverse k is differentiable, and $k'(x) = 1/(g'(t))$ at the corresponding values of x, again by the inverse-function theorem. Then the composite function $F(x) = h(k(x))$ is differentiable, by the chain rule. This completes the proof of Theorem 4. ∎

Theorem 2 is a corollary of Theorem 4 for, along an interval I containing no path critical points, both derivatives $g'(t)$ and $h'(t)$ are different from zero. Since $g' \neq 0$, the path runs along the graph of a function F. Since $h' \neq 0$, the formula $F'(x) = h'(t)/g'(t)$ shows that $F' \neq 0$, and therefore that F is monotone. Thus, between successive critical points, a smooth path runs along a monotone function graph, as claimed in Theorem 2.

3. PATH LENGTH

We saw in the last chapter that we could define the length of a smooth function graph as the limit of lengths of inscribed polygons, and that this limit had a natural formulation as a definite integral.

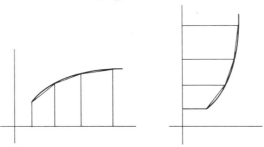

If we rotate the plane over the $y = x$ diagonal, such a function graph rotates to a nonstandard function graph along which x is a function of y. But the inscribed polygonal lengths remain the same and they necessarily have the same limit, so the *length* of a piece of a graph remains unchanged by such a rotation. However, the description of the length calculation for the new graph proceeds with the roles of x and y interchanged, so the integral formula for the length is now

$$\int_a^b \sqrt{1 + \left(\frac{dx}{dy}\right)^2}\, dy.$$

We have seen that a smooth path can be broken up into a succession of arcs, each of which is one of the above two types. The length of a smooth path is therefore a sum of lengths, each of which can be computed by one of the above formulas. But this is a piecemeal way to deal with path length, and we clearly would like a single integral formula with the parameter as variable.

The most straightforward way to approach such a path length formula would be to start over again with inscribed polygons and express their lengths in terms of the parameter t. This works all right, but it involves an extra complication, namely, that the resulting finite t sums are not quite Riemann sums.

A second possibility is to change variables in each of the function-graph integrals, transforming each type to a t-integral, and then amalgamating to a single integral.

Instead, we shall rework the derivative approach. We are better off than the first time we took this approach, in Chapter 8, because the basic geometric assumption is now an established fact. We know that for small arcs and their chords,

$$\frac{\text{arc length}}{\text{chord length}} \to 1$$

as chord length $\to 0$. (See Section 4 in Chapter 10.)

So let s be the length of the smooth path $(x,y) = (g(t), h(t))$ from the fixed point (x_0, y_0) at $t = t_0$. Then:

Theorem 5. *The path length s is a differentiable function of t, and*

$$\left(\frac{ds}{dt}\right)^2 = \left(\frac{dx}{dt}\right)^2 + \left(\frac{dy}{dt}\right)^2.$$

Proof. We assume that $t > t_0$, so that s is an increasing function of t and $\Delta s/\Delta t$ is positive. If we give t a positive increment Δt, then the chord length l for the corresponding arc increment is

$$l = \sqrt{(\Delta x)^2 + (\Delta y)^2}.$$

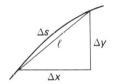

Therefore,

$$\frac{\Delta s}{\Delta t} = \frac{\Delta s}{l} \cdot \frac{l}{\Delta t} = \frac{\Delta s}{l} \sqrt{\left(\frac{\Delta x}{\Delta t}\right)^2 + \left(\frac{\Delta y}{\Delta t}\right)^2}.$$

We know that $\Delta s/l \to 1$ as $\Delta t \to 0$. Therefore $ds/dt = \lim_{\Delta t \to 0}(\Delta s/\Delta t)$ exists and

$$\frac{ds}{dt} = \sqrt{\left(\frac{dx}{dt}\right)^2 + \left(\frac{dy}{dt}\right)^2}.$$

If Δt is negative, then we have to keep track of minus signs in the above calculation, but the result is the same. However, if $t < t_0$, so that s is a decreasing function of t, then the length of the increment in arc is $-\Delta s$ for positive Δt, and this extra minus sign leads to

$$\frac{ds}{dt} = \lim_{\Delta t \to 0} \frac{\Delta s}{\Delta t} = -\sqrt{\left(\frac{dx}{dt}\right)^2 + \left(\frac{dy}{dt}\right)^2}.$$

In every case, however,

$$\left(\frac{ds}{dt}\right)^2 = \left(\frac{dx}{dt}\right)^2 + \left(\frac{dv}{dt}\right)^2. \quad \blacksquare$$

REMARK. In terms of differentials, this equation becomes

$$(ds)^2 = (dx)^2 + (dy)^2,$$

which is easy to remember by thinking of the Pythagorean theorem for an "infinitesimal" right triangle.

REMARK. A function graph traced normally is a particular path with parameter $t = x$:

$$x = x,$$

$$y = f(x).$$

For this special path, our new path formula reduces to the original function formula

$$\frac{ds}{dx} = \sqrt{1 + \left(\frac{dy}{dx}\right)^2}.$$

The new integral formula is of course

$$s = \int_{t_1}^{t_2} \sqrt{\left(\frac{dx}{dt}\right)^2 + \left(\frac{dy}{dt}\right)^2}\, dt.$$

EXAMPLE 1. The path $(x, y) = (\cos\theta, \sin\theta)$ traces the unit circumference once as θ runs from 0 to 2π. The length formula should therefore give the answer 2π. We check that it does:

$$\int_0^{2\pi} \sqrt{\left(\frac{dx}{d\theta}\right)^2 + \left(\frac{dy}{d\theta}\right)^2}\, d\theta = \int_0^{2\pi} \sqrt{\sin^2\theta + \cos^2\theta}\, d\theta$$

$$= \int_0^{2\pi} 1 \cdot d\theta = \theta\Big]_0^{2\pi} = 2\pi.$$

EXAMPLE 2. The path $(x, y) = (\cos(\theta^2), \sin(\theta^2))$ traces the unit circumference once as θ runs from 0 to $\sqrt{2\pi}$. The answer should again be 2π.

$$\int_0^{\sqrt{2\pi}} \sqrt{\left(\frac{dx}{d\theta}\right)^2 + \left(\frac{dy}{d\theta}\right)^2}\, d\theta = \int_0^{\sqrt{2\pi}} \sqrt{4\theta^2 \sin^2\theta^2 + 4\theta^2 \cos^2\theta^2}\, d\theta$$

$$= \int_0^{\sqrt{2\pi}} 2\theta\, d\theta = \theta^2\Big]_0^{\sqrt{2\pi}} = 2\pi.$$

EXAMPLE 3. The path

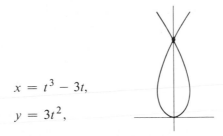

$$x = t^3 - 3t,$$

$$y = 3t^2,$$

is the same as our original example of a path except that the y-coordinates are multiplied by 3. This makes a big difference in the length computation. There is no obvious way of doing it for the original path, but it is very easy for this modified path. Here it is:

$$s = \int \sqrt{\left(\frac{dx}{dt}\right)^2 + \left(\frac{dy}{dt}\right)^2}\, dt = \int \sqrt{9(t^2 - 1)^2 + 36t^2}\, dt$$

$$= \int \sqrt{9(t^2 + 1)^2}\, dt = \int 3(t^2 + 1)\, dt = t^3 + 3t + c.$$

The closed loop in the path runs from $t = -\sqrt{3}$ to $t = \sqrt{3}$, and its length is

$$\int_{-\sqrt{3}}^{\sqrt{3}} = (t^3 + 3t)\Big]_{-\sqrt{3}}^{\sqrt{3}} = 12\sqrt{3}.$$

PROBLEMS FOR SECTION 3

Find the derivative of the path length (ds/dt) for each of the following paths.

1. $x = t, y = t^2$.
2. $x = \cos t, y = \sin t$.
3. $x = 3 \sin t, y = \cos t$.
4. $x = t^3, y = t^2$.

Compute the lengths of the following paths over the given parameter intervals.

5. $3x = 4t^{3/2}, 2y = t^2 - 2t$; $[0, a]$.
6. $x = a \cos^3 t, y = a \sin^3 t$; $[0, \pi/2]$.
7. $x = t^3 + 3t^2, y = t^3 - 3t^2$;   a) $[0, 1]$ b) $[0, 2]$
8. $x = t^3, y = 2t^2$; $[0, 1]$.
9. $x = e^t \cos t, y = e^t \sin t$; $[0, 1]$.
10. $x = t^2 \cos t, y = t^2 \sin t$; $[0, 1]$.
11. $x = \cos t + t \sin t, y = \sin t - t \cos t$; $[0,1]$.
12. A smooth path $(x,y) = (g(t), h(t))$ runs along the graph of a function

$y = f(x)$. Show that if x is increasing with t, then the formulas

$$\frac{ds}{dx} = \sqrt{1 + \left(\frac{dy}{dx}\right)^2}, \quad \frac{ds}{dt} = \sqrt{\left(\frac{dx}{dt}\right)^2 + \left(\frac{dy}{dt}\right)^2}$$

are equivalent, by the chain rule.

13. Continue the line of reasoning begun in the above problem, and thus give an alternative proof of the parametric path-length formula

$$s = \int_a^b \sqrt{\left(\frac{dx}{dt}\right)^2 + \left(\frac{dy}{dt}\right)^2}\, dt.$$

14. Show that the formula

$$s = \int_c^d \sqrt{1 + \left(\frac{dx}{dy}\right)^2}\, dy$$

for the length of a graph $x = g(y)$ is a special case of the parametric formula for path length.

4. ANGLE AS PARAMETER

When a curve loops around the origin, it is frequently convenient to express it as a path with an angle θ as the parameter. However, the angle θ may or may not be the angular coordinate of the path point.

EXAMPLE 1. The central ellipse

$$\frac{x^2}{a^2} + \frac{y^2}{b^2} = 1$$

has the important parametric representation

$$x = a \cos \theta$$
$$y = b \sin \theta.$$

But θ is not the angular coordinate of the path point $(x, y) = (a \cos \theta, b \sin \theta)$ on this ellipse. The geometric meaning of θ here is shown in the accompanying figure.

The angular coordinate θ is the natural parameter when the polar graph of a function f is interpreted as a path, for along the polar graph the x and y coordinates,

$$x = r \cos \theta,$$
$$y = r \sin \theta,$$

satisfy the equation

$$r = f(\theta),$$

so that

$$x = f(\theta) \cos \theta,$$
$$y = f(\theta) \sin \theta$$

along the graph. But these are the equations of a path with the angular coordinate of the path point as parameter.

EXAMPLE 2. The spiral $r = \theta$ is traced by the path

$$x = \theta \cos \theta,$$
$$y = \theta \sin \theta.$$

EXAMPLE 3. The cardioid $r = 1 + \cos \theta$ is traced by the path

$$x = \cos \theta + \cos^2 \theta,$$
$$y = \sin \theta + \sin \theta \cos \theta.$$

The expression of the polar graph

$$r = f(\theta)$$

as the path

$$x = f(\theta)\cos\theta \quad \text{and} \quad y = f(\theta)\sin\theta$$

gives us an easy derivation of the arc-length formula in polar coordinates. We have

$$\frac{dx}{d\theta} = f'(\theta)\cos\theta - f(\theta)\sin\theta,$$

$$\frac{dy}{d\theta} = f'(\theta)\sin\theta + f(\theta)\cos\theta,$$

and when we square each equation and add, we find that

$$\left(\frac{dx}{d\theta}\right)^2 + \left(\frac{dy}{d\theta}\right)^2 = [f'(\theta)]^2 + [f(\theta)]^2$$

$$= r^2 + \left(\frac{dr}{d\theta}\right)^2.$$

(The details of this calculation are left as an exercise.) The Cartesian arc-length formula

$$s = \int_a^b \sqrt{\left(\frac{dx}{d\theta}\right)^2 + \left(\frac{dy}{d\theta}\right)^2}\, d\theta$$

becomes the formula

$$s = \int_a^b \sqrt{[f(\theta)]^2 + [f'(\theta)]^2}\, d\theta$$

$$= \int_a^b \sqrt{r^2 + \left(\frac{dr}{d\theta}\right)^2}\, d\theta$$

for the length of the polar graph of f from $\theta = a$ to $\theta = b$.

EXAMPLE 4. Find the length of the cardioid $r = 1 + \cos\theta = 2\cos^2(\theta/2)$.

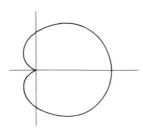

Solution

$$s = \int_0^{2\pi} \sqrt{r^2 + \left(\frac{dr}{d\theta}\right)^2}\, d\theta$$

$$= \int_0^{2\pi} \sqrt{4\cos^4(\theta/2) + 4\cos^2(\theta/2)\sin^2(\theta/2)}\, d\theta$$

$$= 2\int_0^{2\pi} \sqrt{\cos^2\theta/2}\, d\theta = 2\int_0^{2\pi} |\cos(\theta/2)|\, d\theta$$

$$= 4\int_0^{\pi} \cos(\theta/2)\, d\theta = 8\,\sin(\theta/2)\Big]_0^{\pi}$$

$$= 8.$$

EXAMPLE 5. Find the length of the first revolution of spiral $r = \theta$.

Solution. By the formula above, this length is

$$s = \int_0^{2\pi} \sqrt{r^2 + \left(\frac{dr}{d\theta}\right)^2}\, d\theta = \int_0^{2\pi} \sqrt{\theta^2 + 1}\, d\theta.$$

This is a difficult integral. Substituting first $\theta = \tan u$ and then $\sin u = v$, we get:

$$\int \sqrt{\theta^2 + 1}\, d\theta = \int \sec^3 u\, du = \int \frac{\cos u\, du}{(1 - \sin^2 u)^2}$$

$$= \int \frac{dv}{(1 - v^2)^2}.$$

We then express $1/(1 - v^2)^2$ as the sum of its partial fractions. Skipping the details, the answer is:

$$\frac{1}{(1 - v^2)^2} = \frac{1}{4}\left[\frac{1}{(1 - v)^2} + \frac{1}{(1 + v)^2} + \frac{1}{1 - v} + \frac{1}{1 + v}\right].$$

Therefore,

$$\int \frac{dv}{(1 - v^2)^2} = \frac{1}{4}\left[\frac{1}{1 - v} - \frac{1}{1 + v} - \log(1 - v) + \log(1 + v)\right]$$

$$= \frac{1}{4}\left[\frac{2v}{1 - v^2} + \log\left(\frac{1 + v}{1 - v}\right)\right].$$

Since $v = \sin u = \theta/\sqrt{1 + \theta^2}$, we end up with

$$\int \sqrt{\theta^2 + 1}\, d\theta = \frac{1}{2}\theta\sqrt{\theta^2 + 1} + \frac{1}{4} \log \frac{\sqrt{\theta^2 + 1} + \theta}{\sqrt{\theta^2 + 1} - \theta}$$

$$= \frac{1}{2}\theta\sqrt{\theta^2 + 1} + \frac{1}{2} \log(\sqrt{\theta^2 + 1} + \theta).$$

Finally,

$$s = \int_0^{2\pi} \sqrt{\theta^2 + 1}\, d\theta = \pi\sqrt{4\pi^2 + 1} + \frac{1}{2} \log[\sqrt{4\pi^2 + 1} + 2\pi].$$

PROBLEMS FOR SECTION 4

Find the lengths of the following polar graphs.

1. The first revolution of the spiral $r = \theta^2$.

2. $r = \cos \theta$, from $\theta = 0$ to $\theta = \pi/2$.

3. $r = \sec \theta$, from $\theta = 0$ to $\theta = \pi/4$.

4. $r = e^\theta$, from $\theta = 0$ to $\theta = a$.

5. $r = 1/\theta$, from $\theta = \pi/2$ to $\theta = \pi$.

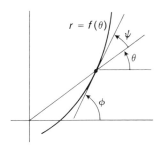

It seems reasonable to specify the direction of a polar graph $r = f(\theta)$ by the angle ψ (Greek psi) it makes with the polar ray, as in the above figure. That this is the natural angle to use is shown by the simplicity of the formula:

$$\cot \psi = \frac{f'(\theta)}{f(\theta)} = \frac{1}{r}\frac{dr}{d\theta}.$$

The following problems center around this formula.

6. Derive the identity

$$\cot(\alpha - \beta) = \frac{1 + \tan \alpha \tan \beta}{\tan \alpha - \tan \beta}$$

from the trigonometric addition formulas (Chapter 4, Section 3).

7. Referring to the above figure again, the slope of the polar graph is $\tan \phi$, where $\phi = \psi + \theta$. Thus,

$$\tan \phi = \frac{dy}{dx} = \frac{dy/d\theta}{dx/d\theta} = \frac{f'(\theta)\sin \theta + f(\theta)\cos \theta}{f'(\theta)\cos \theta - f(\theta)\sin \theta},$$

where the values on the right have been copied from our earlier calculations in the text. Now expand

$$\cot \psi = \cot(\phi - \theta)$$

by the formula in Problem 6, substitute the above value for $\tan \phi$ and simplify. Your answer should be the simple formula for $\cot \psi$ stated earlier (after Problem 5).

8. Prove the identities

$$\tan x = \frac{\sin 2x}{1 + \cos 2x} \quad \text{and} \quad \cot x = \tan\left(\frac{\pi}{2} - x\right).$$

from the trigonometric addition formulas.

Compute $\cot \psi$ for each of the following polar graphs. In each case, try to interpret the answer geometrically. (Problem 8 can help.)

9. $r = e^\theta$ 　　　　　　　　　　　　　10. $r = \sec \theta$

11. $r = \cos \theta$ 　　　　　　　　　　　12. $r = 1 + \cos \theta$

13. Show that the path $x = a \cos \theta$, $y = b \sin \theta$ does lie along the ellipse $x^2/a^2 + y^2/b^2 = 1$, as claimed in the text.

14. Use the parametric representation

$$x = a \cos \theta,$$
$$y = b \sin \theta$$

for the ellipse

$$x^2/a^2 + y^2/b^2 = 1,$$

to prove the following theorem.

Theorem. *Let F and F' be the points* $(c,0)$ *and* $(-c,0)$ *on the x-axis, where* $c = \sqrt{a^2 - b^2}$. *Let* $P = (x,y)$ *be any point on the ellipse. Then the segments PF and PF' have the lengths*

$$\overline{PF} = a - ex \qquad \text{and} \qquad \overline{PF'} = a + ex,$$

where $e = c/a$.

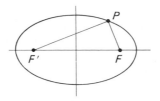

This shows that $\overline{PF} + \overline{PF'} = 2a$; the ellipse is the locus of a point P moving in such a way that the sum of its distance from the two fixed points F and F' is constant and equal to $2a$.

15. Show that the hyperbola

$$\frac{x^2}{a^2} - \frac{y^2}{b^2} = 1$$

has the parametric representation

$$x = a \sec \theta,$$

$$y = b \tan \theta.$$

16. Let F and F' be the points $(c, 0)$ and $(-c, 0)$, where $c = \sqrt{a^2 + b^2}$. Let P be any point on the hyperbola of Problem 15. Using the above parametric representation, show that the segments PF and PF' have the lengths

$$PF = |ex - a|, \qquad PF' = |ex + a|,$$

where $e = c/a$. (*Hint.* Set $\tan^2\theta = \sec^2\theta - 1$ at an appropriate moment.)

17. Assuming the above equations show that

$$|PF - PF'| = 2a.$$

(Consider separately the cases $x > 0$, $x < 0$. Note that $e > 1$ and that $|x| \geq 1$ on the graph.) Restate this equation in words as a locus characterization of the hyperbola, similar to that for the ellipse in Problem 14.

18. Carry out the computation showing that

$$\left(\frac{dx}{d\theta}\right)^2 + \left(\frac{dy}{d\theta}\right)^2 = r^2 + \left(\frac{dr}{d\theta}\right)^2,$$

where $x = r \cos \theta$, $y = r \sin \theta$, and r is a function of θ.

5. CURVATURE

Definition. *The curvature of a curve is the amount it bends per unit length, or its change in direction per unit length, where the change in direction between two points is the angle between the two tangent lines. The curvature is thus the rate of change of the angle ϕ with respect to the arc length s,*

$$\text{curvature} = \frac{d\phi}{ds},$$

where ϕ is the angle from the x-axis to the tangent line.

We shall first investigate the curvature of a function graph. We saw earlier that the sign of the second derivative of a function f tells us which way the graph of f is turning. The magnitude of the second derivative ought to give us an indication of how fast the graph is changing direction, i.e.; of its curvature. But d^2y/dx^2 is not exactly what we want, because $dy/dx = \tan \phi$ and

$$\frac{d^2y}{dx^2} = \frac{d}{dx}\left(\frac{dy}{dx}\right) = \frac{d \tan \phi}{dx},$$

whereas we want $d\phi/ds$. The circle is a good test of suitability. The circle is certainly turning at a constant rate, i.e., has constant curvature. However, from the equation $y = \sqrt{1 - x^2}$ for the upper half of the unit circle we get

$$\frac{dy}{dx} = -\frac{x}{\sqrt{1 - x^2}} \quad \text{and} \quad \frac{d^2y}{dx^2} = -\frac{1}{(1 - x^2)^{3/2}},$$

which is not constant.

The formula for the curvature $d\phi/ds$ depends on the formula for the arc-length derivative ds/dx and the equations

$$\phi = \arctan \frac{dy}{dx}, \qquad \left(\text{from } \tan \phi = \frac{dy}{dx}\right)$$

$$\frac{d\phi}{dx} = \frac{d\phi}{ds} \cdot \frac{ds}{dx}.$$

Then,

$$\frac{d\phi}{ds} = \frac{\dfrac{d\phi}{dx}}{\dfrac{ds}{dx}} = \frac{\dfrac{d}{dx}\left(\arctan \dfrac{dy}{dx}\right)}{\sqrt{1 + \left(\dfrac{dy}{dx}\right)^2}} = \frac{\dfrac{1}{1 + \left(\dfrac{dy}{dx}\right)^2} \cdot \dfrac{d^2y}{dx^2}}{\sqrt{1 + \left(\dfrac{dy}{dx}\right)^2}}.$$

Thus,

$$\text{curvature} = \frac{d\phi}{ds} = \frac{\dfrac{d^2 y}{dx^2}}{\left[1 + \left(\dfrac{dy}{dx}\right)^2\right]^{3/2}}.$$

There is still a sign problem. If you visualize a point running along a curve, you will see that if the tangent line rotates counterclockwise as the curve is traced in one direction, then it rotates clockwise when the curve is traced in the opposite direction. In other words, the curvature $d\phi/ds$ can have either sign, depending on which way we measure arc length along the curve. In our explicit formula above, this ambiguity lurks in the ambiguous sign of ds/dx. We have derived the formula for the case when s increases with x, but if s happened to be measured in the opposite direction we would have acquired a minus sign. What this means is that curvature is a property of a *directed*, or *oriented*, curve. In order to get a definite sign for the curvature, we must be told which way to trace the curve. For undirected curves, we use the *absolute* curvature

$$\kappa = \left|\frac{d\phi}{ds}\right| = \frac{\left|\dfrac{d^2 y}{dx^2}\right|}{\left[1 + \left(\dfrac{dy}{dx}\right)^2\right]^{3/2}}.$$

EXAMPLE. The function $f(x) = x^2$ has a constant second derivative. But its graph is a parabola, and we are familiar enough with its general shape to realize that it flattens out, or becomes less curved, as $|x|$ increases. We thus expect the curvature to approach 0 as $x \to \infty$. In fact

$$\kappa = \frac{\left|\dfrac{d^2 y}{dx^2}\right|}{\left(1 + \left(\dfrac{dy}{dx}\right)^2\right)^{3/2}} = \frac{2}{(1 + 4x^2)^{3/2}}.$$

As a matter of fact, any smooth function of constant concavity whose domain extends to $+\infty$ must of necessity flatten out in this way. If you visualize a graph concave up all the way out, or a graph concave down all the way out, you may see that this has to happen. A general proof would be quite sophisticated, but we can at least check that it happens in every particular case that we look at.

The curvature of a path is computed in the same way. This time we start with

$$\tan \phi = \text{slope of path} = \frac{\dfrac{dy}{dt}}{\dfrac{dx}{dt}}.$$

The proof then follows exactly the same pattern as before, although the computations are a little more complicated because of the quotient in the expression for $\tan \phi$. The result is

$$\text{curvature} = \frac{\dfrac{d^2y}{dt^2} \cdot \dfrac{dx}{dt} - \dfrac{d^2x}{dt^2} \cdot \dfrac{dy}{dt}}{\left[\left(\dfrac{dx}{dt} \right)^2 + \left(\dfrac{dy}{dt} \right)^2 \right]^{3/2}}.$$

Newton's Dot Notation

Newton used the following very efficient notation for derivatives with respect to time:

$$\dot{x} = \frac{dx}{dt},$$

$$\ddot{x} = \frac{d^2x}{dt^2}, \qquad \text{etc.}$$

In particular, velocities and accelerations were always expressed this way.

The dot notation is also useful for paths where the parameter isn't necessarily interpreted as time. For example, in this notation, the path curvature formula is

$$\text{curvature} = \frac{\ddot{y}\dot{x} - \ddot{x}\dot{y}}{[\dot{x}^2 + \dot{y}^2]^{3/2}},$$

which is much simpler and neater than the Leibniz form given above.

REMARK. The computation of path curvature that we outlined above is incomplete, in that it fails for a path point at which the tangent line is vertical ($dx/dt = 0$, $\tan \phi = \infty$). Near such a point, we have to use the alternate formula

$$\cot \phi = \frac{dx/dt}{dy/dt}.$$

The calculation is then substantially the same, and gives the same curvature formula.

The trouble here is that in each case we are working with a slope formula, and there is always one direction in which a slope becomes infinite and the slope formula becomes useless. Later we shall learn to work with a variable pair $(\Delta x, \Delta y)$ rather than a variable quotient $\Delta y / \Delta x$. This new "vector" technique avoids these "blowing up" difficulties and provides a more satisfactory way to approach such geometric notions as direction and curvature. Chapter 14 contains a very brief introduction to such "vector calculus."

PROBLEMS FOR SECTION 5

Compute the curvature of the graph of each of the following functions.

1. $y = x^2$
2. $y = x^{1/2}$
3. $y = \log x$
4. $y = \log(\cos x)$ $(-\pi/2 < x < \pi/2)$
5. $y = mx + b$
6. $y = \sqrt{1 - x^2}$
7. Show that the curvature of $y = \cos x$ oscillates between $+1$ and -1, with period 2π.
8. Find the maximum value of the absolute curvature κ of the graph $y = e^x$.
9. Find the minimum value of the absolute curvature κ of the path

$$x = a \cos^3 t, \qquad y = a \sin^3 t,$$

   over the t interval $[0, \pi/2]$.
10. Show that the circle

$$(x, y) = (r \cos \theta, \quad r \sin \theta)$$

   has constant curvature $\kappa = 1/r$.
11. Prove that a function graph of zero curvature is a straight line.
12. Consider the function $f(x) = x^a$ on the positive x-axis $(0, \infty)$, where the exponent a can be any real number. Show that the curvature of the graph of f has a finite limit as $x \to 0$ if and only if

$$a \leq \frac{1}{2} \qquad \text{or} \qquad a \geq 2.$$

13. Compute the curvature of the ellipse

$$x = a \cos t, \qquad y = b \sin t.$$

   Find its minimum and maximum values.
14. Carry out the proof of the parametric curvature formula.
15. The formula for the absolute curvature κ in polar coordinates is

$$\kappa = \frac{r^2 + 2\left(\dfrac{dr}{d\theta}\right)^2 - r\dfrac{d^2r}{d\theta^2}}{\left[r^2 + \left(\dfrac{dr}{d\theta}\right)^2\right]^{3/2}}.$$

Prove this formula, starting from the parametric form of the formula given in the text, and using the expressions for $dx/d\theta$ and $dy/d\theta$ from the last section.

Using the above formula, compute κ for the following polar graphs.

16. $r = 2a \cos \theta$ 17. $r = a(1 + \cos \theta)$

18. $r = e^{a\theta}$ 19. $r = a\theta$

6. THE PARAMETRIC MEAN-VALUE THEOREM

There is a generalization of the mean-value theorem that seems just as obvious geometrically. Suppose we have a smooth path $(x,y) = (g(t), h(t))$ running from (x_0, y_0) to (x_1, y_1). Then the secant line through these two points must be parallel to the tangent line at some in-between point. If the lines are not vertical, this is equivalent to their slopes being equal:

$$\frac{y_1 - y_0}{x_1 - x_0} = \text{slope of tangent line}$$

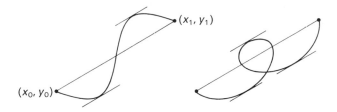

And we saw in Section 1 that the tangent-line slope is $h'(t)/g'(t)$. The following theorem is thus intuitively obvious on geometric grounds.

Theorem 6. *Let the functions g and h be continuous on the closed interval $[a,b]$ and differentiable on its interior. Suppose that $g(a) \neq g(b)$, and that $g'(t)$ and $h'(t)$ are never simultaneously zero. Then there is at least one point T strictly between a and b at which*

$$\frac{h(b) - h(a)}{g(b) - g(a)} = \frac{h'(T)}{g'(T)}.$$

The proof will be given in the next section. Here we shall illustrate its usefulness by proving a theorem called l'Hôpital's rule, and by proving a formula for the tangent-line error E in Chapter 5. Then, in Chapter 13, it will be used to prove a much more general error formula. First, we state a special case of the theorem that better fits these applications.

Theorem 7. *Let g and h be differentiable functions on an interval I containing $x = a$, and suppose that $g(a) = h(a) = 0$ and that $g'(x) \neq 0$ when $x \neq a$. Then for each x different from a, there is a point ξ lying strictly between a and x such that*

$$\frac{h(x)}{g(x)} = \frac{h'(\xi)}{g'(\xi)}.$$

As a first application of Theorem 7, we let $E(x)$ be the error in the tangent-line approximation, as discussed in Chapter 6:

$$E(x) = f(x) - f(a) - f'(a)(x - a).$$

Note that $E(a)$ and $E'(a)$ are both zero, and that $E''(x) = f''(x)$. Let $h(x) = (x - a)^2$. Then $h(a)$ and $h'(a)$ are also both zero. We can then apply Theorem 7 twice in a row, and conclude that

$$\frac{E(x)}{(x - a)^2} = \frac{E'(\xi)}{2(\xi - a)} = \frac{E''(X)}{2} = \frac{f''(X)}{2},$$

where ξ lies between a and x, and X lies between a and ξ. Thus,

Corollary. *Let f be twice differentiable on an interval I containing the point a. Then for every other point x in I, the error*

$$E = f(x) - [f(a) + f'(a)(x - a)]$$

in the tangent-line approximation to f at a can be written in the form

$$E = \frac{f''(X)}{2}(x - a)^2,$$

where X is some number lying strictly between a and x.

It follows from the above formula that if f'' is bounded by K on an interval I about a, then

$$|E(x)| \leq \frac{K}{2}(x - a)^2$$

for each x in I. We thus have a new proof of Theorem 6 of Chapter 5.

Theorem 7 can also be used to answer a natural question about paths, and, in the process, yields a result that is useful in contexts having nothing to do with paths. We saw earlier that a smooth path always has tangent line, with slope $m = h'(t)/g'(t)$. The tangent line could be vertical, in which case $g'(t) = 0$, $h'(t) \neq 0$, and m is infinite. Now suppose the path is smooth except for one (singular) point $t = a$, where $h'(a) = g'(a) = 0$. The slope formula then takes the meaningless form $0/0$, and we say that $h'(t)/g'(t)$ is *indeterminate* at $t = a$. But the path may still have a tangent line there.

EXAMPLE 1. If

$$x = g(t) = t^3,$$

$$y = h(t) = t - \sin t,$$

then $g'(0) = h'(0)$, and $t = 0$ is a singular point where the slope formula $h'(t)/g'(t)$ is indeterminate. Yet the path has slope $1/6$ there, as we shall soon see.

Let $(x, y) = (g(t), h(t))$ be any path through the origin, and suppose that $(0, 0) = (g(a), h(a))$. The path slope there is

$$m = \lim_{\Delta t \to 0} \frac{\Delta y}{\Delta x} = \lim_{\Delta t \to 0} \frac{h(a + \Delta t) - h(a)}{g(a + \Delta t) - g(a)} = \lim_{t \to a} \frac{h(t)}{g(t)}.$$

We can't evaluate this limit by the quotient limit law (the limit of a quotient is the quotient of the limits), because each of the function g and h has the limit 0 at $t = 0$, so the quotient of the limits is the indeterminate $0/0$. The *quotient limit law is valid only when the limit of the denominator is not zero.* We avoided this problem for smooth paths by Theorem 3, which proved that the limit is then $h'(a)/g'(a)$. But what if the path is singular at $t = a$, so that this quotient is also the indeterminate $0/0$?

The solution to this dilemma lies in Theorem 7 which tells us that for any $t \neq a$, there is a number T lying strictly between a and t such that

$$\frac{h(t)}{g(t)} = \frac{h'(T)}{g'(T)}.$$

Suppose that $h'(T)/g'(T)$ approaches a limit l as T approaches a, and consider what happens in the above equation as t approaches a. Then T must also approach a, since T lies between a and t. The righthand side thus approaches the limit l, as $t \to a$, and hence so must the lefthand side. We have proved the following corollary of Theorem 7.

Theorem 8. *Suppose that the functions g and h are differentiable on an interval I containing $x = a$, and suppose that $g(a) = h(a) = 0$ and $g'(x)$ is not equal to zero on I except possibly at $x = a$. Then*

$$\lim_{x \to a} \frac{h(x)}{g(x)} = \lim_{x \to a} \frac{h'(x)}{g'(x)},$$

in the sense that if the derivative limit exists then the function limit exists and has the same value.

This formula is called l'Hôpital's rule. (Rather, it is one of a *collection* of similar results that all go under this name.)

EXAMPLE 2. We return to our original problem, where

$$g(x) = x^3, \quad \text{and} \quad h(x) = x - \sin x.$$

By the theorem,

$$\lim_{x \to 0} \frac{x - \sin x}{x^3} = \lim_{x \to 0} \frac{1 - \cos x}{3x^2},$$

provided the righthand limit exists. We are still in trouble, however, because the righthand quotient is indeterminate at $x = 0$, so we cannot apply the quotient-limit law. However, the numerator and denominator functions in the righthand quotient satisfy the requirements for g and h in the theorem, so we can apply the theorem once more, and conclude that

$$\lim_{x \to 0} \frac{1 - \cos x}{3x^2} = \lim_{x \to 0} \frac{\sin x}{6x},$$

provided the righthand limit exists. This new quotient is again indeterminate at $x = 0$, but we can apply the theorem a third time, to conclude that

$$\lim_{x \to 0} \frac{\sin x}{6x} = \lim_{x \to 0} \frac{\cos x}{6},$$

provided the righthand limit exists. And finally it *does* exist because we now have a quotient of two continuous functions with the denominator nonzero. Thus

$$\lim_{x \to 0} \frac{\cos x}{6} = \frac{\cos 0}{6} = \frac{1}{6},$$

and, backtracking through the above steps in reverse order, we see that each lefthand limit now does exist and has the value $1/6$, by the theorem.

The above example was complicated because of its origin as a path problem to which Theorem 3 could not be applied. It is perhaps typical of the limits evaluated by the theorem that the theorem must be applied more than once, but this doesn't always occur.

EXAMPLE 3. Does the limit of $\sin x / x$ exist as $x \to 0$?

. . . .

Solution. We know, of course, that it does, because we had to establish the limit of this quotient, which is simply the difference quotient for $\sin'(0)$, before we could even prove that $\sin x$ is differentiable. However, it is interesting to note that we can apply the theorem and conclude that

$$\lim_{x \to 0} \frac{\sin x}{x} = \lim_{x \to 0} \frac{\cos x}{1} = \frac{\cos 0}{1} = 1.$$

PROBLEMS FOR SECTION 6

Evaluate the following limits, mostly by l'Hôpital's rule.

1. $\displaystyle\lim_{t\to0} \frac{\cos t - 1}{t}$

2. $\displaystyle\lim_{x\to0} \frac{e^x - 1}{x}$

3. $\displaystyle\lim_{\theta\to0} \theta \csc \theta$

4. $\displaystyle\lim_{x\to0} x \cot x$

5. $\displaystyle\lim_{x\to0} \frac{x - \log(1 + x)}{x^2}$

6. $\displaystyle\lim_{t\to0} \frac{1 - \cos t}{t^2}$

7. $\displaystyle\lim_{\theta\to0} \frac{1 - \cos \theta}{\sin^2\theta}$

8. $\displaystyle\lim_{t\to0} \frac{1 - \cos^2 t}{t^2}$

9. $\displaystyle\lim_{x\to0} \frac{x - \sin x}{x - \tan x}$

10. $\displaystyle\lim_{x\to0} \frac{e^x - x - 1}{x^2}$

11. $\displaystyle\lim_{t\to0} \frac{2\sqrt{1 + t} - t - 2}{t^2}$

12. $\displaystyle\lim_{x\to0} \frac{2 \cos x + x^2 - 2}{x^4}$

13. $\displaystyle\lim_{x\to0} \frac{\log(1 + x)}{x}$

14. $\displaystyle\lim_{x\to0} \log(1 + x)^{1/x}$

15. $\displaystyle\lim_{x\to0} x \log x$

16. $\displaystyle\lim_{x\to0} x \log(\sin x)$

17. $\displaystyle\lim_{t\to0} \frac{\cos t}{1 + t^2}$

18. $\displaystyle\lim_{x\to0} \frac{\cos x}{x^2}$

19. $\displaystyle\lim_{x\to0} x \sin \frac{1}{x}$

20. Prove the weighted-integral mean-value theorem (Theorem 2 in Chapter 9, Section 8) from the parametric mean-value theorem. (Let F be an antiderivative of fg and let G be an antiderivative of g. Consider the quotient $[F(b) - F(a)]/[G(b) - G(a)]$.)

21. Show that

$$\lim_{x\to0} \frac{f(x) - f(-x)}{x} = 2f'(0).$$

22. Show that

$$\lim_{x\to0} \frac{f(x) + f(-x) - 2f(0)}{x^2} = f''(0).$$

23. Prove that

$$\lim_{x\to0} (1 + x)^{1/x} = e.$$

(Look first at Problem 14.)

24. Theorem 6 in Chapter 5 actually proved the stronger assertion that if $K \le f''(x) \le L$ for all the values of x in question, then the error E in the tangent-line approximation satisfies the inequality

$$\frac{K}{2}(x - a)^2 \leq E \leq \frac{L}{2}(x - a)^2.$$

Show that this stronger result is also a consequence of the Corollary to Theorem 7.

25. Prove L'Hôpital's second rule:

If $\lim_{x \to \infty} g(x) = \lim_{x \to \infty} h(x) = 0$, *and if* $g'(x) \neq 0$ *for all sufficiently large* x, *then*

$$\lim_{x \to \infty} \frac{h(x)}{g(x)} = \lim_{x \to \infty} \frac{h'(x)}{g'(x)},$$

in the same sense as before.
[*Hint*. Set $H(y) = h(1/y)$, $G(y) = g(1/y)$, and try to apply Theorem 8 to H/G.]

7. THE GEOMETRIC PRINCIPLES

The parametric mean-value theorem is the last of the important principles that are suggested to us by geometric intuition, so this seems an appropriate moment to gather together these geometric principles and spend a little time clarifying their relationships to each other and their status in the scheme of things. Here is a list of the principles of this sort that we have used.

A. *If f is differentiable on an interval I and if f ′ is never zero on I, then f ′ has a constant sign on I. That is, f′(x) can change sign only by going through 0.*

B. *If f is continuous on I and if f ′ exists and is positive at every interior point of I, then f is increasing on I.*

C. *If f is continuous on I and if f ′ exists and is zero at every interior point of I, then f is a constant function on I.*

D (**Mean value-theorem**). *If f is continuous on* [a,b] *and differentiable at every interior point, then there is at least one interior point X at which*

$$f_{\bullet}'(X) = \frac{f(b) - f(a)}{b - a}.$$

E (**Parametric mean-value theorem**). *Let g and h be continuous on* [a,b] *and differentiable at every interior point. Suppose that* $g(a) \neq g(b)$, *and that* $g'(t)$ *and* $h'(t)$ *are never simultaneously zero. Then there is at least one interior point X at which*

$$\frac{h'(X)}{g'(X)} = \frac{h(b) - h(a)}{g(b) - g(a)}.$$

In contrast to the above principles, the intermediate-value property that we needed in Chapter 7 does not involve differentiability.

F. *If f is continuous on the closed interval* [a,b], *then the range of f includes the whole closed interval between f(a) and f(b). That is, if k lies between f(a) and f(b), then there must be at least one number X in* [a,b] *at which f(X) = k.*

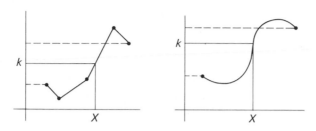

Another equally plausible geometric principle that we have assumed is that a continuous function on a closed interval has a maximum value and a minimum value.

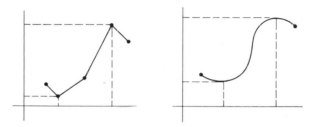

G (**Extreme-value property**). *If f is continuous on a closed interval I, then there are points x_0 and x_1 in I such that $f(x_0)$ is the smallest value of f on I and $f(x_1)$ is the largest value of f on I.*

Unfortunately, the geometric plausibility of these principles does not guarantee their correctness, because geometric intuition isn't always right. For example, considerations like those leading to Principle B, above, might in the beginning have led us to conjecture conversely that if f is strictly increasing over an interval I, then f' is positive everywhere in I. But the example $f(x) = x^3$ showed that this isn't so. Here f is a strictly increasing function, and yet $f'(x) = 3x^2$ is zero at the origin.

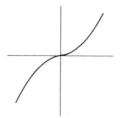

This conjecture, and the "counterexample" showing the conjecture to be false, together form a good illustration of why geometric intuition cannot be trusted absolutely. Something may seem to be true principally because we have not yet noticed, and cannot imagine, a situation in which it is false. If we can't imagine how something might fail to happen, we are tempted to conclude that it must always happen. Of course, according to this principle of reasoning, the poorer our imagination the more facts we could establish!

Once we realize how geometric intuition can be unreliable, we see why any geometric principle that is important to us should be verified analytically. The geometric principles A through G thus do need proofs. However, the inadequacy of intuition as a foundation should not lead you to downgrade it. Intuition of some sort or other is the principal source of ideas, and you should exploit it whenever you can. It ultimately needs analytic support, that's all.

Principles A through G do not all have to be proved from scratch. They form a closely interconnected "family tree," and in the course of investigating their interrelationships it becomes apparent that there are one or two "trunk" properties supporting them all. Then it is only these basic properties that have to be established from the ground up. In fact, we shall prove below that the properties A through E are all logical consequences of the intermediate- and extreme-value properties. Thus, the intermediate-value property and the extreme-value property together form the basis for the geometric principles of calculus.

We start the reduction of the list by eliminating B and C, both of which are immediate consequences of the mean-value property D, as we already have seen (Theorem 7 in Chapter 3 and Theorem 2 in Chapter 5). Here again is the proof of B.

Proof. Let x_1 and x_2 be any two points of I such that $x_1 < x_2$. By the mean-value property, there is some point X between x_1 and x_2 at which

$$f(x_2) - f(x_1) = f'(X)(x_2 - x_1).$$

Since $f'(X) > 0$, by hypothesis, we conclude that $f(x_1) < f(x_2)$. Thus, $f(x_1) < f(x_2)$ whenever $x_1 < x_2$ in I, and f is increasing on I. ∎

The proof of C is essentially the same.

The Mean-Value Property

Consider next the mean-value property. We can visualize obtaining the upper tangent line shown here by starting with a line parallel to the secant and lying wholly above the piece of graph we are considering, and then sliding the line downward, keeping it always parallel to the secant, until it first touches the graph. For a lower tangent line, we would start below and move up.

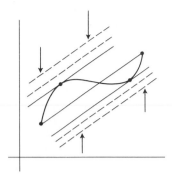

If m is the slope of the secant line, what we have done is to consider all the lines with slope m and to select from among them the highest (or lowest) one that touches the graph. The point of contact is then the highest (or lowest) point on the graph *as measured by these lines*, and it seems clear geometrically that we have a tangent line at that point.

When we begin to consider how we might back up these ideas analytically, or how we might connect them analytically with other fundamental principles, our "highest point" remark above suggests the extreme-value property. In fact, if we were to rotate the whole configuration until the parallel lines become horizontal, then the highest touching line would actually be at the maximum point on the new curve. Such a rotation is simple geometrically, but it is relatively hard to handle analytically. The change-of-coordinate formulas for a rotation are relatively complicated (we haven't even investigated them). Worse, the rotated graph may not be a function graph at all.

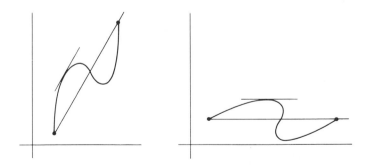

A simpler thing to do analytically is to make the slanting lines horizontal by subtracting mx from all the functions whose graphs appear, and in particular to consider

$$\phi(x) = f(x) - mx.$$

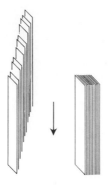

Geometrically, this is not a rotation but a "shearing" motion of the plane, like squaring off a pack of cards. All vertical lines slide along themselves, the line $x = a$ sliding downward a distance ma. Under this motion, all lines of slope m become horizontal and the new configuration looks like the figure at the right. In particular, the secant line has become horizontal, i.e., $\phi(a) = \phi(b)$.

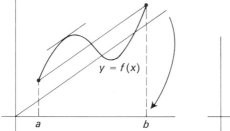

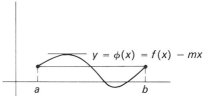

Now we expect that the horizontal line through the highest or lowest value of $\phi(x)$ will be tangent to the graph.

When the above steps are expressed analytically, they become a logical proof of the mean-value property from the extreme-value property. We first prove Rolle's Theorem, which is the above special case with a horizontal secant.

Theorem 9 (Rolle's Theorem). *Suppose that f is continuous on $[a,b]$, and differentiable on (a,b), and that $f(a) = f(b)$. Then the extreme-value property implies that there is an interior point X at which $f'(X) = 0$.*

Proof. If the maximum value of f on $[a, b]$ is not the common endpoint value $f(a) = f(b)$, then a point X, where the maximum value is assumed, must necessarily be an interior point, and so $f'(X) = 0$, by Theorem 1 in Chapter 6. In the same way, if the minimum value of f is not the common endpoint value, then we get an interior point W where $f'(W) = 0$. The only remaining possibility is that the common endpoint value $f(a) = f(b)$ is both the maximum and minimum value of f on $[a, b]$. In this case, f is the constant function $f(x) = f(a)$, and $f'(x) = 0$ at *all* interior points.

Theorem 10. *Rolle's Theorem implies the mean-value theorem.*

Proof. The mean-value theorem reduces to Rolle's Theorem by the device we discussed earlier. We set

$$\phi(x) = f(x) - mx,$$

where m is the slope of the secant segment joining $(a, f(a))$ to $(b, f(b))$,

$$m = \frac{f(b) - f(a)}{b - a}.$$

It seemed clear geometrically that then $\phi(a) = \phi(b)$, but we now have to check this analytically. We have

$$(*) \qquad \begin{aligned} \phi(x) &= f(x) - \frac{f(b) - f(a)}{b - a} x \\ &= \frac{f(x)(b - a) - x(f(b) - f(a))}{b - a}; \end{aligned}$$

and putting x equal to either a or b gives the common value

$$\phi(b) = \phi(a) = \frac{bf(a) - af(b)}{b - a}.$$

Therefore, $\phi'(X) = 0$ for some X interior to $[a, b]$ by Rolle's theorem. But $\phi(x) = f(x) - mx$, so $\phi'(X) = f'(X) - m = 0$. That is,

$$f'(X) = m = \frac{f(b) - f(a)}{b - a}.$$

We have thus proved that (D) follows logically from (G). ∎

The Parametric Mean-Value Theorem

The calculation above also gives us the clue to the proof of the parametric mean-value theorem. In order to use this clue, we first consider a path that runs along the graph of a function $y = f(x)$, from $x = a$ to $x = b$. Such a path has the form

$$x = g(t),$$

$$y = h(t) = f(g(t)).$$

If the parameter t runs across $I = [c, d]$, then $a = g(c)$ and $b = g(d)$. For this path, the numerator in the above formula $(*)$ for $\phi(x)$ becomes

$$h(t)[g(c) - g(d)] - g(t)[h(c) - h(d)].$$

(It was really only the numerator that we made use of.) Call this function $\theta(t)$. Since its formula involves only the path coordinate functions g and h, we can drop the special hypothesis that the path runs along a function graph, and consider $\theta(t)$ along any path with parameter interval $I = [c, d]$. Just as before,

$$\theta(c) = \theta(d) = g(c)h(d) - g(d)h(c),$$

and Rolle's Theorem says that therefore $\theta'(T) = 0$ for some T strictly between c and d. That is,

$$h'(T)[g(c) - g(d)] - g'(T)[h(c) - h(d)] = 0$$

for this T. If $g'(T) = 0$ and $g(c) - g(d) \neq 0$, then $h'(T) = 0$, which contradicts the hypothesis of the theorem. In other words, if $g(c) - g(d)$ is not zero, then $g'(T)$ must also be nonzero. Then the equation above can be rewritten in the form

$$\frac{h'(T)}{g'(T)} = \frac{h(c) - h(d)}{g(c) - g(d)},$$

which is the parametric mean-value theorem. We have proved:

Theorem 11. *Rolle's theorem implies the parametric mean-value theorem* (E).

The Zero Crossing Property for f'

Consider now the zero-crossing principle (A). It seemed geometrically obvious that a graph could change from sloping upward to sloping downward only by passing through a point where its slope is zero. Now, try to examine your visualization of the situation to see *why* this conclusion is obvious. There seem to be two ways to think about it. First, you can visualize a varying tangent line and watch its inclination changing as the point of tangency moves along the curve. If it starts with a positive slope and rotates, as it slides along, to a position with negative slope, then it had to go through a position of zero slope. When we think this way, we are visualizing the tangent line changing position *continuously*, and our conclusion is that if $f'(x)$ changes continuously from a positive value to a negative value, then it has to go through a zero value in the process. The underlying principle is thus the intermediate-value property (F), applied to the continuous function f'.

There is another way to think about what is happening. When we start with a graph sloping upward and end with the graph sloping downward it seems clear that somewhere in between the graph had to go over a highest point, and that at this peak point the slope must be zero.

Such reasoning can be developed into a formal proof of the zero-crossing property for (f') from the extreme value property for f. This is a more general result than that given by our first line of argument (using the intermediate-value property of f'), since it doesn't require the continuity of f'. In practice, this is an empty generality, because in practice a differentiable function will always have a

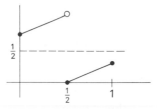

continuous derivative. But in theory, the stronger proof is needed; there *do exist* functions having discontinuous derivatives. (See Problem 4.)

We are now down to the intermediate-value and extreme-value properties of a continuous function. You may feel that these properties are inherently a part of our notion of continuity, and are less in need of analytic justification than the other properties we have discussed. Indeed, until about a hundred years ago they were so taken for granted that they weren't even separately formulated. To Newton and other early workers in calculus, a continuous graph was the path traced by a moving particle, and that such a graph had these properties went without saying.

Nevertheless, this was another case of equating intuitive plausibility with proof. Something was accepted as being true because it was unimaginable that it be false. And if we have agreed that a weak imagination is not a suitable basis for mathematical truth, then the intermediate- and extreme-value properties need analytic proof, just as everything else does.

These proofs were discovered only relatively recently (during the last hundred years or so). They turn out to involve more sophisticated techniques than are needed elsewhere in beginning calculus, and they really belong to the foundations of a larger discipline, called analysis, in which calculus is imbedded. Chapter 20 gives a first look at some of these questions of analysis, and contains a proof of the intermediate-value theorem.

The critical nature of these two properties becomes clearer if we note that even one discontinuity may cause them both to fail. The function graphed above is continuous everywhere on $[0,1]$ except at $x = 1/2$, but it has no maximum value, nor does it assume the intermediate value of $1/2$. Thus we have to know that f is continuous *at every single point* of the interval $[a,b]$ in order to apply these properties.

Moreover, we can't drop an endpoint from $[a,b]$ without losing, possibly, an extreme value. The function $f(x) = x$ loses its maximum value on $[0,1]$ if we drop off $x = 1$, and it loses its minimum if we drop $x = 0$. It has neither a maximum value nor a minimum value on the *open* interval $(0,1)$.

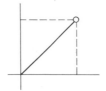

PROBLEMS FOR SECTION 7

Along with (B), there is of course its "change-of-sign" counterpart:

B′ *If f is continuous on I and f′ is negative at every interior point of I, then f is decreasing on I.*

1. Prove B′ from the mean-value theorem.

2. Prove B′ directly from B.

3. Prove the mean-value theorem (D) from the zero-crossing property for f' (A) and the two monotone properties (B) and (B′). [*Hint.* Suppose there is no point X, as claimed in the mean-value theorem. Then concoct a function g whose derivative is never zero in $I = (a, b)$, and apply A and B or B′ to g, ending up with a contradiction.]

4. Let f be the function defined as follows:

    $$f(x) = x^2 \sin(1/x), \quad \text{if } x \neq 0;$$

    $$f(0) = 0.$$

    a) Find the formula for $f'(x)$ when $x \neq 0$, by the differentiation rules.
    b) Show that $f'(x)$ does not have a limit as x approaches 0.
    c) Show that the difference quotient at 0,

    $$\frac{f(0 + k) - f(0)}{h} = h \sin(1/h),$$

    does have a limit as $h \to 0$ (use the squeeze limit law), and, consequently, that $f'(0)$ exists and is zero.

    Thus, f is an everywhere differentiable function whose derivative f' is not a continuous function.

5. Prove the zero-crossing property of f' (A) from the extreme-value property of f (G). [*Hint.* Suppose f' does change sign on I, and choose two points c, d, in I, where f' has opposite signs, with $c < d$. Apply Theorem 4 from Chapter 5 to show that if $f'(c) > 0 > f'(d)$, then the maximum value of f on $[c,d]$ cannot be assumed at either c or d. Now apply Theorem 1 in Chapter 6, etc.]

6. Earlier we came to the shape principle by combining (A) and (B). Prove it now directly from the extreme-value property of a continuous function and Theorem 1, Chapter 6. Here is what you are to prove:

    If f is continuous on an interval I and has no critical points in the interior of I, then f is monotone on I.

[*Hint.* Suppose f is not monotone. Then there must be three points $a < b < c$ in I such that $f(b)$ is either the maximum or the minimum of the values of f at these three points. Then $f'(x) = 0$ at some point between a and c.]

We know from algebra that a polynomial $p(x)$ of degree n,

$$p(x) = x^n + a_{n-1} x^{n-1} + \cdots + a_0,$$

has at most n real roots. That is, there are *at most* n distinct real numbers that are solutions of the equation $p(x) = 0$. There may very well be *fewer* than n. Thus, $p(x) = x^2 + 1$ is a polynomial of degree 2 having *no* real roots.

7. Prove that a polynomial of *odd degree* has at least one real root.

8. Call a polynomial p of degree n *simple* if it does have n distinct real roots. Prove that if p is simple, then so is its derivative $q = p'$.

infinite sequences and series

In Chapter 13 it will be shown that the derivative sum rule remains true for certain "infinitely long polynomials":

$$\text{If } f(x) = \sum_{n=0}^{\infty} a_n x^n, \quad \text{then } f'(x) = \sum_{n=1}^{\infty} n a_n x^{n-1}.$$

Before we can consider the calculus of such infinite polynomials, however, we must learn something about the nature of the infinite sum operation and its arithmetic. The present chapter lays this groundwork.

We start with the infinite decimal representation of a positive real number, a familiar and important description of a real number that exhibits some of the phenomena we have to analyze. We then turn to the notion of a convergent infinite sequence (which makes it possible to define the infinite-sum operation) and to the fundamental property of the real number system called completeness, (which will guarantee that an apparently convergent infinite sum really does represent a number). With these preliminaries out of the way, we begin the study of convergent infinite series and their arithmetic in Section 4.

1. DECIMAL REPRESENTATIONS

In this section we shall describe, mostly without proof, some important properties of real numbers that have only been alluded to up to now.

To begin with, there is the dichotomy between rational and irrational. A *rational* number is a number that can be expressed as a fraction, i.e., as a quotient of two integers, such as $5/3$, $-2/7$, $1/100$, $4 = 4/1$, and $3.1 = 31/10$. A real number that is *not* rational is called *irrational*. For example, it can be shown that $\sqrt{2}, \sqrt{3}, \pi$, and e are all irrational. Proofs of irrationality for these numbers are of varying degrees of difficulty. If p is a positive prime integer, such as 2, 3, 5, or 7, then it is relatively easy to show that $\sqrt{p}$ is irrational. But the proofs that π and e are irrational are very much harder.

A rational number that can be written as a fraction with a power of 10 as denominator is called a *decimal fraction*, and is normally expressed as a *finite decimal*. For example,

$$\frac{31}{10} = 3.1,$$

$$\frac{141}{100} = 1.41,$$

$$\frac{2}{1000} = 0.002.$$

Finite decimals are simpler to handle than general rational numbers because the rules for adding and multiplying fractions reduce, in this case, to the simpler rules for integers.

The crucial property we wish to discuss is that *every real number can be represented by an infinite decimal expansion*. We shall see shortly that a decimal fraction has two such decimal expansions, but for any other number the infinite decimal representation is unique.

We can find the decimal expansion of a rational number by continued long division. For example, to find the expansion of 27/88, we start dividing 88 into 27, and get

$$\frac{27}{88} = 0.306818181\cdots,$$

where the three dots indicate that the sequence of digits goes on forever. This is a very simple infinite decimal expansion because the two-digit block 81 repeats forever.

The decimal expansion of $\sqrt{2}$ can be obtained by a process (algorithm) for extracting square roots that is often taught in school. It is more complicated than long division, but similar in being nonterminating (unless the beginning number happens to be a finite decimal that is a perfect square, such as $\sqrt{1.44} = 1.2$). If we apply this process to compute $\sqrt{2}$, we get

$$\sqrt{2} = 1.414\cdots,$$

where again the three dots indicate that the sequence of digits goes on forever. This time, though, there isn't any block of digits that forever repeats. Moreover, there is no way of determining what comes next except by carrying the computation far enough to find out.

The decimal expansion representing π begins

$$\pi = 3.14159\cdots.$$

The calculation of this expansion is harder still, but there are various ways of going about it.

Whether a number is rational or irrational shows up in its infinite decimal expansion. An expansion of a rational number is *always* a repeating decimal, as in the example above. After a certain point it consists entirely of repetitions of a certain block of digits. This is probably seen more clearly from examples than from a general discussion, and there are several exercises on these repeating decimals in the problem set at the end of the section. It is convenient to indicate the repeated block by an overhead bar, and in this notation the first example is

$$\frac{27}{88} = 0.30\overline{81}.$$

Conversely, we shall see that any repeating decimal represents a rational number. Thus, *the irrational numbers are exactly those numbers having nonrepeating infinite decimal expansions.*

EXAMPLE 1. Assuming that obvious arithmetic operations on infinite decimals are correct, we see that:

If $x = .\overline{81}$, then $100x = 81.\overline{81}$, and

$$99x = 100x - x = 81.\overline{81} - 0.\overline{81} = 81.$$

Therefore, $x = 81/99 = 9/11$. That is,

$$\overline{.81} = \frac{9}{11}.$$

EXAMPLE 2. Assuming the above conclusion, we now have:

$$.306\overline{81} = \frac{306.\overline{81}}{1,000} = \frac{306 + .\overline{81}}{1,000} = \frac{306 + 9/11}{1,000}$$

$$= \frac{3366 + 9}{11,000} = \frac{3375}{11,000} = \frac{27 \cdot 125}{88 \cdot 125} = \frac{27}{88}.$$

When we expand a decimal fraction by long division, we get nothing but zeros at the end, as in

$$\frac{31}{10} = 3.1000 \cdots = 3.1\overline{0}.$$

This fits the finite decimal notation, which can now be viewed as resulting from simply dropping the repeated zero in the infinite decimal expansion. Thus,

$$\frac{31}{10} = 3.1\overline{0} = 3.1.$$

However, each decimal fraction is *also* represented by a *different* decimal expansion having repeated nines at the end. The real meaning of this will appear in Section 4, as a consequence of the meaning of the sum of an infinite series.

EXAMPLE 3. If $x = 0.999 \cdots = 0.\overline{9}$, then $10x = 9.\overline{9}$, and

$$9x = 10x - x = 9.\overline{9} - 0.\overline{9} = 9.$$

Therefore, $x = 9/9 = 1 = 1.\overline{0}$. That is, $0.\overline{9} = 1.\overline{0}$.

EXAMPLE 4. Assuming the above,

$$3.0\overline{9} = \frac{30.\overline{9}}{10} = \frac{30 + .\overline{9}}{10} = \frac{30 + 1}{10} = \frac{31}{10} = 3.1\overline{0}.$$

The above examples show the only way in which a number can have two different decimal representations. *If x is a positive real number that is not equal to* *a decimal fraction, then x has a unique infinite decimal expansion. Every positive decimal fraction has exactly two decimal expansions, one corresponding to its expression as a finite decimal, and the other ending with repeated 9's.*

The one-place decimals divide the number line into intervals, each having a length 1/10. Note that we have written the integers 1 and 2 as the one-place decimals 1.0 and 2.0 in the above diagram. The two-place decimals divide each of the above intervals into 10 intervals, each of length 1/100. For example, [1.60,1.61] is the first of the ten subintervals of [1.6,1.7], and the fifth is [1.64,1.65]. Then the three-place decimals mark off intervals of length $1/1000 = 1/10^3$, and so on. It is apparent that the finite decimals in their totality are scattered along the number line in a very dense way, and we shall see that any *other* number x is exactly located by where it "sits" among the finite decimals. In fact, this is precisely the meaning of the infinite decimal expansion of x.

It is important to understand how this works, so let us return to the expansion of π and interpret its instructions for locating π among the finite decimals. The expansion starts out $3.14\cdots$. This means that π lies between the finite decimals 3.14 and 3.15. Similarly, its expansion through four decimal places is 3.1415, and this means that $3.1415 < \pi < 3.1416$. Thus, the unending decimal expansion $\pi = 3.14159\cdots$ is equivalent to the unending sequence of inequalities

0) $3 < \pi < 4$

1) $3.1 < \pi < 3.2$

2) $3.14 < \pi < 3.15$

3) $3.141 < \pi < 3.142$
$$\vdots \qquad \vdots \qquad \vdots$$

(We have called $3 < \pi < 4$ the zeroth inequality so that the nth inequality will concern n-place decimals.) The first inequality tells us that π lies in the open interval (3.1, 3.2), an interval length of 1/10. The second inequality tells us which tenth part of this first interval contains π; it places π within an interval of length $1/100 = 1/10^2$.

Continuing, the nth inequality locates π on an interval of length $1/10^n$, and since these lengths 10^{-n} become arbitrarily small, the inequalities in their totality locate π exactly.

Since the finite decimal 3.14 can be obtained by cutting off the decimal expansion of π after two places, we call 3.14 the two-place *truncation* of the exansion of π. The following examples illustrate the "positioning" roles of such truncations. Their solutions depend on the laws for inequalities, which are discussed in Appendix 1.

EXAMPLE 5. The inequality interpretation of a truncation of a decimal expansion applies to the finite decimals themselves. For example,

$$8.3 < 8.367 < 8.4.$$

This can be seen in two ways. The procedure for handling inequalities between quotients (cross-multiplying to obtain a common denominator) reduces it to the true inequality

$$8{,}300 < 8{,}367 < 8{,}400.$$

Or we can just note that 8.3 is the one-place truncation of the decimal expansion $8.367\overline{0}$.

EXAMPLE 6. A classical approximation to π is 22/7. Assuming that the expansion of π begins $3.141 \cdots$, show that $\pi < 22/7$ and that $22/7 - \pi < 0.002$.

. . . .

Solution. Dividing out 22/7, we see that its decimal expansion begins $3.142 \cdots$ Therefore,

$$3.141 < \pi < 3.142 < \frac{22}{7} < 3.143.$$

This shows that π is less than 22/7 and that

$$\frac{22}{7} - \pi < 3.143 - 3.141 = 0.002.$$

EXAMPLE 7. The decimal expansion of e begins $2.71 \cdots$. Show from this that $\sqrt{e} < 5/3$.

. . . .

Solution. Since $(5/3)^2 = 25/9 = 2.77 \cdots$, we have

$$(\frac{5}{3})^2 > 2.77 > 2.72 > e,$$

and hence

$$\frac{5}{3} > \sqrt{e}.$$

EXAMPLE 8. Knowing that the decimal expansions of π and $\sqrt{2}$ start 3.14 and 1.41 respectively, what can we say about $\sqrt{2}\,\pi$?

. . . .

Solution. We are assuming that

$$3.14 < \pi < 3.15,$$

$$1.41 < \sqrt{2} < 1.42.$$

Multiplying these two inequalities together, we find that

$$4.4274 < \sqrt{2}\,\pi < 4.4730.$$

In particular,

$$4.4 < \sqrt{2}\,\pi < 4.5,$$

so the expansion of $\sqrt{2}\,\pi$ is 4.4, to one decimal place. (But we don't know from this calculation which of 4.4 and 4.5 is the *closest* one-place decimal to $\sqrt{2}\,\pi$.)

EXAMPLE 9. Show that the inequalities

$$4.43 < x < 4.49 \quad \text{and} \quad |x - 4.46| < .03$$

are equivalent.

. . . .

Solution. We can rewrite the first inequality as

$$4.46 - .03 < x < 4.46 + .03,$$

or

$$-.03 < x - 4.46 < .03,$$

which is equivalent to

$$|x - 4.46| < .03.$$

EXAMPLE 10. Show that an inequality of the form

$$a < x < b$$

can always be rewritten in the form

$$|x - c| < e.$$

. . . .

Solution. Let c be the midpoint of the interval $[a,b]$, and let e be half the interval width $b - a$.

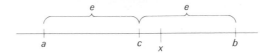

Then it should seem geometrically clear that the inequalities

$$a < x < b \qquad \text{and} \qquad |x - c| < e$$

are equivalent. Algebraically, the proof is as follows. We have defined c and e as

$$c = \frac{b + a}{2},$$

$$e = \frac{b - a}{2}.$$

It follows that $c + e = b$ and $c - e = a$. We thus have, in order,

$$a < x < b,$$

$$c - e < x < c + e,$$

$$-e < x - c < e,$$

$$|x - c| < e,$$

where at each step an inequality is replaced by an equivalent one.

   When we are viewing a number c as an approximation to a number x, the difference $x - c$ is called the *error* in the approximation. The inequality

$$|x - c| < e$$

can thus be read: *c approximates x with an error less than e in magnitude.*

PROBLEMS FOR SECTION 1

Find the repeating infinite decimal expansions of each of the following rational numbers.

1. $\dfrac{1}{3}$ 2. $\dfrac{1}{6}$ 3. $\dfrac{1}{7}$ 4. $\dfrac{2}{7}$

5. $\dfrac{5}{7}$ 6. $\dfrac{1}{9}$ 7. $\dfrac{2}{9}$ 8. $\dfrac{7}{9}$

9. $\dfrac{1}{11}$ 10. $\dfrac{1}{12}$ 11. $\dfrac{1}{13}$ 12. $\dfrac{2}{13}$

13. $\dfrac{3}{13}$ 14. $\dfrac{4}{13}$ 15. $\dfrac{1}{77}$ 16. $\dfrac{1}{88}$

17. $\dfrac{1}{37}$ 18. $\dfrac{1}{17}$ 19. $\dfrac{1}{139}$

20. The number $1/2$ was omitted from the list above because it equals a decimal fraction: $1/2 = 5/10 = .5$. Why were $1/4$, $1/5$, and $1/8$ omitted?

Find the rational number represented by each of the following repeating infinite decimal expansions.

21. $.\overline{5}$ 22. $.\overline{09}$ 23. $.\overline{03}$ 24. $.\overline{04}$ 25. $.\overline{11}$

26. $.\overline{135}$ 27. $.\overline{37}$ 28. $.\overline{36}$ 29. $.0\overline{5}$ 30. $.0\overline{2}$

31. $.3\overline{5}$ 32. $.3\overline{2}$ 33. $.3\overline{30}$ 34. $.3\overline{27}$ 35. $.7\overline{27}$

36. $.8\overline{63}$ 37. $.5\overline{63}$

38. If two numbers have the same decimal expansion through three decimal places, how far apart can they be?

39. Suppose $a = 1.21\cdots$ and $b - 1.25\cdots$. What can be said about the magnitude of $b - a$? $b + a$?

40. Assuming that $\pi = 3.14\cdots$, show that $\pi < \sqrt{10}$.

41. Assuming that $e = 2.7\cdots$ and $\sqrt{2} = 1.4\cdots$, show that $e^2 < 8$. (There is no need to square 2.7 or 2.8.)

42. Assuming that $e = 2.71\cdots$ and $\sqrt{2} = 1.41\cdots$, show that $\sqrt{2}e \approx 3.84$, with an error less than 0.03 in magnitude.

Express each of the following inequalities as a double inequality without absolute value. That is, find a and b so that the inequality is equivalent to $a < x < b$.

43. $|x - 3.141| < .03$ 44. $|x - 3.091| < .05$

45. $|x - 1.99| < .005$ 46. $|x - 7.36| < .02$

47. $|x - 2.113| < .005$ 48. $|x - 1.39| < .002$

Express each of the following double inequalities as a single inequality involving absolute value.

49. $1.404 < x < 1.428$ 50. $3.136 < x < 3.146$

51. $2.717 < x < 2.719$ 52. $1.701 < x < 1.717$

53. $2.186 < x < 2.286$ 54. $4.880 < x < 4.908$

55. If n is a positive integer, show that $1/n$ can be represented as a finite decimal if and only if n is a producer of 2's and 5's, $n = 2^p 5^q$ for some integers p and q.

2. CONVERGENT SEQUENCES

Let $a_0, a_1, a_2, \ldots$ be the infinite sequence of finite decimals obtained by truncating the decimal expansion of π. That is,

$$a_0 = 3,$$

$$a_1 = 3.1,$$

$$a_2 = 3.14,$$

$$a_3 = 3.141,$$

$$\vdots$$

In general, a_n is obtained by cutting off the expansion of π after n places. We have already discussed the manner in which a_n approaches π as n increases, and it is natural to call π the *limit* of the sequence $\{a_n\}$ and write

$$a_n \to \pi \quad \text{as } n \to \infty,$$

or

$$\lim_{n \to \infty} a_n = \pi.$$

Consider next the infinite sequence

$$1, \frac{1}{2}, \frac{1}{3}, \frac{1}{4}, \ldots,$$

where the nth term is $1/n$ for every positive integer n. Then

$$\frac{1}{n} \to 0 \quad \text{as } n \to \infty,$$

in exactly the same way that

$$\frac{1}{x} \to 0 \quad \text{as } x \to \infty.$$

It is clear that sequences have limits in much the same way that functions do, the analogy being with the limit of $f(x)$ as x tends to infinity.

Definition. *If $\{a_n\}$ is an infinite sequence of numbers, then we say that a_n approaches the limit l as n approaches infinity if we can make the difference $l - a_n$ as small as we wish merely by taking n large enough.*

We then write

$$a_n \to l \quad \text{as } n \to \infty,$$

or

$$\lim_{n \to \infty} a_n = l.$$

EXAMPLE 1. If l is positive and if a_n is the n-place decimal truncation of the decimal expansion of l, then

$$a_n \to l \qquad \text{as } n \to \infty.$$

In fact,

$$a_n \le l \le a_n + \frac{1}{10^n}, \qquad \text{so} \quad 0 \le l - a_n \le \frac{1}{10^n}.$$

For example, a_n will be closer to l than $1/1,000,000 = 1/10^6$ for all n larger than 6. In general, if we want $l - a_n < \varepsilon$, where ε is some preassigned "tolerable error," then we only have to take n larger than the number N of places *to the left* of the decimal point in the decimal expansion of $1/\varepsilon$. Then

$$10^N > \frac{1}{\varepsilon},$$

so for all n beyond N,

$$l - a_n \le \frac{1}{10^n} \le \frac{1}{10^N} < \varepsilon.$$

EXAMPLE 2. If $a_n = 1/n$, then obviously

$$a_n \to 0 \qquad \text{as } n \to \infty.$$

Technically, if we want a_n closer to 0 than ε, i.e., if we want

$$\frac{1}{n} < \varepsilon,$$

then we only have to take n larger than $1/\varepsilon$.

A sequence having the limit l is said to *converge* to l. A sequence that doesn't converge to any l is said to *diverge*. Divergence can occur in two "pure" forms: *oscillatory divergence*, as in

$$1, \quad -1, \quad 1, \quad -1, \quad 1, \quad \cdots,$$

where $a_n = (-1)^n$, and *divergence to* ∞, as in

$$1, \quad 4, \quad 9, \quad 16, \quad 25, \quad \cdots,$$

where $a_n = n^2$. Generally a divergent sequence would exhibit a mixture of these two types of divergent behavior.

The computation of sequential limits is governed by the same algebraic limit laws that control function limits:

A. *If $a_n \to a$ and $b_n \to b$ as $n \to \infty$ then*

$$a_n + b_n \to a + b,$$

$$a_n b_n \to ab,$$

$$\frac{a_n}{b_n} \to \frac{a}{b} \qquad \left(\begin{array}{c} \textit{if the denominators} \\ \textit{are nonzero} \end{array} \right).$$

B. *If f is continuous at $x = a$ and if $x_n \to a$ as $n \to \infty$, then*

$$f(x_n) \to f(a) \qquad \textit{as } n \to \infty.$$

C. (Squeeze limit law). *If $0 \le a_n \le b_n$ for all n, and if $b_n \to 0$ as $n \to \infty$, then $a_n \to 0$ as $n \to \infty$.*

Finally, there is a useful "sign-preserving" property of a convergent sequence.

D. *If $a_n \to a$ as $n \to \infty$ and if $a_n \ge 0$ for all n, then $a \ge 0$.*

All of these limit laws should seem to be correct on purely intuitive grounds. For example, if a_n is very close to a and b_n is very close to b, then surely $a_n + b_n$ is very close to $a + b$. But the ultimate proof of a sequential limit law stems from ε estimation and the effective definition of sequential convergence, which will be given in Chapter 20.

EXAMPLE 3. Let a_n again be the n-place decimal obtained by cutting off the expansion of π after n places. We know that $a_n \to \pi$ as $n \to \infty$. Then (B) tells us that

$$a_n^2 \to \pi^2 \qquad \text{and} \qquad e^{a_n} \to e^\pi.$$

Also, $\cos a_n \to -1$. (Why?) If b_n is the n-place truncation of the expansion of $\sqrt{2}$, then

$$a_n + b_n \to \pi + \sqrt{2},$$

$$a_n b_n \to \pi\sqrt{2},$$

and

$$a_n/b_n \rightarrow \pi/\sqrt{2}, $$

by various rules in (A).

EXAMPLE 4. Starting with the fact that $1/n \rightarrow 0$ as $n \rightarrow \infty$, and applying rules from (A), we see that

$$\lim_{n\to\infty} \left(\frac{4-n}{1+3n} \right) = \lim_{n\to\infty} \left(\frac{\frac{4}{n} - 1}{\frac{1}{n} + 3} \right) = \frac{0-1}{0+3} = -\frac{1}{3}.$$

Also,

$$\lim_{n\to\infty} \frac{n^3 - 5n^2}{1 + n^3} = \lim_{n\to\infty} \frac{1 - \frac{5}{n}}{\frac{1}{n^3} + 1} = \frac{1+0}{0+1} = 1.$$

EXAMPLE 5. $\sin \frac{1}{n} \rightarrow 0$ as $n \rightarrow \infty$ by the squeeze limit law (C), because

$$0 < \sin \frac{1}{n} < \frac{1}{n}$$

and $1/n \rightarrow 0$.

We use the words *increasing* and *decreasing* in a slightly weaker sense when we are talking about sequences. A sequence $\{a_n\}$ is said to be *increasing* if

$$a_{n+1} \geq a_n \qquad \text{for all } n.$$

If $a_{n+1} > a_n$ for all n, then we say that $\{a_n\}$ is *strictly increasing*. The meaning of decreasing is similarly modified.

EXAMPLE 6. $a_n = 1/n$ is strictly decreasing.

EXAMPLE 7. If a_n is the n-place decimal truncation of the expansion of π, so that $a_1 = 3.1$ and $a_5 = 3.14159$, then $\{a_n\}$ is an increasing sequence. However, we can't claim that it is strictly increasing because presumably there will be a zero somewhere along the infinite-decimal expansion of π and if the zero occurs in the jth place we will have

$$a_{j-1} = a_j.$$

Here a_n is the lefthand endpoint of the nth closed interval in the sequence of closed intervals that we first looked at in studying the expansion of π. The righthand endpoints form the sequence

$$b_n = a_n + \frac{1}{10^n}.$$

Thus, $b_1 = 3.2$ and $b_4 = 3.1416$. This is a *decreasing sequence*, with the same limit π. This is probably intuitively obvious, but it is also an exercise in the algebraic limit laws:

$$\lim b_n = \lim (a_n + 10^{-n})$$
$$= \lim a_n + \lim 10^{-n}$$
$$= \pi + 0$$
$$= \pi.$$

In the above example, the intervals $[a_n, b_n]$ themselves form a *nested* sequence, in accordance with the following definition.

Definition. *A sequence of closed intervals $I_n = [a_n, b_n]$ is nested if each interval I_n includes the next interval I_{n+1} (and hence includes all later intervals I_{n+j}).*

This just means that $a_n \le a_{n+1}$ and $b_n \ge b_{n+1}$ for each n (and, of course, $a_n \le b_n$), so it is a simple way of saying simultaneously that $\{a_n\}$ is an increasing sequence, that $\{b_n\}$ is a decreasing sequence, and that $a_n \le b_n$ for all n.

You should be able to visualize such a configuration quite clearly on the number line. It is a very useful notion.

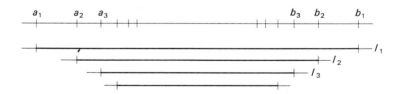

PROBLEMS FOR SECTION 2

Evaluate the limit as $n \to \infty$ of the following sequences.

1. $2 - \dfrac{3}{n} + \dfrac{4}{n^2}$

2. $\dfrac{n+1}{n-1}$

3. $\dfrac{n}{n^2 + 1}$

4. $\dfrac{100n - n^2}{2n^2 + 1}$

5. $\dfrac{5n + 2}{3 - n}$

6. $\dfrac{(2 - n)(3 - n)}{(1 + 2n)(1 + 3n)}$

7. $\dfrac{2n}{\sqrt{4 + n^2}}$

8. $\sqrt{n + 1} - \sqrt{n}$

9. $\sqrt{n^2 + n} - n$

10. $\dfrac{e^n - e^{-n}}{e^n + e^{-n}}$

Here is a principle related to (B) (it can be proved from (B)).

$$\text{If } \lim_{x \to 0} f(x) = l, \qquad \text{then} \qquad \lim_{n \to \infty} f\left(\frac{1}{n}\right) = l.$$

11. Use the new principle stated above to prove that $\lim n \sin(1/n) = 1$.

Use the limit laws, including the new principle above, to find the limits of the following sequences.

12. $n^2\left(1 - \cos\dfrac{1}{n}\right)$

13. $n \log\left(1 + \dfrac{1}{n}\right)$

14. $\dfrac{\log n}{n}$

15. $n(1 - e^{1/n})$

16. Use an algebraic limit law from (A) and the squeeze limit law (C) to prove the more general squeeze limit law:

If $a_n \le b_n \le c_n$ for all n, and if $a_n \to l$ and $c_n \to l$ as $n \to \infty$, then $b_n \to l$ as $n \to \infty$.

3. THE COMPLETENESS OF THE REAL NUMBER SYSTEM.

We saw earlier that an infinite decimal expansion gives instructions for squeezing down on a unique spot on the number line. The expansion defines a sequence of intervals, each being one-tenth of the preceding one, and altogether defining an exact location on the number line. The explicit assumption that such a squeezing-down process always produces a point on the line is called the *axiom of continuity* in geometry.

   The reason for this terminology can be seen by imagining the state of affairs if we were to squeeze down and find nothing there. We would then have located a "crack" in the line, a place where there ought to be a point but there isn't. We could then visualize such a crack as fracturing the line into two unconnected pieces, contradicting our feeling about the continuous nature of the line.

   When we convert the geometric line to the number line, the point in question acquires a coordinate, namely, the number whose infinite decimal expansion we began with. Thus, when we state the axiom of continuity as a property of the real number system, we are asserting that *every infinite decimal is the expansion of a real number.* This property is called the *completeness* of the real number system. It says that every "position" amongst the finite decimals is occupied by a number.

For flexibility, we need a more general formulation of this property. For example, if we compute in base 2 arithmetic, then we end up defining a real number by a different type of interval sequence, where each interval is one of the two *halves* of the preceding interval.

We shall state the general principle in three useful forms.

I. The nested interval property

Suppose that $\{I_n\}$ is a nested sequence of closed intervals whose lengths approach 0 as n tends to infinity. Then there is a unique real number x lying in them all.

For example, if I_n is the *nth* closed interval associated with the decimal expansion of π,

$$I_1 = [3.1, 3.2],$$

$$I_2 = [3.14, 3, 15],$$

$$\vdots$$

$$I_5 = [3.14159, 3.14160],$$

$$\vdots$$

then the I_n form a nested sequence with lengths 10^{-n} tending to 0, and the unique real number common to all the intervals I_n is π.

If $I_n = [a_n, b_n]$ in the nested-interval principle, then of course $a_n \to x$ as $n \to \infty$. The formal proof would first convert the inequality $a_n \le x \le b_n$ to the form

$$0 \le x - a_n \le b_n - a_n.$$

Then, since $b_n - a_n \to 0$, it follows that $x - a_n \to 0$ by the squeeze limit law. Therefore, $a_n \to x$. It is also clear that $b_n \to x$ as $n \to \infty$, and this also has a simple formal proof.

Even more flexible is the following version of completeness.

II. The monotone limit property

An increasing sequence that is bounded above necessarily converges. More explicitly, if $a_n \le a_{n+1}$ for all n, and if there is a number b that is larger than all the numbers a_n, then the sequence $\{a_n\}$ converges.

This should seem intuitively clear if you visualize a_n moving to the right on the number line as n increases, but blocked from ever going beyond b. Then the numbers a_n have to "pile up" at some point short of b, and that is the limit point.

This monotone limit formulation of completeness can be proved from the nested interval formulation by logical reasoning, and the converse is also a theorem. We shall omit these proofs.

The third version of completeness, the completeness axiom, will be discussed in Section 6 of Chapter 20.

PROBLEMS FOR SECTION 3

1. Use the monotone limit property (and the limit laws) to prove the following variant:

   A decreasing sequence that is bounded below necessarily converges.

2. Let $I_n = [a_n, b_n]$ be a nested sequence of closed intervals. Use the monotone limit property and its alternative form above, to prove that both endpoint sequences $\{a_n\}$ and $\{b_n\}$ converge. If their limits are a and b, respectively, prove that $a \le b$.

3. Prove the nested interval property from the monotone limit property. (The problem above is part of this proof, and your job here is to finish it.)

The following four problems outline a proof of the monotone limit property from the nested interval property. You are supposed to finish off each step, assuming the earlier results.

4. Suppose that $\{a_n\}$ is an increasing sequence lying entirely in $[0,1]$. Fix the positive interger m and consider the m-place decimals in $[0,1]$. Show that among these m-place decimals, there is a smallest one, b_m, that is an upper bound of the sequence $\{a_n\}$.

5. We now have an upper bound b_m for each m. Show that this new sequence $\{b_m\}$ is decreasing.

6. Show next that

$$b_n - a_n \to 0 \qquad \text{as } n \to \infty.$$

   [*Hint.* Fix N and show that $b_N - 1/10^N$ is less than some a_n, with $n > N$. Then picture the four numbers

$$b_N, \qquad b_N - \frac{1}{10^N}, \qquad a_n, \qquad \text{and} \qquad b_n$$

   on the line.]

7. Finally, prove that $\{a_n\}$ converges, by the nested-interval property.

8. Suppose that $f(x)$ is an increasing function on $(0,\infty)$, and that $f(x) < B$ for all x. Arguing informally, show that $\lim_{x\to\infty} f(x)$ exists and is at most B. (First consider the sequence $a_n = f(n)$.)

4. THE SUM OF AN INFINITE SERIES

It is surprising and very important that we can sometimes calculate what the sum of an infinite collection of numbers ought to be even though we can't actually carry out the infinite number of additions seemingly required. For example, we shall see below that there is a sense in which the sum of all the powers of $1/2$, starting from $1 = (1/2)^0$, is exactly equal to 2,

$$1 + \frac{1}{2} + \frac{1}{4} + \frac{1}{8} + \cdots = 2,$$

even though the actual calculation of the infinitely many additions indicated on the left is impossible.

This is an example of a *geometric series*, a fundamental type of infinite sum that is sometimes taken up in high school. In general, we are given a number t between -1 and 1, and we want to calculate the infinite sum of all the powers of t,

$$1 + t + t^2 + \cdots + t^n + \cdots$$

We clearly can't perform all the additions that are indicated above, but we *can* start adding and see what happens as we keep on going. The sum of the terms from 1 through t^n is called the *n*th *partial sum*, and is designated s_n. Thus,

$$s_1 = 1 + t,$$

$$s_2 = 1 + t + t^2,$$

$$\vdots \qquad \vdots$$

$$s_n = 1 + t + t^2 + \cdots + t^n.$$

If there is an exact infinite sum s, it should seem reasonable that, as we keep on adding, this accumulating partial sum s_n comes closer and closer to s. That is, we ought to be able to compute the exact infinite sum s as the limit:

$$s = \lim_{n \to \infty} s_n.$$

What makes this computation possible is that we can find a simple explicit formula for the partial sum s_n. It comes from the factorization formula,

$$1 - t^{n+1} = (1 - t)(1 + t + t^2 + \cdots + t^n).$$

We can write this as

$$1 - t^{n+1} = (1 - t)s_n,$$

so

$$s_n = \frac{1 - t^{n+1}}{1 - t}.$$

But we know that $t^{n+1} \to 0$ as $n \to \infty$, because $|t| < 1$. Therefore,

$$s_n = \frac{1 - t^{n+1}}{1 - t} \to \frac{1}{1 - t}$$

as $n \to \infty$. Thus, supposing that there is an exact infinite sum s of the geometric series $1 + t + \cdots + t^n + \cdots$, we can calculate it as the limit of the accumulating partial sum s_n:

$$s = \lim_{n \to \infty} s_n = \frac{1}{1 - t}.$$

EXAMPLE 1.

$$1 + \frac{3}{7} + \frac{9}{49} + \cdots + \left(\frac{3}{7}\right)^n + \cdots = \frac{1}{1 - (3/7)} = \frac{7}{4}.$$

A slight change now completes the picture. Further investigation will not turn up any other way to calculate the exact infinite sum than as the limit $s = \lim s_n$, and we therefore *define* the infinite sum to be the number calculated in this way.

The sum of any other infinite series

$$a_1 + a_2 + \cdots + a_n + \cdots$$

is defined and calculated in the same way. We form the sum s_n of the terms through a_n, and then see what happens as we keep on adding, i.e., as n increases. *If the "lengthening" partial sum s_n approaches a limit s, then we define s to be the exact sum of the infinite series, and we say that the series converges to s.* Our conclusion about the geometric series is thus:

Theorem 1. *If $|t| < 1$, then the geometric series $1 + t + t^2 + \cdots + t^n + \cdots$ converges, and its sum is $1/(1 - t)$.*

EXAMPLE 2. The finite decimals 3.1, 3.14, 3.141, $\cdots$, obtained by truncating the infinite decimal expansion of π, are just the partial sums of the infinite series

$$3 + \frac{1}{10} + \frac{4}{(10)^2} + \frac{1}{(10)^3} + \frac{5}{(10)^4} + \frac{9}{(10)^5} + \cdots.$$

Therefore, the fact that π is given exactly by its infinite decimal expansion can now be reinterpreted as the fact that this series converges to π.

Any other infinite decimal expansion represents a convergent infinite series in the same way. However, these examples are entirely different from the geometric series. There we were able to compute the sum $s = \lim s_n$ by an explicit formula. Here we know the sum exists because of the general property of completeness for the real numbers, but the sum may not have any description other than that provided by the series, i.e., by the infinite decimal expansion. This difference will be clearer after we have looked at the comparison test for convergence (Theorem 4).

When an infinite series fails to have a sum we say that it *diverges*. This can happen in two quite different ways. Suppose that we consider the infinite series

$$1 + 1 + \cdots + 1 + \cdots,$$

i.e., the series

$$a_1 + a_2 + \cdots + a_n + \cdots$$

for which all the terms a_n have the value 1. For this series, the partial sum s_n is just the integer n, which becomes arbitrarily large and doesn't approach any number s. We say that s_n tends to $+\infty$ (plus infinity), and that the series *diverges* to $+\infty$.

A more interesting example of a series that diverges to $+\infty$ is the so-called *harmonic* series,

$$1 + \frac{1}{2} + \frac{1}{3} + \frac{1}{4} + \cdots + \frac{1}{n} + \cdots.$$

Here, the fact that the series diverges isn't so obvious, principally because we don't have a simple formula for the partial sum s_n that we can examine. However, by gathering the terms into groups we see that

$$1 + \frac{1}{2} + \frac{1}{3} + \frac{1}{4} + \frac{1}{5} + \frac{1}{6} + \frac{1}{7} + \frac{1}{8} + \cdots$$
$$> 1 + \frac{1}{2} + \frac{1}{4} + \frac{1}{4} + \frac{1}{8} + \frac{1}{8} + \frac{1}{8} + \frac{1}{8} + \cdots$$
$$= 1 + \frac{1}{2} \quad + \frac{1}{2} \qquad\quad + \frac{1}{2} \qquad + \cdots.$$

Thus, by going out along the harmonic series far enough, we can make the partial sum s_n larger than $1 + m(1/2)$, no matter how large we have taken the integer m. Therefore, the harmonic series diverges to $+\infty$.

The other kind of divergent behavior is illustrated by taking $a_n = (-1)^n$ so that the series is

$$1 + (-1) + 1 + (-1) + \cdots + (-1)^n + \cdots.$$

After we have added n terms, we have either 0 or 1, depending on whether n is even or odd. The partial sum doesn't grow arbitrarily large, but neither does it settle down near one number. Instead it oscillates forever.

These two "pure" forms of divergence can be combined, as in

$$1 + (-2) + 3 + (-4) + \cdots + n(-1)^{n+1} + \cdots.$$

The following theorem states a simple condition that is occasionally useful.

Theorem 2. *If the series $a_0 + a_1 + \cdots$ converges, then $a_n \to 0$ as $n \to \infty$. Therefore, if a_n does not have the limit zero, then the series $a_0 + a_1 + \cdots$ diverges.*

Proof. Suppose that the series converges, with sum s. That is, the partial sum s_n approaches the limit s as $n \to \infty$. Arguing informally, we can see that since s_n and s_{n-1} are both very close to s, their difference, which is a_n, must be close to zero.

The formal argument would be that since we have both $s_n \to s$ and $s_{n-1} \to s$ as $n \to \infty$, then $a_n = s_n - s_{n-1} \to s - s = 0$. ∎

The harmonic series shows that this theorem cannot be reversed: it is a divergent series and yet the general term $1/n$ has the limit zero.

In working with the sums of convergent series we need to use a few principles that will probably seem intuitively obvious. For example, we shall see later on that

$$e = 1 + 1 + \frac{1}{2} + \frac{1}{3!} + \frac{1}{4!} + \cdots + \frac{1}{n!} + \cdots,$$

and that

$$1/e = 1 - 1 + \frac{1}{2} - \frac{1}{3!} + \frac{1}{4!} + \cdots + (-1)^n \frac{1}{n!} + \cdots.$$

Here 4! (read "4 factorial") is $4 \cdot 3 \cdot 2 \cdot 1$, and

$$n! = n(n - 1)(n - 2) \cdots 3 \cdot 2 \cdot 1.$$

Assuming these two equations, we would certainly expect to be able to add the two series term by term, and to conclude that

$$e + 1/e = 2 + 2\left(\frac{1}{2}\right) + 2\left(\frac{1}{4!}\right) + \cdots + 2\left(\frac{1}{(2m)!}\right) + \cdots.$$

Then we would like to be able to multiply throughout by $1/2$, ending up with

$$\frac{e + 1/e}{2} = 1 + \frac{1}{2!} + \frac{1}{4!} + \cdots + \frac{1}{(2m)!} + \cdots$$

We would also like to be able to prove that $e < 3$ by the following argument:

$$e = 1 + 1 + \frac{1}{2} + \frac{1}{3!} + \frac{1}{4!} + \cdots + \frac{1}{n!} + \cdots$$

$$= 1 + 1 + \frac{1}{2} + \frac{1}{2 \cdot 3} + \frac{1}{2 \cdot 3 \cdot 4} + \cdots$$

$$< 1 + 1 + \frac{1}{2} + \frac{1}{2 \cdot 2} + \frac{1}{2 \cdot 2 \cdot 2} + \cdots$$

$$= 1 + \left[1 + \frac{1}{2} + \left(\frac{1}{2}\right)^2 + \left(\frac{1}{2}\right)^3 + \cdots + \left(\frac{1}{2}\right)^n + \cdots \right]$$

$$= 1 + \frac{1}{1 - 1/2} = 1 + 2 = 3.$$

The rules that let us do things like this have to be derived from the rules about sequential limits, because a series sum *is* a sequence limit. However, before looking at these laws we shall introduce the sigma (Σ) notation for infinite series. Remember that $\sum_{j=1}^{n} a_j$ is the sum of the terms a_j from $j = 1$ to $j = n$:

$$\sum_{j=1}^{n} a_j = a_1 + a_2 + \cdots + a_n.$$

For example,

$$\sum_{j=1}^{5} \frac{1}{j^2} = 1 + \frac{1}{2^2} + \frac{1}{3^2} + \frac{1}{4^2} + \frac{1}{5^2} = 1 + \frac{1}{4} + \frac{1}{9} + \frac{1}{16} + \frac{1}{25}.$$

It is then natural to denote the sum of a convergent infinite series by

$$\sum_{j=1}^{\infty} a_j \qquad \text{or} \qquad \sum_{j=0}^{\infty} a_j.$$

If we start at the *m*th term, the sum is $\sum_{j=m}^{\infty} a_j$. We sometimes designate the series itself by

$$\sum a_j \qquad \text{or} \qquad \sum_{j=1}^{\infty} a_j,$$

the rationale being that since we are not indicating the sum to any particular point we are looking at the "unsummed" series.

The general laws for combining series can be discussed conveniently in the Σ-notation, although this is by no means essential. In the statements of these laws, any equation of the form $\sum_{1}^{\infty} x_j = X$ is to be read:

The infinite series $\sum_{1}^{\infty} x_j$ is convergent and its sum is X.

Theorem 3. *If $\sum_{1}^{\infty} a_j = A$ and $\sum_{1}^{\infty} b_j = B$, then $\sum_{1}^{\infty} (a_j + b_j) = A + B$. Also, for any constant c, $\sum_{1}^{\infty} (ca_j) = cA$.*

Proof. We are assuming that the partial sums $\sum_{1}^{n} a_j$ and $\sum_{1}^{n} b_j$ have the limits A and B, respectively, as n tends to ∞. Since we can rearrange a finite sum in any way, it then follows that

$$\sum_{1}^{n} (a_j + b_j) = \sum_{1}^{n} a_j + \sum_{1}^{n} b_j \to A + B$$

as $n \to \infty$, by the law for the limit of the sum of two sequences. Since an infinite sum is by definition the limit of its partial sums, this shows that

$$\sum_{j=1}^{\infty} (a_j + b_j) = A + B.$$

Similarly,

$$\sum_{1}^{n} ca_j = c \sum_{1}^{n} a_j \to cA$$

as $n \to \infty$, so that

$$\sum_{j=1}^{\infty} ca_j = cA.$$

This completes the proof of the theorem.   ∎

Whether a series converges or not is independent of any given initial block of terms, because we always have the identity

$$\sum_{1}^{\infty} a_j = \sum_{1}^{N} a_j + \sum_{N+1}^{\infty} a_j,$$

in the sense that, if either of the two infinite series converges, then so does the other, with the above equation then holding.

This can be considered a special case of the theorem if each of the two incomplete sums on the right is considered to be a full infinite series with zeros in all the missing positions.

The first term on the right above is the partial sum s_N for the given series. The second term, the series sum from the $(N + 1)$st term on, is called the Nth *remainder*, and is frequently designated r_N. Thus, if the series converges, then the equation

$$s = s_N + r_N$$

expresses the infinite sum s as the sum of the first N terms plus the remainder. Note that then

$$r_N \to 0$$

as $N \to \infty$, since $r_N = s - s_N$ and $s_N \to s$.

EXAMPLE 3. If we assume the formula

$$1 + t + \cdots + t^n + \cdots = \frac{1}{1 - t},$$

then we can recover the partial-sum formula algebraically, as follows:

$$\frac{1}{1 - t} = s_n + r_n = s_n + [t^{n+1} + t^{n+2} + \cdots]$$

$$= s_n + t^{n+1}[1 + t + \cdots]$$

$$= s_n + t^{n+1} \cdot \frac{1}{1 - t},$$

and

$$s_n = \frac{1}{1 - t} - \frac{t^{n+1}}{1 - t} = \frac{1 - t^{n+1}}{1 - t}.$$

EXAMPLE 4. The evaluation of a repeating decimal can now be justified, because it is essentially a geometric series from a certain point on. Consider our early example, $0.306\overline{81}$. We group together the terms in the repeating block, and write the infinite decimal as the series sum,

$$\frac{306}{1000} + \frac{1}{1000}\left[\frac{81}{100} + \frac{81}{(100)^2} + \cdots\right]$$

$$= \frac{306}{1000} + \frac{1}{1000} \cdot \frac{81}{100} \cdot \sum_{k=0}^{\infty} \frac{1}{100^k}$$

$$= \frac{306}{1000} + \frac{1}{1000} \cdot \frac{81}{100} \cdot \frac{100}{99} = \frac{306}{1000} + \frac{1}{1000} \cdot \frac{9}{11}$$

$$= \frac{306 \cdot 11 + 9}{11,000} = \frac{3,375}{11,000} = \frac{27}{88}.$$

EXAMPLE 5. This geometric series interpretation also explains why each finite decimal has two decimal expansions. Thus,

$$0.1999\cdots = 0.1\overline{9} = \frac{1}{10} + \frac{9}{(10)^2} + \frac{9}{(10)^3} + \cdots$$

$$= \frac{1}{10} + \frac{9}{100}\sum_{0}^{\infty} \frac{1}{10^j}$$

$$= \frac{1}{10} + \frac{9}{100} \cdot \frac{10}{9} = \frac{2}{10} = .2\overline{0}.$$

PROBLEMS FOR SECTION 4

Find the sum of each of the following series.

1. $1 + \dfrac{1}{4} + \dfrac{1}{16} + \dfrac{1}{64} + \cdots + (\dfrac{1}{4})^n + \cdots$

2. $1 - \dfrac{1}{6} + \dfrac{1}{36} - \dfrac{1}{216} + \cdots + (-\dfrac{1}{6})^n + \cdots$

3. $1 - \dfrac{2}{3} + \dfrac{4}{9} - \dfrac{16}{27} + \cdots + (-\dfrac{2}{3})^n + \cdots$

4. $1 + \dfrac{3}{4} + \dfrac{9}{16} + \dfrac{27}{64} + \cdots + (\dfrac{3}{4})^n + \cdots$

5. $2 + \dfrac{2}{3} + \dfrac{2}{9} + \dfrac{2}{27} + \cdots + 2(\dfrac{1}{3})^n + \cdots$

6. $3 + \dfrac{6}{5} + \dfrac{12}{25} + \dfrac{24}{125} + \cdots + 3(\dfrac{2}{5})^n + \cdots$

7. $\dfrac{1}{4} - \dfrac{1}{8} + \dfrac{1}{16} + \cdots + \dfrac{1}{4}(-\dfrac{1}{2})^n + \cdots$

8. $\dfrac{7}{6} + \dfrac{35}{36} + \dfrac{175}{216} + \dfrac{875}{1296} + \cdots + \dfrac{7}{6}(\dfrac{5}{6})^n + \cdots$

9. $\dfrac{1}{4} + \dfrac{1}{6} + \dfrac{1}{9} + \dfrac{2}{27} + \cdots + \dfrac{1}{4}(\dfrac{2}{3})^n + \cdots$

10. $\sum_{n=0}^{\infty} \dfrac{2^n}{3^n}$

11. $\sum_{n=3}^{\infty} \dfrac{3^n}{4^n}$

12. $\sum_{0}^{\infty} \dfrac{2^n}{5^{n/2}}$

Suppose that each of the following partial sums continues as a geometric series, possibly multiplied by a constant k, so that the nth term is kr^n. Find its sum.

13. $3 + \dfrac{9}{4} + \cdots$

14. $\dfrac{1}{9} + \dfrac{2}{27} + \cdots$

15. $\dfrac{1}{5} - \dfrac{1}{25} + \cdots$

16. $\dfrac{1}{2} + \dfrac{1}{3} + \cdots$

17. Show that a geometric series $\sum r^n$ diverges if $|r| \geq 1$. (Apply Theorem 2.)

18. Suppose that $a_n \geq 1/n$ for every n. Assuming that the harmonic series diverges to $+\infty$, show that the series $\sum a_n$ diverges to $+\infty$.

Assuming the above fact (and possibly using Theorem 3), show that each of the following series is divergent to $+\infty$.

19. $\dfrac{1}{2} + \dfrac{1}{4} + \cdots + \dfrac{1}{2n} + \cdots$ (Suppose it converges and multiply by 2.)

20. $1 + \dfrac{1}{3} + \dfrac{1}{5} + \cdots + \dfrac{1}{2n-1} + \cdots$

21. $\dfrac{1}{4} + \dfrac{1}{7} + \dfrac{1}{10} + \cdots + \dfrac{1}{3n-1} + \cdots$

22. Prove the following divergence criteria from Theorem 3.
    a) If $\sum a_n$ diverges and $c \neq 0$, then $\sum ca_n$ diverges.
    b) If $\sum (a_n + b_n)$ diverges, then either $\sum a_n$ diverges or $\sum b_n$ diverges.

Express each of the following repeating decimals as an infinite series and compute its sum.

23. $0.232323\cdots$

24. $0.6\overline{3}$

25. $0.3\overline{15}$

26. $0.012012012\cdots$

27. $2.0\overline{1}$

28. $4.162162162\cdots$

29. Show that $s_n = 1 - 1/(n+1)$ if s_n is the nth partial sum of the following series:

$$\dfrac{1}{2} + \dfrac{1}{6} + \dfrac{1}{12} + \cdots + \dfrac{1}{n(n+1)} + \cdots$$

[Hint: $(1/6) = (1/2) - (1/3)$.]

30. Consider the series

$$\dfrac{1}{4} + \dfrac{1}{9} + \dfrac{1}{16} + \cdots + \dfrac{1}{(n+1)^2} + \cdots$$

a) Show that its sequence of partial sums s_n is strictly increasing.
b) Show that $s_n < 1$ for each n, assuming the result in the above problem.
c) Conclude that the present series is convergent and that its sum is at most 1.

Show that each of the following series converges, and find its sum.

31. $\dfrac{1}{1 \cdot 3} + \dfrac{1}{3 \cdot 5} + \dfrac{1}{5 \cdot 7} + \cdots + \dfrac{1}{(2n-1)(2n+1)} + \cdots$

    [Hint: $2/(3 \cdot 5) = (1/3) - (1/5)$.]

32. $\dfrac{1}{2 \cdot 4} + \dfrac{1}{4 \cdot 6} + \dfrac{1}{6 \cdot 8} + \cdots$

33. $\dfrac{1}{1 \cdot 3} + \dfrac{1}{2 \cdot 4} + \dfrac{1}{3 \cdot 5} + \cdots + \dfrac{1}{n(n+2)}$

34. Find the sum of the series

$$\frac{1}{1 \cdot 4} + \frac{1}{2 \cdot 5} + \frac{1}{3 \cdot 6} + \cdots + \frac{1}{n(n+3)}.$$

35. A ball bearing is dropped from 10 feet onto a heavy metal plate. Being very elastic, but not *perfectly* elastic, the ball bounces each time to a height that is 19/20 of its preceding height. How far does the bouncing ball travel altogether?

36. A man walks a mile, from A to B, at a steady pace of 3 miles per hour. According to an ancient paradox attributed to the Greek Zeno, he will never get to B, because he first must pass the half-way point, then the 3/4 point, then the 7/8 point, and so on, so that he must pass an infinite number of distance markers on the way, which is clearly impossible. Our modern answer to Zeno is that the man does get to B, in 20 minutes, because two different geometric series each have a finite sum. Explain.

5. THE COMPARISON TEST

The next theorem is the crux of the whole subject. It allows us to establish the convergence of a series by comparing its terms with those of a known convergent series, and is called the *comparison test for convergence.*

Theorem 4. *If* $0 \leq a_n \leq b_n$ *for all n, and if* $\sum b_j$ *is convergent, then the series* $\sum a_j$ *is convergent, and*

$$\sum_{j=1}^{\infty} a_j \leq \sum_{j=1}^{\infty} b_n.$$

EXAMPLE 1. We saw earlier that the series

$$1 + 1 + \frac{1}{2!} + \frac{1}{3!} + \cdots + \frac{1}{n!} + \cdots$$

is term-by-term dominated, in the sense of the theorem, by a geometric series whose sum is 3. Therefore, the displayed series is convergent and its sum is at most 3.

Proof of theorem. Let s_n be the nth partial sum of the series $\sum a_j$, and set $B = \sum_1^\infty b_j$. The hypotheses of the theorem tell us that

$$s_n = \sum_1^n a_j \le \sum_1^n b_j \le B$$

for all n, and that $\{s_n\}$ is an increasing sequence. Therefore the sequence $\{s_n\}$ converges to a limit S that is at most B, by the monotone convergence form of the completeness property. That is, the series $\sum a_j$ has the sum S and $S \le B = \sum_1^\infty b_j$. ∎

If a number s is the sum of an infinite series, we may be able to use the partial sums s_n to estimate s. The error $s - s_n$ in the approximation of s by s_n is the sum of the nth remainder series

$$r_n = a_{n+1} + a_{n+2} + \cdots,$$

and we may be able to show that $s - s_n$ is small by comparing this remainder series with a convergent series whose sum is known.

EXAMPLE 2. This is the ultimate explanation of our original interpretation of an infinite decimal expansion. We now regard the equation

$$\pi = 3.14159 \cdots$$

as meaning that π is the sum of the infinite series

$$\pi = 3 + \frac{1}{10} + \frac{4}{10^2} + \frac{1}{10^3} + \frac{5}{10^4} + \frac{9}{10^5} + \cdots.$$

The second remainder series,

$$r_2 = \frac{1}{10^3} + \frac{5}{10^4} + \frac{9}{10^5} + \cdots,$$

is term-by-term dominated by the series

$$\frac{9}{10^3} + \frac{9}{10^4} + \frac{9}{10^5} + \cdots = \frac{9}{10^3}\left[\sum_{k=0}^\infty \frac{1}{10^k}\right] = \frac{9}{10^3} \cdot \frac{10}{9} = \frac{1}{10^2}.$$

Therefore $r_2 \le .01$, and

$$3.14 = s_2 < \pi = s_2 + r_2 \le 3.14 + 0.01 = 3.15.$$

That is, the bracketing inequality

$$3.14 < \pi < 3.15$$

is now seen to be a consequence of the comparison test for series convergence, applied to the second remainder series r_2.

EXAMPLE 3. Assuming that e is the sum of the series

$$e = 1 + 1 + \frac{1}{2!} + \cdots + \frac{1}{n!} + \cdots,$$

we can compute e by comparing its remainder series with the geometric series. Suppose, for example, that we compare the remainder series r_6 term-by-term with the geometric series that its first terms suggest. We have

$$r_6 = \frac{1}{7!} + \frac{1}{8!} + \frac{1}{9!} + \cdots$$

$$= \frac{1}{7!}\left[1 + \frac{1}{8} + \frac{1}{8\cdot 9} + \frac{1}{8\cdot 9\cdot 10} + \cdots\right]$$

$$< \frac{1}{7!}\left[1 + \frac{1}{8} + \frac{1}{8^2} + \frac{1}{8^3} + \cdots\right] = \frac{1}{7!}\left(\frac{1}{1 - (1/8)}\right) = \frac{1}{7!}\frac{8}{7}$$

$$< \frac{1}{7!}\frac{7}{6} = \frac{1}{6}\cdot\frac{1}{6!}.$$

Thus,

$$s_6 < e = s_6 + r_6 < s_6 + \frac{1}{6}\cdot\frac{1}{6!},$$

and we have e pinned down to an interval of width $(1/6)(1/6!)$. Therefore, each endpoint of the interval, s_6 and $s_6 + (1/6)\cdot(1/6!)$, approximates e with an error less than the interval width $(1/6)(1/6!)$. A quick check shows this to be of the order of magnitude of $1/4,000 = 0.00025$, so s_6 approximates e with 3-place accuracy.

The computation of s_6 goes very simply as follows. (We start with 7 decimal places in order to be ready for a later improvement.)

1	1.0000000
1	1.0000000
$\frac{1}{2}$	0.5000000
$\frac{1}{3!}$	0.1666666 $\cdots$
$\frac{1}{4!}$	0.0416666 $\cdots$
$\frac{1}{5!}$	0.0083333 $\cdots$
$\frac{1}{6!}$	0.0013888 $\cdots$
s_6	= 2.718055 $\cdots$

divide by 3

divide by 4

divide by 5

divide by 6

The interval width $1/6 \cdot 6!$ is $1/6$ of the last line before adding, so

$$r_6 < \frac{1}{6 \cdot 6!} = 0.000231 \cdots < 0.000232.$$

Since

$$s_6 < e < s_6 + \frac{1}{6 \cdot 6!},$$

we have

$$2.718055 < e < 2.718056 + 0.000232 = 2.718288.$$

Thus, $e \approx 2.718$, correct to three decimal places.

Actually, the upper bound $s_6 + 1/6 \cdot 6! = 2.718287 \cdots$ is much closer to e than is the partial sum s_6. This is because

$$s_6 + \left(\frac{1}{7!} + \frac{1}{8!} + \frac{1}{9!}\right) = s_9 < e < s_6 + \frac{1}{6 \cdot 6!},$$

so the error in using $s_6 + 1/6 \cdot 6!$ as the estimate is less than

$$\frac{1}{6 \cdot 6!} - \left(\frac{1}{7!} + \frac{1}{8!} + \frac{1}{9!}\right) = \frac{1}{6!}\left[\frac{1}{6} - \frac{1}{7} - \frac{1}{7 \cdot 8} - \frac{1}{7 \cdot 8 \cdot 9}\right]$$

$$= \frac{1}{6!}\left[\frac{1}{6 \cdot 7} - \frac{1}{7 \cdot 8} - \frac{1}{7 \cdot 8 \cdot 9}\right]$$

$$= \frac{1}{6!}\left[\frac{2}{6 \cdot 7 \cdot 8} - \frac{1}{7 \cdot 8 \cdot 9}\right] = \frac{1}{6!}\left[\frac{2 \cdot 9 - 6}{6 \cdot 7 \cdot 8 \cdot 9}\right]$$

$$= \frac{1}{6!}\left[\frac{12}{6 \cdot 7 \cdot 8 \cdot 9}\right] = \frac{2}{9!} < 6 \cdot 10^{-6}.$$

Thus 2.718288 is larger than e, but by less than 8 in the sixth decimal place, so

$$e = 2.71828 \cdots.$$

We can show in exactly the same way that $s_n + 1/(n \cdot n!)$ approximates e from above, with an error less than

$$\frac{1}{(n + 3)!}\left(1 + \frac{6}{n}\right).$$

The general proof starts out with the inequality,

$$s_n + \frac{1}{(n + 1)!} + \frac{1}{(n + 2)!} + \frac{1}{(n + 3)!} = s_{n+3} < e < s_n + \frac{1}{n \cdot n!},$$

and continues exactly as above. We thus have a very efficient procedure for computing e. The only problem is finding how large n must be for the error bound $(n + 6)/n \cdot (n + 3)!$ to be less than a preassigned value. In any given problem, we would just compute the sequence of factorials until we reach the right value of n.

It may be necessary to multiply a comparison series by a constant before making the comparison.

EXAMPLE 4. Show that the series

$$\frac{1}{3} + \frac{1}{8} + \frac{1}{15} + \cdots + \frac{1}{n(n-2)} + \cdots$$

is convergent by comparing with the known convergent series (Problem 30, Section 1)

$$\frac{1}{9} + \frac{1}{16} + \frac{1}{25} + \cdots + \frac{1}{n^2} + \cdots,$$

where each series starts with $n = 3$.

. . . .

Solution. We can't compare directly because the inequalities go the wrong way. However, we can find a positive constant k such that

$$\frac{1}{n(n-2)} \le k\frac{1}{n^2}$$

for all n from some point on. You may be able to see by inspection that $k = 2$ will do, but in any case we can work it out. Cross-multiplying, we want

$$n^2 \le k\, n(n-2) = kn^2 - 2kn,$$

or

$$2kn \le (k-1)n^2,$$

or

$$\frac{2k}{k-1} \le n.$$

If $k = 2$, this holds for all $n \ge 4$.

Retracing the steps, we see that the given series, starting from its second term, is term-by-term dominated by the known convergent series $2 \sum 1/n^2$, and hence itself converges, by Theorem 4.

PROBLEMS FOR SECTION 5

1. Show that the following result is an immediate corollary of Theorem 4.

Theorem 4'. *If* $0 \leq a_n \leq b_n$ *for all n and if* $\sum a_n$ *diverges, then* $\sum b_n$ *diverges.*

Use the comparison test, including the above corollary, to determine whether the following series converge or diverge. For convergence, compare with a geometric series or the series $\sum 1/n^2$ (Problem 30, Section 1). For divergence, compare with the harmonic series $\sum 1/n$. In some instances, you may have to multiply by a constant before comparing.

2. $\sum \dfrac{1}{n^3}$

3. $\sum \dfrac{1}{\sqrt{n}}$

4. $\sum \dfrac{1}{n^2 + 1}$

5. $\sum \dfrac{1}{n^2 - 1}$

6. $\sum \dfrac{n - 1}{n^2 + 1}$

7. $\sum \dfrac{1}{3n + 5}$

8. $\sum \dfrac{3n + 5}{n^3}$

9. $\sum \dfrac{\log n}{n^3}$

10. $\sum \dfrac{1}{2^n - 1}$

11. $\sum \sin \dfrac{1}{n^2}$

12. $\sum \dfrac{1}{n!}$

13. $\sum \dfrac{n!}{n^n}$

14. Using the infinite-series interpretation of the decimal expansion π = 3.1415 $\cdots$, prove that

$$3.141 < \pi < 3.142.$$

15. Carry out the proof of the error formula

$$E < \frac{1}{(n + 3)!}\left(1 + \frac{6}{n}\right)$$

for the approximation of e by $s_n + 1/(n \cdot n!)$, following the calculation given in the text for the special case $n = 6$.

16. Carry out the computation of e given in Example 3 in Section 5 through three more factorials (i.e., through 1/9!). Compute $S_9 + 1/(9 \cdot 9!)$ through ten decimal places, and use the error formula given in Problem 15 to show that this estimation of e is accurate through eight decimal places.

6. THE RATIO TEST; ABSOLUTE CONVERGENCE

We shall need two particular corollaries of the comparison theorem. The first is the "ratio test," which describes a class of series that can be shown to be convergent by comparison with the geometric series. First an example.

EXAMPLE 1. In the series

$$1\left(\frac{1}{2}\right) + 2\left(\frac{1}{2}\right)^2 + 3\left(\frac{1}{2}\right)^3 + \cdots + n\left(\frac{1}{2}\right)^n + \cdots,$$

a pair of successive terms a_n and a_{n+1} have the ratio

$$\frac{a_{n+1}}{a_n} = \frac{(n+1)\left(\frac{1}{2}\right)^{n+1}}{n\left(\frac{1}{2}\right)^n} = \left(\frac{n+1}{n}\right)\left(\frac{1}{2}\right) = \left(1 + \frac{1}{n}\right)\frac{1}{2}.$$

This "test ratio" a_{n+1}/a_n varies with n, but it has the limit $1/2$ as $n \to \infty$. Therefore, if we choose a number between $1/2$ and 1, say $3/4$, then we will have

$$\frac{a_{n+1}}{a_n} < \frac{3}{4}$$

for all integers n beyond some value N. We don't have to know what N is, but in this example we can easily find its value by solving the inequality

$$\frac{a_{n+1}}{a_n} = \left(1 + \frac{1}{n}\right)\frac{1}{2} < \frac{3}{4},$$

obtaining

$$1 + \frac{1}{n} < \frac{3}{2}, \quad \frac{1}{n} < \frac{1}{2}, \quad n > 2.$$

Thus,

$$\frac{a_{n+1}}{a_n} < \frac{3}{4} \quad \text{or} \quad a_{n+1} < \frac{3}{4}a_n,$$

for all $n \geq 3$. So

$$a_4 < \frac{3}{4}a_3,$$

$$a_5 < \frac{3}{4}a_4 < \left(\frac{3}{4}\right)^2 a_3,$$

$$a_6 < \frac{3}{4}a_5 < \left(\frac{3}{4}\right)^3 a_3,$$

$$\vdots$$

$$a_n < \frac{3}{4}a_{n-1} < \left(\frac{3}{4}\right)^{n-3} a_3.$$

Therefore, if we sum starting with the third term, the comparison test shows that $\sum a_j$ is convergent and that

$$\sum_{n=3}^{\infty} a_n < a_3 \sum_{k=0}^{\infty} \left(\frac{3}{4}\right)^k = a_3\left(\frac{1}{1 - 3/4}\right) = 4a_3.$$

Here is the theorem.

Theorem 5. *Suppose that $\sum a_n$ is a series of positive terms such that*

$$\frac{a_{n+1}}{a_n} \to l$$

as $n \to \infty$. *Then* $\sum a_n$ *converges if* $l < 1$ *and diverges if* $l > 1$.

Proof. Suppose that $l < 1$ and choose a number t between l and 1: $l < t < 1$. Since $a_{n+1}/a_n \to l$, we will have $a_{n+1}/a_n < t$, i.e.,

$$a_{n+1} < ta_n$$

for all indices n beyond some value N. Thus

$$a_{N+1} < ta_N,$$

$$a_{N+2} < ta_{N+1} < t^2 a_N,$$

$$a_{N+3} < ta_{N+2} < t^3 a_N,$$

$$\vdots$$

$$a_{N+m} < ta_{n+m-1} < t^m a_N.$$

Therefore, if we start at the Nth term, the series $\sum a_n$ converges by comparison with the geometric series $a_N \sum t^m$.

If $l > 1$, then it follows as above that $a_{n+1} > a_n$ for all n beyond some value N. Thus the series terms a_j form an increasing sequence from N on. In particular the sequence $\{a_j\}$ does not converge to 0, so $\sum a_j$ diverges, by Theorem 2.

EXAMPLE 2. Consider the series

$$\frac{1}{2} + \frac{2^2}{2^2} + \frac{3^2}{2^3} + \cdots + \frac{n^2}{2^n} + \cdots.$$

The test ratio is

$$\frac{a_{n+1}}{a_n} = \frac{(n+1)^2/2^{n+1}}{n^2/2^n} = \left(\frac{n+1}{n}\right)^2 \frac{1}{2} = \left(1 + \frac{1}{n}\right)^2 / 2.$$

Since $1/n \to 0$ as $n \to \infty$, the test ratio approaches the limit $l = (1 + 0)^2/2 = 1/2$. The series therefore converges, by the theorem.

The theorem says nothing about the case $l = 1$. This is because nothing can be said: the harmonic series is a *divergent* series for which the test ratio has the limit 1, whereas the series $\sum 1/n^2$ is a *convergent* series for which $l = 1$. (Section 7 will treat the general case $\sum 1/n^p$.)

We also have to know something about the convergence of a series $\sum a_n$ whose terms have both signs. If the sum $\sum |a_n|$ of their magnitudes (absolute values) converges, it should seem probable that the series $\sum a_n$ itself must converge, and to a smaller sum, because of the cancelling that goes on as we add together the positive and negative terms.

EXAMPLE 3. Consider the geometric series with ratio $t = -(1/2)$. Theorem 1 shows that

$$1 - \frac{1}{2} + \frac{1}{4} - \frac{1}{8} + \cdots + \left(-\frac{1}{2}\right)^n + \cdots$$

has the sum

$$\frac{1}{1 - \left(-\frac{1}{2}\right)} = \frac{1}{1 + 1/2} = \frac{2}{3}.$$

This is much smaller than the sum of the powers of $1/2$,

$$1 + \frac{1}{2} + \frac{1}{4} + \cdots = 2,$$

and the reason is the cancelling that goes on as we add. We can see exactly the cancelling that occurs by dividing up the series into its positive and negative parts:

$$P = 1 + \frac{1}{4} + \frac{1}{16} + \cdots = \frac{1}{1 - (1/4)} = \frac{4}{3},$$

$$N = -\left(\frac{1}{2} + \frac{1}{8} \cdots\right)$$

$$= \left(-\frac{1}{2}\right)\left[1 + \frac{1}{4} + \cdots\right] = \left(-\frac{1}{2}\right)P = -\frac{2}{3},$$

$$\Sigma = P + N = \frac{4}{3} + \left(-\frac{2}{3}\right) = \frac{2}{3}.$$

Theorem 6. *If $|a_n| \leq b_n$ for all n and if $\sum b_n$ converges, with sum B, then $\sum a_n$ converges, and*

$$\left|\sum_1^\infty a_n\right| \leq B.$$

Proof. We simply divide up $\sum a_n$ into two series, one containing all the positive (and zero) terms a_n, and the other containing all the negative (and zero) terms. Specifically, we define a_n^+ as a_n if $a_n > 0$, and as zero otherwise. Then $0 \leq a_n^+ \leq b_n$ for all n, and so $\sum a_n^+$ converges, with sum $\leq B$, by Theorem 4. Next we define a_n^- as $-a_n$ if $a_n < 0$ and as 0 otherwise, and get, similarly, $0 \leq \sum_1^\infty a_n^- \leq B$. But $a_n = a_n^+ - a_n^-$ for all n. Therefore, $\sum a_n$ converges, with $\sum_1^\infty a_n$

$= \sum_1^\infty a_n^+ - \sum_1^\infty a_n^-$, by Theorem 3. And since each sum on the right lies between 0 and B, their difference lies between $-B$ and B. That is, $|\sum_1^\infty a_n| \leq B$. ∎

Definition. *When we have a series $\sum a_n$ such that $|\sum a_n|$ converges, we say that $\sum a_n$ converges absolutely.*

Theorem 6. can then be restated as follows:

If a series $\sum a_n$ converges absolutely, then it converges, and

$$\left| \sum_1^\infty a_n \right| \leq \sum_1^\infty |a_n|.$$

Theorems 5 and 6 have the following useful corollary:

Corollary. *Suppose $|a_{n+1}/a_n| \to l$ as $n \to \infty$. Then $\sum a_n$ converges absolutely if $l < 1$, and diverges if $l > 1$.*

PROBLEMS FOR SECTION 6

Use the ratio test to conclude what you can about the convergence or divergence of each of the following series.

1. $\sum \dfrac{n+4}{n2^n}$ 2. $\sum nx^n$ 3. $\sum \dfrac{n^2}{2^n}$

4. $\sum n(n-1)x^{n-2}$ 5. $\sum \dfrac{1}{n^3}$ 6. $\sum \dfrac{n^4}{n!}$

7. $\sum \dfrac{1}{\sqrt{n}}$ 8. $\sum \dfrac{n!}{n^n}$ $\left(\text{Use: } \left(1 + \dfrac{1}{n}\right)^n \to e. \right)$

9. $\sum \dfrac{2^n+n}{3^n}$

10. Prove the following corollary of the comparison theorem.

Theorem. *Suppose that $\sum a_n$ and $\sum b_n$ are two series with positive terms, and suppose that the ratio of the nth terms, a_n/b_n, has a limit as $n \to \infty$. Then*
   a) $\sum a_n$ *converges if $\sum b_n$ converges;*
   b) $\sum b_n$ *diverges if $\sum a_n$ diverges.* [*Hint: If $a_n/b_n \to l$ as $n \to \infty$, then $a_n/b_n < l + 1$ for all n beyond some value N.*]

11. Use the above theorem to prove that the series $\sum (3n+5)/n^3$ converges. [*Hint.* For large n, 5 is insignificant compared to $3n$, so $(3n+5)/n^3$ acts like $3/n^2$. So take $b_n = 1/n^2$ and apply the theorem.]

Use the above theorem to establish the convergence or divergence of the following series.

12. $\sum \dfrac{1}{n(n-2)}$ 13. $\sum \dfrac{1}{3n+5}$ 14. $\sum \dfrac{\log n}{n^3}$

15. $\sum \sin \dfrac{1}{n^2}$ 16. $\sum \dfrac{1}{\sqrt{n^2+n}}$

Test for convergence or divergence the series whose general term (from a certain point on) is:

17. $\dfrac{1}{\sqrt{n^4 - 1}}$ 18. $\dfrac{1}{n + \sqrt{n}}$ 19. $\dfrac{1}{\sqrt{n!}}$

20. $\dfrac{n^{10}}{10^n}$ 21. $\dfrac{10^n}{n!}$ 22. $\sin \dfrac{1}{n}$

23. $\dfrac{1}{100n + 10^6}$ 24. $\dfrac{1}{\log n}$ 25. $\dfrac{(n!)^2}{(2n)!}$

26. $\dfrac{2^n (n!)^2}{(2n)!}$

27. Prove that $n!/n^n \to 0$ as $n \to \infty$.   28. Prove that $2^n/n! \to 0$ as $n \to \infty$.

29. Prove Theorem 6 as follows. Set $c_n = a_n + b_n$ and show that $0 \le c_n \le 2b_n$. Then $\Sigma\, c_n$ converges, and $\le \Sigma\, c_n \le 2\, \Sigma\, b_n$, etc.

7. THE INTEGRAL TEST

At the moment we cannot say anything about the convergence of the series

$$1 + \frac{1}{2\sqrt{2}} + \frac{1}{3\sqrt{3}} + \cdots + \frac{1}{n\sqrt{n}} + \cdots .$$

However, look at the following figure. We see that the sum of the above series from 2 to n is the sum of the areas of rectangles lying under the graph of $y = x^{-3/2}$ between $x = 1$ and $x = n$. The area of the region under the graph itself between these limits is

$$\int_1^n x^{-3/2} dx = -2x^{-1/2} \Big]_1^n = 2 - \frac{2}{\sqrt{n}} .$$

Therefore,

$$\sum_{k=2}^n k^{-3/2} < 2 - \frac{2}{\sqrt{n}} < 2.$$

The partial sums s_n of the series thus form an increasing sequence that is bounded above by 2, and therefore convergent, by the monotone limit property. That is, the series $\Sigma\, k^{-3/2}$ converges, and when we add on the first term that was missing in the above inequality we end up with

$$\sum_{k=1}^{\infty} k^{-3/2} \le 3.$$

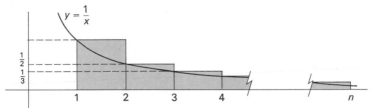

In the same way, but using upper rectangles rather than lower rectangles, we obtain a new proof that the harmonic series $\sum 1/n$ diverges. The figure above shows that

$$\log n = \int_1^n \frac{dx}{x} < \sum_{k=1}^{n-1} \frac{1}{k},$$

where both sides are interpreted as areas. Since $\log n \to \infty$ as $n \to \infty$, it follows that the partial sums of the harmonic series form a sequence diverging to ∞, and hence that $\sum 1/n$ is divergent.

In order to state the theorem convering these two situations, we recall that, by definition,

$$\int_a^{\infty} f(x)\,dx = \lim_{b \to \infty} \int_a^b f(x)\,dx.$$

If the limit exists (as a finite number), we say that $\int_a^{\infty} f$ is a *convergent* improper integral. If the limit does not exist the improper integral is divergent.

Theorem 7. *Let f be a positive decreasing continuous function defined on the interval* $[1,\infty)$*. Then:*

a) *If the improper integral $\int_1^{\infty} f$ converges, then the series $\sum f(n)$ converges, and*

$$\sum_2^{\infty} f(n) \le \int_1^{\infty} f(x)\,dx;$$

b) *If the series $\sum f(n)$ converges, then the improper integral $\int_1^{\infty} f(x)\,dx$ converges, and*

$$\int_1^{\infty} f(x)\,dx \le \sum_1^{\infty} f(n).$$

Proof. Left to the reader. Just write down in general terms the arguments from the two examples.

REMARK. Since a series converges if and only if a remainder series $\sum_N a_k$ converges, we can restate the theorem in terms of the improper integral $\int_N^{\infty} f$.

PROBLEMS FOR SECTION 7

Determine the convergence or divergence of the series whose general term is:

1. $\dfrac{1}{n^p}(p > 1)$ 　　　　　 2. $\dfrac{1}{n^p}(0 < p \le 1)$ 　　　　3. ne^{-n^2}

4. ne^{-n} 　　　　　　　　　 5. $\dfrac{1}{n \log n}$ 　　　　　　6. $\dfrac{1}{n(\log n)^p}(p > 1)$

7. Show that $\sum_1^\infty 1/(n^2) < 2$. (Apply Theorem 7)

8. Show that $\sum_1^\infty 1/(n^2 + 1) < \dfrac{\pi}{2}$.

9. Show that $\sum_1^\infty 1/(n^3) < \dfrac{3}{2}$.

We have been comparing series with series, and series with integrals. Prove the following inequalities by comparing integrals with integrals. (The integral form of the monotone limit property is stated in Problem 8 of Section 3.)

10. $\int_0^\infty \dfrac{dx}{x^2 + x + 1} < \dfrac{\pi}{2}$. 　　　　　11. $\int_2^\infty \dfrac{x\,dx}{x^3 + 1} < \dfrac{1}{2}$.

12. $\int_2^\infty \dfrac{\sqrt{x}\,dx}{x^2 + 1} < \sqrt{2}$. 　　　　　13. $\int_1^\infty e^{-x^2}dx < \dfrac{1}{2e}$.

14. The finite sum $\log(n!) = \log 2 + \log 3 + \cdots + \log n$ can be compared with $\int \log x\,dx$ between appropriate limits, as in the proof of the integral test. Show in this way that

$$\log(n - 1)! < n \log n - n + 1 < \log(n!)$$

for $n \ge 2$. Conclude that

$$(n - 1)! < en^n e^{-n} < n!$$

15. Using the trapezoidal approximation in a similar way, show that

$$\sqrt{n}\,(n - 1)! < en^n e^{-n},$$

or

$$n! < en^n e^{-n} n^{1/2}.$$

16. Finally, interpret $\log k$ as the area of a trapezoid below the tangent line to the graph of $\log x$ at $x = k$, the trapezoid extending from $x = k - 1/2$ to $x = k + 1/2$. Thus $\log(n!)$ is a sum of trapezoidal areas that *include* the area under the graph of $\log x$, except for sticking out too far by $1/2$ on the right, and not reaching quite far enough (by $1/2$) on the left. Show therefore that

$$\log n! + \frac{1}{4} - \frac{1}{2}\log n > n \log n - n + 1,$$

and conclude that

$$n! > e^{7/8} n^n e^{-n} n^{1/2}.$$

power series and

polynomial approximation

We can now show that the elementary transcendental functions e^x, $\sin x$, etc., can be expressed as "infinite polynomials," in the sense that each is an infinite sum of terms $a_n x^n$. For example, the series that we have been using for e will be established by showing more generally that

$$e^x = 1 + x + \frac{x^2}{2!} + \frac{x^3}{3!} + \cdots + \frac{x^n}{n!} + \cdots$$

for all x. Such a series is called a *power series* because its nth term is a constant times x^n. This remarkable equation for e^x, and the others like it that we shall establish, all depend on a few simple facts about power series.

Related to the above infinite series for e^x is the approximation of e^x by the nth partial sum:

$$e^x \approx 1 + x + \cdots + \frac{x^n}{n!}.$$

This is an approximation of e^x by a polynomial. In the second half of the chapter we shall see that such polynomial approximations are important in their own right, quite apart from their relationship to series. As one application, we shall be able to vastly improve the trapezoidal approximation to a definite integral.

This whole subject is important for computation. For example, most of the numerical tables for transcendental functions, such as the tables for the natural logarithm, sine and cosine, are computed from the power-series expansions of these functions.

1. POWER SERIES.

A power series is an infinite series whose nth term is a constant a_n times x^n:

$$a_0 + a_1 x + \cdots + a_n x^n + \cdots$$

Such a series cannot be said to converge or diverge as it stands, since it is not an infinite series of numbers. However, it becomes a numerical series if we give x a numerical value, and we can expect that it may converge for some values of x and diverge for other values. Here is the basic fact.

Theorem 1. *If a power series* $\sum a_n x^n$ *converges for a particular value of x, say the positive number c, then the series converges absolutely for every x in the interval* $(-c, c)$.

Proof. We are assuming that $\sum a_n c^n$ converges. Then $a_n c^n \to 0$ as $n \to \infty$, by Theorem 2 of Chapter 12, so $|a_n c^n| \leq 1$ from some point on, say from the integer N on. We can then see that $\sum a_n x^n$ converges absolutely for all x in the interval

$(-c, c)$, by comparison with the geometric series. For suppose that $|x| < c$ and set $t = |x/c|$. Then $0 \leq t < 1$, and

$$|a_n x^n| = |a_n c^n t^n| = |a_n c^n| \cdot t^n \leq t^n,$$

when $n \geq N$. Therefore the remainder series

$$\sum_{N}^{\infty} |a_n x^n|$$

converges by comparison with the geometric series $\sum_N t^n$. This proves that $\sum a_n x^n$ converges absolutely for every x in the interval $(-c, c)$. ∎

It can be proved, as a corollary of Theorem 1 and the completeness property of the real numbers, that a power series $\sum a_n x^n$ converges exactly on an interval centered at the origin. This is a theoretical result, and does not provide any way of determining the interval of convergence. However, the ratio test often provides a direct computation of the convergence interval, and thus bypasses all such theoretical considerations.

EXAMPLE 1. In order to determine whether the series

$$x + 2x^2 + 3x^3 + \cdots + nx^n + \cdots$$

converges for a given number x we try the ratio test. The test ratio for $\sum |nx^n|$ is

$$\left| \frac{(n+1)x^{n+1}}{nx^n} \right| = \left(\frac{n+1}{n} \right)|x| = \left(1 + \frac{1}{n} \right)|x|,$$

which has the limit $l = |x|$ as n approaches infinity. Therefore, we know (from the corollary of Theorem 6 in Chapter 12) that $\sum nx^n$ will converge absolutely if $|x| < 1$ and diverge if $|x| > 1$. The domain of convergence is thus the interval $(-1, 1)$, with the possible addition of one or both endpoints. Since the ratio test fails when the limit $|x|$ has the value 1, we don't yet know what happens at the endpoints. We shall put off looking at this problem until the end of the chapter.

EXAMPLE 2. The series

$$(x - 3) + 2(x - 3)^2 + \cdots + n(x - 3)^n + \cdots$$

is not of the type we are interested in here, but it has a feature worth noticing.

We apply the ratio test exactly as in the above example:

$$\left| \frac{(n+1)(x-3)^{n+1}}{n(x-3)^n} \right| = \left(1 + \frac{1}{n} \right) |x-3| \to |x-3|.$$

The series therefore converges absolutely if $|x-3| < 1$ and diverges if $|x-3| > 1$, so the interval of convergence is defined by the inequality

$$-1 < x - 3 < 1, \qquad \text{or} \qquad 2 < x < 4.$$

It is thus the interval $(2,4)$, centered about the point $x = 3$.

EXAMPLE 3. For the series

$$1 + x + \frac{x^2}{2!} + \frac{x^3}{3!} + \cdots + \frac{x^n}{n!} + \cdots,$$

we have the test ratio

$$\left| \frac{x^{n+1}/(n+1)!}{x^n/n!} \right| = \frac{|x|}{n+1},$$

which approaches 0 as $n \to \infty$, no matter what value x has. The series therefore converges absolutely for all x, and the interval of convergence is $(-\infty, \infty)$.

If the power series $\sum a_n x^n$ converges on the interval $(-c, c)$, then its sum depends on x and hence is a function f of x,

$$f(x) = a_0 + a_1 x + a_2 x^2 + \cdots + a_n x^n + \cdots$$

The major question we have to answer is this: Does the law that the derivative of a sum is equal to the sum of the derivatives hold for *infinite* sums? In other words, must the function f above necessarily have a derivative f' that is given by

$$f'(x) = a_1 + 2a_2 x + \cdots + na_n x^{n-1} + \cdots$$

on the interval $(-c, c)$?

It is really remarkable that the answer to this question is *yes*, and that we can therefore apply calculus to the study of such infinite-sum functions. We first show that the sum of the derivatives exists.

Theorem 2. *If the power series $\sum a_n x^n$ converges for each x in the open interval $(-c, c)$, then it and its term-by-term differentiated series $\sum na_n x^{n-1}$ both converge absolutely for every x in the interval $(-c, c)$.*

Proof. Consider any x in $(-c, c)$. Then choose a positive number u such that $|x| < u < c$. This number u will now play the role that c played in Theorem 1. Thus, since $\sum a_n u^n$ converges by hypothesis, it follows from Theorem 1 that $\sum a_n x^n$ is absolutely convergent.

For the differentiated series we have to go back to the line of reasoning we used in Theorem 1. Since $\sum a_n u^n$ converges, it follows that $|a_n u^n| \leq 1$ for all n from some integer N on. Set $t = |x/u|$. Then

$$|na_n x^n| = n|a_n u^n| \cdot t^n \leq nt^n$$

for $n \leq N$. Since $0 < t < 1$, the series $\sum nt^n$ converges by the ratio test. Then $\sum na_n x^n$ converges absolutely by comparison. If $x \neq 0$ we can multiply this series by $1/x$ and conclude that $\sum na_n x^{n-1}$ is absolutely convergent (Theorem 3 of Chapter 12). Of course, every power series converges trivially at $x = 0$, so the proof is complete. ∎

This brings us to the major theorem.

Theorem 3. *If $\sum a_n x^n$ converges on the interval $(-c, c)$, then its sum function*

$$f(x) = a_0 + a_1 x + a_2 x^2 + \cdots + a_n x^n + \cdots$$

is differentiable and

$$f'(x) = a_1 + 2a_2 x + \cdots + na_n x^{n-1} + \cdots$$

We shall give the proof of this theorem later on. It is not really hard, but it takes longer than anything we have met so far.

PROBLEMS FOR SECTION 1

In each problem below determine the *open* interval of convergence by the ratio test. The test will fail at every endpoint, so we have to leave the question of endpoint convergence to a later discussion.

1. $1 + x + \dfrac{x^2}{2} + \dfrac{x^3}{3} + \cdots + \dfrac{x^n}{n} + \cdots$

2. $1 + \dfrac{x}{2} + \dfrac{x^2}{4} + \cdots + \dfrac{x^n}{2^n} + \cdots$

3. $1 + x + \dfrac{x^2}{2} + \dfrac{x^3}{2 \cdot 3} + \dfrac{x^4}{2 \cdot 3 \cdot 4} + \cdots + \dfrac{x^n}{n!} + \cdots$

4. $1 + 3x + 9x^2 + \cdots + 3^n x^n + \cdots$

5. $1 + 3x + 5x^2 + \cdots + (2n + 1)x^n + \cdots$

6. $1 + \dfrac{x^2}{2} + \dfrac{x^4}{4} + \dfrac{x^6}{8} + \cdots + \dfrac{x^{2n}}{2^n} + \cdots$

7. $x - \dfrac{x^3}{3} + \dfrac{x^5}{5} - \dfrac{x^7}{7} + \cdots + (-1)^n \dfrac{x^{2n+1}}{2n+1} + \cdots$

8. $1 - \dfrac{x}{1 \cdot 2} + \dfrac{x^2}{2 \cdot 4} - \dfrac{x^3}{3 \cdot 8} + \cdots + (-1)^n \dfrac{x^n}{n2^n} + \cdots$

9. $\displaystyle\sum \dfrac{(x-2)^n}{2^n}$

10. $\displaystyle\sum \dfrac{(x-1)^n}{2^n}$

11. $\displaystyle\sum 2^n (x-1)^n$

12. $\displaystyle\sum \dfrac{(x-3)^n}{n2^n}$

13. $\displaystyle\sum \dfrac{(x-3)^n}{n^2 3^n}$

The following four problems center around the geometric series

$$\frac{1}{1-x} = 1 + x + x^2 + \cdots + x^n + \cdots$$

14. a) Using Theorem 3, find a power series whose sum is $1/(1-x)^2$.
    b) Compute the exact sum of the series $\sum_1^\infty n/2^n$.

15. Compute the exact sum of the series $\sum_1^\infty n^2/2^n$ by some device like that used in the problem above.

16. Prove that

$$\log(1-x) = -\left[x + \frac{x^2}{2} + \frac{x^3}{3} + \cdots + \frac{x^n}{n} + \cdots \right]$$

for x in the interval $(-1, 1)$. [*Hint.* Show first that the series on the right converges on $(-1, 1)$, and let $f(x)$ be its sum. Then compute $f'(x)$ by Theorem 3 and see what you have.]

17. Prove that

$$\arctan x = x - \frac{x^3}{3} + \frac{x^5}{5} - + \cdots + (-1)^n \frac{x^{2n+1}}{2n+1} + \cdots$$

for all x in the interval $(-1, 1)$.

18. Assuming Theorem 3, prove the analogous theorem about integrating a power series term by term. (Show that if the power series

$$a_0 + a_1 x + a_2 x^2 + \cdots + a_n x^n + \cdots$$

converges on an interval I, then so does its term-by-term integrated series

$$a_0 x + \frac{a_1 x^2}{2} + \frac{a_2 x^3}{3} + \cdots + \frac{a_n x^{n+1}}{n+1} + \cdots$$

Then apply Theorem 3.)

Use Theorem 3, and its integration analog discussed above, to identify the following functions.

19. $f(x) = 1 + \dfrac{x}{2} + \dfrac{x^2}{3} + \cdots + \dfrac{x^n}{n-1} + \cdots$

20. $f(x) = 1 + 2x + 3x^2 + \cdots + (n+1)x^n + \cdots$

21. $f(x) = 2 + 3x + 4x^2 + 5x^3 + \cdots + (n+2)x^n + \cdots$

22. $f(x) = 1 + 4x + 9x^2 + \cdots + (n+1)^2 x^n + \cdots$

23. $f(x) = 1 + x + \dfrac{x^2}{1 \cdot 2} + \dfrac{x^3}{2 \cdot 3} + \cdots + \dfrac{x^n}{(n-1)n} + \cdots$

24. The series for $\log(1 - x)$ in Problem 16 above is not useful for computation because it converges too slowly. Show that it can be modified to give the following improved version:

$$\log \frac{1+x}{1-x} = 2\left[x + \frac{x^3}{3} + \frac{x^5}{5} + \cdots + \frac{x^{2n-1}}{2n-1} + \cdots \right].$$

(First write down the $\log(1 + x)$ series and then combine.)

25. In the above series, show that the error after n terms is at most

$$\frac{2}{2n+1} x^{2n+1}\left(\frac{1}{1 - x^2} \right).$$

(Compare the remainder with a geometric series.)

26. Compute $\log 3/2$ to four decimal places by using the above series. (First show that the error after the first three series terms, i.e., through the x^5 term, is less than 10^{-5}.) Note how much more efficient this computation is than the Riemann-sum method of Chapter 10.

27. Compute $\log 3/2$ to eight decimal places.

28. Compute $\log 2$ to five decimal places.

2. THE MACLAURIN SERIES OF A FUNCTION

It is difficult to overemphasize the importance of Theorem 3. To begin with, it determines the series coefficients a_n in terms of the function derivatives $f^{(n)}(0)$.

Theorem 4. *If*

$$f(x) = a_0 + a_1 x + a_2 x^2 + \cdots + a_n x^n + \cdots$$

over an interval $(-r, r)$, then f has derivatives of all orders in this interval and

$$a_n = \frac{f^{(n)}(0)}{n!}$$

for every n, where $f^{(n)}$ is the nth derivative of f. The series can thus be rewritten

$$f(x) = f(0) + f'(0)x + \frac{f''(0)x^2}{2!} + \cdots + \frac{f^{(n)}(0)x^n}{n!} + \cdots$$

REMARK. In order to interpret $f(0)$ as $f^{(0)}(0)/0!$ we adopt the convention that

$$0! = 1.$$

Proof of theorem. By Theorems 2 and 3, we can differentiate the above series as often as we wish. Thus

$$
\begin{aligned}
f(x) = {} & a_0 + & a_1 x + {} & a_2 x^2 + a_3 x^3 + \cdots & +a_n x^n + \cdots \\
f'(x) = {} & a_1 + & 2a_2 x + {} & 3a_3 x^2 + \cdots & +na_n x^{n-1} + \cdots \\
f''(x) = {} & 2a_2 + 3 \cdot {} & 2a_3 x + \cdots & & +n(n-1)a_n x^{n-2} + \cdots \\
f'''(x) = {} & 3!\, a_3 + \cdots & & & +\,n(n-1)(n-2)a_n x^{n-3} + \cdots \\
& \vdots & & \vdots & \\
f^{(n)}(x) = {} & n!\, a_n + \cdots & & &
\end{aligned}
$$

Setting $x = 0$ in these equations reduces them to

$$f(0) = a_0$$
$$f'(0) = a_1$$
$$f''(0) = 2a_2$$
$$f'''(0) = 3!\, a_3$$
$$\vdots$$
$$f^{(n)}(0) = n!\, a_n.$$

Thus $a_j = f^{(j)}(0)/j!$ for all j. ∎

Definition. *The series*

$$\sum \frac{f^{(n)}(0)}{n!} x^n$$

is called the Maclaurin *series for f.*

Let us be clear about what we have proved here. We have proved that *if f is the sum of a power series on an interval* $(-r, r)$, *then f is infinitely differentiable and the series is necessarily the Maclaurin series for f.* We have *not* proved that if *f* is infinitely differentiable, then it is the sum of its Maclaurin series. Two things can go wrong for such a function:

a) Its Maclaurin series $\sum f^{(n)}(0)x^n/n!$ may not converge for any x except $x = 0$;

b) The Maclaurin series may converge to a function different from *f.*

Let us explore the second possibility a little further. Suppose that *f* is infinitely differentiable on $(-r, r)$ and that its Maclaurin series $\sum f''(0)x^n/n!$ converges to a function *g* on $(-r, r)$, with *g* different from *f*. Then the series also has to be the Maclaurin series for *g*, by Theorem 4. That is, its coefficients must also have the form $g^{(n)}(0)/n!$, so

$$f^{(n)}(0) = g^{(n)}(0)$$

for all *n*. Then $(f - g)^{(n)}(0) = 0$ for all *n*. Thus $f - g$ is an infinitely differentiable function on $(-r, r)$ that is not the zero function but has the property that all of its derivatives are zero at $x = 0$. This may seem paradoxical, and yet we can concoct just such a function. If we define *h* by

$$h(x) = e^{-1/|x|} \qquad \text{if } x \neq 0,$$

$$h(0) = 0,$$

then it can be proved that *h* is infinitely differentiable and that all the derivatives of *h* are 0 at $x = 0$. Yet the graph of *h* is like this:

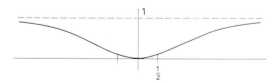

It is decreasing on $(-\infty, 0]$, increasing on $[0, \infty)$, concave up on $[-\frac{1}{2}, \frac{1}{2}]$ and concave down elsewhere.

The Maclaurin series for *h* has only 0 coefficients and its sum is the 0 function, not *h*.

We now establish the power-series representations of the elementary transcendental functions.

Theorem 5. *For every real number* x,

$$e^x = 1 + x + \frac{x^2}{2!} + \frac{x^3}{3!} + \cdots + \frac{x^n}{n!} + \cdots = \sum_{j=0}^{\infty} \frac{x^j}{j!}.$$

Proof. Since $f(x) = e^x$ is its own derivative, we see that $f(0)$, $f'(0)$, $f''(0)$, $\cdots$ all have the same value $e^0 = 1$. The above series is therefore the Maclaurin series for e^x, and hence is the only series that can possibly represent e^x. But does it?

We noted earlier that this series converges for all x, by the ratio test, and therefore defines *some* function g,

$$g(x) = 1 + x + \frac{x^2}{2!} + \frac{x^3}{3!} + \cdots$$

By Theorem 3, $g(x)$ is everywhere differentiable and $g'(x)$ is the sum of the term-by-term differentiated series. But here the differentiated series turns out to be the original series:

$$0 + 1 + \frac{2x}{2!} + \frac{3x^2}{3!} + \frac{4x^3}{4!} + \cdots = 1 + x + \frac{x^2}{2!} + \frac{x^3}{3!} + \cdots,$$

and so $g'(x) = g(x)$. Also $g(0) = 1$. But we know from Chapter 4 (page 158), that e^x is the unique function f satisfying the differential equation $f' = f$ and the "initial condition" $f(0) = 1$. Therefore $g(x) = e^x$. ∎

Theorem 6. *For every real number* x,

$$\cos x = 1 - \frac{x^2}{2!} + \frac{x^4}{4!} - \frac{x^6}{6!} + \cdots + (-1)^m \frac{x^{2m}}{(2m)!} + \cdots$$

Proof. This proof goes exactly like the one above, and we shall only list the steps, leaving their verification to the reader.

a) The series is the Maclaurin series for $\cos x$ and so is the only power series that can possibly represent $\cos x$.

b) The series converges for all x (by the ratio test) and so defines a function g.

c) The function g satisfies the differential equation $g'' = -g$ and the initial conditions $g(0) = 1$, $g'(0) = 0$.

d) But $\cos x$ is the unique function satisfying this differential equation and these initial conditions (Chapter 4, Section 4).

Theorem 7. *For every real number* x,

$$\sin x = x - \frac{x^3}{3!} + \frac{x^5}{5!} - \frac{x^7}{7!} + \cdots + (-1)^m \frac{x^{2m+1}}{(2m+1)!} + \cdots$$

Proof. Just the same as for Theorem 6. ∎

Theorem 3 suggests that perhaps we also can integrate a power series term by term. As a matter of fact, if

$$f(x) = a_0 + a_1 x + \cdots + a_n x^n + \cdots$$

on $(-r, r)$, then the integrated series

$$a_0 x + \frac{a_1 x^2}{2} + \cdots + \frac{a_n x^{n+1}}{n+1} + \cdots$$

also converges on $(-r, r)$, because, after we factor x out, its terms are smaller and we can apply the comparison test. Thus

$$F(x) = a_0 x + \frac{a_1 x^2}{2} + \cdots + \frac{a_n x^{n+1}}{n+1} + \cdots$$

is defined on $(-r, r)$, and $F'(x) = f(x)$, by Theorem 3. Since obviously $F(0) = 0$, we have proved:

Theorem 8. *If the power series*

$$f(x) = a_0 + a_1 x + \cdots + a_n x^n + \cdots$$

converges on $(-r, r)$, *then so does its term-by-term integrated series, and the sum* $F(x)$ *of the integrated series is the antiderivative of* $f(x)$ *which has the value 0 at 0.*

Corollary 1. *For every* x *in the interval* $(\,1, 1)$,

$$\log(1 + x) = x - \frac{x^2}{2} + \frac{x^3}{3} + \cdots + (-1)^{n+1} \frac{x^n}{n} + \cdots$$

Proof. Since

$$1 - x + x^2 + \cdots + (-1)^n x^n + \cdots = \frac{1}{1+x} \left(= \frac{1}{1 - (-x)} \right)$$

on $(-1, 1)$, the sum of the integrated series

$$x - \frac{x^2}{2} + \frac{x^3}{3} + \cdots + (-1)^n \frac{x^{n+1}}{n+1} + \cdots$$

is the antiderivative of $1/(1 + x)$ which has the value 0 at $x = 0$, and so is just $\log(1 + x)$, without any constant of integration.

Corollary 2. *For every x in the interval $(-1, 1)$,*

$$\arctan x = x - \frac{x^3}{3} + \frac{x^5}{5} - \frac{x^7}{7} + \cdots + (-1)^m \frac{x^{2m+1}}{2m + 1} + \cdots$$

Proof. Left to the reader. [*Hint.* Set $y = -x^2$ in $1/(1 + x^2)$ and use the geometric series expansion.] ∎

Theorem 9. *For any fixed real number α,*

$$(1 + x)^\alpha = 1 + \alpha x + \frac{\alpha(\alpha - 1)}{2!} x^2 + \cdots + \frac{\alpha(\alpha - 1)\cdots(\alpha - n + 1)}{n!} x^n + \cdots$$

on the interval $(-1, 1)$.

This proof is left as an exercise. The steps are:

a) Use the ratio test to find the interval of convergence.

b) Show that if g is the sum of the series, then

$$g'(x) = \frac{\alpha g(x)}{(1 + x)}, \qquad g(0) = 1.$$

c) Show that the unique solution of this initial-value problem is the function $(1 + x)^\alpha$.

PROBLEMS FOR SECTION 2

Compute the Maclaurin series for each of the following functions f by evaluating its derivatives $f^{(n)}(0)$.

1. $1/(1 - x)$

2. $\log(1 - x)$

3. x^3

4. $1/(1 + x)$

5. $x^3 - 2x + 5$

6. $\sqrt{1 + x}$

Assuming Theorems 4 through 9, express each of the following functions as a sum of a power series. Also give the interval of convergence in each case.

7. e^{-x^2}

8. $\cos \sqrt{x}$

9. $\arctan x^3$

10. $\int_0^x \cos \sqrt{t}\, dt$

11. $\int_0^x e^{-t^2/2}$

12. Complete the proof of Theorem 6 by giving the details for steps (a), (b) and (c).

13. Prove Theorem 7.

14. Show that if $g'(x) = \alpha g(x)/(1 + x)$, then $g(x) = c(1 + x)^\alpha$ for some constant c. (How can you show that $g(x)/(1 + x)^\alpha$ is a constant?)

15. a) Show that the infinite series defined in Theorem 9 converges on $(-1, 1)$.
     b) Show that its sum $g(x)$ has the property that $g'(x)(1 + x) = \alpha g(x)$.
     c) Assuming the result of Problem 14, show that $g(x) = (1 + x)^{\alpha}$.

16. We know that $x^n e^{-x} \to 0$ as $x \to \infty$ for any positive integer n (see Chapter 7, Section 3, Problem 33). Show therefore that

$$\frac{e^{-1/x^2}}{x^m} \to 0$$

as $x \to 0$, for any positive integer m. It follows then that

$$p\left(\frac{1}{x}\right)e^{-1/x^2} \to 0$$

as $x \to 0$ for any polynomial in $1/x$:

$$p\left(\frac{1}{x}\right) = \frac{a_n}{x^n} + \frac{a_{n-1}}{x^{n-1}} + \cdots + \frac{a_1}{x} + a_0 .$$

17. Show that the derivative of a function of the form

$$f(x) = p\left(\frac{1}{x}\right)e^{-1/x^2}$$

is again of this form, with a different polynomial. It is sufficient to show this for one of the terms of $f(x)$, that is, for a "monomial" term $ae^{-1/x^2}/x^m$.

18. Now define $F(x)$ by

$$F(x) = e^{-1/x^2} \qquad \text{if } x \neq 0,$$

$$F(0) = 0.$$

     a) Show that F is everywhere differentiable, with $F'(0) = 0$ and with $F'(x)$ of the form $p(1/x)e^{-1/x^2}$, when $x \neq 0$, where p is a polynomial. (Use Problems 16 and 17.)
     b) Then show, continuing to use Problems 16 and 17, that all the derivatives of F exist and that $F^{(n)}(0) = 0$ for every n.

The function F is thus an infinitely differentiable function such that $F(x) \neq 0$ when $x \neq 0$, and yet its Maclaurin-series coefficients are all 0.

3. PROOF OF THEOREM 3

We come finally to the postponed proof of the basic theorem on differentiating power series. It does not involve any new principles, but is a little more complicated than anything we have done so far in this chapter.

Theorem 3. *If*

$$f(x) = a_0 + a_1 x + a_2 x^2 + \cdots + a_n x^n + \cdots$$

on an open interval $I = (-c, c)$, *then* f *is differentiable and*

$$f'(x) = a_1 + 2a_2 x + \cdots + na_n x^{n-1} + \cdots$$

on I.

Proof. We know from Theorem 2 that the term-by-term differentiated series $\sum na_n x^{n-1}$ converges on the same interval $(-c, c)$. Call its sum $g(x)$. We want to prove that $g(x) = f'(x)$, that is, that

$$\frac{f(x + \Delta x) - f(x)}{\Delta x} - g(x)$$

approaches 0 as Δx approaches 0. According to Theorem 3, Chapter 12, this algebraic combination is the sum of the series

$$\sum_{n=1}^{\infty} a_n \left[\frac{(x + \Delta x)^n - x^n}{\Delta x} - nx^{n-1} \right].$$

We shall show below that the nth term of this series is less in absolute value than Δx times the nth term of the twice-differentiated series $\sum n(n-1)a_n r^{n-2}$ (for a suitable value of r). We know that this twice-differentiated series is absolutely convergent, by Theorem 2 again, and if we set

$$s = \sum_{n=2}^{\infty} n(n-1)|a_n|r^{n-2},$$

then it will follow directly from Theorem 6 of Chapter 12 that the two sums satisfy the inequality

$$\left| \frac{f(x + \Delta x) - f(x)}{\Delta x} - g(x) \right| \le s|\Delta x|.$$

The left side, therefore, has the limit zero as $\Delta x \to 0$. That is,

$$g(x) = \lim_{\Delta x \to 0} \frac{f(x + \Delta x) - f(x)}{\Delta x},$$

or $g(x) = f'(x)$, as we claimed.

Our proof will thus be complete when we have shown that

$$\left| \frac{(x + \Delta x)^n - x^n}{\Delta x} - nx^{n-1} \right| \le n(n-1)r^{n-2} \cdot |\Delta x|,$$

provided that r is larger than both $|x|$ and $|x + \Delta x|$. But this is a direct consequence of the tangent-line error formula in Chapter 5, Section 3, which tells us that

$$(x + \Delta x)^n - x^n - nx^{n-1}\Delta x = \frac{n(n-1)X^{n-2}(\Delta x)^2}{2}$$

for some number X between x and $x + \Delta x$. Then $|X| < r$, and we obtain the above inequality upon dividing by $|\Delta x|$. ∎

4. ALTERNATING SERIES

The partial sums of the sine and cosine series are much better approximations to $\sin x$ and $\cos x$ than we are able to show at the moment, because we have no test that takes into account the cancelling that goes on in a series whose terms alternate in sign. We now fill this gap.

Theorem 10. *If $\{a_n\}$ is a strictly decreasing sequence of positive numbers such that $a_n \to 0$ as $n \to \infty$, then the series*

$$a_0 - a_1 + a_2 - \cdots + (-1)^n a_n + \cdots$$

is convergent. Moreover, its sum s is approximated by its nth partial sum s_n with an error less than the next term a_{n+1}:

$$|s - s_n| < a_{n+1}.$$

EXAMPLE 1. The Maclaurin series for $\sin x$ is an alternating series in this sense if $|x| \le 2$. Therefore $x - x^3/3!$ approximates $\sin x$ with an error less than $x^5/5!$; for example,

$$\sin 0.2 \approx 0.2 - 0.008/6$$

$$= 0.2 - 0.001333 \cdots$$

$$= 0.198666 \cdots$$

with an error less than

$$\frac{(0.2)^5}{120} = \frac{0.00032}{120} < 0.000003.$$

It will turn out in the proof that the partial sums are alternately too large and too small, so here

$$0.198666 < \sin 0.2 < 0.198670,$$

and

$$\sin 0.2 \approx 0.19867$$

is the closest five-place decimal approximation to $\sin 0.2$.

Proof of theorem. The odd partial sums s_{2m+1} form an *increasing* sequence, as shown by the following bracketing:

$$s_{2m+1} = (a_0 - a_1) + (a_2 - a_3) + \cdots + (a_{2m} - a_{2m+1}).$$

Each term in parentheses is positive, and the odd partial sums s_{2m+1} are the partial sums of this bracketed series. Similarly, the even partial sums s_{2m} form a *decreasing* sequence:

$$s_{2m} = a_0 - (a_1 - a_2) - \cdots - (a_{2m-1} - a_{2m}).$$

Moreover,

$$s_{2m} > s_{2m-1}$$

for all m, since $s_{2m} - s_{2m-1} = a_{2m} > 0$. Therefore, the intervals

$$I_m = [s_{2m-1}, s_{2m}]$$

form a nested sequence of closed intervals with lengths a_{2m} tending to 0. It follows from the nested-interval principle that the two endpoint sequences converge as $m \to \infty$, and to the same limit s. That is, $s_n \to s$ as n runs to ∞ through the even integers, and also as n runs to ∞ through the odd integers. Therefore, $s_n \to s$ as $n \to \infty$.

Finally, since

$$s_{2m-1} \le s \le s_{2m},$$

we see that

$$s - s_{2m-1} \le s_{2m} - s_{2m-1} = a_{2m}.$$

Also $s_{2m+1} \le s \le s_{2m}$, so

$$s_{2m} - s \le s_{2m} - s_{2m+1} = a_{2m+1}.$$

Thus in all cases

$$|s - s_n| \leq a_{n+1}. \quad \blacksquare$$

EXAMPLE 2. Determine a range of x over which the first three terms in the Maclaurin series for $\sin x$ are accurate as an approximation to $\sin x$ to within an error of 1 in the fifth decimal place.

. . . .

Solution. The error is less than $x^7/7!$ so it will be sufficient to restrict x so that

$$\frac{x^7}{7!} < 10^{-5},$$

or

$$x^7 < (7!)10^{-5} = 5040 \cdot 10^{-5} = 0.0504$$

or

$$x < (0.0504)^{1/7}.$$

Solving this requires logarithm tables (or a slide rule) but you can check that $(0.6)^7 < 0.03$ so the desired approximation holds at least over the interval $(0, 0.6)$.

EXAMPLE 3. The series

$$1 - \frac{1}{2} + \frac{1}{3} - \frac{1}{4} + \cdots + (-1)^{n+1}\frac{1}{n} + \cdots$$

is convergent, by the theorem. However it is not absolutely convergent because the harmonic series

$$1 + \frac{1}{2} + \frac{1}{3} + \cdots + \frac{1}{n} + \cdots$$

diverges.

A series that is convergent but *not* absolutely convergent is said to converge *conditionally*. Conditionally convergent series are of little interest to us here. They converge so slowly that their partial sums are of little use in estimating their sums. Moreover, we are not allowed to handle them algebraically with freedom. For example, we cannot express the above sum as the difference of the sums of the positive and negative parts,

$$s = (1 + \frac{1}{3} + \frac{1}{5} + \cdots) - (\frac{1}{2} + \frac{1}{4} + \frac{1}{6} + \cdots),$$

because each of these series diverges to infinity and we have the nonsensical

$$s = \infty - \infty.$$

So for us the importance of alternating series is the error estimate $|s - s_n| < a_{n+1}$ in situations where we do have rapid, absolute convergence.

The alternating-series test lets us determine the exact domain of convergence for many power series.

EXAMPLE 4. Consider again the series $\sum x^n/n$. The ratio test shows that it converges if $|x| < 1$ and diverges if $|x| > 1$, but it leaves up in the air the behavior of the series at $x = \pm 1$. We can now complete the picture.

At $x = 1$ the series becomes the harmonic series, which we know to be divergent.

At $x = -1$ it is the *alternating harmonic* series

$$-1 + \frac{1}{2} - \frac{1}{3} + \frac{1}{4} + \cdots + \frac{(-1)^n}{n} + \cdots,$$

which we now know converges, by Theorem 16.

The domain of convergence is thus exactly the set of numbers x such that $-1 \le x < 1$, that is, the interval $[-1, 1)$, closed on the left and open on the right.

PROBLEMS FOR SECTION 4

Determine completely the domain of convergence of the power series whose general term is:

1. $\dfrac{(-1)^n x^n}{n}$

2. $\dfrac{x^n}{n2^n}$

3. $\dfrac{x^n}{\sqrt{n}}$

4. $\dfrac{(x - 3)^n}{n^2}$

5. $(-1)^n \dfrac{(x - 2)^n}{2^n}$

6. $\dfrac{(-1)^n (x - 2)^n}{n2^n}$

7. $\dfrac{n(x - 1)^n}{2^n}$

8. $\dfrac{(x - 1)^n}{n3^n}$

9. $\dfrac{(x - 1)^n}{n^2 3^n}$

10. Show that for any value of x the Maclaurin series for $\sin x$ is alternating from some term on.

11. Compute $\sin 2$ to eight decimal places. (First show that the error after the first three terms is less than 3×10^{-9}.)

12. Compute $\sin 5$ to three decimal places.

13. Prove that

$$\int_0^x e^{-t^2/2} = x - \frac{x^3}{3 \cdot 2} + \frac{x^5}{5 \cdot 2^2 \cdot 2!} - \frac{x^7}{7 \cdot 2^3 \cdot 3!} + \cdots$$

$$+ \frac{(-1)^n x^{2n+1}}{(2n + 1)2^n n!} + \cdots$$

Use the above series to compute $\int_0^1 e^{-t^2/2}$ to two decimal places.

14. The arctan x series

$$\arctan x = x - \frac{x^3}{3} + \frac{x^5}{5} + \cdots + (-1)^n \frac{x^{2n+1}}{2n + 1} + \cdots$$

provides the standard method of computing π, but here again a little artistic modification pays enormous dividends. Prove first that, if $\tan \theta = 1/5$, then

$$\tan\left(4\theta - \frac{\pi}{4}\right) = \frac{1}{239}.$$

(Use the formulas for $\tan(x + y)$ and $\tan(x - y)$ from Chapter 4. Compute $\tan 2\theta$, $\tan 4\theta$, and $\tan(4\theta - \pi/4)$, in that order.) Conclude that

$$\pi = 16 \arctan \frac{1}{5} - 4 \arctan \frac{1}{239}.$$

This remarkable identity has been known for nearly 300 years.

15. We start by seeing what we can do with only the first term of the series for arctan $1/239$. Using just the crude estimate $239 > 200$ in the error term, show that $4/239$ approximates $4 \arctan 1/239$ with an error less than $2(10)^{-7}$.

16. The decimal expansion of $1/239$ (and hence that of $4/239$) repeats in blocks of 7. Carry out the long division far enough to determine this expansion. Then use the result from the above problem, and the inequality arctan $x < x$, to show that

$$0.01673620 < 4 \arctan \frac{1}{239} < 0.01673641.$$

17. Now compute 16 arctan 0.2 through the $x^9/9$ term, and show that

$$3.15832895 < 16 \arctan 0.2 < 3.15832899.$$

18. Combine the above two results using Problem 14, and so prove that

$$3.1415925 < \pi < 3.1415928.$$

Thus $\pi = 3.141593$ to the nearest six decimal places, and the expansion of π begins $3.141592 \cdots$

19. Show that if arctan $1/239$ is replaced by its series expansion through the second term (the $x^3/3$ term) in the formula

$$\pi = 16 \arctan \frac{1}{5} - 4 \arctan \frac{1}{239},$$

and if arctan $1/5$ is replaced by its expansion through the eighth term (the $x^{15}/15$ term), then the right side approximates π with an error less than 5×10^{-12}. (Use the crude estimates $239 > 200$ and $2^{13} = 8192 < 10^4$.)

5. MACLAURIN POLYNOMIALS

If a function f is the sum of a power series,

$$f(x) = a_0 + a_1 x + \cdots + a_n x^n + \cdots = \sum_0^\infty a_j x^j$$

on $(-r, r)$, then $f(x)$ is *approximated* by the nth partial sum

$$s(x) = a_0 + a_1 x + \cdots + a_n x^n.$$

We can get an idea of the nature of this approximation from the infinite series. We can factor out x^{n+1} from the remainder series $r(x) = f(x) - s(x)$,

$$r(x) = a_{n+1} x^{n+1} + \cdots + a_{n+j} x^{n+j} + \cdots$$

$$= x^{n+1}[a_{n+1} + a_{n+2} x + \cdots + a_{n+j} x^{j-1} + \cdots]$$

$$= x^{n+1} h(x),$$

where h is the sum of a power series converging on $(-r, r)$. In particular, h is continuous, and on any *closed* subinterval I, say $I = [-r/2, r/2]$, $|h|$ has a maximum value K. Thus

$$|f(x) - s(x)| \le K|x|^{n+1}$$

on I. This shows the general nature of the approximation of f by its partial-sum polynomial $s(x)$; as x becomes small, the error becomes small like a multiple of x^{n+1}. But unless we know that K is small, we can only be sure that this approximation is good locally around the origin, i.e., for sufficiently small x.

The polynomial $s(x)$ above is very special, being the nth partial sum of the Maclaurin series for $f(x)$. We shall call it the nth *Maclaurin polynomial* of $f(x)$, or the *Maclaurin polynomial of degree n*. Its formula is

$$p_n(x) = f(0) + f'(0)x + \frac{f''(0)}{2} + \cdots + \frac{f^{(n)}(0)}{n!} x^n.$$

EXAMPLE 1. The fourth Maclaurin polynomial of e^x is

$$p(x) = 1 + x + \frac{x^2}{2} + \frac{x^3}{6} + \frac{x^4}{24}.$$

EXAMPLE 2. The fifth Maclaurin polynomial of $\cos x$ is

$$p(x) = 1 - \frac{x^2}{2!} + \frac{x^4}{4!}.$$

In a power series the nth partial sum is generally understood to be the sum through the term $a_n x^n$, regardless of how many preceding coefficients a_n are zero.

We can also form Maclaurin polynomials for functions that are not sums of power series. All that is required is that the function in question have derivatives up through some order n around the origin.

EXAMPLE 3. The function

$$f(x) = x^{7/2} + x^3 - 2x^2 + x + 3$$

has derivatives through the order 3 around the origin, and you can check that its second-degree Maclaurin polynomial is

$$f(0) + \frac{f'(0)}{1}x + \frac{f''(0)}{2!}x^2 = 3 + x - 2x^2.$$

We could also form its third-degree Maclaurin polynomial. Note, however, that $f^{(4)}(0)$ does not exist, because

$$\frac{d^4 x^{7/2}}{dx^4} = \frac{7}{2} \cdot \frac{5}{2} \cdot \frac{3}{2} \cdot \frac{1}{2} x^{-1/2},$$

which is undefined at $x = 0$. Therefore, f does not have any further Maclaurin polynomials.

EXAMPLE 4. The function $\tan x$ can be proved to be the sum of its Maclaurin series near the origin, but we don't know any way of obtaining the series coefficients except to compute them by the Maclaurin formula. We can obtain the first few coefficients in this manner, but the differentiations required become more and more cumbersome, and after a while we give up. Here we shall compute the third Maclaurin polynomial. We have

$$f(0) = 0;$$

$$\frac{d}{dx}\tan x = \sec^2 x = \tan^2 x + 1, \qquad f'(0) = 1;$$

$$\frac{d^2}{dx^2}\tan x = 2\tan x \sec^2 x$$

$$= 2\tan^3 x + 2\tan x, \qquad f''(0) = 0;$$

$$\frac{d^3}{dx^3}\tan x = 6\tan^2 x \sec^2 x + 2\sec^2 x$$

$$= 6\tan^4 x + 8\tan^2 x + 2, \qquad f'''(0) = 2.$$

The third Maclaurin polynomial of tan x is thus

$$p(x) = x + \frac{x^3}{3}.$$

This is also the fourth Maclaurin polynomial. Can you see why this is so without explicitly making the next calculation?

The question now is whether a Maclaurin polynomial of f *always* approximates f in the local way that we discovered above. That is, if f has derivatives through order n at the origin, so that we can form its nth-degree Maclaurin polynomial, p, does it necessarily follow that

$$|f(x) - p(x)| \le K|x|^{n+1}$$

on some interval I about the origin?

Let us see to what extent we can repeat, or at least imitate, the reasoning that we used in the power-series case. The important fact there was that the remainder $r(x) = f(x) - p(x)$ had x^{n+1} as a factor. We can state this property of $r(x)$ in other ways. One is that the power series whose sum is $r(x)$ has only zero coefficients before the $(n+1)$st coefficient. That is, the nth Maclaurin polynomial for $r(x)$ is just the zero polynomial. That is, the derivatives

$$r(0), \qquad r'(0), \qquad \cdots, \qquad r^{(n)}(0)$$

are all zero. These latter two formulations can be checked in the general situation.

Theorem 11. *If*

$$p(x) = a_0 + a_1 x + \cdots + a_n x^n$$

is a Maclaurin polynomial for $f(x)$, and if we set $r(x) = f(x) - p(x)$, then

$$0 = r(0) = r'(0) = \cdots = r^{(n)}(0).$$

That is, the nth Maclaurin polynomial of $f - p$ is zero.

Proof. By assumption, the derivatives of f exist up through the order n at the origin and the coefficients a_k of the polynomial p have the values

$$a_k = \frac{f^{(k)}(0)}{k!}$$

for $k = 0, 1, \ldots, n$. If we now apply Theorem 4 directly to the polynomial $p(x)$

itself, we see that the coefficients a_k also have the values

$$a_k = \frac{p^{(k)}(0)}{k!},$$

for $k = 0, 1, \ldots, n$. Therefore, $p^{(k)}(0) = f^{(k)}(0)$ for $k = 0, 1, \ldots, n$, and since $r^{(k)}(x) = f^{(k)}(x) - p^{(k)}(x)$ for all the derivatives that exist, it follows that $r^{(k)}(0) = 0$ for $k = 0, 1, \ldots, n$. ∎

Our approximation question can now be rephrased: If $g(x)$ is a function defined on an interval I about the origin, and if the derivatives $g'(0), g''(0), \ldots, g^{(n)}(0)$ all exist and all have the value 0, that is, if the nth Maclaurin polynomial of g exists and is identically zero, must it necessarily follow that

$$|g(x)| \leq K|x|^{n+1}$$

on some interval about the origin?

One should always try out general conjectures on simple special cases, and here even the case $n = 0$ breaks down. In this case we are asking whether the hypothesis that $g(0) = 0$ implies that

$$|g(x)| \leq K|x|.$$

A counterexample is provided by $g(x) = x^{1/3}$: the inequality $|x^{1/3}| \leq K|x|$ is equivalent to

$$|x^{-2/3}| \leq K,$$

which contradicts the fact that

$$x^{-2/3} \to \infty \qquad \text{as} \quad x \to 0.$$

So our first conjecture fails.

On the other hand, if g' exists on some interval I about the origin, then the mean-value theorem tells us that

$$g(x) = g(x) - g(0) = g'(X) \cdot x$$

for some X between 0 and x. Then, if we know that $|g'(x)|$ assumes a maximum value M on the interval I, we can conclude that

$$|g(x)| = |g'(X)| \cdot |x| \leq M|x|$$

for all x in I. Moreover, in the power-series case, the derivatve g' does exist and is continuous. These considerations suggest a more "educated" conjecture which turns out to be correct. We now guess that if g has continuous derivatives up through the order $n + 1$, and if g and its first n derivatives all vanish at the origin, then

$$|g(x)| \leq K|x|^{n+1}$$

in some interval about the origin. This desired conclusion can be written

$$\left| \frac{g(x)}{x^{n+1}} \right| \leq K,$$

which might lead us to study the quotient function $g(x)/x^{n+1}$. In any event, the parametric mean-value theorem can be applied repeatedly to this quotient, and here is what we get:

Theorem 12. *Suppose that $g(x)$ has derivatives up through the order $n + 1$ in some interval I about the origin, and that*

$$0 = g(0) = g'(0) = \cdots = g^{(n)}(0).$$

Then for each x in I, there is a number X lying strictly between 0 and x, such that

$$g(x) = \frac{g^{(n+1)}(X)}{(n + 1)!} x^{n+1}.$$

Proof. We apply the parametric mean-value theorem in the form of Theorem 7, Chapter 11, repeatedly to the quotient $g(x)/x^{n+1}$. First

$$\frac{g(x)}{x^{n+1}} = \frac{g'(x_1)}{(n + 1)x_1^n}$$

for some x_1 between 0 and x. Then

$$\frac{g'(x_1)}{(n + 1)x_1^n} = \frac{g''(x_2)}{(n + 1)nx_2^{n-1}}$$

for some point x_2 between 0 and x_1. We can continue in this way right up through the nth derivative, because all the derivatives $g^{(k)}(0)$ are 0. Thus,

$$\frac{g(x)}{x^{n+1}} = \frac{g'(x_1)}{(n + 1)x_1^n} = \frac{g''(x_2)}{(n + 1)nx_2^{n-1}} = \cdots = \frac{g^{(n)}(x_n)}{(n + 1)! \, x_n} = \frac{g^{(n+1)}(X)}{(n + 1)!},$$

and this final value for the quotient gives us the theorem.

Theorems 11 and 12 combine as follows:

Theorem 13. *If f has derivatives up through the order $n + 1$ in some interval I about* the origin, and if $s(x)$ is the nth Maclaurin polynomial of f, then

$$f(x) - s(x) = \frac{f^{(n+1)}(X)}{(n+1)!} x^{n+1},$$

where X is a number depending on x and lying between 0 and x.

Proof. Theorems 11 and 12 show that if $r(x) = f(x) - s(x)$, then

$$r(x) = \frac{r^{(n+1)}(X)}{(n+1)!} x^{n+1}.$$

Moreover, $r^{(n+1)} = f^{(n+1)}$, because $s(x)$ is a polynomial of degree at most n and so its $(n+1)$st derivative is zero. The equation in the theorem is thus established upon replacing $r^{(n+1)}(X)$ by $f^{(n+1)}(X)$, and r by $f - s$. ∎

The expression of $f(x)$ in the form $f(x) = s(x) + r(x)$ now becomes a formula for $f(x)$ made up of its nth Maclaurin polynomial plus an explicit formula for the remainder $r(x)$:

$$f(x) = f(0) + \frac{f'(0)}{1} x + \frac{f''(0)}{2!} x^2 + \cdots + \frac{f^{(n)}(0)}{n!} x^n + \frac{f^{(n+1)}(X)}{(n+1)!} x^{n+1}.$$

This is called the *Maclaurin formula with remainder.*

The explicit formula for the remainder $r(x) = f(x) - s(x)$ is very useful. For example, if we can determine that $|f^{(n+1)}(x)|$ assumes the maximum value M on the interval I, then

$$|f(x) - s(x)| \leq \frac{M}{(n+1)!} |x|^{n+1}$$

for all x in I, and we have the error-estimating inequality we were looking for.

Beyond this, if f is infinitely differentiable, then we can fix x and vary n, and possibly show that the nth remainder $r_n(x) = f(x) - s_n(x)$ approaches 0. We would then have another way of proving that f is the sum of its Maclaurin series.

EXAMPLE 5. For $f(x) = e^x$, the Maclaurin formula with remainder is

$$e^x = 1 + \frac{x}{1} + \frac{x^2}{2!} + \cdots + \frac{x^n}{n!} + e^X \frac{x^{n+1}}{(n+1)!}$$

for some X between 0 and x. If x is positive then $e^X \leq e^x$, and

$$r_n(x) \leq \frac{e^x x^{n+1}}{(n+1)!}.$$

(If x is negative, then $e^x \leq 1$ and we omit the e^x factor.) We know that $x^n/n! \to 0$ as $n \to \infty$. Therefore $r_n(x) \to 0$ for all x. That is

$$s_n(x) = e^x - r_n(x) \text{ approaches } e^x,$$

so

$$e^x = \lim_{n \to \infty} s_n(x) = \lim_{n \to \infty} \sum_{k=0}^{n} \frac{x^k}{k!} = \sum_{k=0}^{\infty} \frac{x^k}{k.}$$

We thus have a second proof that e^x is the sum of its Maclaurin series. However when x is positive the above estimate of the remainder $r_n(x)$ is not is good as the one we obtained in Chapter 12 by comparing the remainder series with the geometric series.

PROBLEMS FOR SECTION 5

Find the nth Maclaurin polynomial $p_n(x)$ of the function $f(x)$, where:

1. $f(x) = \arcsin x, \quad n = 3$
2. $f(x) = \log(1 + x), \quad n = 3$
3. $f(x) = \sec x, \quad n = 3$
4. $f(x) = \log(\cos x), \quad n = 4$
5. $f(x) = \sin(x + \frac{\pi}{4}), \quad n = 4$
6. $f(x) = (1 + x)^a, \quad n = 4$
7. $f(x) = \tan x, \quad n = 4$

Find the nth Maclaurin formula with remainder, $f(x) = p_n(x) + r_n(x)$, for each function f below. Also determine a constant K such that

$$|f(x) - p_n(x)| \leq Kx^{n+1}$$

on the interval $I = [-\frac{1}{2}, \frac{1}{2}]$.

8. $f(x) = \arcsin x, \quad n = 2$
9. $f(x) = \sec x, \quad n = 2$
10. $f(x) = \tan x, \quad n = 2$
11. $f(x) = e^{-x}, \quad n = 4$
12. $f(x) = \sin(x - \frac{\pi}{4}), \quad n = 4$
13. $f(x) = (1 + x)^{3/2}, \quad n = 3$
14. $f(x) = \int_0^x e^{-t^2} dt, \quad n = 3$

15. Prove that every Maclaurin polynomial of an odd (even) function is odd (even).

16. The second Maclaurin polynomial of $f(x)$ is $2 + (x/2) - (x^2/8)$. Show that the second Maclaurin polynomial of $1/f(x)$ is

$$\frac{1}{2} - \frac{x}{8} + \frac{x^2}{16},$$

by computing the derivatives of $g = 1/f$ and evaluating them at the origin.

17. Let the function $g(x)$ in the above problem have the second Maclaurin polynomial $A + Bx + Cx^2$. Supposing that f and g both have continuous third derivatives, we can write

$$f(x) = 2 + \frac{x}{2} - \frac{x^2}{8} + x^3 F(x),$$
$$g(x) = A + Bx + Cx^2 + x^3 G(x),$$

where F and G are continuous. Use the fact that $f(x)g(x) = 1$ to solve for the coefficients A, B, and C.

18. The second Maclaurin polynomial of $f(x)$ is $1 + x + (x^2/6)$. Find the second Maclaurin polynomial of $g = 1/f$:

a) by the method of Problem 16;
b) by the method of Problem 17.

19. Find the third Maclaurin polynomial for $\tan x$ by the method of Problem 17, using this time the equation $\cos x \tan x = \sin x$ and the known polynomials for sine and cosine.

20. Prove that $\sin x$ is the sum of its Maclaurin series by using the Maclaurin formula with remainder, as in Example 5 in the text.

21. Prove that $1/(1 - x)$ is the sum of its Maclaurin series over the interval $(-1, 1/2)$ by using the Maclaurin formula with remainder.

22. Suppose that f has derivatives up through order $n - 1$ on an interval I about the origin, with

$$f(0) = f'(0) = \cdots = f^{(n-1)}(0) = 0.$$

Suppose furthermore that $f^{(n-1)}(x)$ is differentiable at $x = 0$, so that $f^{(n)}(0)$ exists. Show that $f(x)/x^n \to f^{(n)}(0)/n!$ as $x \to 0$. (Apply l'Hôpital's rule.)

23. If f has derivatives of order $n - 1$ on an interval about the origin and if $f^{(n-1)}(x)$ is differentiable at $x = 0$, then

$$f(0), \quad f'(0), \quad \cdots, \quad f^{(n)}(0)$$

all exist, and we can write down the nth Maclaurin polynomial $p_n(x)$ for f. Show that

$$\lim_{x \to 0} \frac{f(x) - p_n(x)}{x^n} = 0.$$

(Assume and apply the conclusion in the preceding problem.)

24. Let p and q be two polynomials of degree n at most, and suppose each of

them is an nth-order approximation of f around the origin, in the sense that

$$\lim_{x \to 0} \frac{f(x) - p(x)}{x^n} = 0,$$

$$\lim_{x \to 0} \frac{f(x) - q(x)}{x^n} = 0.$$

Show that the two polynomials p and q must necessarily be the same. In view of the above problem, we see that this property uniquely characterizes the nth Maclaurin polynomial of f.

25. Use Problem 24 to prove that every Maclaurin polynomial of an even (odd) function is even (odd).

6. TAYLOR EXPANSIONS

Suppose that g is the sum of a power series,

$$g(t) = a_0 + a_1 t + a_2 t^2 + \cdots + a_n t^n + \cdots$$

on an interval $(-c, c)$, and suppose that we set $t = x - a$ and $f(x) = g(x - a)$. Then the above series turns into a series expansion of $f(x)$:

$$f(x) = a_0 + a_1(x - a) + a_2(x - a)^2 + \cdots + a_n(x - a)^n + \cdots$$

As t runs across the interval $(-c, c)$, then $x = t + a$ runs across $(a - c, a + c)$. For any x in this interval, the series $\sum a_n(x - a)^n$ is the same as the series $\sum a_n t^n$ for $t = x - a$ in the interval $(-c, c)$, Thus, $f(x)$ is the sum of the x series on $(a - c, a + c)$.

Conversely, if we can express a function $f(x)$ as the sum of a convergent power series of the form

$$f(x) = a_0 + a_1(x - a) + \cdots + a_n(x - a)^n + \cdots$$

on an interval centered at $x = a$, say $J = (a - c, a + c)$, then we have

$$f(t + a) = a_0 + a_1 t + \cdots + a_n t^n + \cdots$$

for t in the interval $I = (-c, c)$, and this is the Maclaurin series for $g(t) = f(t + a)$ there.

What we are in effect doing is "sliding," or translating, the graph of f to a new position to obtain the graph of g. The equation $g(t) = f(t + a)$ says that the behavior of g around $t = 0$ is identical to the behavior of f around $t = a$.

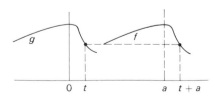

The coefficients a_n are, of course, the Maclaurin coefficients for the function g,

$$a_n = \frac{g^{(n)}(0)}{n!},$$

but they now can be reevaluated in terms of the derivatives of f at the point $x = a$. This follows from the relationship $f(x) = g(x - a)$. Differentiating repeatedly, we see that $f'(x) = g'(x - a)$, $f''(x) = g''(x - a)$, etc., and therefore

$$f'(a) = g'(0), \qquad f''(a) = g''(0), \qquad \text{etc.}$$

Thus, $f^{(n)}(a) = g^{(n)}(0)$ for all n, and so

$$a_n = \frac{f^{(n)}(a)}{n!}$$

for all n. The series for f can thus be rewritten

$$f(x) = f(a) + \frac{f'(a)}{1!}(x - a) + \frac{f''(a)}{2!}(x - a)^2 + \cdots + \frac{f^{(n)}(a)}{n!}(x - a)^n + \cdots$$

This is called the *Taylor series for f around* $x = a$.

A function f that is infinitely differentiable on an interval $(a - c, a + c)$ may or may not be the sum of its Taylor series there, just as a function g that is infinitely differentiable on $(-c, c)$ may or may not be the sum of its Maclaurin series.

The nth partial sum in the Taylor series for f around $x = a$,

$$s_n(x) = f(a) + \frac{f'(a)}{1}(x - a) + \cdots + \frac{f^{(n)}(a)}{n!}(x - a)^n,$$

is called the nth *Taylor polynomial for f at* $x = a$. It can be formed for any function having n derivatives at $x = a$.

EXAMPLE 1. The third Taylor polynomial for $f(x) = \log x$ at $x = 1$ is

$$p(x) = f(1) + \frac{f'(1)}{1}(x - 1) + \frac{f''(1)}{2!}(x - 1)^2 + \frac{f'''(1)}{3!}(x - 1)^3$$

$$= 0 + \frac{1}{1}(1 - x) + \frac{(-1)}{2!}(x - 1)^2 + \frac{2}{3!}(x - 1)^3$$

$$= (x - 1) - \frac{1}{2}(x - 1)^2 + \frac{1}{3}(x - 1)^3.$$

If f has $n + 1$ derivatives in an interval about $x = a$, then the Taylor remainder

$$r_n(x) = f(x) - s_n(x)$$

can be written in the form

$$r_n(x) = \frac{f^{(n+1)}(X)}{(n + 1)!}(x - a)^{n+1},$$

where X is some number between x and a. This formula is obtained directly from the Maclaurin remainder for $g(t) = f(t + a)$ upon setting $t = x - a$. Then

$$f(x) = f(a) + \frac{f'(a)}{1}(x - a) + \cdots + \frac{f^{(n)}(a)}{n!}(x - a)^n + \frac{f^{(n+1)}(X)}{(n + 1)!}(x - a)^{n+1}.$$

This is called the *Taylor formula with remainder*.

EXAMPLE 2. The second Taylor polynomial for $f(x) = \sin x$ around $a = \pi/4$ is

$$f(\pi/4) + \frac{f'(\pi/4)}{1}(x - \pi/4) + \frac{f''(\pi/4)}{2!}(x - \pi/4)^2$$

$$= \sin(\pi/4) + [\cos(\pi/4)](x - \pi/4) - [\sin(\pi/4)](x - \pi/4)^2/2$$

$$= \frac{\sqrt{2}}{2}\left[1 + \left(x - \frac{\pi}{4}\right) - \frac{(x - \pi/4)^2}{2}\right].$$

This polynomial approximates $\sin x$ with the error (remainder)

$$\frac{-\cos(X)}{3!}(x - \frac{\pi}{4})^3,$$

where X is some number between x and $\pi/4$. Since $|\cos X| \le 1$, the error is at most

$$\frac{\left|x - \frac{\pi}{4}\right|^3}{3!}.$$

For values of x near $\pi/4$, this approximation to $\sin x$ will have an accuracy that could be obtained from the Maclaurin series only by using many more terms. However, this advantage is offset in part by the fact that here we have to contend with decimal approximations to $\pi/4$.

EXAMPLE 3. Any polynomial $p(x)$ of degree n is equal to its own nth Taylor polynomial at any point $x = a$. We can see this from the Taylor formula because the remainder is then 0. Thus, if $p(x) = x^3 - x$, then

$$p'(x) = 3x^2 - 1, \qquad p''(x) = 6x, \qquad p'''(x) = 6;$$

$$p(2) = 6, \qquad p'(2) = 11, \qquad p''(2) = 12, \qquad p'''(2) = 6;$$

$$p(x) = 6 + \frac{11}{1}(x - 2) + \frac{12}{2!}(x - 2)^2 + \frac{6}{3!}(x - 2)^3.$$

Therefore,

$$x^3 - x = 6 + 11(x - 2) + 6(x - 2)^2 + (x - 2)^3.$$

EXAMPLE 4. The fact that a polynomial in x can be rewritten as a polynomial of the same degree in $(x - a)$ is really an algebraic fact rather than a calculus fact. We simply set $x = (x - a) + a$ and expand all the powers of x. Thus,

$$x^3 - x = [(x - 2) + 2]^3 - [(x - 2) + 2]$$

$$= (x - 2)^3 + 3(x - 2)^2 \cdot 2 + 3(x - 2) \cdot 2^2 + 8 - (x - 2) - 2$$

$$= (x - 2)^3 + 6(x - 2)^2 + 11(x - 2) + 6.$$

In order to prove that an infinitely differentiable function f is the sum of its Taylor series at $x = a$, we can try the "initial-value problem" approach that we used in Section 4, or we can try to show that the remainder approaches 0 in the Taylor formula with remainder. Nothing here is new, since the Taylor series for f is equivalent to the Maclaurin series for $g(t) = f(t + a)$.

PROBLEMS FOR SECTION 6

Find the nth Taylor polynomial of the function f around the point a, where

1. $f(x) = \log x$, $n = 3$, $a = 1$
2. $f(x) = \sin x$, $n = 4$, $a = \pi/4$
3. $f(x) = \sin x$, $n = 4$, $a = \pi/6$
4. $f(x) = \sin x$, $n = 4$, $a = \pi/3$
5. $f(x) = e^x$, $n = 3$, $a = a$
6. $f(x) = \tan x$, $n = 3$, $a = \pi/4$

Find the nth Taylor formula with remainder, $f(x) = p_n(x) + r_n(x)$, about the point $x = a$, where:

7. $f(x) = \cos x$, $n = 4$, $a = \pi/4$
8. $f(x) = \cos x$, $n = 4$, $a = \pi/3$
9. $f(x) = \sin x$, $n = 4$, $a = \pi/6$
10. $f(x) = e^x$, $n = n$, $a = a$
11. $f(x) = x^{1/3}$, $n = 3$, $a = 1$
12. $f(x) = x^{-1/4}$, $n = 3$, $a = 1$
13. $f(x) = x^r$, $n = 3$, $a = 1$
14. $f(x) = \tan x$, $n = 2$, $a = \pi/4$
15. $f(x) = \log \cos x$, $n = 3$, $a = \pi/4$

Write out the Taylor series for:

16. e^x about $x = a$
17. $\sqrt{x}$ about $x = 1$
18. $1/x$ about $x = 1$
19. $\sin x$ about $x = \pi/2$
20. $\log x$ about $x = 1$
21. $\sin x$ about $x = \pi$
22. $\sin x$ about $x = \pi/4$
23. Carry out the details of the calculation of the Taylor remainder formula from the Maclaurin remainder formula.
24. Show that

$$\frac{\sqrt{2}}{2}\left[1 - \frac{\pi}{20} - \frac{\pi^2}{800}\right]$$

approximates $\sin \pi/5$ with an error less than 0.001.

25. Express $p(x) = x^3 + 2x^2 + 1$ as its Taylor polynomial about $a = 1$.
26. Express $p(x) = x^4 - 5x^3 + 10x$ as its Taylor polynomial about $a = 2$.
27. Prove that $\sin x$ is the sum of its Taylor series about $\pi/2$ by using the Taylor formula with remainder and proving that the remainder goes to zero.
28. Show that the Taylor series for $\sin x$ about $a = \pi/6$ expresses the identity

$$\sin x = \sin\left(x - \frac{\pi}{6}\right)\cos\frac{\pi}{6} + \cos\left(x - \frac{\pi}{6}\right)\sin\frac{\pi}{6}.$$

29. Prove that e^x is the sum of its Taylor series about $x = a$, by evaluating the remainder and showing that it goes to zero.

7. THE MODIFIED TRAPEZOIDAL RULE

In this section we shall go back and pick up the estimation of definite integrals where we abandoned it at the end of Chapter 10. The trapezoidal rule can be vastly improved by including a correction that takes into account the exact directions of the graph at the two endpoints. Here is the result.

Theorem 14. *If the fourth derivative $f^{(4)}$ exists and is bounded by K on the interval $[a, b]$, then*

$$\int_a^b f \approx \text{trapezoidal estimate} - (\Delta x)^2 \frac{f'(b) - f'(a)}{12},$$

with an error at most

$$\frac{K(b-a)}{720}(\Delta x)^4.$$

Before taking up the proof of the theorem, let us see how it improves things with our test calculation $\int_1^{3/2} dx/x = \log(3/2)$. We use the same four-fold subdivision of $[1, (3/2)]$, but now add the endpoint-derivative correction to the trapezoidal estimate that we looked at earlier. Then

$$\log\frac{3}{2} = \int_1^{3/2}\frac{dx}{x} \approx \frac{1}{8}\left[\frac{1}{2} + \frac{8}{9} + \frac{4}{5} + \frac{8}{11} + \frac{1}{3}\right] - \frac{1}{64}\left(\frac{1 - 4/9}{12}\right)$$

$$= \left\{\frac{1}{16} + \frac{1}{9} + \frac{1}{10} + \frac{1}{11} + \frac{1}{24}\right\} - \frac{5}{6912}$$

$$= + \begin{cases} 0.06250000\cdots \\ 0.11111111\cdots \\ 0.10000000\cdots \\ 0.09090909\cdots \\ \underline{0.04166666\cdots} \end{cases}$$

$$ \quad 0.40618686\cdots$$

$$- \quad \underline{0.00072338\cdots}$$

$$= \quad 0.4054634\cdots$$

Here $f^{(4)}(x) = 24/x^5$, which is bounded by $K = 24$ over the interval $[1, 3/2]$. The error in the improved estimate is therefore at most

$$\frac{K(b-a)}{720}(\Delta x)^4 = \frac{24}{720} \cdot \frac{1}{2} \cdot \left(\frac{1}{8}\right)^4 = \frac{1}{245{,}760} < \frac{10^{-5}}{2} = 0.000005.$$

Thus, we have calculated that

$$\log 3/2 = 0.405463 \cdots$$

with an error less than 0.000005 in magnitude, and have five-place accuracy. (However, we don't know whether 0.40546 or 0.40547 is the *closest* five-place decimal. The table gives the latter.)

We now turn to the proof of Theorem 14.

Lemma 1. *If q is a polynomial of degree at most two then*

$$q(b) - q(a) = \frac{q'(b) + q'(a)}{2}(b - a)$$

for any two points a and b.

Proof. If $q(x) = Ax^2 + Bx + C$, then

$$q(b) - a(a) = A(b^2 - a^2) + B(b - a)$$

$$= [A(b + a) + B](b - a)$$

$$= \frac{[(2Ab + B) + (2Aa + B)]}{2}(b - a)$$

$$= \frac{q'(b) + q'(a)}{2}(b - a). \quad \blacksquare$$

There are similar (but more complicated) identities for higher-degree polynomials. We shall need the next one. Note that it includes Lemma 1 as a special case.

Lemma 2. *If p is a polynomial of degree at most four, then*

$$p(b) - p(a) = \frac{p'(b) + p'(a)}{2}(b - a) - \frac{p''(b) - p''(a)}{12}(b - a)^2$$

for any two points a and b.

Proof. We can write p as its own Taylor polynomial at a and also as its Taylor polynomial at b. Thus,

$$p(x) = p(a) + p'(a)(x - a) + \frac{p''(a)}{2}(x - a)^2 + \frac{p'''(a)}{6}(x - a)^3 + \frac{p^{(4)}(a)}{24}(x - a)^4,$$

$$p(x) = p(b) + p'(b)(x - b) + \frac{p''(b)}{2}(x - b)^2 + \frac{p'''(b)}{6}(x - b)^3 + \frac{p^{(4)}(b)}{24}(x - b)^4.$$

Now evaluate the first equation at $x = b$, the second at $x = a$, and subtract the second from the first. The resulting equation can be written

$$2[p(b) - p(a)] = [p'(b) + p'(a)](b - a) - \frac{p''(b) - p''(a)}{2}(b - a)^2$$
$$+ \frac{p'''(b) + p'''(a)}{6}(b - a)^3.$$

The fourth-degree terms cancel because the fourth derivative of p is a constant. We can now replace the third-order term by a second-order term, by applying Lemma 1 to the second-degree polynomial $q(x) = p''(x)$. The new numerical coefficient on the second-order term is

$$-\left(\frac{1}{2}\right) + \left(\frac{1}{3}\right) = -\frac{1}{6},$$

and the lemma now follows upon dividing by 2. ∎

Lemma 3. *If the fourth derivative of f exists and is bounded by K over the interval $[a, b]$, then the difference between $\int_a^b f$ and the number*

$$\frac{1}{2}(b - a)(f(b) + f(a)) - \frac{1}{12}(b - a)^2(f'(b) - f'(a))$$

is at most

$$\frac{K}{720}(b - a)^5$$

in magnitude.

Proof. Let $p(x)$ be the polynomial of degree at most 3 determined by the conditions

$$p(a) = f(a), \qquad p(b) = f(b), \qquad p'(a) = f'(a), \qquad p'(b) = f'(b).$$

Problems 4 and 5 develop the existence and uniqueness of such a polynomial p. Then the difference $g(x) = f(x) - p(x)$ is zero at a and b, as is its derivative $g'(x)$. Moreover, $g^{(4)}(x) = f^{(4)}(x)$, since $p^{(4)} = 0$, so $g^{(4)}$ is bounded by K on the interval $[a, b]$. It follows from Exercise 33 in Section 8 of Chapter 9 that

$$\left|\int_a^b g\right| \le \frac{K}{720}(b - a)^5.$$

But

$$\int_a^b g = \int_a^b f - \int_a^b p$$
$$= \int_a^b f - \left[\frac{1}{2}(b - a)(p(b) + p(a)) - \frac{1}{12}(b - a)^2(p'(b) - p'(a))\right]$$

by Lemma 2. (Note that if $q(x)$ is an antiderivative of $p(x)$, then q is a polynomial of degree at most 4.) Since

$$p(b) = f(b), \qquad \cdots, \qquad p'(a) = f'(a),$$

we end up with the assertion of Lemma 3. ∎

Proof of Theorem 14. Let the points $a = x_0, x_1, \ldots, x_n = b$ subdivide the interval $[a, b]$ into n equal subintervals of common length $\Delta x = (b - a)/n$. We apply Lemma 3 to each subinterval, and have the approximations:

$$\int_{x_{n-1}}^{x_n} f \approx \Delta x \frac{f(x_n) + f(x_{n-1})}{2} - (\Delta x)^2 \frac{f'(x_n) - f'(x_{n-1})}{12},$$
$$\vdots$$
$$\int_{x_1}^{x_2} f \approx \Delta x \frac{f(x_2) + f(x_1)}{2} - (\Delta x)^2 \frac{f'(x_2) - f'(x_1)}{12},$$
$$\int_{x_0}^{x_1} f \approx \Delta x \frac{f(x_1) + f(x_0)}{2} - (\Delta x)^2 \frac{f'(x_1) - f'(x_0)}{12},$$

where each approximation is in error by at most

$$\frac{K}{720}(\Delta x)^5.$$

Now we add these approximate equalities. The sum of the integrals is $\int_{x_0}^{x_n} f = \int_a^b f$. The sum of the first terms on the right is the trapezoidal approximation to $\int_a^b f$. Finally, the sum of the second terms is

$$(\Delta x)^2 \frac{f'(x_n) - f'(x_0)}{12} = (\Delta x)^2 \frac{f'(b) - f'(a)}{12},$$

since all the intermediate evaluations $f'(x_k)$ cancel out.

We thus end up with

$$\int_a^b f \approx \text{trapezoidal estimate} + (\Delta x)^2 \frac{f'(a) - f'(b)}{12}.$$

Since the error at each step is at most

$$\frac{K}{720}(\Delta x)^5,$$

the total error is at most

$$n \frac{K}{720}(\Delta x)^5 = \frac{K(b - a)}{720}(\Delta x)^4,$$

where we have again used the fact that $n\Delta x = b - a$. This proves the theorem. ∎

PROBLEMS FOR SECTION 7

1. What would the error be in the estimate of log 2 provided by Theorem 14 if [1,2] were divided into 10 subintervals? 20 subintervals?

2. Estimate log 2 by the formula of Theorem 14, using a six-fold subdivision of [1,2]. Compute the error estimate as the first step, and use it as a guide in deciding how far to carry out the decimal expansions.

3. Carry out the details of the proof of Lemma 2.

4. a) Show that the most general polynomial p of degree at most 3, such that $p(a) = p'(a) = 0$, can be written

$$p(x) = A(x - a)^3 + B(x - a)^2.$$

   [Hint Write p as its own Taylor polynomial at a.]

   b) Show that if $b \neq a$, then we can arrange to give $p(b)$ and $p'(b)$ any values we wish by choosing the coefficients A and B suitably. Show, in fact, that if C and D are any numbers, then the equations

$$p(b) = C, \qquad p'(b) = D$$

   can be *uniquely* solved for A and B in terms of C and D.

5. Assuming the conclusions of the preceding problem, show that we can find a unique polynomial $p(x)$, of degree at most 3, such that

$$p(a), \qquad p'(a), \qquad p(b), \qquad \text{and} \qquad p'(b)$$

   have any four assigned values.

6. Strengthen Theorem 14 by showing that if the fourth derivative $f^{(4)}$ does not change sign on $[a, b]$, then the error in the modified trapezoidal approximation has the sign of $f^{(4)}$. (Look back at Problem 33 in Section 8, Chapter 9. It gives more information than we have used.)

   Now look again at the $\log(3/2)$ example in the text and show, furthermore, that the decimal expansion of $\log(3/2)$ begins $0.40546 \cdots$

7. Show that the modified trapezoidal estimate gives the exact answer if f is a polynomial of degree at most 3.

8. Show that the error in Theorem 14 is exactly $(\Delta x)^4 (b - a)/720$ times the constant value of $f^{(4)}(x)$ if f is a fourth-degree polynomial. (This will require the more precise result in Problem 33, Section 8, Chapter 9.)

9. Assume the above problem, and apply Theorem 14 to a fourth-degree polynomial using the trivial subdivision of $[a, b]$ (no subdividing points). State the result as a lemma about fifth-degree polynomials, analogous to Lemma 2.

The identity

$$\pi = 4 \arctan 1 = 4 \int_0^1 \frac{dx}{1 + x^2}$$

can be used to calculate π. In the following problems we consider the 10-fold subdivision of $[0,1]$ by the one-place decimals 0.0, 0.1, ..., 0.9, 1.0. The calculation of the higher derivatives of $f(x) = 1/(1 + x^2)$ becomes rather complicated, and you may assume that $|f^{(4)}(x)| \leq 24$ on the interval $[0,1]$. The correct beginning of the decimal expansion of π is

$$\pi = 3.141592653589 \cdots$$

10. Show that the error in the approximation of $\pi = 4 \int_0^1 dx/(1 + x^2)$ is:
    a) less than 0.0067 when the unmodified trapezoidal sum is used (Chapter 10, Theorem 2);
    b) less than $0.000014 = 1.4(10)^{-5}$ when the modified trapezoidal sum is used (Theorem 14 above).

A modern desk calculator gave the following values for $1/(1 + x^2)$ at the subdividing points:

1/1.01	0 . 9 9 0 0 9 9 0 0 9 9 0 0
1/1.04	0 . 9 6 1 5 3 8 4 6 1 5 3 8
1/1.09	0 . 9 1 7 4 3 1 1 9 2 6 6 0
1/1.16	0 . 8 6 2 0 6 8 9 6 5 5 1 7
1/1.25	0 . 8 0 0 0 0 0 0 0 0 0 0 0
1/1.36	0 . 7 3 5 2 9 4 1 1 7 6 4 7
1/1.49	0 . 6 7 1 1 4 0 9 3 9 5 9 7
1/1.64	0 . 6 0 9 7 5 6 0 9 7 5 6 0
1/1.81	0 . 5 5 2 4 8 6 1 8 7 8 4 5

It is clear from the top line that the calculator did not round off at the last place, and the above values are therefore the 12-place truncations in the decimal expansions. So each approximates its entry with an error lying between 0 and 10^{-12}. Note that one row is exact.

11. Compute the sum of the above 12-place decimals, and modify it to correspond to the trapezoidal sum. (Two entries are missing, as well as the factor Δx.) Then multiply by 4. The result is an approximation of π. Show that the error is approximately 0.0017. Show also that this error is due almost entirely to the trapezoidal sum, and not to the above approximation of this sum. That is, show that the error in your approximation of the exact trapezoidal sum is insignificant as compared to the error in the trapezoidal sum itself.

12. Now modify the approximation by the endpoint-derivative correction term of Theorem 14, and compare with the expansion of π. Your answer should be correct through 8 decimal places.

13. The actual estimate of π obtained above is much better than theoretically predicted error (Problem 10). Go over the proof of Theorem 14 with the present case in mind and try to see why this might be.

8. SIMPSON'S RULE

Simpson's rule is a good compromise between accuracy and simplicity. It is not as accurate as the modified trapezoidal rule, by a factor of 4, but it avoids the two endpoint-derivative evaluations. The basic estimate can be established by a rather complicated integration by parts, similar to the derivations in the last section but less straightforward. We shall omit the proof.

Lemma 4. *The "weighted" Riemann-sum estimate*

$$\int_{a-h}^{a+h} f(x)\,dx \approx \frac{h}{3}[f(a-h) + 4f(a) + f(a+h)]$$

is accurate to within an error of

$$\frac{K}{90}h^5,$$

if the fourth derivative of f exists and is bounded by K over the interval $[a-h, a+h]$.

Theorem 15 (Simpson's rule). *If the points $a = x_0, x_1, \ldots, x_n = b$ subdivide the interval $[a, b]$ into an even number of equal subintervals of length $\Delta x = (b-a)/n$ each, then*

$$\int_a^b f \approx \frac{\Delta x}{3}[f(x_0) + 4f(x_1) + 2f(x_2) + 4f(x_3) + \cdots + 4f(x_{n-1}) + f(x_n)].$$

If $f^{(4)}(x)$ is bounded by K over the interval $[a, b]$, then the error in the approximation is at most

$$\frac{K(b-a)}{180}(\Delta x)^4.$$

Note that the coefficient pattern in this estimating sum, including the outside factor $(1/3)$, is

$$\frac{1}{3}, \frac{4}{3}, \frac{2}{3}, \frac{4}{3}, \frac{2}{3}, \frac{4}{3}, \quad \cdots \quad \frac{2}{3}, \frac{4}{3}, \frac{1}{3}.$$

Since there are $n + 1$ terms, the sum of these coefficients is n, the number of intervals in the subdivision. The right side is a "weighted" average of the values $f(x_i)$, where $i = 0, 1, \ldots, n$.

Proof of Theorem 15. We apply the estimate of the lemma to successive pairs of subintervals:

$$\int_{x_0}^{x_2} f \approx \frac{\Delta x}{3}[f(x_0) + 4f(x_1) + f(x_2)]$$

$$\int_{x_2}^{x_4} f \approx \frac{\Delta x}{3}[f(x_2) + 4f(x_3) + f(x_4)]$$

$$\vdots$$

$$\int_{x_{n-2}}^{x_n} \approx \frac{\Delta x}{3}[f(x_{n-2}) + 4f(x_{n-1}) + f(x_n)].$$

The sum of these approximate equalities is Simpson's rule:

$$\int_a^b f \approx \frac{\Delta x}{3}[f(x_0) + 4f(x_1) + 2f(x_2) + 4f(x_3) + \cdots + 4f(x_{n-1}) + f(x_n)].$$

If $f^{(4)}(x)$ is bounded by K over the whole interval $[a, b]$, then the error in each application of Lemma 4 is at most $K(\Delta x)^5/90$, and since there are $(n/2)$ such approximations, the total accumulated error is at most

$$\frac{n}{2} \cdot \frac{K}{90} \cdot (\Delta x)^5 = \frac{K(b - a)}{180}(\Delta x)^4$$

(since $n\Delta x = b - a$). ∎

EXAMPLE 1. We consider our canonical $\log(3/2)$ problem for the last time. Now the estimate is

$$\log \frac{3}{2} = \int_1^{3/2} \frac{dt}{t} \approx \frac{\Delta x}{3}[x_0 + 4x_1 + 2x_2 + 4x_3 + x_4]$$

$$= \frac{1}{24}\left[1 + \frac{32}{9} + \frac{8}{5} + \frac{32}{11} + \frac{2}{3}\right]$$

$$= \frac{1}{24} + \frac{4}{27} + \frac{1}{15} + \frac{4}{33} + \frac{1}{36}$$

$$= \begin{cases} .0416666 \cdots \\ .1481481 \cdots \\ .0666666 \cdots \\ .1212121 \cdots \\ \underline{.0277777 \cdots} \\ .4054711 \cdots \end{cases}$$

Since $f^{(4)}(x) = 24/x^5$ is bounded by $K = 24$ over the interval $[1, \frac{3}{2}]$, the error in the above estimate is at most

$$\frac{K(b-a)}{180}(\Delta x)^4 = \frac{24}{180} \cdot \frac{1}{2} \cdot \left(\frac{1}{8}\right)^4$$

$$= \frac{1}{61,440} < \frac{1}{6}(10)^{-4} < 0.000017.$$

Thus,

$$0.40545 < \log(3/2) < 0.40549$$

and

$$\log \frac{3}{2} \approx 0.4055$$

to the nearest four decimal places.

PROBLEMS FOR SECTION 8

1. Show that Simpson's rule is exact for a polynomial of degree 3 or less.

2. Apply Simpson's rule to $f(x) = x^4$ over the interval $[-a, a]$ with $\Delta x = a$. Show that the error is exactly

    $$\frac{K(b-a)}{180}(\Delta x)^4$$

    in magnitude, where K is the constant value of $f^{(4)}$. This shows that the error formula cannot be improved.

3. Estimate $\pi = 4\int_0^1 dx/(1+x^2)$ by applying Simpson's rule to the data provided for Problem 11 in the last section. This is just a heavy dose of arithmetic, but the use of calculus in applied problems generally ends this way. Your error should be approximately $2.6(10)^{-5}$.

4. Now compute the expected error from the statement of Simpson's rule (Theorem 15), and compare with the actual error obtained above.

chapter 14

vectors in the plane

The interaction between vectors and calculus leads to a huge increase in the range of problems to which calculus can be successfully applied, and such vector calculus is one of the principal topics of advanced courses in calculus. In this book we can only introduce the subject and give a few applications.

This chapter develops the algebra of vectors in the plane and its geometric interpretations, and then treats the differential calculus for vector functions of a parameter t.

Chapters 16 and 17 take up vector interpretations of differentiation that arise in the study of functions of two or three independent variables. Chapter 15 is an introduction to such functions, containing no vector material.

1. THE POINT OPERATIONS

There are some simple operations on number pairs that have geometric meanings in the coordinate plane, and the resulting algebra of number pairs is a valuable new tool both in geometry and in calculus.

In discussing these operations, we shall use capital letters for points in the coordinate plane, i.e., for ordered pairs of numbers, the convention being that X is the pair (x,y), and

$$A = (a, b), \qquad U = (u, v), \qquad X' = (x', y'), \qquad X_0 = (x_0, y_0),$$

$$\Delta X = (\Delta x, \Delta y), \qquad \text{etc.}$$

This convention rules out using adjacent capital letters, and it seems a rather poor scheme, but there is nothing better unless the whole coordinate notation is changed over to "vector" notation. This new notation will be described briefly in Section 4 of Chapter 17.

We start with the natural definition of the sum $X + A$ of two points in terms of the sums of their coordinates:

Definition

$$(x, y) + (a, b) = (x + a, y + b).$$

Thus the first coordinate of the sum $X + A$ is the sum of the two first coordinates, and similarly for the second coordinates. For example,

$$(1, 2) + (3, -1) = (1 + 3, 2 - 1) = (4, 1).$$

The point equation

$$X_1 = X + \Delta X$$

is thus the equation

$$(x_1, y_1) = (x, y) + (\Delta x, \Delta y).$$

By the definition, this reduces to

$$(x_1, y_1) = (x + \Delta x, y + \Delta y),$$

and hence is equivalent to the two numerical equations

$$x_1 = x + \Delta x,$$
$$y_1 = y + \Delta y,$$

for the first and second coordinates, respectively.

Along with addition we need the product tX of a number t and a point X. Again, the definition is the obvious one.

Definition

$$t(x, y) = (tx, ty).$$

Thus

$$U = 3A$$

means that $(u, v) = 3(a, b) = (3a, 3b)$, and hence that

$$u = 3a \quad \text{and} \quad v = 3b.$$

The following laws of algebra for these two operations are immediate consequences of the corresponding laws for numbers.

(a1) $\qquad\qquad X + A = A + X \qquad$ for all A and X.

(a2) $\qquad (U + X) + A = U + (X + A) \qquad$ for all U, X and A.

(a3) There is a unique point 0 such that $X + 0 = 0 + X = X \qquad$ for all X.

(a4) For each X there is a unique point, called the *negative of X*, and designated $-X$, such that

$$X + (-X) = 0.$$

The above laws concern addition only. *Multiplication* of points by numbers satisfies the further laws:

(m1) $\qquad\qquad (st)X = s(tX).$

(m2) $\qquad\qquad (s + t)X = sX + tX.$

(m3) $\qquad\qquad t(X + A) = tX + tA.$

(m4) $\qquad\qquad 1X = X.$

These laws are verified by using the definitions of the operations and the laws of algebra for numbers. For example, (m3) is proved as follows:

$$t[X + A] = t[(x,y) + (a,b)] = t(x + a, y + b) = (t(x + a), t(y + b))$$

$$= (tx + ta, ty + tb) = (tx, ty) + (ta, tb)$$

$$= t(x,y) + t(a,b) = tX + tA.$$

The proof uses the definitions of the two operations together with the distributive law for numbers.

You probably can see that the zero point 0 of (a3) must be the zero pair (0,0), that is the origin of coordinates, and that $-X$ is the pair $(-x, -y)$.

The geometric interpretation of the two point operations is given by the following theorem. In its statement we use the notation $\overrightarrow{AX}$ for the directed line segment, or *arrow, from* the point A *to* the point X.

Theorem 1. *Two arrows $\overrightarrow{AX}$ and $\overrightarrow{A'X'}$ have the same length and direction if and only if*

$$X' - A' = X - A.$$

More generally, the arrows $\overrightarrow{AX}$ and $\overrightarrow{A'X'}$ are parallel if and only if

$$X' - A' = t(X - A)$$

for some real number t. In this situation, furthermore:
a) the two directions are the same if t is positive and opposite if t is negative;
b) the length of $\overrightarrow{A'X'}$ is $|t|$ times the length of $\overrightarrow{AX}$.

EXAMPLE 1. Show that the arrows in the left figure below have the same length and direction.

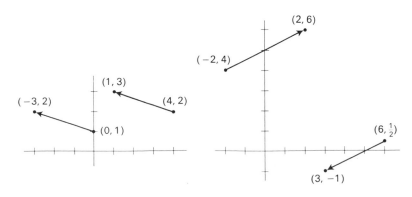

. . . .

Solution

$$X' - A' = (1, 3) - (4, 2) = (-3, 1),$$

$$X - A = (-3, 2) - (0, 1) = (-3, 1).$$

Since these differences are the same number pair, the theorem says that the arrows have the same length and direction.

EXAMPLE 2. The arrows in the right figure above look parallel. This conjecture can be checked by applying the theorem. We compute:

$$X - A = (3, -1) - (6, 1/2) = (-3, -3/2),$$

$$X' - A' = (2, 6) - (-2, 4) = (4, 2)$$

$$= -\frac{4}{3}(-3, -\frac{3}{2}) = -\frac{4}{3}(X - A).$$

Thus, the arrows are parallel. Furthermore, they point in opposite directions ($t = -4/3$ is negative), and the ratio of their lengths, "primed to unprimed," is $4/3$.

Proof of theorem. For directions pointing into a given quadrant, say the first quadrant, as shown in the figure, we can prove the theorem by considering similar triangles. But a general argument that is valid in all cases requires using $\cos \theta$ and $\sin \theta$, where θ is the angle from the direction of the positive x-axis to the direction of the arrow. We noted in Chapter 4 that θ is not quite uniquely determined by the direction, but that θ_1 and θ_2 have the same direction if and only if $\theta_1 - \theta_2$ is a multiple of 2π, and therefore if and only if

$$(\cos \theta_1, \sin \theta_1) = (\cos \theta_2, \sin \theta_2).$$

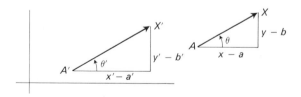

If l is the length of the arrow $\overrightarrow{AX}$

$$l = \sqrt{(x - a)^2 + (y - b)^2},$$

then

$$\cos \theta = \frac{x - a}{l}, \qquad \sin \theta = \frac{y - b}{l}.$$

(See Section 3 in Chapter 4.) For the arrow $\overrightarrow{A'X'}$ we have, similarly,

$$\cos \theta' = \frac{x' - a'}{l'}, \qquad \sin \theta' = \frac{y' - b'}{l'}.$$

The two arrows are parallel if and only if

$$(\cos \theta', \sin \theta') = \pm(\cos \theta, \sin \theta),$$

for this equation just says that the two directions are the *same* (+), or *opposite* (−). We substitute into this equation the above values of $\cos \theta$, etc., and obtain, successively,

$$\frac{1}{l'}(x' - a', y' - b') = \pm\frac{1}{l}(x - a, y - b),$$

$$\frac{1}{l'}(X' - A') = \pm\frac{1}{l}(X - A),$$

and

$$X' - A' = t(X - A),$$

where $t = \pm l'/l$. We have already noted that the directions are the same when t is positive and opposite when t is negative. And by definition, $|t| = l'/l$ is the ratio of the lengths. The theorem is thus proved. ∎

PROBLEMS FOR SECTION 1

For each pair of points X and A given below, find the requested algebraic combination.

1. $X = (1,4)$, $A = (2,1)$; find $X - A$.

2. $X = (3/2, 1/4)$, $A = (7/3, 1/6)$; find $X + 6A$.

3. $X = (3/2, 1/4)$, $A = (7/3, 1/6)$; find $6(X + A)$.

4. $X = (1,4)$, $A = (2,7)$; find $\frac{1}{2}X - 2A$.

Prove the following laws from the definitions of the point operations and the laws of algebra for numbers.

5. $A + X = A$ if and only if $X = (0,0)$.

6. $A + U = (0,0)$ if and only if $u = -a$ and $v = -b$.

7. $(U + X) + A = U + (X + A)$.

8. $(s + t)X = sX + tX$.

In each case below apply Theorem 1 to show that $\overrightarrow{AX}$ is parallel to $\overrightarrow{A'X'}$. Also give the ratio of their lengths, and state whether their directions are the same or opposite.

9. $A = (1, 3)$ $X = (2, 5)$ $A' = (6, 17)$ $X' = (5, 15)$.

10. $A = (2, 3)$ $X = (4, 1)$ $A' = (0, -8)$ $X' = (-5, -3)$.

11. $A = (-1, 2)$ $X = (2, -2)$ $A' = (0, 0)$ $X' = (-3/2, 2)$.

12. $A = (3, 0)$ $X = (3, 5)$ $A' = (0, 9)$ $X' = (0, 4)$.

13. $A = (6, 2)$ $X = (3, -1)$ $A' = (0, -1)$ $X' = (1, 0)$.

14. A theorem of geometry says that if two sides of a quadrilateral are parallel and equal, then the other two sides are also parallel and equal, so the figure is a parallelogram. Formulate and prove this as a theorem about arrows and the point operations.

2. THE PARAMETRIC EQUATION OF A LINE

Two arrows $\overrightarrow{X_0 X}$ and $\overrightarrow{X_0 X_1}$, that stem from the same initial point X_0, are parallel just when they lie along the same line. Therefore, by Theorem 1, a point X will lie on the line through the distinct points X_0 and X_1 if and only if

$$X - X_0 = t(X_1 - X_0)$$

for some real number t. Solving for X we can rewrite this equation as

$$X = X_0 + t(X_1 - X_0),$$

or

$$X = tX_1 + (1 - t)X_0.$$

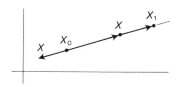

A more symmetric form is

$$X = sX_0 + tX_1,$$

where $s + t = 1$.

As the real number t runs from $-\infty$ to $+\infty$, the varying point X sweeps out the line through the two points X_0 and X_1. In this situation, t is called a *parameter*, and the above equation is a *parametric equation* for the line. This is consistent with the use of the word *parameter* in Chapter 11. The parameter t is not "part of" the point X, but t determines X.

EXAMPLE 1. The line through the points $X_0 = (1,2)$ and $X_1 = (4,1)$ has the parametric equation

$$X = X_0 + t(X_1 - X_0)$$
$$= (1,2) + t(3,-1).$$

This is the equation

$$(x,y) = (1 + 3t, 2 - t),$$

and is equivalent to the pair of numerical equations

$$x = 1 + 3t,$$
$$y = 2 - t.$$

Note that these are parametric equations for the line in exactly the sense of Section 1 in Chapter 11. The parameter can easily be eliminated to obtain an x,y-equation for the line. For instance, $t = 2 - y$ from the second equation, so

$$x = 1 + 3t = 1 + 3(2 - y) = 7 - 3y.$$

The line has the equation

$$x + 3y = 7.$$

Going back to the original equation

$$X - X_0 = t(X_1 - X_0),$$

Theorem 1 says that X is t times as far from X_0 as X_1 is. For example, X will be the midpoint of the segment X_0X_1 if $t = 1/2$, and solving for X with this value of t yields the midpoint formula

$$X = \frac{X_0 + X_1}{2}.$$

When $t = 2/3$, we get the point $2/3$ of the way from X_0 to X_1, and its formula comes out to be

$$X = \frac{1}{3}X_0 + \frac{2}{3}X_1.$$

In general,

Theorem 2. *If $s + t = 1$ and both numbers are positive, then*

$$X = sX_0 + tX_1$$

is the point on the segment $X_0 X_1$ that divides it in the ratio t/s.

Note that the dividing point is nearer the endpoint having the *larger coefficient.*

EXAMPLE 2. The midpoint of the segment from $X_0 = (1, 2)$ to $X_1 = (4, 4)$ is

$$X = \frac{X_0 + X_1}{2} = \frac{(1, 2) + (4, 4)}{2}$$

$$= \frac{1}{2}(5, 6) = \left(\frac{5}{2}, 3\right).$$

EXAMPLE 3. The point $(3/4)$ of the way from X_0 to X_1 on the above segment is

$$X = \frac{1}{4}X_0 + \frac{3}{4}X_1$$

$$= \frac{(1, 2) + 3(4, 4)}{4} = \frac{(1, 2) + (12, 12)}{4}$$

$$= \frac{(13, 14)}{4} = \left(\frac{13}{4}, \frac{7}{2}\right).$$

We can already solve some problems in geometry by this new algebra.

EXAMPLE 4. Prove that the medians of a triangle are concurrent.

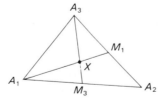

Solution. We start off by computing the point X where the two medians shown above intersect. Since X lies on the segment $M_3 A_3$, it can be written

$$X = tM_3 + (1 - t)A_3$$

for some number t between 0 and 1. Moreover, $M_3 = (A_1 + A_2)/2$. Thus,

$$X = \frac{t}{2}A_1 + \frac{t}{2}A_2 + (1 - t)A_3.$$

But X also lies on the segment $A_1 M_1$, so that, similarly,

$$X = (1 - s)A_1 + \frac{s}{2}A_2 + \frac{s}{2}A_3$$

for some number s between 0 and 1. These two representations must give the same point X, and this will happen if $s = t$ and $(t/2) = (1 - t)$. This gives $s = t = (2/3)$, and

$$X = \frac{A_1 + A_2 + A_3}{3}.$$

Since this representation is symmetric in the letters A_1, A_2, and A_3, the point X lies on all three medians. (In more detail, if we had started with another pair of medians, the calculation of their point of intersection would be the same as the above calculation except that the indices 1, 2, and 3 would be permuted. The resulting formula for the point of intersection would thus be the same as the above formula except for this permutation of the indices. But the above formula remains the same under any permutation of the indices, and the two points of intersection are thus the same point.)

PROBLEMS FOR SECTION 2

In each of the following problems, write the parametric equation of the line through the given points. Also give the equivalent pair of scalar equations.

1. $X_0 = (1, 4)$, $X_1 = (2, -2)$ 2. $X_0 = (1, 1)$, $X_1 = (3, 0)$
3. $X_0 = (1, 1)$, $X_1 = (4, 1)$ 4. $X_0 = (2, 0)$, $X_1 = (2, 1)$
5. $X_0 = (-1, 1/2)$, $X_1 = (1/2, 3/4)$

In each of the following problems, find the point X which divides the line segment into the given ratio.

6. $X_0 = (1, 1)$, $X_1 = (-6, 0)$; X is $1/3$ of the way from X_1 to X_0.
7. $X_0 = (0, 1)$, $X_1 = (10, 4)$; X is $1/4$ of the way from X_0 to X_1.
8. $X_0 = (1/2, 1/3)$, $X_1 = (4/3, 1/2)$; X is $2/5$ of the way from X_0 to X_1.
9. $X_0 = (-1, -4)$, $X_1 = (-6, -2)$; X is $7/8$ of the way from X_1 to X_0.
10. $X_0 = (4, -3)$, $X_1 = (-1, 6)$; X is $5/6$ of the way from X_0 to X_1.
11. Prove that the diagonals of a parallelogram bisect one another.
12. Prove that the line segment determined by the midpoints of two sides of a triangle is parallel to the third side and half its length.

13.   Let X_1, X_2, X_3 and X_4 be the vertices of an arbitrary quadrilateral. Show that the midpoints of the sides are the vertices of a parallelogram.

14.   A theorem of geometry says that if the opposite sides of a quadrilateral are parallel, then the opposite sides are also equal, and the figure is a parallelogram. However, this clearly can be false if all four points lie on a straight line.

Prove the following correct algebraic version of the theorem.

Theorem. *If* $(A_2 - X_2) = s(A_1 - X_1)$ *and* $(A_2 - A_1) = t(X_2 - X_1)$, *then either* $s = t = 1$ *or the four points are collinear.*

15.   Prove the following geometric theorem.

Theorem. *A line parallel to one side of a triangle cuts the other two sides proportionately.* (*Take one vertex at the origin.*)

3. VECTORS

The operations $A + X$ and tX are called the *vector operations* on the points of the coordinate plane, and their use to prove geometric theorems is called *vector geometry*. We call a point X a *vector* when we intend to use it or think about it in connection with the vector operations. (This may seem a rather odd definition of a vector, but it agrees with modern usage. If you ask a mathematician what a vector is he will answer in all seriousness that "A vector is just an element of a vector space." He means that a vector is just something that is *used* in a vectorial way.)

If $U = X - A$, we call U *the vector of the arrow* $\overrightarrow{AX}$. Then, according to Theorem 1, two arrows $\overrightarrow{AX}$ and $\overrightarrow{A'X'}$ have the same vector U if and only if they have the same direction and length. A vector U is thus equivalent to a direction and length, the common direction and length of all the arrows having U as vector. In particular, this is the length and direction of $\overrightarrow{OU}$, since $U = U - O$.

Any arrow $\overrightarrow{AX}$ displays the length and direction of its vector $U = X - A$, and is thus a geometric representation of U. It is the representation of U that is "laid off from A" or "located at A," and we shall frequently call arrows *located vectors*.

The *parallelogram law* is an appealing interpretation of vector addition, but it is flawed by an exceptional case that it doesn't cover. It says:

If the vectors A and U are not parallel, then the sum X = A + U is the fourth vertex of the parallelogram having the segments OA and OU as two of its sides.

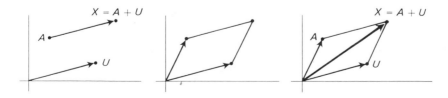

The geometric line is one-dimensional and the geometric plane is two-dimensional. What this means can be described in various ways, the simplest being that a point on the coordinate line is specified by one coordinate and a point in the coordinate plane is specified by two coordinates. We turn now to the vector characterization of the dimension of the plane. The term *scalar* is used in the following theorem; in vector language this just means a number.

Theorem 3. *If A_1 and A_2 are any two nonparallel vectors in the plane, then each vector X is expressible in a unique way as a linear combination of A_1 and A_2. That is, X has the form*

$$X = sA_1 + tA_2,$$

where s and t are uniquely determined scalars.

Proof. The figure below shows how to find s and t.

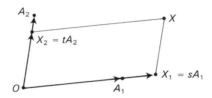

We draw the two lines through X parallel to the vectors A_1 and A_2, obtaining a parallelogram as shown. This figure expresses X as a sum $X = X_1 + X_2$ of a vector X_1 parallel to A_1 and a vector X_2 parallel to A_2. Then X_1 has to be of the form sA_1, and X_2 of the form tA_2, so $X = sA_1 + tA_2$, as claimed.

The uniqueness of s and t is implicit in the above construction, and it can also be checked algebraically, as follows. Suppose there were a different representation

$$X = s'A_1 + t'A_2,$$

with, say, $s' \neq s$. Subtracting the second representation from the first gives

$$0 = (s - s')A_1 + (t - t')A_2.$$

But then

$$A_1 = cA_2,$$

where $c = -(t - t')/(s - s')$, and A_1 is thus parallel to A_2. Since this contradicts our hypothesis that the two vectors are not parallel, we see that a different representation is impossible. ∎

Because the vectors in the plane are specified by pairs of numbers (s,t) in this way, we say that the coordinate plane is a *two-dimensional vector space*. The fixed pair of vectors A_1 and A_2 is called a *basis* for the vector space, and the representation

$$X = sA_1 + tA_2$$

is the *expansion* of the vector X with respect to this basis. The coefficients s and t are the *components* of X with respect to the basis (A_1, A_2).

There is an alternative, purely algebraic, proof of Theorem 3, consisting of solving the equations

$$sa_1 + ta_2 = x,$$

$$sb_1 + tb_2 = y,$$

for s and t. This process will be illustrated in the example below.

EXAMPLE 1. Find the components of the vector $(2,1)$ with respect to the basis $A_1 = (1, 3)$, $A_2 = (2, -1)$.

. . . .

Solutions. By definition, the components are the unique numbers s and t such that

$$(2, 1) = sA_1 + tA_2$$

$$= s(1, 3) + t(2, -1)$$

$$= (s + 2t, 3s - t).$$

This vector equation is equivalent to the two scalar equations

$$s + 2t = 2 \quad \text{and} \quad 3s - t = 1,$$

and we find s and t by solving these equations simultaneously. Multiplying the second equation by 2 and adding it to the first gives

$$7s = 4.$$

Thus, $s = 4/7$, and then, since $(4/7) + 2t = 2$, we see that $t = 5/7$. The component pair is thus $(4/7, 5/7)$.

Now consider $E_1 = (1, 0)$ and $E_2 = (0, 1)$, the special pair of "unit vectors" along the coordinate axes. Then for any vector $X = (x, y)$, we have

$$X = (x, y) = x(1, 0) + y(0, 1)$$

$$= xE_1 + yE_2.$$

Thus the *coordinates* of X are also its *components* with respect to the "standard" basis E_1, E_2 associated with the coordinate system.

Physicists often use the notation

$$\mathbf{i} = (1,0)$$

$$\mathbf{j} = (0,1)$$

for the standard basis vectors. Then

$$X = (x,y) = x\mathbf{i} + y\mathbf{j}$$

is the standard basis expansion of X.

We conclude this section with the notion of a *vector field*. Wind velocity is an important vector in meteorology. A steady wind is specified by its direction and magnitude. But usually, at any given moment, wind velocity varies from point to point, and these varying located vectors make up a vector field. *A vector field is a collection of located vectors, one at each point in the plane.* We can get an idea of such a field of vectors by plotting a few of its arrows, as below.

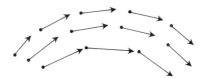

PROBLEMS FOR SECTION 3

1. Show that if $A \neq 0$, then the line through the point X_0 parallel to the vector A has the parametric equation

$$X = X_0 + tA.$$

Write out the following examples of the parametric equation in Problem 1, and reduce each equation to the equivalent pair of scalar equations.

2. $X_0 = (1,2)$, $A = (-3,0)$.   3. $X_0 = (-2,3)$, $A = (1,1)$.

4. $X_0 = (0,0)$, $A = (3,4)$.   5. $X_0 = (0,0)$, $A = (a,b)$.

6. $X_0 = (x_0,y_0)$, $A = (1,1)$.   7. $X_0 = (2,-1)$, $A = (1,4)$.

8. Show what goes wrong in the parallelogram law in the exceptional case.

9. Find the point of intersection of the lines $X = (1,2) + s(1,1)$ and $X = (2,2) + t(3,-1)$.

10. Show that if A_1 and A_2 are not parallel, then the lines $X = X_1 + sA_1$, $X = X_2 + tA_2$, have a unique point of intersection. (Apply Theorem 3.)

Find the components of each of the following vectors with respect to the basis $(1,2)$, $(-1,-1)$

11. $X = (7, 1)$ 　　　　　　　　　　　　12. $X = (-3, 4)$

13. $X = (6, 1)$ 　　　　　　　　　　　　14. $X = (-1, -3)$

15. $X = (0, 4)$ 　　　　　　　　　　　　16. $E_1 = (1, 0)$

17. $E_2 = (0, 1)$

18. Let $S_1 = (s_1, t_1)$ and $S_2 = (s_2, t_2)$ be the component pairs of

$$E_1 = (1, 0) \quad \text{and} \quad E_2 = (0, 1)$$

with respect to a basis $\{A_1, A_2\}$. Show that the component pair of any vector $X = (x, y)$ is $xS_1 + yS_2$.

19. Now go back to Problems 11 through 17. Use the components of E_1 and E_2, as determined in Problems 16 and 17, to write out the answers to Problems 11 through 15 by the formula just proved in Problem 18.

20. Prove Theorem 3 by algebra. That is, solve the equation

$$X = sA_1 + tA_2$$

for s and t in terms of x and y, by the usual technique for solving two equations in two unknowns. You will find that this solution process works only if A_1 and A_2 satisfy a certain algebraic condition, and that this condition is equivalent to their not being parallel.

21. The component pair $S = (s, t)$ of a vector X with respect to a basis $\{A_1, A_2\}$ is just another point in the coordinate plane. Show that the operation taking X to S (also referred to as the *mapping* $X \to S$) is linear. That is, show that if S_1 and S_2 are the component pairs of X_1 and X_2, then

$$aS_1 + bS_2$$

is the component pair of $aX_1 + bX_2$, for any scalars a and b.

If we let ϕ be the function that assigns to each vector X its component pair S, $S = \phi(X)$, then the conclusion you are to prove can be stated as:

$$\phi(aX_1 + bX_2) = a\phi(X_1) + b\phi(X_2).$$

That is, ϕ "preserves" linear combinations, and for this reason we call ϕ a linear mapping.

4. THE DOT PRODUCT

We shall use the ordinary absolute-value symbol $|X|$ for the magnitude of the vector $X = (x, y)$:

$$|X| = \sqrt{x^2 + y^2}.$$

Geometrically, $|X|$ is the *distance* from the origin to the point X and also the *length* of the arrow $\overrightarrow{OX}$. According to Theorem 1,

$$|tA| = |t| \cdot |A|.$$

A *unit vector* is a vector of length one. Any nonzero vector X determines a unique unit vector U in the same direction, that is, a vector U such that

$$|U| = 1,$$

$$U = tX, \quad t > 0.$$

For these requirements lead to

$$1 = |tX| = |t| \cdot |X|, \qquad \text{and} \qquad |t| = \frac{1}{|X|}.$$

But t is positive, so $t = 1/|X|$. Thus U is uniquely determined as

$$U = \frac{X}{|X|}.$$

The fact that U defined this way is a unit vector can of course be checked directly. It is called the *normalization* of X.

EXAMPLE 1. The unit vector in the direction of $X = (1,2)$ is $U = X/|X|$ $= (1,2)/\sqrt{5} = (1/\sqrt{5}, 2/\sqrt{5})$.

Since there is exactly one unit vector in a given direction, we sometimes identify a direction with its unit vector.

The trigonometric identity

$$\cos^2\theta + \sin^2\theta = 1$$

can be interpreted as saying that

$$(\cos\theta, \sin\theta)$$

is a unit vector. And every unit vector U can be expressed this way, since U is just a point (u,v) on the unit circle. Thus

$$U = (\cos\theta, \sin\theta),$$

where θ is the angle from the direction of the positive x-axis to the direction of U.

In the general situation, where $U = X/|X|$ is the unit vector (direction) of X,

$$(\cos \theta, \sin \theta) = U = \frac{X}{|X|} = \left(\frac{x}{r}, \frac{y}{r}\right),$$

where $r = |X| = \sqrt{x^2 + y^2}$. The cross-multiplied forms

$$X = |X|U,$$

$$(x, y) = r(\cos \theta, \sin \theta),$$

can be called the polar-coordinate representation of the vector $X = (x, y)$, that is, its representation as the product of its length times its direction.

Now let A be a second nonzero vector, with unit vector

$$(\cos \alpha, \ \sin \alpha) = \frac{A}{|A|} = \left(\frac{a}{l}, \frac{b}{l}\right),$$

where $l = |A| = \sqrt{a^2 + b^2}$. Then $\phi = \theta - \alpha$ is the angle from A to X and

$$\cos \phi = \cos(\theta - \alpha) = \cos \theta \cos \alpha + \sin \theta \sin \alpha$$

$$= \frac{x}{r} \cdot \frac{a}{l} + \frac{y}{r} \cdot \frac{b}{l}$$

$$= \frac{ax + by}{rl}.$$

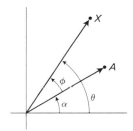

We could also compute $\sin \phi$. But it turns out that the $\sin \phi$ formula doesn't generalize to three dimensions, whereas the $\cos \phi$ formula does. This reflects the fact that there is no natural way to define a *signed* angle *from* one direction *to* another in three dimensions. We can only look at the *undirected* angle *between* two vectors, and such an angle is specified by its cosine (since $\cos \phi = \cos(-\phi)$).

The numerator in the above formula for $\cos \phi$ comes up so often that we give it a special name. It is called the *dot product* of the vectors A and X, and is designated $A \cdot X$.

Definition. *The dot product $A \cdot X$ of the vectors $A = (a, b)$ and $X = (x, y)$ is defined by*

$$A \cdot X = ax + by.$$

EXAMPLE 2. The dot product of the vectors $(1,4)$ and $(-3,2)$ is

$$(1,4) \cdot (-3,2) = 1(-3) + 4 \cdot 2 = -3 + 8 = 5.$$

EXAMPLE 3. The length $|X|$ of the vector X is

$$|X| = (X \cdot X)^{1/2},$$

since $|X|^2 = x^2 + y^2 = x \cdot x + y \cdot y = X \cdot X$.

The formula for $\cos \phi$ that we computed above is an important property of the dot product.

Theorem 4. *If ϕ is the angle between two nonzero vectors A and X, then*

$$\cos \phi = \frac{A \cdot X}{|A||X|}.$$

Corollary. *The vectors A and X are perpendicular if and only if $A \cdot X = 0$.*

Proof. The angle ϕ between the vectors is $90°$ $(\pi/2)$ if and only if $\cos \phi = 0$, and this is equivalent to $A \cdot X = 0$, by the formula of the theorem. ∎

EXAMPLE 4. In order to check the vectors $A = (1,2)$ and $X = (4,-2)$ for perpendicularity, we compute

$$A \cdot X = ax + by = 1 \cdot 4 + 2(-2) = 0.$$

Thus, they are perpendicular $(A \perp X)$.

EXAMPLE 5. Find the angle ϕ between the vectors $(1,-1)$ and $(3,2)$.
. . . .

Solution

$$\cos \phi = \frac{A \cdot X}{|A||X|} = \frac{1 \cdot 3 + (-1)2}{\sqrt{1+1}\sqrt{9+4}} = \frac{1}{\sqrt{26}},$$

and $\phi = \arccos 1/\sqrt{26}$.

EXAMPLE 6. Show that the straight line

$$ax + by + c = 0$$

is perpendicular to the vector $A = (a,b)$.
. . . .

Solution Let X_1 and X_0 be any two distinct points on the line. Then

$$ax_1 + by_1 + c = 0, \qquad ax_0 + by_0 + c = 0;$$

and subtracting these two equations gives the equation

$$a(x_1 - x_0) + b(y_1 - y_0) = 0.$$

But this just says that

$$A \cdot (X_1 - X_0) = 0$$

and hence that $A \perp (X_1 - X_0)$. Since the direction of $X_1 - X_0$ is the direction of the arrow $\overrightarrow{X_0 X_1}$ along the line, we see that the vector A is perpendicular to the line.

Reversing the above steps gives a new derivation of the equation of a line. Given a line l in the coordinate plane, choose a fixed point $X_0 = (x_0, y_0)$ on l and a fixed vector $A = (a, b)$ perpendicular to the direction of l. Then a point X will be on l if and only if the arrow $\overrightarrow{X_0 X}$ is perpendicular to the vector A. In view of Theorem 4, this is exactly the condition

$$A \cdot (X - X_0) = 0.$$

So

$$a(x - x_0) + b(y - y_0) = 0,$$

or

$$ax + by + c \qquad (c = -cx_0 - by_0),$$

is an equation of the line l.

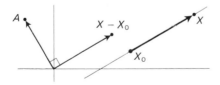

We now do for the dot product what we did in the beginning for vector addition and multiplication by scalars: We list the algebraic properties that let us use $X \cdot A$ directly, without recourse to its component formula. They are:

(s1) $$X \cdot X = |X|^2,$$

(s2) $$X \cdot A = A \cdot X,$$

(s3) $$(X_1 + X_2) \cdot A = X_1 \cdot A + X_2 \cdot A,$$

(s4) $$(tX) \cdot A = t(X \cdot A).$$

As before, the proofs depend on properties of numbers via components. Thus,

$$(X_1 + X_2) \cdot A = ((x_1,y_1) + (x_2,y_2)) \cdot (a,b)$$
$$= (x_1 + x_2, y_2 + y_2) \cdot (a,b)$$
$$= (x_1 + x_2)a + (y_1 + y_2)b$$
$$= (x_1 a + y_1 b) + (x_2 a + y_2 b)$$
$$= (x_1,y_1) \cdot (a,b) + (x_2,y_2) \cdot (a,b)$$
$$= X_1 \cdot A + X_2 \cdot A,$$

which is (s3).

Note that, because of the commutative law (s2), both (s3) and (s4) can be turned around:

$$A \cdot (X_1 + X_2) = A \cdot X_1 + A \cdot X_2,$$
$$A \cdot (tX) = t(A \cdot X).$$

These formal properties of the dot product will be basic in the discussion of the three-dimensional dot product in Chapter 17. Here we make only one application.

If B_1 and B_2 are perpendicular unit vectors, we say that they form an *orthonormal basis* for the coordinate plane. The components of a vector X with respect to an orthonormal basis can be computed as dot products, without having to solve simultaneous linear equations.

Theorem 5. *If the pair of vectors B_1,B_2 is an orthonormal basis for the plane, then the components of any vector X with respect to this basis are the numbers*

$$x_1 = X \cdot B_1, \qquad x_2 = X \cdot B_2.$$

Proof. The components are the unique numbers x_1 and x_2 such that

$$X = x_1 B_1 + x_2 B_2.$$

Theorem 4 guarantees the existence of these numbers. But then

$$X \cdot B_1 = (x_1 B_1 + x_2 B_2) \cdot B_1$$
$$= x_1(B_1 \cdot B_1) + x_2(B_2 \cdot B_1) \qquad \text{(by (s3) and (s4))}$$
$$= x_1 \cdot 1 + x_2 \cdot 0 = x_1,$$

by the definition of an orthonormal basis. Similarly, $X \cdot B_2 = x_2$. ∎

EXAMPLE 7. The vectors $A_1 = (3,4)$ and $A_2 = (8,-6)$ are perpendicular. Normalize them, so that they will form an orthonormal basis, and then compute the components of $(1,0)$ with respect to this basis.

Solution

$$|A_1| = \sqrt{9 + 16} = 5,$$

$$|A_2| = \sqrt{64 + 36} = 10.$$

The basis is

$$B_1 = \frac{A_1}{|A_1|} = \frac{1}{5}(3,4) = \left(\frac{3}{5}, \frac{4}{5}\right),$$

$$B_2 = \frac{A_2}{|A_2|} = \frac{1}{10}(8,-6) = \left(\frac{4}{5}, -\frac{3}{5}\right).$$

Then

$$(1,0) = c_1 B_1 + c_2 B_2,$$

where

$$c_1 = (1,0) \cdot B_1 = (1,0) \cdot \left(\frac{3}{5}, \frac{4}{5}\right) = \frac{3}{5}$$

and

$$c_2 = (1,0) \cdot B_2 = (1,0) \cdot \left(\frac{4}{5}, \frac{3}{5}\right) = \frac{4}{5}.$$

EXAMPLE 8. Find the components of $(2,1)$ with respect to the above basis.

Solution

$$c_1 = (2,1) \cdot \left(\frac{3}{5}, \frac{4}{5}\right) = \frac{6}{5} + \frac{4}{5} = 2$$

$$c_2 = (2,1) \cdot \left(\frac{4}{5}, -\frac{3}{5}\right) = \frac{8}{5} - \frac{3}{5} = 1.$$

Note that these are also the components of $(2,1)$ with respect to the standard basis. This implies that there is some unusual symmetry in the configuration, and the diagram below shows what it is.

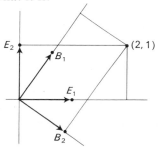

We saw in the theorem that the first component c_1 of a vector X with respect to an orthonormal basis depends only on the first basis vector U_1, according to the formula

$$c_1 = X \cdot U_1.$$

Moreover, then $c_1 U_1$ is "the part of X parallel to U_1", in the sense that after subtracting $c_1 U_1$ from X, the remainder

$$X - c_1 U_1 = c_2 U_2$$

is perpendicular to U_1.

This view of the theorem focuses attention on the relation of a vector X to a single unit vector U. If $c = X \cdot U$, then cU is always the part of X parallel to U in just the above sense. For the remainder

$$X - cU$$

is always perpendicular to U, as appears on dotting with U:

$$(X - cU) \cdot U = X \cdot U - c(U \cdot U) = c - c \cdot 1 = 0.$$

With these facts in mind, we call $X \cdot U$ *the component of X in the direction of the unit vector U*.

PROBLEMS FOR SECTION 4

Determine the length of each of the following vectors, and give the unit vector in the same direction. Draw each vector as an arrow from the origin and also as an arrow from the point $(-1,2)$.

1. $V = (3,4)$ 2. $X = (4,-1)$ 3. $A = (-3,1)$

4. $X = (1/2, 1/2)$ 5. $V = (1,-2)$

Find $A \cdot X$ and the angle ϕ between A and X ($\phi = $ arc cos (something)). Draw A and X as arrows from the origin, and label ϕ.

6. $A = (2,1)$, $X = (1,4)$ 7. $A = (1/2,-2)$, $X = (1,-1)$

8. $A = (3,-1/2)$, $X = (-1/2, 2/3)$.   9. $A = (1/4, -1/3)$, $X = (1/3, 1/4)$.

10. $A = (2,-1)$, $X = (2,2)$.

11. U is a unit vector perpendicular to $A = (1,2)$. Determine U (two answers).

12. U is a unit vector making the angle $\pi/3$ with $A = (1,0)$. Determine U (two answers).

13. U is a unit vector making the angle $\pi/3$ with $A = (1,1)$. Determine U.

14. The same question when $A = (1,2)$.

15. Determine the collection of all vectors X making the angle $\pi/3$ with $A = (1,0)$.

16. Find the equation of the line through $X_0 = (1,-5)$ perpendicular to the vector $A = (-2,1)$.

17. Find the equation of the line through the point $X_0 = (3,2)$ perpendicular to the vector $(-2,3)$.

18. Let l be the line through the point X_0 perpendicular to the nonzero vector A. Show that l goes through the origin if and only if X_0 is perpendicular to A.

19. Let l be the line through the point X_0 perpendicular to the nonzero vector A. Show that its distance from the origin is $|A \cdot X_0|/|A|$.

20. Prove the law (s4): $t(X \cdot A) = (tX) \cdot A$.

The vectors $B_1 = (2/\sqrt{5}, 1/\sqrt{5})$ and $B_2 = (-1/\sqrt{5}, 2/\sqrt{5})$ form an orthonormal basis. For each of the following vectors X, compute its components x_1 and x_2 with respect to the basis $\{B_1, B_2\}$, by Theorem 5. Then directly verify the correctness of the equation

$$X = x_1 B_1 + x_2 B_2.$$

21. $X = (3,4)$ 22. $X = (-1,1)$

23. $X = (0,1)$ 24. $X = (1/2, 1/2)$

25. $X = (1,0)$ 26. $X = (\sqrt{5}, \sqrt{5})$

27. $X = (6/\sqrt{5}, -2/\sqrt{5})$ 28. $X = (-1/\sqrt{5}, 1/\sqrt{5})$

29. $X = (2/\sqrt{5}, -1/\sqrt{5})$ 30. $X = (\sqrt{5}, -2/\sqrt{5})$

31. Let B_1 and B_2 be perpendicular unit vectors. Prove that $\{B_1, B_2\}$ is a basis directly from the dot-product component formulas $x_i = X \cdot B_i$, $i = 1, 2$, without resorting to Theorem 3. (Assume and use the fact that if U is perpendicular to both B_1 and B_2, then $U = 0$.)

32. Let $\{B_1, B_2\}$ be an orthonormal basis. Prove that if x_1 and x_2 are the components of X and u_1 and u_2 are the components of U (all with respect to the basis B_1, B_2), then

$$X \cdot U = x_1 u_1 + x_2 u_2.$$

For each pair A and X given below, find the component of X in the direction of A.

33. $A = (1, 1)$, $X = (2, 1)$.   34. $A = (1, -2)$, $X = (2, 1)$.

35. $A = (1, -2)$, $X = (1, 2)$.   36. $A = (10, 1)$, $X = (1, 0)$.

37. $A = (1,0)$, $X = (10,1)$.   38. $A = (10,1)$, $X = (0,1)$.

39. If A is a given nonzero vector and k is a given constant, we know that $A \cdot X = k$ is the equation of a line perpendicular to A. Now reinterpret this locus as the collection of all vectors X having a certain component property.

5. THE TANGENT VECTOR TO A PATH

In the same way that a pair of numbers x and y combine to form a single vector

$$X = (x,y),$$

a pair of functions

$$x = g(t),$$
$$y = h(t),$$

can be combined to form a single *vector-valued* function

$$(x,y) = (g(t), h(t))$$

or

$$X = G(t).$$

This gives us a new way to look at paths. To be sure, in Chapter 11 we required that a path be made up of a pair of *smooth* coordinate functions, but we didn't have to. We can consider a path that is only continuous, for instance. However, a discontinuous vector-valued function wouldn't be considered to be a path.

In order to discuss the continuity of a path (vector-valued function) we fix a point on the path,

$$X_0 = (x_0, y_0) = (g(t_0), h(t_0)),$$

and then give t an increment Δt to get a new point

$$X_1 = (x_0 + \Delta x, y_0 + \Delta y) = (g(t_0 + \Delta t), h(t_0 + \Delta t)).$$

The *path increment* is then the vector

$$\Delta X = X_1 - X_0 = (\Delta x, \Delta y).$$

It is the vector of the arrow $\overrightarrow{X_0 X_1}$ from the fixed path point X_0 to the varying path

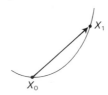

point X_1. Then the path is *continuous* at X_0 if $\Delta X \to 0$ as $\Delta t \to 0$. This means that the magnitude of ΔX approaches 0,

$$|\Delta X| = \sqrt{(\Delta x)^2 + (\Delta y)^2} \to 0,$$

and this occurs just when both Δx and Δy approach 0 as Δt approaches 0. That is:

A path $X = (g(t), h(t))$ *is continuous at* $t = t_0$ *if and only if* both *coordinate functions g and h are continuous there.*

In the same way we find that a vector function $X = (g(t), h(t))$ has the limit $L = (l, m)$ as t approaches t_0 if and only if $g(t) \to l$ and $h(t) \to m$ as $t \to t_0$.

The discussion so far has been preliminary; our main concern is the *derivative* of a vector-valued function X of a parameter t. There is nothing new in the definition except the use of vector limits. We form the difference quotient

$$\frac{\Delta X}{\Delta t} = \frac{1}{\Delta t}(\Delta x, \Delta y) = \left(\frac{\Delta x}{\Delta t}, \frac{\Delta y}{\Delta t}\right),$$

and take the limit as $\Delta t \to 0$. Since

$$\frac{\Delta x}{\Delta t} \to \frac{dx}{dt} = g'(t)$$

and

$$\frac{\Delta y}{\Delta t} \to \frac{dy}{dt} = h'(t),$$

it follows that the vector

$$\frac{\Delta X}{\Delta t} = \left(\frac{\Delta x}{\Delta t}, \frac{\Delta y}{\Delta t}\right)$$

approaches the vector

$$\left(\frac{dx}{dt}, \frac{dy}{dt}\right) = (g'(t), h'(t))$$

as Δt approaches 0. We state this conclusion as a theorem.

Theorem 6. *The vector function*

$$X = (x, y) = (g(t), h(t))$$

is differentiable if and only if its component functions g and h are both differentiable, and then

$$\frac{dX}{dt} = \left(\frac{dx}{dt}, \frac{dy}{dt}\right) = (g'(t), h'(t)).$$

We can also write

$$X = G(t) \quad \text{and} \quad \frac{dX}{dt} = G'(t).$$

Now consider the geometric meaning of this limit. If $\Delta t > 0$, then the vector $\Delta X/\Delta t$ has the same direction as the increment vector $\Delta X = X_1 - X_0$. So if we locate the $\underline{\text{vector}}$ $\Delta X/\Delta t$ at X_0, then it will lie along the arrow $\overrightarrow{X_0 X_1}$. It will be longer than $\overrightarrow{X_0 X_1}$ if $\Delta t < 1$, the multiplying scalar $1/\Delta t$ then being greater than 1.

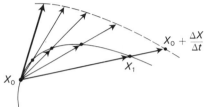

The direction of the limit vector

$$\frac{dX}{dt} = \lim_{\Delta t \to 0} \frac{\Delta X}{\Delta t}$$

is thus the limit of the direction of the secant arrow $\overrightarrow{X_0 X_1}$ as X_1 approaches X_0 along the path. But the limit of the direction of the secant arrow is what we *mean* by the tangent direction. Therefore:

The path derivative dX/dt can be interpreted geometrically as a vector tangent to the path. It is called *the path tangent vector.*

We normally draw it located at the path point, as the arrow from X to $X + dX/dt$.

EXAMPLE 1. Find the tangent vector to the path

$$X = (x, y) = (t^3 - 3t, t^2)$$

at the point corresponding to $t = -1$. The same for $t = 0$, $t = \frac{3}{2}$, and $t = 2$.

. . . .

Solution. At the general point $X = (x, y) = (t^3 - 3t, t^2)$, the path has the tangent vector

$$\frac{dX}{dt} = \left(\frac{dx}{dt}, \frac{dy}{dt} \right) = (3t^2 - 3, 2t).$$

At $t = -1$, the path point X is $(2,1)$ and the tangent vector is

$$\frac{dX}{dt}\bigg|_{t=-1} = (3(-1)^2 - 3, 2(-1)) = (0, -2).$$

When $t = 0$,

$$X = (0,0), \qquad \frac{dX}{dt} = (-3, 0).$$

When $t = \frac{3}{2}$,

$$X = \left(-\frac{9}{8}, \frac{9}{4}\right), \qquad \frac{dX}{dt} = \left(\frac{15}{4}, 3\right).$$

When $t = 2$,

$$X = (2,4), \qquad \frac{dX}{dt} = (9.4).$$

This path is the same one we began our path discussion with in Chapter 11. If we locate the vector dX/dt at the path point X where it is tangent, i.e., if we consider the arrow from X to $X + dX/dt$, then the four tangent vectors computed above look like this.

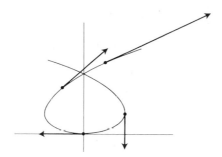

The direction of dX/dt shows the path direction. (The magnitude of dX/dt shows "how fast the path point X is changing with respect to t", but we haven't yet examined this aspect of dX/dt.)

EXAMPLE 2. Show that a path whose tangent vector is constant and not zero must necessarily be a straight line.
. . . .

Solution. Let $X = (g(t), h(t))$ be the path and let A be the nonzero constant tangent vector. Our hypothesis is that

$$\frac{dX}{dt} = (g'(t), h'(t)) = A = (a, b),$$

for all t. That is,

$$g'(t) = a,$$

$$h'(t) = b,$$

for all t. Integrating these two equations, we see that

$$x = g(t) = at + c,$$

$$y = h(t) = bt + d,$$

or

$$X = At + C.$$

This is the parametric equation for a line in vector form.

EXAMPLE 3. Find the line tangent to the curve $X = (x, y) = (2 \cos t, \sin t)$ at the point corresponding to $t_0 = \pi/6$.

. . . .

Solution. At $t_0 = \pi/6$,

$$X_0 = (2 \cos \pi/6, \sin \pi/6) = (\sqrt{3}, 1/2),$$

$$\left.\frac{dX}{dt}\right|_{\pi/6} = (-2 \sin t, \cos t)_{\pi/6} = (-1, \sqrt{3}/2).$$

The line through X_0 in the direction of the vector A has the parametric equation

$$X = X_0 + tA.$$

Here the direction is given by the tangent vector, and the equation of the line is

$$X = (\sqrt{3}, 1/2) + t(-1, \sqrt{3}/2)$$

$$= (\sqrt{3} - t, 1/2 + t\sqrt{3}/2).$$

The component equations are

$$x = \sqrt{3} - t,$$

$$y = \frac{1 + \sqrt{3}\, t}{2}.$$

EXAMPLE 4. Show that if the vector variable $X = (x, y)$ is a differentiable function of the parameter t, then $|X| = \sqrt{x^2 + y^2}$ is a constant if and only if X is always perpendicular to dX/dt.

. . . .

Solution. The assumption of constant magnitude can be written

$$x^2 + y^2 = |X|^2 = \text{const},$$

where x and y are differentiable functions of t. Differentiating with respect to t gives

$$2x\frac{dx}{dt} + 2y\frac{dy}{dt} = 0,$$

or

$$2(X \cdot \frac{dX}{dt}) = 0.$$

Thus X is perpendicular to dX/dt. Conversely, if these vectors are perpendicular, then the above argument can be reversed, to give

$$\frac{d}{dt}(x^2 + y^2) = 0,$$

$$x^2 + y^2 = \text{const}.$$

EXAMPLE 5. If $X_0 = (g(t_0), h(t_0))$ is the point closest to the origin on the path $X = (g(t), h(t))$, then the tangent vector to the path at X_0 is necessarily perpendicular to the vector X_0.

This is because $|X|^2 = g^2(t) + h^2(t)$ has its minimum value at X_0, so its derivative is zero there. But

$$\frac{d|X|^2}{dt} = 2\left(X \cdot \frac{dX}{dt}\right),$$

as we saw in Example 4, so

$$X_0 \cdot \frac{dX}{dt}\bigg|_0 = 0,$$

as claimed.

A tangent vector

$$\frac{dX}{dt} = (g'(t), h'(t))$$

can perfectly well be the zero vector $(0,0)$, in which case it does not specify a unique direction. However, this is exactly what we ruled out when we defined a smooth path, so a smooth path has a continuously varying, nowhere-zero, tangent vector. In particular, a smooth path has a continuously changing direction. This is the ultimate justification of Theorem 1 in Chapter 11.

PROBLEMS FOR SECTION 5

Find the derivative dX/dt for each of the following vector functions. Also, determine the points where dX/dt is perpendicular to X.

1. $F(t) = (t, 1/t)$ 2. $X = (e^{2t}, e^{-t})$

3. $\Phi(t) = (t^2, t + 3)$ 4. $X = (2t, t + 1)$

5. $X = (\cos t, \sin t)$ 6. $U = (\sqrt{1 - t^2}, t)$

7. $G(t) = (3 \cos t, 2 \sin t)$ 8. $X = (t^2, (1 + t)^2)$

Find the vector tangent to each of the paths below at the given point. Then write down a parametric equation for the tangent line at the point.

9. $X = (40t, 30t - 16 t^2)$; when $t = 1$

10. $X = (4 \cos t, 3 \sin t)$; when $t = \pi/3$

11. $X = (e^t \cos t, e^t \sin t)$; when $t = 0$

12. $X = (t^2, t^3)$; when $t = 2$

13. $X = (e^t + e^{-t}, e^t - e^{-t})$; when $t = 0$

14. $X = (1/t, 1/(t^2 + 1))$; when $t = -1$

15. $X = (\log t, 2/t)$; when $t = 1$

16. $X = ((t^2 + 1)/(t^2 + 2), 2t)$; when $t = 0$

17. The ellipse

$$\frac{x^2}{a^2} + \frac{y^2}{b^2} = 1$$

has the parametric equation

$$X = (a \cos \theta, b \sin \theta).$$

Use this representation to find a parametric equation of the tangent line to the ellipse at $(x_0, y_0) = (a \cos \theta_0, b \sin \theta_0)$.

18. The path $X = (a \sec \theta, b \tan \theta)$ runs along the hyperbola $(x/a)^2 - (y/b)^2 = 1$. Use this parametric representation of the hyperbola to find the equation of its tangent line at $(x_0, y_0) = (a \sec \theta_0, b \tan \theta_0)$.

19. Redraw the figure showing the meaning of Theorem 6 for the case when Δt is negative.

20. Show that the line tangent to the path $X = G(t)$ at the point $X_0 = G(t_0)$ has the parametric equation

$$X - X_0 = G'(t_0)(t - t_0).$$

21. Show that a path whose tangent vector has a constant *direction* must lie along a straight line. (The hypothesis is that $dX/dt = f(t)U$, where U is a constant unit vector.)

22. Show that a path $X = G(t)$ is uniquely determined by its derivative $G'(t)$ and an initial value $X_0 = G(t_0)$. That is, if

$$G(t) = (g(t), h(t)) \qquad \text{and} \qquad L(t) = (l(t), m(t))$$

are two paths such that $G'(t) = L'(t)$ for all t and $G(t_0) = L(t_0)$, then

$$G(t) = L(t) \quad \text{for all } t.$$

23. Let $X = G(t) = (g(t), h(t))$ and $X = P(s) = (p(s), q(s))$ be two paths that do not intersect. Let d be the shortest distance between them, and suppose that we can find a closest pair of points,

$$X_1 = G(t_0) \quad \text{and} \quad X_2 = L(s_0),$$

i.e., a pair such that

$$|X_2 - X_1| = d.$$

Prove that the segment $X_1 X_2$ is perpendicular to both curves.

24. If one path is a straight line l in the above problem, then the tangent to the other path must be parallel to l at the point closest to l. Use this principle to find the point on $X = (t, t^2 - 1)$ closest to the line $X = (-s - 1, 2s)$.

/

6. VECTOR FORMS OF THE DERIVATIVE RULES

In the beginning sections of the chapter we were sometimes able to prove things about vectors without referring explicitly to their components. For example, we proved in this way that the medians of a triangle are concurrent. Instead of explicitly calculating with components we were able to use the vector laws.

In a similar way, derivative calculations involving vector functions of t become more transparent and frequently more efficient if we use vector forms of the derivative rules, as given below.

1. If X and U are differentiable vector functions of t, then so is $X + U$ and

$$\frac{d}{dt}(X + U) = \frac{dX}{dt} + \frac{dU}{dt}.$$

2. If w is a differentiable scalar function of t and X is a differentiable vector function of t, then wX is a differentiable vector function of t and

$$\frac{d}{dt}(wX) = w\frac{dX}{dt} + \frac{dw}{dt}X.$$

3. If X and U are differentiable vector functions of t, then $X \cdot U$ is a differentiable scalar function of t and

$$\frac{d}{dt}(X \cdot U) = X \cdot \frac{dU}{dt} + \frac{dX}{dt} \cdot U.$$

Rules (2) and (3) are simply the product rule for the two products we have for vectors.

4. If X is a differentiable vector function of the parameter r, and if $r = f(t)$ is differentiable, then X is a differentiable vector function of t, and

$$\frac{dX}{dt} = \frac{dX}{dr}\frac{dr}{dt}.$$

This is the rule for changing the path parameter; it is another chain rule.

5. If dX/dt is identically zero, then X is a constant vector.

These reformulations all follow from the original derivative rules and Theorem 6. For example,

$$\frac{d}{dt}[X \cdot U] = \frac{d}{dt}[(x,y) \cdot (u,v)] = \frac{d}{dt}(xu + yv)$$

$$= x\frac{du}{dt} + \frac{dx}{dt}u + y\frac{dv}{dt} + \frac{dy}{dt}v$$

$$= (x,y) \cdot \left(\frac{du}{dt}, \frac{dv}{dt}\right) + \left(\frac{dx}{dt}, \frac{dy}{dt}\right) \cdot (u,v)$$

$$= X \cdot \frac{dU}{dt} + \frac{dX}{dt} \cdot U.$$

EXAMPLE. We reconsider Examples 2 and 4 from the last section. In the second example we were given that dX/dt is the constant vector A. Now one such vector function of t is At, since

$$\frac{d}{dt}(At) = A,$$

by rule (2) above. Thus

$$\frac{d}{dt}(X - At) = 0$$

(rule (1)), and so

$$X - At = \text{constant vector } C,$$

$$X = At + C$$

(rule (5)). This was the conclusion of Example 2.

In the fourth example we considered a path $X = G(t)$ for which $|X|$ is constant. We can now compute

$$0 = \frac{d}{dt}|X|^2 = \frac{d}{dt}(X \cdot X) = X \cdot \frac{dX}{dt} + \frac{dX}{dt} \cdot X$$

$$= 2X \cdot \frac{dX}{dt},$$

by rule (3), and so $X \perp dX/dt$, as before.

PROBLEMS FOR SECTION 6

1. Let $X = F(t)$ be a twice-differentiable path such that $d^2X/dt^2 \equiv 0$. Prove that $X = At + C$, so the path is simply the parametric representation of a straight line.

2. Prove the rule (4) by a component calculation, the way (3) was proved in the text.

3. Prove the rule (5).

4. Let $X = F(t)$ be a path whose tangent vector has a constant direction. Prove that the path lies along a straight line (without resorting to component calculations).

5. The tangent vector of the path $X = F(t)$ is everywhere perpendicular to X. Prove that the path runs along a circle about the origin.

6. The tangent vector of the path $X = F(t)$ is everywhere perpendicular to the constant vector A ($A \neq 0$). Prove that the path runs along a line perpendicular to A.

7. Let $X = F(t)$ be a path whose tangent vector dX/dt is always in the direction of X. Prove that X runs along a line through the origin. (Write $X = |X|U$, where U is unit vector, presumably varying with t. Apply rule (2) and remember that dU/dt is necessarily perpendicular to U, as in Example 1.)

8. If X is a differentiable vector function of t, prove from rule (3) that

$$\frac{d}{dt}|X| = \frac{X \cdot \dfrac{dX}{dt}}{|X|}.$$

9. If $0 < c < a$, and $F = (c,0)$ and $G = (\ c,0)$, we know (Chapter 11, Section 4, Problem 14) that the set of points X such that

$$|X - F| + |X - G| = 2a$$

is the ellipse

$$\frac{x^2}{a^2} + \frac{y^2}{b^2} = 1,$$

where $b^2 = a^2 - c^2$. We can consider the above ellipse to be parametrized in some way, so that X traces the elliptical path with parameter t. The actual parametrization doesn't matter. Now differentiate the locus identity

$$|X - F| + |X - G| = 2a$$

with respect to t, using Problem 8, and so prove that the arrows $\overrightarrow{FX}$ and $\overrightarrow{GX}$ make equal angles with the tangent to the ellipse at X.

This is the so-called *optical property* of the ellipse, because it shows that the rays from a point source of light placed at G reflect off the ellipse and all come together (are "focused") at F (and vice versa). Each of the points F and G is called a *focus* of the ellipse.

10. It was shown in Chapter 5, Section 6, that the polar graph of the equation $r(1 - \varepsilon \cos \theta) = 1$ is an ellipse if $0 < \varepsilon < 1$. An important additional fact is that this ellipse has one focus at the origin. Prove this now from the values of a and b given there. (This is just algebra.)

7. ARC LENGTH AS PARAMETER

We saw in Chapter 11 that a smooth plane path $X = (x, y) = (g(t), h(t))$ has length, and that its arc length s measured *forward* from some fixed point is determined by the equation

$$\frac{ds}{dt} = \sqrt{\left(\frac{dx}{dt}\right)^2 + \left(\frac{dy}{dt}\right)^2} .$$

The right side is now recognizable as the magnitude of the path tangent vector

$$\frac{dX}{dt} = \left(\frac{dx}{dt}, \frac{dy}{dt}\right).$$

The equation can thus be rewritten

$$\left|\frac{dX}{dt}\right| = \frac{ds}{dt},$$

and says:

> *The magnitude of the path tangent vector is the rate of change of the path arc length s with respect to the parameter t.*

Note that if the parameter t happens to be equal to the arc length s, or more generally, if $s = t + c$, then

$$\left|\frac{dX}{dt}\right| = \frac{ds}{dt} = 1$$

and the tangent vector is always a unit vector. The same argument works backward and shows that if the tangent vector is always a unit vector then $ds/dt = 1$ and $s = t +$ constant. Thus

Theorem 7. *The parameter of a smooth plane path is equal to the path arc length (plus a constant) if and only if the tangent vector is always a unit vector.*

In general the parameter t will not be the arc length, but we can always reparametrize the path by its arc length. The argument goes as follows. Since the tangent vector to a smooth path is never zero (the definition of smoothness), $ds/dt = |dX/dt|$ is always positive. Therefore s is an increasing function of t which has an everywhere-differentiable inverse, by the inverse function theorem. That is, t is a differentiable function of s, say

$$t = \phi(s).$$

Then

$$x = g(t) = g(\phi(s)) = l(s),$$

$$y = h(t) = h(\phi(s)) = m(s)$$

is a new smooth parametrization of the path, with the path length s as the new parameter. Strictly speaking, this change of parameter gives us a new path. In any event, we can now write the tangent vector dX/dt in the form

$$\frac{dX}{dt} = \frac{dX}{ds} \frac{ds}{dt},$$

by rule (4) from the last section. This is the "polar coordinate" representation for the tangent vector, ds/dt being its magnitude and dX/ds being its direction, i.e. the unit tangent vector T that we think of as its direction.

We have noted in several contexts that a vector function of constant magnitude is necessarily perpendicular to its derivative. When the vector function is the unit tangent vector T to a path we can say more.

Theorem 8. *Let* $X = G(s)$ *be a path with arc length as parameter, and suppose that the unit tangent vector* T *is a differentiable function of* s. *Then the vector* dT/ds *is perpendicular to the path, and points in the direction in which the path is turning. Moreover, its magnitude is the absolute curvature of the path. Thus* dT/ds *has the "polar coordinate" representation*

$$\frac{dT}{ds} = \kappa N,$$

where κ *is the absolute path curvature and* N *is the unit vector normal to the path in the direction in which the path is turning.*

Proof. The first step is "old hat." We differentiate the identity

$$|T|^2 = T \cdot T = 1$$

by the product law for dot products:

$$T \cdot \frac{dT}{ds} + \frac{dT}{ds} \cdot T = 0.$$

Therefore,

$$2\left(T \cdot \frac{dT}{ds}\right) = 0$$

and the vectors T and dT/ds are perpendicular. Since T is tangent to the path, dT/ds is perpendicular to the path.

For the second assertion we can only give an intuitive argument. Consider the approximate equality

$$\frac{\Delta T}{\Delta s} \approx \frac{dT}{ds}.$$

Taking Δs positive, this says, in particular, that the direction of dT/ds is approximately the same as the direction of ΔT. But this is approximately the direction in which the curve is turning. Therefore the direction of dT/ds is approximately the direction in which the curve is turning, and since the approximation can presumably be made arbitrarily good by taking Δs sufficiently small, the two directions must in fact be equal. (It would be more appropriate to conclude from the above plausibility argument that it seems correct to *define* the direction in which the path is bending to be the direction of the vector dT/ds.)

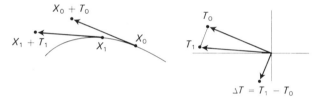

Now write the unit vector T in the form

$$T = (\cos \phi, \sin \phi),$$

where ϕ is the angle from the positive x-axis to the direction of the path. Then

$$\frac{dT}{ds} = \frac{dT}{d\phi}\frac{d\phi}{ds} = (-\sin \phi, \cos \phi)\frac{d\phi}{ds}.$$

But $d\phi/ds$ was our definition of the *curvature* of a path. And $(-\sin \phi, \cos \phi)$ is another unit vector. Therefore

$$\left|\frac{dT}{ds}\right| = |(-\sin \phi, \cos \phi)|\left|\frac{d\phi}{ds}\right| = 1 \cdot \kappa = \kappa,$$

where $\kappa = |d\phi/ds|$ is the absolute curvature of the path. The theorem is thus proved. ∎

EXAMPLE. A path that doesn't curve should lie along a straight line. This conjecture is easy to prove using the above formula. For if $\kappa = 0$, then

$dT/ds = 0$ and T is a constant, say, $T = A$. Since $T = dX/ds$ we conclude, as in Example 1, that

$$X = X_0 + sA.$$

Here s is arc length. If the path had been given in terms of some other parameter t, then the arc length s is a function of t, say $s = f(t)$, and then the path runs along the above line according to the equation

$$X = X_0 + f(t)A.$$

PROBLEMS FOR SECTION 7

1. a) Reparametrize the path

$$(x,y) = (e^t \cos t, e^t \sin t)$$

with the arc length s as parameter. (This involves finding s as a function of t, then t as a function of s, then substituting.)
    b) Then compute $T = dX/ds$ and dT/ds, and verify that they are orthogonal.
    c) Compute $\kappa = |dT/ds|$.
2. Do the same for the path $(x,y) = \dfrac{2}{3}(\cos^3 t, \sin^3 t)$.
3. The same for the path

$$(x,y) = (\frac{4}{3}t^{3/2}, \frac{1}{2}t^2 - t).$$

8. VELOCITY AND ACCELERATION

Now let the parameter t be time, so that the path $X = G(t)$ is the trajectory of a moving particle, and ds/dt is its speed. We could define the velocity of the particle as the time rate of change of its position X, that is, as the tangent vector dX/dt. But it is intuitively more satisfying to define the velocity as the vector V whose direction is the direction in which the particle is moving, and whose magnitude is the speed ds/dt at which the particle is moving. It then *follows* that

$$V = \frac{dX}{dt},$$

because dX/dt meets both of these requirements.

Acceleration is the *time rate of change of velocity*. It is thus the vector

$$A = \frac{dV}{dt} = \frac{d^2X}{dt^2}.$$

What is interesting about vector acceleration is that it is due *both* to change in speed *and* to change in direction. If *either* the speed is changing *or* the direction is changing, then the particle is accelerating.

Think how you are affected by changes in velocity when you ride in a car. If the car speeds up while going down a straight road the back of the seat presses harder against you, and you attribute this extra pressure to the straight-ahead acceleration of the car. Now consider what happens when the car travels with *constant* speed along a winding road. When you ride around a curve, you are thrown sidewise. This is a consequence of the *sidewise* acceleration of the car, and the greater or lesser extent that you feel this sidewise pressure is a rough measure of the magnitude of the sidewise acceleration. On this basis you can reach some conclusions about this perpendicular acceleration. You are thrown more to the side when you go at the same speed around a sharper curve, or if you go at a higher speed around the same curve. Thus the acceleration must increase in magnitude if *either* the speed or the curvature of the path is increased.

The exact formula for these different acceleration effects is contained in the following theorem.

Theorem 9. *If* $X = G(t)$ *is the path of a moving particle, with the time t as parameter, then the acceleration A has the formula*

$$A = \frac{dv}{dt} T + \kappa v^2 N,$$

where v is the speed ds/dt, T and N are the unit tangent and normal vectors, and κ is the absolute curvature of the path.

Proof. We start with

$$V = \frac{dX}{dt} = \frac{dX}{ds}\frac{ds}{dt} = vT$$

and differentiate again:

$$A = \frac{dV}{dt} = v\frac{dT}{dt} + \frac{dv}{dt} T.$$

But

$$\frac{dT}{dt} = \frac{dT}{ds}\frac{ds}{dt} = \kappa Nv$$

from Theorem 8, and substituting this value into the formula for A gives the formula of the theorem. ∎

It follows from the theorem that the component of the acceleration vector A normal to the curve is

$$A \cdot N = \kappa v^2,$$

in the direction in which the curve is bending. This shows explicitly how the "sidewise" acceleration varies with curvature κ and speed v.

EXAMPLE. A circle of radius r has curvature

$$\kappa = \frac{1}{r}.$$

Thus a particle that moves along this circle with *constant* speed v is nevertheless accelerating, according to the formula

$$A = \kappa v^2 N = \left(\frac{v^2}{r}\right)N.$$

That is, its acceleration is directed toward the center of the circle and is of magnitude v^2/r.

PROBLEMS FOR SECTION 8

Find the velocity and acceleration vectors, the speed v, and the tangential and normal components (dv/dt and κv^2) of the acceleration, for each of the following motions:

1. $(x,y) = (\cos 3t, \sin 3t)$.

2. $(x,y) = (3 \cos t, 3 \sin t)$

3. $(x,y) = (3 \cos t, 2 \sin t)$

4. $(x,y) = (t, \sin t)$

5. $(x,y) = (t, t^2)$

6. $(x,y) = (t^2, t^3)$

7. $(x,y) = (2 \cos 3t, 3 \sin 2t)$

8. $(x,y) = (\cos t \sin 3t, \sin t \sin 3t)$

9. A road lies along the parabola $100y = x^2$. A truck moving along the road is so loaded that the normal component of its acceleration must not exceed 25. What limitation does this impose on its speed as it rounds the vertex of the parabola?

10. A particle moves along the ellipse $(x,y) = (3 \cos t, 2 \sin t)$ with constant speed 5. What are the maximum and minimum values of the magnitude of its acceleration?

11. Is it possible for a particle to move along a curving path without experiencing any sidewise acceleration? Consult the formula of Theorem 9.

12. If you travel along a road on which the curvature is never zero, how must you adjust your speed so that your sidewise acceleration has a constant magnitude?

13. There are two circular tracks, the large one having twice the radius of the smaller. If you are comfortable driving 50 mph on the smaller track, how much faster can you drive on the larger track without experiencing a greater radial (sidewise) acceleration?

14. Suppose that the parameter in the standard parametric equations of the ellipse

$$\frac{x^2}{a^2} + \frac{y^2}{b^2} = 1$$

is the time t, so that the parametric equations

$$x = a \cos t \quad \text{and} \quad y = b \sin t$$

describe the motion of a particle moving along the ellipse. Show that the acceleration vector

$$A = \frac{d^2 X}{dt^2}$$

is always directed to the center of the ellipse (the origin).

9. KEPLER'S LAWS AND NEWTON'S UNIVERSAL LAW OF GRAVITATION

In the last section, the acceleration vector of a moving particle was resolved into its components parallel and perpendicular to the velocity vector $X'(t)$. It is also possible to resolve the acceleration vector into components parallel and perpendicular to the path position vector $X(t)$, and this resolution of the acceleration vector has a spectacular application. Using it, we shall be able to derive Newton's universal law of gravitation from Kepler's laws of planetary motion as an almost routine exercise in calculus. Newton did this in the seventeenth century, and it was one of his greatest achievements.

There was a third man whose work was essential to this enterprise. He was Tycho Brahé, an astronomer in the employ of the Emperor Rudolph of Bohemia in Prague, who had made thousands of observations of the positions of the planets, and who was engaged in the early years of the seventeenth century in completing and organizing his observations into tables of planetary motions.

Kepler was an astronomer, more speculatively inclined than Tycho, who strove to discover the secrets of the motions of the planets. He knew that Tycho had the best data available, and asked permission to visit and study Tycho's observations. He was welcomed to Prague and given the title Imperial Mathematician, in the year 1601. There, during 16 years of analysis, conjecture, false starts, and failures, he discovered, one by one, his three laws.

Kepler's laws

The first two laws concern the orbit of a single planet, while the third law relates the various planetary orbits to each other.

I. *A planet moves about the sun in such a way that the vector from the sun to the planet sweeps out equal areas in equal times. That is, the vector sweeps out area at a constant rate.*

II. *The path of a planet is an ellipse with the sun at one focus.*

III. *The* **square** *of the period T of a planetary orbit is proportional to the* **cube** *of its semimajor axis a. That is, the ratio T^2/a^3 is the same for all the planets.*

The new resolution of the acceleration vector $\ddot{X}(t)$ that we need for Kepler's laws can be obtained most easily by using polar coordinates. We write X in its polar coordinate form,

$$X = r(\cos\theta, \sin\theta) = rU,$$

where U is the unit vector of X, $r = |X|$, and r and θ are functions of t. Then

$$\dot{U} = \frac{d}{dt}(\cos\theta, \sin\theta) = (-\sin\theta, \cos\theta)\dot{\theta} = \dot{\theta}P,$$

where $P = (-\sin\theta, \cos\theta)$ is the unit vector perpendicular to U and leading U by $\pi/2$ (since $(-\sin\theta, \cos\theta) = (\cos(\theta + \pi/2), \sin(\theta + \pi/2))$). Similarly,

$$\dot{P} = \frac{d}{dt}(-\sin\theta, \cos\theta) = -\dot{\theta}U.$$

Therefore, the velocity vector $V = \dot{X}$ and acceleration vector $A = \dot{V}$ can be written

$$V = \dot{X} = \frac{d}{dt}(rU) = \dot{r}U + r\dot{U} = \dot{r}U + r\dot{\theta}P,$$

$$A = \dot{V} = (\ddot{r}U + \dot{r}\dot{\theta}P) + (\dot{r}\dot{\theta}P + r\ddot{\theta}P - r(\dot{\theta})^2 U).$$

Thus,

$$A = (\ddot{r} - r\dot{\theta}^2)U + (r\ddot{\theta} + 2\dot{r}\dot{\theta})P,$$

and this is the formula we have been looking for.

We can now analyze the consequences of Kepler's first law. If polar coordinates are introduced in the plane containing the planetary orbit, with the sun as the origin, then the area formula in Chapter 8, Section 7, says that the derivative of the area swept out with respect to the polar angle θ is $r^2/2$. Therefore,

$$\frac{d}{dt}(\text{area swept out}) = \frac{r^2}{2}\frac{d\theta}{dt} = \frac{\dot{\theta}r^2}{2}.$$

Kepler's law says that area is being swept out at a constant rate, so

$$\frac{\dot{\theta}r^2}{2} = \text{constant},$$

and

$$\frac{d}{dt}\left(\frac{\dot{\theta}r^2}{2}\right) = \frac{\ddot{\theta}r^2 + 2\dot{\theta}r\dot{r}}{2} = 0.$$

In particular, $\ddot{\theta}r + 2\dot{\theta}\dot{r} = 0$, so the new formula obtained above for the path acceleration vector reduces to

$$A = (\ddot{r} - r\dot{\theta}^2)U,$$

which is a vector directed along the line through the origin. We have proved the following theorem.

Theorem 10. *If a path $X(t)$ is traced in such a way that the arrow $\overrightarrow{OX}(t)$ sweeps out area at a constant rate, then the acceleration vector $\ddot{X}(t)$ is always directed along the line through $X(t)$ and O (i.e., $\ddot{X}(t)$ is always a scalar multiple of $X(t)$).*

Note that this result is *independent of the shape of the path*, and that it includes, as a special case, the result in the last section about uniform motion around a circle.

Note also in this situation that the magnitude of the acceleration vector is just the magnitude of its radial component, which by the component formula is

$$|A| = |\ddot{r} - r\dot{\theta}^2|.$$

Kepler's second law says that a planet moves in an elliptical path with one focus at the sun. If we take the origin at the sun, and the major axis of the ellipse along the x-axis, the elliptical path has the polar equation

$$r = \frac{k}{1 - \varepsilon \cos \theta}.$$

where $0 < \varepsilon < 1$ and k is positive. (See Example 4 in Section 6 of Chapter 5 and Problem 10 in Section 6 of this chapter.) We can now take the major step toward Newton's inverse-square law of gravitational attraction.

Theorem 11. *Kepler's first and second laws together imply that the acceleration of a planet is directed toward the sun, and is inversely proportional to the square of the distance from the sun in magnitude.*

Proof. We have already seen, in Theorem 10, that the acceleration of the planet is directed along the line to the sun, and that its magnitude is therefore the absolute value of its radial component, which is

$$|A| = |\ddot{r} - r\dot{\theta}^2|,$$

by the new resolution formula for $A = \ddot{X}$. Here

$$r = \frac{k}{1 - \varepsilon \cos \theta},$$

by Kepler's second law (and our discussion above). We differentiate this equation, and use the fact that $r^2\dot\theta$ is a constant C, by Kepler's first law, obtaining

$$\dot r = \frac{-k\varepsilon \sin \theta}{(1 - \varepsilon \cos \theta)^2}\dot\theta$$

$$= \frac{-\varepsilon \sin \theta}{k}r^2\dot\theta$$

$$= -\frac{C}{k}\varepsilon \sin \theta.$$

Differentiating again yields the equation

$$\ddot r = -\frac{C}{k}\varepsilon \cos \theta \cdot \dot\theta.$$

The radial component of the acceleration can now be computed:

$$\ddot r - r\dot\theta^2 = -\frac{C}{k}\varepsilon(\cos \theta)\dot\theta - \frac{C\dot\theta}{r}$$

$$- -C\dot\theta\left[\frac{1}{r} + \frac{\varepsilon \cos \theta}{k}\right]$$

$$= -C\dot\theta\left[\frac{1 - \varepsilon \cos \theta}{k} + \frac{\varepsilon \cos \theta}{k}\right] = -\frac{C}{k}\dot\theta$$

$$= -\left(\frac{C^2}{k}\right)\frac{1}{r^2}.$$

This shows the acceleration to be inversely proportional to r^2 in magnitude, and to be directed toward the origin, completing the proof of the theorem. ∎

Theorem 12. *Kepler's third law implies that the constant of proportionality C^2/k is the same for all the planets.*

Proof. This calculation will be left as an exercise. It depends on the following facts.

1. The semimajor and semiminor axes of the ellipse $r = k/(1 - \varepsilon \cos \theta)$ are

    $$a = \frac{k}{1 - \varepsilon^2} \quad \text{and} \quad b = \sqrt{1 - \varepsilon^2}\, a$$

    (See Chapter 5, Section 6).

2. The area of an ellipse is πab (Chapter 9, Problem 1 in the problem set at the end of the chapter).

3. If T is the period of the planet, that is, the time it takes the planet to make one complete revolution, then the area enclosed by its elliptical orbit is

$CT/2$. This is because $C/2$ is the constant rate at which $\overrightarrow{OX}(t)$ sweeps out area, by definition.

According to Newton's second law, the (vector) acceleration A of a moving body is related to the (vector) force F (which causes the acceleration) by the equation

$$F = mA,$$

where m is the mass of the body. Therefore, our three theorems above can be summarized as follows:

The motion of each planet is governed by a force directed toward the sun, of magnitude

$$k\frac{m}{r^2},$$

where r is the distance from the planet to the sun, m is the mass of the planet, and the constant k is the same for all planets.

It is easy to perform a mental experiment showing that the force is also directly proportional to the mass of the attracting body (the sun); for if the sun were divided into two equal parts, then the total force exerted by the sun on the planet would surely be the sum of the two equal forces exerted by its two equal parts, so each part would exert half of the total force. Reasoning in this way, we are led at once to Newton's universal law of gravitation:

If two bodies of mass m and M are at a distance r apart, then they exert on each other an attractive (gravitational) force of magnitude

$$\gamma\frac{mM}{r^2},$$

where γ is a universal constant (independent of m, M, and r).

PROBLEMS FOR SECTION 9

1. Prove Theorem 12.

2. Prove the converse of Theorem 10 (by reversing the chain of reasoning used there). That is, prove that if the acceleration vector $A = \ddot{X}(t)$ is always parallel to $\overrightarrow{OX}(t)$, then $\overrightarrow{OX}(t)$ sweeps out area at a constant rate.

The next three problems are aimed at proving the following theorem.

Theorem 13. *If a particle moves in the plane in such a way that its acceleration is directed toward the origin and is proportional to $1/r^2$ in magnitude, then the particle moves along a conic section (ellipse, parabola, or hyperbola) with the origin as a focus.*

We already know from the second problem that then $r^2\dot\theta$ is a nonzero constant C, which we shall assume to be positive. In particular, $\dot\theta$ is positive, so θ is an increasing function of t, and therefore t is a function of θ. That is, there is a uniquely determined time t_0 at which θ reaches any given value θ_0. this means, in particular, that r is a function of θ along the path. Our problem is to determine the nature of this function.

3. Use the chain rule $\dot u = (du/d\theta)\dot\theta$ to show that

$$\frac{d\left(\dfrac{1}{r}\right)}{d\theta} = -\frac{\dot r}{C}$$

(where C is the constant value of $r^2\dot\theta$), and hence that

$$\frac{d^2\left(\dfrac{1}{r}\right)}{d\theta^2} = \frac{-\ddot r}{C\dot\theta}.$$

4. The theorem hypothesis amounts to the equation

$$\ddot r - r(\dot\theta)^2 = -\frac{k}{r^2},$$

where k is a positive constant. Why is this? Use this equation and the results of the problem above to show that

$$\frac{d^2\left(\dfrac{1}{r}\right)}{d\theta^2} + \frac{1}{r} = \frac{k}{C^2}.$$

5. In Chapter 19 it will be shown that the differential equation $\ddot y + y = C$ has the general solution $y = A \cos t + B \sin t + C$. Assuming this, show that the general solution of the above differential equation can be written in the form

$$kr = \frac{C^2}{(1 - \varepsilon \cos(\theta - \theta_0))},$$

where ε is a positive constant. According to Chapter 5, Section 6, the graph of this equation is an ellipse if $\varepsilon < 1$, a parabola if $\varepsilon = 1$, and a hyperbola if $\varepsilon > 1$.

6. Suppose that $0 < \varepsilon < 1$, and that the ellipse

$$kr = \frac{C^2}{1 - \varepsilon \cos\theta}$$

is the path of a particle moving in such a way that $r^2\dot{\theta} = C$. Show that

$$\frac{T^2}{a^3} = \frac{4\pi^2}{k},$$

where T is the period of the motion (the time required for one complete revolution) and a is the semimajor axis of the ellipse. (This is Kepler's third law.)

chapter 15
functions of
two variables

More often than not the dependency relationships that arise in mathematics and its applications involve several variables. The volume of a pyramid with a square base depends on its base edge b and altitude h, according to the formula

$$V = \frac{1}{3}b^2 h.$$

Thus, V is a function of the two independent variables b and h.

The temperature T indicated on a weather map varies with position, and is a function of latitude and longitude. If we include the variation of T with time also, then T is a function of three variables, say, x, y, and t.

The state of a small homogeneous body of gas is determined by its pressure p, volume V, and temperature T. According to Boyle's law, these state variables are related by

$$pV = k(T + C),$$

where k and C are constants. This equation determines any one of the state variables as a function of the other two.

We investigate the variation of such a function by a "divide-and-conquer" strategy. We first study the variation that occurs when only one of the independent variables is permitted to change. The tool here is the *partial derivative*, which is just the ordinary derivative with respect to the one changing variable. Then we learn how to combine the information provided by the several partial derivatives in order to understand the way in which the function varies. This "multidimensional" calculus is an enormous subject that we will barely touch upon in this book.

In this chapter we shall consider only the separate partial derivatives, and two relatively simple applications, to maximum-minimum problems and to the equation of a tangent plane to a graph in three dimensions. In Chapters 16 and 17 we consider the combined action of the partial derivatives, and investigate the simplest consequences of their vector interpretation.

1. FUNCTIONS OF TWO VARIABLES; PARTIAL DERIVATIVES

Functions of several variables involve the same language and the same conventions that are used for functions of one variable. Thus, each of the expressions $x^2 + y^2$ and $x \sin y$ defines a function $f(x, y)$ of the two independent variables x and y, since each of them has a uniquely determined value for each pair of numbers x and y. Similarly,

$$g(x, y, z, w) = (x^2 + y^2)zw$$

is a function of four variables. Except for an occasional remark, we shall consider only functions of two variables in this chapter.

In general, a function $f(x,y)$ will not be defined for all pairs of numbers (x,y). For example,

$$f(x,y) = y \log x$$

is defined only when $x > 0$. Thus:

Definition. *A function of two variables is an operation that determines a unique output number z when applied to any given input pair of numbers (x,y) chosen from a certain domain.*

Each of the equations

$$z = x^2 + y,$$

$$xyz = 1,$$

$$z^3 = x/y,$$

determines z as a function of x and y. Thus a function of *two* variables can be defined by an equation in *three* variables, just as a function of one variable can be defined by an equation in two variables. Similarly,

$$e^w = xyz$$

determines w as a function of the *three* variables x, y, and z.

The domain of a function of two variables $f(x,y)$ will generally be some sort of region in the plane bounded by one or more curves. For example, $f(x,y) = \sqrt{1 - x^2 - y^2}$ is defined when $1 - x^2 - y^2 \geq 0$, that is, when

$$x^2 + y^2 \leq 1.$$

The domain of f is the "unit circular disk," and it is bounded by the unit circle $x^2 + y^2 = 1$.

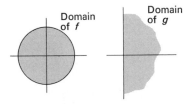

Domain
of *f*

Domain
of *g*

The domain of

$$g(x,y) = y \log x$$

is the right half-plane, and its boundary is the y-axis.

A point (a,b) is *in the interior* of a domain D if some small circular disk about (a,b) lies entirely in D. Any other point of D is a *boundary point*. Thus a point is a boundary point of D if every circle about it contains both points in D and points not in D. As stated above, the boundary points of a domain normally lie along one or more boundary curves.

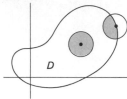

If $z = x^2 y^3 + xy + y^2$, and if we hold y fixed, then z becomes a function of x alone and we can calculate its derivative. This derivative is called the *partial derivative of z with respect to x*. In Leibniz notation it is written

$$\frac{\partial z}{\partial x},$$

where ∂x is read 'del x'. Thus, treating y as a constant and computing the derivative of z with respect to x, we have

$$\frac{\partial z}{\partial x} = 2xy^3 + y.$$

Similarly, we obtain $\partial z/\partial y$ by regarding x as a constant and computing the derivative of z with respect to y:

$$\frac{\partial z}{\partial y} = 3x^2 y^2 + x + 2y.$$

This procedure is followed for any number of independent variables. For example, if

$$w = x^2 - xy + y^2 + 2yz + 2z^2 + z,$$

then

$$\frac{\partial w}{\partial x} = 2x - y,$$

$$\frac{\partial w}{\partial y} = -x + 2y + 2z,$$

$$\frac{\partial w}{\partial z} = 2y + 4z + 1.$$

The partial derivatives $\partial z/\partial x$ and $\partial z/\partial y$ have geometric interpretations as slopes, but these slopes have to be pictured on the surface S that is the graph of $z = f(x,y)$ in three dimensions. Three-dimensional coordinate systems and graphs will be discussed in Sections 3 and 4, and the remarks below may not be clear until then.

Holding y fixed at the value $y = b$ restricts the point (x, y, z) to a plane, parallel to the xz-plane, that slices the surface S in the curve shown in the lefthand figure below. The slope of this curve is the derivative of z with respect to x while y is held fixed at the value $y = b$, and this is exactly the partial derivative $\partial z / \partial x$. That is, $\partial z / \partial x$ is the slope of the surface S in the direction parallel to the x-axis. Similarly $\partial z / \partial y$ is the slope in the direction parallel to the y-axis, as shown in the righthand figure.

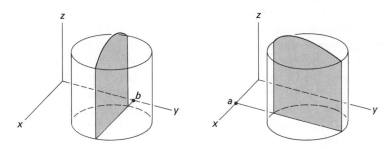

In addition to the Leibniz notation for partial derviatives, we shall need function notation, analogous to the f' notation for functions of one variable. We shall use

$$D_1 f \qquad \text{and} \qquad D_2 f$$

for the partial derivative of f with respect to its first and second variables, respectively. Thus, if $z = f(x, y)$, then

$$\frac{\partial z}{\partial x} = D_1 f(x, y),$$

$$\frac{\partial z}{\partial y} = D_2 f(x, y).$$

As in the case of functions of one variable, the function notation is superior to the Leibniz notation when substitutions have to be shown. For example, the value of $\partial z / \partial x$ at $(x, y) = (a, b)$ is given in the two notations by

$$D_1 f(a, b) = \frac{\partial z}{\partial x}\bigg|_{(a,b)}.$$

EXAMPLE. If $f(x, y) = xy + x^2 - y$, then

$$D_1 f(x, y) = y + 2x, \qquad D_2 f(x, y) = x - 1,$$
$$D_1 f(2, 3) = 3 + 2 \cdot 2 = 7, \qquad D_2 f(2, 3) = 2 - 1 = 1.$$

At a couple of points we shall have to mention continuity, in order to state things correctly. This notion extends naturally to a function of two variables $f(x,y)$:

The function f is continuous at the point (a,b) if its value $f(x,y)$ can be made to be as close to $f(a,b)$ as we wish by taking the point (x,y) suitably close to (a,b).

In the plane, the closeness of two points is measured by the distance between them, so being suitably close to (a,b) means lying within a suitably small circle about (a,b). This is a simultaneous condition on x and y, and the notion of continuity thus involves simultaneous variation in x and y.

Since this chapter is only a preview of functions of more than one variable, we shall not look into whether there are reasonable ways in which such a function can be discontinuous. Every function that will come up will be continuous wherever it is defined. The reason lies along the lines of the Chapter 3 theorem, that a differentiable function is automatically continuous. But when two independent variables are involved, the argument from derivatives to continuity is more subtle, and we shall just state the appropriate theorem from Chapter 20.

Theorem. *Suppose that both partial derivatives of $f(x,y)$ exist everywhere inside a circle C, and are everywhere less in magnitude than a fixed constant k. Then f is continuous inside of C.*

PROBLEMS FOR SECTION 1

1. Write out the definitions of $\partial z/\partial x$ and $\partial z/\partial y$ (where $z = f(x,y)$) as limits of difference quotients, using the increment notation.

For each of the following functions, find $\partial z/\partial x$ and $\partial z/\partial y$.

2. $z = 2x - 3y$ 3. $z = 3x^2 - 2xy + y$

4. $z = \sqrt{x^2 + y^2}$ 5. $z = e^{ax+by}$

6. $z = e^{xy}$ 7. $z = \sin(2x + 3y)$

8. $z = \cos((x^2 - y^2)/2)$ 9. $z = xye^{-x^2/2}$

10. $z = y\log(x + y)$ 11. $z = y\log(xy)$

12. $z = e^{-x}\cos y$ 13. $z = \sin ax \cos by$

14. $z = \dfrac{x}{y} - \dfrac{y}{x}$ 15. $z = \arctan \dfrac{x}{y}$

16. $z = x^y$ 17. $z = \dfrac{x + y}{x - y}$

In the problems below, compute the partial derivatives using function notation, and determine the indicated values.

18. $f(u,v) = 3u^2 + 2uv + v^3$, $D_1 f(2, -3)$, $D_2 f(1, 1)$.

19. $f(s,t) = se^{st}$, $D_1 f(2, 3)$, $D_2 f(2, -1)$.

20. $f(x,y) = \sin(3x + y^2)$, $D_1 f(\pi, 0)$, $D_2 f(0, \sqrt{\pi})$.

21. $f(x,y,z) = xy^2z^3$, $D_1 f(0,2,-2)$, $D_2 f(0,2,-2)$, $D_3 f(1,2,-2)$.

22. $f(x,u,t) = x^2 + ut^3$, all partials at $(2,0,1)$.

23. $f(s,t) = s^2 \sin t$, $D_1 f(a,b)$, $D_2 f(a,0)$.

A function can have a property called "homogeneity of degree n" that we won't define here, but which can be shown to be equivalent to the condition

(*) $$x D_1 f(x,y) + y D_2 f(x,y) = nf(x,y).$$

24. Show that $z = \arctan(y/x)$ is homogeneous of degree 0. That is, show that

$$0 = x \frac{\partial z}{\partial x} + y \frac{\partial z}{\partial y}.$$

Each of the following functions is homogeneous. In each case verify (*) and determine the degree of homogeneity.

25. $z = x^4 + y^4 - x^2y^2$ 26. $z = (x + y)/(x - y)$

27. $z = x/(x^2 + y^2)$

28. Polar coordinates and rectangular coordinates are related by the equations

$$x = r \cos \theta, \qquad r = \sqrt{x^2 + y^2},$$

$$y = r \sin \theta, \qquad \theta = \arctan \frac{y}{x}.$$

Find all the partial derivatives of these functions.

29. The boundary of a plane region G need not consist of one or more curves. What is the domain of the function $f(x,y) = 1/(x^2 + y^2)$? What is its boundary?

30. Find a function whose domain boundary consists of just the two points $(1,0)$ and $(-1,0)$.

31. Find a function whose domain boundary consists of the unit circle $x^2 + y^2 = 1$ plus the origin $(0,0)$.

2. MAXIMUM–MINIMUM PROBLEMS

Suppose that the function $z = f(x,y)$ assumes a maximum (or minimum) value at a point (x_0, y_0) in the interior of its domain. If we hold y constant at the value y_0, then $f(x, y_0)$ is a function of the one variable x having its maximum value at $x = x_0$, and so its derivative must be zero there, as in Chapter 6. That is,

$\partial z/\partial x = 0$ at the point $(x,y) = (x_0, y_0)$. Similarly, $\partial z/\partial y = 0$ at this point. The equations

$$\frac{\partial z}{\partial x} = 0, \qquad \frac{\partial z}{\partial y} = 0$$

are thus two equations in two unknowns that are satisfied by the maximum point (x_0, y_0), and we may be able to solve them simultaneously and so determine (x_0, y_0).

In analogy with the earlier definition for a function of one variable, we call such a point (x_0, y_0) where both partial derivatives are zero a *critical point* of f.

EXAMPLE 1. Find the dimensions of the rectangular box with open top and one-cubic-foot capacity that has the smallest surface area.

. . . .

Solution. The box is shown in the figure.

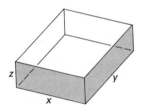

Its volume is xyz and this is required to be 1:

$$xyz = 1.$$

The total surface area is

$$A = xy + 2xz + 2yz.$$

We can eliminate one variable, say z, by solving for z in the first equation and substituting in the second, giving $z = 1/xy$ and

$$A = xy + \frac{2}{y} + \frac{2}{x}.$$

Now A is a function of x and y. Its domain is the first quadrant, because x and y have to be positive, but no other limitation is placed on their values. If A has a minimum value, it has to be at a critical point, that is, a point where

$$\frac{\partial A}{\partial x} = y - \frac{2}{x^2} = 0,$$

$$\frac{\partial A}{\partial y} = x - \frac{2}{y^2} = 0.$$

So we solve these equations simultaneously. First,

$$yx^2 = 2,$$

$$xy^2 = 2.$$

Dividing gives $x/y = 1$ and $x = y$. Therefore, $x^3 = 2$ and $x = y = 2^{1/3}$. Then

$$z = \frac{1}{xy} = 2^{-2/3}.$$

The *proportions* of the box are given by

$$x = y = 2z.$$

The box has a square bottom, with edge length twice the height of the box.

REMARK. In Chapter 6 we had a theorem guaranteeing that in a certain situation a function assumes a maximum value, and telling how to find it. Here we have no such theoretical guarantee that a maximum exists. We have only argued that *if f* has a maximum value at an interior point, *then* its partial derivatives must both be zero there. This means that, in any given problem, you should convince yourself that it is reasonable to expect a maximum or minimum value somewhere. For example, in the problem above you probably feel that among all rectangular boxes of a fixed capacity, there has to be one with minimal surface area. But this is tricky business, and we ought to nail down the existence of extreme values in a manner analogous to (but of necessity more complicated than) the argument in Chapter 6. (See Examples 3 and 4.)

EXAMPLE 2. If we restrict the domain of

$$z = f(x, y) = y^3 + xy - x^3$$

to the unit circular disk $x^2 + y^2 \leq 1$, does f have a maximum value at an interior point?

. . . .

Solution. An interior maximum point (x,y) has to be a critical point of f. That is,

$$\frac{\partial z}{\partial x} = y - 3x^2 = 0,$$

$$\frac{\partial z}{\partial y} = 3y^2 + x = 0,$$

at such a point. We can solve these equations simultaneously by substituting from the first into the second, to get

$$27x^4 + x = 0.$$

Factoring this equation,

$$x(27x^3 + 1) = 0,$$

shows its solutions to be $x = 0$, $x = -1/3$. The corresponding y values are then obtained from the critical point equations. They are $y = 0$ and $y = 1/3$, respectively, and the critical points are thus $(0,0)$ and $(-1/3, 1/3)$. Since

$$f(0,0) = 0,$$

$$f\left(-\frac{1}{3}, \frac{1}{3}\right) = -\frac{1}{27},$$

while

$$f(0, 1) = 1,$$

we see that neither critical point yields a maximum value for f. Therefore f does not assume a maximum value at an interior point of its restricted domain.

This example can be pursued further. We call a domain *closed* if it includes its boundary and *finite* if it lies entirely inside some circle. The restricted domain in the above example is both closed and finite. We shall now assume the following theoretical maximum principle:

If a function f is continuous on a finite, closed domain D, then f assumes a maximum value and a minimum value on D.

EXAMPLE 3. It follows from this maximum principle that the function f of Example 2 has a maximum value somewhere on its domain $x^2 + y^2 \le 1$. We saw there that the maximum value could not be at an interior point, and it therefore must be somewhere on the boundary circle $x^2 + y^2 = 1$. We thus want the maximum of $z = y^3 + xy - x^3$ subject to the auxiliary condition $x^2 + y^2 = 1$. But this is now a one-variable problem. We can either solve for y in $x^2 + y^2 = 1$ and substitute in the equation for z, or we can use the method of auxiliary variables. This remaining one-variable problem is rather complicated, and we shall not go any further with it. Instead, we shall work out a simpler problem of the same type.

EXAMPLE 4. Find the maximum value of

$$z = x^2 + 4x^2y^2 - y^2$$

on the closed circular disk $x^2 + y^2 \le 1$.

Solution. The maximum value occurs either at an interior point or at a boundary point. The only possible interior location is a critical point, determined by the critical point equations

$$\frac{\partial z}{\partial x} = 2x + 8xy^2 = 0,$$

$$\frac{\partial z}{\partial y} = 8x^2y - 2y = 0.$$

These equations reduce to

$$x(1 + 4y^2) = 0,$$

$$y(4x^2 - 1) = 0.$$

The first shows that $x = 0$, and then the second says that $y = 0$. Thus the only critical point is the origin $(x, y) = (0, 0)$. The value of z there is zero, and this is not the maximum value of z since we see by inspection that $z = 1$ at $(1, 0)$.

The maximum therefore occurs on the boundary curve $x^2 + y^2 = 1$, where the problem reduces to a one-variable problem. In this example, the reduction is very simple. On the boundary $y^2 = 1 - x^2$, and so

$$z = x^2 + 4x^2y^2 - y^2 = x^2 + (1 - x^2)(4x^2 - 1)$$

$$= 6x^2 - 4x^4 - 1.$$

This is over the x-interval $[-1, 1]$. We saw in Chapter 6 that the maximum value of z on this interval must occur at an endpoint or at a point where $dz/dx = 0$. At the two endpoints $z = 1$. On the other hand, the equation

$$\frac{dz}{dx} = 12x - 16x^3 = 0$$

has the roots $x = 0, \pm\sqrt{3}/2$, and the corresponding values of z are $z = -1$ and $z = 6(3/4) - 4(3/4)^2 - 1 = 5/4$. This second value is larger than the common endpoint value, so z has its maximum value $5/4$ at the boundary points determined by $x = \pm\sqrt{3}/2$. At these points $y^2 = 1 - x^2 = 1/4$ and $y = \pm1/2$. There are thus four maximum points, symmetric with respect to the two axes, and the one in the first quadrant is $(\sqrt{3}/2, 1/2)$.

PROBLEMS FOR SECTION 2

Each of the following functions is defined on the whole plane and has a minimum or maximum value. Find this extreme value and the point (or points) where it occurs.

1. $z = x^2 + 2xy + 2y^2 - 6y$ 2. $z = 4x + 6y - x^2 - y^2$

3. $z = x^4 - x^2y^2 + y^4 + 4x^2 - 6y^2$ 4. $z = x^4 + 2xy^2 + y^4$

5. $z = 3x^4 - 4x^3y + y^6$

Let D be the first quadrant. Each of the following functions has a maximum or a minimum on D. Find this extreme value and the point where it occurs.

6. $z = xy + \dfrac{1}{x} + \dfrac{1}{y}$ 7. $z = 8y^3 + x^3 - 3xy$

8. $z = y + 2x - \log xy$

9. Find the minimum distance between the parabola $y = x^2$ and the line $y = x - 1$.

10. A rectangular box with capacity one cubic foot is to be made with an open top. The material for the bottom costs half again as much as the material for the sides. Find the dimensions of the least expensive box.

11. Find the maximum and minimum values of

$$f(x,y) = x^2 + y^2 - xy - x$$

on the unit square $-1 \le x \le 1, \quad -1 \le y \le 1$.

12. Find the maximum and minimum values of

$$f(x,y) = 2x^4 - 3x^2y^2 + 2y^4 - x^2$$

on the closed unit disk $x^2 + y^2 \le 1$.

13. Find the minimum distance between the path $(x,y) = (t, t^2 + 1)$ and the path $(x,y) = (-s - 1, 2s)$.

14. The temperature in degrees Fahrenheit at each point (x,y) in the region $0 \le x, y \le 1$ is given by $T = 48xy - 32x^3 - 24y^2$. Find the points of maximum and minimum temperature and the temperature at each of these points.

15. Find positive numbers x, y, and z such that $x + 3y + 2z = 18$ and xyz is a maximum. [*Hint.* Use the technique of implicit differentiation covered in Section 4.7. The use of this technique for partial differentiation is analogous.]

3. COORDINATES IN SPACE

The graph of a three-variable equation such as $xyz = 1$ or $z = f(x,y)$ requires a three-dimensional coordinate system, with three mutually perpendicular lines in space through a common origin O as axes. We generally draw and label such space axes as shown below. If you have difficulty visualizing this configuration as three perpendicular lines in space, imagine looking down into a corner of a room.

You see one horizontal plane, the floor, and two mutually perpendicular vertical planes, the two walls. The lines in the coordinate diagram are the edges where two planes intersect. They are mutually perpendicular, and form the positive axes of our space axis system.

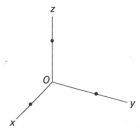

Each axis is given a coordinate system with common origin at O and a common unit of distance. The plane through the y- and z-axes is then a coordinate plane, called the yz-coordinate plane. It is viewed as a vertical "back wall" plane. The xy-plane is a horizontal "floor" plane, and it has the normal xy-plane appearance when viewed from a point on the positive z-axis. The xz-plane is a vertical "left wall" plane perpendicular to the yz-plane. Figures in these coordinate planes have to be distorted when they are drawn on the page. In particular, the fact that their axes are perpendicular cannot be shown and must be visualized.

The three coordinate planes divide space into 8 regions, called *octants*, coming together at O. This total coordinate plane configuration can be visualized, and a corner of a room again is helpful. This time imagine a room corner inside a building. Then the floor extends into other rooms and the walls extend down to the floor below. We can indicate the extended edges that we can't see by dotted lines or light lines. There are eight rooms sharing the corner point O, four on our floor having O as a floor point, and four on the floor below having O as a ceiling corner point.

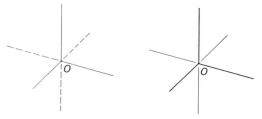

Consider now an arbitrary point p in space. Its coordinates can be obtained by dropping perpendiculars to the axes, just as in the plane. However, the resulting figure, shown at the left at the top of the next page, is hard to visualize. We therefore describe the process differently.

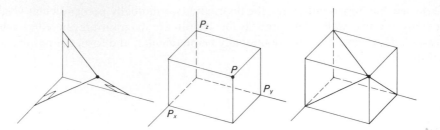

First, pass the plane through p perpendicular to the x-axis. Then define the x-coordinate of p to be the coordinate on the x-axis of the point p_x where the plane and axis intersect. If the other two coordinates of p are defined in the same way, then the three planes through p and the three coordinate planes altogether determine a rectangular "coordinate box" for p. It can be described as the rectangular box having three of its edges along the coordinate axes and having p as the vertex diagonally opposite to the origin O. The middle figure above shows this box, and it should look three-dimensional to you. In the right figure the three original perpendicular lines are now seen as lying on faces of the box.

Just as we did in one and two dimensions, we identify the geometric point p with its coordinate triple (x, y, z) and speak of the point (x, y, z).

If the line segment joining the points

$$p_1 = (x_1, y_1, z_1) \qquad \text{and} \qquad p_2 = (x_2, y_2, z_2)$$

is not parallel to any coordinate plane, then p_1 and p_2 are diagonally opposite vertices of a certain rectangular box, just as were p and the origin O in the above coordinate box.

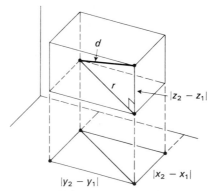

The edge lengths are now given by the coordinate differences, as shown, and the distance formula is again a consequence of the Pythagorean theorem:

$$d^2 = r^2 + (z_2 - z_1)^2 = (x_2 - x_1)^2 + (y_2 - y_1)^2 + (z_2 - z_1)^2.$$

Thus:

> *The distance d between the points (x_1, y_1, z_1) and (x_2, y_2, z_2) in coordinate three-space is given by*

$$d = \sqrt{(x_2 - x_1)^2 + (y_2 - y_1)^2 + (z_2 - z_1)^2}.$$

Supposing that one's geometric imagination is good enough to visualize the coordinate-box drawing as a figure in space, there is still a practical problem of how to draw space configurations on a two-dimensional page in such a way that they can be reinterpreted *accurately* as three-dimensional arrays. Here are three conventions that overcome most difficulties.

I. *Parallel lines in space are always drawn parallel.* We wouldn't draw the end of a room the way it would look to us from inside the room, as below on the left, but rather as on the right. This convention amounts to assuming that we are always outside of anything that we draw, and far enough away so that we have the same orientation to all of its parts, and parallel edges appear parallel to our eyes.

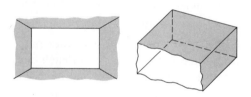

II. *We interpret parallel lines in a figure as representing parallel lines in space.* This is part of a more general "convention of general position." If a space configuration has a pair of lines that are not parallel, then as we view it from various far-away positions, these lines in general will not look parallel. It takes a special orientation for them to look parallel, and we suppose that we have avoided all such special viewpoints and have drawn the figure in "general position."

III. Similarly, if a point P is not on a line l in space, then from most points of view P will be seen to be off l. The assumption of general position therefore lets us conclude that *if a point P is shown on a line l in a figure, then P is on l in space.*

There is still the problem that if we indicate a space point P by a dot in the usual way, then we don't really know where P is, because there is a whole line in space that looks "end on" like a single point. The simplest way to fix the location of P is to show also the point Q directly below it in the xy-plane. If you can visualize the point Q, shown below, as a point in the floor plane, and P as a point directly over it, then you should be able to "see" P in space.

We can also convey the space location of P by other auxiliary lines, the coordinate box giving probably the strongest impression of where P is.

EXAMPLE. We draw the coordinate box for the point $(3, 4, 2)$ in one of the several possible ways. The steps, shown in order below, are:

1.   Mark the three coordinate points on the axes;

2.   Through each marked point draw the two lines parallel to the other two axes, thus obtaining three rectangles, one in each coordinate plane. These are the three faces of the coordinate box that have the origin as a vertex;

3.   Each of the three rectangles just drawn has one new vertex. Through it draw a line representing a perpendicular to its coordinate plane, that is, a line parallel to the axis not in its coordinate plane. These three lines will meet at the point representing the space point $(3, 4, 2)$.

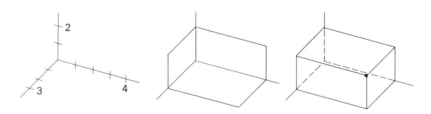

PROBLEMS FOR SECTION 3

Draw a three-dimensional coordinate system and locate the following points on it.

1. $P_1(6, 2, 1)$ 2. $P_2(-6, 3, 1)$ 3. $P_3(1, -2, 5)$

4. $P_4(2, 3, 4)$ 5. $P_5(2, 1, -3)$ 6. $P_6(-1, -2, 3)$

7. $P_7(2, -1, -4)$ 8. $P_8(-2, 3, -1)$ 9. $P_9(-6, -2, -1)$

10. $P_{10}(2, 1, -4)$

11. Find the distance between P_1 and P_4.

12. Find the distance between P_4 and P_9.

13. Find the distance between P_{10} and P_1.

14. Find the distance between P_6 and P_5.

15. Find the distance between P_3 and P_2.

What is the locus of a point:

16. Whose x-coordinate is always 0?

17. Whose y-coordinate is always 3?

18. Whose x- and z-coordinates are always 0?

4. GRAPHS IN COORDINATE SPACE

The graph of an equation in x, y, and z is the space configuration made up of all the points (x, y, z) satisfying the equation, and it will generally be some kind of two-dimensional surface.

EXAMPLE 1. The graph of

$$x^2 + y^2 + z^2 = 25$$

is the sphere about the origin of radius 5. For, taking square roots and replacing x by $x - 0$, etc., we get the equivalent equation

$$\sqrt{(x - 0)^2 + (y - 0)^2 + (z - 0)^2} = 5,$$

which says that the distance from $(0, 0, 0)$ to (x, y, z) is 5, by the distance formula. The graph consists of all such points (x, y, z) and is thus the sphere. Here is a drawing of the part of the sphere in the first octant.

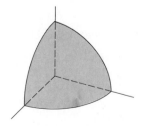

EXAMPLE 2. In three dimensions, the graph of

$$x = a$$

is a *plane*, namely, the plane through the point $(a, 0, 0)$, parallel to the *yz*-coordinate plane (perpendicular to the *x*-axis). This is just a restatement of the definition of *x*-coordinates.

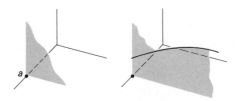

Note that this plane is itself a coordinate plane with y and z as coordinates. The graph *in this plane* of the equation

$$z = \sqrt{y + 1}$$

is partly shown at the right above. It consists of the space points of the form $(a, y, z) = (a, y, \sqrt{y + 1})$.

Although each plane $x = a$ is a *yz*-plane in this way, only the special plane $x = 0$ is called the *yz-coordinate plane*, since it is the one containing the *y*- and *z*-axes.

Similarly, $y = b$ is (the equation of) the plane through $(0, b, 0)$ parallel to the *xz*-coordinate plane, and $z = c$ is the plane through $(0, 0, c)$ parallel to the *xy*-coordinate plane.

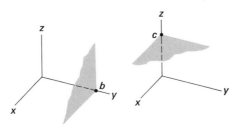

The points satisfying *both equations*

$$x = a \qquad \text{and} \qquad y = b$$

are the points lying on both of the planes $x = a$ and $y = b$, i.e., on the vertical line through the point $(a, b, 0)$.

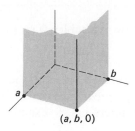

$(a, b, 0)$

The three-dimensional graph of an equation in two variables will be a surface parallel to the axis of the missing variable.

EXAMPLE 3. If a point P lies on the graph of $2z = 4 - y^2$ then so does any other point Q having the same y- and z-coordinates. The graph thus contains the line through p parallel to the x-axis.

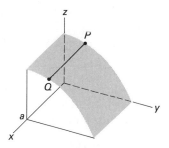

A piece of the graph in the first octant is shown above. Every plane $x = a$ is a yz-plane which intersects the surface in the same parabola $2z = 4 - y^2$. Similarly, the graph of $2z = 4 - x^2$ is a "parabolic cylinder" parallel to the y-axis, as shown below at the left. These two parabolic cylinders, together with the coordinate planes, bound a solid figure in the first octant, as sketched at the right. It has square cross sections parallel to the xy-plane.

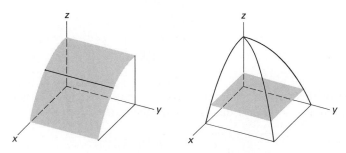

The graph of a function of two variables $f(x, y)$ is the graph of the equation

$$z = f(x, y).$$

If we interpret the domain D of f as a subset of the xy-coordinate plane, then the graph of f is a surface S that is *spread out* over D, in the sense that each vertical line through a point of D intersects S exactly once, in the point $(x, y, z) = (x, y, f(x, y))$.

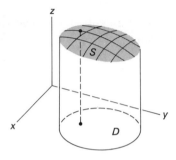

It is very helpful in visualizing such a surface S to consider the *curves* in which it intersects several vertical planes

$$x = a \qquad \text{and} \qquad y = b.$$

Its intersection with the plane $x = a$ is the graph in this yz-plane of the function of one variable $f(a, y)$, that is, the graph of the equation

$$z = f(a, y).$$

It is the *slice* in which the plane $x = a$ cuts S.

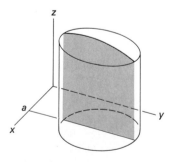

The slope of this curve is the derivative of z with respect to y, while x is held fixed at the value $x = a$, and this is exactly the partial derivative

$$\left. \frac{\partial z}{\partial y} \right|_{x=a} = D_2 f(a, y).$$

Thus we can interpret $\partial z / \partial y$ as *the slope of the graph of $z = f(x,y)$ in the y-direction*.

Similarly, the intersection of S with the plane $y = b$ is the graph in that xz-plane of the equation $z = f(x, b)$.

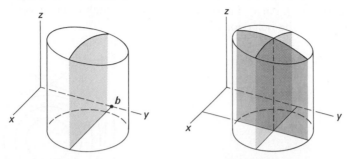

Its slope is $\partial z / \partial x$, evaluated along $y = b$, that is, $D_1 f(x, b)$. Thus, $\partial z / \partial x$ can be interpreted as *the slope of the graph $z = f(x, y)$ in the x-direction*.

If (a, b) is in the domain of f, then these two slice curves necessarily intersect over $(a, b, 0)$, the point of intersection being

$$(a, b, z) = (a, b, f(a, b)).$$

If we sketch two or three curves of each type, even very roughly, we get an impression of how the surface is curving. We did this in the figure on page 620.

Finally, the intersection of S with the horizontal plane $z = c$ is the graph in that xy-plane of the equation

$$c = f(x, y).$$

This horizontal curve may or may not intersect a given vertical slice, say the one in the plane $y = b$. It will just in case

$$z = f(x, b)$$

assumes the value c. An example where the curves do not intersect is shown below for the parabolic cap $z = 4 - x^2 - y^2$.

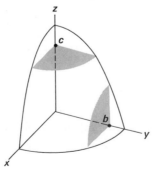

When we draw the graphs

$$c = f(x,y)$$

in the xy-coordinate plane, that is, in the domain D of f, it is natural to call them the level curves of f. An xy-figure showing several such level curves is an alternate way of picturing the behavior of $f(x,y)$.

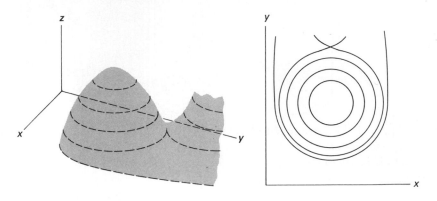

EXAMPLE 4. The level curves of the parabolic cap

$$5z = 25 - (x^2 + y^2)$$

are just circles. Nevertheless, their spacing can indicate the shape of the space graph of the equation. This is the principle of a contour map.

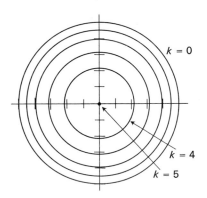

PROBLEMS FOR SECTION 4

Describe and sketch (as best you can) the graphs of the following equations.

1. $x^2 + y^2 + z^2 = 36$ 2. $x + y = 1$ 3. $y + z = 1$

4. $x + z = 1$ 5. $z = 1 - y^2, x = 2$ 6. $z = 1 - y^2$

7. $x^2 + y^2 = 4, z = 1$ 8. $x^2 + y^2 = 4$ 9. $y^2 + z^2 = 4$

10. $2x + y + z = 4$ (We shall see in the next section that this is a plane. It can be pictured by marking its intercepts on the axes and then drawing the lines joining pairs of these intercepts in the three coordinate planes. These are the lines in which the given plane intersects the coordinate planes.)

11. $4x^2 + 9y^2 = 36$ 12. $4x^2 + 4z^2 + 9y^2 = 36$

13. $x^2 + y^2 + z^2 - 4x - 2y - 6z = 36$ (Complete the squares.)

Draw the figure representing the solid in the first octant bounded by the coordinate planes and the following surfaces.

14. $x + y + z = 2$ (See comment with Problem 10.)

15. $y + z = 1, x = 2.$ 16. $x + z = 1, y = 2.$

17. $x + y = 1, z = 2.$ 18. $x^2 + y^2 = 1, z = 2.$

19. $x^2 + y^2 = 1, x + z = 1.$ 20. $x^2 + y^2 + z^2 = 25, y = 4.$

21. $y^2 + z^2 = 16, x^2 + y^2 = 16.$

22. $x^2 + y^2 = 16, x + y + z = 8$ (See comment with Problem 10.)

23. $4x^2 + 4y^2 + z^2 = 64, y + z = 4.$ 24. $x^2 + y^2 = z^2 + 9, z = 4.$

5. PLANES AND TANGENT PLANES

The graph of

$$z = 6 - x - 2y$$

intersects every vertical plane $x = a$ or $y = b$ in a straight line. For example, it intersects the yz-plane $x = 2$ in the line $z = 4 - 2y$.

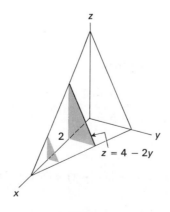

Pretty clearly the graph of $z = 6 - x - 2y$ is a plane, and a proof of this fact could be obtained by reasoning about these straight line intersections. In a similar way, it could be shown that the graph of

$$z = ax + by + c$$

is always a plane. However, a more elegant proof will be given in Chapter 17, and we shall just assume this fact for now. If (x_0, y_0, z_0) is a point on the plane, so that

$$z_0 = ax_0 + by_0 + c,$$

then subtracting the two equations gives the equation

$$z - z_0 = a(x - x_0) + b(y - y_0),$$

analogous to the point–slope equation of a line.

EXAMPLE 1. Find the plane through the point $(1, 1, 1)$ that intersects the yz-coordinate plane in the line $z = 2y + 1$.

....

Solution. We can start from either form of the plane equation. If we use the second form, with $(x_0, y_0, z_0) = (1, 1, 1)$, then the equation of the plane is

$$z - 1 = a(x - 1) + b(y - 1).$$

This must reduce to $z = 2y + 1$ when $x = 0$. Setting $x = 0$ and solving for z gives

$$z = by + (1 - a - b).$$

Comparing this with

$$z = 2y + 1$$

and equating coefficients, we see that

$$\left. \begin{array}{r} b = 2 \\ 1 - a - b = 1 \end{array} \right\} \quad a = -2.$$

Therefore,

$$z - 1 = -2(x - 1) + 2(y - 1),$$

so

$$z = -2x + 2y + 1$$

is the plane in question.

Note that the plane

$$z = ax + by + c$$

has the constant slope

$$\frac{\partial z}{\partial x} = a$$

in the direction parallel to the x-axis, and the constant slope

$$\frac{\partial z}{\partial y} = b$$

in the direction parallel to the y-axis.

EXAMPLE 2. We shall find the equation of the plane π that is tangent to the surface $4z = 25 - 3x^2 - y^2$ at the point $(x_0, y_0, z_0) = (2, 1, 3)$.

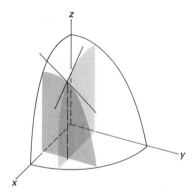

(You should check that this point does lie on the surface.) We assume that there *is* a tangent plane and that our only problem is to find its equation. In Chapter 17 we will convince ourselves that the plane that we get meets every criterion of tangency.

Since the tangent plane contains the point $(2, 1, 3)$, its point–slope equation is

$$z - 3 = a(x - 2) + b(y - 1),$$

The surface is the graph of the function

$$f(x,y) = \frac{25 - 3x^2 - y^2}{4}.$$

At the point of tangency the surface and plane must have the same slope in the x-direction. That is, we must have $D_1 f(2,1) = a$. But

$$D_1 f(x,y) = -\frac{3x}{2},$$

and

$$D_1 f(2,1) = -3.$$

Therefore, $a = -3$.

Similarly, the surface and tangent plane must have the same slope in the y-direction at the point of tangency, so that

$$b = D_2 f(2,1) = -\frac{2y}{4}\bigg|_{(2,1)} = -\frac{1}{2}.$$

The equation of the tangent plane is therefore

$$z - 3 = -3(x - 2) - \frac{1}{2}(y - 1)$$

or

$$2z = -6x - y + 19.$$

The same argument shows that the plane

$$z - z_0 = a(x - x_0) + b(y - y_0)$$

will be tangent to the surface

$$z = f(x,y)$$

at the point $(x_0, y_0, z_0) = (x_0, y_0, f(x_0, y_0))$ if

$$a = D_1 f(x_0, y_0),$$

and

$$b = D_2 f(x_0, y_0),$$

Thus the equation of the tangent plane to the graph of

$$z = f(x,y)$$

at the point over (x_0, y_0) is

$$z - z_0 = [D_1 f(x_0, y_0)](x - x_0) + [D_2 f(x_0, y_0)](y - y_0),$$

where $z_0 = f(x_0, y_0)$.

EXAMPLE 3. Find the equation of the plane tangent to the surface $z = xy^2$ at the point over $(x_0, y_0) = (-3, 2)$.

. . . .

Solution. We compute the remaining coefficients for the above form of the tangent plane equation:

$$z_0 = f(-3, 2) = (-3)(2)^2 = -12,$$

$$D_1 f(-3, 2) = y^2 \Big|_{(-3,2)} = 2^2 = 4,$$

$$D_2 f(-3, 2) = 2xy \Big|_{(-3,2)} = 2(-3)(2) = -12.$$

Substituting these values in the general form above gives

$$z + 12 = 4(x + 3) - 12(y - 2),$$

or

$$z = 4x - 12y + 24,$$

as the equation of the tangent plane.

PROBLEMS FOR SECTION 5

Sketch the following planes in the first octant.

1. $x + z + 2y = 4$ 2. $z - y = 0$

3. $z - y + x = 0$ 4. $z = x + y - 1$

5. $y = x + z + 1$

The following planes do not cross the first octant. Extend the coordinate axes by dotted lines, and then sketch each plane in the same manner in some other octant.

6. $z = -1$ 7. $x = -2$

8. $y = -1$ 9. $x + y + z = -1$

10. $x + y + z = 0$

11. The planes

$$z = ax + by + c,$$

$$z = Ax + By + C$$

are parallel if either they are the same plane or if they never intersect. In the second case, the two equations have no simultaneous solutions. Show by algebra that the planes are parallel if and only if

$$A = a$$

and

$$B = b.$$

Assume the condition given in Problem 11 in doing Problems 12 through 14 below.

12. Determine the equation of the plane through the point $(1, -2, 3)$ and parallel to the plane $x - 3y + 2z = 0$.

13. Determine the equation of the plane passing through $(6, 2, 1)$ parallel to the xy-plane.

14. Determine the equation of the plane parallel to the z-axis with x-intercept 2 and y-intercept -3.

Find the equation of the plane containing the following points:

15. $(2, 1, 3)$, $(-1, -2, 4)$ $(4, 2, 1)$.   16. $(1, 1, 1)$, $(-2, 2, 2)$, $(1, 1, 0)$.

17. $(4, 2, 1)$, $(-1, -2, 2)$, $(0, 4, 5)$.

18. Determine the equation of the plane having the intercepts 3, 5, -2 on the three coordinate axes.

In the following problems, determine the equation of the tangent plane to the surface at the point indicated.

19. $xy + yz + zx = 11, (1, 2, 3)$ 20. $xyz = x + y + z, (1/2, 1/3, -1)$

21. $z = \log \sqrt{x^2 + y^2}, (-3, 4, \log 5)$ 22. $z = e^x \sin y, (1, \pi/2, e)$

23. $z = xy, (2, -1, -2)$ 24. $z = 3x^2 + 2y^2 - 11, (2, 1, 3)$

25. Find the point on the plane $x + 2y + 3z = 4$ closest to the origin.

26. Find the point on the surface $xyz = 4$ closest to $(1, 0, 0)$.

6. WHAT ABOUT HIGHER DIMENSIONS?

By now it is clear that visualizing and representing equation graphs in three dimensions is much more complicated than in two dimensions, and the corresponding problems for equations in four or more variables are hopeless. It is natural to wonder whether the effort spent on the three-dimensional situation was worth it, and to ask how one can go about trying to understand the behavior of functions of three or more variables.

What we do is to proceed more or less by analogy with the two- and three-dimensional cases, keeping in mind both their similarities and their differences. For example, we consider the set of all quadruples of numbers such as $(1, 3, -4, 2)$ as forming a four-dimensional coordinate space, and we consider the graph of an equation in four variables as forming a three-dimensional "surface" in this 4-space. The graph of a function of three variables $f(x, y, z)$ is the graph of the equation

$$w = f(x, y, z).$$

We view the three-dimensional domain D of f as lying in the three-dimensional coordinate "plane" consisting of all points of the form $(x, y, z, 0)$. Then the graph of f is a three-dimensional "surface" spread out "over" the domain D. If (a, b, c) is in the domain of f, we picture it as the point

$$(a, b, c, 0) \quad \text{in 4 space}$$

and the graph point over it is

$$(a, b, c, f(a, b, c)).$$

In the same way we view the graph of a function of n variables as forming an n-dimensional "surface" in $(n + 1)$-dimensional coordinate space. We thus discuss higher-dimensional situations in geometric language suggested by our experience in two and three dimensions, and we still reason geometrically, but in a looser, less concrete way.

As our geometric grip loosens, and we reason more by geometric analogy than by direct geometric visualization, we have to bolster our weaker intuition by paying more attention to its algebraic and analytic counterparts. It is here that the theory of vector spaces comes to our aid. The algebraic study of vector spaces gives us very sharp and clear notions of planes and lines and perpendicularity in higher dimensions, and we can then use these flat objects to approximate curving configurations near points of tangency in almost exactly the same way that a tangent line approximates a curve and a tangent plane approximates a surface.

We took a first look at vectors in Chapter 14, but a whole course in linear algebra is needed before one can feel confident about the geometry of $(n + 1)$-dimensional space and the behavior of functions of n variables.

chapter 16

the chain rule

The chain rule for functions of several variables has an importance beyond anything suggested by the one-variable case. One could say that the chain rule takes on new dimensions. The present chapter explores some of this new territory in two dimensions, and the next chapter contains the analogous results in three dimensions. More advanced courses pursue the same ideas in their full generality.

The last two sections of the chapter take up higher-order partial derivatives.

1. THE CHAIN RULE

Since a partial derivative of a function of several variables is nothing but the ordinary derivative of a suitable restriction of that function, it follows that, by and large, the rules for computing partial derivatives are the same as for ordinary derivatives. Thus the sum, product, and quotient rules are all the same.

The chain rule acquires a new form, however, since now it concerns the composition of an outside function of *two or more variables* with *two or more inside functions*.

EXAMPLE 1. If $z = x^2 + y^2$ and if $x = 2t$ and $y = t^3$, then

$$z = (2t)^2 + (t^3)^2 = t^2(4 + t^4).$$

Here is the Leibniz form of the chain rule for a function of two variables. We call $f(x, y)$ *smooth* if it has continuous partial derivatives.

Chain rule. *If z is a smooth function of x and y, and if x and y are differentiable functions of t, then z is a differentiable function of t, and*

$$\frac{dz}{dt} = \frac{\partial z}{\partial x}\frac{dx}{dt} + \frac{\partial z}{\partial y}\frac{dy}{dt}.$$

Note that the formula is "dimensionally" correct in the same way the old formula was. The positions of x cancel symbolically, one being up and one down, as do the positions of y, so that symbolically the righthand side reduces to z up and t down, which agrees with the left.

Of course, z will be a function of t only if the domains and ranges overlap. We can picture this overlapping in the xy-plane by considering the two functions

$$x = g(t) \qquad \text{and} \qquad y = h(t)$$

as together defining a path

$$(x, y) = (g(t), h(t)).$$

Then this path must run, for a while at least, in the region D that is the domain of $z = f(x,y)$. The path could very well meander in and out of D, and in this case the domain of the composite function

$$z = f(x,y) = f(g(t), h(t))$$

is made up of two or more separate t-intervals, corresponding to the separate pieces of the path lying in D. Such diagrams lead us to interpret the composite-function derivative dz/dt as the derivative of $f(x,y)$ *along the path.*

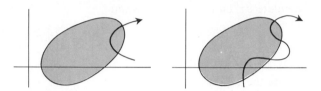

The chain rule will be proved in the next section, and its meaning will become clearer in terms of vectors in Section 4. Here we shall merely perform some computations with it.

EXAMPLE 2. We verify the correctness of the chain rule in the situation of Example 1. We had

$$z = x^2 + y^2, \qquad x = 2t, \qquad \text{and} \qquad y = t^3.$$

By the chain rule,

$$\frac{\partial z}{\partial t} = \frac{\partial z}{\partial x}\frac{dx}{dt} + \frac{\partial z}{\partial y}\frac{dy}{dt}$$

$$= 2x \cdot 2 + 2y \cdot 3t^2$$

$$= 2(2t) \cdot 2 + 2t^3 \cdot 3t^2 = 8t + 6t^5.$$

On the other hand,

$$z = (x^2 + y^2) = (2t)^2 + (t^3)^2 = 4t^2 + t^6,$$

from which the same answer,

$$dz/dt = 8t + 6t^5,$$

follows by direct computation.

EXAMPLE 3. What is the formula for $\partial z/\partial u$ if

$$z = f(x,y), \qquad x = g(u,v), \qquad y = h(u,v) ?$$

Solution. Here z has become a function of u and v, and we find $\partial z/\partial u$ by holding v constant and taking the ordinary derivative with respect to u. This reduces the present situation to the chain rule, and we have

$$\frac{\partial z}{\partial u} = \frac{\partial z}{\partial x}\frac{\partial x}{\partial u} + \frac{\partial z}{\partial y}\frac{\partial y}{\partial u}.$$

This formula is the chain rule for partial derivatives.

EXAMPLE 4. Calculate $\partial z/\partial u$ in two ways when

$$z = x^2 + y^2, \qquad x = uv^2, \qquad y = u^2v.$$

. . . .

Solution 1. Substitute the expressions for x and y:

$$z = x^2 + y^2 = (uv^2)^2 + (u^2v)^2$$

$$= u^2v^4 + u^4v^2.$$

Then compute $\partial z/\partial u$:

$$\frac{\partial z}{\partial u} = 2uv^4 + 4u^3v^2.$$

. . . .

Solution 2. Use the chain rule for partial derivatives:

$$\frac{\partial z}{\partial u} = \frac{\partial z}{\partial x}\frac{\partial x}{\partial u} + \frac{\partial z}{\partial y}\frac{\partial y}{\partial u}$$

$$= 2xv^2 + 2y\,2uv$$

$$= 2(uv^2)v^2 + 2(u^2v)2uv$$

$$= 2uv^4 + 4u^3v^2.$$

These Leibniz expressions treat z ambiguously. Two quite different functions were being differentiated when we wrote $\partial z/\partial x$ and $\partial z/\partial u$.

EXAMPLE 5. The following very common situation is a good example of how bad things can be. If z is a function of x and y, and if y is a function of x, then z is a function of x and

$$\frac{dz}{dx} = \frac{\partial z}{\partial x} + \frac{\partial z}{\partial y}\frac{dy}{dx}.$$

Correctly interpreted this is a valid and important special case of the chain rule. But the correct interpretation requires us to keep in mind that dz/dx and $\partial z/\partial x$ are derivatives of two entirely different functions! In simple situations we can put up with such loose talk, but we must always be ready to come to the rescue with more precise formulations of the chain rule involving function notation.

Good enough for most purposes is the intermediate form, part Leibniz notation and part function notation, as follows:

Chain rule. *If* $z = f(x,y)$ *with* f *smooth, and if* x *and* y *are differentiable functions of* t*, then* z *is a differentiable function of* t*, and*

$$\frac{dz}{dt} = D_1 f(x,y)\frac{dx}{dt} + D_2 f(x,y)\frac{dy}{dt}.$$

In this form the Leibniz notation is permitted for the derivatives with respect to the ultimate independent variable or variables, but the function notation is used for the partials with respect to the intermediate variables.

Consider again the important situation of Example 5.

EXAMPLE 6. If $z = f(x,y)$ and y is a function of x, then we have the special case $t = x$ in the above formula: z becomes a function of x alone, and the formula becomes

$$\frac{dz}{dx} = D_1 f(x,y) + D_2 f(x,y)\frac{dy}{dx}.$$

Note how this formulation avoids the ambiguity of the pure Leibniz formula.

Most precise, but also most ungainly, is the pure function form. Leaving out the differentiability hypotheses, it is:
   If $\phi(t) = f(g(t), h(t))$*, then*

$$\phi'(t) = D_1 f(g(t), h(t))g'(t) + D_2 f(g(t), h(t))h'(t).$$

PROBLEMS FOR SECTION 1

Find dz/dt in each of the following problems.
1. $z = x^2 - 2xy + y^2$, $x = (t + 1)^2$, $y = (t - 1)^2$.
2. $z = x \sin y$, $x = 1/t$, $y = \arctan t$.
3. $z = \log \sqrt{(x - y)/(x + y)}$, $x = \sec t$, $y = \tan t$.
4. $z = x^2 y^3$, $x = 2t^3$, $y = 3t^2$.
5. $z = \log(x^2 + y^2)$, $x = e^{-t}$, $y = e^t$.

6. The altitude of a right circular cone is 8 inches and is increasing at 0.5 in/min. The radius of the base is 10 inches and is decreasing at the rate of 0.2 in/min. How fast is the volume changing?

Calculate $\partial z/\partial v$ and $\partial z/\partial u$ in two ways in the following problems.

7. $z = x^2 + y^2$, $x = u + v$, $y = u - v$.

8. $z = x^2 + 3xy + y^2$, $x = \sin u + \cos v$, $y = \sin u - \cos v$.

9. $z = e^{xy}$, $x = u^2 + 2uv$, $y = 2uv + v^2$.

10. $z = \sin(4x + 5y)$, $x = u + v$, $y = u - v$.

11. Show that $z = f(y/x)$ is homogeneous of degree 0; that is, show that

$$x\frac{\partial z}{\partial x} + y\frac{\partial z}{\partial y} = 0.$$

12. Suppose that

$$z = f(u, v), \qquad u = g(x, y), \qquad \text{and} \qquad v = h(x, y),$$

where f is homogeneous of degree m and g and h are each homogeneous of degree n. Show that z is homogeneous of degree mn as a function of x and y. That is,

$$z = f(g(x, y), h(x, y))$$

is homogeneous of degree mn. (Recall from the last chapter that we call $z = f(x, y)$ *homogeneous of degree n* if

(*) $$x\frac{\partial z}{\partial x} + y\frac{\partial z}{\partial y} = nz.)$$

13. The condition (*) above is not the usual definition of homogeneity. Show that a function $z = f(x, y)$ will satisfy (*) if it has the property

(* *) $$f(tx, ty) = t^n f(x, y).$$

(Differentiate with respect to t.)

14. Show, conversely, that if a function satisfies (*), then it satisfies (**). You will need the following fact:

If $du/dt = n(u/t)$, then $u = ct^n$.

(See Problem 58 in Section 1 of Chapter 4.)

15. If x and y are independent variables, then their differentials dx and dy are new independent variables, wholly independent of x and y in their values, but used along with x and y. The definition of the differential of a *dependent* variable is analogous to that given earlier for a function of one independent variable (at the end of Chapter 4).

Definition. *If* $z = f(x,y)$, *then*

$$dz = D_1 f(x,y)dx + D_2 f(x,y)dy.$$

Show that this relationship between dx, dy, and dz remains true even if x and y are functions of other variables, say $x = g(u,v)$ and $y = h(u,v)$, so that the underlying independent variables are u and v throughout.

If u and v are both functions of x and y, show that:

16. $d(u + v) = dy + dv.$

17. $d(uv) = udv + vdu.$

18. $d\left(\dfrac{u}{v}\right) = \dfrac{v\,du - u\,dv}{v^2}.$

19. If $w = f(u)$, then $dw = f'(u)du.$

2. PROOF OF THE CHAIN RULE

The proof of this theorem will depend on the fact that if we hold one variable fixed in $f(x,y)$, then the mean-value theorem can be applied to the remaining function of the other variable. For example, if we set $g(x) = f(x,b)$, then the mean-value property

$$g(x + \Delta x) - g(x) = g'(X) \cdot \Delta x$$

becomes

$$f(x + \Delta x, b) - f(x, b) = D_1 f(X, b) \cdot \Delta x.$$

With this observation we can prove the theorem.

Theorem 1. *If the functions* $x = g(t)$ *and* $y = h(t)$ *are differentiable at* t_0, *and if* $z = f(x,y)$ *is smooth inside some small circle about the point*

$$(x_0, y_0) = (g(t_0), h(t_0)),$$

then the composite function

$$k(t) = f(g(t), h(t))$$

is differentiable at t_0, *and*

$$k'(t_0) = D_1 f(x_0, y_0) \cdot g'(t_0) + D_2 f(x_0, y_0) \cdot h'(t_0).$$

Proof. The proof is almost the same as the earlier proof for functions of one variable (Chapter 4, Theorem 5′). If we give t an increment Δt, then x and y acquire increments Δx and Δy, and $z = f(x,y)$ acquires the increment

$$\Delta z = f(x + \Delta x, y + \Delta y) - f(x,y).$$

(We have dropped the zero subscripts to simplify the notation.) What is new in this situation is that the intermediate variables x and y together form a *variable point* (x, y) with new value $(x + \Delta x, y + \Delta y)$. In order to bring in the partial derivatives of f, we make this point change in two steps, first changing x by Δx and *then* y by Δy. That is, we take one step horizontally off the path and then a step vertically back onto the path. Then Δz is the sum of its two partial changes:

$$\Delta z = [f(x + \Delta x, y + \Delta y) - f(x + \Delta x, y)] + [f(x + \Delta x, y) - f(x, y)].$$

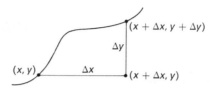

Each bracket involves a change in only one variable, and can be rewritten by the mean-value theorem for functions of that variable. Thus,

$$\Delta z = D_2 f(x + \Delta x, Y) \cdot \Delta y + D_1 f(X, y) \cdot \Delta x,$$

where Y lies between y and $(y + \Delta y)$, and X lies between x and $(x + \Delta x)$. Now divide by Δt,

$$\frac{\Delta z}{\Delta t} = D_2 f(x + \Delta x, Y) \frac{\Delta y}{\Delta t} + D_1 f(X, y) \frac{\Delta x}{\Delta t},$$

and let Δt approach 0. The limit on the right exists and has the value on the right below. Therefore, dz/dt exists and

$$\frac{dz}{dt} = D_2 f(x, y) \frac{dy}{dt} + D_1 f(x, y) \frac{dx}{dt},$$

which is the chain rule. Note that we have needed the continuity of the partial derivative $D_2 f$ in order to conclude that

$$D_2 f(x + \Delta x, Y) \to D_2 f(x, y)$$

as $\Delta t \to 0$. The continuity of D_1 was used, too.

PROBLEMS FOR SECTION 2

The Leibniz form for the chain rule for a function of three variables is:

If w is a smooth function of x, y, and z, and if x, y, and z are differentiable functions of t, then w is a differentiable function of t, and

$$\frac{dw}{dt} = \frac{\partial w}{\partial x} \frac{dx}{dt} + \frac{\partial w}{\partial y} \frac{dy}{dt} + \frac{\partial w}{\partial z} \frac{dz}{dt}.$$

1. State the pure function form of this rule.

2. Prove this three-variable chain rule, following the plan of the two-variable proof in the text. You will have to express Δw as the sum of *three* "partial" increments.

3. State the chain rule for a function of four variables.

4. State the chain rule for a function of n variables $x_1, x_2, \ldots, x_n$.

3. IMPLICIT FUNCTIONS

The chain rule for functions of two variables lets us discuss implicit functions in a general context that was unavailable in Chapter 4. The goal of the discussion is not further technique but conceptual consolidation. You should at least read the material and try to understand what is being said; we shall have to apply it later on.

The graph of an equation of the form

$$f(x,y) = k$$

is called a *level set* of the function f, as we saw in the last chapter. For example, the level sets of the function

$$f(x,y) = x^2 + y^2$$

are the circles about the origin:

$$x^2 + y^2 = k.$$

Generally, as in this example, a level set of a smooth function will be some sort of a smooth curve, although probably not a function graph. But it may be made up of two or more separate function graphs $y = \phi(x)$, just as the above level set $x^2 + y^2 = k$ consists of the graphs of the two functions

$$y = \phi_1(x) = \sqrt{k - x^2} \qquad \text{and} \qquad y = \phi_2(x) = -\sqrt{k - x^2}.$$

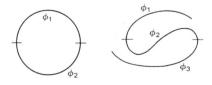

By hypothesis, $z = f(x,y)$ has the constant value k along the graph of any such function $y = \phi(x)$,

$$z = f(x, \phi(x)) = k,$$

so if f and ϕ are smooth functions then

(*)
$$\frac{dz}{dx} = D_1 f + D_2 f \frac{dy}{dx} = 0.$$

The middle term is just the chain-rule evaluation of dz/dx when $z = f(x, y)$ and y is a function of x. The result is zero because z is constant as a function of x. If $D_2 f$ is not zero, the above equation can then be solved for dy/dx:

$$\frac{dy}{dx} = \frac{-D_1 f(x, y)}{D_2 f(x, y)},$$

or

$$\phi'(x) = \frac{-D_1 f(x, \phi(x))}{D_2 f(x, \phi(x))}.$$

In the language of Chapter 4, a function $y = \phi(x)$, such that

$$f(x, \phi(x)) = k$$

is a *function solution* of the equation

$$f(x, y) = k,$$

and is defined *implicitly* by this equation. So we have obtained above a general formula for the derivative of an implicitly defined function.

Note that if $D_2 f = 0$ in (*), then also $D_1 f = 0$, and the point $(x, y) = (x, \phi(x))$ is a critical point of f. So if the graph of ϕ contains no critical point of f, then $D_2 f \neq 0$ along the graph and we can solve for $\phi'(x)$ as above.

Altogether we have shown that:

Theorem 2. *If $f(x, y)$ is a smooth function and if $y = \phi(x)$ is a differentiable function defined implicitly by the equation $f(x, y) = k$, and if there are no critical points of f on the graph of ϕ, then*

$$\phi'(x) = -\frac{D_1 f(x, \phi(x))}{D_2 f(x, \phi(x))}.$$

This whole analysis works in the same way when we consider a piece of the level set $f(x, y) = k$, along which x is a smooth function of y,

$$x = \psi(y).$$

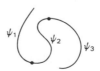

Such a piece is not a standard function graph, because its independent variable runs vertically, but we would still call ψ an implicitly defined function. This time we have

$$z = f(\psi(y), y) = k,$$

$$\frac{dz}{dy} = (D_1 f)\frac{dx}{dy} + D_2 f = 0,$$

$$\psi'(y) = \frac{dx}{dy} = -\frac{D_2 f(x, y)}{D_1 f(x, y)} = -\frac{D_2 f(\psi(y), y)}{D_1 f(\psi(y), y)},$$

again provided that $(x, y) = (\psi(y), y)$ is not a critical point of f.

So far, the condition that there be no critical points of f along the piece of the level set we are looking at has been a requirement tacked on at the last minute to permit dividing by $D_2 f$ or $D_1 f$ in the derivative formula. If (x_0, y_0) is a critical point on the level set $f = k$, then we don't get a derivative formula for an implicitly defined function through that point. This raises the possibility that near a critical point the level set may not even *be* a smooth function graph, and simple examples bear out these forebodings.

EXAMPLE 1. The function

$$f(x, y) = y^2 - x^2$$

has a critical point at the origin. Its level set through the origin, the graph of $y^2 - x^2 = 0$, is the pair of straight lines $y = x$ and $y = -x$, so in the neighborhood of the origin the level set is not a function graph at all.

EXAMPLE 2. The function

$$f(x, y) = y^3 - x^2$$

has a critical point at the origin. Its level set through the origin is the graph of the function $y = \phi(x) = x^{2/3}$, but $\phi'(0)$ does not exist. The level set has a cusp at the origin.

EXAMPLE 3. The level set $y^2 - x^3 = 0$ is the graph of $x = \psi(y) = y^{2/3}$. It is a nonstandard function graph, and is the image of the graph in Example 2 under the rotation of the plane over the 45° line $y = x$. Again, the origin is a bad point: the graph has a cusp there, ψ' fails to exist there, and

$$f(x,y) = y^2 - x^3$$

has a critical point there.

Such examples suggest the true significance of the "no critical point" condition: It is what guarantees that, in the short run, a level set necessarily lies along a smooth function graph, either standard or nonstandard. Near a noncritical point (x_0, y_0), the equation $f(x, y) = k$ does define y as a smooth function of x, or x as a smooth function of y, depending on whether

$$D_2 f(x_0, y_0) \neq 0 \quad \text{or} \quad D_1 f(x_0, y_0) \neq 0.$$

This theorem is called the *implicit function theorem*. Here it is again, stated somewhat more exactly.

Theorem 3. *Suppose that $f(x,y)$ is smooth, and that $D_2 f(x_0, y_0) \neq 0$, where (x_0, y_0) is a point on the level set $f(x, y) = k$. Then the level set runs along a smooth function graph as it goes through (x_0, y_0). That is, there is a rectangle about (x_0, y_0) within which the equation*

$$f(x, y) = k$$

determines y as a smooth function of x. Similarly, if $D_1(x_0, y_0) \neq 0$, then $f(x, y) = k$ defines x as a smooth function of y in a suitably small rectangle about (x_0, y_0).

Suppose, then, that the level set $f(x, y) = k$ contains no critical point of f. Then locally the level set is a smooth function graph, either standard or nonstandard.

That is, along a sufficiently short stretch of the level set through any given point P_0, either y is a smooth function of x,

$$y = \phi(x),$$

or x is a smooth function of y,

$$x = \psi(y).$$

Now this is exactly the local characterization of the locus of a smooth path, as we saw in Chapter 11. Doesn't this mean that the whole level set is (the locus of) a smooth path? Not quite, because of what the following example shows.

EXAMPLE 4. The level set

$$5(x^2 + y^2) - (x^2 + y^2)^2 = 4$$

is the pair of concentric circles

$$x^2 + y^2 = 1, \qquad x^2 + y^2 = 4.$$

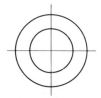

So what we *can* say is that such a smooth level set is made up of one or more smooth-path loci, and we can discuss the local properties of the level set in terms of a smooth path

$$(x, y) = (g(t), h(t))$$

that runs along it. The advantage of this "path" language is that it permits us to discuss a situation in which locally one variable is a function of the other without having to say which is which.

The proof of the implicit-function theorem is frequently omitted from a first course. In later years, a much more general form of the theorem is proved, by much more powerful methods than are available to us now. However, Theorem 3 has an elementary and very instructive proof. We shall give it below, but star it because it is long.

If $D_2 f(x_0, y_0)$ is not zero, we can suppose it is positive (otherwise, look at $-f$), and then by continuity we can choose a square about (x_0, y_0) on which $D_2 f$ has a positive lower bound m. Also, by continuity, we can suppose that $|D_1 f|$ has an upper bound M on the square. The crux of the matter is then the following theorem.

Theorem 4. *Suppose that $z = f(x,y)$ is a smooth function defined throughout a square S of side $2b$ about a point (x_0, y_0), and suppose that on S the partial derivatives satisfy the inequalities*

$$\frac{\partial z}{\partial y} > m, \qquad \left| \frac{\partial z}{\partial x} \right| < M,$$

where m and M are positive constants. We trim the square down to a rectangle R about (x_0, y_0) of width $2a$, where

$$a = \frac{bm}{M},$$

and we take k as the value $f(x_0, y_0)$ at the center of R. Then in R the level set

$$f(x,y) = k$$

is the graph of a smooth function $y = \phi(x)$, with domain the whole closed interval $I = [x_0 - a, x_0 + a]$ forming the base of R.

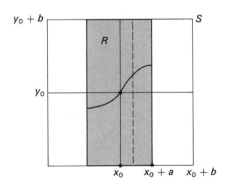

Proof. We shall prove that $f(x,y) > k$ along the top edge of the rectangle R. It will then follow, in exactly the same way, that $f(x,y) < k$ along the bottom edge of R. Therefore, along any vertical segment $x = x_1$ crossing R, $f(x_1, y)$ is an increasing function of y that starts with a value less than k and ends with a value greater than k. So $f(x_1, y)$ assumes the value k exactly once, by the intermediate-value theorem. That is, the level set $f(x,y) = k$ intersects each vertical segment of R exactly once. That is, in R, the level set $f = k$ is the graph of a function ϕ with domain

$$I = [x_0 - a, x_0 + a].$$

So the fact that in R the level set is a function graph follows from the fact that $f(x,y) > k$ along the top edge of the rectangle R. To prove this we go vertically from the center (x_0,y_0). Along this vertical segment $z = f(x_0,y)$ is an increasing function of y, and its total increase is at least mb since its derivative is everywhere larger than m. (This is by the mean-value theorem or its corollary, Theorem 3 of Chapter 5.) So, at the center of the top edge, we have

$$f(x_0,y_0 + b) > k + mb.$$

Along the top edge we have a function of x,

$$f(x,y_0 + b),$$

with derivative bounded by M. In particular, its derivative is bounded below by $-M$, so for the same reason it cannot decrease from its central value by more than Ma:

$$f(x,y_0 + b) \geq f(x_0,y_0 + b) - Ma.$$

Combining this with the preceding inequality, and remembering that

$$mb - Ma = 0,$$

we see that

$$f(x,y_0 + b) > k.$$

The proof that $f(x,y_0 - b) < k$ goes in exactly the same way and we omit it. So we have our implicitly defined function ϕ.

It remains to be shown that ϕ is smooth. To do this, we go back to the proof of the chain rule and copy out the formula

$$\Delta z = f(x + \Delta x, y + \Delta y) - f(x, y)$$
$$= \Delta x \, D_1 f(X,y) + \Delta y \, D_2 f(x + \Delta x, Y),$$

where X is some number between x and $(x + \Delta x)$, and Y is some number between y and $(y + \Delta y)$. Here the two points (x,y) and $(x + \Delta x, y + \Delta y)$ are on the level set $f = k$, so $\Delta z = k - k = 0$ and

$$\frac{\Delta y}{\Delta x} = -\frac{D_1 f(X,y)}{D_2 f(x + \Delta x, Y)}.$$

It then follows from the derivative bounds that

$$\left| \frac{\Delta y}{\Delta x} \right| < \frac{M}{m},$$

or

$$|\Delta y| < \frac{M}{m}|\Delta x|,$$

so ϕ is continuous. But now, as $\Delta x \to 0$, we have

$$X \to x, \qquad \Delta y \to 0, \qquad \text{and } Y \to y.$$

Since the partial derivatives $D_1 f$ and $D_2 f$ are continuous, we can take limits in the above expression for $\Delta y/\Delta x$ and conclude that $y = \phi(x)$ is differentiable and

$$\frac{dy}{dx} = -\frac{D_1 f(x,y)}{D_2 f(x,y)}.$$

In terms of ϕ, this is the formula

$$\phi'(x) = -\frac{D_1 f(x, \phi(x))}{D_2 f(x, \phi(x))}.$$

PROBLEMS FOR SECTION 3

Use Theorem 3 to determine for each of the following cubic equations $f(x,y) = 0$, whether or not the equation determines y as a function of x locally about each point (x_0, y_0) on the graph. Determine any exceptional points (x_0, y_0) near which the equation may *not* determine y as a function of x. (According to the theorem these are the simultaneous solutions of the two equations $f(x,y) = 0$ and $D_2 f(x,y) = 0$.)

1. $x^3 + y + y^3 = 0$.   2. $1 - xy + y^3 = 0$.

3. $1 + xy^2 + y^3 = 0$.   4. $x^3 + xy + y^3 = 0$.

5. $x^3 + xy^2 + y^3 = 0$.

6. Study the local solvability of the equation

$$1 - xy + x^2 \log y = 0$$

for y in terms of x.

7. In the proof of the implicit function theorem, we used the fact that if $f'(t) > m$ on an interval $[c, d]$, then

$$f(d) - f(c) > m(d - c).$$

Give a careful proof of this fact.

8. In the notation of the theorem, prove that

$$f(x, y_0 - b) < k$$

for every x in the interval $[-a, a]$. (Imitate the proof given for the inequality $f(x, y_0 + b) > k$, with suitable modifications.)

9. What inequality law (or laws) is needed for the conclusion $|\Delta y/\Delta x| < M/m$ in the proof of the differentiability of the implicitly defined function ϕ?

10. Give all the details of the final limit evaluation that yields the formula for dy/dx.

4. THE GRADIENT OF A FUNCTION AND
THE VECTOR FORM OF THE CHAIN RULE

The two partial derivatives of f can be viewed as the components of a vector. This vector is called the *gradient* of f and is designated grad f. Thus,

$$\text{grad } f = \left(\frac{\partial z}{\partial x}, \frac{\partial z}{\partial y} \right) = (D_1 f, D_2 f),$$

and

$$[\text{grad } f](a, b) = \left(\frac{\partial z}{\partial x}, \frac{\partial z}{\partial y} \right)\bigg|_{(x,y)=(a,b)} = (D_1 f(a, b), D_2 f(a, b)).$$

Note that a critical point of f is just a point where the vector grad f is zero.

EXAMPLE 1. If $z = f(x, y) = (x^2 + 2y)/4$, then

$$(\text{grad } f)(x, y) = (\partial z/\partial x, \partial z/\partial y) = (x/2, 1/2).$$

Thus,

$$(\text{grad } f)(-3, 0) = (x/2, 1/2)_{(-3,0)} = (-3/2, 1/2).$$

When we view (x, y) in the coordinate plane as a vector

$$X = (x, y),$$

then a function of two variables $z = f(x, y)$ can be viewed as a function of one vector variable X, and so we shall often write

$$f(x, y) = f(X).$$

For example, grad $f(a, b) = \text{grad } f(A)$.

We often view grad $f(A)$ as a vector located at the point A. Then, for each point X, we have associated the located vector grad $f(X)$, and grad f is a *field of vectors*, or a *vector field*. We can sketch the vector field grad f by drawing some of its arrows.

EXAMPLE 2. If $f(x,y) = (x^2 + y^2)/4$, then

$$(\text{grad } f)(x,y) = \left(\frac{\partial}{\partial x}\left(\frac{x^2 + y^2}{4} \right), \frac{\partial}{\partial y}\left(\frac{x^2 + y^2}{4} \right) \right) = \left(\frac{x}{2}, \frac{y}{2} \right) = \frac{1}{2}(x,y).$$

Thus,

$$\text{grad } f(1,2) = \frac{1}{2}(1,2) = \left(\frac{1}{2}, 1 \right),$$

$$\text{grad } f(0,-1) = \frac{1}{2}(0,-1) = \left(0, -\frac{1}{2} \right), \qquad \text{etc.}$$

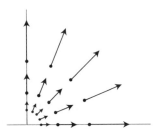

EXAMPLE 3. In Example 1 we had

$$(\text{grad } f)(x,y) = \left(\frac{x}{2}, \frac{1}{2} \right),$$

and we can directly draw some of these vectors as arrows based at the corresponding points (x,y).

The chain rule

$$\frac{dz}{dt} = \frac{\partial z}{\partial x}\frac{dx}{dt} + \frac{\partial z}{\partial y}\frac{dy}{dt}$$

$$= (D_1 f)\frac{dx}{dt} + (D_2 f)\frac{dy}{dt}$$

can now be given the following very important reformulation:

$$\frac{dz}{dt} = \text{grad } f \cdot \frac{dX}{dt}$$

$$= \text{grad } f \cdot \text{path tangent vector.}$$

This dot-product formula for the derivative of a function along a path has many consequences.

To begin with, consider the curve in the plane defined by

$$f(x,y) = k,$$

where k is a constant. In Chapter 15 we called this a *level curve* of f, since it is defined geometrically by the intersection of the graph of $z = f(x,y)$ with the horizontal plane $z = k$. We saw in the last section that such a level curve is necessarily a smooth path, provided grad f is never zero. More explicitly, if (x_0, y_0) is a point of the level set $f(x, y) = k$ at which grad f is not zero, then there is a small rectangle R about (x_0, y_0) whose intersection with the level set is the locus of a smooth path.

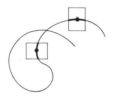

Suppose, then, that $X = G(t)$ is a smooth path running along the level curve $f(x,y) = k$. Then $z = f(x,y)$ has the constant value k along the path, and $dz/dt = 0$. Therefore

$$0 = \frac{dz}{dt} = \text{grad } f \cdot \frac{dX}{dt},$$

so the vector grad f is perpendicular to the path tangent vector. Since the path traces the level curve, its tangent vector is tangent to the curve. We have thus proved the following theorem.

Theorem 5. *If the function $f(x,y)$ is smooth and $X_0 = (x_0, y_0)$ is not a critical point of f, then* grad $f(X_0)$ *is perpendicular to the level curve of f through X_0.*

A vector perpendicular to a curve is said to be *normal* to the curve.

EXAMPLE 4. According to the theorem, a vector normal to the ellipse

$$\frac{x^2}{6} + \frac{4y^2}{6} = \frac{25}{6}$$

at the point $(x_0, y_0) = (3, 2)$ is given by

$$\text{grad}\left(\frac{x^2}{6} + \frac{4y^2}{6}\right)\Bigg|_{(3,2)} = \left(\frac{2}{6}x, \frac{8}{6}y\right)\Bigg|_{(3,2)} = \left(1, \frac{8}{3}\right).$$

In the figure below, we draw this normal vector located at X_0.

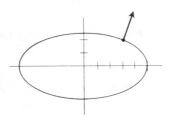

The theorem gives us an easy way to write down the equation of the tangent line to a level curve at a point X_0 where grad $f \neq 0$. Since it is perpendicular to grad $f(X_0)$, its equation is

$$\text{grad } f(X_0) \cdot (X - X_0) = 0,$$

as we saw in Section 4 of Chapter 14.

If the gradient of f is zero at X_0, this equation will not give a line because both coefficients will be zero. In fact, we know from examples in the last section that the level curve $f(x, y) = k$ may not *have* a tangent line at a critical point.

EXAMPLE 5. Find the line tangent to the curve

$$x^3 + xy + y^3 = 5$$

at the point $X_0 = (2, -1)$.
. . . .

Solution. First verify that $(2, -1)$ is on the graph:

$$2^3 + 2(-1) + (-1)^3 = 8 - 2 - 1 = 5.$$

Then compute grad $f(2, -1)$, where $f(x, y) = x^3 + xy + y^3$:

$$\text{grad } f(2, -1) = (3x^2 + y, x + 3y^2)_{(2, -1)} = (11, 5).$$

Substituting these values in the above formula for the tangent line gives

$$\text{grad } f(X_0) \cdot (X - X_0) = (11, 5) \cdot (x - 2, y + 1)$$

$$= 11(x - 2) + 5(y + 1) = 0$$

or

$$11x + 5y - 17 = 0.$$

This is the equation of the line tangent to the curve $x^3 + xy + y^5 = 5$ at the point $(2, -1)$.

EXAMPLE 6. Show that the general tangent-line formula above agrees with our earlier calculations for the line tangent to the graph of a function $y = g(x)$.

. . . .

Solution. This graph is the level curve

$$g(x) - y = 0,$$

and the gradient of this special function $f(x, y) = g(x) - y$ is

$$\text{grad } f(x, y) = (g'(x), -1).$$

The dot-product formula

$$\text{grad } f(X_0) \cdot (X - X_0) = 0$$

reduces in this case to

$$g'(x_0)(x - x_0) - 1(y - y_0) = 0,$$

or

$$y - y_0 = g'(x_0)(x - x_0),$$

which is the original tangent-line formula.

There is a converse to Theorem 5, as follows.

Theorem 6. *Suppose the path* $X = (g(t), h(t))$ *and the function* $z = f(x, y)$ *are so related that at every point of the path the path direction is perpendicular to* grad f. *Then the path runs along a level curve of* f.

Proof. The derivative of $z = f(x, y)$, along the path

$$\frac{dz}{dt} = \text{grad } f \cdot \frac{dX}{dt},$$

is everywhere zero, by hypothesis, so z, as a function of t, must therefore be a constant k. That is,

$$z = f(x, y) = k$$

along the path $(x, y) = (g(t), h(t))$, which is another way of saying that the path runs along the level curve $f(x, y) = k$.

PROBLEMS FOR SECTION 4

For each of the following determine grad f at the given point.

1. $f(x,y) = x^2 + y^2$; $(3,4)$
2. $f(x,y) = \arctan \dfrac{x}{y}$; $(1,2)$
3. $f(x,y) = \cos(xy)$; $(1, \pi/6)$
4. $f(x,y) = e^x \tan y$; $(0, \pi/4)$
5. $f(x,y) = x^3 + 2x^2y - 3xy + 4xy^2 - y^3$; $(1,1)$
6. $f(x,y) = x^{2y}$; $(2,1)$
7. $f(x,y) = 2x - y$; (a,b)

In the following problems the functions are divided by a constant, just to make the vectors short enough so they won't clutter up the page. You can use a different constant if you wish. Sketch the gradient field of $f(x,y)$ when:

8. $f(x,y) = \dfrac{x+y}{4}$.

9. $f(x,y) = \dfrac{(y-x^2)}{4}$.

10. $f(x,y) = -\dfrac{x}{4y}$.

Find a vector, other than the zero vector, that is normal to each curve at the given point.

11. $\sqrt{x^2 - y^2} = 1$; $(\sqrt{2}, 1)$.
12. $xe^{xy} = 2e$; $(2, 1/2)$.
13. $\arcsin \sqrt{x^2 + y^2} = \pi/6$; $(1/4, \sqrt{3}/4)$.
14. $\arctan xe^y = \pi/4$; $(1/e, 1)$.

Using gradients, find the tangent line to each curve below at the given point.

15. $xy = 2$; $(1,2)$.
16. $xy^2 = 1$; $(1/4, 2)$.
17. $x^2 + y^2 = 25$; $(3,4)$.
18. $x^2/9 + y^2/4 = 1$; $(2, (2/3)\sqrt{5})$.
19. $x^2 - 2xy - 3y^2 = 5$; $(2, -1)$.
20. $x^2 + xy + y^2 - 2x + y = 0$; $(2, -3)$.
21. $x^2 - 3xy^2 + 2y^3 = 16$; $(0,2)$.
22. $xe^{xy} = 2$; $(2,0)$.
23. $(x+y)\arctan xy = (5/8)\pi$; $(2, 1/2)$.
24. $xy \log x = 2e$; $(e,2)$.
25. Show that the level curves of $f(x,y) = x^2 + y^2$ are everywhere perpendicular to the level curves of $g(x,y) = y/x$. Sketch a few curves from each family.

26. Show that the level curves of $f(x,y) = x^2 - y^2$ are everywhere perpendicular to the level curves of $g(x,y) = xy$. Sketch a few curves from each family.

27. Let $f(x,y) = k$ be a level curve not containing the origin, and let (x_0, y_0) be a point on this level set that is closest to the origin. Prove that

$$X_0 = (x_0, y_0) \quad \text{and} \quad \text{grad } f(X_0)$$

are parallel vectors.

28. Let f be a smooth function, and let the domain D of $f(x,y)$ be in one piece, in the sense that a smooth path can be drawn in D between any two points of D. Show that f is a constant function if and only if grad f is identically 0. (This depends on, and generalizes, Theorem 7 in Chapter 3 (Section 10).)

29. Show that

$$\text{grad}(f + g)(X) = \text{grad } f(X) + \text{grad } g(X)$$

$$\text{grad } cf(X) = c \text{ grad } f(X).$$

30. Let D be a domain in one piece, as in Problem 28, and suppose f and g are two functions with domain D such that

$$\text{grad } f = \text{grad } g$$

on D. Use Problems 28 and 29 to show that there is a constant C such that

$$f(x,y) = g(x,y) + C$$

for all (x,y) in D.

31. Find a function f such that

$$\text{grad } f(X) \equiv X.$$

What is the most general such function?

32. Show that

$$\text{grad } fg(X) = f(X) \text{ grad } g(X) + g(X) \text{ grad } f(X).$$

5. DIRECTIONAL DERIVATIVES

We are now in a position to compute the rate at which $f(x,y)$ changes as the point (x,y) moves through (x_0, y_0) in the direction of a unit vector U. This rate of change will be called *the derivative of f in the direction U at the point* (x_0, y_0), and will be designated

$$D_U f(x_0, y_0).$$

If we go a distance t from the point X_0 in the direction U, we end up at the point $X = X_0 + tU$. This is true because U is a unit vector: the distance from X_0 to X is

$$|X - X_0| = |(X_0 + tU) - X_0| = |tU| = |t||U|,$$

and this is t only because $|U| = 1$ and t is positive. Anyway, the change in the value of f per unit distance in that direction, or the average rate of change of f in that direction, is

$$\frac{f(X_0 + tU) - f(X_0)}{t}.$$

As usual, the true, or "instantaneous," rate of change is defined as the limit of the average rate of change, so:

Definition

$$D_U f(X_0) = \lim_{t \to 0} \frac{f(X_0 + tU) - f(X_0)}{t}.$$

In order to evaluate this *directional derivative*, note that it is the ordinary derivative of a certain function of t: if

$$\phi(t) = f(X_0 + tU) = f(x_0 + tu, y_0 + tv),$$

then

$$D_U f(X_0) = \lim_{t \to 0} \frac{\phi(t) - \phi(0)}{t} = \phi'(0).$$

But ϕ is a composite function. It is the restriction of $f(x, y) = f(X)$ to the straight line path $X = X_0 + tU$. So the derivative $\phi'(t)$ is calculated by the chain rule, and has the value

$$\phi'(t) = \operatorname{grad} f(X_0 + tU) \cdot \frac{dX}{dt}$$

$$= \operatorname{grad} f(X_0 + tU) \cdot U.$$

We only want $\phi'(0)$. Thus:

Theorem 7. *The directional derivative $D_U f(X_0)$ is the value at X_0 of the derivative of f along the straight-line path through X_0 in the direction U. Its chain-rule evaluation is*

$$D_U f(X_0) = \operatorname{grad} f(X_0) \cdot U.$$

EXAMPLE 1. Find the derivative of $z = x^2 y$ at the point $X_0 = (2, -3)$ in the direction of the vector $A = (3, 4)$.

Solution. Since $A = (3, 4)$ is not a unit vector, we first have to find the unit vector having the direction of A. This is

$$U = \frac{A}{|A|} = \frac{(3, 4)}{\sqrt{3^2 + 4^2}} = \left(\frac{3}{5}, \frac{4}{5}\right).$$

Next, we compute

$$(\text{grad } f)(2, -3) = \left(\frac{\partial z}{\partial x}, \frac{\partial z}{\partial y}\right)_{(2, -3)}$$

$$= (2xy, x^2)_{(2, -3)} = (-12, 4).$$

Then,

$$D_U f(X_0) = \text{grad } f(X_0) \cdot U$$

$$= (-12, 4) \cdot \left(\frac{3}{5}, \frac{4}{5}\right)$$

$$= -12 \cdot \frac{3}{5} + 4 \cdot \frac{4}{5} = \frac{-36 + 16}{5} = -4.$$

If $U = (1, 0)$, then $D_U f$ is simply the first partial derivative of f:

$$D_{(1,0)} f = D_1 f = \frac{\partial z}{\partial x}$$

(where $z = f(x, y)$). Similarly $D_{(0,1)} f$ is the second partial derivative $D_2 f = \partial z / \partial y$. These facts can be read off from the chain-rule formula for $D_U f$ by substituting the special values for u and v, or they can be seen directly from the original limit definition of $D_U f$, by comparing it with the limit definitions of $\partial z / \partial x$ and $\partial z / \partial y$. The symbol D has been used somewhat ambiguously in the contexts D_1 and D_U, but no harm results.

Now recall from Section 4 of Chapter 14 that if U is a unit vector, then the dot product $A \cdot U$ can be interpreted as the component of the vector A in the direction U. Also, $A \cdot U = |A| \cos \phi$, where ϕ is the angle between the direction of A and U. The formula

$$D_U f(X) = \text{grad } f(X) \cdot U$$

thus shows that:

The directional derivative $D_U f(X)$ is the component of the vector grad $f(X)$ *in the direction U, and*

$$D_U f(X) = |\text{grad } f(X)| \cdot \cos \phi,$$

where ϕ is the angle between U and the vector grad $f(X)$.

If X is held fixed, but the direction U of the derivative is allowed to vary, then $\cos \phi$ varies between -1 and $+1$ in this formula. In particular, $D_U f(X)$ has its maximum value $|(\text{grad } f)(X)|$, when U is in the direction of $(\text{grad } f)(X)$, for then $\cos \phi$ has its maximum value 1. Also

$$D_U f(X) = 0$$

whenever the direction U is perpendicular to $\text{grad } f(X)$. This can be thought of as the "local" expression of the fact that f is constant along any path that runs perpendicular to the vector field $\text{grad } f$. (See Theorem 3.)

EXAMPLE 2. If $f(x,y) = xy^2$, find the maximum value of the directional derivatives of f at $X_0 = (1,1)$ and the direction U in which f has this maximum rate of change.

. . . .

Solution. First,

$$\text{grad } f(1,1) = (y^2, 2xy)_{(1,1)} = (1,2).$$

Then, as we saw above, $D_U f(1,1)$ will have the maximum value

$$|\text{grad } f(1,1)| = |(1,2)| = \sqrt{5},$$

when U is the direction of $\text{grad } f(1,1)$,

$$U = \frac{\text{grad } f(1,1)}{|\text{grad } f(1,1)|} = \frac{(1,2)}{\sqrt{5}} = \left(\frac{1}{\sqrt{5}}, \frac{2}{\sqrt{5}}\right).$$

We can interpret the relation between $\text{grad } f(X)$ and the directions in which the directional derivative $D_U f(X)$ has its maximum, zero, and minimum values in terms of our experiences on a hilly countryside.

Two straight highways intersecting at right angles can be used as xy-axis for the neighboring countryside, and then the terrain itself is the graph of a certain function $z = f(x,y)$. If we start at a point on a hillside and walk levelly along the side of the hill, we are walking along a level curve of f. (Strictly speaking we are walking *over* a level curve. The level curve itself is normally drawn in the xy-base plane.) If we now turn and start walking upward at right angles to the level curve we are necessarily moving in the direction of *greatest steepness* of the hill, for our new direction, specified as a direction in the base plane, is the direction of the *gradient of f*, and we have just seen that this is the direction in which $z = f(x,y)$ has the maximum rate of change. A stream of water flows down the hillside along the shortest path to the bottom, and this means that it is always running in the direction of greatest steepness downward, which is the direction of $-\text{grad } f$.

The countryside $z = f(x,y)$ can be pictured in the xy-base plane by a series of level curves. These are the contour lines of a geodetic map. We get from one contour line to the next most quickly along the perpendicular direction. This is pretty clear geometrically, and is a geometric interpretation of the fact that the maximum rate of change of $z = f(x,y)$ is in the direction of the gradient of f. As we move perpendicularly to the contour lines, the number of lines we cross per unit distance is a rough measure of the magnitude of grad f. When the contour lines bunch together they indicate a steep hillside and a large rate of change of $z = f(x,y)$. With a little experience interpreting geodetic contour lines, one gets a direct three-dimensional image from looking at the two-dimensional map.

PROBLEMS FOR SECTION 5

1. Establish the formula

$$\frac{\text{grad } f(X_0) \cdot A}{|A|}$$

for the derivative of f in the direction of the vector A at the point X_0.

For each of the functions below, find the derivative at the given point and in the direction of the given vector A. (Either use the formula of Problem 1, or work directly from Theorem 4 as in Example 1 in the text.)

2. $f(x,y) = x^2 y$; $(1,2)$; $A = (3/5, 4/5)$.

3. $f(x,y) = x^2 - y^2$; $(1/2, -1)$; $A = (\sqrt{3}/2, 1/2)$.

4. $f(x,y) = xe^{xy}$; $(2,1)$; $A = (12/13, 5/13)$.

5. $f(x,y) = \log[(x+y)/(x-y)]$; $(1/2, 2)$; $A = (1/4, 3/4)$.

6. $f(x,y) = \cos x \sin y$; $(\pi/3, \pi/3)$; $A = (1/2, 1/3)$.

7. $f(x,y) = e^{x \tan y}$; $(1, \pi/4)$; $A = (1/e, 2/e)$.

8. $f(x,y) = \sqrt{x^2 + y^2}$; $(3,4)$; $A = (1/4, 1/3)$.

9. $f(x,y) = \arctan x/y$; $(1/2, 1/3)$; $A = (1/5, 1/12)$.

10. $f(x,y) = x^y$; $(e,0)$; $A = (0,1)$.

11. $f(x,y) = x^{xy}$; $(1,2)$; $A = (2,0)$.

Find the maximum value of the derivative of the given function at the point indicated, and the direction of the maximum rate of change.

12. $g(x,y) = xy^2 + yx^2$; $(1/2, 1)$.

13. $g(x,y) = x^2 y^2 - 2xy^3 + x$; $(1/2, 2)$.

14. $g(x,y) = \tan x \sec y$; $(\pi/4, \pi/3)$.

15. $g(x,y) = \log \dfrac{x^2 - y^2}{x^2 + y^2}$; $(1/2, 1/4)$.

16. $g(x,y) = ye^{x/y}$; $(0,2)$.

17. $g(x,y) = \dfrac{x-y}{x+y}$; $(1/4, 3/4)$.

18. $g(x,y) = \sqrt{xy}$; $(2,8)$.

19. $g(x,y) = \arctan x/y$; $(1, 1/2)$.

20. Write out the definition of the first partial derivative $D_1 f$ as the limit of a difference quotient, and show that it agrees with the definition of the directional derivative $D_U f$ when $U = E_1 = (1,0)$.

21. Suppose that $z = f(x,y)$ and $(x,y) = (g(t), h(t))$. In Section 1 we interpreted the chain-rule derivative dz/dt as the derivative of f along the path $X = G(t)$. Show now that:

    The derivative of f along the path $X = G(t)$ is equal to the directional derivative of f in the direction of the path, multiplied by the magnitude $|dX/dt|$ of the path tangent vector.

22. Given $f(x,y) = x^3 y$, find the directions U for which

$$D_U f(2,1) = 10.$$

23. If we know the derivative of f at X_0 in two independent directions, then we can solve for grad $f(X_0)$. Suppose that $D_U f(X_0)$ has the value 6 in the direction of the vector $(3,4)$ and the value $-\sqrt{5}$ in the direction of the vector $(2, -4)$. Find grad $f(X_0)$.

24. Find grad $f(X_0)$ if $D_U f(X_0) = 1$ in the direction of $(1, -\sqrt{3})$ and $D_U f(X_0) = -1$ in the direction of $(1, \sqrt{3})$.

Write down the derivative of f at $A = (a,b)$ in the direction of the unit vector $U = (u,v)$, for each of the following functions.

25. $f(x,y) = x + y$;   26. $f(x,y) = (x^2 + y^2)/2$.

27. $f(x,y) = xy$.

28. A smooth path of length L runs from the point X_0 to the point X_1. A function f, defined on a region containing the path, has both partial derivatives bounded by a constant K. Show that

$$|f(X_1) - f(X_0)| \le 2KL.$$

6. HIGHER-ORDER PARTIAL DERIVATIVES

Another new situation that arises in the study of partial derivatives is the occurrence of *mixed* higher-order derivatives, like $\partial^2 z/(\partial x\,\partial y)$. The notation means what you would expect:

$$\frac{\partial^2 z}{\partial y \, \partial x} = \frac{\partial}{\partial y}\left(\frac{\partial z}{\partial x}\right),$$

$$\frac{\partial^2 z}{\partial x \, \partial y} = \frac{\partial}{\partial x}\left(\frac{\partial z}{\partial y}\right).$$

For example, if

$$z = x^2 y^3 + x^4 y,$$

then

$$\frac{\partial z}{\partial x} = 2xy^3 + 4x^3 y,$$

$$\frac{\partial^2 z}{\partial y \, \partial x} = \frac{\partial}{\partial y}\left(\frac{\partial z}{\partial x}\right) = 6xy^2 + 4x^3,$$

$$\frac{\partial z}{\partial y} = 3x^2 y^2 + x^4,$$

$$\frac{\partial^2 z}{\partial x \, \partial y} = \frac{\partial}{\partial x}\left(\frac{\partial z}{\partial y}\right) = 6xy^2 + 4x^3.$$

You will note in the above example that

$$\frac{\partial^2 z}{\partial x \, \partial y} = \frac{\partial^2 z}{\partial y \, \partial x}.$$

Such equality is not accidental, for this "commutative" law is a theorem. In proving this theorem, we shall have to evaluate the mixed partial derivatives at explicit points, which is most easily done in function notation. We set

$$D_{21} f = D_1(D_2 f),$$

$$D_{12} f = D_2(D_1 f).$$

Note the reversal in the order of the indices in these definitions. The idea is to think of D_{21} as $(D_2)_1$, with the index nearest D representing the derivative that is first computed. This is obviously a possible source of confusion, but the theorem saves the day, since it shows that the order doesn't matter.

Theorem 8. *If $f(x,y)$ has continuous partial derivatives of the first and second order, then*

$$D_{12} f(x,y) = D_{21} f(x,y).$$

Proof. We shall show that both mixed partials are approximately equal to the "second-order difference quotient"

(*) $$\frac{f(x + h, y + k) - f(x + h, y) - f(x, y + k) + f(x, y)}{hk}.$$

Consider first the function

$$\phi(x) = f(x, y + k) - f(x, y),$$

where y and $(y + k)$ are both held fixed. The numerator in the above expression (*) is then $\phi(x + h) - \phi(x)$, and hence can be rewritten

$$\phi(x + h) - \phi(x) = h\phi'(X) = h[D_1 f(X, y + k) - D_1 f(X, y)]$$

by the mean-value theorem. Here, X is some number lying between x and $(x + h)$. Since $D_1 f(X, y)$ is differentiable as a function of y, we can apply the mean-value theorem once more:

$$h[D_1 f(X, y + k) - D_1 f(X, y)] = hk\, D_{12} f(X, Y),$$

where Y lies between y and $y + k$. Thus the double difference quotient (*) has the values $D_{12} f(X, Y)$.

We now start over again with the function

$$\psi(y) = f(x + h, y) - f(x, y),$$

where x and $x + h$ are both held fixed. The numerator in (*) is then $\psi(y + k) - \psi(y)$, and, proceeding exactly as before, through two applications of the mean-value theorem, we find this time that the double difference quotient (*) has the value $D_{21} f(X', Y')$, where Y' lies between y and $y + k$ and X' lies between x and $x + h$.

Altogether we have evaluated (*) in two different ways. In particular, we have proved that there are numbers X and X' both lying between x and $x + h$, and numbers Y and Y' both lying between y and $y + k$, such that

$$D_{12} f(X, Y) = D_{21} f(X', Y').$$

Now, let h and k both approach zero. Then X and X' both approach x, and Y and Y' both approach y, and since the partial derivatives $D_{12} f$ and $D_{21} f$ are continuous, the above equation becomes

$$D_{12} f(x, y) = D_{21} f(x, y)$$

in the limit. This proves the theorem.

PROBLEMS FOR SECTION 6

Find *all* of the second partial derivatives of the following functions.

1. $z = 3x^2 - 2xy + y$ 2. $z = x \cos y - y \cos x$

3. $z = e^{x^2 y}$ 4. $z = \sin(2x^2 + 3y)$

5. $z = x^2 + 3xy + y^2$ 6. $z = y \log x$

7. $z = xy$ 8. $z = \sin 3x \cos 4y$

9. $z = \dfrac{x}{y^2} - \dfrac{y}{x^2}$ 10. $z = \arctan y/x$

11. Let f be a function of one variable and set

$$z = f(x - ct).$$

Show that

$$\frac{\partial^2 z}{\partial t^2} = c^2 \frac{\partial^2 z}{\partial x^2}.$$

12. Show, more generally, that

$$z = f(x - ct) + g(x + ct)$$

satisfies the above partial differential equation.

13. If $z = \log(x^2 + y^2)$, show that

$$\frac{\partial^2 z}{\partial x^2} + \frac{\partial^2 z}{\partial y^2} = 0.$$

14. If $z = \arctan \dfrac{y}{x}$, show that

$$\frac{\partial^2 z}{\partial x^2} + \frac{\partial^2 z}{\partial y^2} = 0.$$

15. If $z = \dfrac{e^{-x^2/4y}}{\sqrt{y}}$, show that

$$\frac{\partial z}{\partial y} = \frac{\partial^2 z}{\partial x^2}.$$

7. THE CLASSIFICATION OF CRITICAL POINTS

In Chapter 5 we saw that if x_0 is a critical point of f at which the second derivative $f''(x_0)$ exists and is nonzero, then f has a relative maximum or minimum at x_0, depending on the sign of $f''(x_0)$. On the other hand, if $f''(x_0) = 0$, then nothing can be concluded.

A similar result holds for a function $z = f(x, y)$ of two variables, the requirement that $f''(x_0) \neq 0$ being replaced by $D(x_0, y_0) \neq 0$, where the "discriminant" function $D(x, y)$ is defined by

$$D(x, y) = \left(\frac{\partial^2 z}{\partial x \, \partial y}\right)^2 - \frac{\partial^2 z}{\partial x^2} \frac{\partial^2 z}{\partial y^2}.$$

However, a new possibility occurs now, namely, that f may have a maximum at (x_0,y_0) along one line through this point, and a minimum at (x_0,y_0) along a different line. The resulting configuration of the graph of f around (x_0,y_0) is called a *saddle point*, a term that the figure below shows to be appropriate.

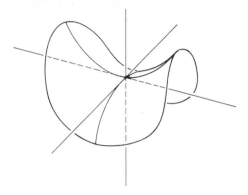

In this situation, the level set through $X_0 = (x_0,y_0)$ is a pair of intersecting curves, dividing the neighborhood of X_0 into four quadrant-like pieces on which $f(x,y) - f(x_0,y_0)$ is alternately positive and negative.

We wish, then, to analyze the possible behavior of $f(x,y)$ around a critical point (x_0,y_0), at which $D(x_0,y_0) \neq 0$, and we shall start with the case $D(x_0,y_0) < 0$. Since $D(x,y)$ is continuous, we can restrict our attention to a small circular disk R about X_0 on which $D(x,y)$ is everywhere negative. Inspection of the formula for $D(x,y)$ shows that then $\partial^2 z/\partial x^2$ and $\partial^2 z/\partial y^2$ can never be zero and must have the same sign throughout R. We shall suppose that both are positive.

So our total hypotheses are that $D(x,y)$ is negative throughout R, $\partial^2 z/\partial x^2$, and $\partial^2 z/\partial y^2$ are both positive throughout R, while the center point (x_0,y_0) in R is a critical point of f. Under these assumptions we shall show that $f(x_0,y_0)$ is the minimum value of f on R. That is, we shall show that if (x_1,y_1) is any other point of R, then $f(x_0,y_0) < f(x_1,y_1)$.

We first dispose of the case where $x_0 = x_1$, that is, where the two points

$$X_0 = (x_0,y_0) \qquad \text{and} \qquad X_1 = (x_1,y_1)$$

are on the same vertical line. Then the function $\phi(y) = f(x_0, y)$ satisfies

$$\phi(y_0) = f(x_0, y_0) \qquad \text{and} \qquad \phi(y_1) = f(x_1, y_1).$$

And since its second derivative $\phi''(y) = D_{22}f(x_0, y)$ is everywhere positive by hypothesis, it follows that ϕ is concave up, and has its minimum value at its critical point y_0, by the results in Chapter 5. In particular

$$f(x_1, y_1) = \phi(y_1) > \phi(y_0) = f(x_0, y_0).$$

If X_0 and X_1 are not on a vertical line, the same argument works but the details are more complicated. Let

$$y - y_0 = m(x - x_0)$$

be the nonvertical line through X_0 and X_1. Along this line $f(x, y)$ becomes a function of x alone,

$$\phi(x) = f(x, y_0 + m(x - x_0)),$$

and we can compute the first and second derivatives of ϕ by the chain rule:

$$\phi' = D_1 f + (D_2 f)\frac{dy}{dx} = D_1 f + m D_2 f,$$

$$\phi'' = D_1(D_1 f + m D_2 f) + D_2(D_1 f + m D_2 f)\frac{dy}{dx}$$

$$= D_{11}f + m D_{21}f + m D_{12}f + m^2 D_{22}f$$

$$= m^2 D_{22}f + 2m D_{12}f + D_{11}f,$$

where all the partial derivatives are evaluated at the point $(x, y_0 + m(x - x_0))$. If $\phi''(x)$ were ever zero, then the slope m would have to be related to the second partial derivatives of f by the well-known quadratic formula

$$m = \frac{-b \pm \sqrt{b^2 - 4ac}}{2a},$$

and this can be shown to be impossible because of our hypotheses. An equivalent procedure is to complete the square on the right in the above formula for ϕ'', giving

$$\phi'' = D_{22}f\left(m + \frac{D_{12}f}{D_{22}f}\right)^2 + \frac{D_{11}f D_{22}f - (D_{12}f)^2}{D_{22}f}.$$

This displays ϕ'' directly in terms of quantitatives whose signs we know (by hypothesis), and shows that ϕ'' is always positive. So ϕ is concave up, as before, and assumes its minimum value at its critical point x_0. Thus

$$f(x_0, y_0) = \phi(x_0) < \phi(x_1) = f(x_1, y_1),$$

and we have now proved that $f(x_0, y_0)$ is smaller than the value of f at *any* other point (x_1, y_1) in R. That is, f has a relative minimum at (x_0, y_0).

Going back to the beginning, if $Df = (D_{12}f)^2 - D_{11}f D_{22}f$ is negative at (x_0, y_0) but the common sign of D_{11} and D_{22} is negative, then the above argument repeats almost verbatim to show that ϕ'' is always negative; hence ϕ is concave down, and hence $f(x_0, y_0)$ is larger than $f(x_1, y_1)$ for every other point (x_1, y_1) in R. So f has a relative maximum at (x_0, y_0) in this case, and the following theorem has been proved.

Theorem 9. *Suppose that $f(x, y)$ has continuous partial derivatives of the first and second order, and suppose that $X_0 = (x_0, y_0)$ is a critical point of f at which the discriminant $D(x, y)$ is negative. Then f has a relative minimum at X_0 or a relative maximum at X_0, depending on whether $D_{11}f(x_0, y_0)$ and $D_{22}f(x_0, y_0)$ are both positive or both negative.*

There remains the case where (x_0, y_0) is a critical point of $f(x, y)$ at which the discriminant is positive. Suppose that $D_{22}f(x_0, y_0) \neq 0$, and consider the quadratic polynomial

$$q(m) = am^2 + 2bm + c,$$

where

$$a = D_{22}f(x_0, y_0),$$

$$b = D_{12}f(x_0, y_0),$$

$$c = D_{11}f(x_0, y_0).$$

The quadratic formula for the roots of $q(m)$ is

$$m = \frac{-2b \pm \sqrt{(2b)^2 - 4ac}}{2a} = \frac{-b \pm \sqrt{b^2 - ac}}{a},$$

where $b^2 - ac$ is the discriminant $Df(x_0, y_0)$, which we are now assuming to be positive. The quadratic formula therefore gives two roots:

$$m_1 = (-b - \sqrt{D})/a, \qquad m_2 = (-b + \sqrt{D})/a,$$

and $q(m)$ factors as follows:

$$q(m) = a(m - m_1)(m - m_2).$$

In particular, $q(m)$ changes sign as m crosses m_1, say from $-$ to $+$. Then $q(m)$ is positive for any m between m_1 and m_2. Now $q(m)$ is the value we found for $\phi''(x_0)$, where $\phi(x)$ is the restriction of $f(x, y)$ to the line through (x_0, y_0) of slope m. So, for any m between m_1 and m_2, $f(x, y)$ is concave up as $X = (x, y)$ moves through $X_0 = (x_0, y_0)$ along the line of slope m. Similarly, $f(x, y)$ is concave down as X moves through X_0 along a line whose slope m is either less than m_1 or greater than m_2. This means that X_0 is a saddle point for f (although the proof is not complete).

We have given above a partial proof of the following theorem.

Theorem 10. *If $f(x,y)$ has continuous partial derivatives of the first and second order, and if (x_0,y_0) is a critical point of f at which the discriminant of f is positive, then (x_0,y_0) is a saddle point for f.*

EXAMPLE 1. It is clear on the face of it that $z = x^2 + y^2$ is minimum at the origin, but let us disregard the obvious and see what the machinery of critical-point analysis has to say. The first partial derivatives $\partial z/\partial x = 2x$ and $\partial z/\partial y = 2y$ vanish simultaneously only at the origin, which is therefore the only critical point. The second partial derivatives are all constants, with the values

$$\frac{\partial^2 z}{\partial x^2} = 2, \qquad \frac{\partial^2 z}{\partial y^2} = 2, \qquad \frac{\partial^2 z}{\partial x \, \partial y} = 0.$$

The discriminant thus has the constant value

$$0^2 - 2 \cdot 2 = -4.$$

So Theorem 9 tells us that $f(x,y) = x^2 + y^2$ has a relative minimum at the origin.

EXAMPLE 2. Classify the critical point or points of $z = x^2 - y^2$.

. . . .

Solution. We find that $\partial z/\partial x = 2x$ and $\partial z/\partial y = -2y$ are simultaneously zero only at the origin, so that is the only critical point. The second partial derivatives are again all constant,

$$\frac{\partial^2 z}{\partial x^2} = 2, \qquad \frac{\partial^2 z}{\partial y^2} = -2, \qquad \frac{\partial^2 z}{\partial x \, \partial y} = 0,$$

and the discriminant $D(x,y)$ has the constant value

$$\left(\frac{\partial^2 z}{\partial x \, \partial y} \right)^2 - \frac{\partial^2 z}{\partial x^2} \frac{\partial^2 z}{\partial y^2} = 0^2 - 2(-2) = 4.$$

This time the origin is a saddle point, by Theorem 10.

EXAMPLE 3. Study the critical points of the function

$$z = x^3 y + 3x + y$$

by applying Theorem 9 or Theorem 10.

. . . .

Solution. The equations

$$\frac{\partial z}{\partial x} = 3x^2y + 3 = 0,$$

$$\frac{\partial z}{\partial y} = x^3 + 1 = 0$$

determine the solutions: $x = -1$, from the second equation, and then $y = -1$ from the first. Therefore $(-1, -1)$ is the only critical point. The discriminant

$$\left(\frac{\partial^2 z}{\partial x \, \partial y}\right)^2 - \frac{\partial^2 z}{\partial x^2}\frac{\partial^2 z}{\partial y^2} = (3x^2)^2 - (6xy)\cdot 0 = 9x^4$$

is positive at $(-1, -1)$, so this point is a saddle point, by Theorem 10.

PROBLEMS FOR SECTION 7

Classify the critical points of each of the following functions of x and y.

1. $x^2 + 2y^2 + 2x - y$ 2. $x - y + xy$

3. $x^3 - xy + y^2$ 4. $x^2 + 3xy + y^2$

5. xy (Draw a few level curves.)   6. $x^2y^2 + 2x - 2y$

7. $x^2y + y^2 + x$ 8. $x^3 - 3xy + y^3$

9. $x^2 - xy + y^3 - x$ 10. $x^4 + y^4$

11. $x^4 + y^4 + 4x - 4y$ 12. $x^4 - y^4 + 4x - 4y$

13. $x^2 + y^2 + \dfrac{2}{xy}$ 14. $\dfrac{1}{x^2} + \dfrac{4}{y^2} + xy$

15. $e^x + e^y - e^{x+y}$ 16. $x \sin y$

17. Prove the following corollary of (part of) Theorem 9.

> **Theorem.** *Suppose that the discriminant of $f(x,y)$ is negative at the point (x_0, y_0), and that the second partial derivatives $\partial^2 z/\partial x^2$ and $\partial^2 z/\partial y^2$ are both positive there. Prove that, for points (x,y) sufficiently close to (x_0, y_0), the graph of $z = f(x,y)$ lies entirely above its tangent plane at (x_0, y_0).*

vectors in space

So far we have looked only at plane vectors and some of their applications to calculus. When we turn to three dimensions, we discover that things are pretty much the same. We obtain the same kinds of results by essentially the same methods, and this means that our attention in this chapter will be largely directed toward taking note of the things that are different.

1. VECTORS IN SPACE

Space vectors can be introduced by repeating the earlier development practically word for word, so we shall just mention the highlights.

Since a point X in coordinate space is a coordinate triple $X = (x, y, z)$, the definitions of the sum $(X + A)$ and the product tX are now

$$(x, y, z) + (a, b, c) = (x + a, y + b, z + c),$$

$$t(x, y, z) = (tx, ty, tz).$$

The magnitude $|X|$ of the vector $X = (x, y, z)$ is the distance from the origin to X, and its formula is

$$|X| = \sqrt{x^2 + y^2 + z^2}.$$

Then the formula for the distance d between the space points X_1 and X_2 can be rewritten, in terms of vector subtraction, as

$$d = |X_1 - X_2|.$$

The *direction* of $A = (a, b, c)$ is again presented by its unit vector

$$U = \frac{A}{|A|} = \left(\frac{a}{r}, \frac{b}{r}, \frac{c}{r}\right),$$

where $r = |A| = \sqrt{a^2 + b^2 + c^2}$. But the geometric interpretation of the components of U necessarily changes slightly from what it was in the two-dimensional case. There we could write

$$U = (\cos\theta, \sin\theta).$$

where θ is the angle from the positive x-axis to the direction U. This clearly does not generalize to three dimensions, but the new interpretation of U is nevertheless very close. If a coordinate box is drawn with vertices at A, and if the *face* diagonals are drawn from A (two are shown), then it is apparent that

$$\frac{a}{r} = \cos\alpha, \quad \frac{b}{r} = \cos\beta, \quad \frac{c}{r} = \cos\gamma,$$

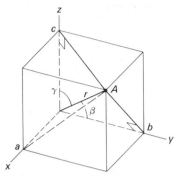

where α, β, and γ are the angles between the direction of A and the positive x-, y-, and z-axes. Thus,

$$U = \frac{A}{|A|} = (\cos \alpha, \cos \beta, \cos \gamma).$$

U is the triple of so-called "direction cosines" of A.

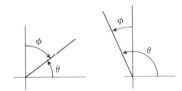

If direction cosines are introduced in the plane, as shown above, then $\theta - \phi = \pi/2$, and

$$\cos \phi = \cos(\theta - \frac{\pi}{2}) = \sin \theta.$$

Thus,

$$(\cos \theta, \cos \phi) = (\cos \theta, \sin \theta),$$

and the direction cosines reduce to the cosine and sine of the single angle θ. Remember, however, that when the plane direction is presented as $(\cos \theta, \sin \theta)$, then θ must be the *signed* angle *from* the positive x-axis *to* the vector direction. On the other hand, if the direction is written $(\cos \theta, \cos \phi)$, then θ and ϕ can be taken to be the positive *unsigned* angles *between* the vector direction and the axes, because of the identity $\cos \alpha = \cos(-\alpha)$.

The translation-of-axes theorem (page 40) is proved for space just as it was for the plane. Because of our present convention about capital letters, we have to state the theorem in slightly different notation.

Theorem 1. *If a translated $x'y'z'$-coordinate system is chosen with its origin O' at the point A having old coordinates (a, b, c), then the new coordinates (x', y', z') of a point*

P are obtained from its old coordinates (x, y, z) *by the change-of-coordinates equations*

$$x' = x - a, \quad y' = y - b, \quad z' = z - c.$$

We use the new axes at A to compute the cosines of the direction of the arrow $\overrightarrow{AX}$, and get

$$\cos \alpha = \frac{x'}{r} = \frac{x - a}{r}, \qquad \cos \beta = \frac{y'}{r} = \frac{y - b}{r}, \qquad \cos \gamma = \frac{z'}{r} = \frac{z - c}{r},$$

where

$$r = \sqrt{(x')^2 + (y')^2 + (z')^2} = \sqrt{(x - a)^2 + (y - b)^2 + (z - c)^2}$$
$$= |X - A|.$$

Except for the change from $(\cos \theta, \sin \theta)$ to the triple of direction cosines, the proof of Theorem 1 in Chapter 14 is now valid for three dimensions as well as for two. Thus;

Theorem 2.

a) *Two arrows* $\overrightarrow{AX}$ *and* $\overrightarrow{A'X'}$ *have the same length and direction if and only if*

$$X' - A' = X - A,$$

that is, if and only if their vectors

$$U' = X' - A' \qquad and \qquad U = X - A$$

are the same.

b) *Two vectors* U *and* U' *are parallel if and only if*

$$U' = tU$$

for some scalar t. In this case the directions are the same if t is positive and opposite if t is negative, and $|U'| = |t| \, |U|$.

The parametric equation of the line through the distinct points X_0 and X_1 comes out to be

$$X = X_0 + t(X_1 - X_0),$$

just as before.

EXAMPLE 1. The line through $X_0 = (1, 2, 3)$ and $X_1 = (3, -2, 1)$ has the parametric equation

$$X = (1, 2, 3) + t(2, -4, -2).$$

This is the equation

$$(x, y, z) = (1 + 2t, 2 - 4t, 3 - 2t),$$

and is equivalent to the three scalar equations

$$x = 1 + 2t, \qquad y = 2 - 4t, \qquad z = 3 - 2t.$$

Here the parameter t *cannot* be eliminated, and the parametric equations are *not* equivalent to a single linear equation in x, y, and z. In fact, we know that such an equation has a *plane* as its graph.

EXAMPLE 2. The midpoint of the segment from $X_1 = (1, 2, 3)$ to $X_2 = (3, -4, 5)$ is

$$\frac{X_1 + X_2}{2} = \frac{(1, 2, 3) + (3, -4, 5)}{2} = \frac{(4, -2, 8)}{2} = (2, -1, 4).$$

EXAMPLE 3. We saw earlier that the three medians of the triangle $A_1 A_2 A_3$ are concurrent in the point $(A_1 + A_2 + A_3)/3$. This point is actually the *centroid* of the triangle. It is algebraically the average value of the three vectors A_1, A_2, and A_3. The proof given for this median result is valid for a triangle in space. Now consider four space points A_1, A_2, A_3, and A_4 not lying in any one plane. They are the vertices of a *tetrahedron*. Through each vertex we draw a line to the centroid of the opposite side, and we ask whether these four lines are concurrent.

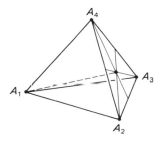

Here we shall just guess the answer and then verify that it is correct. We guess that

$$\frac{A_1 + A_2 + A_3 + A_4}{4},$$

the *average* of the four vertex vectors, is the point we want. Is it on the segment joining the vertex A_1 to the centroid $(A_2 + A_3 + A_4)/3$? Yes, because the equation

$$\frac{A_1 + A_2 + A_3 + A_4}{4} = \frac{1}{4}A_1 + \frac{3}{4}\left(\frac{A_2 + A_3 + A_4}{3}\right)$$

shows it to be the point 3/4 of the way along the segment from the vertex to the centroid.

The dot product in three dimensions has essentially the same definition,

$$(x, y, z) \cdot (r, s, t) = xr + ys + zt,$$

and it has the same properties (s1) through (s4), for the same reasons. In view of this second example of a dot product, we can take these properties as the *axioms* for a dot product. Here they are again, for ready reference:

(s1) $$X \cdot X = |X|^2,$$

(s2) $$X \cdot A = A \cdot X,$$

(s3) $$(X_1 + X_2) \cdot A = X_1 \cdot A + X_2 \cdot A,$$

(s4) $$(tX) \cdot A = t(X \cdot A).$$

The formula relating the dot product to $\cos \phi$ cannot be proved as before. Instead, we shall see that it is a consequence of the dot-product axioms. Consider first the special case of perpendicularity.

Theorem 3. *The vectors A and X are perpendicular if and only if* $A \cdot X = 0$.

Proof. By the Pythagorean theorem, A and X are perpendicular if and only if

$$|A|^2 + |X|^2 = |A - X|^2.$$

Using (s1) and then the two versions of (s3), this becomes

$$A \cdot A + X \cdot X = (A - X) \cdot (A - X)$$
$$= A \cdot (A - X) - X \cdot (A - X)$$
$$= A \cdot A - A \cdot X - X \cdot A + X \cdot X.$$

Since $X \cdot A = A \cdot X$, this equation reduces to $2(A \cdot X) = 0$, and hence to $A \cdot X = 0$, proving the theorem.

EXAMPLE 4. We see that $A = (-1, 2, 3)$ and $X = (3, 3, -1)$ are perpendicular, by computing

$$A \cdot X = (-1)3 + 2 \cdot 3 + 3 \cdot (-1) = -3 + 6 - 3 = 0.$$

EXAMPLE 5. Find the most general vector X perpendicular to

$$A_1 = (1, 2, 3) \quad \text{and} \quad A_2 = (2, 3, 1).$$

· ·

Solution. $X = (x, y, z)$ will be perpendicular to these two vectors if and only if $A_1 \cdot X = 0$ and $A_2 \cdot X = 0$, that is, if and only if

$$x + 2y + 3z = 0,$$

$$2x + 3y + z = 0.$$

Eliminating z (by multiplying the second equation by 3 and subtracting), we see that

$$5x + 7y = 0 \quad \text{and} \quad y = -(5/7)x.$$

Similarly, we find that $z = x/7$. Thus a vector X is perpendicular to both $A_1 = (1, 2, 3)$ and $A_2 = (2, 3, 1)$ if and only if it can be written in the form

$$X = (x, -\frac{5}{7}x, \frac{1}{7}x) = x(1, -\frac{5}{7}, \frac{1}{7}).$$

Next we formally repeat an argument given in Chapter 14.

Theorem 4. *If X and A are any two vectors, with $A \neq 0$, then there is a uniquely determined scalar c such that $X - cA$ is perpendicular to A.*

Proof. This is obvious geometrically from a diagram, but we want to deduce it just from the axioms. If $X - cA \perp A$, then

$$0 = (X - cA) \cdot A = X \cdot A - c(A \cdot A)$$

$$= X \cdot A - c|A|^2,$$

by (s3), (s4), and (s1). Therefore,

$$c = \frac{X \cdot A}{|A|^2}.$$

Conversely, with this value of c, the calculation above can be reversed and $X - cA$ is perpendicular to A.

The $\cos \phi$ law is implicit in this result, for we now have a right triangle as shown below at the left if $c > 0$, and at the right if $c < 0$. In particular, the sign of c is the sign of $\cos \phi$, and since $\cos \phi = $ adjacent/hypotenuse except for sign, it follows that

$$\cos \phi = \frac{c|A|}{|X|} = \frac{X \cdot A}{|A|^2} \cdot \frac{|A|}{X} = \frac{X \cdot A}{|X||A|}.$$

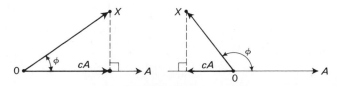

We have proved:

Theorem 5. *If ϕ is the angle between the nonzero vectors A and X, then*

$$\cos \phi = \frac{A \cdot X}{|A||X|}.$$

EXAMPLE 6. Find the angle between $(1, 2, 1)$ and $(1, -1, 2)$.

Solution

$$\cos \phi = \frac{A \cdot X}{|A||X|} = \frac{1 \cdot 1 + 2(-1) + 1 \cdot 2}{\sqrt{6}\sqrt{6}} = \frac{1}{6},$$

so

$$\phi = \arccos\left(\frac{1}{6}\right).$$

PROBLEMS FOR SECTION 1

In the following, find the unit vector in the same direction as the given vector.

1. $(2, 1, 2)$ 　　　　　　　　　　　　　　2. $(-3, 6, -6)$

3. $(7, -6, -6)$ 　　　　　　　　　　　　4. $(-3, 4, 12)$

5. $(1/2, 1/2, 1/4)$

Given X and A, find $X \cdot A$ and the angle ϕ between X and A.

6. $X = (2, 1, 2)$, $A = (1, 1, 0)$ 　　　　7. $X = (1/2, 0, -1/2)$, $A = (2, 1, 2)$

8. $X = (3, 1, \sqrt{2})$, $A = (1, 0, 0)$ 　　　9. $X = 1(1, 1/2, -1)$, $A = (0, 1, 1/2)$

10. $X = (1, 1, 1)$, $A = (2, -2, 1)$ 　$(\phi = \arctan(\ ))$

Write the parametric equation of the line through the two given points.

11. $(1, -2, 1)$, $(2, 1, -1)$ 　　　　　　12. $(1/2, 0, 1)$, $(1/2, -1, 2)$

13. $(-2, -1, 2)$, $(1, 3, 4)$ 　　　　　　14. $(4, 1/4, -1)$, $(-1, 1/2, 1)$

15. $(-1, -2, 1)$, $(2, 0, 2)$

Write the scalar equations of the line through the two given points.

16. $(-3, 1, 2)$, $(2, -1, -1)$ 　　　　　17. $(1, -1, 4)$, $(1, 1/2, 0)$

18. $(-1, 1/2, -1/3)$, $(-2, -2, 1)$ 　　　19. $(-2, -1, 0)$, $(1/2, -1, -2)$

20. $(1/4, 1/2, 1/3)$, $(1, 0, 1)$

Write the parametric equation of the line through X_0 in the direction of the vector A, where:

21. $X_0 + (0,0,0)$, $A = (1,-2,-1)$. 22. $X_0 = (4,-2,2)$, $A = (-2,1,1)$.

23. $X_0 = (1,1,1)$, $A = (2,0,-1)$. 24. $X_0 = (-3,1,2)$, $A = (5,-2,-3)$.

25. Let A and L be two unit vectors, making angles, α, β, γ, and λ, μ, ν, respectively, with the coordinate axes. Interpret the formula

$$\cos \phi = \cos \alpha \cos \lambda + \cos \beta \cos \mu + \cos \gamma \cos \nu.$$

Find the most general vector X perpendicular to A_1 and A_2 in the following problems.

26. $A_1 = (1,2,3)$, $A_2 = (0,1,2)$. 27. $A_1 = (1,0,1)$, $A_2 = (1,1,0)$.

28. $A_1 = (2,-1,1)$, $A_2 = (0,2,2)$.

29. Let A_1 and A_2 be any pair of independent vectors. A vector X is perpendicular to both A_1 and A_2 if and only if

$$a_1 x + b_1 y + c_1 z = 0,$$

$$a_2 x + b_2 y + c_2 z = 0.$$

We can solve for X as in Example 5 in the text and in the above problems. If we multiply the first equation by c_2 and the second equation by c_1 and then subtract, this will eliminate z, and leave one equation in x and y. (Etc.) Show that this procedure leads to the following result.

Theorem. *A vector X is perpendicular to both A_1 and A_2 if and only if X is a scalar multiple of the vector*

$$(b_1 c_2 - c_1 b_2, c_1 a_2 - a_1 c_2, a_1 b_2 - b_1 a_2).$$

The vector in the above theorem is called the *cross product* of A_1 and A_2 and is designated $A_1 \times A_2$. Thus:

Definition

$$A_1 \times A_2 = (b_1 c_2 - c_1 b_2, c_1 a_2 - a_1 c_2, a_1 b_2 - b_1 a_2).$$

Like the dot product, the cross product has a collection of algebraic properties that we can establish from the above definition, and which let us then use the cross product without necessarily referring back to its components. Prove the following properties of the cross product.

30. $A_1 \times A_2 = 0$ if and only if A_1 and A_2 are dependent. [Hint. Suppose $a_1 \neq 0$ and set $k = (a_2/a_1)$. Show that if $A_1 \times A_2 = 0$, then $A_2 = kA_1$.]

31. $A_1 \times A_2 = -(A_2 \times A_1)$.

32. $(A_1 \times A_2) \cdot A_3 = A_1 \cdot (A_2 \times A_3)$.

33. $A_1 \times A_2$ is perpendicular to both A_1 and A_2. (Use Problems 32 and 30.)

34. $A \times (U_1 + U_2) = (A \times U_1) + (A \times U_2)$.

35. $|A_1 \times A_2|^2 = |A_1|^2 |A_2|^2 - |A_1 \cdot A_2|^2$.

36. $|A_1 \times A_2| = |A_1||A_2|\sin \phi$, where ϕ is the angle between the directions of A_1 and A_2 (from Problem 35 and the dot-product cosine formula).

37. $|A_1 \times A_2|$ is the area of the paralellogram having A_1 and A_2 for sides.

38. Rework Problem 26, using the cross product.

39. Do the same for Problem 27.

40. Do the same for Problem 28.

2. PLANES

We saw earlier that the dot-product perpendicularity condition led to an easy derivation of the equation of a line. In three dimensions, the same proof leads to the equation of a *plane*.

Let $X_0 = (x_0, y_0, z_0)$ be any fixed point and let $A = (a, b, c)$ be a fixed nonzero vector. Let π be the plane that passes through the point X_0 and is perpendicular to the direction of A. A point X will lie on the plane π if and only if the arrow $\overrightarrow{X_0 X}$ is perpendicular to A. Thus, X is on π if and only if

$$A \cdot (X - X_0) = 0,$$

or

$$a(x - x_0) + b(y - y_0) + c(z - z_0) = 0.$$

This equation is therefore the "point–direction" equation of the plane.

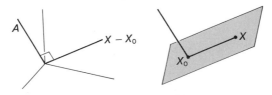

If we set

$$(ax_0 + by_0 + cz_0) = d,$$

the equation takes the general linear form

$$ax + by + cz = d.$$

Conversely, any such first-degree equation is the equation of a plane, provided that at least one of the coefficients a, b, and c is really there (nonzero). For, if $X_0 = (x_0, y_0, z_0)$ is any fixed triple satisfying the equation

$$ax_0 + by_0 + cz_0 = d,$$

then subtracting the two equations gives the equivalent form

$$a(x - x_0) + b(y - y_0) + c(z - z_0) = 0,$$

or

$$A \cdot (X - X_0) = 0.$$

Thus, the points $X = (x, y, z)$ whose coordinates satisfy the equation $ax + by + cz - d = 0$ are exactly the points X such that $X - X_0$ is perpendicular to A, that is, the points lying on the plane π through X_0 and perpendicular to A. Altogether we have proved:

Theorem 6. *The plane through the point (x_0, y_0, z_0) and perpendicular to the nonzero vector $A = (a, b, c)$ has the equation*

$$A \cdot (X - X_0) = 0,$$

or

$$ax + by + cz = d,$$

where $d = A \cdot X_0$. Conversely, the graph of the equation

$$ax + by + cz = d$$

is a plane perpendicular to $A = (a, b, c)$, provided that A is not the zero vector.

The planes

$$ax + by + cz = d_1$$

and

$$ax + by + cz = d_2$$

have the same normal vector $A = (a, b, c)$ and are parallel. The plane parallel to them through the origin has the equation

$$ax + by + cz = 0.$$

If X_0 is a fixed point on the plane

$$ax + by + cz = d,$$

then Theorem 6 says that a point X is on this plane if and only if $P = X - X_0$ is on the parallel plane through the origin,

$$ax + by + cz = 0.$$

(This was the condition $A \cdot (X - X_0) = 0$.) In other words, as P runs over a plane through the origin, the point $X = P + X_0$ runs over the parallel plane through X_0.

There is another way that we can look at planes. The geometric proof given for Theorem 3 in Chapter 14 lifts verbatim to the present context and shows that:

Theorem 7. *If π is a plane through the origin, and if A_1 and A_2 are two nonparallel vectors in π, then π consists exactly of the set of all linear combinations*

$$X = sA_1 + tA_2$$

of A_1 and A_2.

In this situation we call the plane π the *linear span* of the vectors A_1 and A_2, or the *plane spanned by A_1 and A_2.*

The formula

$$X = sA_1 + tA_2$$

can be called a *parametric representation* for the plane π, with parameters s and t. It is exactly analogous to the parametric equation for a line through the origin

$$X = tA.$$

We noted above that the parallel plane π_0 through the point X_0 is obtained by adding X_0 to all the points of π, so π_0 has the parametric representation

$$X = sA_1 + tA_2 + X_0.$$

We now have two views of a plane through the origin. On the one hand it is the *linear span* of any two of its vectors, A_1 and A_2 (provided they are independent). On the other hand, it is the collection of vectors perpendicular to a (normal) vector A. These two viewpoints must, of course, be algebraically consistent. It must be true that:

1. An independent pair of vectors A_1 and A_2 determine a mutually normal vector A, uniquely to within scalar multiples. That is, A_1 and A_2 determine a unique mutually normal *direction*. The problem of finding A from A_1 and A_2 was illustrated in Example 5 of the last section. (An explicit formula for A is given by the cross product, $A = A_1 \times A_2$. See Problems 29 and 33 in the last section.)

2. Given a nonzero vector A, we can find an independent pair A_1 and A_2 both perpendicular to A. This calculation is illustrated below.

EXAMPLE 1. Find an independent pair of vectors A and X such that each is perpendicular to the vector $(1, -3, 2)$.

. . . .

Solution. Let us first find X. The perpendicularity requirement is

$$x - 3y + 2x = 0.$$

This is one condition on the three unknowns x, y, and z, and determines one of them as a function of the other two. For example, x and y can be chosen arbitrarily, say

$$(x, y) = (2, 0),$$

in which case z is determined as -1. Thus,

$$(x, y, z) = (2, 0, -1)$$

is one vector perpendicular to $(1, -3, 2)$. A second vector, $A = (a, b, c)$, can be found in exactly the same way, by choosing its first two components (a, b) arbitrarily and then determining c from the perpendicularity condition

$$a - 3b + 2c = 0,$$

obtaining

$$c = \frac{(3b - a)}{2}.$$

In order to ensure that A will not be a scalar multiple of X, it is sufficient to guard against this in its first two components. Thus we choose any pair (a, b) *not* of the form $k(x, y) = (2k, 0)$. If we take $(a, b) = (1, 1)$, then

$$c = \frac{(3b - a)}{2} = \frac{(3 - 1)}{2} = 1$$

and

$$A = (1, 1, 1).$$

Another way of finding A is to solve the *first* problem starting from $(1, -3, 2)$ and $X = (2, 0, -1)$, that is, to find A as a vector mutually perpendicular to these two vectors. Then $(1, -3, 2)$, X, and A, is a trio of *mutually perpendicular vectors*.

PROBLEMS FOR SECTION 2

Write the equation of the plane through the given point and perpendicular to the given vector.

1. Point $(0,0,0)$; vector $(1,2,1)$.
2. Point $(-1,1,1/2)$; vector $(3,-1,-2)$.
3. Point $(4,1,1)$; vector $(1/2,1,0)$.
4. Point $(1,-1,4)$; vector $(-1,-2,1)$.
5. Point $(6,4,-6)$; vector $(1/3,1/4,-1/6)$.

In Problems 6 through 9 find an independent pair of vectors A and X each perpendicular to the given vector.

6. $(1,1,0)$ 7. $(1,2,-1)$

8. $(2,2,2)$ 9. $(5,-1,3)$

10. Find the point of intersection of the line

$$X = (1,2,-1) + t(2,-2,3)$$

and the plane

$$3x + y - z = 2.$$

11. Let π be the plane through the origin and the two points $A_1(2,-1,1)$ and $A_2(1,3,1)$, that is, the linear span of the vectors A_1 and A_2. Prove that its intersection with the xy-coordinate plane is the line

$$X = t(1,-4,0).$$

Eliminate the parameter and find the xy-equation of the line.

12. Find the parametric representation of the plane π through the two lines

$$X = (1,2,-1) + t(2,-2,3),$$

$$X' = (1,2,-1) + s(5,1,0).$$

13. Use the cross product to write the equation of the plane in Problem 11 in the form $A \cdot (X - X_0) = 0$. (See Problems 29 and 33 in the last section.)

14. The same question for the plane in Problem 12.

15. Find the line of intersection of the planes

$$x + 2y - z = 0 \quad \text{and} \quad 3x - y + 2z = 0.$$

(Read off vectors A_1 and A_2 perpendicular to the planes and again use the cross product.)

16. Find the plane through the origin perpendicular to each of the planes

$$x + y + z = 3 \quad \text{and} \quad x - 2y - z = 1.$$

17. Find the plane through $(1, 2, -1)$ perpendicular to each of the planes

$$2x - y + z = 0 \quad \text{and} \quad x + 3y - 4z = 1.$$

3. DIMENSION

The analogue of Theorem 3 in Chapter 14 shows Cartesian space to be a *three-dimensional vector space*.

Theorem 8. *If A_1, A_2, and A_3 are independent, in the sense that they do not all lie in a single plane through the origin, then each vector X_0 is expressible in a unique way as a linear combination of A_1, A_2, and A_3. That is, X_0 has the form*

$$X_0 = rA_1 + sA_2 + tA_3,$$

where r, s, and t are uniquely determined scalars.

Proof. In the last section it was shown that the plane through X_0 parallel to the two vectors A_1 and A_2 consists exactly of the vectors of the form

$$X = pA_1 + qA_2 + X_0,$$

where p and q are any scalars. Since A_3 is not parallel to this plane, the line

$$X = tA_3$$

meets the plane in a unique point. That is, there is a unique scalar t such that the point $X = tA_3$ can also be expressed in the form $X = pA_1 + qA_2 + X_0$, with p and q uniquely determined. So

$$tA_3 = pA_1 + qA_2 + X_0,$$

and

$$X_0 = rA_1 + sA_2 + tA_3$$

(where $r = -p, s = -q$), all the scalars being uniquely determined. ∎

The above proof is partly geometric, being based on the unique intersection point of a plane and a line not parallel to the plane. A purely algebraic proof consists in solving the equation $X_0 = rA_1 + sA_2 + tA_3$ algebraically for r, s, t.

This is equivalent to the set of three scalar equations:

$$ra_1 + sb_1 + tc_1 = x_0,$$
$$ra_2 + sb_2 + tc_2 = y_0,$$
$$ra_3 + sb_3 + tc_3 = z_0,$$

and with three equations in three unknowns we expect that the usual solution by elimination procedure will work out, and yield a unique solution triple (r, s, t) for each triple (x_0, y_0, z_0). There is, however, an algebraic requirement that must be satisfied, namely, that the *determinant*

$$\begin{vmatrix} a_1 & b_1 & c_1 \\ a_2 & b_2 & c_2 \\ a_3 & b_3 & c_3 \end{vmatrix}$$

must be *nonzero*. This is the algebraic expression of our hypothesis that the three vectors A_1, A_2, and A_3 do not all lie in a plane. But we shall not pursue these algebraic versions.

Theorem 8 shows Cartesian space to be a three-dimensional vector space in the same sense that Theorem 3 of Chapter 14 showed the Cartesian plane to be two-dimensional. As in the two-dimensional case, the independent spanning triple A_1, A_2, A_3 of Theorem 8 is called a *basis* for the vector space, and if

$$X = rA_1 + sA_2 + tA_3,$$

then r, s, and t are the *components* of X with respect to the basis. The triple

$$E_1 = (1, 0, 0), \qquad E_2 = (0, 1, 0), \qquad E_3 = (0, 0, 1)$$

is called the *standard basis* of Cartesian space. Since

$$X = (x, y, z)$$
$$= x(1, 0, 0) + y(0, 1, 0) + z(0, 0, 1)$$
$$= xE_1 + yE_2 + zE_3,$$

we see that the coordinates of X can be reinterpreted as its components with respect to the standard basis. Three mutually perpendicular unit vectors form a basis of a very special kind, called an *orthonormal basis*. As in the plane, the components of a vector with respect to an orthonormal basis can be directly computed as dot products.

Theorem 9. *The components of a vector X with respect to an orthonormal basis B_1, B_2, B_3 are just the dot products*

$$X \cdot B_1, \qquad X \cdot B_2, \qquad \text{and} \qquad X \cdot B_3.$$

Proof. If

$$X = rB_1 + sB_2 + tB_3,$$

then

$$
\begin{aligned}
X \cdot B_1 &= r(B_1 \cdot B_1) + s(B_2 \cdot B_1) + t(B_3 \cdot B_1) \\
&= r1 + s0 + t0 \\
&= r.
\end{aligned}
$$

Similarly $X \cdot B_2 = s$ and $X \cdot B_3 = t$. ∎

Now consider again a plane π through the origin. It follows from Theorem 7 that if the vectors X and P lie in π, then so does their sum $(X + P)$, and so does tX for any scalar t. We indicate this property of π by saying that π is *closed under the vector operations.* Thus, π is exactly like the Cartesian plane from the point of view of the vector operations. (Of course, they are also identical from the point of view of geometry.)

Going on with the analogy, because the vectors in π are specified in the above way by pairs of numbers (s, t), we say that π is a two-dimensional vector space with A_1 and A_2 as a basis. This time, though, we have a vector space imbedded in a larger vector space, the plane π imbedded in Cartesian space, which is a three-dimensional vector space. We therefore call π a (two-dimensional) vector *subspace* of Cartesian space.

Returning, finally, to a line L through the origin, with parametric equation

$$X = tA,$$

we note that L also is closed under the vector operations and hence is a vector subspace of Cartesian space. It has the single vector A as basis, and each of its points $X = tA$ has the single component t. That is, L is a *one-dimensional* vector subspace.

PROBLEMS FOR SECTION 3

1. The vectors $A_1 = (2, 1, 1)$ and $A_2 = (1, -1, -1)$ are orthogonal (perpendicular). Find an orthonormal basis $\{B_1, B_2, B_3\}$ such that B_1 and B_2 are the directions of A_1 and A_2, respectively.

2. Same problem for the vectors $A_1 = (1, 2, 3)$ and $A_2 = (3, -3, 1)$.

3. Same problem for $A_1 = (1, 1, 1)$ and $A_2 = (1, -1, 0)$.

4. Check that

$$B_1 = \frac{1}{3}(2, 1, 2), \qquad B_2 = \frac{1}{3}(2, -2, -1), \qquad \text{and} \qquad B_3 = \frac{1}{3}(1, 2, -2)$$

form an orthonormal basis.

Using the above basis and Theorem 9, express each of the following vectors X in the form

$$X = x_1 B_1 + x_2 B_2 + x_3 B_3.$$

Then verify directly the correctness of each expansion.

5. $X = E_1 = (1,0,0)$.   6. $X = E_2 = (0,1,0)$.

7. $X = E_3 = (0,0,1)$.   8. $X = (1,2,3)$.

9. $X = (4,0,-2)$.   10. $X = (1,-1,1)$.

11. $X = (-2,-2,1)$.   12. $X = (3,1,-2)$.

13. We have seen, in the examples and problems, that two nonparallel vectors A_1 and A_2 determine a nonzero vector A perpendicular to them both, and uniquely up to scalar multiples. That is, if X is any other vector perpendicular to both A_1 and A_2, then $X = tA$ for some scalar t. Use this fact to prove the following result:

If X is a vector perpendicular to A_1, A_2, and A, then $X = 0$.

In particular, if A_1, A_2, and A_3 are any three mutually perpendicular nonzero vectors, then the only vector perpendicular to them all is 0.

14. The result above is one way of saying that space is three-dimensional, because it lets us show directly, without using Theorem 8, that an orthonormal triple of vectors B_1, B_2, B_3 is a basis. Prove this. (Show that if $x_i = X \cdot B_i$, where $i = 1, 2, 3$, then

$$X - (x_1 B_1 + x_2 B_2 + x_3 B_3)$$

is perpendicular to all three vectors B_i.)

15. Let $\{B_1, B_2, B_3\}$ be an orthonormal basis and let the vectors X and A have components (x_1, x_2, x_3) and (a_1, a_2, a_3) with respect to this basis. Prove that

$$A \cdot X = a_1 x_1 + a_2 x_2 + a_3 x_3 = \sum_1^3 a_i x_i.$$

4. THE VECTOR COORDINATE NOTATION

This notation reverses the role of letters and indices. Instead of labeling the axes x, y, and z, we label them 1, 2, and 3. The different coordinates of a point are then labeled with the same letter and different subscripts, such as

$$X = (x_1, x_2, x_3),$$

$$A = (a_1, a_2, a_3), \qquad \text{etc.}$$

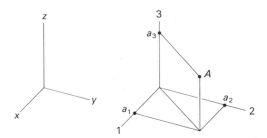

The common letter now represents the point, and different points are designated by different letters. The scalar product is now

$$X \cdot A = x_1 a_1 + x_2 a_2 + x_3 a_3 = \sum_{i=1}^{3} x_i a_i.$$

The equation of the plane through the point C perpendicular to the vector A is

$$0 = A \cdot (X - C) = \sum_{i=1}^{3} a_i(x_i - c_i).$$

There would not be much gain in going over to this notation in two and three dimensions, although it would avoid the necessity of our inelegant convention that $A = (a, b)$, $X = (x, y)$, etc.; and the above summation formulas are perhaps a little nicer than their older forms. However, vector space theory is important in part because it works as well for large dimensions as it does for dimensions two and three; the change to vector notation is a lifesaver when we have to consider functions of a large number of variables. Thus, the formula

$$A \cdot X = a_1 x_1 + a_2 x_2 + \cdots + a_n x_n$$

$$= \sum_{i=1}^{n} a_i x_i$$

for the dot product of two n-dimensional vectors $A = (a_1, a_2, \ldots, a_n)$ and $X = (x_1, x_2, \ldots, x_n)$ is about as easy to write down and discuss as the two- and three-dimensional special cases. But this formula would be virtually impossible to handle if we hadn't gone over to the new notation.

PROBLEMS FOR SECTION 4

1. Write the formula for the cross product $A \times X$ in vector notation. (See Problem 29 in Section 1.)

5. PATHS AND GRADIENTS IN SPACE

We shall not generally graph space paths, because drawing three-dimensional configurations is so complicated. However, three coordinate functions of a parameter t,

$$x = g(t), \qquad y = h(t), \qquad z = k(t),$$

can be viewed as defining a path in space exactly as two functions do in the plane. If t is interpreted as time, the path can be interpreted as the trajectory of a moving particle. The path will be called smooth if the three coordinate functions all have continuous derivatives and if the three derivatives are never simultaneously zero.

A space path can be viewed as a single vector-valued function of t,

$$X = (x, y, z) = (g(t), h(t), k(t)),$$

and, exactly as in the plane, the derivative of this vector-valued function is the vector function

$$\frac{dX}{dt} = (\frac{dx}{dt}, \frac{dy}{dt}, \frac{dz}{dt}) = (g'(t), h'(t), k'(t)).$$

As before, dX/dt can be interpreted as the tangent vector to the path at the point $X = (g(t), h(t), k(t))$.

EXAMPLE 1. The path $X = (\cos t, \sin t, t/2)$ is a spiral rising over the unit circle in the xy-plane. Its tangent vector is

$$\frac{dX}{dt} = (-\sin t, \cos t, 1/2).$$

When $t = \pi/2$, the path point is $(0, 1, \pi/4)$ and the tangent vector is $(-1, 0, 1/2)$.

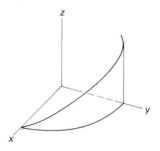

EXAMPLE 2. Show that a space path $X = G(t)$ lies entirely on the surface of a sphere about the origin if and only if

$$X \perp dX/dt \qquad \text{for all } t.$$

. . . .

Solution. To say that the path lies on the surface of a sphere about O is to say that $|X|^2 = x^2 + y^2 + z^2$ has a constant value, say

$$x^2 + y^2 + z^2 = r^2,$$

along the path. Differentiating with respect to t gives

$$2x\frac{dx}{dt} + 2y\frac{dy}{dt} + 2z\frac{dz}{dt} = 0,$$

or

$$2X \cdot \frac{dX}{dt} = 0.$$

Thus $X \perp dX/dt$ for such a path. Conversely, if X and dX/dt are always perpendicular, then this argument can be reversed to show that $|X|$ is constant.

If f is a function of three variables each of which is a function of t, then there is a corresponding chain rule that you probably can guess.

Theorem 10. *If $w = f(X) = f(x, y, z)$ is smooth, and if*

$$X = (x, y, z) = (g(t), h(t), k(t))$$

is a differentiable path running through the domain of f, then along the path w is a differentiable function of t and

$$\frac{dw}{dt} = \frac{\partial w}{\partial x}\frac{dx}{dt} + \frac{\partial w}{\partial y}\frac{dy}{dt} + \frac{\partial w}{\partial z}\frac{dz}{dt}$$

$$= D_1 f\frac{dx}{dt} + D_2 f\frac{dy}{dt} + D_3 f\frac{dz}{dt}.$$

The proof is just like the two-dimensional proof, the only change being that Δw is now broken down into the sum of three partial increments corresponding to separate changes in x, y, and z.

The gradient of a function of three variables $f(x, y, z)$ is the three-dimensional vector

$$\text{grad } f = (D_1 f, D_2 f, D_3 f),$$

and the three-dimensional chain rule has the same dot-product formula as before,

$$\frac{dw}{dt} = \text{grad } f \cdot \frac{dX}{dt},$$

where, of course, the three-dimensional dot product is involved.

The directional derivatives of a function of three variables $f(x, y, z)$ are defined the same way as in the plane, with the same final formulation:

For any space direction U, that is, any three-dimensional unit vector U, the derivative of f in the direction U at the point X is

$$D_U f(X) = \text{grad } f(X) \cdot U.$$

Thus $D_U f(X_0)$ is the value at X_0 of the derivative of f along the straight line $X = Ut + X_0$ through X_0 in the direction U.

EXAMPLE 3. Find the derivative of $f(x, y, z) = xy^2z^3$ at the point $(1, 1, 1)$ in the direction of the vector $(2, 1, -2)$.

. . . .

Solution. The gradient of f at $(1, 1, 1)$ is

$$\text{grad } f(1, 1, 1) = (y^2z^3, 2xyz^3, 3xy^2z^2)\Big|_{(1,1,1)} = (1, 2, 3).$$

The unit vector U in the direction of $A = (2, 1, 2)$ is

$$U = \frac{A}{|A|} = \frac{(2, 1 - 2)}{\sqrt{4 + 1 + 4}} = (\frac{2}{3}, \frac{1}{3}, -\frac{2}{3}).$$

Then

$$D_U f(1, 1, 1) = \text{grad } f(1, 1, 1) \cdot U$$

$$= (1, 2, 3) \cdot (\frac{2}{3}, \frac{1}{3}, -\frac{2}{3})$$

$$= (\frac{2}{3} + \frac{2}{3} - \frac{6}{3}) = -\frac{2}{3}.$$

PROBLEMS FOR SECTION 5

Find the tangent vector to each of the following paths at the given point.

1. $X = (t, t^2, t^3)$; $t = -1$.
2. $X = (2 \cos t, 3 \sin t, t^2)$; $t = \pi/6$.
3. $X = (3t, t - 2/3, \sin t)$; $t = \pi/3$.
4. $X = (\sec t, \tan t, \sin t)$; $t = \pi/3$.
5. $X = (\log t, \sqrt{t}, 1/t)$; $t = 4$.

Determine the gradient vector of each of the following functions at the given point.

6. $f(x, y, z) = \cos xy + \sin xz$; $P = (0, 2, -1)$.
7. $f(x, y, z) = ax + by + cz$; $P = (x_0, y_0, z_0)$.
8. $f(x, y, z) = x^2 + 3xy + y^2 + z^2$; $P = (1, 0, 2)$.
9. $f(x, y, z) = x \cos y + y \cos z + z \cos x$; $P = (\pi/2, \pi/2, 0)$.
10. $f(x, y, z) = \log(x^2 + y^2) + e^{yz}$; $P = (3, 4, 0)$.

Find the derivative of each of the following functions at the given point in the direction of the given vector.

11. $g(x, y, z) = xyz$; $P = (1, 1/2, 2)$; $A = (3, -1, 2)$.

12. $g(x, y, z) = x^2 + 2xy - y^2 + xz + z^2$; $P = (2, 1, 1)$; $A = (1, 2, 2)$.

13. $g(x, y, z) = x^2 y + xye^z - 2xze^y$; $P = (1, 2, 0)$; $A = (2, -3, 6)$.

14. $g(x, y, z) = \log \sqrt{x^2 + y^2 + z^2}$; $P = (1, 2, -1)$; $A = (1, -2, 2)$.

15. $g(x, y, z) = xe^{yz} + yze^x$; $P = (-4, 2, -2)$; $A = (2, 2, -1)$.

16. The directional derivative $D_U f(X_0)$ has the value 3, 1, and -1 in the directions of the positive coordinate axes. Find grad $f(X_0)$.

17. The directional derivative $D_U f(X_0)$ has the value 3, 1, and -1 in the directions of the vectors

$$(0, 1, 1), \qquad (1, 0, 1), \qquad \text{and} \qquad (1, 1, 0),$$

respectively. Find grad $f(X_0)$.

18. Find the direction of *maximum rate of change* of

$$f(x, y, z) = x^2 + xy - 2y^2 + 4xz + z^2$$

at the point $(2, 3, -2)$. What is this maximum rate of change? In what direction is the rate of change at f at this point *minimum?*

19. Find the maximum rate of change and the direction in which it occurs for

$$f(x, y, z) = \log[(x + y)/(x + z)]$$

at the point $(1, 2, 1)$.

Prove the following derivative rules for space paths.

20. If U and X are differentiable vector functions of t, then so is $aU + bX$ for any constants a and b, and

$$\frac{d}{dt}(aU + bX) = a\frac{dX}{dt} + b\frac{dU}{dt}.$$

21. If w is a differentiable scalar function of t, and X is a differentiable vector function of t, then wX is a differentiable vector function of t, and

$$\frac{d}{dt}wX = w\frac{dX}{dt} + \frac{dw}{dt}X.$$

22. If X and U are differentiable vector functions of t, then $X \cdot U$ is a differentiable scalar function of t, and

$$\frac{d}{dt}X \cdot U = X \cdot \frac{dU}{dt} + \frac{dX}{dt} \cdot U.$$

23. If X is a differentiable vector function of t and if $t = \phi(r)$ is differentiable, then X is a differentiable vector function of r, and

$$\frac{dX}{dr} = \frac{dt}{dr}\frac{dX}{dt}.$$

Try to use the above rule in proving the remaining problems.

24. a) Find the smallest distance between the lines

$$X = t(1, 2, -1) \quad \text{and} \quad U = (1, 1, 1) + s(3, -2, 2).$$

b) If X_0 and U_0 are the points on the two lines obtained in part (a), show that $X_0 - U_0$ is perpendicular to the directions of both lines.

25. If X_0 and U_0 are the points on any two space paths at which the paths are closest together, show that the vector $X_0 - U_0$ is perpendicular to both paths.

26. Let $X = F(t)$ be a space path that does not pass through the origin, and let $X_0 = F(t_0)$ be a point on the path that is closest to the origin. Show that the path tangent vector at X_0 is perpendicular to X_0.

27. Let $X = F(t)$ be a twice-differentiable path such that

$$\frac{d^2 X}{dt^2} \equiv 0.$$

Prove that $X = At + K$, so that the path is simply the standard parametric representation of a straight line.

28. Let $X = F(t)$ be a path whose tangent vector has a constant direction. Prove that the path lies along a straight line.

29. Let $X = F(t)$ be a path whose tangent vector is everywhere perpendicular to X. Prove that the path lies on a sphere about the origin. Also prove the converse.

30. Let $X = F(t)$ be a path whose tangent vector is everywhere perpendicular to the constant vector A $(A \neq 0)$. Prove that the path lies in a plane perpendicular to A.

31. Let X be a path whose tangent vector is always parallel to the vector X. Prove that X runs along a line through the origin.

6. LEVEL SURFACES AND TANGENT PLANES

We don't try to draw the graph of a function of three variables

$$w = f(x, y, z),$$

because it would involve trying to picture a four-dimensional configuration.

However, the equation

$$f(x,y,z) = c$$

defines (in general) a two-dimensional surface in three-dimensional space, called a *level surface* of f, and such surfaces *can* be visualized. For example, the level surfaces of

$$f(x,y,z) = x^2 + y^2 + z^2$$

are simply spheres about the origin, such as

$$x^2 + y^2 + z^2 = 4.$$

Although it is possible to draw pictures of such surfaces by the techniques discussed in Chapter 15, it is hard to do, and we shall continue to argue without using pictures.

Now suppose that

$$X = (x,y,z) = (g(t), h(t), k(t))$$

is any differentiable path lying in the level surface

$$f(x,y,z) = c.$$

Then $w = f(x,y,z)$ has the constant value c along this path, and its path derivative dw/dt is zero. That is,

$$\text{grad } f \cdot \frac{dX}{dt} = 0.$$

We therefore have the space analogue of Theorem 5 of Chapter 16.

Theorem 11. *At every point $X_0 = (x_0, y_0, z_0)$ on the level surface $f(x,y,z) = k$, the vector* grad $f(X_0)$ *is perpendicular to the surface, in the sense that* grad $f(X_0)$ *is perpendicular to every differentiable curve lying in the surface and passing through X_0.*

A direction perpendicular to a surface is said to be *normal* to the surface.

EXAMPLE 1. Find a vector normal to the surface $xyz = 1$ at the point $(1,2,1/2)$.

Solution. This is a level surface of $f(x,y,z) = xyz$ and grad f is a normal vector, by the theorem. Therefore,

$$\text{grad } f(1,2,1/2) = (yz, xz, xy)_{(1,2,1/2)}$$
$$= (1, 1/2, 2)$$

is a normal vector.

EXAMPLE 2. Find a parametric equation for the line normal to the surface

$$x^2 + y^2 + z^2 = 14$$

at the point $X_0 = (1, 2, 3)$.

. . . .

Solution. This is the level surface

$$f(x, y, z) = 14,$$

where $f(x, y, z) = x^2 + y^2 + z^2$. Since grad $f(X_0)$ is normal to the surface, it is parallel to the line we want. A vector parametric equation of the line is thus

$$X - X_0 = t \text{ grad } f(X_0),$$

or

$$(x, y, z) - (1, 2, 3) = t(2x_0, 2y_0, 2z_0) = t(2, 4, 6).$$

This can be written in the form

$$(x, y, z) = (2t + 1)(1, 2, 3)$$

or

$$(x, y, z) = s(2, 4, 6),$$

where $s = 2t + 1$.

EXAMPLE 3. Show that the line normal to the surface

$$x^2 + y^2 + z^2 = k$$

at any point $X_0 = (x_0, y_0, z_0)$ passes through the origin. (This is the general form of the special result in Example 2. We expect it to be true because this level surface is a sphere about the origin, and any line perpendicular to a sphere should be a diameter.)

. . . .

Solution. If $f(x, y, z) = x^2 + y^2 + z^2$, then

$$\text{grad } f(X_0) = (2x, 2y, 2z)_{X_0} = (2x_0, 2y_0, 2z_0) = 2X_0.$$

The parametric equation for the line is thus

$$X - X_0 = t \text{ grad } f(X_0) = 2t X_0,$$

or

$$X = (2t + 1)X_0 = s X_0,$$

where $s = 2t + 1$. The points X on the line are the scalar multiples of X_0, and $X = 0$ when $s = 0$.

EXAMPLE 4. Suppose that the space path

$$X = (x, y, z) = (g(t), h(t), k(t))$$

is everywhere perpendicular to the gradient of the function $F(x, y, z)$. Show that the path lies entirely on a level surface of F.

. . . .

Solution. We simply have to show that F is constant along the curve, that is, that the composite function

$$s = F(x, y, z) = F(g(t), h(t), k(t))$$

is a constant function of t. By the chain rule,

$$\frac{dw}{dt} = \text{grad } F \cdot \frac{dX}{dt},$$

and this is everywhere 0, by hypothesis. Since $dw/dt = 0$, w is constant, say

$$w = F(g(t), h(t), k(t)) = k,$$

and this just says that the path lies on the level surface

$$F(x, y, z) = k.$$

We can now give a satisfactory account of the tangent plane to a surface.

Definition. *If $f(x, y, z)$ is smooth, and X_0 is a point on the level surface $f(x, y, z) = k$ at which $\text{grad } f$ is not zero, then the tangent plane to the surface at X_0 is the plane through X_0 that is tangent to every differentiable path passing through X_0 and lying in the surface.*

According to Theorem 11, the tangent plane is perpendicular to $\text{grad } f(X_0)$, and its equation is therefore

$$\text{grad } f(X_0) \cdot (X - X_0) = 0,$$

or

$$[D_1 f(X_0)](x - x_0) + [D_2 f(X_0)](y - y_0) + [D_3 f(X_0)](z - z_0) = 0.$$

EXAMPLE 5. Find the plane tangent to the surface $xe^y \sin z = 1$ at the point $X_0 = (2, 0, \pi/6)$.

. . . .

Solution. First verify that $(2, 0, \pi/6)$ is on the surface:

$$2e^0 \sin\left(\frac{\pi}{6}\right) = 2 \cdot 1 \cdot \frac{1}{2} = 1.$$

Then compute grad $f(2, 0, \pi/6)$, where $f(x, y, z) = xe^y \sin z$:

$$\text{grad } f(2, 0, \pi/6) = (e^y \sin z, xe^y \sin z, xe^y \cos z)_{(2,0,1/2)}$$
$$= (1/2, 1, \sqrt{3}).$$

Substituting these values in the above formula for the tangent plane gives

$$\text{grad } f(X_0) \cdot (X - X_0) = (1/2, 1, \sqrt{3}) \cdot (x - 2, y, z - \pi/6)$$
$$= 1/2(x - 2) + y + \sqrt{3}(z - \pi/2) = 0,$$

or

$$x + 2y + 2\sqrt{3}z - (4 + \sqrt{3}\pi) = 0.$$

This is the equation of the plane tangent to the surface $xe^y \sin z = 1$ at the point $(2, 0, \pi/6)$.

EXAMPLE 6. Show that the above formula for a tangent plane agrees with the calculation in Chapter 15 for the plane tangent to the graph of the function $z = g(x, y)$.

. . . .

Solution. This graph is the level surface $g(x, y) - z = 0$, and the gradient of this special function

$$f(x, y, z) = g(x, y) - z$$

is

$$\text{grad } f(x, y, z) = (D_1 g(x, y), D_2 g(x, y), -1).$$

The dot-product formula

$$\text{grad } f(X_0) \cdot (X - X_0) = 0$$

thus reduces in this case to

$$[D_1 g(x_0, y_0)](x - x_0) + [D_2 g(x_0, y_0)](y - y_0) - (z - z_0) = 0,$$

or

$$z - z_0 = [D_1 g(x_0, y_0)](x - x_0) + [D_2 g(x_0, y_0)](y - y_0).$$

This is the Chapter 15 formula.

PROBLEMS FOR SECTION 6

In Problems 1 through 5, find a vector normal to the level surface at the given point.

1. $-2x^2 - 6y^2 + 3z^2 = 4$; $(1, -1, -2)$.
2. $x^2y - y^2z + z^2x = 33$; $(2, 1/2, 4)$.
3. $xe^{xy} - ze^{yz} = 0$; $(-1, 2, -1)$.
4. $\sin(x^2yz) = 1/2$; $(-1, 1/2, \pi/2)$.
5. a) $\log xyz = 0$; $(1/2, 1/3, 6)$.
   b) $xyz = 1$; $(1/2, 1/3, 6)$.

Find a parametric equation for the line normal to the surface at the given point.

6. $z = xy$; $(2, -1, -2)$.
7. $z = e^x \sin y$; $(1, \pi/2, e)$.
8. $z = x^2y^2$; $(-2, 2, 16)$.
9. $xy + yz + xz = 1$; $(2, 3, -1)$.
10. $\log \sqrt{x^2 + y^2 - z^2} = 0$; $(1, -1, 1)$.

Find the plane tangent to the given surface at the given point.

11. $-2x^2 - 6y^2 + 3z^2 = 4$; $(1, -1, 2)$.
12. $x^2y - y^2z + z^2x = 33$; $(2, 1/2, 4)$.
13. $xe^{xy} - ze^{yz} = 0$; $(-1, 2, -1)$.
14. $z = e^x \sin y$; $(1, \pi/2, e)$.
15. $\log \sqrt{x^2 + y^2 - z^2} = 0$; $(1, -1, 1)$.
16. Find the point(s) on the level surface

$$\frac{x^2}{9} + \frac{y^2}{4} + z^2 = 1$$

at which the tangent plane is perpendicular to $(1, 1, \sqrt{3})$

17. The surface

$$z = -(x^{2/3} + y^{2/3})$$

has a "spike" at the origin and has no tangent plane there. Given any nonvertical plane

$$z = ax + by$$

through the origin; show that there is a circle $x^2 + y^2 = r^2$ within which the plane lies above the surface. (That is, any nonvertical plane can be supported on the point of a vertical pin.)

Prove the following properties of grad f.

18. $\operatorname{grad}(f + g) = \operatorname{grad} f + \operatorname{grad} g$.

19. $\operatorname{grad} fg(X) = f(X)\operatorname{grad} g(X) + g(X)\operatorname{grad} f(X)$.

20. $\operatorname{grad}(f \circ g)(X) = f'(g)(X))\operatorname{grad} G(X)$. (Here f is a function of one varia-
     ble and $f \circ g$ is the composition product of f and g, so that $f \circ g(X) = f(g(X))$.)

21. We often have to find the maximum value of a function $f(X) = f(x, y, z)$
     on a level surface S defined by $g(x, y, z) = 0$. That is, we want to maximize
     $f(X)$ subject to the "constraint" $g(X) = 0$. Suppose that f attains its
     maximum value on S at a point X_0 not at the boundary of S. Show that
     grad $f(X_0)$ and grad $g(X_0)$ are necessarily parallel:

$$\operatorname{grad} f(X_0) = k \operatorname{grad} g(X_0),$$

for some constant k. (Let $X = P(t)$ be any path lying in S and passing
through X_0: $X_0 = P(t_0)$. Then $f(P(t))$ has its maximum value at t_0 and its
derivative must be zero there. Use the vector form of the chain rule to
conclude that grad $f(X_0)$ is perpendicular to every such path, etc.)

According to the above problem, the point X_0 where $f(X)$ has its maximum value
on the level surface $g(X) = 0$ satisfies the equations

$$g(X_0) = 0,$$

$$\operatorname{grad}(f - kg)(X_0) = 0,$$

for some constant k. The second equation is a vector equation, so altogether we
have four equations in the four unknowns

$$x_0, \quad y_0, \quad z_0, \quad k,$$

and we may be able to solve them to determine the maximum point X_0. This is
the method of *Lagrange multipliers*.

For example, suppose we want to find the maximum value of $f(x) = x + 2y$
$- 2z$ on the sphere $x^2 + y^2 + z^2 = 4$, using the Lagrange multiplier k. This
means solving the equations

$$x^2 + y^2 + z^2 = 4,$$

$$(1, 2, -2) - k(2x, 2y, 2z) = 0,$$

simultaneously. The vector equation is equivalent to the three scalar equations

$$2kx = 1, \quad 2ky = 2, \quad 2kz = -2,$$

and substituting from them into the first equation gives

$$\frac{1}{4k^2} + \frac{1}{k^2} + \frac{1}{k^2} = 4,$$

and $k = \pm\frac{3}{4}$. Therefore

$$X = (x, y, z) = \pm\frac{4}{3}(\frac{1}{2}, 1, -1).$$

These are the *critical points* of f on the level surface. We can see that the plus sign given f is its maximum value and the minus sign its minimum value.

22.   Find the maximum and minimum values of

$$f(x, y, z) = x - 2y + 5z$$

on the sphere $x^2 + y^2 + z^2 = 30$.

23.   Find the minimum value of $x^2 + y^2 + z^2$ on the plane $x - 2y + 5z = 1$.

24.   Find the point of the plane $2x - 3y - z = 10$ closest to the origin.

25.   Find the volume of the largest rectangular solid with sides parallel to the coordinate planes that can be inscribed in the ellipsoid

$$\frac{x^2}{9} + \frac{y^2}{4} + z^2 = 1.$$

26.   Same problem for the ellipsoid

$$\frac{x^2}{a^2} + \frac{y^2}{b^2} + \frac{z^2}{c^2} = 1.$$

27.   Find the maximum and minimum values of $f(x, y, z) = (8x - y + 27z)/2$ on the surface

$$x^4 + y^4 + z^4 = 1.$$

7. IMPLICIT FUNCTIONS AGAIN

In the examples we have looked at, the graph of an equation of the form

$$f(x, y, z) = k$$

has always turned out to be a surface in space. The question is whether this necessarily has to happen. The situation turns out to be exactly analogous to our two-variable discussion in the last chapter. The surface $f(x, y, z) = k$ is a level set of the function of three variables $w = f(x, y, z)$, and we can show, exactly as before, that if the point (x_0, y_0, z_0) on this level set is not a critical point of f, then near (x_0, y_0, z_0) the equation

$$f(x, y, z) = k$$

either determines z as a smooth function of x and y,

$$z = \phi(x, y),$$

or it determines y as a smooth function of x and z,

$$y = \psi(x, z),$$

or it determines x as a smooth function of y and z. That is, we can view the level set

$$f(x, y, z) = k$$

near (x_0, y_0, z_0) either as a surface spread out over a region in the xy-coordinate plane, *or* as a surface spread out over a region in the xz-coordinate plane, *or* as a surface spread out over a region in the yz-coordinate plane. These possibilities are not mutually exclusive. Two, or even all three, can occur simultaneously. For example, the linear equation

$$x + 2y - 3z = 5$$

can be solved for each of the variables in terms of the other two, so we can view it in all three ways.

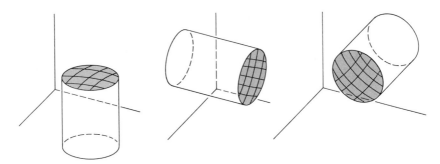

The basic theorem has exactly the structure of Theorem 4 in the last chapter.

Theorem 12. *Suppose that* $w = f(x, y, z)$ *is a smooth function defined throughout a cube* C *of side* $2b$ *about a point* (x_0, y_0, z_0), *and suppose that in* C *the partial derivatives satisfy the inequalities*

$$\frac{\partial w}{\partial z} > m, \qquad |\frac{\partial w}{\partial x}| < M, \qquad |\frac{\partial w}{\partial y}| < M,$$

where m and M are positive constants. We thin the cube down to a rectangular box R about (x_0, y_0, z_0) with a square base S of side $2a$, where

$$a = \frac{bm}{2M},$$

and we take k as the value $f(x_0, y_0, z_0)$ at the center of R. Then in R the level set

$$f(x, y, z) = k$$

is the graph of a smooth function $z = \phi(x, y)$, with domain the whole square S forming the base of R.

The proof is identical to that of the earlier theorem except for a couple of points, and we shall mention only the changes.

In the proof that $f > k$ on the top square of R, that is, that $f(x, y, z_0 + b) > k$, we go from the center point $(x_0, y_0, z_0 + b)$ to the general point $(x, y, z_0 + b)$ in *two* steps, an x-change step and then a y-change step. Each of these steps permits a maximum decrease in the value of f of amount Ma, so the total decrease is less than $2Ma$. This explains the new factor 2 in the definition of a.

The other change is in notation. In the proof that the implicitly defined function ϕ is smooth, we hold either x or y fixed and show that the *partial* derivative of ϕ with respect to the other variable exists, concluding, say, that

$$\frac{\partial z}{\partial y} = -\frac{D_2 f(x, y, z)}{D_3 f(x, y, z)}.$$

The implicit-function theorem lets us fill a logical gap in the derivation of the tangent-plane formula in the last section. We assumed there that grad $f(X_0)$ has the *unique* direction normal to the level surface $f(X) = k$, in the sense of Theorem 11. It is intuitively obvious that a smooth surface has a unique normal direction at each of its points, but we ought to be able to prove this analytically, and it is the implicit-function theorem that lets us do it. It tells us that if grad $f(X_0) \neq 0$, then the level surface $f(X) = k$ is a smooth function graph near X_0. Which variable is the dependent variable only affects the figure we draw, and doesn't matter analytically. So suppose that the level surface S near X_0 is the graph of

$$z = g(x, y),$$

where g is smooth and $z_0 = g(x_0, y_0)$. The xz-plane $y = y_0$ intersects S in the curve $z = g(x, y_0)$. This is the path

$$x = x, \qquad y = y_0, \qquad z = g(x, y_0),$$

with parameter x, and its tangent vector at X_0 is

$$(1, 0, D_1 g(x_0, y_0)).$$

Similarly, the yz-plane $x = x_0$ intersects S in a path with tangent vector

$$(0, 1, D_2 g(x_0, y_0))$$

at X_0. These two vectors are not parallel, so there is a unique direction perpendicular to them both, as we noted in Section 2. Thus the direction normal to S at X_0 is uniquely determined.

The above discussion is based on the hypothesis that grad $f(X_0) \neq 0$. If X_0 is a critical point of f, then the level surface may not have a tangent plane at that point.

EXAMPLE 7. The level surface

$$f(x, y, z) = x^2 + y^2 - z^2 = 0$$

is a cone with vertex at the origin. A cone does not have a tangent plane at its vertex, so the origin must be a critical point of f. We see that it is:

$$\text{grad } f(0) = (2x, 2y, -2x)_{(0,0,0)} = 0.$$

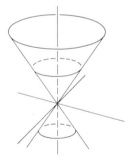

8. ARC LENGTH IN SPACE

So far we have discussed length for plane paths only. When we turn to space paths, a sin of omission that we committed in Chapters 11 and 14 catches up with us. The straightforward definition of path length, as the limit of the lengths of inscribed polygons, involves approximation by finite sums that are not quite Riemann sums. It can be shown, nevertheless, that these sums do converge to the path-length integral,

$$\int_a^b \sqrt{\left(\frac{dx}{dt}\right)^2 + \left(\frac{dy}{dt}\right)^2}\, dt$$

in the plane, and

$$\int_a^b \sqrt{\left(\frac{dx}{dt}\right)^2 + \left(\frac{dy}{dt}\right)^2 + \left(\frac{dz}{dt}\right)^2} \, dt$$

in space. We managed to avoid having to go into this in the plane, but the problem remains with us in space. So we shall simply assume without proof that a smooth space path has length, and that

$$\frac{\text{chord length}}{\text{arc length}} \to 1$$

as chord length approaches 0. Then the argument that we gave in Section 3 of Chapter 11 applies practically verbatim, and proves that

$$\left(\frac{ds}{dt}\right)^2 = \left(\frac{dx}{dt}\right)^2 + \left(\frac{dy}{dt}\right)^2 + \left(\frac{dz}{dt}\right)^2.$$

Thus the magnitude of the path tangent vector dX/dt is again

$$\left|\frac{dX}{dt}\right| = \frac{ds}{dt}.$$

Theorem 7 of Chapter 14 therefore holds for a space path, with the same proof, as does Theorem 8 down to the final statement

$$\left|\frac{dT}{ds}\right| = \kappa.$$

There we had to use a plane argument that does not generalize to space, because our definition of curvature is tied to the plane and does not generalize. However, the formula

$$\kappa = \left|\frac{dT}{ds}\right|$$

would have been a reasonable alternative definition for the curvature of a path. That is, it is reasonable to consider the curvature of a path to be the magnitude of the rate of change of the unit tangent vector T with respect to the arc length s. Then the last part of Theorem 8 in Chapter 14 proves the two possible curvature definitions in the plane to be equivalent. But, for a space curve, the only possible definition of the absolute curvature κ is our new one,

$$\kappa = \left|\frac{dT}{ds}\right|.$$

With this definition, Theorem 8 shows that a space path satisfies the same law

$$\frac{dT}{ds} = \kappa N,$$

where the unit vector N is normal to the path.

The proof of Theorem 9 in Chapter 14 is also the same. Here are the statements of these three theorems again.

Theorem 13. *The parameter of a smooth space path is equal to the path arc length (plus a constant) if and only if the tangent vector is always a unit vector.*

Theorem 14. *Let $X = G(s)$ be a twice-differentiable space path with arc length as parameter and unit tangent vector T. Then*

$$\frac{dT}{ds} = \kappa N,$$

where N is the unit vector normal to the path in the direction in which the path is turning, and κ is (by definition) the absolute curvature of the path.

Theorem 15. *If $X = G(t)$ is the path of a moving particle, with time t as parameter, then the tangential and normal components of the acceleration A are given by*

$$A = \frac{dv}{dt} T + \kappa v^2 N,$$

where v is the speed $ds/dt = |dX/dt|$, and T, κ, and N are as in the above theorem.

EXAMPLE 1. Consider again the spiral

$$X = (\cos t, \sin t, t/2)$$

of Example 1 in Section 5. We see that the tangent vector

$$\frac{dX}{dt} = (-\sin t, \cos t, 1/2)$$

has constant magnitude

$$\left| \frac{dX}{dt} \right| = \frac{\sqrt{5}}{2}.$$

Thus $ds/dt = \sqrt{5}/2$, and if we start measuring s at the point when $t = 0$, then

$$s = \frac{\sqrt{5}}{2} t.$$

Changing to the parameter s, the path is

$$X = (\cos \frac{2}{\sqrt{5}} s, \sin \frac{2}{\sqrt{5}} s, \frac{s}{\sqrt{5}}),$$

Then the unit tangent vector is

$$T = \frac{dX}{ds} = \frac{2}{\sqrt{5}}(-\sin \frac{2}{\sqrt{5}} s, \cos \frac{2}{\sqrt{5}} s, \frac{1}{2}),$$

so

$$\frac{dT}{ds} = \frac{4}{5}(-\cos \frac{2}{\sqrt{5}} s, -\sin \frac{2}{\sqrt{5}} s, 0) = \frac{4}{5} N,$$

where N is a unit vector. Comparing this with

$$\frac{dT}{ds} = \kappa N,$$

we see that the absolute curvature has the constant value

$$\kappa = \frac{4}{5}.$$

PROBLEMS FOR SECTION 8

1. Let $X = F(t)$ be a differentiable vector function of t, and set $\dot{X} = dX/dt$ (Newton's dot notation). Prove the rule

$$\frac{d}{dt}|X| = \frac{X \cdot \dot{X}}{|X|}.$$

    [*Hint.* See Problem 22 in Section 5.]

2. The absolute curvature $\kappa = |dT/ds|$ has the following formula in terms of derivatives with respect to the parameter t.

$$\kappa^2 = \frac{|\dot{X}|^2 |\ddot{X}|^2 - (\dot{X} \cdot \ddot{X})^2}{|\dot{X}|^6}.$$

    Prove this formula, in the following steps:
    a) $\dot{X} = |\dot{X}| T$
    b) $\ddot{X} = |\dot{X}|^2 \frac{dT}{ds} + \frac{(\dot{X} \cdot \ddot{X})}{|\dot{X}|} T$
    c) Compute $\ddot{X} \cdot \ddot{X}$ from (b) and solve for $\kappa^2 = (\frac{dT}{ds} \cdot \frac{dT}{ds})$.

Compute the absolute curvature κ of the following space paths from the formula for κ in Problem 2.

3. $X = (\cos t, \sin t, t/2)$. (Compare with Example 2.)

4. $X = (t, t^2/2, t^3/3)$.

5. $X = (\cos t, \sin t, \sin t)$.

6. The formula in Problem 2 applies to plane curves as well as to space curves. Show that, in the plane, it reduces to our earlier formula

$$\kappa = \frac{|\ddot{y}\dot{x} - \ddot{x}\dot{y}|}{[(\dot{x})^2 + (\dot{y})^2]^{3/2}}.$$

(This requires looking at components.)

7. Let $X = F(t)$ be a space path of zero curvature. Prove that it necessarily runs along a straight line. (Use the formula in Problem 2, and consider two cases. If $\ddot{X}$ is identically zero, then the path is the standard parametric representation of a straight line. (This was an earlier problem, but work it out again.) If $\ddot{X} \neq 0$, then show that $\cos \phi = \pm 1$, where ϕ is the angle between $\dot{X}$ and $\ddot{X}$. Finally, apply Problem 31 of Section 5.)

chapter 18

double and
triple integrals

A function of two variables $f(x,y)$ can be integrated over a plane region G in much the same way that a function of one variable is integrated over an interval. The number we get is called the *double integral* of f over G, and is designated

$$\iint_G f(x,y)\,dx\,dy.$$

Similarly, a continuous function of three variables $f(x,y,z)$, defined over a finite region G in space, has a *triple integral*

$$\iiint_G f(x,y,z)\,dx\,dy\,dz,$$

and a continuous function of n variables $f(x_1,\ldots,x_n)$ has an n-tuple integral over an analogous domain in "n-dimensional space." Such multiple integrals are as important as partial derivatives for the more advanced theory and applications of calculus. The present chapter is intended as an introduction to this large subject, and is restricted to some properties and applications of double and triple integrals.

Like the one-variable integral $\int_a^b f$, the double integral $\iint_G f$ has an extensive array of interpretations and applications. Here are a few.

1   If f is positive, then $\iint_G f(x,y)\,dx\,dy$ is the volume of the region that lies between the surface $z = f(x,y)$ and the xy-plane, with the plane region G as its base.

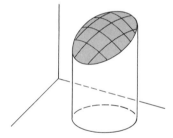

2   For any continuous f,

$$\frac{\iint_G f(xy)\,dx\,dy}{\text{area of } G} \quad \text{is the average value of } f \text{ over } G.$$

3   If G is a plane lamina with variable mass density $\rho(x,y)$, then

$$\iint_G \rho(x,y)\,dx\,dy$$

is the mass of G.

4   If $\rho(x,y)$ is a probability density function defined over the plane, then $\iint_G \rho$ is the probability that a randomly chosen point will land in the region G.

5   The surface area of the piece of the graph of $z = f(x,y)$ that lies over G is given by

$$\iint_G \sqrt{1 + \left(\frac{\partial z}{\partial x}\right)^2 + \left(\frac{\partial z}{\partial y}\right)^2}\,dx\,dy.$$

This is only a partial list. In general, if Q is any quantity that can be interpreted as being spread out over the plane, and if the distribution of Q has a continuous density $\rho(x,y)$, then the amount of Q lying over the plane region G is

$$\iint_G \rho(x,y)\,dx\,dy.$$

Like the definite integral $\int_a^b f$, the double integral $\iint_G f$ is ultimately defined as the limit of certain Riemann sums. But the purely analytic theory will be postponed, and we shall proceed provisionally on the basis of geometric reasoning. We shall assume certain geometrically obvious properties of the volume of a solid body in space, and shall develop the double integral on the basis of these assumptions. This is analogous to our earlier treatment of the integral of a function of one variable, which was based on geometrically obvious properties of the areas of plane regions.

1. VOLUMES BY ITERATED INTEGRATION

Suppose that the region G in the xy-plane is bounded by the graphs of continuous functions g and h, and lies between the x values $x = a$ and $x = b$, as in either figure below.

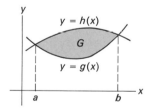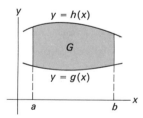

We consider the solid based on G and bounded above by the piece of the surface $z = f(x,y)$ lying over G, where f is a positive continuous function of two variables.

This solid is the intersection of the region under the surface $z = f(x,y)$ and the cylinder having G as its base. We shall often refer to it as the solid *based* on G and *capped* by the surface $z = f(x,y)$.

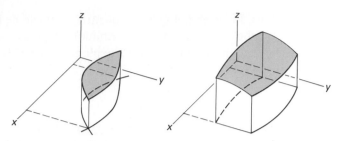

The cross section of this solid in the plane $x = x_0$ is the plane region under the graph of $z = f(x_0, y)$, from $y_1 = g(x_0)$ to $y_2 = h(x_0)$. Its area is

$$A(x_0) = \int_{y_1}^{y_2} f(x_0, y)\, dy = \int_{g(x_0)}^{h(x_0)} f(x_0, y)\, dy,$$

by the area considerations in Chapter 8. As we vary the slicing plane, the cross section changes continuously, in the manner discussed in Section 9 of Chapter 8.

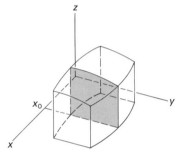

(The general proof of this fact for an arbitrary positive continuous function $z = f(x, y)$ is beyond our present capabilities, but it is easy enough to prove for the types of functions we actually meet in practice. See Problems 30 and 31.) Therefore,

$$V = \int_a^b A(x)\, dx,$$

by Theorem 10 of Chapter 8. Substituting the above value for $A(x)$, we end up with a new volume formula:

$$V = \int_a^b \left[\int_{g(x)}^{h(x)} f(x, y)\, dy \right] dx.$$

This evaluation of V is called an *iterated* integral. We first integrate $f(x,y)$ with respect to y, holding x fixed. The limits of integration will depend on the fixed value of x, and so will the resulting value of the integral. It was called $A(x)$ above. Then we integrate this function of x between fixed limits a and b, and obtain the number that is called the iterated integral. All in all, we start with a function of two variables $f(x,y)$, and first "integrate y out," leaving a function of the one variable x, and then "integrate x out," ending up with a number.

Here is the theorem we have proved (supposing that the x cross section varies continuously with x):

Theorem 1. *Let S be the solid based on the region G in the xy-plane and capped by the surface $z = f(x,y)$, where f is continuous and nonnegative on G. Suppose, furthermore, that in the xy-plane G is bounded below and above by the graphs of the continuous functions g and h, and left and right by the straight lines $x = a$ and $x = b$. Then the volume V of S is given by the iterated integral*

$$V = \int_a^b \left[\int_{g(x)}^{h(x)} f(x,y)\,dy \right] dx.$$

EXAMPLE 1. Evaluate the iterated integral

$$\int_0^1 \left(\int_{x^2}^x xy^2\,dy \right) dx.$$

Sketch the "domain of integration."

. . . .

Solution. We first compute the inside y integral, treating x as a constant:

$$\int_{x^2}^x xy^2\,dy = \frac{xy^3}{3} \bigg]_{x^2}^x = \frac{1}{3}[x^4 - x^7].$$

Then we compute the outside x integral:

$$\frac{1}{3}\int_0^1 (x^4 - x^7)\,dx = \frac{1}{3}\left[\frac{x^5}{5} - \frac{x^8}{8} \right]_0^1 = \frac{1}{3}\left(\frac{1}{5} - \frac{1}{8} \right) = \frac{1}{40}.$$

Thus,

$$\int_0^1 \left(\int_{x^2}^x xy^2\,dy \right) dx = \frac{1}{40}.$$

For each fixed x_0 between $x = 0$ and $x = 1$ the inside function of y is integrated from $y = x_0^2$ to $y = x_0$. The domain of integration is thus the plane region between the graphs $y = x^2$ and $y = x$, as shown below.

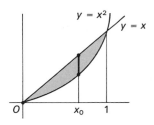

EXAMPLE 2. Find the volume of the space region under the graph of $z = x + y$ based on the triangle in the xy-plane that is cut off from the first quadrant by the line $x + y = 1$.

. . . .

Solution. The figure is shown below. Such a figure should usually be drawn. We don't really need it here, because when the domain of integration is provided explicitly, as it is here, the volume can be computed by just plugging into the iterated integration formula. The triangular domain is bounded above and below by the graphs $y = 1 - x$ and $y = 0$, and runs left to right from $x = 0$ to $x = 1$. The volume integral is therefore

$$V = \int_0^1 \left[\int_0^{1-x} (x + y)\, dy \right] dx.$$

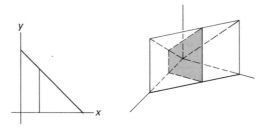

For the inside integral we have

$$\int_0^{1-x} (x + y)\, dy = \left[xy + \frac{y^2}{2} \right]_0^{(1-x)} = x(1 - x) + \frac{(1 - x)^2}{2}$$

$$= \frac{1}{2}(1 - x^2).$$

Therefore,

$$V = \int_0^1 \frac{1 - x^2}{2}\, dx = \frac{1}{2}\left[x - \frac{x^3}{3} \right]_0^1 = \frac{1}{3}.$$

REMARK. The brackets or parentheses setting off the inside integral of an iterated integral are generally omitted, it being understood that the inside integral is computed first. We would thus write the above iterated integral as follows:

$$V = \int_0^1 \int_0^{1-x} (x + y)\, dy\, dx.$$

EXAMPLE 3. Find the volume of the region in the first octant under the graph of $z = f(x, y) = 1 - y - x^2$.

. . . .

Solution. The domain is the region in the xy-plane bounded by the coordinate axes and the level curve $f(x, y) = 0$, that is, the parabola $y = 1 - x^2$.

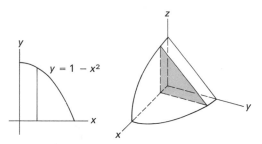

So the inside y integral runs from $y = 0$ to $y = 1 - x^2$, and the volume formula is

$$V = \int_0^1 \int_0^{1-x^2} (1 - y - x^2)\, dy\, dx.$$

The figure for the solid at the right above also shows the edge curves in the other two coordinate planes. The edge curve in the xz-coordinate plane is obtained by setting $y = 0$ in the equation

$$z = 1 - y - x^2,$$

and is the parabola $z = 1 - x^2$. In the zy-coordinate plane it is the straight line $z = 1 - y$. These edge curves, together with the cross section at x, are enough to give a good visual image of the solid.

Here is the volume computation:

$$V = \int_0^1 \int_0^{1-x^2} (1 - y - x^2)\, dy\, dx$$

$$= \int_0^1 \left[(1 - x^2)y - \frac{y^2}{2} \right]_0^{1-x^2} dx$$

$$= \int_0^1 \frac{(1 - x^2)^2}{2}\, dx = \frac{1}{2}\left[x - \frac{2x^3}{3} + \frac{x^5}{5} \right]_0^1 = \frac{4}{15}.$$

PROBLEMS FOR SECTION 1

Compute the following iterated integrals. In each case sketch the region that is the domain of integration in the xy-plane.

1. $\int_0^1 \int_0^1 xy\, dy\, dx$

2. $\int_0^1 \int_0^1 dy\, dx$

3. $\int_0^{\pi/2} \int_0^x \cos y\, dy\, dx$

4. $\int_0^{\pi/2} \int_0^{\sin x} dy\, dx$

5. $\int_{-1}^1 \int_x^1 xy\, dy\, dx$

6. $\int_0^1 \int_0^{\sqrt{1-x^2}} x\, dy\, dx$

7. $\int_{-1}^1 \int_{-\sqrt{1-x^2}}^{\sqrt{1-x^2}} y\, dy\, dx$

8. $\int_1^2 \int_x^{2x} y\, dy\, dx$

9. $\int_0^1 \int_{x^2}^x (x + y)\, dy\, dx$

10. $\int_{1/4}^{3/2} \int_{x^2}^x (x + y)\, dy\, dx$

Use iterated integration to find the volume of each of the space regions described in Problems 11 through 21 below. Sketch each configuration.

11. The unit cube $0 \le x \le 1, 0 \le y \le 1, 0 \le z \le 1$.

12. The solid based on the rectangle $0 \le x \le a, 0 \le y \le b$, in the xy-plane, and capped by the plane $z = c$.

13. The solid based on the unit square and capped by the plane $z = x + y$.

14. The solid based on the first quadrant of the unit circle $x^2 + y^2 \le 1$ and capped by the plane $z = y$.

15. The piece of the first octant cut out by the plane $x + y = 1$ and the parabolic cylinder $z = 1 - y^2$.

16. The piece of the first octant cut out by the plane $x + y = 1$ and the parabolic cylinder $z = 1 - x^2$.

17. The piece of the first octant cut out by the plane $z = x$ and the cylinder $x^2 + y^2 = 1$.

18. The piece of the first octant cut out by the plane

$$\frac{x}{a} + \frac{y}{b} + \frac{z}{c} = 1,$$

where a, b, and c are all positive.

19. The interior of the ellipsoid

$$\frac{x^2}{a^2} + \frac{y^2}{b^2} + \frac{z^2}{c^2} = 1.$$

20. The solid bounded by the parabolic cylinders $z = 1 - x^2$, $x = 1 - y^2$, $x = y^2 - 1$, and the plane $z = 0$. (Divide the base region into two pieces and compute the volume as the sum of two integrals.)

21. The solid bounded by the parabolic cylinder $x = 2 - y^2$ and the planes $y = x, z = 0$, and $z = 1 - y$. (Here also the base region will have to be divided into two pieces.)

22. If $f_1(x,y) \le f_2(x,y)$ over the region G, then the solid cut out from the cylinder through G by the graphs of f_2 and f_1 has the volume

$$V = \iint_G (f_2(x,y) - f_1(x,y)) \, dy \, dx.$$

Justify this.

23. Find the volume of the solid lying over the unit square $0 \le x \le 1$, $0 \le y \le 1$, and bounded above and below by the surface $z = y$ and $z = y^3$. (Apply the formula in Problem 22.)

24. Find the volume of the solid lying over the quarter of the unit circle in the first quadrant and bounded above and below by the surface $z = y$ and $z = y^3$.

25. Find the volume of the piece of the first octant lying between the surfaces $z = y$ and $z = 1 - x^2$.

26. For some base regions G, it may be convenient and even necessary to interchange the role of x and y and apply the volumes-by-slicing formula to varying y-slices. Draw the base region G for which the following iterated integral is a correct volume formula, and verify its correctness by giving an argument based on volumes-by-slicing, as we did in the text:

$$V = \int_a^b \left(\int_{g(y)}^{h(y)} f(x,y)\,dx \right) dy.$$

27. Redo Problem 17 by this method.

28. Redo Problem 20 by this method. Now the base region can be left in one piece.

29. Redo Problem 21 by this method. This time the base region can be left intact.

30. Let S be the solid based on the rectangle $a \le x \le b$, $c \le y \le d$, and capped by the surface $z = f(x,y)$. Suppose that the partial derivative $\partial z/\partial x$ exists and is bounded by B ($|\partial z/\partial x| \le B$) on this rectangle. Let R_{x_0} be the cross section of S in the plane $x = x_0$. Thus, R_{x_0} is the region in this plane under the graph $z = f(x_0, y)$, from $y = c$ to $y = d$. We want to show that R_{x_0} varies continuously as x_0 varies near a fixed $x_0 = k$, and for this purpose we draw all these regions R_{x_0} in a single yz-plane. (See left below.)

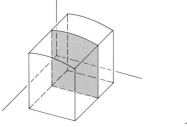 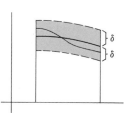

We choose any positive number δ and draw the band of vertical width 2δ centered on the curve $z = f(k,y)$, as shown in the righthand figure. (The band runs from the bottom curve ($z = f(k,y) - \delta$) up to the top curve $z = (f(k,y) + \delta)$.)

Show, then, that the graph $z = f(x_0, y)$ lies in this band for every x_0 in the interval $(k - \delta/B, k + \delta/B)$. Since the band width 2δ can be taken as small as desired, this shows that R_{x_0} varies continuously with x_0.

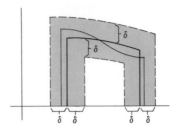

31. Consider the situation of the above problem, but with the base rectangle replaced by the base region G lying between the graphs $y = g(x)$ and $y = h(x)$, from $x = a$ to $x = b$. Assume that the derivatives g' and h' are bounded by the same B. Now the cross section R_{x_0} has varying width. Prove, however, that R_{x_0} is squeezed between the minimum and maximum regions of the type shown above, for every x_0 in the interval $(k - \delta/B, k + \delta/B)$.

2. RIEMANN SUMS AND THE DOUBLE INTEGRAL.

Let $f(x,y)$ be a continuous function defined on a plane region G. Then we can define what we mean by a Riemann sum for f over G, and we shall see that the relationship between the Riemann sums for f and the definite integral of f is the same for such a function of two variables as it is for a function of one variable.

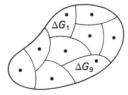
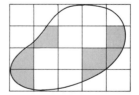

In order to define a Riemann sum for $f(x,y)$ over G, we first subdivide G in any manner into a finite number of small subregions ΔG_i where $i = 1, \ldots, n$, as indicated in the figures above. The simplest way to do this is to impose a lattice of rectangles over G, as at the right, but it is desirable for flexibility to allow more general subdivisions. Next, we choose an arbitrary "evaluation point" (x_i, y_i) in the subregion ΔG_i, for each i. Then the Riemann sum for f associated with this subdivision and evaluation set is

$$\sum_{i=1}^{n} f(x_i, y_i)\Delta A_i,$$

where ΔA_i is the area of the subregion ΔG_i.

EXAMPLE 1. We consider $f(x,y) = y/x$ over the unit square G having its lower left corner at $(1,0)$. We subdivide the square into four congruent subsquares, and use the lower left corner of each as its evaluation point. Then the Riemann sum for y/x over G associated with this subdivision and this evaluation set has the value

$$f\left(1,0\right)\Delta A_1 + f\left(\frac{3}{2},0\right)\Delta A_2 + f\left(1,\frac{1}{2}\right)\Delta A_3 + f\left(\frac{3}{2},\frac{1}{2}\right)\Delta A_4$$

$$= \frac{0}{1}\cdot\frac{1}{4} + \frac{0}{3/2}\cdot\frac{1}{4} + \frac{1/2}{1}\cdot\frac{1}{4} + \frac{1/2}{3/2}\cdot\frac{1}{4}$$

$$= \frac{1}{4}\left[0 + 0 + \frac{1}{2} + \frac{1}{3}\right]$$

$$= \frac{1}{4}\cdot\frac{5}{6} = \frac{5}{24}.$$

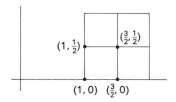

EXAMPLE 2. If we use the *upper* right corners as evaluation points in the above situation, then the Riemann sum has the value

$$f\left(\frac{3}{2},\frac{1}{2}\right)\Delta A_1 + f\left(2,\frac{1}{2}\right)\Delta A_2 + f\left(\frac{3}{2},1\right)\Delta A_3 + f\left(2,1\right)\Delta A_4$$

$$= \frac{1/2}{3/2}\cdot\frac{1}{4} + \frac{1/2}{2}\cdot\frac{1}{4} + \frac{1}{3/2}\cdot\frac{1}{4} + \frac{1}{2}\cdot\frac{1}{4}$$

$$= \frac{1}{4}\left[\frac{1}{3} + \frac{1}{4} + \frac{2}{3} + \frac{1}{2}\right] = \frac{1}{4}\cdot\frac{7}{4} = \frac{7}{16}.$$

EXAMPLE 3. Find the Riemann sum for the above function f and Region G if G is divided into nine equal squares and the lower left corners are taken as evaluation points.

. . . .

Solution. The points on the x-axis again contribute zero terms, and the remaining 6 terms are

$$f\left(1,\frac{1}{3}\right)\Delta A_4 + f\left(\frac{4}{3},\frac{1}{3}\right)\Delta A_5 + f\left(\frac{5}{3},\frac{1}{3}\right)\Delta A_6 + f\left(1,\frac{2}{3}\right)\Delta A_7 + f\left(\frac{4}{3},\frac{2}{3}\right)\Delta A_8$$

$$+ f\left(\frac{5}{3},\frac{2}{3}\right)\Delta A_9$$

$$= \frac{1/3}{1}\cdot\frac{1}{9} + \frac{1/3}{4/3}\cdot\frac{1}{9} + \frac{1/3}{5/3}\cdot\frac{1}{9} + \frac{2/3}{1}\cdot\frac{1}{9} + \frac{2/3}{4/3}\cdot\frac{1}{9} + \frac{2/3}{5/3}\cdot\frac{1}{9}$$

$$= \frac{1}{9}\left[\frac{1}{3} + \frac{1}{4} + \frac{1}{5} + \frac{2}{3} + \frac{2}{4} + \frac{2}{5}\right] = \frac{1}{9}\left[\frac{47}{20}\right] = \frac{47}{180}.$$

When we vary the subdivision and evaluation set, we get a different value for the Riemann sum. We shall see that if the function f and region G are reasonably well-behaved, then this varying Riemann sum Σ approaches a limit I as the subdivision is taken finer and finer. More exactly, we can make the error in the approximation $\Sigma \approx I$ as small as we wish simply by using a subdivision all of whose subregions ΔG_i have suitably small size. This limit I is called the *double integral* of f over G. Thus, when we say that the double integral exists and is a certain number, we mean that the varying Riemann sum approaches this number as its limit as the maximum size of its subregions approaches zero.

Definition. *The double integral of f over G, designated*

$$\iint_G f(x,y)\,dA,$$

is the limit (if it exists) of the Riemann sums for f over G in the above sense.

The basic theorem is then:

Theorem 2. *If $f(x,y)$ is continuous on a bounded closed region G, then the double integral*

$$\iint_G f(x,y)\,dA$$

exists.

Moreover, when the region G is simple enough, the double integral of f over G can be evaluated by iterated integration.

Theorem 3. *Let $f(x,y)$ be a continuous function defined over the closed plane region G, and suppose that G is bounded above and below by the graphs of $y = h(x)$ and*

$y = g(x)$, *and runs from* $x = a$ *to* $x = b$. *Then*

$$\iint_G f(x,y)\,dA = \int_a^b \int_{g(x)}^{h(x)} f(x,y)\,dy\,dx.$$

If, instead, the left and right parts of the boundary of G lie along the graphs $x = p(y)$ *and* $x = q(y)$, *respectively, and G extends from* $y = c$ *to* $y = d$, *then*

$$\iint_G f(x,y)\,dA = \int_c^d \int_{p(y)}^{q(y)} f(x,y)\,dx\,dy.$$

In the long run Theorems 2 and 3 have to be given purely analytic proofs. Although such proofs are beyond our present capabilities, we can prove Theorem 2 subject to natural assumptions about volume, and using one final property of a continuous function. On this basis we shall show that if V is the volume of the solid S based on G and capped by the graph of f, then V is approximated arbitrarily closely by the Riemann sums for f over G. That is, the Riemann sums approach V as their limit, so we have a geometric proof of the existence of the double integral $\iint_G f(x,y)\,dA$. Theorem 3 is then simply the statement that this same volume can be computed by iterated integration, by Theorem 1 and its alternate form with variables reversed. The geometric proof of Theorem 2 will be given in the next section.

Theorems 2 and 3 have important applications, most of which come about as follows. We want to compute some quantity Q from geometry or physics, or from some other discipline. We first show that Q can be estimated by a certain type of finite sum. We then observe that this finite sum can also be interpreted as a Riemann sum for a certain function $\rho(x,y)$ over a region G. But this means that

$$Q = \iint_G \rho(x,y)\,dA,$$

since both Q and the double integral are approximated arbitrarily closely by the same estimating number. Finally, we *compute* Q by iterated integration, by virtue of Theorem 3.

In more detail, a quantity Q can be calculated in this way if it can be interpreted as a "distribution" over the plane with a continuous "density" $\rho(x,y)$. The procedure is exactly like that discussed in Chapter 10 for quantities distributed over the line. We first show that if ΔQ is the amount of the quantity lying over an incremental region ΔG, and if (x_0, y_0) is an arbitrary "evaluation point" in ΔG, then

$$\Delta Q \approx \rho(x_0, y_0)\Delta A,$$

with an error that is small in comparison with the area ΔA of ΔG. When we sum these estimates over a subdivision of a region G, we see that the amount Q of the quantity over G is given approximately by

$$Q \approx \sum \rho(x_i, y_i)\Delta A_i,$$

the error being small in comparison with the area A of G. Since the sum on the right is also a Riemann sum for ρ over G, we conclude that Q is given exactly by the double integral

$$Q = \iint_G \rho(x,y)\,dA.$$

There are several applications of this procedure in the remaining sections of this chapter.

PROBLEMS FOR SECTION 2

1. Evaluate the Riemann sum for $f(x,y) = x/(1+y)^2$ over the unit square $0 \leq x \leq 1$, $0 \leq y \leq 1$, using the subdivision obtained by dividing the square into nine congruent subsquares and choosing the *upper right* vertex of each square as evaluation point.

2. Evaluate the Riemann sum for the same function and subdivision but with the *lower left* vertices as evaluation points.

3. Prove that if the double integrals of f and g exist over a region G, then so does the double integral of $af(x,y) + bg(x,y)$ and

$$\iint_G [af(x,y) + bg(x,y)]\,dA = a\iint_G f(x,y)\,dA + b\iint_G g(x,y)\,dA.$$

Assume that the usual limit laws hold here: The limit of a sum is the sum of the limits, etc.

4. Prove that if $f(x,y)$ has the constant value k over the region G, then its double integral over G exists and

$$\iint_G f = kA,$$

where A is the area of G.

5. The proof to be given shortly that a continuous function has a double integral over a bounded closed region G works only for positive functions. Prove from this that it is nevertheless true that an arbitrary continuous function $f(x,y)$ has a double integral over G. [Hint: Let k be any number less than the minimum value of f on G. Then $g(x,y) = f(x,y) - k$ is positive on G. Now apply the above two problems.

6. Prove that if the base region G is divided into two nonoverlapping subregions G_1 and G_2, then

$$\iint_G f\,dA = \iint_{G_1} f\,dA + \iint_{G_2} f\,dA.$$

(This should again reduce to the assumption that a certain limit law remains valid for this type of limit.)

A corollary of Theorem 3 is that if the base region G can be described in *both* ways specified there, then the two iterated integrals are necessarily equal:

$$(*) \qquad \int_a^b \int_{g(x)}^{h(x)} f(x,y)\, dy\, dx = \int_c^d \int_{p(y)}^{q(y)} f(x,y)\, dx\, dy.$$

In Problems 7 through 12, an iterated integral is given that specifies a base region G in one of the above ways, and you are to work out the correct limits of integration for the other integral. (This may require that the region G be divided into two or more parts, with the given integral expressed as the sum of two or more integrals.) In each case *draw* the region G from the given limits of integration and then describe G the other way.

7. $\int_0^1 \int_0^x f(x,y)\, dy\, dx$

8. $\int_0^1 \int_0^{x^2} \cdots\, dy\, dx$

9. $\int_0^1 \int_{x^2}^x \cdots\, dy\, dx$

10. $\int_0^1 \int_{1-x}^{1+x} \cdots\, dy\, dx$

11. $\int_{-1}^1 \int_0^{y^2} \cdots\, dx\, dy$

12. $\int_0^1 \int_{y^2}^1 \cdots\, dx\, dy$

The iterated integral identity $(*)$ can be used to evaluate either integral in terms of the other. This is called *reversing the order of integration*. Sometimes a difficult integration can be made easier by reversing the order of integration. Compute the following iterated integrals by reversing the order of integration.

13. $\int_0^1 \int_y^1 e^{-x^2}\, dx\, dy$

14. $\int_0^1 \int_0^{y^2} y \cos(1 - x)^2\, dx\, dy$

15. $\int_0^1 \int_x^1 \sqrt{1 - y^2}\, dy\, dx$

16. Suppose we choose N points $(x_1, y_1), \ldots, (x_N, y_N)$ that are scattered in a fairly uniform way throughout a plane region G. That is, there must be roughly the same number of chosen points "per unit area" wherever we look in G. If the area of G is A, give some kind of an argument to support the claim that

$$\frac{A}{N} \sum_{i=1}^N f(x_i, y_i)$$

is approximately equal to the double integral of f over G.

17. Let G be the unit square $0 \le x \le 1, 0 \le y \le 1$.
    a) Compute $\iint_G xy\, dA$ by iterated integration.
    b) Compute the above estimate of this double integral for the nine vertices obtained by dividing the unit square into four equal subsquares.

18. Work out the above problem for $f(x,y) = x + y$.

19. The same for $f(x,y) = x^2 + y^2$.

20. The same for $f(x,y) = x^2 y^2$.

21. The estimates in Problems 19 and 20 were not very good. They would improve with a finer subdivision, of course, but there is something else

wrong. In a sense a vertex is surrounded by only 1/4 as much area as an interior point and an edge point that is not a vertex is surrounded by 1/2 as much area as an interior point. We are therefore not "weighting" the effect of these boundary points correctly. In order to adjust for this edge effect, suppose we count each of the four corner values $f(x_i, y_i)$ once, each of the four edge midpoint values twice, and the middle value $f(1/2, 1/2)$ four times. Then we have effectively increased the number of terms in the sum from 9 to 16, so we now divide by 16 rather than by 9. Recompute the estimate for $f(x, y) = x^2 + y^2$, using this weighted sum.

22. Do the above problem for $x^2 y^2$.

23. a) If the unit square is divided into 9 equal subsquares and if their 16 vertices are weighted as in Problem 21, show that we must now divide by $N = 36$.

   b) Recompute the estimate for the double integral of $x^2 y^2$ over the unit square using this new weighted sum.

   c) Compute the double integral again and compare answers.

3. PROOF OF THEOREM 2

We assume, to begin with, that any solid body has a uniquely determined volume. By a solid body we mean a region in space bounded by (pieces of) a finite number of surfaces. For example, a circular cylinder is a solid body bounded by three surfaces, its two bases and its curved lateral surface. A cube is also a cylinder, but it is a solid body bounded by six surfaces, its six faces.

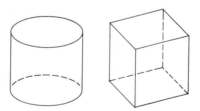

Furthermore, we assume:

1. The volume of a cylinder is given by

$$V = Ah$$

where A is the area of its base and h is its altitude.

2. If a solid body is divided into two nonoverlapping pieces, then its volume is the sum of the volumes of its two constituents

$$V = V_1 + V_2$$

 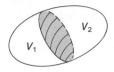

Note that:

2'. If a solid S includes a solid S_1, then the volume of S is greater than the volume of S_1.

This follows from principle (2), because

$$V = V_1 + V_2 > V_1,$$

where V is the volume of S, V_1 is the volume of S_1, and V_2 is the volume of the solid S_2 obtained by removing S_1 from S.

We are interested in the volume of the solid S based on a region G in the xy-plane and capped by the graph $z = f(x,y)$, where f is a positive continuous function. We consider, to begin with, a small incremental closed subregion ΔG of G, and the piece ΔS of the solid S lying over ΔG. Let m and M be the minimum and maximum values of f on the subregion ΔG, and let C_m and C_M be the cylinders based on ΔG with altitudes m and M respectively. Then C_m is included in ΔS, which is included in C_M, so

$$\text{volume } (C_m) \leq \text{volume } (\Delta S) \leq \text{volume } (C_M),$$

by the volume principle (2'). By (1), this is the inequality

$$m\Delta A \leq \Delta V \leq M\Delta A,$$

where ΔA is the area of ΔG.

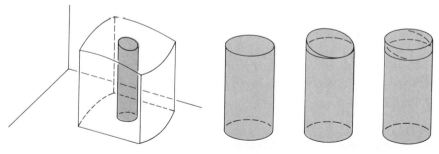

Now let (x_0, y_0) be an arbitrary "evaluation point" in ΔG. Then

$$m \leq f(x_0, y_0) \leq M,$$

so

$$m\Delta A \leq f(x_0, y_0)\Delta A \leq M\Delta A.$$

Thus the numbers ΔV and $f(x_0,y_0)\Delta A$ both lie on the closed interval $[m\Delta A, M\Delta A]$, so they differ by less than the length of this interval. That is,

$$\Delta V \approx f(x_0,y_0)\Delta A,$$

with an error less than $(M - m)\Delta A$ in magnitude. If the values of f on the incremental subregion ΔG vary by less than a number e, that is, if $M - m < e$, then the error in the above estimate is less than $e\Delta A$.

Now suppose that we are given a positive number e, and suppose that we can then subdivide the region G into a finite number of subregions ΔG_i, where $i = 1, \ldots, n$, that are small enough so that the values of f vary by less than e over each subregion ΔG_i. (That is, if M_i and m_i are the maximum and minimum values of f on ΔG_i, then $M_i - m_i < e$, for each subregion ΔG_i.) Let (x_i,y_i) be an arbitrary evaluation point in the subregion ΔG_i, for each i. Then, for each i, we know from the above reasoning that the volume ΔV_i based on ΔG_i can be estimated as the product $f(x_i,y_i)$ times the area ΔA_i of ΔG_i,

$$\Delta V_i \approx f(x_i,y_i)\Delta A_i,$$

with an error less than $e\Delta A_i$. The sum of these errors is less than

$$\sum e\Delta A_i = e \sum \Delta A_i = eA,$$

where A is the area of G. Also $V = \sum \Delta V_i$ is the volume of the solid S based on G. So when we sum the above estimates, we end up with the estimate

$$V \approx \sum_{i=1}^{n} f(x_i,y_i)\Delta A_i,$$

with an error less than eA.

We have thus shown that the volume V that we are interested in can be estimated by a Riemann sum for f over G. Moreover, we can make the error bound eA as small as we wish by first taking e small and then choosing a corresponding sufficiently fine subdivision. That is, V is the limit of the Riemann sums of f over G. In particular, the double integral

$$\iint_G f(x,y)\,dA$$

exists, proving Theorem 2, and its value is the volume V of the space region based on G and capped by the graph of f.

The only thing that remains to be examined is our assumption that if we are given any positive number e, then any suitably "fine" subdivision will consist of subregions ΔG_i on each of which the values of f vary by less than e.

We define the *diameter* of a closed region to be the maximum distance apart of any two of its points. Thus the diameter of a square of side s is $\sqrt{2}s$, the length of its diagonal, and the diameter of a triangle is the longest of its side lengths. When we say that a subdivision is "suitably fine," we just mean that the diameters of its subregions ΔG_i are all less than a suitably small number d. The fact that we can find such a number d for a given e is a new property of a continuous function on a closed domain. Here is a direct statement of this property without any reference to subdivisions.

Theorem 4. *If $f(x,y)$ is continuous on a bounded closed region G, and if we are given any positive number e, then we can find a positive number d that is suitably small in the following sense: Whenever two points (x,y) and (x',y') in G are closer together than the distance d, then*

$$|f(x,y) - f(x',y')| < e.$$

This property says that we have *uniform* control over the continuous variation of $f(x,y)$ over the whole of G. The theorem states that any function $f(x,y)$ that is continuous on a bounded closed region G is necessarily *uniformly* continuous in the above sense. The proof of this theorem is beyond the level of a first course, but we can verify it for the situations that usually arise in practice. (See Problems 1 and 2.) In any event, if we are given e we can now proceed as follows. First we find the above uniform control number d. Then we subdivide G into subregions ΔG_i each of which has diameter less than d. Then the variation of f is less than e on each subregion ΔG_i and the proof of Theorem 2 is complete.

PROBLEMS FOR SECTION 3

We have been using the diameter of a closed region G as a measure of how big G is, in connection with Riemann sum subdivisions. There is a refinement of this measure that we might call the *length* of G. We say that

$$\text{length of } G \leq l$$

if any two points of G can be connected by a path that lies in G and has length less than or equal to l. If, in addition, there is a pair of points in G for which the shortest connecting path has length l, then the length of G is exactly l.

1. Now let G be a region whose length is less than l, and let $f(x,y)$ be continuously differentiably on G, with both partial derivatives bounded by a constant K. Prove that

    $$|f(x_1,y_1) - f(x_2,y_2)| \leq \sqrt{2}Kl$$

    for any two points (x_1,y_1) and (x_2,y_2) in G.

2. Let f be as in the problem above and let G be a convex region. This means that if X_1 and X_2 are any two points of G, then the line segment joining them lies entirely in G. Prove that f is uniformly continuous on G.

3. Suppose that G is a region that can be subdivided into subregions ΔG_i having arbitrarily small length. Let $f(x,y)$ be a positive continuously differentiable function on G, with both partial derivatives bounded by K.

    a) Show first that if G is divided into subregions ΔG_i all having length less than e, and if (x_i, y_i) is an arbitrary evaluation point in ΔG_i for each i, then the volume V of the solid based on G and capped by the surface $z = f(x,y)$ is given approximately by

    $$V = \Sigma f(x_i, y_i) \Delta A_i,$$

    with an error less than $\sqrt{2}KAe$. (A is the area of G and ΔA_i is the area of ΔG_i.)

    b) Conclude that the double integral $\iint_G f$ exists, where the limit of the Riemann sum is taken in a sense slightly different from before.

4. THE MASS OF A PLANE LAMINA

In Chapter 8 we defined the density of a lamina at the point (x_0, y_0) as the limit

$$\rho(x_0, y_0) = \lim \frac{\Delta m}{\Delta A},$$

where Δm and ΔA are, respectively, the mass and area of an incremental region ΔG containing the point (x_0, y_0), and where the limit is taken as ΔG shrinks down on this point, that is, as the diameter of ΔG approaches 0. Thus

$$\rho(x_0, y_0) \approx \frac{\Delta m}{\Delta A},$$

where the error can be made as small as desired by taking the diameter of ΔG suitably small.

Density is thus a sort of derivative of mass with respect to area. We shall now see that the mass m can be recovered from its "two-dimensional derivative" $\rho(x,y)$ by two-dimensional integration. That is, the mass of the piece of the lamina occupying a region G is given by the double integral of ρ over G,

$$\iint_G \rho(x,y)\,dA.$$

To begin with, we subdivide the lamina G into small sublaminas ΔG_i, and choose an evaluation point (x_i, y_i) in ΔG_i for each i. Then,

$$\frac{\Delta m_i}{\Delta A_i} \approx \rho(x_i, y_i)$$

for each i. We assume that we can make all the errors simultaneously less than a preassigned positive number e by choosing the subdivision fine enough (i.e., by taking the diameters of the subregions ΔG_i all less than some suitably small number). Then, for each index i,

$$\Delta m_i \approx \rho(x_i, y_i) \Delta A_i,$$

with an error less than $e \Delta A_i$. The sum of the errors is thus less than eA, where A is the area of G. Therefore, the total mass m of the region G is given approximately by

$$m \approx \sum_{i=1}^{n} \rho(x_i, y_i) \Delta A_i,$$

with an error at most eA. And we can make this approximation as good as we wish, since in the beginning we can take e to be *any* positive number, however small.

So far we have been looking at the sum $\sum \rho(x_i, y_i) \Delta A_i$ as an estimate of the total mass m of the lamina. But this sum is also a Riemann sum approximating the double integral $\iint_G \rho(x, y) \, dA$. So the double integral and the mass m are both estimated with arbitrarily small error by the same number, and it follows that m is equal to the double integral. Thus:

Theorem 5. *If G is a plane lamina with variable density $\rho(x, y)$, then the total mass m of G is*

$$m = \iint_G \rho(x, y) \, dA.$$

REMARK. We assumed without proof that the estimates $\Delta m_i / \Delta A_i \approx \rho(x_i, y_i)$ can be made good simultaneously, and to this extent the proof of the theorem is incomplete.

EXAMPLE. The density of a square lamina of side length $2a$ is proportional to the square of the distance from the center. Find the mass of the lamina.

. . . .

Solution: If we center the square at the origin then the density $\rho(x, y)$ is given by

$$\rho(x, y) = k(x^2 + y^2),$$

where k is the constant of proportionality. This was our hypothesis. The mass of the lamina is therefore

$$m = k \int_{-a}^{a} \int_{-a}^{a} (x^2 + y^2) \, dx \, dy = k \int_{-a}^{a} \left[\frac{x^3}{3} + xy^2 \right]_{-a}^{a} dy$$

$$= 2k \int_{-a}^{a} \left(\frac{a^3}{3} + ay^2 \right) dy = 2k \left[\frac{a^3 y}{3} + \frac{ay^3}{3} \right]_{-a}^{a}$$

$$= \frac{8}{3} ka^4.$$

PROBLEMS FOR SECTION 4

In each of the following problems find the mass of the plane lamina G having the given shape and density.

1. G is the unit square with lower left vertex at the origin, and $\rho(x,y) = xy$.

2. G is the piece of the unit circle $x^2 + y^2 \leq 1$ in the first quadrant and $\rho(x,y) = xy$.

3. G is the lamina between the graphs of $y = x$ and $y = x^2$ and $\rho(x,y) = \sqrt{x}$.

4. G is the parabolic cap bounded by the parabola $y = 4 - x^2$ and the x-axis. The density is proportional to the distance from the x-axis.

5. G is the same as in Problem 4, but the density is proportional to the distance from the y-axis.

6. G is the unit circle $x^2 + y^2 \leq 1$ and the density is proportional to the distance from the y-axis.

7. G is the unit circle and the density is proportional to the distance from the x-axis.

8. G is the unit circle and the density is proportional to the sum of the distances from the two coordinate axes.

9. G is the square symmetric about the origin with one vertex at (a,a). Its density is proportional to the distance from the line $x + y = 0$ (that is, $\rho(x,y) = k|x + y|$.)

10. G is the piece of the first quadrant under the line $x + y = 1$ and the density is e^{x+y}.

11. G is the lamina between the graph of $y = \sin x$ and the x-axis, from $x = 0$ to $x = \pi$. The density is proportional to the distance from the y-axis.

12. G is the above lamina and the density is proportional to the distance from the x-axis.

13. G is the lamina between the graph $y = e^{-x}$ and the x-axis, from $x = 0$ to infinity; $\rho(x,y) = x$.

5. MOMENTS AND CENTER OF MASS OF A PLANE LAMINA

The proof of the following theorem is similar in spirit to the proof of the mass formula in the last section.

Theorem 6. *If a plane lamina with variable density $\rho(x,y)$ occupies the region G, then the moment M of the lamina about the axis $x = r$ is given by*

$$M = \iint_G (x - r)\rho(x,y)\,dA.$$

By definition, the center of mass $(\bar{x}, \bar{y})$ of the lamina is its balancing point. This means that the lamina is required to balance on the axis $x = \bar{x}$, so the moment of the lamina about the axis $x = \bar{x}$ must be zero, and similarly for the axis $y = \bar{y}$. We therefore find $\bar{x}$ from the equation $M = 0$, which is

$$0 = \iint_G (x - \bar{x})\rho(x,y)\,dA = \iint_G x\rho(x,y)\,dA - \bar{x}\iint_G \rho(x,y)\,dA.$$

Solving for $\bar{x}$ in this equation, and for $\bar{y}$ in the corresponding moment equation about the axis $y = \bar{y}$, we have the corollary:

Corollary: *The center of mass $(\bar{x}, \bar{y})$ of the above lamina has the coordinates*

$$\bar{x} = \frac{\iint_G x\rho(x,y)\,dA}{\iint_G \rho(x,y)\,dA} = \frac{\text{moment about the } y\text{-axis}}{\text{total mass } m},$$

$$\bar{y} = \frac{\iint_G y\rho(x,y)\,dA}{\iint_G \rho(x,y)\,dA} = \frac{\text{moment about the } x\text{-axis}}{\text{total mass } m}.$$

EXAMPLE Find the center of mass of the square G having lower left vertex at the origin and upper right vertex at (a,a), if its density is proportional to the square of the distance from the origin.

. . . .

Solution. If k is the constant of proportionality for the density,

$$\rho(x,y) = k(x^2 + y^2),$$

then the calculation in the example in the last section shows that the mass of the square lamina is $2ka^4/3$.

For the moment about the y-axis, we have

$$\int_0^a \int_0^a x \cdot k(x^2 + y^2)\, dx\, dy = k \int_0^a \left(\frac{x^4}{4} + \frac{x^2 y^2}{2} \right)_0^a dy$$

$$= k \int_0^a \left(\frac{a^4}{4} + \frac{a^2 y^2}{2} \right) dy = k \left[\frac{a^4 y}{4} + \frac{a^2 y^3}{6} \right]_0^a$$

$$= k \left(\frac{a^5}{4} + \frac{a^5}{6} \right) = \frac{5}{12} k a^5.$$

Thus

$$\bar{x} = \frac{\text{moment about } y\text{-axis}}{\text{total mass}} = \frac{\frac{5}{12} k a^5}{\frac{2}{3} k a^4} = \frac{5}{8} a.$$

Finally, since the whole configuration is symmetric in x and y, we also have $\bar{y} = 5a/8$.

We turn now to the proof of Theorem 6.

According to our general discussion of the moments of a plane lamina (§5 of Chapter 10), the moment about a line parallel to the y-axis will decrease if the mass of the lamina is shifted to the left, and will increase if the mass is moved to the right. Therefore, if ΔG is an incremental piece of the lamina lying over an interval $[x', x'']$ on the x-axis, and if the mass Δm on ΔG is slid left and accumulated at a point on the line $x = x'$, then the moment of ΔG about the axis $x = r$ is decreased to the moment of this point mass, which is $(x' - r)\Delta m$. That is

$$(x' - r)\Delta m \leq \Delta M,$$

where ΔM is the moment of ΔG.

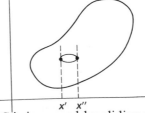

Similarly, the moment of ΔG is increased by sliding the mass of ΔG to the right and accumulating it at a point on the line $x = x''$. So, altogether, we have the moment ΔM of ΔG squeezed between these two point mass moments

$$(x' - r)\Delta m \leq \Delta M \leq (x'' - r)\Delta m.$$

Therefore, if (x_0, y_0) is any point of ΔG, then

$$\Delta M \approx (x_0 - r)\Delta m,$$

with an error at most $(x'' - x')\Delta m$ in magnitude. (Because the numbers ΔM and $(x_0 - r)\Delta m$ both lie on the interval $[(x' - r)\Delta m, (x'' - r)\Delta m]$, they differ from each other by less than the width of this interval, which is $(x'' - x')\Delta m$.) Since $x'' - x'$ is the width of the incremental lamina ΔG, we can restate the above estimate as follows:

If the width of the incremental lamina ΔG is less than d, then

$$\Delta M \approx (x_0 - r)\Delta m$$

with an error less than $d\Delta m$.

We now proceed just as we did when studying the mass distribution over G. We subdivide G into small incremental sublaminas ΔG_i, and choose an arbitrary evaluation point (x_i, y_i) in each piece ΔG_i. We suppose that the widths of the sublaminas ΔG_i are all less than some number d. Then we can apply the above estimate of the moment ΔM_i of ΔG_i in terms of its mass Δm_i and its evaluation point (x_i, y_i), and have

$$\Delta M_i \approx (x_i - r)\Delta m_i,$$

with an error less than $d\Delta m_i$, for each under i. The total moment M of the lamina is the sum $\sum \Delta M_i$ of its incremental moments, and its total mass m is the sum $\sum \Delta m_i$ of its incremental masses. Finally, the sum of the errors in the above approximations is less than

$$\sum d\Delta m_i = d \sum \Delta m_i = dm.$$

So when we add up these estimates of the incremental moments we end up with the approximation

$$M \approx \sum (x_i - r)\Delta m_i,$$

with an error less than dm. Note that this error can be made as small as we wish by taking the maximum width d of the subregions ΔG_i suitably small.

We now recall that $\Delta m_i \approx \rho(x_i, y_i)\Delta A_i$, with an error less than $e\Delta A_i$ (when the diameter of ΔG_i is suitably small), so

$$(x_i - r)\Delta m_i \approx (x_i - r)\rho(x_i, y_i)\Delta A_i,$$

with an error less than $|x_i - r|e\Delta A_i$. Let D be the maximum value $|x_i - r|$ can have. Thus, D is the maximum distance from the axis $x = r$ to a point of the lamina G. Then the error in the last estimate is less than $De\Delta A_i$. The sum of these errors is thus less than $\sum De\Delta A_i = De \sum \Delta A_i = DeA$, so

$$\sum (x_i - r)\Delta m_i \approx \sum (x_i - r)\rho(x_i, y_i)\Delta A_i,$$

with an error at most DeA. Here again, the error can be made as small as we wish by choosing e suitably small. And when we combine this estimate with the one for M above, we see that

$$M \approx \sum (x_i - r)\rho(x_i, y_i)\Delta A_i,$$

with an error that can be made arbitrarily small (by choosing the subdividing regions ΔG_i small enough).

Since the sum on the right is also a Riemann sum for the double integral $\iint_G (x - r)\rho(x, y)\, dA$, we conclude as before that M is given exactly by this double integral. (The double integral and the moment M are each approximated arbitrarily closely by the same estimating number, so they must be equal.) This completes the proof of the theorem.

PROBLEMS FOR SECTION 5

In each of the following problems find the center of mass of the lamina G having the given shape and density. Assume the given value of the mass.

1. G is the unit square with lower left vertex at the origin, and $\rho(x, y) = xy$. Its mass is $1/4$.

2. G is the piece of the unit circle $x^2 + y^2 \le 1$ in the first quadrant and $\rho(x, y) = xy$. Its mass is $1/8$.

3. G is the lamina between the graphs of $y = x$ and $y = x^2$ and $\rho(x, y) = \sqrt{x}$. The mass is $4/35$.

4. G is the parabolic cap bounded by the parabola $y = 4 - x^2$ and the x-axis. The density is proportional to the distance from the x-axis. The mass is $256k/15$, where k is the constant of proportionality for the density.

5. G is the same as in Problem 4, but the density is proportional to the distance from the y-axis. The mass is $8k$.

6. G is the right half of the unit circle $x^2 + y^2 \le 1$ and the density is proportional to the distance from the y-axis. The mass is $2k/3$.

7. G is the right half of the unit circle and the density is proportional to the distance from the x-axis. The mass is $2k/3$.

8. G is the first quadrant of the unit circle and the density is proportional to the sum of the distances from the two coordinate axes. The mass is $2k/3$.

9. G is the square symmetric about the origin with one vertex at (a,a). Its density is proportional to the distance from the line $x + y = 0$ (that is, $\rho(x,y) = k|x + y|$.) The mass is $8ka^3/3$.

10. G is the piece of the first quadrant under the line $x + y = 1$ and the density is e^{x+y}. The mass is 1.

11. G is the lamina between the graph of $y = \sin x$ and the x-axis, from $x = 0$ to $x = \pi$. The density is proportional to the distance from the y-axis. The mass is πk.

12. G is the above lamina and the density is proportional to the distance from the x-axis. The mass is $\pi k/4$.

13. G is the lamina between the graph $y = e^{-x}$ and the x-axis, from $x = 0$ to infinity; $\rho(x,y) = x$. The mass is 1.

6. THE DOUBLE INTEGRAL IN POLAR COORDINATES

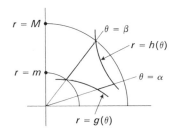 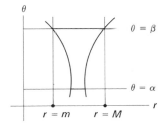

Let G be a region in the xy-plane lying between the polar graphs

$$r = g(\theta) \quad \text{and} \quad r = h(\theta),$$

from $\theta = \alpha$ to $\theta = \beta$, as at the left above. With θ restricted to lie between α and β, each point (x,y) in G determines unique polar coordinates r and θ, where $r = (x^2 + y^2)^{1/2}$ and θ is the unique angle between α and β such that $(\sin \theta, \cos \theta) = (x/r, y/r)$. Thus each point (x,y) in G determines a unique point (r,θ) in the Cartesian plane (with axes labelled r and θ). These (r,θ)-points fill out the region G' bounded by the graphs of $r = g(\theta)$ and $r = h(\theta)$, and the horizontal lines $\theta = \alpha$ and $\theta = \beta$, as shown in the righthand figure above. We say that G' is the image of G under the mapping $(x, y) \rightarrow (r, \theta)$. The areas of G and G' are different, but we know how to compute them both (Sections 6 and 7 in Chapter 8). They are

$$A = \int_\alpha^\beta \frac{[h^2(\theta) - g^2(\theta)]}{2} d\theta \quad \text{and} \quad A' = \int_\alpha^\beta [h(\theta) - g(\theta)] d\theta.$$

There is a simple comparison between these two areas that seems very crude, but it becomes more precise when applied to the incremental subregions of a subdivision.

Lemma 1. *There is a point (r_0, θ_0) in the region G' having the ratio A/A' as its first coordinate. That is,*

$$A = r_0 A'.$$

where r_0 is the r-coordinate of some point (r_0, θ_0) in the region G'.

Proof. The mean-value theorem for integrals (Theorem 2 of Chapter 9), applied to the A integral, guarantees the existence of a number θ_0 between α and β, such that

$$\int_\alpha^\beta \frac{h^2(\theta) - g^2(\theta)}{2} \, d\theta = \frac{h(\theta_0) + g(\theta_0)}{2} \int_\alpha^\beta [h(\theta) - g(\theta)] \, d\theta.$$

But this is the identity

$$A = \frac{h(\theta_0) + g(\theta_0)}{2} A',$$

or

$$A = r_0 A',$$

where $r_0 = [h(\theta_0) + g(\theta_0)]/2$. Moreover, the point

$$(r_0, \theta_0) = ([h(\theta_0) + g(\theta_0)]/2, \theta_0)$$

is in G'; in fact, it is the midpoint of the segment crossing G' from the left boundary point $(g(\theta_0), \theta_0)$ to the right boundary point $(h(\theta_0), \theta_0)$. The lemma is thus proved. ∎

The role played by the conversion factor r_0 in passing from A' to A becomes clear when a small region near the origin is compared with one further away (see figure).

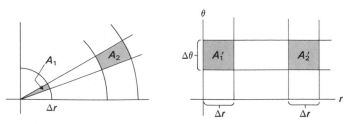

The above lemma is the heart of the following theorem.

Theorem 7. *If $f(x,y)$ is continuous on the bounded closed region G, and if we define $F(r,\theta)$ by*

$$F(r,\theta) = f(r \cos \theta, r \sin \theta),$$

then

$$\iint_G f(x,y)\, dA = \iint_{G'} F(r,\theta) r\, dA',$$

where G' is the image of G under the mapping $(x,y) \to (r,\theta)$.

Proof. If we subdivide G into small subregions ΔG_i, then their images $\Delta G_i'$ under the mapping $(x,y) \to (r,\theta)$ form a corresponding subdivision of G'. For each i, we choose an evaluation point (r_i, θ_i) in $\Delta G_i'$ such that $\Delta A_i = r_i \Delta A_i'$ (by the lemma), and we take (x_i, y_i) as the corresponding point in ΔG_i,

$$(x_i, y_i) = (r_i \cos \theta_i, r_i \sin \theta_i).$$

Then,

$$\iint_G f(x,y)\, dA \approx \sum f(x_i, y_i)\Delta A_i,$$

$$\iint_{G'} F(r,\theta) r\, dA' \approx \sum F(r_i, \theta_i) r_i \Delta A_i',$$

where each Riemann-sum approximation can be made as good as desired by choosing the subdivision fine enough. But the two Riemann sums are equal, for

$$f(x_i, y_i) = f(r_i \cos \theta_i, r_i \sin \theta_i) = F(r_i, \theta_i),$$

and

$$\Delta A_i = r_i \Delta A_i'$$

by the choice of r_i. The two double integrals are thus both approximated as closely as desired by the same number, so they must be equal. ∎

The theorem above is the simplest case of a "change of variables" in a multiple integral, analogous to the substitution law for single integrals. The general theory of such transformations of multiple integrals is studied in advanced calculus. It is an important subject, with many applications. For example, the present theorem provides us with a new way to evaluate a double integral by iterated integration.

Theorem 8. *Let $f(x, y)$ be continuous on a bounded closed region G, and suppose that the boundary of G lies along the polar graphs $r = g(\theta)$, and $r = h(\theta)$, and the rays*

$\theta = \alpha$ *and* $\theta = \beta$, *where* $\alpha < \beta$ *and* $g(\theta) < h(\theta)$ *for all* θ *between* α *and* β. *Then*

$$\iint_G f(x,y)\,dA = \int_\alpha^\beta \left[\int_{g(\theta)}^{h(\theta)} F(r,\theta)r\,dr \right] d\theta,$$

where $F(r,\theta) = f(r \cos \theta, r \sin \theta)$.

Proof. This is an immediate corollary of Theorems 3 and 7. ∎

EXAMPLE 1. Compute the double integral

$$\int_G e^{-(x^2+y^2)}\,dA,$$

where G is the circular disk $x^2 + y^2 \le a^2$.

. . . .

Solution. In polar coordinates G can be described by the conditions $0 \le r \le a$, $0 \le \theta \le 2\pi$. Moreover,

$$e^{-(x^2+y^2)} = e^{-r^2}.$$

Thus

$$\int_G e^{-(x^2+y^2)}\,dA = \int_0^{2\pi} \left[\int_0^a e^{-r^2} r\,dr \right] d\theta$$

$$= \int_0^{2\pi} \left[-e^{-r^2}/2 \right]_0^a d\theta$$

$$= \int_0^{2\pi} \frac{1 - e^{-a^2}}{2} d\theta = \pi(1 - e^{-a^2}).$$

This is an example of a double integral that is easier to compute than its one-dimensional analogue. In fact, the integral

$$\int_0^a e^{-x^2}\,dx$$

cannot by computed at all in terms of elementary functions.

EXAMPLE 2. Compute the volume of a ball of radius a.

. . . .

Solution. The ball is bounded by the spherical surface

$$x^2 + y^2 + z^2 = a^2.$$

Moreover, the total volume of the ball is eight times the volume of the piece of the ball in the first octant. Thus

$$V = 8 \iint_G \sqrt{a^2 - (x^2 + y^2)} \, dA,$$

where G is the part of the circular disk $x^2 + y^2 \le a^2$ lying in the first quadrant. By Theorem 8, we can compute this double integral as the following iterated integral in polar coordinates:

$$V = 8 \int_0^{\pi/2} \left[\int_0^a \sqrt{a^2 - r^2} \, r \, dr \right] d\theta$$

$$= 8 \int_0^{\pi/2} \left[-\frac{1}{3}(a^2 - r^2)^{3/2} \right]_0^a d\theta$$

$$= \frac{8a^3}{3} \int_0^{\pi/2} d\theta = \frac{8a^3}{3} \cdot \frac{\pi}{2} = \frac{4}{3}\pi a^3.$$

Here again we are helped by the extra r in the integrand of the polar coordinate integral.

EXAMPLE 3. Find the center of mass of a homogeneous semicircular disk of radius R.

. . . .

Solution. Assuming that the lamina occupies the right half of the circle $x^2 + y^2 \le R^2$ and has constant density 1, its moment M about the y-axis is $\iint_G x \, dA$, where G can be described in terms of polar coordinates by the conditions that r runs from 0 to R and θ from $-\pi/2$ to $\pi/2$. Also $x = r \cos \theta$, so by Theorem 8,

$$M = \iint_G x \, dA = \int_{-\pi/2}^{\pi/2} \left[\int_0^R (r \cos \theta) r \, dr \right] d\theta$$

$$= \int_{-\pi/2}^{\pi/2} \left[\frac{r^3 \cos \theta}{3} \right]_{r=0}^{r=R} d\theta = \frac{R^3}{3} \int_{-\pi/2}^{\pi/2} \cos \theta \, d\theta$$

$$= \frac{R^3}{3} \left[\sin \theta \right]_{-\pi/2}^{\pi/2} = \frac{2}{3} R^3.$$

The x-coordinate of the center of mass is the moment about the y-axis divided by the area (= the mass):

$$\bar{x} = \frac{M}{A} = \frac{\frac{2}{3} R^3}{\frac{\pi R^2}{2}} = \frac{4}{3} \frac{R}{\pi}.$$

The disk will balance on the x-axis, by symmetry, so $\bar{y} = 0$.

PROBLEMS FOR SECTION 6

1. Find the volume of the solid in the half-space $z \geq 0$ based on the circular disk $x^2 + y^2 \leq 9$ and capped by the sphere $x^2 + y^2 + z^2 = 25$.

2. Find the volume of the solid cut out of the spherical ball $x^2 + y^2 + z^2 \leq b^2$ by the cylinder $x^2 + y^2 = a^2$, where $0 < a \leq b$.

3. Find the volume of the cone $0 \leq z \leq 1 - r$.

4. Find the volume of the cone $0 \leq z \leq (a - r)h/a$.

5. Find the volume of the solid based on the interior of the cardioid $r = 1 + \cos \theta$ and capped by the cone $z = 2 - r$.

6. Find the volume of the solid based on the interior of the circle $r = \cos \theta$ and capped by the cone $z = 1 - r$.

7. Find the volume of the solid based on the interior of the circle $r = \cos \theta$ and capped by the sphere $r^2 + z^2 = 1$.

8. Find the volume of the solid based on the interior of the circle $r = \cos \theta$ and capped by the plane $z = x$.

9. Show that the mass of a circular lamina of radius R whose density is proportional to the distance from the center is $2\pi k R^3/3$, where k is the constant of proportionality.

10. Compute the center of mass of the right half of the above lamina.

11. Compute the mass and center of mass of the semicircular lamina $0 \leq r \leq 1, 0 \leq \theta \leq \pi$ if the density is proportional to $r(1 - r)$.

12. Compute the mass and center of mass of the semicircular lamina $0 \leq r \leq 1, 0 \leq \theta \leq \pi$, if its density is proportional to $\sqrt{1 - r^2}$.

13. Find the volume of the solid obtained by intersecting the cylinders

$$x^2 + y^2 = a^2 \quad \text{and} \quad x^2 + z^2 = a^2.$$

7. SURFACE AREA

Let π and P be two planes that intersect at an angle θ. Their line of intersection will be called their axis.

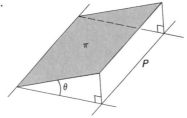

When we project from π vertically down onto P, lengths parallel to the axis remain unchanged, but lengths perpendicular to the axis are multiplied by $\cos \theta$, and hence are decreased by this constant factor.

It follows that if a region R on π is projected vertically downward to P, then it is compressed by the factor $\cos \theta$ in the direction perpendicular to the axis, and its area is therefore decreased by the factor $\cos \theta$:

$$\text{area of image region} = \cos \theta(\text{area of } R).$$

We assume this as self-evident. However, a proof could be constructed by starting with rectangles having one side parallel to the axis, and then approximating more general regions by finite collections of such rectangles, somewhat in the manner of our Riemann sum approximations. We shall use this relationship in reverse.

Lemma 2. *If G is a region in the xy-plane, and if π is the plane $z = ax + by + c$, then the area of the region on π lying over G is*

$$\sqrt{1 + a^2 + b^2}\,(\text{area of } G).$$

Proof. We know that the vector $(-a, -b, 1)$ is normal to π, and that if θ is the angle between this vector and the z-axis, then

$$\cos \theta = \frac{1}{\sqrt{1 + a^2 + b^2}}.$$

(See Section 1 in Chapter 17.) Since θ is also the angle between π and the xy-plane, a region G' on π has its area decreased by the factor $\cos \theta = 1/\sqrt{1 + a^2 + b^2}$ when it is projected vertically downward to the xy-plane, becoming the region G on the xy-plane. That is,

$$\text{area of } G = \frac{1}{\sqrt{1 + a^2 + b^2}}(\text{area of } G'),$$

and the lemma is just the cross-multiplied form of this equation. ∎

Now consider the piece of the surface $z = f(x, y)$ lying over the region G in the xy-plane. Its tangent plane at $z_0 = f(x_0, y_0)$,

$$z - z_0 = \frac{\partial z}{\partial x}\bigg|_{(x_0, y_0)}(x - x_0) + \frac{\partial z}{\partial y}\bigg|_{(x_0, y_0)}(y - y_0),$$

remains very nearly parallel to the surface near the point of tangency. So if ΔG is an incremental subregion containing the evaluation point (x_0, y_0), and if ΔS is

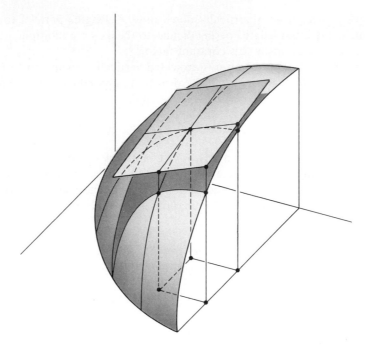

the area of the little piece of surface lying over ΔG, then ΔS is approximately
equal to the corresponding area on the tangent plane, which is

$$\sqrt{1 + \left(\frac{\partial z}{\partial x}\right)_0^2 + \left(\frac{\partial z}{\partial y}\right)_0^2} \Delta A,$$

by the lemma. We have used the abbreviation $(\partial z/\partial x)_0$ for the value of the
derivative at (x_0, y_0). It seems intuitively clear that this approximation improves
as ΔA approaches 0, in the sense that the ratio of the two areas approaches 1.
This is the same thing as saying that

$$\Delta S \approx \sqrt{1 + \left(\frac{\partial z}{\partial x}\right)_0^2 + \left(\frac{\partial z}{\partial y}\right)_0^2} \Delta A,$$

with an error that is small compared to ΔA. Therefore if we subdivide G into
suitably small subregions ΔG_i and add up the above estimates, we see that the
area S of the surface over G is given approximately by

$$S \approx \Sigma \sqrt{1 + \left(\frac{\partial z}{\partial x}\right)_i^2 + \left(\frac{\partial z}{\partial y}\right)_i^2} \Delta A_i,$$

with an error that is small compared to the area A of G. The partial derivatives
are evaluated at arbitrarily chosen points (x_i, y_i) in ΔG_i.

The discussion above is not a real proof, but it should seem plausible and it is the best we can do for surface area.

Since the sum on the right above is a Riemann sum for the function $\sqrt{1 + (\partial z/\partial x)^2 + (\partial z/\partial y)^2}$, we conclude as usual that S is given exactly by the double integral of this function over G.

We have thus shown that the following theorem is a consequence of plausible geometric assumptions.

Theorem 9. *If $f(x,y)$ is a continuously differentiable function whose domain includes the region G, then the area of the piece of the surface $z = f(x,y)$ lying over G is*

$$\iint_G \sqrt{1 + \left(\frac{\partial z}{\partial x}\right)^2 + \left(\frac{\partial z}{\partial y}\right)^2}\, dA.$$

EXAMPLE. Find the formula for the area of the surface of a sphere.

. . . .

Solution. The sphere of radius R about the origin has the equation

$$x^2 + y^2 + z^2 = R^2.$$

In order to compute the partial derivatives $\partial z/\partial x$ and $\partial z/\partial y$, we can solve for z and differentiate, or we can simply differentiate implicitly in the equation as it stands. The latter method gives

$$2x + 2z\frac{\partial z}{\partial x} = 0,$$

so $\partial z/\partial x = -x/z$. Similarly, $\partial z/\partial y = -y/z$. Since the total surface area S is eight times the area in the first octant, we have

$$S = 8 \iint_G \sqrt{1 + \left(\frac{\partial z}{\partial x}\right)^2 + \left(\frac{\partial z}{\partial y}\right)^2}\, dA$$

$$= 8 \iint_G \sqrt{1 + \left(\frac{x}{z}\right)^2 + \left(\frac{y}{z}\right)^2}\, dA,$$

where G is the quarter circle in the first quadrant. If we write 1 as z/z in the integrand, we see that

$$S = 8 \iint_G \frac{R}{z}\, dA.$$

In polar coordinates, $z = \sqrt{R^2 - (x^2 + y^2)} = \sqrt{R^2 - r^2}$, so

$$\iint_G \frac{R}{z} dA = R \int_0^{\pi/2} \left[\int_0^R \frac{r\,dr}{\sqrt{R^2 - r^2}} \right] d\theta$$

$$= R \int_0^{\pi/2} \left[-\sqrt{R^2 - r^2} \right]_0^R d\theta = R^2 \int_0^{\pi/2} d\theta$$

$$= \frac{\pi}{2} R^2.$$

The total area S is eight times this, so the formula for the area of the surface of a sphere of radius R is

$$S = 4\pi R^2.$$

PROBLEMS FOR SECTION 7

Find the area of each of the following surfaces.

1. The piece of the paraboloid $z = 4 - x^2 - y^2$ that lies above the xy-plane. (Polar coordinates are simplest.)

2. The surface cut out of the cone $z = 1 - r$ by the cylinder $r = \cos\theta$.

3. The piece of the sphere $x^2 + y^2 + z^2 = 25$ lying over the circular disk $x^2 + y^2 \le 16$.

4. The piece of the sphere $x^2 + y^2 + z^2 = R^2$ cut out by the cylinder $x^2 + y^2 = a^2$ and lying above the xy-plane.

5. The piece of the sphere $x^2 + y^2 + z^2 = 1$ lying over the interior of the circle $r = \cos\theta$.

6. The piece of the plane $z = 6 - 2x - 3y$ lying over the unit square with lower left corner at the origin.

7. The piece of the above plane that is cut off by the first octant.

8. The piece of the plane $z = ax + by + c$ lying over the region G. (That is, prove that the Lemma is a special case of Theorem 9.)

9. The cone of altitude h and base radius a.

10. When $r \le a$, the equation

$$az = h(a - r)$$

describes the cone whose altitude is h and whose base is the circular disk $r \le a$. Let G be any region lying in this disk. Show that the area of the part of the cone lying over G is

$$\sqrt{1 + \left(\frac{h}{a}\right)^2} \; (\text{area of } G).$$

8. ITERATED TRIPLE INTEGRATION.

Suppose that a solid S, occupying a region G in space, has a variable continuous density $\rho(x,y,z)$. When we consider how to estimate the total mass m of S in terms of its density function ρ, we are led to an approximation of m by a finite sum $\sum \rho(x_i, y_i, z_i)\Delta V_i$ having all the earmarks of a Riemann sum, except for involving incremental volumes ΔV_i rather than the incremental areas ΔA_i (or incremental lengths Δx_i) of our earlier discussions. Then m is given exactly by the limit of this three-dimensional Riemann sum as the incremental volumes ΔV_i approach zero in a suitable manner. This limit is called the *triple integral* of ρ over G, and is designated $\iiint_G \rho(x,y,z)dV$. Thus, the mass m of S is the triple integral of its density function ρ:

$$m = \iiint_G \rho(x,y,z)\,dV.$$

The development outlined above will be given in the final section, and will lead to the notion of the triple integral $\iiint_G f(x,y,z)\,dV$ for any continuous function f. The reasoning will be similar to that used earlier for the double integral. Again, it will turn out that the multiple integral can be evaluated by iterated integration, that is by "integrating out" the variables one at a time. We shall assume this for the moment, and turn to the practical problem of determining the limits of integration. They occur at the boundary of the integration domain G, and a sketch of G is almost essential in order to get them right.

EXAMPLE 1. Use iterated integration to evaluate the triple integral

$$\iiint_G x\,dV,$$

where G is the region cut off from the first octant by the plane $x + y + z = 2$.

. . . .

Solution. The region G is shown in the figure below. Suppose that we decide to integrate first with respect to z. This means holding x and y fixed and integrating the resulting function of z. Geometrically, the domain of this integration is a vertical line segment crossing G from the xy-plane to the plane $x + y + z = 2$. Along this segment, z varies from $z = 0$ to $z = 2 - x - y$.

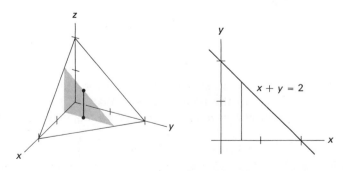

So the first integral is

$$\int_0^{2-x-y} x\,dz = xz\Big]_{z=0}^{z=2-x-y}$$

$$= x(2 - x - y).$$

With z integrated out, we have left a double integral of a function of x and y, over the region R in the xy-plane that is the base of the original space region G. Here R is the triangle cut off in the first quadrant by the line $x + y = 2$. If we choose to integrate next with respect to y, the calculation finishes off as follows:

$$\int_0^2 \int_0^{2-x} x(2 - x - y)\,dy\,dx = \int_0^2 \left[-x\frac{(2 - x - y)^2}{2} \right]_0^{2-x} dx$$

$$= \frac{1}{2} \int_0^2 x(2 - x)^2\,dx$$

$$= \frac{1}{2} \int_0^2 (4x - 4x^2 + x^3)\,dx$$

$$= \frac{1}{2} \left[2x^2 - \frac{4x^3}{3} + \frac{x^4}{4} \right]_0^2$$

$$= 4 - \frac{16}{3} + 2 = \frac{2}{3}.$$

In reducing the evaluation of a triple integral to iterated integration, the crucial step is a *slicing* formula that generalizes our earlier volumes-by-slicing formula.

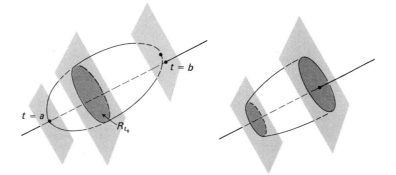

We choose some axis, which we shall call the t-axis for neutrality, and we slice G by planes perpendicular to the t-axis. Suppose that the first and last of these planes to touch G occur at $t = a$ and $t = b$. For each t_0 between a and b, let R_{t_0}

be the cross section of G in the plane $t = t_0$, and let $I(t_0)$ be the double integral of f over the plane region R_{t_0}:

$$I(t_0) = \iint_{R_{t_0}} f\, dA.$$

Then,

$$\iiint_G f\, dV = \int_a^b I(t)\, dt.$$

The slicing formula expresses a triple integral as the "single" integral of a varying double integral. Equally important is a complementary *piercing* formula, expressing a triple integral as a double integral of a varying single integral, as follows:

In the same t-axis setup that we used above, let π be the base plane $t = 0$, and let R be the base of the solid G in π. That is, R is obtained by projecting all the points of G perpendicularly onto π. For each point p in R, we suppose that the line through p perpendicular to π (and therefore parallel to the t-axis) intersects the space region G in an interval $[T_1, T_2]$. This interval will vary with p, so what we really are saying is that G runs from a lower surface $t = T_1(p)$ to an upper surface $t = T_2(p)$. The piercing formula evaluates the triple integral of f over G by first forming the one-dimensional integral of f over the interval $[T_1(p), T_2(p)]$ for each p, $\int_{T_1(p)}^{T_2(p)} f\, dt$, and then computing the double integral over R of the resulting function of p:

$$\iiint_G f\, dV = \iint_R \left[\int_{T_1(p)}^{T_2(p)} f\, dt \right] dA.$$

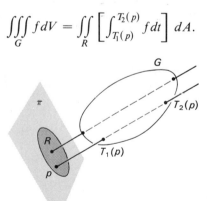

The slicing and piercing formulas are both valid for completely general three-dimensional solids G, but our proofs and applications are restricted to "standard" regions, that is, to regions that can be described in a standard manner by the integration limits of an iterated integral.

We shall prove these formulas in Section 10. By taking the t-axis as the x-axis and evaluating the double integral in the slicing formula as the iterated integral

$\iint f(x, y, z)\, dz\, dy$, we obtain the iteration evaluation procedure we were using in Example 1. In the same way, or by using the piercing formula, we can justify integrating out the variables x,y,z in any other iteration order.

Here is a formal statement of this evaluation identity. It is proved simply by inserting the double-integral evaluation formula of Theorem 3 into the slicing or piercing formula.

Theorem 10. *Let G be the space region based on the region R in the xy-plane and capped by the graph $z = \phi(x,y)$. Suppose also that the plane region R is bounded by the lower graph $y = g(x)$, the upper graph $y = h(x)$, and the lines $x = a$ and $x = b$ in the accompanying figure. Then*

$$\iiint\limits_{G} f(x, y, z)\, dV = \int_a^b \int_{g(x)}^{h(x)} \int_0^{\phi(x,y)} f(x, y, z)\, dz\, dy\, dx.$$

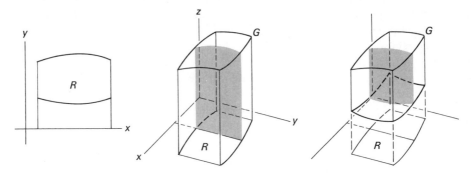

A slightly more general domain is shown in the righthand figure. Here, G lies over the same base region R in the xy-plane, but runs from the lower surface $z = \psi(x,y)$ to the upper surface $z = \phi(x,y)$. Now,

$$\iiint\limits_{G} f\, dV = \int_a^b \int_{g(x)}^{h(x)} \int_{\psi(x,y)}^{\phi(x,y)} f\, dz\, dy\, dx.$$

EXAMPLE 2. Suppose G is the region between the parabolic surfaces $z = x^2 + y^2$ and $z = 2 - (x^2 + y^2)$. The intersection curve of these surfaces is the circle $x^2 + y^2 = 1$ in the plane $z = 1$. The part of G in the first octant is shown below. The base region R in the xy-plane is the interior of the unit circle, with y ranging therefore from $y = -\sqrt{1 - x^2}$ to $y = \sqrt{1 - x^2}$. Thus

$$\iiint\limits_{G} f\, dV = \int_{-1}^{1} \int_{-\sqrt{1-x^2}}^{\sqrt{1-x^2}} \int_{x^2+y^2}^{2-(x^2+y^2)} f\, dz\, dy\, dx.$$

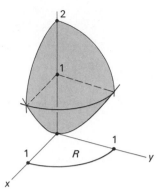

'EXAMPLE 3. Use iterated triple integration to find the volume of the tetrahedral region of Example 1.

. . . .

Solution. The volume of a space region G is simply the triple integral over G of the constant 1,

$$V = \iiint_G 1 \, dV$$

(because the volume equals the mass of a distribution having constant density one). Thus

$$V = \int_0^2 \int_0^{2-x} \int_0^{2-x-y} dz \, dy \, dx$$

$$= \int_0^2 \int_0^{2-x} (2 - x - y) \, dy \, dx$$

$$= \int_0^2 \left[-\frac{(2-x-y)^2}{2} \right]_0^{2-x} dx = \frac{1}{2} \int_0^2 (2-x)^2 \, dx$$

$$= -\frac{1}{6}(2-x)^3 \bigg]_0^2 = \frac{2^3}{6} = \frac{4}{3}.$$

There is nothing sacred about the order in which the variables are integrated when we evaluate a multiple integral by iterated integration, and sometimes a different order will make for an easier calculation, as we have already seen in the double-integral sections. Here is a triple-integral example.

EXAMPLE 4. Compute the volume of the wedge cut from the cylinder $y^2 + z^2 = 1$ in the first octant by the planes $y = x$ and $x = 0$.

. . . .

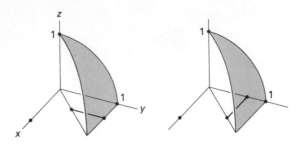

Solution. In the order we have been using so far, this sets up as the iterated integral

$$V = \int_0^1 \int_x^1 \int_0^{\sqrt{1-y^2}} dz\, dy\, dx = \int_0^1 \int_x^1 \sqrt{1-y^2}\, dy\, dx,$$

and now we have to compute the rather nasty integral $\int \sqrt{1-y^2}\, dy$. But if we interchange the order of the x and y integrations, we have

$$V = \int_0^1 \int_0^y \int_0^{\sqrt{1-y^2}} dz\, dx\, dy = \int_0^1 \int_0^y \sqrt{1-y^2}\, dx\, dy$$

$$= \int_0^1 y\sqrt{1-y^2}\, dy = -\frac{1}{2} \int_1^0 u^{1/2}\, du \qquad (u = 1 - y^2)$$

$$= \frac{1}{3} u^{3/2} \Big]_0^1 = \frac{1}{3}.$$

In Section 10 we will be in a position to make a Riemann-sum estimate of turning moments. We shall skip this development, however, since the reasoning proceeds exactly as in the two-dimensional case, with analogous conclusions:

The moment M about the origin of a mass distribution with density $\rho(x,y,z)$ is the vector

$$M = (M_x, M_y, M_z) = \left(\iiint x\rho\, dV, \iiint y\rho\, dV, \iiint z\rho\, dV \right).$$

The center of mass of the distribution is given by

$$(\bar{x}, \bar{y}, \bar{z}) = \frac{M}{m} = \left(\frac{M_x}{m}, \frac{M_y}{m}, \frac{M_z}{m} \right),$$

where $m = \iiint \rho\, dV$ is the total mass of the distribution.

EXAMPLE 5. Find the centroid of the tetrahedron in Example 1.

· ·

Solution. The centroid of a space region G is the center of mass of a distribution in G having a constant density k, and k may as well be taken to have the value $k = 1$, since it factors out and cancels in the quotients defining the center of mass.

The moment of the above tetrahedron about the origin thus has the x-component $M_x = \iiint_G x\, dV$, which we computed in Example 1 to be 2/3. The mass of the tetrahedron is its volume, which we found in Example 3 to be 4/3. Thus,

$$\bar{x} = \frac{M_x}{m} = \frac{2/3}{4/3} = \frac{1}{2}.$$

Moreover, the configuration is symmetric with respect to the variables x, y, and z, so $\bar{y}$ and $\bar{z}$ must also have the value 1/2. Thus

$$(\bar{x}, \bar{y}, \bar{z}) = (\tfrac{1}{2}, \tfrac{1}{2}, \tfrac{1}{2}) = \tfrac{1}{2}(1, 1, 1).$$

EXAMPLE 6. Find the centroid of the piece of the unit ball lying in the first octant.

· ·

Solution. The unit sphere has the equation $x^2 + y^2 + z^2 = 1$, so

$$M_z = \iiint_G z\, dV = \int_0^1 \int_0^{\sqrt{1-x^2}} \int_0^{\sqrt{1-x^2-y^2}} z\, dz\, dy\, dx$$

$$= \int_0^1 \int_0^{\sqrt{1-x^2}} \frac{z^2}{2}\Big]_0^{\sqrt{1-x^2-y^2}} dy\, dz = \frac{1}{2}\int_0^1 \int_0^{\sqrt{1-x^2}} (1 - x^2 - y^2)\, dy\, dz$$

$$= \frac{1}{2}\int_0^1 \left[(1 - x^2)y - \frac{y^3}{3}\right]_0^{\sqrt{1-x^2}} dx = \frac{1}{3}\int_0^1 (1 - x^2)^{3/2}\, dx$$

$$= \frac{1}{3}\int_0^{\pi/2} \cos^4 u\, du \qquad (\text{from } x = \sin u)$$

$$= \frac{1}{3}\int_0^{\pi/2} \left(\frac{1 + \cos 2u}{2}\right)^2 du$$

$$= \frac{1}{12}\int_0^{\pi/2} \left(1 + 2\cos 2u + \left(\frac{1 + \cos 4u}{2}\right)\right) du$$

$$= \frac{1}{12}\int_0^{\pi/2} \frac{3}{2}\, du \qquad (\text{since } \int_0^{\pi/2} \cos 2u\, du = 0, \text{ etc.})$$

$$= \frac{\pi}{16}.$$

Since the volume of one eighth of the unit ball is $\pi/6$, we have

$$\bar{z} = \frac{M_z}{V} = \frac{\pi/16}{\pi/6} = \frac{3}{8}.$$

By symmetry, $\bar{x}$ and $\bar{y}$ have the same value, so the centroid is

$$(\bar{x}, \bar{y}, \bar{z}) = \frac{3}{8}(1, 1, 1).$$

PROBLEMS FOR SECTION 8

Find the volume of each of the following solids by triple (iterated) integration.

1.   The tetrahedron cut from the first octant by the plane $3x + 2y + z = 6$.

2.   The rectangular solid cut off from the first octant by the planes $x = a$, $y = b$, $z = c$.

3.   The tetrahedron cut off from the first octant by the plane

$$\frac{x}{a} + \frac{y}{b} + \frac{z}{c} = 1.$$

4.   The solid cut off from the first octant by the parabolic surfaces $z = 1 - x^2$ and $y = 1 - x^2$.

5.   The solid cut off from the first octant by the parabolic surfaces $z = 1 - x^2$ and $x = 1 - y^2$.

6.   The solid cut off from the first octant by the plane $z = x$ and the cylinder $x^2 + y^2 = 1$.

7.   The solid cut off from the first octant by the plane $z = y$ and the cylinder $x^2 + y^2 = 1$.

8.   The solid cut off from the first octant by the cylinders $x^2 + y^2 = 1$ and $x^2 + z^2 = 1$.

9.   The solid bounded by the cylinders $x^2 + y^2 = 1$ and $y^2 + z^2 = 1$.

10. The solid bounded above by $z = 1 - x^2$ and below by $z = y^2$.

11. Compute the triple integral of $f(x,y,z)=y$ over the region G lying in the first octant between the xy-plane and the surface $z=4-(x^2+y^2)$.

12. Compute the triple integral of $f(x,y,z) = xy$ over the region G cut off from the first octant by the ellipsoid

$$\frac{x^2}{a^2} + \frac{y^2}{b^2} + \frac{z^2}{c^2} = 1.$$

13. Compute the triple integral of $f(x,y,z) = xyz$ over the solid cut off from the first octant by the cylinders $x^2 + z^2 = 4$ and $x^2 + y^2 = 4$.

14. Calculate the centroid of the wedge cut out of the cylinder $x^2 + y^2 = 1$ by the plane $z = y$ above and the plane $z = 0$ below.

15. Calculate the centroid of the solid cut from the first octant by the ellipsoid

$$\frac{x^2}{a^2} + \frac{y^2}{b^2} + \frac{z^2}{c^2} = 1.$$

(Assume that the volume of the ellipsoid is $(4/3)\pi abc$.)

9. CYLINDRICAL AND SPHERICAL COORDINATES.

There is another possibility for evaluating the double integral in the slicing and piercing formulas. If there is some element of symmetry about the t-axis in the integrand f or the domain G, then it may be easier to compute the double integral as an iterated integral in polar coordinates, using Theorems 7 and 8 from Section 6. In the theorem below, the t-axis is taken to be the z-axis.

Theorem 11. *If G' is the region in r,θ,z-space corresponding to the region G in x,y,z-space, and if we define F by $F(r,\theta,z) = f(r \cos \theta, r \sin \theta, z)$, then*

$$\iiint\limits_{G} f(x,y,z)\,dV = \iiint\limits_{G'} F(r,\theta,z)r\,dV'.$$

Proof. The image region G' consists of those points (r,θ,z) for which the point $(x,y,z) = (r \cos \theta, r \sin \theta, z)$ is in G. (Remember that only one value of θ is chosen for each pair (x,y), normally, the unique θ in $[0,2\pi)$ such that $(\cos \theta, \sin \theta) = (x/r, y/r)$. If $R(z_0)$ and $R'(z_0)$ are the cross sections of G and G' in the plane $z = z_0$, then for each z_0, $R'(z_0)$ is the image of $R(z_0)$ under the polar-coordinate mapping, so

$$\iint\limits_{R(z_0)} f(x,y,z_0)\,dA = \iint\limits_{R'(z_0)} F(r,\theta,z_0)r\,dA',$$

by Theorem 7. Letting $I(z_0)$ be the common value of these equal double integrals, we then have

$$\iiint\limits_{G} f(x,y,z)\,dV = \int_a^b I(z)\,dz$$

$$= \iiint\limits_{G'} F(r,\theta,z)r\,dV'$$

by two applications of the slicing formula. ∎

There is an equally short proof of the theorem using the piercing formula. We can now evaluate the triple integral on the right as an iterated integral in the variables r,θ,z (by Theorem 10), so, in combination, the two theorems give us a new evaluation procedure for the triple integral $\iiint_G f\,dV$.

We normally interpret this new procedure in terms of "cylindrical-coordinate" graphs, which are the analogs of polar graphs. In the above theorem, a triple (r_0,θ_0,z_0) is plotted in space by a Cartesian r,θ,z-axis system in the usual way, and the region G' is constructed in this axis system. However, if we set $x = r\cos\theta$, $y = r\sin\theta$, and interpret r and θ as polar coordinates of the point (x,y), then we call the triple (r,θ,z) the "cylindrical" coordinates of the point $(x,y,z) = (r\cos\theta, r\sin\theta, z)$. There is a simple reason for the terminology; in this interpretation the graph of $r = a$ is the cylinder of radius a about the z-axis. This is quite different from the graph of $r = a$ in the Cartesian r, θ, z-axis, where it is a plane parallel to the θz-coordinate plane.

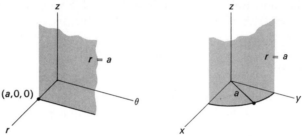

Thus, the *cylindrical graph* of an equation such as $r = f(\theta,z)$, or $z = g(r,\theta)$, or $h(r,\theta,z) = 0$ consists of the points in x,y,z-space whose cylindrical coordinates satisfy the equation. It is quite different from the standard graph of the equation in the Cartesian r,θ,z-axis system.

We can now interpret the combination of Theorems 10 and 11 as the evaluation of a triple integral by an iterated integral in cylindrical coordinates. The variables z, r, and θ can be integrated in any order for which we can set up the corresponding limits of integration. Here is a typical situation.

Theorem 12. *Suppose that the region G is bounded by the cylindrical graphs of $z = q(r,\theta)$, $r = g(\theta)$, and $r = h(\theta)$, and runs from $\theta = \alpha$ to $\theta = \beta$. Then*

$$\iiint_G f(x,y,z)\,dV = \int_\alpha^\beta \int_{g(\theta)}^{h(\theta)} \int_0^{q(r,\theta)} F(r,\theta,z)r\,dz\,dr\,d\theta.$$

EXAMPLE 1. Find the mass of the cylinder C defined by the inequalities $x^2 + y^2 \le 1$, $0 \le z \le 1$, if the density of C is given by

$$\rho(x,y,z) = (1 - z)\sqrt{x^2 + y^2}.$$

Solution. In cylindrical coordinates, the density is $(1 - z)r$, and the cylinder is specified by the inequalities $0 \le r \le 1$, $0 \le z \le 1$ (and $0 \le \theta \le 2\pi$, or $-\pi < \theta < \pi$). (These inequalities describe a rectangular solid C' in the r,θ,z-coordinate system.) Thus,

$$
m = \iiint_G \rho(x, y, z)\,dV = \int_0^1 \int_0^{2\pi} \int_0^1 (1 - z)r^2\,dr\,d\theta\,dz
$$

$$
= \int_0^1 \int_0^{2\pi} (1 - z)\frac{r^3}{3}\bigg]_0^1 d\theta\,dz = \frac{1}{3}\int_0^1 \int_0^{2\pi} (1 - z)\,d\theta\,dz
$$

$$
= \frac{1}{3}\int_0^1 (1 - z)\theta\bigg]_0^{2\pi} dz = \frac{2\pi}{3}\int_0^1 (1 - z)\,dz
$$

$$
= \frac{2\pi}{3}\left[z - \frac{z^2}{2}\right]_0^1 = \frac{\pi}{3}.
$$

The *spherical* coordinates of a point are shown in the accompanying figure. Essentially, they combine the polar angle θ in the xy-plane with polar coordinates ρ,ϕ in the *half-plane* determined by θ and the z-axis. Note that ϕ ranges from 0 to π only. Referring to the figure, we see first that $z = \rho \cos \phi$ and $r = \rho \sin \phi$.

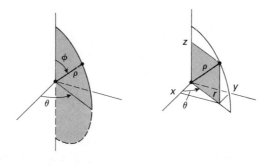

Then

$$
x = r \cos \theta = \rho \sin \phi \cos \theta,
$$

$$
y = r \sin \theta = \rho \sin \phi \sin \theta.
$$

The change-of-coordinate formulas are thus

$$x = \rho \sin \phi \cos \theta,$$

$$y = \rho \sin \phi \sin \theta,$$

$$z = \rho \cos \phi.$$

REMARK. Note that we are using ρ here for the distance to the origin; earlier ρ was our standard symbol for density.

Triple integrals in which the integrand f or the domain of integration G exhibit some elements of spherical symmetry about the origin may be computed most easily by iterated integration in spherical coordinates. Here is the basic theorem.

Theorem 13. *Let G'' be the region in ρ,ϕ,θ-space that is the image of G under the spherical-coordinate mapping $(x, y, z) \to (\rho, \phi, \theta)$. That is, G'' consists of all triples (ρ,ϕ,θ) for which the corresponding points (x,y,z) lie in G. Let $f(x,y,z)$ be a continuous function on G, and define the function F by*

$$F(r, \phi, \theta) = f(\rho \sin \phi \cos \theta, \rho \sin \phi \sin \theta, \rho \cos \phi).$$

Then

$$\iiint\limits_{G} f(x, y, z)\, dV = \iiint\limits_{G''} F(\rho, \phi, \theta)\rho^2 \sin \phi \, dV''.$$

Sketch of Proof. This is just the "square" of Theorem 11. We start with the formula at the end of that theorem,

$$\iiint\limits_{G'} F(r, \theta, z)r\, dV',$$

and apply Theorem 11 to it, this time making the change to polar coordinates ρ,ϕ in the z,r-plane: $z = \rho \cos \phi, r = \rho \sin \phi$. Theorem 13 then follows. ∎

We can now evaluate the integral on the right in the theorem above as an iterated integral in the variables ρ,ϕ,θ (by Theorem 10), where the limits of integration are determined by introducing spherical coordinates in x,y,z-space.

EXAMPLE 2. Compute the volume of the sphere of radius a by using spherical coordinates.

. . . .

Solution. The volume of a body is the same as its mass if its density is constant and equal to one. That is, the volume of G is obtained as the triple integral over G of the constant function 1. Also, the sphere of radius a is described in terms of the spherical-coordinate variables by the inequalities $0 \leq r \leq a$, $0 \leq \phi \leq \pi$, $0 \leq \theta \leq 2\pi$. Therefore,

$$V = \iiint_G 1 \, dV = \iiint_{G''} \rho^2 \sin \phi \, dV'$$

$$= \int_0^{2\pi} \int_0^{\pi} \int_0^a \rho^2 \sin \phi \, d\rho \, d\phi \, d\theta$$

$$= \int_0^{2\pi} \int_0^{\pi} \frac{\rho^3}{3} \sin \theta \Big]_{\rho=0}^a d\phi \, d\theta$$

$$= \frac{a^3}{3} \int_0^{2\pi} \int_0^{\pi} \sin \phi \, d\phi \, d\theta = \frac{a^3}{3} \int_0^{2\pi} [-\cos \phi]_0^{\pi} d\theta$$

$$= \frac{2a^3}{3} \int_0^{2\pi} d\theta = \frac{2a^3}{3} \cdot 2\pi = \frac{4}{3} \pi a^3.$$

PROBLEMS FOR SECTION 9

In Problems 1 through 7 find the volume of the given solid by iterated integration in cylindrical or spherical coordinates.

1. The region between the paraboloid $z = x^2 + y^2$ and the plane $z = 4$.

2. The region cut from the sphere $x^2 + y^2 + z^2 = b^2$ by the cylinder $x^2 + y^2 = a^2$ (where $a \leq b$).

3. The "ice cream cone" cut from the sphere $x^2 + y^2 + z^2 = 4$ by the vertical cone with vertex at the origin and semi-vertex angle equal to $\pi/6$.

4. The region cut from the sphere $x^2 + y^2 + z^2 = a^2$ by the upper half of the cone generated by rotating the line $z = mx$ about the z-axis. Use cylindrical coordinates.

5. The region cut from the sphere $\rho = a$ by the cone $\phi = \alpha$ (spherical coordinates). Interpret your answer for the case $\alpha > \pi/2$, and, in particular for $\alpha = \pi$.

6. The region between the paraboloid $z = x^2 + y^2$ and the plane $z = 2x$.

7. The region bounded by the sphere $x^2 + y^2 + z^2 = 6$ and the paraboloid $z = x^2 + y^2$.

8. Find the mass of a ball of radius a if its density is proportional to the distance from the center.

9. Find the mass of a ball of radius a if its density is inversely proportional to the distance from the center.

10. The density of a ball is $k\rho^\alpha$, where ρ is the distance to the origin and α is a constant. Find what limitation is placed on the exponent α by the requirement that the mass of the ball be finite.

11. The density of a ball centered at the origin is proportional to the distance from the xy-plane. Find the mass of the ball.

12. The density of a ball centered at the origin is proportional to the distance from the z-axis. Find the mass of the ball.

13. A cylinder is bounded by the cylindrical surface $x^2 + y^2 = a^2$ and the planes $z = 0, z = b$. Its density is zr. Find its mass.

14. Find the centroid of the region between the xy-plane and the paraboloid $z = 4 - (x^2 + y^2)$.

15. Find the centroid of the "ice cream cone" region of Problem 3.

16. Find the centroid of the region of Problem 7.

17. Find the centroid of the region cut off from the first octant by the sphere $x^2 + y^2 + z^2 = a^2$.

18. Write out the proof of Theorem 11, using the piercing formula.

10. THE TRIPLE INTEGRAL.

The definition of the triple integral is exactly analogous to our earlier definition of the double integral. Let $f(x, y, z)$ be any continuous function defined over a bounded closed region G in space. A Riemann sum for f over G is defined in the following way: We subdivide G in any manner into a finite number of small subregions ΔG_i, and in each of the subregions ΔG_i we choose an arbitrary evaluation point (x_i, y_i, z_i). Then the Riemann sum for f defined by this subdivision of G and the evaluation set is

$$\sum f(x_i, y_i, z_i)\Delta V_i,$$

where ΔV_i is the volume of G_i, for each i.

If we change to a different subdivision and evaluation set, then the Riemann sum will probably change in value. Suppose that this varying Riemann sum $\sum$ approaches a limit L as the subdivision is taken finer and finer. More precisely, suppose that we can make $\sum - L$ as small as we wish simply by taking a subdivision whose subregions ΔG_i all have diameters less than a suitably small positive number d. If this happens, then the limit L of the varying Riemann sum $\sum$ is called the *triple integral of f over G*, and is designated

$$\iiint\limits_{G} f(x, y, z)\, dV.$$

That is,

Definition

$$\iiint_G f(x,y,z)\,dV = \lim \sum_{i=1}^{n} f(x_i, y_i, z_i)\Delta V_i,$$

provided that the limit exists in the above sense.

The basic theorem is that this limit does exist for any continuous function f.

Theorem 14. *If $f(x,y,z)$ is continuous on the bounded closed region G, then its triple integral over G exists.*

So far we have parroted our earlier double-integral discussion almost verbatim. However, the proof of Theorem 14 has to be different, because the graph of a function of three variables lies in four-dimensional space, and we have no secure intuitions about four-dimensional volume. Instead, we shall prove this theorem by interpreting a positive continuous function as the density $\rho(x,y,z)$ of a mass distribution, and then applying certain intuitively obvious principles that govern the relationship between a mass distribution and its density ρ. We shall then have a "physical" proof of the theorem.

Suppose, then, that the continuous function $\rho(x,y,z)$ is the density of a distribution of mass through a space region G. Let m be the total mass of G and let V be the volume of G. We assume:

I.   If the density $\rho(x,y,z)$ has the constant value k throughout G, then $m = kV$.

II. If $\rho(x,y,z) \leq k$ throughout G, then $m \leq kV$. If $\rho(x,y,z) \geq k$ throughout G, then $m \geq kV$.

These principles let us analyze the relationship between m and ρ. Starting with some given small positive number e, we subdivide G into a finite number of subregions ΔG_i on each of which the variation of ρ is less then e. That is, if ρ_i' and ρ_i'' are the minimum and maximum values of $\rho(x,y,z)$ on the subregion ΔG_i, then $\rho_i'' - \rho_i' < e$, for each i. This restriction on the variation of ρ will be met automatically if we subdivide G into subregions whose diameters are all less than a suitably small number d. (See Section 3.)

According to principle II above, the mass Δm_i occupying the subregion ΔG_i satisfies the inequality

$$\rho_i'\Delta V_i \leq \Delta m_i \leq \rho_i''\Delta V_i,$$

for each i. Now let (x_i, y_i, z_i) be an arbitrary evaluation point in ΔG_i, for each i. Then,

$$\rho_i' \leq \rho(x_i, y_i, z_i) \leq \rho_i'',$$

by the definition of ρ_i' and ρ_i'', so we also have the inequality

$$\rho_i'\Delta V_i \leq \rho(x_i, y_i, z_i)\Delta V_i \leq \rho_i''\Delta V_i.$$

Since two numbers lying in the same interval differ by at most the length of the interval, it follows, for each i, that the numbers Δm_i and $\rho(x_i, y_i, z_i)\Delta V_i$ differ by at most $(\rho''_i - \rho'_i)\Delta V_i$ in magnitude. That is,

$$\Delta m_i \approx \rho(x_i, y_i, z_i)\Delta V_i,$$

with an error at most $(\rho''_i - \rho'_i)\Delta V_i$. Since $\rho''_i - \rho'_i \leq e$ for each i, the sum of these errors is at most

$$\sum e\Delta V_i = e \sum \Delta V_i = eV,$$

and since $m = \sum m_i$. is the total mass of G, we conclude that

$$m \approx \sum \rho(x_i, y_i, z_i)\Delta V_i$$

with an error at most eV.

Note that we can make this error as small as we wish by choosing e suitably small to begin with and then using any subdivision for which the diameters of the subregions ΔG_i are all less than the associated small number d. This shows that m is the limit of the varying Riemann sum as the subdivision of G is taken finer and finer. So the triple integral $\iiint_G \rho(x, y, z)\, dV$ exists (and has the value m).

Since any positive continuous function $f(x,y,z)$ can be interpreted as the density of a mass distribution, we thus have a "physical" proof of Theorem 14, for the case that f is positive. Moreover, any function that is not positive can be reduced to a positive f by a simple device (Problem 6), so we are done. ∎

The above proof of Theorem 14 is based on physical intuitions about density and mass. However, with a little more mathematical sophistication, the reasoning we used can be converted into a sound analytic proof. We shall take this step (in Chapter 20) for the definite integral $\int_a^b f$, and the argument given there will cover double and triple integrals with only slight modifications.

We turn now to the piercing theorem.

Theorem 15. *Let G be the region in space based on the bounded closed plane region R in the xy-plane and capped by the surface $z = \phi(x, y)$, where ϕ is continuous on R. Let f be continuous on G. Then*

$$\iiint_G f\, dV = \iint_R \left(\int_0^{\phi(x, y)} f(x, y, z)\; dz \right) dA.$$

Condensed proof. We can suppose that f is positive, since the general case is easily reduced to this case (Problem 7). It may be helpful, although it is not necessary, to think of f as the density of a mass distribution. Choose any fixed point (x_0, y_0) in R and a small incremental subregion ΔR containing (x_0, y_0). Let m and M be the minimum and maximum values of $\phi(x,y)$ over ΔR, and let B be the maximum value of f on G. We are going to let the diameter of ΔR approach zero. Then m and M both approach $\phi(x_0, y_0)$, and $M - m \to 0$, by the continuity of ϕ. It follows from this that if ΔG is the region based on ΔR and capped by the surface $z = \phi(x, y)$, if C is the cylinder with base ΔR and altitude m, and if ΔA is the area of ΔR, then

$$\frac{1}{\Delta A} \iiint_{\Delta G} f\, dV - \frac{1}{\Delta A} \iiint_{C} f\, dV \to 0.$$

This is because $\Delta G - C$ is included in a cylinder with base ΔR and altitude $M - m$, so its volume is less than $(M - m)\Delta A$. Since the above difference is $1/\Delta A$ times the integral of f over $\Delta G - C$, it is thus less than $B(M - m)\Delta A/\Delta A = B(M - m)$, which approaches zero.

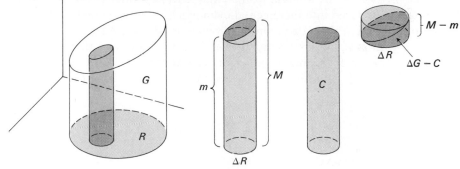

Next, we note that

$$\int_0^m f(x_0, y_0, z)\, dz \to \int_0^{\phi(x_0, y_0)} f(x_0, y_0, z)\, dz,$$

since $m \to \phi(x_0, y_0)$.

Finally, we shall show below that

$$\frac{1}{\Delta A} \iiint_{C} f\, dV - \int_0^m f(x_0, y_0, z)\, dz \to 0.$$

In combination, these three limits show that

$$\frac{1}{\Delta A} \iiint_{\Delta G} f\, dV \to \int_0^{\phi(x_0, y_0)} f(x_0, y_0, z)\, dz,$$

as the diameter of ΔR approaches zero. That is, if we view the triple integral of f over G as a (mass) distribution over the xy-plane, then it has the distribution density

$$\rho(x, y) = \int_0^{\phi(x, y)} f(x, y, z)\, dz.$$

So $\iiint_G f \, dV$ can be computed as the double integral of its density over R (Section 4), and the theorem is proved.

To complete the theorem we have to prove that

$$\frac{1}{\Delta A} \iiint_C f \, dV \approx \int_0^m f(x_0, y_0, z) \, dz$$

with an error that can be made arbitrarily small merely by choosing the diameter of ΔR small enough. To see this, let

$$0 = z_0 < z_1 < \cdots = z_n = m$$

be any subdivision of the interval $[0,m]$. The vertical cylinder (with ΔR as base and altitude m) is sliced into incremental cylinders ΔC_i by the planes $z = z_i$. The altitude of ΔC_i is $\Delta z_i = z_i - z_{i-1}$, and its volume is $\Delta V_i = \Delta A \Delta z_i$.

If we are given a small positive number e, then we can ensure that f varies by less than e over each of the incremental cylinders ΔC_i by taking ΔR small enough and the subdivision $\{z_i\}$ fine enough. (See Section 3.) Then we have Riemann-sum estimates:

$$\iiint_C f \, dV \approx \Sigma f(x_0, y_0, z_i) \Delta V_i,$$

with an error at most $e\Delta V = em\Delta A$, and

$$\int_0^m f(x_0, y_0, z) \, dz \approx \Sigma f(x_0, y_0, z_i) \Delta z_i,$$

with an error at most em. Dividing the first estimate throughout by ΔA, the first Riemann sum becomes the second, so we end up with

$$\frac{1}{\Delta A} \iiint_C f \, dV \approx \int_0^m f(x_0, y_0, z) dz,$$

with an error at most $2em$. Moreover, this holds provided only that the diameter of ΔR is suitably small, since the subdivision $\{z_i\}$ has disappeared in this final approximation. We have thus completely proved the theorem. ∎

PROBLEMS FOR SECTION 10

Prove the following four properties of the triple integral from its definition as the limit of Riemann sums, assuming that the standard limit laws are valid for the type of limit used in this definition.

1. If f is the constant function k and if V is the volume of the space region G, then

$$\iiint_G f\, dV = kV.$$

2. $\iiint_G (af + bg)\, dV = a \iiint_G f\, dV + b \iiint_G g\, dV.$

3. If G is divided into two nonoverlapping subregions G_1 and G_2, then

$$\iiint_G f\, dV = \iiint_{G_1} f\, dV + \iiint_{G_2} f\, dV.$$

4. If $f \geq 0$ on G, then $\iiint_G f\, dV \geq 0.$

5. Now prove from 1, 2, and 4 that, if f is bounded above by the constant k on G, then

$$\iiint_G f\, dV \leq kV,$$

where V is the volume of G.

6. Assuming that the triple integral exists for a positive continuous function (Theorem 14, as proved in the text), prove that it exists for any continuous function (Theorem 14 as stated). (*Hint.* Let k be the maximum value of f on G, write $f = k - (k - f)$ and apply various results already established.)

7. Prove Theorem 15 for an arbitrary continuous function by reducing to a positive function, in the manner suggested in the above problem.

The following three problems refer to the slicing theorem as stated in Section 8.

8. Suppose that G is a cylinder parallel to the z-axis, so that its slices R_z perpendicular to the z-axis are all the same plane region R. Let ΔG be the slab sliced from G by the planes $z = z_0$ and $z = z_0 + \Delta z$, and subdivide ΔG into incremental cylinders ΔG_i by subdividing R into incremental bases ΔR_i. Using Riemann sums related to these subdivisions, show that

$$\frac{1}{\Delta z} \iiint_{\Delta G} f\, dV \approx \iint_{R_{z_0}} f(x, y, z_0)\, dA$$

with an error that can be made arbitrarily small merely by taking Δz suitably small. (*Hint.* Follow the general scheme used in the last part of the proof of Theorem 15.)

9. Now suppose that G is a general region whose cross section R_z perpendicular to the z-axis varies continuously with z in the manner defined in Section 9 of Chapter 8. Thus the slab ΔG sliced from G by the planes $z = z_0$ and

$z = z_0 + \Delta z$ is squeezed between two cylinders C' and C'' whose base areas A' and A'' both approach the area of R_{z_0} as Δz approaches 0. Prove that

$$\frac{1}{\Delta z} \iiint\limits_{\Delta G} f \, dV \rightarrow \iint\limits_{R_{z_0}} f(x, y, z_0) \, dA,$$

as Δz approaches 0, by arguing in the general manner of the proof of Theorem 15, and using the result in Problem 8.

10. Prove the slicing theorem by interpreting the above limit as a statement about the density of a distribution along the z-axis.

some elementary differential equations

The laws governing natural phenomena are often expressed by differential equations. For example, we saw in Chapter 6 that the differential equation

$$\frac{dy}{dt} = ky$$

can be interpreted as the law of *normal population growth*. In the absence of inhibiting or stimulating factors, a population reproduces itself at a rate proportional to its size, which is exactly what the above equation says. Then, solving the equation, we are startled to learn that a population grows exponentially: The solution of the population equation is

$$y = y_0 e^{kt},$$

where the constant y_0 is the size of the initial population, i.e., the size at time $t = 0$. Moreover, we can now answer such quantitative questions as: How frequently will the population double in size? And when will it reach a magnitude that we have chosen, perhaps arbitrarily, as being critical?

Thus the differential equation states our understanding of the growth process as a *rate-of-change* phenomenon, and its solution shows us the growth pattern and predicts the future.

Elastic vibration is another example. We studied a simple idealized elastic system at the end of Chapter 6. It was instructive about frequency, amplitude, and phase, but it was not a very good model of reality because it did not take resistance into account. This topic will come up again at the end of the present chapter. By that time we shall be able to handle the effects due to resistance and to an external driving force; and we shall be in a position to analyze a wide variety of vibratory phenomena.

As these examples suggest, differential equations lead to important applications of calculus; many mathematicians feel that they are the most important applications. However, the most we can try to do here is to carry the subject far enough to finish up our discussion of the simple elastic system. With that in mind we turn to a few beginning topics of the theory.

1. SEPARATION OF VARIABLES.

The equation

$$3\frac{dy}{dx} + \frac{x}{y^2} = 0$$

states a relationship between an unknown function $y = f(x)$ and its derivative $dy/dx = f'(x)$, and is thus a differential equation for $y = f(x)$. It is a differential equation of the *first order* because it involves only the first derivative of f. A *solution* of the differential equation is a function $y = f(x)$ making the equation true, i.e., a function f such that

$$3f'(x) + \frac{x}{[f(x)]^2} \equiv 0.$$

Now imagine that y represents a solution in the original equation and let us see what we can learn about this solution by manipulating the equation into different but equivalent forms. If we multiply by y^2, the equation becomes

$$3y^2 \frac{dy}{dx} + x = 0.$$

We now recognize the first term as dy^3/dx, and the equation can be rewritten

$$\frac{d}{dx}\left[y^3 + \frac{x^2}{2}\right] = 0.$$

Since $du/dx = 0$ only if $u = $ constant, we conclude that

$$y^3 + \frac{x^2}{2} = C,$$

so

$$y = (C - x^2/2)^{1/3}.$$

Alternatively, after multiplying by y^2, we could write the equation as

$$\frac{d}{dx}y^3 = -x.$$

Therefore,

$$y^3 = \int -x\,dx = \frac{-x^2}{2} + C,$$

as before. We have thus "integrated" the equation and have found the most general solution function, $y = f(x)$. Note that the general solution contains a constant of integration.

The only tricky step above was recognizing $3y^2\,dy/dx$ as dy^3/dx. This step can be made more obvious by writing the differential equation in *differential* form,

$$3y^2\,dy + x\,dx = 0.$$

We then just integrate across, obtaining

$$\int 3y^2\,dy + \int x\,dx = C,$$

or

$$y^3 + \frac{x^2}{2} = C,$$

which we know to be the right answer.

In general:

Theorem 1. *A differentiable function $y = f(x)$ satisfies the differential equation*

$$g(y)\,dy + h(x)\,dx = 0$$

if and only if it satisfies the equation

$$\int g(y)\,dy + \int h(x)\,dx = C$$

for some constant C. That is, if $G(y)$ and $H(x)$ are antiderivatives of $g(y)$ and $h(x)$, respectively, then the solutions of the differential equation are exactly the differentiable functions defined implicitly by the equation

$$G(y) + H(x) = C,$$

for all values of the constant C.

Proof. We argue backward. The function

$$G(f(x)) + H(x)$$

is constant if and only if its derivative is 0, i.e., if and only if

$$G'(f(x))f'(x) + H'(x) = 0.$$

That is, a differentiable function $y = f(x)$ satisfies the equation

$$G(y) + H(x) = \text{constant}$$

if and only if

$$G'(y)\frac{dy}{dx} + H'(x) = 0$$

or

$$g(y)\frac{dy}{dx} + h(x) = 0,$$

$$(\text{or} \quad g(y)\,dy + h(x)\,dx = 0).$$

Thus the solutions of the differential equation are exactly the differentiable functions defined implicitly by the equations $G(y) + H(x) = C$ for various constants C. ∎

A differential equation of the form

$$g(y)\,dy + h(x)\,dx = 0$$

is said to have *separated variables*; putting an equation into this form is *separating the variables*.

EXAMPLE 1. The equation

$$\frac{dy}{dx} + \frac{x}{y} = 0$$

can be put into this form by multiplying by $y\,dx$, the equation then becoming

$$y\,dy + x\,dx = 0.$$

By Theorem 1, this differential equation has the general solution

$$\int y\,dy + \int x\,dx = C,$$

or

$$\frac{y^2}{2} + \frac{x^2}{2} = C,$$

or

$$x^2 + y^2 = C$$

(replacing $2C$ by C). As curves in the coordinate plane, the solutions are the *circles about the origin*. A *function* is a solution if and only if its graph lies along one of these curves. The function solutions divide up into the upper half-circle functions

$$y = \sqrt{C - x^2},$$

and the lower half-circle functions

$$y = -\sqrt{C - x^2}.$$

Viewing a circle as the union of upper and lower half circles is geometrically unappealing, and in this sense it may be more satisfying to view the solutions of

$$h(x)\,dx + g(y)\,dy = 0$$

as the curves

$$H(x) + G(y) = C,$$

rather than as the functions whose graphs lie along these curves.

EXAMPLE 2. If we solve the above equation using the derivative form of the separated equation, then the steps could be shown like this:

$$\frac{dy}{dx} + \frac{x}{y} = 0,$$

$$y\frac{dy}{dx} + x = 0,$$

$$\frac{d}{dx}\left[\frac{y^2 + x^2}{2}\right] = 0,$$

$$y^2 + x^2 = \text{constant}.$$

EXAMPLE 3. Solve the differential equation

$$\frac{dy}{dx} + xy = 0.$$

. . . .

Solution. First separate the variables,

$$\frac{dy}{y} + x\,dx = 0,$$

and then integrate, obtaining

$$\log|y| + \frac{x^2}{2} = c.$$

Therefore,

$$\log|y| = c - \frac{x^2}{2},$$

$$|y| = e^{c - x^2/2}.$$

The right side is never 0, so y can never change sign. Therefore,

$$y = Ce^{-x^2/2},$$

where $C > 0$ if y is always positive and $C < 0$ if y is always negative. Note that the constant of integration appears here as the multiplicative constant C.

It may happen that the process of separating the variables loses a solution of the original equation. For example, we divided by y to separate variables in the above differential equation. This introduces the new requirement in the separated equation that $y \neq 0$, and warns us to check back to see whether the added restriction loses a solution. We see by inspection that the constant function $y = 0$ is indeed such a lost solution.

In this example, the lost solution is recovered when we change the form of the constant of integration from c to $C = e^c$. The solutions of the separated equation correspond to values of C different from 0, but we can try putting $C = 0$, and we find that we then recover the lost solution of the original equation.

Lost solutions may or may not be recovered in this way in the final form of the general solution. There can also be other types of "extraneous" solutions that are not covered by the "general" solution. However, we shall see in Section 4 that *linear* differential equations have no extraneous solutions; the general solution contains every solution.

EXAMPLE 4. A more realistic view (model) of population growth will take into account other factors besides the size of the population. For example, there may be some compelling reason why the population x can never grow beyond a limiting size L, due perhaps to a limitation on food or living space. In that case we would expect the growth rate to lessen as the population size increases, and to approach 0 as x approaches L. This hypothesis does not by itself determine the new growth law. The simplest way for such an inhibition to express itself is for the growth rate dx/dt to be proportional both to the population size x and to its remaining possible room for growth $L - x$. Then

$$\frac{dx}{dt} = kx(L - x).$$

There might, on the other hand, be a reason why the limiting size L is "doubly depressing" on the growth rate, so that

$$\frac{dx}{dt} = kx(L - x)^2.$$

In order to solve the first equation we separate variables and integrate:

$$\int \frac{dx}{x(L - x)} = \int k\, dt,$$

$$\frac{1}{L} \log \frac{x}{L - x} = kt + c,$$

$$\frac{x}{L - x} = Ce^{Lkt},$$

where the integration constant $C = e^c$ is positive. This equation can be solved for x, and the solutions are the functions

$$x = \frac{CLe^{Lkt}}{1 + Ce^{Lkt}}.$$

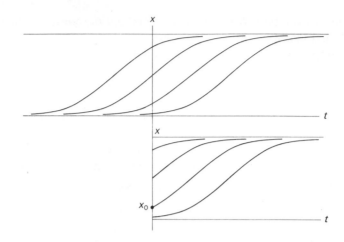

The solution curves fill up the strip $0 < x < L$, as indicated above. If we don't project these growth curves into the past, i.e., if we don't use *negative time*, then the solution curves are confined to the half strip $0 < x < L$, where $t \geq 0$. For any x_0 between 0 and L, there is a growth curve emanating from $(0,x_0)$, showing the evolution of a population that starts at size x_0. That is, $x = x_0$ at $t = 0$ is an *initial condition* that picks out a unique solution curve. (19-1)

We lose the solutions $x = 0$ and $x = L$ when we separate variables in the above equation. The first lost solution is recovered in the general solution by giving the integration constant C the value of 0, but the solution $x = L$ is not part of the general solution. (In a sense, it corresponds to setting $C = \infty$.)

Note also that the lost solutions form the boundary of the region that is filled up by the solution curves of the general solution.

PROBLEMS FOR SECTION 1

Solve the following differential equations.

1. $\dfrac{dy}{dx} = 2y$

2. $y\dfrac{dy}{dx} = 1$

3. $\dfrac{dy}{dx} = ky$

4. $\dfrac{dy}{dx} = y^2$

5. $\dfrac{dy}{dx} = y^2 x$

6. $\dfrac{dy}{dx} = e^y$

7. $\dfrac{dy}{dx} = xe^y$

8. $\dfrac{dy}{dx} = ye^x$

9. $\dfrac{dy}{dx} = \sqrt{1 - y^2}$

10. $\dfrac{dy}{dx} = x^2$

11. $\dfrac{dy}{dx} = \dfrac{2x}{y}$

12. $\dfrac{dy}{dx} = \dfrac{2y}{x}$

13. $x^3 \dfrac{dy}{dx} = y^2(x - 4)$ 14. $4xy\,dx + (x^2 + 1)\,dy = 0$

15. $dy/dx = f(x)$ (Show that separation of variables amounts merely to integrating $f(x)$ for this type of equation.)

16. $\dfrac{dy}{dx} = y^2 \cos x$ 17. $x\,dy + \cos^2 y\,dx = 0$

18. $\sqrt{1 - x^2}\,dy + \sqrt{1 - y^2}\,dx = 0$ 19. $\dfrac{dy}{dx} = \cos^2 y \sin x$

20. $dy/dx = (y/x) + (y/x)^2$.   Here the variables cannot be separated. Show, however, that the substitution $y = vx$ reduces the given equation to one in x and v, in which the variable *can* be separated, and hence solve the equation. (Note that dy/dx becomes $d(xv)\,dx = v + x\,dv/dx$.)

21. Show that the equation

$$\frac{dy}{dx} = f\!\left(\frac{y}{x}\right)$$

reduces to a variables-separable equation in v and x under the substitution $y = vx$.

22. $(x^2 + y^2)\,dx - 2xy\,dy = 0$.

23. A population grows at a rate proportional to its size, but, because of a steadily deteriorating environment, its growth rate is also *inversely proportional to time*. Find the law of population growth.

24. The growth rate of a population of size P is proportional to P, but is also inversely proportional to $P^{1/2}$, because of difficulties in maintaining the food supply. Find the law of population growth.

25. The velocity of water flowing from a pipe at the bottom of a cylindrical tank is proportional to the square root of the depth of the water in the tank. If the water level drops from 10 feet to 9 feet in the first hour, how long will it take to reach the level of x feet?

26. A chemical reaction converts a substance A into a substance B at a rate that is proportional to the amount of A present and inversely proportional to the amount of B. Find a formula relating the amount of B to the time t. (If x is the amount of A and y is the amount of B, assume that x and y have the initial values $x = c, y = 0$.)

2. ORTHOGONAL TRAJECTORIES

We have considered a number of examples of differential equations, all in the general form

$$\frac{dy}{dx} = F(x, y).$$

In each case, the general solution contained a *constant* of *integration* C. For example, the general solution of

$$x\frac{dy}{dx} = 2y$$

is

$$y = Cx^2.$$

Each value of the constant C determines a solution curve. As C varies the solution curves vary, and in their totality they fill up some region G in the plane. In the above example the solution curves fill up the whole plane except for the y-axis.

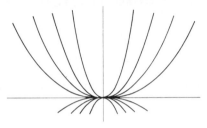

In the last example in the preceding section they fill up a strip or a half-strip. In such situations we call the constant of integration C a *parameter*, and we say that the solutions of the differential equation form a *one-parameter* family of curves filling up G. This use of the word "parameter" is different from, but related to, its use in connection with paths.

We can often recapture a differential equation from its general solution by differentiating and eliminating the parameter.

EXAMPLE 1. Along each curve of the one-parameter family

$$y = Cx^2,$$

we have

$$\frac{dy}{dx} = 2Cx.$$

Since y/x^2 has the constant value C along the curve, we can set $C = y/x^2$ in the second equation and thus recapture the differential equation

$$\frac{dy}{dx} = \frac{2y}{x}.$$

EXAMPLE 2. The parameter C in

$$y = x^2 + C$$

disappears upon differentiation. The differential equation of this one-parameter family is thus

$$\frac{dy}{dx} = 2x.$$

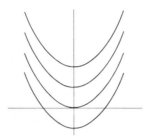

EXAMPLE 3. Sketch the one-parameter family

$$y = \sin(x + C),$$

where $x + C$ is restricted to lie in the interval $[-\pi/2, \pi/2]$. Find its differential equation.

. . . .

Solution. For each value of C, x lies between $-\pi/2 - C$ and $\pi/2 - C$, and $y = \sin(x + C)$ runs from -1 to $+1$. The curves fill up the strip $-1 \le y \le 1$, as shown below.

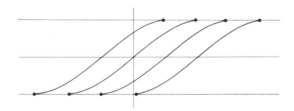

In principle, we can eliminate the parameter C by solving for C either in the given equation for y or in the differential equation for dy/dx, and then substituting its value in the other equation. We can describe the procedure as eliminating C *between* the two equations, ending up with a single equation relating y to dy/dx. Here we can first replace $y = \sin(x + C)$ by the equivalent equation

$$\arcsin y = x + C,$$

and then C drops out upon differentiating, giving the differential equation of the family at once:

$$\frac{1}{\sqrt{1 - y^2}} \frac{dy}{dx} = 1,$$

or

$$\frac{dy}{dx} = \sqrt{1 - y^2}.$$

Another procedure is to differentiate the given equation $y = \sin(x + C)$,

$$\frac{dy}{dx} = \cos(x + C),$$

and then use the fact that $\cos u = \sqrt{1 - \sin^2 u}$ for all u in the interval $[-\pi/2, \pi/2]$. Thus,

$$\frac{dy}{dx} = \cos(x + C) = \sqrt{1 - \sin^2(x + C)}$$

$$= \sqrt{1 - y^2},$$

which is the same equation as before.

An *orthogonal trajectory* of a family of curves is a curve that crosses each curve of the family at right angles. The orthogonal trajectories of a one-parameter family of curves will themselves form a second one-parameter family, and they can be found by solving a differential equation.

EXAMPLE 4. Find the orthogonal trajectories of the family of parabolas $y = Cx^2$.

. . . .

Solution. We have seen that the differential equation of this family of curves is

$$\frac{dy}{dx} = \frac{2y}{x}.$$

A curve cutting each of these parabolas orthogonally will have a slope that is everywhere the *negative reciprocal* of the above slope, so it will be a solution of the equation

$$\frac{dy}{dx} = -\frac{x}{2y},$$

or

$$2y \, dy + x \, dx = 0.$$

These orthogonal trajectories are thus the curves

$$y^2 + \frac{x^2}{2} = C,$$

a one parameter family of *ellipses*.

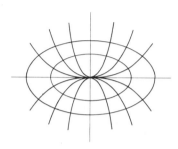

PROBLEMS FOR SECTION 2

In each of the following problems, find the differential equation of the given one-parameter family of curves. Sketch a few curves of the family, and indicate the region of the plane that they cover.

1. $y = x + C$ 2. $y = Cx$
3. $x^2 + y^2 = C$ 4. $y = \sqrt{x + C}$
5. $y = \sqrt{1 - x^2} + C$ 6. $y = Ce^x/(1 + Ce^x)$
7. The one-parameter family of curves

$$y = (x + C)^2$$

has the added complication that *two* curves of the family pass through each point of its region G. Show that this is so, and sketch a few of the curves. Also find a differential equation of the family, and show from the differential equation why one would expect two solutions through each point.

8. Do the same for the one-parameter family

$$(y + C)^2 + x^2 = 1.$$

Find the orthogonal trajectories of each of the following families of curves. Sketch a few curves of the given family and a few of the orthogonal trajectories.

9. The straight lines $y = Cx$
10. The hyperbolas $y = \dfrac{1}{x} + C$
11. The parabolas $y^2 = x + C$
12. The hyperbolas $y = C/x$
13. The circles $x^2 + y^2 = C$
14. The curves $y = C - \arctan x$
15. The circles $x^2 + Cx + y^2 = 0$.

16. If $y = (x + C)/(1 - Cx)$, show that

$$\frac{dy}{dx} = \frac{1 + y^2}{1 + x^2}.$$

17. Solve the above differential equation, and show that its general solution can be expressed in the above form.

3. DIRECTION FIELDS

The differential equation

$$\frac{dy}{dx} = F(x, y)$$

itself has a geometric interpretation that is vivid and very useful. It assigns the slope $dy/dx = F(x, y)$ to the point (x,y) and hence determines a *direction* at each point. We say that the equation defines a *direction field* or a *field of directions* in the plane. We can sketch the direction field by drawing a short line segment (having the given slope) through each of a few points.

EXAMPLE 1. Sketch the direction field for the equation

$$\frac{dy}{dx} + \frac{x}{y} = 0.$$

. . . .

Solution. Here the slope at (x,y) is given by

$$\frac{dy}{dx} = -\frac{x}{y}.$$

Since $-x/y$ is the negative reciprocal of y/x, which is the slope of the line segment from the origin to (x,y), the direction field consists of the directions perpendicular to these segments.

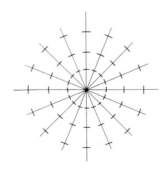

A solution of the differential equation is a curve having the direction $dy/dx = F(x,y)$ at each point (x,y) on the curve, i.e. a curve tangent at each of its points to the little segment showing the field directions at that point. With this in mind, it is practically obvious from the above figure that the solutions of

$$\frac{dy}{dx} = -\frac{x}{y}$$

are the circles about the origin.

EXAMPLE 2. The equation

$$\frac{dy}{dx} = y$$

has a feature similar to the first example, in that the slope dy/dx has the same value at each point along a line, this time a horizontal straight line.

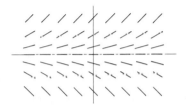

In practice we may meet differential equations of the form $dy/dx = F(x,y)$ where the function $F(x,y)$ is so complicated that we cannot explicitly solve the equation. However, we can still draw the direction field to see what the solutions look like, and this suggests a procedure for constructing approximate solutions. We start at some initial point (x_0,y_0) and proceed along the direction segment at that point to a new point (x_1,y_1). We then change directions and go along the direction segment at (x_1,y_1) to a third point (x_2,y_2), etc. We thus construct a "polygonal" function p, whose graph is linear over each of the intervals

$$[x_0, x_1], \quad [x_1, x_2], \quad [x_2, x_3], \quad \cdots,$$

and such that

$$p'(x_0) = F(x_0, p(x_0)), \qquad p'(x_1) = F(x_1, p(x_1)), \qquad \text{etc.}$$

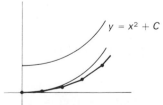

$y = x^2 + C$

The figure suggests that if the points x_0, x_1, x_2, ... are taken close together, then the polygonal function $p(x)$ computed as above will be a good approximation to an actual solution $f(x)$, at least in the short run. The figure also suggests that the error $|f(x) - p(x)|$ does grow as x moves away from x_0. However, there is a computation that one can make that shows how close together the points x_0, x_1, x_2, ... should be taken in order to ensure that the error remains less than some preassigned tolerance over a given interval I.

PROBLEMS FOR SECTION 3

1. Sketch the direction field of the differential equation $(dy/dx) = x$. Use the direction field to sketch the solution curve through the point $(1,1)$. Now find the equation of the solution curve passing through this point by solving the differential equation and evaluating the constant of integration.

2 Do the same for the equation $dy/dx = y/x$.

3. Do the same for the equation $dy/dx = xy$.

4. THE FIRST-ORDER LINEAR EQUATION

A first-order linear differential equation is an equation of the form

$$\frac{dy}{dx} + a(x)y + b(x) = 0,$$

where the coefficients $a(x)$ and $b(x)$ are any functions of x. Superficially, the reason for the word *linear* is that we call $ay + b$ a linear expression in y. The deeper meaning of linearity will be discussed a little later.

A linear equation cannot generally be solved by separating the variables. However, if $b(x)$ is the zero function we have the special case

$$\frac{dy}{dx} + a(x)y = 0,$$

called the linear *homogeneous* equation, in which the variables can be separated. Proceeding as usual, we obtain the separated equation

$$\frac{dy}{y} + a(x)\,dx = 0,$$

which has the general solution

$$\int \frac{dy}{y} + \int a(x)\,dx = c,$$

or

$$\log|y| + A(x) = c,$$

where $A(x)$ is any particular antiderivative of $a(x)$. Then

$$|y| = e^c e^{-A(x)}.$$

And since the right side is never zero, y can never change sign and the general solution can be written

$$y = Ce^{-A(x)}.$$

Note that $y = 0$ is a solution of the homogeneous equation $dy/dx + a(x)y = 0$ that is lost in the separated equation $dy/y + a(x)dx = 0$, but is recovered in the general solution by giving the parameter C the value 0.

Note also that the growth equation $dy/dx - ky = 0$ is a special case of the linear homogeneous equation.

The solution of the inhomogeneous linear equation requires a new method, associated with a device that is very important in the theory and practice of differential equations.

What we do is to look for a solution y in the form $y = uv$, where v is a function that we already have in hand for some reason or other, and u is an unknown function to be determined. Then requiring y to be a solution of the given equation turns into requiring u to be a solution of a different equation, and we hope that the new equation for u will be easier to solve than the old equation for y. It often will be if the multiplier function v is somehow related to the beginning equation.

As a first application of this procedure, we take v to be a nonzero solution of the linear homogeneous equation

$$\frac{dy}{dx} + a(x)y = 0.$$

(We know a formula for v, but there is no need to write it down. We do need the fact that v is never zero, so that we can write down $1/v$.) Then we try to find a solution y of the *inhomogeneous* equation

$$\frac{dy}{dx} + a(x)y = r(x)$$

in the form $y = vu$, where u is some function of x to be determined. Setting $y = vu$, we see that the inhomogeneous equation becomes

$$v\frac{du}{dx} + u\frac{dv}{dx} + a(x)uv = r(x)$$

or

$$u\left[\frac{dv}{dx} + a(x)v\right] + v\frac{du}{dx} = r(x)$$

or

$$v \frac{du}{dx} = r(x),$$

since, by assumption, the bracket is zero. Thus

$$\frac{du}{dx} = \frac{r(x)}{v}$$

and

$$u = \int \left(\frac{r(x)}{v} \right) dx + C.$$

Therefore,

$$y = uv = v \int \left(\frac{r(x)}{v} \right) dx + Cv.$$

Moreover, if y is *any* solution of the inhomogeneous equation, then we can set $u = y/v$, proceed as above, and conclude that y is given by the above formula. So there are no extraneous solutions. We have thus found a formula for all the solutions of the inhomogeneous linear equation, without having to know explicitly what the homogeneous solution v is. Of course, $v = e^{-A(x)}$, from the separation-of-variables calculation given earlier, so all in all we have proved the following theorem.

***Theorem* 2**. *The general solution of the linear differential equation*

$$\frac{dy}{dx} + a(x)y = r(x)$$

is given by

$$y = Ce^{-A(x)} + e^{-A(x)} \int e^{A(x)} r(x) \, dx,$$

where $A(x)$ is any particular antiderivative of the coefficient function $a(x)$, and C is the constant of integration. Moreover, every solution is of this form; there are no extraneous solutions.

The solution formula in Theorem 2 seems complicated and awkward at the moment, but its general form can be understood once we have looked into the nature of linearity. This will be the topic of the next section. Meanwhile, the formula should not be memorized. It is better to solve any particular inhomogeneous equation by retracing the steps in the proof of the theorem. This is called the *variation-of-parameter* method, for the following reason. The general solution of the homogeneous equation is a one-parameter family of solutions of the form $y = Cv(x)$. Here we replace the parameter C by an unknown function $u(x)$—

hence the phrase "variation of parameter"—and try for a solution of the inhomogeneous equation in the form $y = u(x)v(x)$.

EXAMPLE 1. Solve the linear inhomogeneous equation

$$\frac{dy}{dx} - y = \cos x,$$

by the variation-of-parameter method.

. . . .

Solution. We first find one solution of the homogeneous equation

$$\frac{dv}{dx} - v = 0.$$

You probably can mentally separate variables here and see that $v = e^x$ is a solution. Or you may just happen to remember that e^x is its own derivative. We then set $y = e^x u$ in the given equation, and get

$$ue^x + e^x\frac{du}{dx} - ue^x = \cos x,$$

$$e^x\frac{du}{dx} = \cos x,$$

$$\frac{du}{dx} = e^{-x}\cos x.$$

The integral of $e^{-x}\cos x$ is worked out by integrating by parts, and is

$$u = \int e^{-x}\cos x = \frac{1}{2}[e^{-x}\sin x - e^{-x}\cos x] + C.$$

Therefore,

$$y = e^x u = \frac{1}{2}[\sin x - \cos x] + Ce^x$$

is the general solution of the equation

$$\frac{dy}{dx} - y = \cos x.$$

PROBLEMS FOR SECTION 4

Solve the following equations by the variation-of-parameter method.

1. $\dfrac{dy}{dx} - 2y = e^x$ 2. $\dfrac{dy}{dx} - 2y = e^{2x}$

3. $\dfrac{dy}{dx} + ay = e^x$ 4. $\dfrac{dy}{dx} + ay = e^{bx}$

5. $\dfrac{dy}{dx} + y = x$ 6. $\dfrac{dy}{dx} - y = x^2$

7. $\dfrac{dy}{dx} + \dfrac{y}{x} = x$ 8. $\dfrac{dy}{dx} - \dfrac{y}{x} = x$

9. $\dfrac{dy}{dx} + \dfrac{y}{x} = \dfrac{1}{x}$ 10. $\dfrac{dy}{dx} - \dfrac{y}{x} = \dfrac{1}{x}$

11. $\dfrac{dy}{dx} + ay = \sin x$ 12. $\dfrac{dy}{dx} + y \sin x = \sin x$

13. $\dfrac{dy}{dx} + 2xy = x^3$ 14. $\dfrac{dy}{dx} - \dfrac{y}{x} = \log x$

15. $\dfrac{dy}{dx} - y \tan x = 1$ 16. $\dfrac{dy}{dx} + y \tan x = 1$

5. LINEARITY

We shall now examine some important properties of a linear differential equation that were implicit in the results of the last section. These properties will explain the general form of the rather complicated solution formula we obtained there, and they will provide a shortcut to the solutions of an important type of inhomogeneous equation. Finally, with the general nature of linearity in mind, we will be in a better position to understand the more complicated second-order equations that we shall turn to next.

A new notation is helpful here: We set

$$L(y) = \frac{dy}{dx} + a(x)y,$$

so that the linear equations

$$\frac{dy}{dx} + a(x)y = 0, \qquad \frac{dy}{dx} + a(x)y = r(x)$$

abbreviate to

$$L(y) = 0, \qquad L(y) = r(x).$$

EXAMPLE 1. If the equation is

$$\frac{dy}{dx} + xy = 0,$$

then $L(y) = \dfrac{dy}{dx} + xy$. In particular,

$$L(x^3) = \frac{d}{dx}(x^3) + x(x^3) = 3x^2 + x^4.$$

Similarly,

$$L(-x) = -1 - x^2,$$

$$L(x^3 - x) = -1 + 2x^2 + x^4$$

$$= L(x^3) - L(x);$$

$$L(e^x) = e^x(x + 1),$$

$$L(e^{-x^2}) = -xe^{-x^2},$$

$$L(e^{-x^2/2}) = 0.$$

Theorem 3. *For any two differentiable functions* $y_1 = f_1(x)$ *and* $y_2 = f_2(x)$, *and any two constants* c_1 *and* c_2,

$$L(c_1 y_1 + c_2 y_2) = c_1 L(y_1) + c_2 L(y_2).$$

Proof. We just compute.

$$L(c_1 y_1 + c_2 y_2) = \frac{d}{dx}(c_1 y_1 + c_2 y_2) + a(x)(c_1 y_1 + c_2 y_2)$$

$$= c_1 \left[\frac{dy_1}{dx} + a(x)y_1 \right] + c_2 \left[\frac{dy_2}{dx} + a(x)y_2 \right]$$

$$= c_1 L(y_1) + c_2 L(y_2). \quad \blacksquare$$

The theorem says that if

$$z = c_1 y_1 + c_2 y_2,$$

then

$$L(z) = c_1 L(y_1) + c_2 L(y_2).$$

That is, when we apply L to a linear combination of two functions y_1 and y_2, we get *that same* linear combination of the functions $L(y_1)$ and $L(y_2)$. Applying L *preserves linear combination relationships*. For this reason we call L a *linear differential operator*, and we call the homogeneous equation

$$L(y) = 0,$$

or

$$\frac{dy}{dx} + a(x)y = 0,$$

a homogeneous *linear* differential equation. Similarly,

$$\frac{dy}{dx} + a(x)y = r(x)$$

or

$$L(y) = r(x),$$

is an *inhomogeneous linear* differential equation.

We already have explicit formulas for the solutions of these equations, but now we can see that the form these formulas take is due to the linearity of the differential operator L. That is, we can show that the solutions of $L(y) = 0$ and $L(y) = r(x)$ must have a certain general form simply because L is linear, and independently of any prior knowledge of these solutions.

Theorem 4. *If $v = \phi(x)$ is one nonzero solution of the linear homogeneous equation*

$$L(y) = 0,$$

then the constant multiples of ϕ, $Cv = C\phi(x)$, form the family of all solutions.
    If $u = \psi(x)$ is any particular solution of the linear inhomogeneous equation

$$L(y) = r(x),$$

then its general solution is

$$y = Cv + u = C\phi(x) + \psi(x).$$

That is, the general solution of the inhomogeneous equation is the sum of the general solution of the homogeneous equation and any particular solution of the inhomogeneous equation.

Proof. The linearity of L shows that if $L(y) = 0$, then $L(Cy) = CL(y) = 0$. Thus if $v = \phi(x)$ is one solution, then its constant multiples $Cv = C\phi(x)$ are all solutions. However, we do need Theorem 2 to tell us that there are no other solutions.

Now suppose that we have found, somehow, one solution $u = \psi(x)$ of the inhomogeneous equation

$$L(u) = r(x).$$

Let y be any other solution. Then

$$L(y - u) = L(y) - L(u) = r(x) - r(x) = 0,$$

by Theorem 3, so $y - u$ is a solution of the homogeneous equation, and hence of the form

$$y - u = Cv = C\phi(x),$$

by the first part of the theorem above. Thus, every solution y of the inhomogeneous equation is of the form

$$y = u + Cv = \psi(x) + C\phi(x).$$

Finally, these functions are all solutions, for

$$L(y) = L(u + Cv) = L(u) + CL(v)$$
$$= r(x) + C \cdot 0$$
$$= r(x).$$

This completes the proof of the theorem. ∎

Note that the above proof depends almost entirely on the linearity of L as stated in Theorem 3.

If we now reexamine the explicit general solution that we obtained earlier,

$$y = Ce^{-A(x)} + e^{-A(x)} \int e^{A(x)} r(x) \, dx,$$

where $A(x)$ is an integral of $a(x)$, we can see the structure promised by the above theorem. It is the sum

$$y = Cv + u,$$

where

$$Cv = Ce^{-A(x)}$$

is the general solution of the homogeneous equation $L(y) = 0$, and

$$u = e^{-A(x)} \int e^{A(x)} r(x) \, dx$$

is a particular solution of the inhomogeneous equation $L(u) = r(x)$.

If a is a constant, we see by inspection that $u = e^{-ax}$ is a solution of the homogeneous equation

$$\frac{dy}{dx} + ay = 0.$$

The general solution of the inhomogeneous equation

$$\frac{dy}{dx} + ay = r(x)$$

then reduces, by virtue of Theorem 4, to finding a particular solution. The variation-of-parameter method is always available, but it can lead to a difficult integration, as we saw in the examples and problems in the last section. There is another, very simple, procedure, called the method of *undetermined coefficients*, that often works here. We illustrate it by some examples.

EXAMPLE 2. Solve the differential equation

$$\frac{dy}{dx} - 2y = 3 \sin x + \cos x.$$

. . . .

Solution. Here the function $r(x) = 3 \sin x + \cos x$, and each of its derivatives, is a linear combination,

$$c_1 \sin x + c_2 \cos x,$$

of the two functions $\sin x$ and $\cos x$. In this situation, with one exception to be mentioned later, the differential equation has a uniquely determined particular solution of the same form. In order to find it, we note that

$$L(\sin x) = \cos x - 2 \sin x,$$

$$L(\cos x) = -\sin x - 2 \cos x,$$

so, if $u = c_1 \sin x + c_2 \cos x$, then

$$\begin{aligned} L(u) &= L(c_1 \sin x + c_2 \cos x) \\ &= c_1 L(\sin x) + c_2 L(\cos x) \\ &= c_1(\cos x - 2 \sin x) + c_2(-\sin x - 2 \cos x) \\ &= (c_1 - 2c_2)\cos x + (-2c_1 - c_2)\sin x. \end{aligned}$$

We want $L(u)$ to be equal to $3 \sin x + \cos x$. That is, we want

$$-2c_1 - c_2 = 3 \quad \text{and} \quad c_1 - 2c_2 = 1.$$

Solving these equations simultaneously, we find that $c_1 = c_2 = -1$. Therefore, $-\sin x - \cos x$ is a particular solution. Theorem 4 now tells us that the general solution of the inhomogeneous equation is the sum of this particular solution and the general solution of the homogeneous equation $dy/dx - 2y = 0$, which is Ce^{2x}. Thus

$$y = Ce^{2x} - \sin x - \cos x$$

is the general solution of the inhomogeneous equation

$$\frac{dy}{dx} - 2y = 3 \sin x + \cos x.$$

EXAMPLE 3. Find the general solution of the differential equation

$$\frac{dy}{dx} - 2y = e^{-x}.$$

. . . .

Solution. Here $r(x)$ and each of its derivatives is of the form ce^{-x}, so we look for a particular solution u of the same form: $u = ce^{-x}$. We see by inspection that $L(e^{-x}) = -3e^{-x}$, so

$$L(u) = L(ce^{-x}) = cL(e^{-x}) = -3ce^{-x},$$

and since this is required to be the function e^{-x}, the coefficient c must be $-1/3$. Thus, $-e^{-x}/3$ is a particular solution of the differential equation

$$\frac{dy}{dx} - 2y = e^{-x},$$

and its general solution is then

$$y = Ce^{2x} - \frac{1}{3}e^{-x},$$

by Theorem 4.

EXAMPLE 4. Find a general solution of the differential equation

$$\frac{dy}{dx} - 2y = e^{2x}.$$

. . . .

Solution. Here we run into the difficulty alluded to earlier. We note that $L(e^{2x}) = 0$, so if $u = ce^{2x}$, then

$$L(u) = L(ce^{2x}) = cL(e^{2x}) = 0,$$

and it is impossible to find c so that $L(u) = e^{2x}$.

In general, if the function $r(x)$ on the right in the inhomogeneous equation

$$L(y) = r(x)$$

happens to be a solution of the homogeneous equation $L(y) = 0$, then the undetermined-coefficient method cannot be directly applied.

But, just in this case, the variation-of-parameter method reduces to the completely trivial integration $v = \int dx = x + C$, and the general solution is

$$(x + C)r(x) = Cr(x) + xr(x).$$

So $xr(x)$ is the particular solution we were looking for.

Returning to Example 4, the particular solution is xe^{2x} and the general solution is

$$Ce^{2x} + xe^{2x} = (C + x)e^{2x}.$$

Problems 11 and 13 will show a couple of other ways of looking at this special case.

PROBLEMS FOR SECTION 5

In each of the following problems, find the general solution by using the method of undetermined coefficients.

1. $\dfrac{dy}{dx} - 2y = e^x$

2. $\dfrac{dy}{dx} - ay = e^{bx} \qquad (a \neq b)$

3. $\dfrac{dy}{dx} + y = x$

4. $\dfrac{dy}{dx} + ay = x^2$

5. $\dfrac{dy}{dx} + ay = 1$

6. $\dfrac{dy}{dx} + 3y = \sin x$

7. $\dfrac{dy}{dx} + 2y = xe^x$

8. $\dfrac{dy}{dx} - y = xe^x$

9. $\dfrac{dy}{dx} + y = x \sin x$

10. $\dfrac{dy}{dx} + ay = e^x \sin x$

11. Show that the differential equation

$$\frac{dy}{dx} + ay = r(x)$$

is equivalent to

$$\frac{d}{dx}(e^{ax}y) = e^{ax}r(x).$$

Then find a solution by inspection in the case $r(x) = e^{-ax}$.

12. Show that the solution formula in Theorem 2 is an immediate consequence of the reformulated differential equation in Problem 11.

13. Show that for any $L(y) = dy/dx + a(x)y$,

$$L(xy) = xL(y) + y,$$

so

$$L(xy) = y \qquad \text{if} \qquad L(y) = 0.$$

14. Find the general solution of

$$\frac{dy}{dx} - y = e^x.$$

(Use Problem 11 or Problem 13 for the particular solution.)

15. Find the general solution of

$$\frac{dy}{dx} + 4y = e^{-4x}.$$

16. Show that the general solution of the differential equation

$$L(y) = b_1(x) + b_2(x)$$

is the sum of a particular solution of $L(y) = b_1(x)$, a particular solution of $L(y) = b_2(x)$, and the general solution of $L(y) = 0$.

17. Find the general solution of

$$\frac{dy}{dx} + y = 1 + e^x$$

by following the above prescription.

18. Solve

$$\frac{dy}{dx} - y = x + \sin x.$$

19. Find the general solution of

$$\frac{dy}{dx} - y = x + e^x.$$

20. Find the general solution of

$$\frac{dy}{dx} + y = e^x + e^{-x}.$$

21. If $L(y) = dy/dx - y/x$, compute
    a) $L(x^4)$ \hspace{2cm} b) $L(x^2)$
    c) $L(x)$ \hspace{2.2cm} d) $L(xe^x)$.

22. If $L(y) = (dy/dx) + a(x)y$, show that

$$uL(v) - vL(u) = u^2 \frac{d}{dx}\left(\frac{v}{u}\right).$$

23. If $L(y) = (dy/dx) + a(x)y$, show that
    a) $L(uv) = uL(v) + v\dfrac{du}{dx}$,
    b) $2L(uv) = uL(v) + vL(u) + \dfrac{d}{dx}(uv)$.

24. Let L and M be any two first-order linear differential operators with constant coefficients, say

$$L(y) = y' + ay, \qquad M(y) = y' + by.$$

Show that L and M commute:

$$L[M(f(x))] = M[L(f(x))]$$

for any twice-differentiable function f.

6. THE SECOND-ORDER LINEAR EQUATION

We shall now turn our attention to certain important *second-order* differential equations, i.e., equations involving the derivatives of f up through the order two. In the prime notation,

$$y' = \frac{dy}{dx}, \qquad y'' = \frac{d^2y}{dx^2}, \qquad \text{etc.,}$$

the equations we shall study are

$$y'' + a(x)y' + b(x)y = 0,$$

$$y'' + a(x)y' + b(x)y = r(x).$$

These equations are *linear*, as we shall see. This property leads to a relatively simple classification for the family of solution functions, much like the first-order situation that we summarized in Theorem 4. On the other hand, these equations have very important applications to natural phenomena associated with oscillation and vibration, resonance and damping, growth and decay. So our mathematical understanding of linearity helps us to describe and predict these natural phenomena.

Given the coefficient functions $a(x)$ and $b(x)$, we set

$$L(y) = y'' + a(x)y' + b(x)y$$

$$= f''(x) + a(x)f'(x) + b(x)f(x),$$

for any function $y = f(x)$ having two derivatives. Thus, if $a(x) = 2$ and $b(x) = 3$, then

$$L(y) = y'' + 2y' + 3y,$$

$$L(f(x)) = f''(x) + 2f'(x) + 3f(x),$$

$$L(\sin x) = -\sin x + 2 \cos x + 3 \sin x$$

$$= 2(\sin x + \cos x),$$

$$L(e^x) = e^x + 2e^x + 3e^x = 6e^x,$$

$$L(e^{-x}) = e^{-x} - 2e^{-x} + 3e^{-x} = 2e^{-x}, \quad \text{etc.}$$

Theorem 5. *L is a linear differential operator. That is, if f_1 and f_2 are any functions having at least two derivatives, and if c_1 and c_2 are any two constants, then*

$$L(c_1 f_1 + c_2 f_2) = c_1 L(f_1) + c_2 L(f_2).$$

Proof. The proof is just a computational verification, like the proof of Theorem 3. We leave it as an exercise. ∎

The two differential equations we want to study are the linear homogeneous equation

$$L(f(x)) = 0$$

and the linear inhomogeneous equation

$$L(f(x)) = r(x).$$

The general properties of their solution families are stated in the following theorem.

Theorem 6. *If y_1 and y_2 are solutions of the homogeneous equation $L(y) = 0$, then so is any linear combination $y = c_1 y_1 + c_2 y_2$. Moreover, if y_1 and y_2 are independent, in the sense that neither is a constant multiple of the other, then*

$$y = c_1 y_1 + c_2 y_2$$

is the general solution of $L(y) = 0$.
If u is any particular solution of the inhomogeneous equation $L(u) = r(x)$, then

$$y = (c_1 y_1 + c_2 y_2) + u$$

is the general solution of the inhomogeneous equation (supposing y_1 and y_2 to be independent). That is, the general solution of the inhomogeneous equation is the sum of a particular solution and the general solution of the homogeneous equation.

Partial proof. The proof of this theorem has an easy part and a hard part, and we are ready now only for the easy part.
If $L(y_1) = 0$ and $L(y_2) = 0$, then

$$L(c_1 y_1 + c_2 y_2) = c_1 L(y_1) + c_2 L(y_2) = 0,$$

by Theorem 5. Thus the family of solutions of the homogeneous equation $L(y) = 0$ is closed under forming linear combinations. What we cannot prove here is that if y_1 and y_2 are independent solutions, then the functions $c_1 y_1 + c_2 y_2$ form the whole solution family.

Now suppose that u is some particular solution of the inhomogeneous equation $L(u) = r(x)$. If y is any other solution, then

$$L(y - u) = L(y) - L(u) = r - r = 0,$$

and so $y - u$ is a solution of the homogeneous equation. Since $y = (y - u) + u$, we see that if we have one solution u of the inhomogeneous equation, then any other solution y can be written as u plus a solution of the homogeneous equation. ∎

According to Theorem 5, the general solution of the inhomogeneous equation depends on finding

1. the general solution of the homogeneous equation,   and

2. a particular solution of the inhomogeneous equation.

The variation-of-parameter method is the basic theoretical tool in reducing (2) to (1). It now starts off like this: First, we find the general solution of the homogeneous equation in the form

$$y = c_1 v_1 + c_2 v_2.$$

Then we try for a particular solution of the inhomogeneous equation $L(y) = r(x)$, in the form

$$y = u_1 v_1 + u_2 v_2,$$

where u_1 and u_2 are unknown functions that have to be determined. Since we have replaced the parameters c_1 and c_2 by unknown functions u_1 and u_2, as in the first-order case, this method is called *variation of parameters*. But it is now more complicated than in our first-order examples, and it is best understood in the context of linear algebra. Moreover, we may not even be able to get started, because solving the homogeneous equation can be much harder in the second-order case. We shall therefore break off at this point, and leave variation of parameters to a future course.

The situation is much simpler if the coefficient functions $a(x)$ and $b(x)$ are both constants, and from now on this will be assumed. In this case, we can solve the homogeneous equation completely, and will do so in the next section. Moreover, independently of the homogeneous solution, we can find a particular solution of the inhomogeneous equation if the function $r(x)$ on the right lets us use the method of undetermined coefficients. This will be taken up now for the simplest context of the problem: to find a particular solution of the equation

$$L(y) = r(x),$$

when $r(x)$ is *not* itself a solution of the homogeneous equation.

EXAMPLE 1. Find a particular solution of the equation

$$L(y) = \sin x,$$

where

$$L(y) = y'' + 3y' + 2y.$$

. . . .

Solution. We try for a solution in the form

$$u = a \sin x + b \cos x.$$

That is, we try for a solution u in the form of a linear combination of $r(x)$ and its derivatives. Remember that, for this to succeed, it is essential that $r(x)$ and its derivatives involve only a finite number of functions. So the device cannot work for a function like $r(x) = 1/x$. Anyway, we see by inspection that

$$L(\sin x) = \sin x + 3 \cos x,$$

$$L(\cos x) = \cos x - 3 \sin x;$$

so

$$\begin{aligned} L(u) &= L(a \sin x + b \cos x) \\ &= aL(\sin x) + bL(\cos x) \\ &= a(\sin x + 3 \cos x) + b(\cos x - 3 \sin x) \\ &= (a - 3b)\sin x + (3a + b)\cos x. \end{aligned}$$

Thus, $L(u) = \sin x$ if and only if

$$a - 3b = 1, \qquad 3a + b = 0.$$

The solutions of these equations are $a = 1/10$, $b = -3/10$; so

$$u = \frac{1}{10} \sin x - \frac{3}{10} \cos x$$

is a particular solution of the inhomogeneous equation.

EXAMPLE 2. In order to obtain a particular solution of

$$L(y) = y'' + 3y' + 2y = e^t,$$

we try a linear combination of e^t and its derivative. But they are all the same function, so we just try

$$u = ce^t.$$

Since $L(e^t) = 6e^t$, we then have

$$L(u) = L(ce^t) = 6ce^t,$$

and the equation

$$L(u) = e^t$$

reduces to

$$6ce^t = e^t.$$

Thus, $6c = 1$, and a particular solution is $u = e^t/6$.

EXAMPLE 3. It is always possible to find a solution of

$$L(y) = r_1(x) + r_2(x),$$

by finding a solution u_1 of $L(y) = r_1(x)$, a solution u_2 of $L(y) = r_2(x)$, and then adding:

$$L(u_1 + u_2) = L(u_1) + L(u_2) = r_1(x) + r_2(x).$$

For an equation like

$$y'' - 2y' + y = \sin x - \cos x,$$

this would be inefficient, since each separate solution would have to be of the form

$$a \sin x + b \cos x,$$

and there would be duplication of effort.

In the case of

$$y'' + 2y' - y = e^x + x,$$

however, x and its derivatives remain completely distinct from e^x and its derivatives, and in such a situation the divide-and-conquer strategy may pay off. We first calculate (mentally) that

$$L(e^x) = 2e^x,$$

so we will have

$$L(ce^x) = e^x$$

if $c = 1/2$. That is, $u_1 = e^x/2$ is a particular solution of $L(y) = e^x$.

Next, we calculate what L does to x and its derivatives,

$$L(x) = 2 - x, \qquad L(1) = -1,$$

so $L(ax + b \cdot 1) = a(2 - x) + b(-1) = -ax + (2a - b)$. This will be the function x if $a = -1$ and $b = 2a = -2$. That is,

$$u_2 = -x - 2$$

is a particular solution of $L(y) = x$. Then

$$u = \frac{e^x}{2} - x - 2$$

is a particular solution of $L(y) = e^x + x$.

PROBLEMS FOR SECTION 6

In each of the following problems find a particular solution by the method of undetermined coefficients.

1. $y'' + y = x$
2. $y'' + 4y = e^x$
3. $y'' - 4y = e^x$
4. $y'' - 4y' = e^{3x}$
5. $y'' + 5y = \cos x$
6. $y'' - y = x^2$
7. $y'' + y' = x^2$
8. $y'' - y = Ax^2 + Bx + C$
9. $y'' + 2y = e^x + 2e^{-x}$
10. $y'' - y' = \cos x + 2x$
11. $y'' + y = xe^x$
12. $y'' + y' - 2y = x$
13. $y'' + y' - 2y = e^{-x}$
14. $y'' + y' - 2y = e^{2x}$
15. $y'' + y' - 2y = \sin x$
16. $y'' + y' = x^3$
17. $y'' + y' = x + \sin x$
18. $y'' + y' = x^2 e^{-x}$
19. $y'' - 2y' + y = x^2 e^x$
20. $y'' - 2y = xe^x + 1$
21. $y'' - 2y = e^x \sin x$
22. $y'' + 4y = x \sin x$
23. $y'' - 2y' + 3y = e^x \sin x$
24. Supposing that the differential operator L has constant coefficients, show that it commutes with differentiation. That is, show that

$$\frac{d}{dx} L(f(x)) = L(f'(x)).$$

7. THE HOMOGENEOUS EQUATION WITH CONSTANT COEFFICIENTS

In view of our first-order experience, it seems reasonable to check whether $y = e^{kx}$ can ever be a solution of the homogeneous equation

$$L(y) = y'' + ay' + by = 0.$$

EXAMPLE 1. Can $y = e^{kx}$ be a solution of

$$L(y) = y'' + 2y' - 3y = 0 ?$$

. . . .

Solution. We have

$$L(e^{kx}) = k^2 e^{kx} + 2k e^{kx} - 3 e^{kx}$$

$$= (k^2 + 2k - 3) e^{kx}.$$

This will be the zero function if and only if

$$k^2 + 2k - 3 = 0.$$

Since

$$k^2 + 2k - 3 = (k + 3)(k - 1),$$

the roots of the equation are

$$k = -3 \quad \text{and} \quad k = 1.$$

Thus, $y = e^{-3x}$ and $y = e^x$ are both solutions of $L(y) = 0$, and they are the only pure exponential solutions.

EXAMPLE 2. Continuing the above example,

$$y = c_1 e^{-3x} + c_2 e^x$$

is a solution for any constants c_1 and c_2, by Theorem 6. Moreover, since we cannot express e^{-3x} as a constant times e^x, Theorem 6 says, furthermore, that every solution of $L(y) = 0$ is of the above form. However, we haven't proved this part of the theorem, so let us see if we can find a direct proof for this example, using the product device again.

Let u be any solution of the equation and write $u = vy$, where v is to be determined and y is a soluton we already know. Here we can take y as e^{-3x} or e^x, and the latter looks simpler. So we set $u = ve^x$. Then

$$u' = ve^x + v'e^x,$$

$$u'' = ve^x + 2v'e^x + v''e^x,$$

and

$$0 = L(u) = u'' + 2u' - 3u$$

$$= v(e^x + 2e^x - 3e^x) + 4v'e^x + v''e^x$$

$$= (v'' + 4v')e^x.$$

Therefore,

$$v'' + 4v' = 0,$$

which is a first-order equation in v'. Its general solution is

$$v' = c_1 e^{-4x},$$

so

$$v = c_1 e^{-4x} + c_2,$$

and finally

$$u = v e^x = c_1 e^{-3x} + c_2 e^x.$$

This proves that there are no other solutions than the ones we already had.

Now let us consider the general constant coefficient equation, which we shall write in the form

$$L(y) = y'' + 2py' + qy = 0.$$

Note that we are writing $2p$ for the coefficient of y'. For example, if $L(y) = y'' + 2y' - 3y$, then

$$p = 1 \quad \text{and} \quad q = -3.$$

This artificial notation simplifies the algebra. Then

$$L(e^{kx}) = k^2 e^{kx} + 2pk e^{kx} + q e^{kx}$$
$$= (k^2 + 2pk + q)e^{kx}.$$

This will be the zero function if and only if

$$k^2 + 2pk + q = 0,$$

which is called the *characteristic equation* of L. The solutions of the characteristic equation are given by the quadratic formula

$$k = \frac{-2p \pm \sqrt{4p^2 - 4q}}{2} = -p \pm \sqrt{p^2 - q}.$$

There are three cases.

Case 1. If $p^2 > q$ and if we set $\Delta = \sqrt{p^2 - q}$, then there are two distinct real roots

$$k_1 = -p + \Delta \quad \text{and} \quad k_2 = -p - \Delta,$$

and the situation is exactly like Example 1. In fact, if we replace -3 and 1 by k_1 and k_2, the argument in Example 2 turns into a proof of the following theorem.

Theorem 7. *If the characteristic equation*

$$k^2 + 2pk + q = 0$$

has distinct real solutions $k = k_1, k_2$, then the solutions of the linear homogeneous equation

$$y'' = 2py' + y = 0$$

are exactly the functions of the form

$$y = c_1 e^{k_1 x} + c_2 e^{k_2 x}. \; + \; c_3 e^{k_3 x}$$

Case II. If $p^2 - q = 0$, then there is only one exponential solution e^{kx} given by $k = -p$. We illustrate this situation with an example.

EXAMPLE 3. For

$$y'' + 2y' + y = 0,$$

the characteristic equation

$$k^2 + 2k + 1 = (k + 1)^2 = 0$$

has only one root, $k = -1$. Thus, e^{-x} is the only pure exponential solution. But this leads to the general solution by the same device as in Example 2. Any other solution u can be written $u = ve^{-x}$, with v to be determined. Then,

$$u' = -ve^{-x} + v'e^{-x},$$

$$u'' = ve^{-x} - 2v'e^{-x} + v''e^{-x},$$

and

$$0 = L(u) = u'' + 2u' + u$$

$$= v(e^{-x} - 2e^{-x} + e^{-x}) + v''e^{-x}$$

$$= v''e^{-x}.$$

Therefore,

$$v'' = 0,$$

$$v' = c_1$$

$$v = c_1 x + c_2,$$

and

$$u = ve^{-x} = c_1 xe^{-x} + c_2 e^{-x}.$$

With $2p$ in place of 2, the above argument proves the following theorem.

Theorem 8. *If the characteristic equation*

$$k^2 + 2pk + q = 0$$

has exactly one real solution, i.e., if $q = p^2$ and the equation is $(k + p)^2 = 0$, then the solutions of

$$y'' + 2py' + y = 0$$

are exactly the functions

$$y = c_1 xe^{-px} + c_2 e^{-px}.$$

Case III. If $p^2 < q$, then neither of the roots

$$k = -p \pm \sqrt{p^2 - q}$$

is real, and there are no pure exponential solutions. Let us, nevertheless, proceed as in Case II, and write an arbitrary solution u in the form

$$u = ve^{-px},$$

with v to be determined. Then,

$$u' = -pve^{-px} + v'e^{-px},$$

$$u'' = p^2 ve^{-px} - 2pv'e^{-px} + v''e^{-px},$$

and

$$0 = L(u) = u'' + 2pu' + qu$$

$$= v(p^2 - 2p^2 + q)e^{-px} + v''e^{-px}.$$

The equation for u thus reduces to a simpler equation for v,

$$v'' = -\beta^2 v,$$

where $\beta = \sqrt{q - p^2}$. ("β" is the Greek letter *beta*.)We know two solutions of this equation, namely

$$v = \sin \beta x \qquad \text{and} \qquad v = \cos \beta x.$$

Therefore, every linear combination

$$v = c_1 \sin \beta x + c_2 \cos \beta x$$

is a solution. We must show that there are no other solutions. The consistent thing to do would be to try the same device once more. We would set $v = w \cos \beta x$, and find w by solving a first-order equation. This would work, but not as well as before, because we have to divide by $\cos \beta x$ in the process and then worry about the points where $\cos \beta x = 0$. Instead, we shall resort to the end of Section 4 in Chapter 4, where the proof for the case $\beta = 1$ was sketched. That proof works just as well for arbitrary β, and takes care of our present dilemma.

Theorem 9. *If the characteristic equation*

$$k^2 + 2pk + q = 0$$

has no real solutions, i.e. if $p^2 - q < 0$, then the solutions of the equation

$$y'' + 2py' + qy = 0$$

are exactly the functions

$$y = e^{-p(x)}[c_1 \cos \beta x + c_2 \sin \beta x]$$

where $\beta = \sqrt{q - p^2}$.

Theorems 7 through 9 show that the general solution of the constant coefficient homogeneous equation

$$y'' + 2py' + qy = 0$$

is one of three types, depending on whether q is less than, equal to, or greater than p^2. If $q < p^2$, then we can write $q = p^2 - m^2$, where $m = \sqrt{p^2 - q}$. Similarly, if $q > p^2$, then q is of the form $q = p^2 + m^2$. We can then summarize the three theorems in the following more harmonious form.

Theorem 10. *The general solution of the constant coefficient homogeneous equation*

$$y'' + 2py' + qy = 0$$

is:

$$e^{-px}[c_1 e^{mx} + c_2 e^{-mx}] \qquad \text{if } q = p^2 - m^2,$$

$$e^{-px}[c_1 x + c_2] \qquad\qquad \text{if } q = p^2,$$

$$e^{-px}[c_1 \cos mx + c_2 \sin mx] \qquad \text{if } q = p^2 + m^2.$$

Finally, Theorem 6 tells us that the general solution of the inhomogeneous equation

$$y'' + ay' + by = r(x)$$

is the sum of the general solution of the homogeneous equation

$$y'' + ay' + by = 0,$$

as determined in the above theorems, and a particular solution of the inhomogeneous equation, determined possibly by the method of undetermined coefficients.

EXAMPLE 4. Find the general solution of

$$y'' + 3y' + 2y = \sin t.$$

. . . .

Solution. The characteristic equation

$$k^2 + 3k + 2 = 0$$

factors into

$$(k + 2)(k + 1) = 0,$$

and hence has the two roots $k = -2, -1$. The general solution of the homogeneous equation

$$y'' + 3y' + 2y = 0$$

is therefore

$$c_1 e^{-x} + c_2 e^{-2x}.$$

On the other hand, we found a particular solution of the inhomogeneous equation to be

$$\frac{1}{10} \sin x - \frac{3}{10} \cos x$$

in the first example of the last section. The general solution is therefore

$$c_1 e^{-x} + c_2 e^{-2x} + \frac{1}{10} \sin x - \frac{3}{10} \cos x,$$

by Theorem 6.

It may happen, when we write down the general solution for the homogeneous equation associated with

$$y'' + ay' + by = r(x),$$

that we discover the righthand function $r(x)$ to be already one of the homogeneous solutions. In that case, the undetermined-coefficient procedure has to be modified. In view of our experience with the first-order equations, it is natural to try a linear combination of $xr(x)$ and its derivatives. We then have to calculate $L(xr(x))$, and it will shorten our work if we know ahead of time what the result of this calculation is.

Lemma 1. *If* $L(y) = y'' + ay' + by$, *then*

$$L(xy) = xL(y) + (2y' + ay).$$

Therefore, if $y = f(x)$ *is a solution of the homogeneous equation;* $L(f) = 0$, *then*

$$L(xf) = 2f' + af.$$

The proof will be left as an exercise.

The peculiar form of the above result will make more sense after some of the problems. (See Problem 29.)

EXAMPLE 5. Find the general solution of

$$y'' - 4y' + 4y = e^{2x}.$$

. . . .

Solution. The characteristic equation is

$$k^2 - 4k + 4 = (k - 2)^2 = 0,$$

which has the single (repeated) root $k = 2$. The general solution of the homogeneous equation is therefore

$$e^{2x}(c_1 x + c_2),$$

by Theorem 8 or Theorem 10.

In this case, since $r(x) = e^{2x}$ and $xr(x) = xe^{2x}$ are both solutions of the homogeneous equation, we apply Lemma 1 to $x(xe^{2x}) = x^2 e^{2x}$:

$$L(x^2 e^{2x}) = xL(xe^{2x}) + 2(xe^{2x})' - 4(xe^{2x}) = 2e^{2x}.$$

So if $u = cx^2 e^{2x}$, then

$$L(u) = L(cx^2 e^{2x}) = cL(x^2 e^{2x}) = 2ce^{2x}.$$

This is required to be equal to $r(x) = e^{2x}$, so $c = \dfrac{1}{2}$ and the particular solution is $x^2 e^{2x}/2$. The general solution is therefore

$$y = e^{2x}\left[\frac{x^2}{2} + c_1 x + c_2\right],$$

by Theorem 6.

EXAMPLE 6. Find a solution of

$$y'' + y = \sin x.$$

. . . .

Solution. We would normally try a linear combination of $\sin x$ and its derivatives, $u = a \sin x + b \cos x$. But we find here that

$$L(\sin x) = 0 = L(\cos x).$$

So we try

$$u = ax \sin x + bx \cos x,$$

instead. By Lemma 1,

$$L(x \sin x) = xL(\sin x) + 2 \cos x = 0 + 2 \cos x$$

$$= 2 \cos x,$$

$$L(x \cos x) = -2 \sin x;$$

so

$$L(u) = L(ax \sin x + bx \cos x) = aL(x \sin x) + bL(x \cos x)$$

$$= 2a \cos x - 2b \sin x.$$

We want $L(u) = \sin x$, so $a = 0$ and $b = -\dfrac{1}{2}$. Thus the particular solution is

$$u = -\frac{1}{2}x \cos x,$$

and the general solution is

$$c_1 \sin x + \left(c_2 - \frac{x}{2}\right)\cos x.$$

PROBLEMS FOR SECTION 7

Find the general solution of each of the following equations.

1. $y'' + 2y = 0$
2. $y'' + 2y = 1$
3. $y'' + 2y = x^2$
4. $y'' + 2y = \sin x$
5. $y'' + 4y = \sin 2x$
6. $y'' - 2y = x + 1$
7. $y'' - y = e^x$
8. $y'' - y' = 1$
9. $y'' - y = e^x \sin x$
10. $y'' - y = xe^x$
11. $y'' + y' - 2y = 0$
12. $y'' + y' - 2y = x$
13. $y'' + y' - 2y = \sin x$
14. $y'' + y' - 2y = e^x$

15. $y'' - 2y' + y = 0$

16. $y'' - 2y' + y = x$

17. $y'' - 2y' + y = \sin x$

18. $y'' - 2y' + y = e^x$

19. $y'' + 2y' + 2y = 0$

20. $y'' + 2y' + 2y = x$

21. $y'' + 2y' + 2y = e^x$

22. $y'' + 2y' + 2y = \sin x$

23. Let D be the operation of differentiation, so that

$$Df = f', \qquad Df(x) = f'(x).$$

Then $L(f) = f'' + af' + bf$ can be thought of as a quadratic polynomial in D applied to f,

$$L(f) = D^2 f + aDf + bf = (D^2 + aD + b)f = p(D)f,$$

where p is the quadratic polynomial

$$p(x) = x^2 + ax + b.$$

Show that if $p(x)$ can be factored,

$$p(x) = (x - k_1)(x - k_2),$$

then

$$p(D)f = (D - k_1)(D - k_2)f,$$

where the right side is interpreted as

$$(D - k_1)[(D - k_2)f].$$

The above formula gives a systematic way of obtaining the solutions of the homogeneous equation. However, the third case, where the roots k_1 and k_2, obtained from the quadratic formula, involve the imaginary number $i = \sqrt{-1}$, carries us beyond the scope of this course.

24. Supposing that

$$D^2 + aD + b = (D - k_1)(D - k_2)$$

in the above sense, with $k_1 \neq k_2$, show that the general solution of the homogeneous equation

$$L(y) = (D^2 + aD + b)y = 0$$

is

$$y = C_1 e^{k_1 x} + C_2 e^{k_2 x}.$$

[Hint: Set $u = (D - k_2)y$. Then the homogeneous second-order equation $L(y) = 0$ reduces to the *first*-order linear equation

$$(D - k_1)u = 0.$$

Solving this gives u, and then $(D - k_2)y = u$ is a first-order inhomogeneous equation for y, that can be solved by variation of parameters, or undetermined coefficients.]

25. Now solve the homogeneous equation by the above scheme in the case where

$$D^2 + aD + b = (D - k)^2$$

(i.e., where $k_1 = k_2 = k$). In this case, use variation of parameters for the last step.

Use the method developed in Problem 24 to find the general solution of each of the following problems, by two applications of the first-order variation-of-parameter calculation.

26. $y'' - y' - 2y = e^x$.   27. $y'' - y' - 2y = e^{2x}$.

28. $y'' - y = 1$.

29. Show that

$$p(D)(xf) = xp(D)f + p'(D)f,$$

so

$$\text{if} \quad p(D)f = 0, \quad \text{then} \quad p(D)(xf) = p'(D)f.$$

(Here $p(x) = x^2 + ax + b$, $p'(x) = 2x + a$.)

30. Show that, if $p(D) = D^2 + aD + b$, then

$$p(D)(fg) = [p(D)f]g + [p'(D)f]g' + \frac{1}{2}[p''(D)f]g''.$$

8. ELASTIC OSCILLATION

Although all sorts of elastic objects vibrate when they are struck or otherwise disturbed, the vibration normally dies away, as when a note is struck on a piano. We ascribe this attenuation to a *resistance* to the motion, air resistance in the case of the piano string.

If the object is continuously "excited," then the vibration can continue indefinitely, despite resistance. Violin tones and organ tones are such sustained vibrations. A continuously excited vibration can even *increase* in intensity. This is the phenomenon of *resonance*, and it may be desirable or catastrophic. A suspension bridge is an elastic object, and a group of people marching in step over a suspension bridge at the natural frequency of the bridge can cause the bridge to vibrate with an increasing amplitude that may eventually destroy it. This is why soldiers are trained to break cadence when crossing a bridge. On the other hand, the current in electric circuits oscillates according to the same principles of vibration, and there resonance can be a desirable phenomenon.

The ideal mass–spring system of Chapter 6 becomes realistic when we add a resistance force proportional to the velocity, $-r\,dx/dt$. Then its motion is governed by Newton's law $ma - F = 0$, in the form

$$m\frac{d^2x}{dt^2} + \left[r\frac{dx}{dx} + sx\right] = 0,$$

or

$$\ddot{x} + \frac{r}{m}\dot{x} + \frac{s}{m}x = 0.$$

If, in addition, the system is continuously excited by a periodic external force of the form $c \sin \beta t$, then its motion is governed by Newton's law in the form

$$m\frac{d^2x}{dt^2} + r\frac{dx}{dt} + kx = c \sin \beta t.$$

(This is also the equation for the motion of electricity in a circuit to which a generator has been attached. The earlier equation governs the flow of electricity in a "free" circuit. We shall not analyze those electrical applications, however.)

Here again we have differential equations that state our understanding of a natural phenomenon, and their solutions will show the possible patterns of elastic behavior and permit quantitative predictions.

The analysis of this situation is broken down into the following problems.

PROBLEMS FOR SECTION 8

1. Consider first the case where the resistance r dominates the combined effect of the mass m and spring stiffness s, in the sense that

$$r^2 > 4ms.$$

   Supposing that the initial velocity (at time $t = 0$) is zero, and the initial position is positive, show that the motion is of the form

$$x = c[ke^{-lt} - le^{-kt}],$$

   where $0 < l < k$, and c is positive.

2. Assuming the above formula, show that the graph of x as a function of t has the form shown in the following diagram, with a single point of inflection at $t = [\log(k/l)]/(k - l)$.

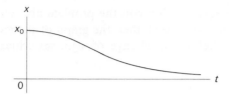

3. In the same situation of a dominant resistance, suppose that the initial position is at $x = 0$ but that the initial velocity is positive. Show that then the motion is of the form

$$x = c[e^{-lt} - e^{-kt}],$$

where $c > 0$ and $k > l > 0$.

4. Show that the graph of the above motion has the general features shown below.

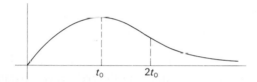

In particular, show that there is one critical point $t = t_0$ at which x has its maximum value, and one point of inflection, located at $2t_0$.

The next two problems analyze in more detail the same situation of a dominant resistance, when the initial position x_0 is positive, say $x_0 = 1$, and the initial velocity v_0 has various values.

5. Show first that if $x_0 = 1$, the motion has the form

$$x = ae^{-lt} + (1 - a)e^{-kt},$$

where $0 < l < k$, Then show that the initial velocity v_0 is less than $-k$ if and only if $a < 0$, and that the graph of the motion in this case has the general form shown below. (One change in sign, one critical point, one point of inflection.)

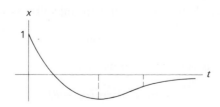

6. Show from the equation for x in the problem above:
   a) If $-k \leq v_0 \leq -k/(k+l)$, then the graph of the motion has the general form shown below (no change of sign, no critical point, no point of inflection).

   b) If $v_0 > -k/(k+l)$, then the graph is always positive and has one point of inflection, as below:

In the next two problems we assume that the resistance r balances the combined effect of the mass m and stiffness s, in the sense that $r^2 = 4ms$.

7. Show that if $r^2 = 4ms$ and if the initial conditions are $(x_0, v_0) = (1, 0)$, then the motion is given by

$$x = (1 + kt)e^{-kt},$$

   where $k = r/2m$. Show also that the graph of this motion has the same features as in Problem 2 (but with a different point of inflection).

8. If the initial conditions are $(x_0, v_0) = (0, 1)$, show that the motion is given by

$$x = te^{-kt},$$

   and that its graph has the same features as in Problem 4.

9. Consider finally the case where the resistance r is small in comparison with the combined effect of the mass m and spring stiffness s, in the sense that

$$r^2 < 4ms.$$

   Show that then the motion is given by

$$x = ae^{-kt}\sin(\omega t + \theta),$$

where the exponential damping rate k has the value $k = r/2m$, and the frequency $\omega/2\pi$ is determined by

$$\omega = \frac{1}{2m}\sqrt{4ms - r^2}.$$

(See Section 5 in Chapter 6 for the source of the amplitude a and phase angle θ.)

10. Show that the graph of x, as a function of t in the above problem, has the general features shown below.

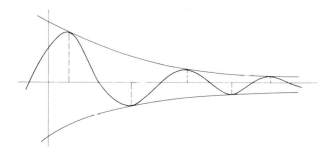

How is this graph affected by a decrease in the resistance r (m and s remaining unchanged)? What is the limiting configuration as r approaches zero?

We consider one more situation, in which an extra periodic "driving force" is impressed on the system, say a "sinusoidal" force, proportional to $\sin \alpha t$. The modified differential equation can be taken to be

$$(*) \qquad \qquad \ddot{x} + \frac{r}{m}\dot{x} + \frac{s}{m}x = \sin \alpha t,$$

where $r^2 < 4ms$, and where we have written $\sin \alpha t$ instead of the more general $c \sin \alpha t$, for simplicity. The motion is now a superposition (sum) of a transient vibration that dies out exponentially as discussed above (a solution of the homogeneous equation) and a "steady-state" sinusoidal vibration

$$A \sin(\alpha t + \beta)$$

(the particular solution of the inhomogeneous equation obtained by undetermined coefficients). Note that the total motion settles down to this steady-state simple harmonic motion as the transient component wears off. If in the beginning the transient and steady-state components tend to cancel one another, then the total effect is a vibration with an amplitude that builds up as time passes, and if the transient and steady-state frequencies are nearly the same, the limiting amplitude can be very large. This is the phenomenon of resonance.

11. Show that the particular solution of the above equation obtained by the method of undetermined coefficients can be written in the form

$$A \sin(\alpha t + \beta),$$

where the amplitude A is given by

$$A = \frac{1}{\sqrt{\left(\dfrac{s}{m} - \alpha^2\right)^2 + \left(\dfrac{\alpha r}{m}\right)^2}} \, .$$

Conclude that A is proportional to $1/r$ if the driving frequency $\alpha/2\pi$ is equal to the natural frequency $\sqrt{s/m}/2\pi$ of the undamped and undriven system.

12. With m, s, and r fixed, but α variable, prove that the maximum value of the amplitude A in the above problem is

$$A_{\max} = \frac{2m^2}{r\sqrt{4ms - r^2}},$$

when $\alpha = \sqrt{ms - r^2/2}\big/m$.

13. If a simple electric circuit contains a total capacitance of amount $1/s$, a total resistance r, and a total inductance m, and if x is the amount of electricity that has moved past a fixed point in a displacement from the equilibrium position, then x fluctuates with time according to the same homogeneous equation

$$m\ddot{x} + r\dot{x} + sx = 0.$$

This is for a "free" circuit, unaffected by any outside influence. If a sinuisoidal generator is attached, then the fluctuation of x is determined by the inhomogeneous equation (*), with $\sin \alpha t$ possibly replaced by $c \sin \alpha t$ for some constant c.

But here we are more interested in the behavior of the *current* $i = \dot{x}$. Show that if x satisfies (*), then $i = \dot{x}$ satisfies

(**) $$m\frac{d^2 i}{dt^2} + r\frac{di}{dt} + si = \alpha \cos(\alpha t).$$

14. Show without calculation, by applying a general result from Section 6 (see Problem 24), that if $A \sin(\alpha t + \beta)$ is a particular solution of (*), then $\alpha A \cos(\alpha t + \beta)$ is a particular solution of (**). Then show, from the result in Problem 11, that the amplitude $A' = \alpha A$ of this oscillation is given by

$$A' = \frac{1}{\sqrt{\left(\dfrac{s}{\alpha m} - \alpha\right)^2 + \left(\dfrac{r}{m}\right)^2}},$$

and hence that A' is maximum (for fixed s, m, and r) when

$$\alpha = \sqrt{\frac{s}{m}}.$$

Conclude that the resonant frequency is equal to the frequency of the undamped $(r = 0)$ freely oscillating circuit.

9. SERIES SOLUTIONS OF DIFFERENTIAL EQUATIONS

Infinite series apply naturally to the solution of linear differential equations, because such a differential equation automatically determines the Maclaurin series of a solution function.

EXAMPLE 1. Consider the differential equation

$$y' - y = 0.$$

If we assume that a solution $y = f(x)$ is the sum of a power series

$$y = a_0 + a_1 x + a_2 x^2 + \cdots + a_n x^2 + \cdots,$$

then

$$y' = a_1 + 2a_2 x + 3a_3 x^2 + \cdots + (n + 1)a_{n+1} x^n + \cdots$$

and

$$y' - y = (a_1 - a_0) + (2a_2 - a_1)x + (3a_3 - a_2)x^2 + \cdots$$
$$+ [(n + 1)a_{n+1} - a_n]x^n + \cdots$$

Since $y' - y$ is the zero function, its Maclaurin series coefficients must all be zero, so

$$a_1 = a_0,$$

$$a_2 = \frac{a_1}{2} = \frac{a_0}{2},$$

$$a_3 = \frac{a_2}{3} = \frac{a_0}{3!},$$

$$\vdots \quad \vdots$$

$$a_{n+1} = \frac{a_n}{n + 1} = \frac{a_0}{(n + 1)!}.$$

and

$$y = a_0 \left[1 + x + \frac{x^2}{2!} + \cdots + \frac{x^n}{n!} + \cdots \right],$$

where a_0 is an arbitrary constant, the "constant of integration." Of course, we recognize this solution to be $a_0 e^x$.

What we are doing here is essentially the reverse of our procedure in Section 2 of Chapter 13. There we noted that the function

$$y = 1 + x + \frac{x^2}{2} + \cdots + \frac{x^n}{n!} + \cdots$$

satisfies the differential equation $y' - y = 0$ (and we used this fact to conclude that $y = e^x$). Here we directly compute that, if the sum of a power series satisfies the differential equation $y' - y = 0$, then the series must be $a \sum_0^\infty x^n/n!$ Note that this calculation demands no prior knowledge of the exponential function, and if we have never seen e^x, we could still solve the differential equation $y' - y = 0$ as the sum of a power series in the above way and then use the series to compute the solution function. In this situation it would be necessary to check that the solution series does in fact converge.

EXAMPLE 2. If $y = f(x)$ is the sum of a power series

$$y = a_0 + a_1 x + \cdots + a_n x^n + \cdots,$$

and satisfies the differential equation

$$y'' + y = 0,$$

then the coefficients are determined by the same kind of recursive procedure as in the first example, except that this time the even coefficients go back to a_0 and the odd coefficients go back to a_1. Thus,

$$y = a_0 + a_1 x + a_2 x^2 + a_3 x^3 + a_4 x^4 + a_5 x^5 + \cdots$$

$$y'' = 2a_2 + 3 \cdot 2a_3 x + 4 \cdot 3a_4 x^2 + 5 \cdot 4a_5 x^3 + 6 \cdot 5a_6 x^4 + 7 \cdot 6a_7 x^5 + \cdots,$$

and, since $y'' + y = 0$, we see that

$$a_2 = -\frac{a_0}{2}, \qquad\qquad a_5 = -\frac{a_3}{5 \cdot 4} = \frac{a_1}{5!}$$

$$a_3 = -\frac{a_1}{3!}, \qquad\qquad a_6 = -\frac{a_4}{6 \cdot 5} = -\frac{a_0}{6!}$$

$$a_4 = -\frac{a_2}{4 \cdot 3} = \frac{a_0}{4!} \qquad a_7 = -\frac{a_5}{7 \cdot 6} = -\frac{a_1}{7!}$$

and so on. Thus

$$y = a_0\left[1 - \frac{x^2}{2} + \frac{x^4}{4!} - \frac{x^6}{6!} + \cdots\right] + a_1\left[x - \frac{x^3}{3!} + \frac{x^5}{5!} - \frac{x^7}{7!} + \cdots\right],$$

which we recognize to be $a_0\cos x + a_2\sin x$.

Again, if we knew nothing about the trigonometric functions, we could check that both series above converge everywhere to some functions $C(x)$ and $S(x)$ respectively, and we could compute these functions by the partial sums of the series and make up tables for them. Then the general solution of the differential equation $y'' + y = 0$ is expressed in terms of these two new functions by

$$y = a_0 C(x) + a_1 S(x),$$

as we saw above. This is only an academic exercise, but it has the following point: We can proceed in exactly this way with more complicated differential equations, computing the new functions that are their solutions as the sums of power series determined by the differential equation, and then studying these new solution functions on the basis of these power-series definitions.

The solutions of a second-order linear homogeneous equation with *variable* coefficients,

$$y'' + a(x)y' + b(x)y,$$

are generally not elementary functions, which means that there cannot be any general procedure for turning out solutions as finite combinations of familiar functions. Yet in some circumstances it is easy to compute the solutions as sums of power series in the manner of the above examples.

EXAMPLE 3. Solve the equation

$$y'' - xy = 0$$

by power series.

. . . .

Solution. If

$$y = c_0 + c_1 x + c_2 x^3 + \cdots + c_n x^n + \cdots,$$

then

$$y'' = 2c_2 + 3\cdot 2c_3 x + 4\cdot 3c_4 x^2 + \cdots + (n + 2)(n + 1)c_{n+2} x^n + \cdots,$$

$$xy = c_0 x + c_1 x^2 + \cdots + c_{n-1} x^n + \cdots,$$

and

$$y'' - xy = 2c_2 + (3 \cdot 2c_3 - c_0)x + (4 \cdot 3c_4 - c_1)x^2 + \cdots$$
$$+ ((n + 2)(n + 1)c_{n+2} - c_{n-1})x^n + \cdots$$

Therefore $y'' - xy = 0$ if and only if

$$2c_2 = 0,$$
$$3 \cdot 2c_3 = c_0,$$
$$4 \cdot 3c_4 = c_1,$$
$$\vdots \qquad \vdots$$
$$(n + 2)(n + 1)c_{n+2} = c_{n-1}.$$

These equations define the coefficients c_n recursively in terms of c_0 and c_1 (and c_2):

$$c_2 = 0, \qquad c_3 = \frac{c_0}{3 \cdot 2}, \qquad c_4 = \frac{c_1}{4 \cdot 3}, \qquad c_5 = \frac{c_2}{5 \cdot 4} = 0,$$
$$c_6 = \frac{c_3}{6 \cdot 5} = \frac{c_0}{6 \cdot 5 \cdot 3 \cdot 2}, \qquad c_7 = \frac{c_4}{7 \cdot 6} = \frac{c_1}{7 \cdot 6 \cdot 4 \cdot 3},$$

and so on. Thus

$$y = c_0 f(x) + c_1 g(x)$$

where

$$f(x) = 1 + \frac{x^3}{3 \cdot 2} + \frac{x^6}{6 \cdot 5 \cdot 3 \cdot 2} + \frac{x^9}{9 \cdot 8 \cdot 6 \cdot 5 \cdot 3 \cdot 2} + \cdots,$$

and

$$g(x) = x + \frac{x^4}{4 \cdot 3} + \frac{x^7}{7 \cdot 6 \cdot 4 \cdot 3} + \frac{x^{10}}{10 \cdot 9 \cdot 7 \cdot 6 \cdot 4 \cdot 3} + \cdots$$

There is no easy way to write down a formula for the nth coefficient in either of these series, but the pattern is clear. Moreover, it is easy to see, from the ratio test, that both series converge for all x, because the recursive relations give us simple explicit formulas for the ratios of successive terms. For example, $f(x)$ is of the form

$$f(x) = \sum_0^\infty a_n x^{3n},$$

where $a_n/a_{n-1} = 1/(3n(3n - 1))$. The ratio of the corresponding series terms is thus $x^3/(3n(3n - 1))$, which approaches 0 as $n \to \infty$, no matter what x is. Therefore, f and g are everywhere defined functions, and each satisfies the differential equation $y'' - xy = 0$, because the coefficients in its Maclaurin expansion satisfy the recursive relations that were equivalent to this differential equation. That $y = c_0 f(x) + c_1 g(x)$ is the general solution of this differential equation follows from Theorem 6.

PROBLEMS FOR SECTION 9

Show that the general solution of each of the following differential equations can be expressed as the sum of convergent power series. In each case, write down the recursion formula for the coefficients, and determine the convergence interval by the ratio test. Write down the first few terms of each series, and, if possible, the general term. (It is also instructive, where possible, to find the solutions by our earlier methods and compare answers.)

1. $y' - xy = 0$ 2. $y'' + xy' = 0$

3. $y' - x^2 y = x$ 4. $(1 - x)y' + y = 0$

5. $(1 - x^2)y' - xy = 0$ 6. $y'' - 2y = 1$

7. $y'' - x^2 y = 0$ 8. $y'' + 2x^2 y - xy = 0$

9. $(1 - x^2)y'' - y = 0$

10. Solve the equation

$$xy'' - y = 0$$

by the power series method. What conclusion can you draw about the possibility of expressing the general solution

$$y = af(x) + bg(x)$$

as a power series?

11. Supposing that $y = f(x)$ is one solution of

$$xy'' - y = 0$$

(the one found above), show that $z = u(x)y$ is another solution if and only if

$$u(x) = C_1 \int \frac{dx}{y^2} + C_2.$$

12. Show that no solution of the equation

$$x^2y'' - y = 0$$

can be represented by a convergent Maclaurin series.

chapter 20

calculating
a real number;

the analytic
basis of calculus

We shall begin this chapter by considering the general nature of a numerical computation. Almost incidentally, we shall improve our understanding of such fundamental notions as continuity, limits, and the intermediate-value theorem. This may seem paradoxical, but the foundations of calculus turn out to have strong computational overtones, and a computational procedure usually contains the seeds of an analytic development. For example, we shall see that an attempt to set up a computational scheme for an "intermediate value" turns into an analytic proof of the intermediate-value theorem.

With this link between computation and theory in mind, let us look back on some of our early development. We accepted the existence and properties of certain fundamental entities on the basis of geometric intuition. For example, we assumed that plane regions have well-defined and well-behaved numerical areas, in order to conclude that every continuous function has an antiderivative, and we gave geometric derivations of the properties of sin x and cos x. Our introduction to e^x was shakier: we simply plucked it out of thin air.

However, in due course, we learned to compute all of these quantities, and we now expect such a computation to provide an analytic foundation. So we should be on the verge of an analytic development of all the material which first appeared under the sponsorship of intuition.

It seems worth while to take this final step. We shall then come full circle, and end our engagement with the serious aspects of calculus more or less where we began it, by providing the analytic structure that supports these intial intuitive points of departure. With all these intuitive props knocked away, we shall find that calculus rests on a single support, the completeness of the real numbers.

1. CALCULATING A NUMBER

It seems reasonable that knowing how to calculate a number like $\sqrt{2}$ or π should mean knowing how to find its infinite decimal expansion. Of course, in practice we can never write out the full infinite sequence of digits, so all we can mean is that we know how to compute the decimal expansion *as far out as we wish to go.* Square roots like $\sqrt{2}$ can be calculated in this sense by a procedure (algorithm) that is often taught in school.

According to this provisional definition, being able to calculate x means being able to find, for any particular positive integer n, the unique n-place decimal $r = m/10^n$ such that $r \leq x < r + 1/10^n$. (Allowing for equality on the left takes care of the remote possibility that the number x being computed might itself turn out to be a decimal fraction.)

In practice the above requirement may be hard to meet. The result of a calculation will not generally be recognizable as the first n places in the expansion of the number being calculated, and if we try to force an answer into this form we may lose the accuracy we have worked hard to obtain. Examples will show this better than any general discussion can.

EXAMPLE 1. Knowing that the decimal expansion of π begins $3.14\ldots$, what can we conclude about π^2?

· ·

Solution. We are assuming that

$$3.14 < \pi < 3.15.$$

Squaring, we find that

$$9.8596 < \pi^2 < 9.9225.$$

In particular,

$$9.85 < \pi^2 < 9.95,$$

$$-0.05 < \pi^2 - 9.9 < 0.05,$$

$$|\pi^2 - 9.9| < 0.05.$$

Thus $\pi^2 \approx 9.9$, with an error less than 0.05 in magnitude. That is, 9.9 is the *closest* one-place decimal to π^2. But we don't know from this calculation whether the decimal expansion of π^2 starts $9.8\ldots$ or $9.9\ldots$.

EXAMPLE 2. We find in a calculation that 4.5376 approximates a number x with an error less than 0.005 in magnitude. What can we say about the decimal expansion of x?

· ·

Solution. By hypothesis

$$4.5376 - 0.005 < x < 4.5376 + 0.005,$$

or

$$4.5326 < x < 4.5426.$$

The only thing definite that we can say about the decimal expansion of x is that it begins with 4.5. But this doesn't do justice to our information. Much better is the conclusion that

$$x \approx 4.54$$

with an error less than 0.01 in magnitude.

EXAMPLE 3. We find in a calculation that 4.539 approximates a number x with an error less than 0.003 in magnitude. What can we say about x?

· ·

Solution. Here our information is that

$$4.536 < x < 4.542,$$

and we can say that

$$x \approx 4.54$$

to the nearest two decimal places, i.e., with an error less than 0.005. (But the decimal expansion of x can start either 4.53 ... or 4.54 ....)

EXAMPLE 4. The decimal expansion of a number x begins 0.547 .... What can we say about x^2?

. . . .

Solution. By assumption

$$0.547 \le x < 0.548.$$

Squaring, we find that

$$.299209 \le x^2 < .300304.$$

Therefore $x^2 \approx .300$, with an error less than 0.001. But we don't know whether the decimal expansion of x^2 begins 0.299 ... or 0.300 ...

These examples show that requiring the first n places in the decimal expansion of a solution imposes an extraneous burden on the primary objective of obtaining a close approximation to the solution. We shall therefore reject our provisional definition in favor of the following more flexible alternative. In it we shall use the traditional Greek epsilon, ε, for a small positive number.

Definition. *To be able to calculate a number x means to be able to find a rational number that approximates x as closely as we wish. That is, no matter how small the error ε that we can tolerate, we know how to find a rational number r such that*

$$|x - r| < \varepsilon.$$

Although we have abandoned the requirement that we be able to find the exact truncated decimal expansion of a number being calculated, we were nevertheless successfully estimating by decimal approximations in our examples, and it is natural to ask to what extent we can compute by finite decimals.

Theorem 1. *We can calculate x if and only if for each positive integer n we can find an n-place decimal d that approximates x to within an error of 1 in the last decimal place*:

$$|x - d| < 10^{-n}.$$

Proof. Suppose we can calculate x. In particular, given n, we can find a rational r such that

$$|x - r| < \frac{10^{-n}}{2}.$$

Then we can compute the decimal expansion of r by long division, and so determine the n-place decimal d that is *closest* to r:

$$|r - d| \leq \frac{10^{-n}}{2}.$$

(This will be either the first n places in the expansion of r, or the number one greater in the last place, whichever is closer to r.) Then

$$|x - d| = |(x - r) + (r - d)|$$

$$\leq |x - r| + |r - d| \leq \frac{10^{-n}}{2} + \frac{10^{-n}}{2} = 10^{-n}. \quad \blacksquare$$

In practice, the following result is extremely useful.

Theorem 2. *We can calculate x if we can approximate x arbitrarily closely by numbers that we can calculate. That is, we can calculate x if no matter what the tolerable error ε, we can find a number y that we can calculate such that $|x - y| < \varepsilon$.*

Proof. Given ε, we want to find a rational number r such that $|x - r| < \varepsilon$. To do this, we first locate a number y that we can calculate such that

$$|x - y| < \frac{\varepsilon}{2}.$$

This is possible by hypothesis. Then we find a rational number r such that

$$|y - r| < \frac{\varepsilon}{2}.$$

We can do this because we can calculate y. Then

$$|x - r| = |(x - y) + (y - r)| \leq |x - y| + |y - r| < \frac{\varepsilon}{2} + \frac{\varepsilon}{2} = \varepsilon,$$

so we have found the required rational approximation to x. $\quad \blacksquare$

We use this device when we compute function values. Consider, for example, the function $f(x) = 10^x$, and the problem of computing $f(\sqrt{2}) = 10^{\sqrt{2}}$.

We shall suppose that we know how to calculate nth roots of integers. Then we can calculate 10^r for any rational number $r = m/n$ since $10^r = (10^m)^{1/n}$. We also suppose that we know how to find a rational number r so that 10^r is as close as we want to $10^{\sqrt{2}}$. This means that we know how to approximate $10^{\sqrt{2}}$ as closely as we wish by calculable numbers, and we can therefore calculate $10^{\sqrt{2}}$.

The actual calculation process would follow the steps in the proof of the theorem. Given the error ε that we can tolerate, we first determine a rational number r so that

$$|10^{\sqrt{2}} - 10^r| < \frac{\varepsilon}{2}.$$

We then calculate 10^r to within $\varepsilon/2$, i.e., find a rational number s such that

$$|10^r - s| < \frac{\varepsilon}{2}.$$

Then

$$|10^{\sqrt{2}} - s| < \varepsilon,$$

and so s is the desired approximation.

In Chapter 10 we studied the estimation of a definite integral $\int_a^b f$ by its Riemann sums and by the more sophisticated trapezoidal sums. It takes just one more small step to see that such an estimate provides a calculation of $\int_a^b f$. Consider first the approximation of $\log(3/2) = \int_1^{3/2} dx/x$ by its Riemann sum S. We found that

$$\left| \int_1^{3/2} \frac{dx}{x} - S \right| < \frac{\Delta x}{4},$$

where Δx is the common length of all the subintervals in the subdivision of $[1, (3/2)]$. Therefore, if we want the error in the approximation to be less than ε, it suffices to make

$$\frac{\Delta x}{4} < \varepsilon,$$

and hence to take an equally spaced subdivision for which

$$\Delta x < 4\varepsilon.$$

Moreover, the Riemann sum S is a rational number in this example. The Riemann sum approximation is thus a method of calculating $\int_1^{3/2} dx/x$.

For the general smooth function f on the closed interval $[a,b]$, we found that

$$\left| \int_a^b f - S \right| < \frac{K(b-a)}{2} \Delta x,$$

where K is a bound for f' over the interval $[a,b]$. In order to make the error in this approximation less than ε, we only have to see to it that

$$\frac{K(b-a)}{2} \Delta x < \varepsilon,$$

and hence to take an equally spaced subdivision of $[a,b]$ for which the common interval width Δx satisfies

$$\Delta x < \frac{2}{K(b-a)} \varepsilon.$$

Here the approximating number S will not in general be rational. But if we can calculate the limits of integration a and b, and the values of f occurring in the Riemann sum S, then S can be calculated. Now the Riemann sum provides a calculation of $\int_a^b f$ by virtue of Theorem 2.

The same thing is true of other estimates that we have made. We know, for example, that

$$\sin x \simeq \sum_{k=0}^{n-1} (-1)^k \frac{x^{2k+1}}{(2k+1)!},$$

with an error less than $x^{2n+1}/(2n+1)!$ in magnitude. This error bound is hard to handle with precision. However, in Problem 14 of Chapter 12, Section 7, it is proved that

$$n! > \left(\frac{n}{e}\right)^n$$

for all values of n, and this inequality makes it easy to find a value of n for which the error bound is less than a preassigned tolerance ε. The Maclaurin polynomials therefore provide a calculation of $\sin x$.

PROBLEMS FOR SECTION 1

1. Archimedes proved that

$$3\frac{10}{71} < \pi < 3\frac{1}{7}.$$

   Show from this that $22/7$ approximates π with an error less than $1/497$.

2. Find what the above double inequality tells us about the decimal expansion of π.

3. From a classical partial-fraction expansion of e, we find that

$$2\frac{334}{465} < e < 2\frac{385}{536}.$$

   Show therefore that either of these rational numbers approximates e with an error less than $1/249, 000$.

4. Compute e from the above data to the nearest five decimal places.

Here are some decimal expansions through two decimal places:

$$\pi = 3.14\cdots, \qquad \sqrt{2} = 1.41\cdots, \qquad e = 2.71\cdots,$$

$$\log 3 = 1.09\cdots, \qquad \sin\sqrt{2} = 0.98\cdots$$

Determine what can be said about the one-place decimal approximations to each of the following products. In each case the answer should be *either* the decimal expansion through one decimal place, *or* the closest one-place decimal approximation (or both).

5. $e\sqrt{2}$ 6. $\pi \log 3$ 7. $\sqrt{2} \sin \sqrt{2}$ 8. e^2

9. $\pi \sin \sqrt{2}$ 10. $e \log 3$ 11. $\log 3 \sin \sqrt{2}$

12. If $|x - 1.4849| < 0.001$, how close must x be to 1.485?

13. If $|x - 1.099| < 0.005$, how close must x be to 1.1?

14. If $|x - 4.473| < 0.001$, how close must x be to 4.47?

15. If $|x - 2.152| < 0.003$, how close must x be to 2.15? 2.16?

16. If $|x - 2.289| < 0.001$, how close must x be to 2.28? 2.29?

17. In the situation of Theorem 1, if we start with a rational r even closer to a positive number x, then we can get a better answer. Show that if we first find a rational r such that

$$|x - r| < \frac{10^{-n}}{4},$$

and then find the closest n-place decimal to r,

$$|r - d| \le \frac{10^{-n}}{2},$$

then we can determine *either* the n-place decimal closest to x (it will be d), *or* the first n places in the decimal expansion of x (it will be d or $(d - 10^{-n})$). [Hint. Consider separately the cases $|r - d| \le 10^{-n}/4$; $d - r > 10^{-n}/4$; $r - d > 10^{-n}/4$.]

18. Suppose that we want to compute sin 1 to ten decimal places. By using the inequality $n! > (n/e)^n$, show that if $p(x)$ is the Maclaurin polynomial for sin x of degree 13 (and having seven terms), then $p(1)$ will approximate sin 1 with an error less than $10^{-10}/3$.

19. Continuing the above problem, show that if we compute $p(1)$ correctly through 11 decimal places, then we will know either the 10-place decimal closest to sin 1, or the 10-place truncation in the decimal expansion of sin 1.

20. The polynomial $p(1)$ in Problem 18 is a somewhat better estimate of sin 1 than was claimed because there is slack in the inequality $n! > (n/e)^n$. Show by a direct estimate of $(15)!$ that $p(1)$ approximates sin 1 with an error less than 10^{-12}. (In computing 15! the arithmetic can be simplified by dropping down to a smaller but much simpler integer after 7!, and again after 13!.)

21. If a series of positive terms $\Sigma\, a_n$ is found to be convergent by comparison with a geometric series $k \,\Sigma\, r^n$, then its sum s can be calculated by its partial sums $s_n = \Sigma_1^n\, a_j$. Supposing that we are using negative powers of 10 as measures of smallness, let p be a fixed positive integer such that $r^p < 1/10$. Show that the error in the approximation of s by s_n is then less than $c10^{-n/p}$, where $c = k/(1 - r)$.

2. CONTINUITY

A question that we needed to answer to complete the calculation of $10^{\sqrt{2}}$ sketched in the last section was: Given the tolerable error ε, how close must we take r to $\sqrt{2}$ in order to ensure that

$$|10^{\sqrt{2}} - 10^r| < \varepsilon\,?$$

A similar problem occurs when we try to calculate π^3, knowing how to calculate π. We have to know how close to take the rational number r to π in order to ensure that

$$|\pi^3 - r^3| < \varepsilon.$$

We can ask this question in neutral terms, and for a general function f, as follows:

Given a positive number ε, how close do we have to take x to a in order to ensure that

$$|f(a) - f(x)| < \varepsilon\,?$$

Or

How small should we take Δx in order to ensure that

$$|\Delta y| < \varepsilon,$$

where $\Delta y = f(x + \Delta x) - f(x)$?

This question concerns the quantitative aspect of continuity. We said, in Chapter 2, that a function f is continuous at $x = a$ if $f(x)$ can be made to be as close to $f(a)$ as we wish by taking x suitably close to a. This was a purely qualitative description. But in making computations we have to find out *how close* is "suitably close."

The smooth functions of calculus give us the simplest answer to this question, because of the following theorem. (It is the general statement of the procedure we were following in the examples in Section 11 of Chapter 3. Afterward, we shall look again at a couple of those earlier examples, with our discussion of decimal approximations in mind.)

Theorem 3. *If $|f'|$ is bounded by the constant k on an interval I, that is, if $|f'(x)| \le k$ for every x in I, then*

$$|\Delta y| \le k |\Delta x|$$

for any two points x and $(x + \Delta x)$ in I.

Proof. By the mean-value theorem,

$$\Delta y = f(x + \Delta x) - f(x) = f'(X) \cdot \Delta x,$$

where X is some number between x and $x + \Delta x$. But then $|f'(X)| \le k$ by hypothesis, and so

$$|\Delta y| = |f'X| \cdot |\Delta x| \le k |\Delta x|,$$

proving the theorem.   ∎

By virtue of this inequality, if now we want to ensure that

$$|\Delta y| < \varepsilon,$$

it is sufficient to see to it that

$$k |\Delta x| < \varepsilon,$$

i.e., to take

$$|\Delta x| < \frac{\varepsilon}{k}.$$

This is the simplest conceivable solution to the continuity problem.

EXAMPLE 1. If we consider $y = f(x) = x^3$ on the interval $I = [-4, 4]$, then

$$f'(x) = 3x^2 \le 3 \cdot 4^2 = 48,$$

and hence

$$|\Delta y| \le 48 |\Delta x|,$$

by the theorem. Therefore, in order to ensure that $|\Delta y| < \varepsilon$, it is sufficient to take

$$48 |\Delta x| < \varepsilon, \quad \text{or} \quad |\Delta x| < \frac{\varepsilon}{48}.$$

In other words, if x and a are both on the interval $I = [-4, 4]$, then

$$|a^3 - x^3| < \varepsilon \quad \text{whenever } |a - x| < \frac{\varepsilon}{48}.$$

EXAMPLE 2. Find a two-place decimal that approximates π^3 with an error less than 0.01.

. . . .

Solution. Following the scheme of Theorem 1, we first look for a rational number x such that

$$|\pi^3 - x^3| < 0.005.$$

According to Example 1, this will hold if

$$|\pi - x| \leq \frac{0.005}{48}.$$

In particular, it is sufficient to take

$$|\pi - x| \leq \frac{0.005}{50} = 10^{-4}.$$

That is,

$$\text{if } |\pi - x| < 10^{-4}, \qquad \text{then } |\pi^3 - x^3| < 0.005.$$

We can therefore take $x = 3.1415$, the 4-place truncation of the expansion of π. Some brutal arithmetic then shows that

$$x^3 = 31.0035 \cdots$$

We know that this approximates π^3 with an error < 0.005, and so $\pi^3 \approx 31.00$, with an error less than 0.01.

With the same amount of work, we can get a better approximation. This is because we know that the expansion of π begins $3.14159 \ldots$, and therefore that

$$x = 3.1416$$

is a four-place decimal that approximates π with an error that is $< 0.00001 = 10^{-5}$. In this case

$$|\pi^3 - x^3| \leq 48|\pi - x| < 48(10^{-5}) < 0.0005.$$

The arithmetic is similar to that in our first estimate above, and this time we find that

$$x^3 = (3.1416)^3 = 31.00649 \cdots$$

Here, we know that x^3 is larger than π^3 but by less than 0.0005. Therefore,

$$\pi^3 \approx 31.006,$$

with an error less than 0.0005.

Although the above theorem furnishes a very simple way to compute, it is important to find ways to make the continuity calculation that are independent of the derivative. There are two reasons for this further investigation.

First, we have to know that some functions are continuous before we can prove that they have derivatives.

EXAMPLE 3. Look back at the calculation of $d\sqrt{x}/dx = 1/2\sqrt{x}$. The difference quotient was

$$\frac{\Delta y}{\Delta x} = \frac{\sqrt{(x + \Delta x)} - \sqrt{x}}{\Delta x} = \frac{1}{\sqrt{x + \Delta x} + \sqrt{x}}.$$

In order to get the answer $1/2\sqrt{x}$, we have to know that $\sqrt{x + \Delta x}$ approaches $\sqrt{x}$ as $\Delta x \to 0$, i.e., that $\sqrt{x}$ is continuous. A prior continuity calculation is thus required, and to avoid circularity in our logic, we cannot base this calculation on the derivative of $\sqrt{x}$ and so can't use Theorem 3.

So what do we do? Generally speaking, we make the same calculation and algebraic simplification of Δy that we did for the difference quotient in Chapter 3, but then we derive an inequality instead of taking a limit. In the above case, where

$$\frac{\Delta y}{\Delta x} = \frac{1}{\sqrt{x + \Delta x} + \sqrt{x}},$$

we note that since $\sqrt{x + \Delta x}$ is of necessity nonnegative, we make the right side larger (if anything) by omitting $\sqrt{x + \Delta x}$. That is,

$$\frac{\Delta y}{\Delta x} \leq \frac{1}{\sqrt{x}}.$$

Multiplying by Δx, and allowing for the possibility that Δx is negative, we get

$$|\Delta y| \leq \frac{1}{\sqrt{x}} |\Delta x|.$$

Now we fix a positive number a and restrict x to the interval $[a, \infty)$. Then $1/\sqrt{x} \leq 1/\sqrt{a}$, and so

$$|\Delta y| \leq k |\Delta x|,$$

where $k = 1/\sqrt{a}$. We have thus obtained an inequality like that of the theorem by a direct calculation.

EXAMPLE 4. Let us see how this direct continuity calculation turns out for our original function $y = f(x) = x^3$ on the interval $[-4,4]$. Here, the factoring procedure works most smoothly because it uses the two numbers r and x that we are restricting to $[-4,4]$. We have

$$f(r) - f(x) = r^3 - x^3 = (r - x)(r^2 + rx + x^2),$$

and so

$$|f(r) - f(x)| \leq |r - x|(16 + 16 + 16) = 48|r - x|.$$

Thus, $|\Delta y| \leq 48|\Delta x|$, as before.

If we try the increment approach, we get

$$\Delta y = f(x + \Delta x) - f(x) = (x + \Delta x)^3 - x^3$$

$$= x^3 + 3x^2\Delta x + 3x(\Delta x)^2 + (\Delta x)^3 - x^3$$

$$= \Delta x(3x^2 + 3x\Delta x + \Delta x^2).$$

In this form, it isn't useful to assume that both x and $(x + \Delta x)$ lie in $[-4,4]$. Let us assume that x lies there, and that $|\Delta x|$ is at most 1; then $|3x^2 + 3x\Delta x + \Delta x^2| \leq 48 + 12 + 1 = 61$, and so

$$|\Delta y| \leq 61|\Delta x|,$$

a different inequality stemming from different assumptions as to where the two points x and $(x + \Delta x)$ lie. In this case, in order to ensure that

$$|\Delta y| < \varepsilon,$$

it is sufficient to see to it that

$$61|\Delta x| < \varepsilon,$$

i.e., to take

$$|\Delta x| < \frac{\varepsilon}{61}.$$

The second reason for making the direct continuity calculation goes back to Chapter 5: We frequently need to establish the continuity of f at points where f' does not exist.

EXAMPLE 5. Consider, again, $y = x^{1/3}$ around the origin. We know that $f'(x) = x^{-2/3}$ blows up at $x = 0$. However, since

$$\Delta y = f(0 + \Delta x) - f(0) = (0 + \Delta x)^{1/3} - 0^{1/3} = (\Delta x)^{1/3},$$

the inequality

$$|\Delta y| < \varepsilon$$

will hold if

$$|\Delta x|^{1/3} < \varepsilon, \quad \text{or} \quad |\Delta x| < \varepsilon^3.$$

Thus Δy does go to 0 with Δx, but much more slowly. For example, in order to make $|\Delta y| < 1/100$, we have to take

$$|\Delta x| < \left(\frac{1}{100}\right)^3 = \frac{1}{1,000,000}.$$

EXAMPLE 6. The upper semicircle $y = f(x) = \sqrt{1 - x^2}$ is another example. It is clear from geometry that this graph has vertical tangents at $x = 1$ and $x = -1$, and that f' fails to exist at these two points. Nevertheless, we can work out by simple algebra that

$$|\Delta y| \leq \sqrt{2}\,|\Delta x|^{1/2}$$

at $x_0 = \pm 1$. Therefore,

$$|\Delta y| < \varepsilon$$

if

$$\sqrt{2}\,|\Delta x|^{1/2} < \varepsilon,$$

i.e., if

$$|\Delta x| < \frac{\varepsilon^2}{2}.$$

Here again, $|\Delta y|$ approaches 0 much more slowly than Δx. For example, in order to ensure that $|\Delta y| < .01$, using the above inequalities, we have to take $|\Delta x| < (0.01)^2/2 = 0.00005$.

Let us summarize our various continuity calculations. In each case we wanted to know how small to take $|\Delta x|$ in order to ensure that $|\Delta y| < \varepsilon$, where ε is some arbitrary preassigned positive number which can be thought of as the tolerable error. In each case, we found a number δ (Greek delta) depending on ε which served as an adequate control for the size of Δx. Our various control numbers δ were

$$\delta = \frac{\varepsilon}{k}, \qquad \delta = \varepsilon^3, \qquad \delta = \frac{\varepsilon^2}{2}.$$

A general description of the continuity calculation would therefore run as follows: We first obtain an inequality which limits $|\Delta y|$ in terms of $|\Delta x|$. From this inequality, we see that in order to ensure that $|\Delta y| < \varepsilon$, it is sufficient to take $|\Delta x|$ less than a certain control number δ depending on ε.

This description of "effective" continuity is the modern *definition* of continuity.

Definition. *The function* $y = f(x)$ *is continuous at the point* $x = x_0$ *in its domain if for each positive "tolerable error"* ε *we can find an adequate "control" number* δ. *By this we mean that we can ensure that* $|\Delta y| < \varepsilon$ *by taking* $|\Delta x| < \delta$:

$$|\Delta x| < \delta \Rightarrow |\Delta y| < \varepsilon,$$

($\Rightarrow$ *means* implies), *or*

$$|x - x_0| < \delta \Rightarrow |f(x) - f(x_0)| < \varepsilon.$$

Let us consider next the "effective" continuity of the multiplication operation $y = uv$. In the examples above, the direct proof of continuity always started off as though its objective were a derivative, by setting up a formula for the dependent increment Δy, and manipulating it so that its dependence on Δx becomes apparent. The same scheme works here for the product function $y = uv$. If we start with fixed values $y_0 = u_0 v_0$ and give u and v increments Δu and Δv, then the new value of y is

$$y_0 + \Delta y = (u_0 + \Delta u)(v_0 + \Delta v).$$

Multiplying this out and subtracting $y = uv$, we see that

$$\Delta y = u_0 \Delta v + v_0 \Delta u + \Delta u \Delta v.$$

This increment formula is not new. It was the basis for our proof of the product rule for derivatives in Chapter 4. Here we use it differently, to get an inequality limiting $|\Delta y|$. First,

$$|\Delta y| \leq |u_0| \cdot |\Delta v| + |v_0| \cdot |\Delta u| + |\Delta u| \cdot |\Delta v|.$$

Then suppose we choose some positive number δ less than 1, and restrict Δu and Δv to be at most δ in magnitude. Substituting the inequalities $|\Delta u| \leq \delta$, $|\Delta v| \leq \delta$, in the right side above, and noting that $\delta^2 < \delta$ because $\delta < 1$, we find

$$|\Delta y| \leq |u_0|\delta + |v_0|\delta + \delta^2$$

$$< [|u_0| + |v_0| + 1]\delta.$$

If we want $|\Delta y| < \varepsilon$, it is thus sufficient to make

$$[|u_0| + |v_0| + 1]\delta = \varepsilon,$$

i.e., to take

$$\delta = \frac{\varepsilon}{|u_0| + |v_0| + 1}.$$

We have shown that the increment in the product function $y = uv$ will be less than ε in magnitude if the increments Δu and Δv in the two factors are each bounded by

$$\delta = \frac{\varepsilon}{|u_0| + |v_0| + 1}.$$

We have thus effectively demonstrated the continuity of multiplication.

Now consider how the continuity calculation would go for the product of two continuous functions. Suppose that $u = f(x)$ and $v = g(x)$ are both continuous at $x = x_0$. Given ε, we want to find how close to take x to x_0 in order to guarantee that

$$|f(x)g(x) - f(x_0)g(x_0)| < \varepsilon,$$

or, more simply,

$$|uv - u_0 v_0| < \varepsilon.$$

We saw above that this inequality will hold if

$$\Delta u = u - u_0 \qquad \text{and} \qquad \Delta v = v - v_0$$

are each less than

$$\varepsilon' = \frac{\varepsilon}{|u_0| + |v_0| + 1}$$

in magnitude. Since $u = f(x)$ is continuous at x_0, we can find a positive control number δ_1 for the inequality

$$|\Delta u| < \varepsilon'.$$

That is, the inequality will hold whenever we restrict $\Delta x = x - x_0$ to be less than δ_1 in magnitude. Similarly, we can find a positive control number δ_2 for the continuous function $v = g(x)$, so that

$$|\Delta v| < \varepsilon'$$

whenever $|\Delta x| < \delta_2$. We want *both* $|\Delta u| < \varepsilon'$ and $|\Delta v| < \varepsilon'$, and this double requirement will be met if $|\Delta x|$ meets both control requirements,

$$|\Delta x| < \delta_1 \qquad \text{and} \qquad |\Delta x| < \delta_2.$$

We can state this final condition more simply by setting δ equal to the *smaller* of the two control numbers δ_1 and δ_2. So if $|\Delta x| < \delta$, then Δx meets both control requirements. Backtracking, we see that then both $|\Delta x| < \varepsilon'$ and $|\Delta v| < \varepsilon'$ hold, and therefore that $|uv - u_0 v_0| < \varepsilon$. That is,

if $|x - x_0| < \delta$, then $|f(x)g(x) - f(x_0)g(x_0)| < \varepsilon$.

We have thus found a suitable control number δ for the product function $f(x)g(x)$ at $x = x_0$.

In this way, by explicitly tracking down how the continuity calculation goes, we have proved:

Theorem 4. *If f and g are continuous at $x = x_0$, then so is the product function fg.*

This is the way all the limit and continuity laws are established. In every case we prove that a function is continuous, or that a limit exists and has a certain value, by exhibiting an explicit scheme that will furnish a control number δ for any given positive number ε. This means, of course, that we use the corresponding "effective" definition of a function limit:

Definition. *The function f has the limit l at $x = x_0$ if for every positive number ε, we can find a positive number δ such that*

$$0 < |x - x_0| < \delta \Rightarrow |f(x) - l| < \varepsilon.$$

Note that we don't allow x to take on the value x_0 when we consider whether or not $f(x)$ has the limit l as x approaches x_0.

The notion of sequential convergence is made precise in the same quantitative way.

Definition. *The sequence $\{a_n\}$ converges to the number l as its limit as n tends to infinity if, given any positive number ε, we can find an integer N such that*

$$n > N \Rightarrow |a_n - l| < \varepsilon.$$

Sooner or later one should work through all of this material in a systematic way. It is probably not appropriate to take the time to do this in a first calculus course, and here we shall just spotlight a few additional results in the exercises.

Finally, we consider the counterpart of Theorem 3 for functions of two variables.

Theorem 5. *Suppose that on a circular disk R both partial derivatives of $z = f(x,y)$ exist and are bounded by k in magnitude. That is,*

$$|D_1 f(x,y)| \le k,$$

$$|D_2 f(x,y)| \le k,$$

for all points (x,y) in R. Then

$$|\Delta z| \leq k(|\Delta x| + |\Delta y|)$$

for any two points (x,y) and $(x + \Delta x, y + \Delta y)$ in R.

Proof. Up to a point we just repeat the proof of the chain rule from Chapter 16 (page 638). We go from the first point (x,y) to the second point $(x + \Delta x, y + \Delta y)$ in two steps, first a horizontal step to $(x + \Delta x, y)$ and then a vertical step. (It may be necessary to make the first step vertical and the second horizontal, as the figure shows.) Then Δz can be written as the sum of the two corresponding partial increments,

$$\Delta z = f(x + \Delta x, y + \Delta y) - f(x,y)$$
$$= [f(x + \Delta x, y + \Delta y) - f(x + \Delta x, y)] + [f(x + \Delta x, y) - f(x,y)],$$

and we can apply the mean-value theorem for functions of one variable to each term on the right, obtaining

$$\Delta z = D_2 f(x + \Delta x, Y) \cdot \Delta y + D_1 f(X,y) \cdot \Delta x.$$

All this is exactly as before. But now we use the assumed inequalities on the partial derivatives and have at once the inequality we want:

$$|\Delta z| \leq k|\Delta y| + k|\Delta x| = k(|\Delta x| + |\Delta y|).$$

It was because of this theorem that we could be sure that the functions we met in Chapter 13 were all continuous.

REMARK. We can restate the theorem in terms of the length $|\Delta X|$ of the vector $\Delta X = (\Delta x, \Delta y)$, because of the fact that

$$a + b \leq \sqrt{2}\sqrt{a^2 + b^2}$$

for any two numbers. (Square both sides and see what happens.) The inequality

$$|\Delta z| \leq k(|\Delta x| + |\Delta y|)$$

thus implies that

$$|\Delta z| \leq \sqrt{2}\,k|\Delta X|.$$

EXAMPLE 7. Theorem 5 provides another way to establish the continuity inequality for the product operation $z = f(x,y) = xy$. The partial derivatives

$$\frac{\partial z}{\partial x} = y, \qquad \frac{\partial z}{\partial y} = x,$$

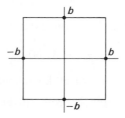

are both less than b in magnitude on the square $|x| \le b$, $|y| \le b$. Therefore,

$$|\Delta z| \le b(|\Delta x| + |\Delta y|)$$

for any two points (x,y) and $(x + \Delta x, y + \Delta y)$ in the square. If we set $(r, s) = (x + \Delta x, y + \Delta y)$, then we can rewrite this new version of the continuity inequality as

$$|rs - xy| \le b(|r - x| + |s - y|).$$

It is closely related to the earlier version, the differences being due to different assumptions about the magnitudes of x, y, and their increments.

PROBLEMS FOR SECTION 2

1. How many places in the decimal expansion

    $$\log 4 = 1.3 \cdots$$

    will be needed to approximate $(\log 4)^2$ with an error less than 0.01? (Use Theorem 3.)

2. How many places in the decimal expansion

    $$\log_{10} \pi = 0.4 \cdots$$

    are needed to approximate $(\log_{10} \pi)^3$ with an error less than 0.005?

3. How many places in the expansion

    $$\log 100 = 4.6 \cdots$$

    are needed to approximate $(\log 100)^4$ with an error less than 0.005?

4. How many places in the decimal expansion

$$e = 2.7 \cdots$$

   are needed to approximate e^5 with an error less than 0.002?

5. Example 2 in the text did not make full use of the inequality

$$3.1415 < \pi < 3.1416.$$

   Show that the conclusion of Example 2 can be strengthened to the fact that

$$\pi^3 = 31.00 \cdots$$

   That is, the decimal expansion of π^3 begins with 31.00.

6. In the second part of Example 2, show, furthermore, that

$$\pi^3 = 31.006 \cdots$$

7. Prove the following counterpart of Theorem 3.

   Theorem 3'. *If*

$$|f'(x)| \geq b > 0$$

   for every x *in an interval* I, *then*

$$|\Delta y| \geq b|\Delta x|$$

   for any two points x *and* $(x + \Delta x)$ *in* I.

8. Show that if $e = 2.7 \cdots$, then $e^3 < 23$, *without* computing $(2.8)^3$. (Use Problem 7 and the obvious inequalities $e > 8/3$, $3 - e > 0.2$.)

Prove the inequalities in Problems 9 through 13 by direct arguments (that is, without using calculus).

9. If $0 < x < a$, then

$$4x^3(a - x) < a^4 - x^4 < 4a^3(a - x).$$

10. If x and y are both greater than the positive number a, then

$$\left| \frac{1}{x} - \frac{1}{y} \right| < \frac{1}{a^2}|x - y|.$$

11. If $0 < x < a$ then $a^n - x^n < na^{n-1}(a - x)$.

12. Assuming that $|\sin x| \leq |x|$ for all x, show that

$$|\cos x - \cos y| \leq |x - y|.$$

13. If $0 < b < y$ then $y^{1/4} - b^{1/4} < \frac{1}{4}b^{-3/4}(y - b)$. (This can be proved directly from Problem 1 by relabeling.)

14. If $y = 1/x$, show by a direct estimation of Δy, as in Examples 3 and 4, that if $|x|$ and $|x + \Delta x|$ are both larger than a positive number b, then

$$|\Delta y| \leq \varepsilon \qquad \text{whenever } |\Delta x| \leq b^2\varepsilon.$$

15. If $y = x^4$, show by a direct algebraic estimate of Δy that if $|x|$ and $|x + \Delta x|$ are both less than a, then

$$|\Delta y| \leq \varepsilon \qquad \text{whenever } |\Delta x| \leq \frac{\varepsilon}{4a^3}.$$

16. Show that if $e = 2.7182 \cdots$, then

$$e^4 - (2.718)^4 < 0.05.$$

(Use the preceding problem or Problem 9.)

17. Show that if $y = \sqrt{1 - x^2}$, then $y \leq \sqrt{2}(1 - x)^{1/2}$. Similarly, show that $y \leq \sqrt{2}(1 + x)^{1/2}$. This completes Example 6 in the text.

In each of the following four problems, find how small to take $\Delta x = x - a$ in order to ensure that $\Delta y = f(x) - f(a)$ will be less than ε in magnitude.

18. $f(x) = (x - 2)^{2/3}$; $a = 2$.   19. $f(x) = (4 - x^2)^{4/5}$; $a = 2$.

20. $f(x) = \sqrt{x - 1}$; $a = 1$.   21. $f(x) = \sqrt{x^2 - 1}$; $a = 1$.

22. Prove that if f is continuous and $f(a) \neq 0$, then $1/f$ is continuous at $x = a$.

23. Show by a direct algebraic argument that $f(x, y) = x + y$ is a continuous function of two variables. (This is much easier than the proof for the product in the text.)

24. Prove that if f and g are continuous at $x = a$, then so is $f + g$.

25. Show that if $x \neq 0$ and if we can calculate x, then we can calculate $1/x$.

26. Show that if we can calculate the numbers x and y, then
    a) we can calculate $x + y$, and
    b) we can calculate xy.

27. Proceeding either on the basis of the above problem or by a direct argument, show that if we can calculate N numbers $x_1, x_2, \ldots, x_n$, then we can calculate
    a) the sum $x_1 + x_2 + \cdots + x_n$, and
    b) the product $x_1 x_2 \cdots x_n$.

28. Suppose that we can calculate the value $f(x)$ of the function f at every rational number x, and suppose that a and b are rational numbers. Show

that we can calculate the Riemann sum

$$\Delta x \sum_{k=0}^{n-1} f(a + k\Delta x),$$

where $\Delta x = (b - a)/n$. (Assume any result from the above two problems.)

29. Give a rigorous ε, N proof of the law: If $a_n \to a$ and $b_n \to b$ as $n \to \infty$, then $a_n b_n \to ab$ as $n \to \infty$. (Imitate the proof of Theorem 4, replacing the control numbers δ_1 and δ_2 by control integers N_1 and N_2, etc.)

30. Prove the sign-preserving law: If $a_n \to a$ as $n \to \infty$, and if $a_n \geq 0$ for all n, then $a \geq 0$. [*Hint:* Suppose, on the contrary, that $a < 0$, and take $\varepsilon = |a| = -a$, in the definition of convergence. You should be able to show that, then,

$$a_n < 0 \qquad \text{for all } n \text{ larger than } N,$$

which is a contradiction.]

31. Suppose that $f(x)$ is an increasing function on $(0, \infty)$ and that $f(x) < B$ for all x. Prove that $\lim_{x \to \infty} f(x)$ exists and is at most B. (First consider the sequence $a_n = f(n)$.)

32. Suppose that f is increasing and bounded on $(0, 1)$. Prove that $\lim_{x \to 1} f(x)$ exists. (This can be thrown back to the above problem or attacked directly.)

3. THE INTERMEDIATE-VALUE THEOREM

This is the theorem:

Theorem. *If f is continuous on the closed interval $[a, b]$ and if l is any number between $f(a)$ and $f(b)$, then there is at least one point X in $[a, b]$ for which $f(X) = l$.*

If we set $g(x) = f(x) - l$, and possibly change sign, then the intermediate-value theorem reduces to the zero-crossing theorem for a continuous function.

Theorem 6. *If g is continuous on the closed interval $[a,b]$, and if $g(a) < 0 < g(b)$, then $g(X) = 0$ for some number X in $[a,b]$.*

Let us see how we might go about computing such a number X. We shall first sketch a "theoretical" computation, called *approximation by double position*, and then look into whether it will really work in practice.

Suppose, to be specific, that $[a, b] = [1, 2]$. As a first step we divide $[1,2]$ into 10 equal subintervals with endpoints

$$1.0, \quad 1.1, \quad 1.2, \quad \cdots, \quad 1.9, \quad 2.0.$$

Then we run through the corresponding values of f:

$$f(1.0), \quad f(1.1), \quad \cdots, \quad f(1.9), \quad f(2.0).$$

We shall suppose that we are never lucky enough to land exactly on a number where f is zero, so each value is either positive or negative. Since we start with the negative value $f(1.0)$ and end with the positive value $f(2.0)$, there will be a first sign change in this sequence of eleven values, say, between 1.2 and 1.3. Then

$$f(1.2) < 0 < f(1.3).$$

Our second step is to divide $[1.2, 1.3]$ into ten equal subintervals, with endpoints

$$1.20, \quad 1.21, \quad \cdots, \quad 1.29, \quad 1.30,$$

and again select the first pair where f changes sign, say,

$$f(1.26) < 0 < f(1.27).$$

Continuing in this way, we construct an infinite decimal, beginning, say, $1.26745\ldots$, such that

$$f(1) < 0 < f(2)$$
$$f(1.2) < 0 < f(1.3)$$
$$f(1.26) < 0 < f(1.27)$$
$$\vdots$$
$$f(1.26745) < 0 < f(1.26746)$$

Now let X be the number having this infinite decimal expansion. Then $f(X)$ has to be 0. Why? Since the lefthand numbers $1, 1.2, 1.26, \ldots$ form an infinite sequence $\{a_n\}$ converging to X, and since f is continuous at X, we have

$$f(X) = \lim_{n \to \infty} f(a_n).$$

But $f(a_n) < 0$ for all n. Therefore $f(X) \leq 0$ by the sign-preserving property of a convergent sequence. Similarly, since the sequence $\{b_n\}$ of righthand numbers converges to X from above, and since $f(b_n) > 0$ for all n, it follows that

$$f(X) = \lim_{n \to \infty} f(b_n) \geq 0.$$

That is, $f(X)$ must be simultaneously ≤ 0 and ≥ 0, and hence must be 0.

The above process actually constitutes a proof of the zero-crossing theorem. Of course, to be called a proof it has to be described in more general terms.

Moreover, it is somewhat more efficient and also more suitable for computing machines to use successive bisections rather than successive division into ten parts. This means that the number being computed is not now described by an infinite decimal, but instead is specified by an infinite sequence of closed intervals, each of which is one half of its predecessor.

Proof of Theorem. Set $I_0 = [a, b]$ and let $I_1 = [a_1, b_1]$ be the left or right half of I_0 depending on whether the value of f at the midpoint $c = (a + b)/2$ is positve or negative. That is, if $f(c) > 0$, we set $a_1 = a$ and $b_1 = c$, and if $f(c) < 0$, we set $a_1 = c$ and $b_1 = b$. In either case,

$$f(a_1) < 0 < f(b_1).$$

(We suppose we are not lucky enough to find that $f(c) = 0$.) We continue bisecting in this way, at the nth step choosing $I_n = [a_n, b_n]$ as that half of I_{n-1} for which

$$f(a_n) < 0 < f(b_n).$$

Since $b_n - a_n = (b - a)/2^n$, the conditions for the nested-interval form of the completeness property are met, and the sequences $\{a_n\}$ and $\{b_n\}$ therefore converge to a common limit x. Then

$$f(x) = \lim_{n \to \infty} f(a_n) \leq 0,$$

$$f(x) = \lim_{n \to \infty} f(b_n) \geq 0,$$

and so $f(x) = 0$. ∎

Let us now consider the actual computation of a solution x of $f(x) = 0$, following the bisection method. We shall suppose that f is smooth and increasing, so that we are, in effect, calculating a value of the smooth inverse function.

First of all, we will be calculating the values of f at the successive midpoints c_n only to within a certain fixed error δ. For example, we may compute every value $f(c_n)$ to five decimal places, with an error at most $\delta = 10^{-5}$. Suppose that we come to a first midpoint c for which our calculated function value is zero (to within the error δ). Do we then know that c approximates a solution point x to within some allowable error ε? If we can't conclude this, our computation is hopeless.

In order to be able to make this essential conclusion, we need an added strong hypothesis: namely, that f' is bounded below by a fixed positive number m:

$$f'(x) \geq m > 0$$

for every x in (a,b). This means that f is everywhere increasing at least as fast as a line of slope m. (It would also be all right if f were decreasing and $f'(x) \leq -m < 0$.) We then have the following theorem.

Theorem 7. *Let f be a differentiable function on the closed interval $[a,b]$ such that $f(a) < 0$, $f(b) > 0$, and $f'(x) \geq m > 0$ for every x in the interval. Suppose also that a point c has been somehow selected. Then c approximates the unique solution of $f(x) = 0$ to within an error of $|f(c)|/m$.*

Proof. Let x_0 be the exact solution. It exists by the zero-crossing theorem. Then by the mean-value theorem

$$f(c) = f(c) - f(x_0) = f'(X)(c - x_0)$$

for some X between c and x_0. Then

$$|f(c)| = |f'(X)||c - x_0| \geq m|c - x_0|,$$

and so

$$|c - x_0| \leq \frac{|f(c)|}{m},$$

as claimed. ∎

It follows from this theorem that if we want to compute the solution of $f(x) = 0$ with an error less than ε, and if $f' \geq m > 0$ everywhere, then it is sufficient to compute the midpoint function values $f(c)$ to within an error $\delta = \varepsilon m$, and stop when one is found whose value is zero to within this error.

EXAMPLE 1. A table gives 1.73 as the value of $\sqrt{3}$ to two decimal places. Check this on the basis of the above theorem, using the value $(1.73)^2 = 2.9929$.

....

Solution. We note that $\sqrt{3}$ is the solution of $f(x) = x^2 - 3 = 0$, and that $\sqrt{3} > 1.5$. Since the derivative $f'(x) = 2x$ is greater than $3 = m$ in this region, 1.73 is closer to $\sqrt{3}$ than

$$\frac{|f(1.73)|}{m} = \frac{3 - 2.9929}{3} = \frac{0.0071}{3} < 0.0025.$$

In particular,

$$1.73 < \sqrt{3} < 1.733.$$

EXAMPLE 2. We find that $(1.3)^3 = 2.197$. How close can we say that 1.3 is to $2^{1/3}$?

....

Solution. Here $2^{1/3}$ is the solution of $f(x) = x^3 - 2 = 0$, and we check that $1.2 < 2^{1/3}$ (since $(1.2)^3 = 1.728 < 2$). Since $f'(x) = 3x^2 > 4$ for $x \geq 1.2$, the error in taking 1.3 as an approximation of $2^{1/3}$ is less than $(2.197 - 2)/4 < 0.05$. Thus, 1.3 is the closest one-place decimal to $2^{1/3}$.

REMARK. These computations are new only in their interpretation. They are similar to examples given in Section 2, and in Chapter 3, Section 11.

The extreme-value property can also be proved from the completeness property, but the proof is a logical existence proof that cannot be made *effective*. That is, we cannot prove that f assumes a maximum value M by, in effect, *computing* a value of x at which $f(x)$ has the sought-for value, the way we did for the intermediate-value theorem. This proof will be omitted, but it is an essential step in deducing calculus theory from the completeness of the real numbers.

PROBLEMS FOR SECTION 3

In the following three problems, use Theorem 7 in the manner of Examples 1 and 2 in the text.

1. In looking for $\sqrt{10}$ by double approximation, we find that $(3.16)^2 = 9.9856$. Show, therefore, that $\sqrt{10} - 3.16 < 0.003$.

2. Given that $(2.08)^3 = 8.9989 \cdots$, show that $9^{1/3} - 2.08 < 0.0001$.

3. Given that $(3.11)^3 = 30.080 \cdots$, show that $|30^{1/3} - 3.11| < 0.003$.

4. There is a gap in the text discussion. We claimed that the successive midpoint computations will eventually yield a midpoint c for which $|f(c)| < \delta$, where δ is the error allowed in the computation of the function value. Show that if f' is bounded above by M, then we must obtain a midpoint c for which $|f(c)| < \delta$ at least by the Nth step, provided that

$$\frac{b-a}{2^N} < \frac{\delta}{M}.$$

4. THE EXPONENTIAL FUNCTION

We turn now to the intuitive "soft spots" mentioned at the beginning of the chapter. Our first job is to give a purely analytic definition of e^x and then to deduce its properties from this analytic definition. We are not allowed to assume anything at all about the function. Everything must be proved.

We start by defining a function $E(x)$ as the sum of the power series

$$E(x) = \sum_{n=0}^{\infty} \frac{x^n}{n!}.$$

The ratio test shows that the series converges for all x, so we are starting with a function that is at least defined for all x in an explicit uniform way. Moreover, its value can be computed directly from its series definition, as we saw in Chapter

12. Then, according to the theorem on the differentiation of power series, $E(x)$ is differentiable, and we find that

$$E'(x) = E(x)$$

when we look at the differentiated series. Also, of course,

$$E(0) = 1.$$

Thus, E is the solution of the initial-value problem

$$\frac{dy}{dx} = y, \qquad y = 1 \quad \text{when } x = 0.$$

We shall now prove that $E(x) = e^x$, where the base e is defined as $E(1)$, i.e., as

$$e = 1 + 1 + \frac{1}{2} + \frac{1}{3!} + \frac{1}{4!} + \cdots$$

$$= \sum_0^\infty \frac{1}{n!}.$$

Theorem 8. *E satisfies the addition law*

$$E(x + y) = E(x)E(y).$$

Proof. We first remark that if f is any other function satisfying the differential equation

$$\frac{dy}{dx} = y,$$

then

$$f(x)E(-x) = \text{constant} .$$

For then

$$\frac{d}{dx}(f(x)E(-x)) = f'(x)E(-x) + f(x)E'(-x)(-1)$$

$$= f(x)E(-x) - f(x)E(-x) = 0,$$

and we know that a function whose derivative is zero must necessarily be a constant.

Next we note that $f(x) = E(x + a)$ is such a function f, for any fixed a. Therefore

$$E(x + a)E(-x) = c.$$

Setting $x = 0$, we see that $c = E(a)$. Thus,

$$E(x + a)E(-x) = E(a),$$

for all numbers x and a. If we set $s = x + a$ and $t = -x$ in this identity, and note that then $a = s + t$, we obtain the addition law

$$E(s + t) = E(s)E(t)$$

for all s and t. ∎

Theorem 9. *If $g(x)$ is a continuous function defined for all x, and satisfies the addition law*

$$g(x + y) = g(x)g(y)$$

and the initial condition $g(0) \neq 0$, then

$$g(x) = a^x,$$

where the base a is defined by $a = g(1)$.

Proof. If $a = g(1)$, then

$$g(2) = g(1 + 1) = g(1) \cdot g(1) = a \cdot a = a^2,$$

$$g(3) = g(2 + 1) = g(2) \cdot g(1) = a^2 \cdot a = a^3,$$

$$\cdot \qquad \cdot$$
$$\cdot \qquad \cdot$$
$$\cdot \qquad \cdot$$

$$g(n) = g(n - 1 + 1) = g(n - 1)g(1) = a^{n-1} \cdot a = a^n,$$

for all positive integers n. (This is really an argument by mathematical induction.) More generally, but in exactly the same way, we see that

$$g(nx) = [g(x)]^n$$

for any x and any positive integer n. Taking $x = 1/n$, we have $[g(1/n)]^n = g(1) = a$, so that

$$g(1/n) = a^{1/n},$$

by the definition of the nth root of a. Then,

$$g\left(\frac{m}{n}\right) = g\left(m\left(\frac{1}{n}\right)\right) = \left[g\left(\frac{1}{n}\right)\right]^m = [a^{1/n}]^m = a^{m/n}.$$

Thus,

$$g(r) = a^r$$

for every positive rational r. This holds for negative rationals as well, for $g(-r)g(r) = g(r - r) = g(0) = 1$, so

$$g(-r) = 1/g(r) = 1/a^r = a^{-r}.$$

Finally, suppose that x is irrational and let r approach x through rational values. Then

$$g(x) = \lim_{r \to x} g(r) = \lim_{r \to x} a^r = a^x.$$

This completes the proof of the theorem. Note, however, that we are assuming continuity to be part of what we mean by an exponential function a^x. ∎

In particular, if we set $e = E(1)$, then $E(x) = e^x$. We have thus deduced the basic properties of e^x from purely analaytic considerations.

PROBLEMS FOR SECTION 4

According to the plan of the text, the logarithm function can now be defined as the inverse of the exponential function, as in Chapter 7. When it comes to the computation of $\log x$, however, we use the definite integral or a power series. Since our present philosophy is to turn computational schemes into definitions, we should consider the possibility of defining $\log x$ in these two ways. The exercises below explore these alternative approaches to the logarithm function.

1. Prove that if $L(x) = \int_1^x dt/t$, then $L(ab) = L(a) + L(b)$ for all positive a and b. (We know that $L'(x) = 1/x$; now compute $L'(ax)$ by the chain rule and compare results.)

2. The preceding problem shows that $L(x)$ is a logarithm, but is it the inverse of e^x? Show now that $L(e^x) = x$. (Differentiate by the chain rule, and use the initial value $\log 1 = 0$.)

3. It can be proved from the above problem that the other cancellation identity, $e^{L(x)} = x$, also holds. In order to show this directly, first show that if $f(x) = xf'(x)$, then $f(x) = Cx$. (Consider the derivative of $f(x)/x$.) Then show that $f(x) = e^{L(x)}$ has this property, and so conclude that $e^{L(x)} = x$. We then know that $L(x) = \int_1^x dt/t$ is the inverse of e^x.

In many texts, the logarithm function is defined first, by $\log x = \int_1^x dt/t$, and then the exponential function is defined as its inverse. The next few problems develop this idea.

4. Show by a substitution change of variable that if $0 < y < 1$, then

$$\int_y^1 \frac{dt}{t} = \int_1^{1/y} \frac{dt}{t}.$$

5. a) Show that $\int_1^x dt/t \to \infty$ as $x \to \infty$. (Review the integral test from Section 7 of Chapter 12.)

    b) Show that $\int_x^1 dt/t \to \infty$ as $x \to 0$. (Use Problem 4.)

6. Use the inverse-function theorem of Chapter 7 and Problem 5 to show that if we define $\log x$ as $\int_1^x dt/t$, for all positive x, then $\log x$ has a differentiable inverse $E(x)$ defined for all real numbers.

7. Problem 1 shows that $\log x$ is a logarithm. Now show from this that $E(x)$ is an exponential function, i.e., that $E(x + y) = E(x)E(y)$.

8. Show from the above definition of $E(x)$ that $E'(x) = E(x)$. What then would be the alternative proof that $E(x)$ is an exponential function?

Finally, we consider the definition of $\log x$ by power series. The straightforward Maclaurin series

$$\log(1 - x) = -\left(x + \frac{x^2}{2} + \frac{x^3}{3} + \cdots + \frac{x^n}{n} + \cdots\right)$$

is unsatisfactory since it converges only on $(1,1)$.

9. The first problem is to determine a possible new definition. So suppose that we know all about $\log x$, especially the material in Chapter 13, and show that then

$$\log x = F\left(\frac{x - 1}{x + 1}\right),$$

where

$$F(x) = 2\left[x + \frac{x^3}{3} + \frac{x^5}{5} + \cdots + \frac{x^{2n+1}}{2n + 1} + \cdots\right].$$

10. Now define $\log x$ as above. That is, define a new function $L(x)$ by

$$L(y) = F\left(\frac{y - 1}{y + 1}\right), \qquad F(x) = 2\sum_0^\infty \frac{x^{2n+1}}{2n + 1}.$$

Show first that the domain of L is the interval $(0,\infty)$.

11. Show next that $F'(x) = 2/(1 - x^2)$ and then that $L'(y) = 1/y$.

12. Assuming Problems 1 through 3, we are on the verge of knowing that $L(x)$ is the inverse of e^x. Complete the job.

5. THE TRIGONOMETRIC FUNCTIONS

We can remove all intuitive props and leave the functions $\sin x$ and $\cos x$ standing on a completely analytic basis by the same general procedure that we used in the last section for e^x. Remember, we must give analytic definitions of the functions and then deduce their properties from these definitions.

We start by defining functions $C(x)$ and $S(x)$ as the sums of the power series

$$C(x) = 1 - \frac{x^2}{2!} + \frac{x^4}{4!} - \cdots = \sum_{n=0}^{\infty} (-1)^n \frac{x^{2n}}{(2n)!},$$

$$S(x) = x - \frac{x^3}{3!} + \frac{x^5}{5!} - \cdots = \sum_{n=0}^{\infty} (-1)^n \frac{x^{2n+1}}{(2n+1)!}.$$

The ratio test shows these series to be everywhere convergent, so we are starting with functions S and C that are analytically defined for all x in an explicit and uniform manner. Moreover, we can compute their values directly from these series definitions, as we saw in Chapter 13. Note that $C(x)$ is an even function and $S(x)$ is odd.

According to the series differentiation theorem, both functions $S(x)$ and $C(x)$ are everywhere differentiable, and we find that

$$S' = C \qquad \text{and} \qquad C' = -S$$

when we look at their term-by-term differentiated series. It follows that each function is a solution of the second-order differential equation

$$(*) \qquad\qquad \frac{d^2 y}{dx^2} = -y.$$

Also, the series definitions show that C and S have the "initial" values

$$C(0) = 1, \qquad S(0) = 0;$$

$$C'(0) = 0, \qquad S'(0) = 1.$$

In Chapter 4, Section 5, it was shown that if $C(x)$ is a solution of the differential equation $(*)$ and satisfies the initial conditions

$$C(0) = 1, \qquad C'(0) = 0,$$

then any other solution f of $(*)$ can be written in the form

$$(**) \qquad\qquad f(x) = f(0)C(x) - f'(0)C'(x).$$

The addition formulas are an immediate consequence.

Theorem 10. *The functions C and S satisfy the addition formulas*

$$C(x + a) = C(a)C(x) - S(a)S(x),$$

$$S(x + a) = C(a)S(x) + S(a)C(x).$$

Proof. We just check that $f(x) = C(x + a)$ is a solution of (∗) and that $f'(x) = -S(x + a)$. The formula (∗∗) then turns into the first addition formula. Similarly, $f(x) = S(x + a)$ is a solution of (∗) and $f'(x) = C(x + a)$, giving the second addition formula. ∎

We have now established all the basic analytic properties of the trigonometric functions. However, before we conclude that $C(x)$ and $S(x)$ really are cos x and sin x, we should show that these analytic properties imply the geometric property by which we originally defined cosine and sine.

Theorem 11. *The path $(C(t), S(t))$ traces the unit circle counterclockwise, and its length from $(1,0)$ to $(C(\theta), S(\theta))$ is θ.*

Proof. Theorem 10 (with $x = -a$) and the initial values of C and S show that

$$S^2(t) + C^2(t) = 1,$$

so the path $X(t) = (C(t), S(t))$ runs along the unit circle. The formula giving the arc length s as a function of the parameter t is

$$s = \int \sqrt{\left(\frac{dx}{dt}\right)^2 + \left(\frac{dy}{dt}\right)^2}\, dt.$$

So the length of the present circular path from 0 to θ is

$$\int_0^\theta \sqrt{[-S(t)]^2 + [C(t)]^2}\, dt = \int_0^\theta 1\, dt = t\,\Big]_0^\theta = \theta.$$

The analytic meaning of counterclockwise motion is that $C(t)$ is decreasing and $S(t)$ is increasing as long as $(C(t), S(t))$ is in the first quadrant, with similar monotone behavior in the other quadrants. This behavior of C and S follows from their derivative relationships $S' = C$, $C' = -S$. For example, S is increasing as long as C is positive, and C is decreasing as long as S is positive, so the point $(C(t), S(t))$ moves "northwestward" across the first quadrant. ∎

Even if we knew nothing about π, we could conclude from the above analysis that the path must reach $(0,1)$ at least by the time the parameter t reaches the value $t = 2$. This is so because, as long as the path is running northwest, the path length from $t = 0$ to $t = \theta$ is at most $S(\theta) + (1 - C(\theta))$. This is visibly so for an inscribed polygon, and therefore holds for path length upon taking the limit. But $S(\theta) + (1 - C(\theta))$ remains less than 2 as far as the path point $(0,1)$. Since the path length is θ, we conclude that $\theta < 2$ as long as the point $(C(\theta), S(\theta))$ remains in the first quadrant, and therefore that $(C(\theta), S(\theta))$ must become $(0,1)$ at least by the time θ reaches the end of the interval $[0,2]$.

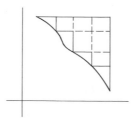

Theorem 11 shows that the functions C and S have the geometric properties by which we originally defined cosine and sine, so we are done.

PROBLEMS FOR SECTION 5

1. Prove that if a function $F(x)$ is defined by $F(x) = \int_0^x dt/(1+t^2)$, then F is a left inverse of the tangent function on the interval $(-\pi/2, \pi/2)$:

$$F(\tan x) = x.$$

(We know that $F'(x) = 1/(1 + x^2)$. Use the chain rule and initial values.)

2. Prove that if a function $G(x)$ is defined by $G(x) = \int_0^x dt/\sqrt{1 - t^2}$, then G is a left inverse of the sine function:

$$G(\sin x) = x.$$

(We know $G'(x)$ from the integral definition. Use the chain rule and initial values.)

Because the arc sine and arc tangent functions are only one-sided inverses of the trigonometric functions, it is not easy to make one of them the starting point for a systematic development of the trigonometric functions. Nevertheless it can be done, and the next few problems take up some of the steps.

3. Show that $\int_0^a dt/\sqrt{1 - t^2}$ has a finite limit as $a \to 1$. (Compare with the integral of $1/\sqrt{1 - t}$; see Problem 32 in Section 2.) We then define $\int_0^1$ as this limit,

$$\int_0^1 \frac{dt}{\sqrt{1 - t^2}} = \lim_{a \to 1} \int_0^a \frac{dt}{\sqrt{1 - t^2}},$$

and say that the *improper integral* $\int_0^1 dt/\sqrt{1 - t^2}$ *converges* to this value.

4. Define $\pi/2$ as the number $\int_0^1 dt/\sqrt{1-t^2}$, and define the function F on the interval $[0,1]$ by

$$F(x) = \int_0^x \frac{dt}{\sqrt{1-t^2}}.$$

a) Show that F has an increasing differentiable inverse $S(x)$, with domain $[0,\pi/2]$ and range $[0,1]$.

b) Then show that

$$[S(x)]^2 + [S'(x)]^2 = 1.$$

(Differentiate the identity $F(S(x)) = x$.)

c) Now show from (b) that

$$S''(x) = -S(x).$$

The principal problem comes when we try to extend $S(x)$ to a larger domain. It is sufficient to show how to add $[\pi/2,\pi]$ to the domain of S, since further extensions can be made in the same way.

5. Show that $G(x) = S(x - \pi/2)$ is defined on $[\pi/2,\pi]$ and satisfies the differential equations

$$[G(x)]^2 + [G'(x)]^2 = 1,$$

$$G''(x) = -G(x),$$

there.

6. Show next that if we extend the domain of S by setting

$$S(x) = G'(x)$$

on $[\pi/2,\pi]$, then S, S', and S'' are all continuous on the enlarged domain $[0,\pi]$, and the differential equation $S' = -S$ continues to hold.

7. By repeatedly extending as above, we end up with $S(x)$ defined on the whole number line $(-\infty,\infty)$, and satisfying

$$S'' = -S, \qquad S(0) = 0, \qquad S'(0) = 1.$$

Discuss how the rest of the theory is now already available.

6. THE COMPLETENESS AXIOM

We discussed two forms of completeness in Section 3 of Chapter 12. Here is our third and most flexible version. We call it the completeness *axiom* because we shall assume it without proof, and then shall deduce the earlier forms of the completeness property from it. It is also the form of completeness that we shall use in discussing the definite integral in the next section.

The Completeness Axiom. *Let A and B be two collections of real numbers such that*:

1. *$a \leq b$ for every a in A and every b in B. That is, A lies entirely to the left of B when the sets are pictured on the number line.*

2. *The difference $b - a$ can be chosen arbitrarily small. That is, if we are given any positive number ε, no matter how small, we can find a in A and b in B such that $b - a < \varepsilon$.*

Then there is a uniquely determined number x separating A from B. That is, there is a number x with the property that

$$a \leq x \leq b$$

for every a in A and every b in B, and x is the only number having this property.

If the numbers in the collection A and B are all rational, or, more generally, if they are all numbers that we can compute, then, in choosing a pair satisfying $b - a < \varepsilon$, we are actually computing the number x. For then

$$0 \leq x - a \leq b - a < \varepsilon,$$

and x is approximated by the computable number a with an error less than ε. So here again we have a method for simultaneously defining and computing a number, like the nested-interval method (but unlike the monotone-limit property.)

It is possible to prove the completeness axiom from the assumption that every infinite decimal represents a number. We shall not give this proof (hence the title "axiom"), but shall show that the other two forms of completeness that we have considered are consequences of the axiom. These proofs have no direct bearing on the next section, and may be omitted.

We first prove that the nested interval property is a consequence of the completeness axiom.

Proof. Let the nested sequence be $I_n = [a_n, b_n]$, and let A be the set of lefthand endpoints a_n and B the set of righthand endpoints b_n. We claim that then A and B satisfy the hypotheses of the completeness axiom.

We first have to show that $a_i \leq b_j$ for any i and j. Suppose that $i \leq j$. Then $a_i \leq a_j$ because $\{a_n\}$ is increasing, and $a_j \leq b_j$ by hypothesis. Therefore $a_i \leq b_j$. If $i > j$, then $a_i \leq b_i \leq b_j$ in a similar way. In either case we have

$$a_i \leq b_j.$$

The second requirement on A and B is met because $b_n - a_n \to 0$.

Therefore, by the completeness axiom, there is a unique number x such that $a_i \leq x \leq b_j$ for all i and j, i.e., a unique number x lying in all the intervals I_n. ∎

The proof of the monotone-limit property is much fussier, but let us see how it goes. We are given an increasing sequence $\{a_n\}$ that has at least one upper bound b, and we want to prove that a_n converges to a limit, by using the completeness axiom. Let A be the collection of sequence terms a_n and let B be the collection of all numbers b such that b is an upper bound for the sequence $\{a_n\}$. Then $a \leq b$ for every a in A and every b in B, by definition. In order to apply the completeness axiom we have to show that, given any positive number ε, we can find particular numbers a in A and b in B such that

$$b - a < \varepsilon.$$

Suppose, for the sake of definiteness, that the sequence $\{a_n\}$ lies entirely in the unit interval $[0,1]$. Thus, 1 is in B but 0 is not in B. Given ε, choose a fixed positive integer K greater than $1/\varepsilon$ and consider the numbers j/K for $j = 0, 1, \ldots, K$. These numbers subdivide the unit interval into subinterval of common length $1/K$. The first of these numbers, $0 = 0/K$, is not in B. The last of these numbers, $1 = K/K$, is in B. Therefore, as j runs through the integers from 0 to K, there will be a first value $J > 0$ such that J/K is in B. The preceding value $(J - 1)/K$ is not in B, and hence is less than at least one number a in A. We thus have a pair a in A and $b = J/K$ in B lying in the interval

$$\left[\frac{J - 1}{K}, \frac{J}{K} \right]$$

of length $1/K$. Therefore

$$b - a \leq \frac{1}{K} < \varepsilon.$$

We can now apply the completeness axiom and conclude that there is a unique number x separating A from B:

$$a \leq x \leq b,$$

for all a in A and all b in B. We claim that then

$$a_n \to x \quad \text{as} \quad n \to \infty,$$

i.e., that for any given ε, all the sequence terms a_m, from some point on, lie within the distance ε of x:

$$0 \leq x - a_m < \varepsilon.$$

But we just saw that we could find *one* sequence term $a = a_n$ and *one* upper bound b in B, such that

$$b - a_n < \varepsilon.$$

Then, for all later terms a_m, we have

$$a_n \leq a_m \leq x \leq b.$$

Therefore

$$0 \leq x - a_m \leq b - a_n < \varepsilon$$

for all m beyond n, as required. We have thus proved the monotone-convergence form of the completeness property from the completeness axiom. ∎

PROBLEMS FOR SECTION 6

The problems below consitute, in their totality, a proof of the completeness axiom from the decimal-expansion property.

1. Prove the completeness axiom from the nested-interval property. (First choose c_n from A and d_n from B, so that $d_n - c_n < 1/n$. Then let a_n be the maximum of the numbers $c_1, c_2, \ldots, c_n$, and let b_n be the minimum of $d_1, \ldots, d_n$. Now complete the proof.)

The proof of the nested-interval property, from the monotone-limit property, was taken up in the Problems in Section 5 of Chapter 12. The final step in this reversed chain of deductions is to prove the monotone-limit property from the decimal-expansion property. Accordingly, let a_n be a strictly increasing sequence of real numbers lying in the interval $[0,1]$. Let $a_{n,m}$ be the m-place decimal obtained by truncating the decimal expansion of a_n. Show the following things.

2. For a fixed m, the sequence of finite decimals $a_{n,m}$ is increasing and constant from a certain point on. Let l_m be this final constant value.

3. If $m < p$, then l_m is the m-place truncation of l_p.

4. Let l be the number whose decimal expansion is made up of the finite decimals l_m. That is, l_m is the m-place truncation of the decimal expansion of l. The number l exists, since by hypothesis every infinite decimal represents a number. Show that l is the limit of the original sequence $\{a_n\}$.

5. Steps 1 through 4 above prove the monotone-limit property for the special case of a strictly increasing sequence lying in $[0,1]$. Show how the monotone-limit property for any other increasing sequence that is bounded above can be reduced to this special case.

7. THE RIEMANN INTEGRAL

Our development of the definite integral was based on geometric intuition. We assumed that any plane region of a certain sort has a definite numerical area, and that areas combine with each other in the appropriate manner. Assuming this and the extreme-value property, we established that every continuous function has an antiderivative, and we were in business. On the other hand, we learned to *calculate* a definite integral by a purely analytic process, the Riemann-sum approximation, having no direct dependency on such geometric assumptions. But according to an informal paraphrase of completeness, any number that we can calculate surely exists. We therefore ought to be able to use completeness to turn the Riemann-sum estimation procedure into an analytic definition of $\int_a^b f$, independent of all geometric assumptions. In particular, if we go back to our first definition of a Riemann sum, where we obtained one sum S_n for each integer n by subdividing $[a,b]$ into n equal subintervals, then we ought to be able to use the completeness property to prove that the sequence $\{S_n\}$ is convergent, so that we can define $\int_a^b f$ as the limit of S_n as n tends to infinity.

The limit definition of $\int_a^b f$ suggested above would seem to offer the simplest possible solution to our problem, but it doesn't work out as well as one would hope. It is too inflexible, in somewhat the same sense that the decimal expansion definition of calculability was too inflexible.

Instead, we shall use the main version of completeness that we called the completeness axiom. We shall see that certain Riemann sums are clearly too large if anything, and we use them to form a collection U of "upper numbers." Other Riemann sums will be clearly too small if anything, and we use them to form a collection L of lower numbers. If we can then show by the completeness axiom that there is exactly one number separating U from L, then we are led to *define* $\int_a^b f$ as this unique number, since it is the only possible value for $\int_a^b f$. Supposing that this all works out, we will have $\int_a^b f$ analytically defined for every continuous function f on a closed interval $[a,b]$. But there will still be work to be done because we still have to prove analytically that $\int_a^b f$ has all the properties that we have come to expect of it. Working this all out is a long tedious process, and we shall omit some of the details.

We consider first the "upper" and "lower" Riemann sums. It will pay us, in the interests of flexibility, to use general Riemann sums, as in Section 3 of Chapter 10. Moreover it is permissible and appropriate to be guided by geometric intuition, as we were there, except that now we must not depend on it. Accordingly, let Δ stand for *any* subdivision of the interval $[a,b]$ into a finite number of subintervals. Then Δ is determined by a finite sequence of subdividing points

$$a = x_0 < x_1 < \cdots < x_n = b.$$

Given such a subdivision Δ and given a continuous function f on $[a,b]$, we construct two extreme Riemann sums, the maximum and minimum of all possible Riemann sums associated with the subdivision Δ.

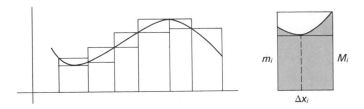

To do this, let m_i and M_i be the minimum and maximum, respectively, of the values of f on the ith subinterval $[x_{i-1}, x_i]$. Then

$$l_\Delta = \sum_{i=1}^{n} m_i \Delta x_i$$

is the smallest of all the Riemann sums associated with the subdivision Δ, and

$$u_\Delta = \sum_{i=1}^{n} M_i \Delta x_i$$

is the largest such Riemann sum. We shall call them the *lower* and *upper* Riemann sums for Δ. The notation l_Δ and u_Δ exhibits Δ but not f, which suggests that we are going to vary the subdivision Δ but keep the function f fixed. The point is that we want to derive the existence of the number $\int_a^b f$ from the completeness property, and this means that we need not just one pair of upper and lower estimates for $\int_a^b f$, but whole collections of such estimates that will squeeze down on a unique number.

So we let L be the collection of all numbers obtained as lower Riemann sums for f over the interval $[a,b]$. That is, L is the set of numbers l_Δ obtained from *all* subdivisions Δ of $[a,b]$. Similarly, let U be the set of all *upper* Riemann sums.

It is clear from our intuition about the area A enclosed between the graph of f and the x-axis that

$$l_{\Delta_1} \leq A$$

for each subdivision Δ_1, and that

$$A \leq u_{\Delta_2}$$

for each subdivision Δ_2, and hence that

$$l_{\Delta_1} \leq u_{\Delta_2}$$

for any two subdivisions Δ_1 and Δ_2. That is, the collection of numbers L lies entirely below the collection of numbers U. We are then well on our way to being able to apply the completeness axiom.

But we are not allowed to depend on geometric reasoning here, so we must find a direct analytic proof of the inequality

$$l_{\Delta_1} \leq u_{\Delta_2}.$$

Let us suppose for the moment that this has been done and go on. Next we must show that for any positive ε we can find a lower sum l and an upper sum u such that

$$u - l < \varepsilon.$$

If we can do this then the completeness axiom says that the two classes U and L squeeze down on a unique number, and we can define $\int_a^b f$ to be this number.

So we have two analytic chores to perform in order to free the definite integral from its dependence on intuition, and base it instead on the completeness of the real numbers. We take them up in reverse order.

We first want to prove that we can make $u - l$ as small as we wish by choosing a suitable subdivision Δ. We shall prove this for two classes of functions.

Lemma 1. *If the derivative f' exists and is bounded by K over the interval $[a,b]$, and if δ is the maximum of the interval widths Δx_i in a subdivision Δ, then*

$$u_\Delta - l_\Delta \leq K(b - a)\delta.$$

Proof. We recall that, when the derivative is bounded by K, then

$$|\Delta y| \leq K|\Delta x|$$

for any two points x and $x + \Delta x$. This will be true in particular for the two points in the ith interval $[x_{i-1}, x_i]$ at which f assumes its minimum and maximum values m_i and M_i. Therefore,

$$M_i - m_i \leq K\Delta x_i \leq K\delta$$

for all i. Therefore

$$u_\Delta - l_\Delta = \sum M_i \Delta x_i - \sum m_i \Delta x_i = \sum (M_i - m_i)\Delta x_i$$
$$\leq \sum K\delta\Delta x_i = K\delta \sum \Delta x_i = K\delta(b - a). \quad \blacksquare$$

Thus $u_\Delta - l_\Delta$ will be less than ε if

$$K\delta(b - a) < \varepsilon$$

i.e., if we use any subdivision whose maximal interval width δ satisfies

$$\delta < \frac{\varepsilon}{K(b - a)}.$$

Lemma 2. *If f is increasing on* [*a,b*], *then*

$$u_\Delta - l_\Delta \leq \delta[f(b) - f(a)],$$

δ *again being the maximum interval width in the subdivision* Δ.

Before giving the algebraic proof, we note that the inequality of the lemma can be directly visualized geometrically. The right side of the figure shows that the sum of the shaded rectangular areas is $u_\Delta - l_\Delta$, and the left side of the figure shows these areas fitted into a rectangle of width δ and height $f(b) - f(a)$.

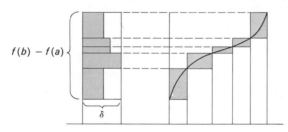

Proof of the lemma. The maximum value of an increasing function occurs at the righthand endpoint and the minimum value at the lefthand endpoint. Thus, over the interval $[x_{i-1}, x_i]$ we have

$$M_i = f(x_i), \qquad m_i = f(x_{i-1}).$$

Then

$$u_\Delta - l_\Delta = \sum_{i=1}^{n} M_i \Delta x_i - \sum m_i \Delta x_i$$

$$= \sum_{i=1}^{n} f(x_i)\Delta x_i - \sum f(x_{i-1})\Delta x_i$$

$$= \sum_{i=1}^{n} [f(x_i) - f(x_{i-1})]\Delta x_i$$

$$\leq \sum_{i=1}^{n} [f(x_i) - f(x_{i-1})]\delta$$

$$= \delta\begin{bmatrix} f(x_1) & - & f(x_0) \\ + & f(x_2) & - & f(x_1) \\ + & f(x_3) & - & f(x_2) \\ & \vdots & \\ + & f(x_n) & - & f(x_{n-1}) \end{bmatrix}$$

$$= \delta[f(x_n) - f(x_0)]$$

$$= \delta[f(b) - f(a)]. \quad \blacksquare$$

Next, we must prove:

Lemma 3. *For any two subdivisions Δ_1 and Δ_2.*

$$l_{\Delta_1} \le u_{\Delta_2}$$

Proof. This depends on the following two observations.

I. *Whenever we* refine *a subdivision, by adding new subdividing points to the ones we already have, then the lower Riemann sum increases (or stays unchanged) and the upper Riemann sum decreases (or stays the same). That is, if Δ_2 is a refinement of Δ_1, then*

$$l_{\Delta_1} \le l_{\Delta_2}, \qquad u_{\Delta_1} \ge u_{\Delta_2}.$$

In order to see this, we note that Δ_2 can be obtained from Δ_1 in a finite number of steps, where we add one new subdividing point at each step. So it suffices to prove these inequalities at each such step. But if the new point added at such a step is x', and if it goes between x_i and x_{i+1}, then the change that occurs in the lower sum is confined to the interval $[x_i, x_{i+1}]$, and can be directly visualized in terms of rectangular areas, as below. Clearly the lower sum increases if anything, and the algebraic verification of this geometrically obvious fact will be left to the reader. Similarly, the upper sum decreases when one new subdividing point is added.

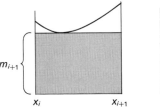 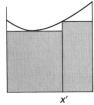

II. *Any two subdivisions Δ_1 and Δ_2 have a common refinement Δ.* We can obtain Δ from Δ_1 by adding those subdividing points of Δ_2 that are not already used in Δ_1. But we also obtain Δ from Δ_2 by adding the subdividing points of Δ_1 not already used in Δ_2.

Using the common refinement Δ and (I), we conclude that

$$l_{\Delta_1} \le l_{\Delta} \le u_{\Delta} \le u_{\Delta_2},$$

where the two outer inequalities follow from (I), and the middle inequality follows directly from the definition of the upper and lower sum of a subdivision. Therefore, $l_{\Delta_2} \le u_{\Delta_2}$, completing the proof of Lemma 3 (and our second step). ∎

We have now proved the two properties of U and L needed for the completeness axiom. We therefore conclude that there is a unique number, that we designate $\int_a^b f$, separating U from L. That is:

Definition. $\int_a^b f$ is defined as the unique number such that

$$l_{\Delta_1} \le \int_a^b f \le u_{\Delta_2}$$

for all subdivisions Δ_1 and Δ_2.

Our final task is to establish analytically that $\int_a^b f$ has the properties that seemed so obvious when it was defined on the basis of areas.

The first lemma below depends on the remark that if l' and l'' are lower Riemann sums for f over the intervals $[a,b]$ and $[b,c]$, respectively, then $l' + l''$ is a lower Riemann sum for f over $[a,c]$. And similarly for upper Riemann sums. (We are assuming that $a < b < c$, here.) This depends on the obvious fact that a subdivision of $[a,b]$ combines with a subdivision of $[b,c]$ to give a subdivision of $[a,c]$.

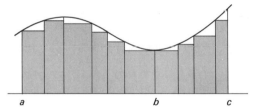

Lemma 4. If $a < b < c$, then

$$\int_a^b f + \int_b^c f = \int_a^c f.$$

Proof. It u' and l' are any upper and lower Riemann sums for f over the interval $[a,b]$, then

$$l' \le \int_a^b f \le u'.$$

Similarly, if u'' and l'' are any upper and lower Riemann sums for f over $[b,c]$, then

$$l'' \le \int_b^c f \le u''.$$

Then

$$l' + l'' \le \int_a^b f + \int_b^c f \le u' + u''.$$

But $u' + u''$ and $l' + l''$ are upper and lower Riemann sums for f over the combined interval $[a,c]$, so also

$$l' + l'' \leq \int_a^c f \leq u' + u''.$$

But when two numbers lie in the same interval, their distance apart is less than the interval width. Therefore

$$\left| \int_a^c f - \left(\int_a^b f + \int_b^c f \right) \right| \leq (u' + u'') - (l' + l'') = (u' - l') + (u'' - l'').$$

By choosing suitable subdivisions of $[a,b]$ and $[b,c]$, the differences $u' - l'$ and $u'' - l''$ can be made as small as desired, so the right side above can be made less than any given positive ε. The left side is therefore a fixed nonnegative number that is less than every positive number ε, and hence is zero. That is,

$$\int_a^c f - \left(\int_a^b f + \int_b^c f \right) = 0,$$

which is what we wanted to prove.   ∎

Lemma 5. *If m and M are the minimum and maximum values of f on $[a,b]$, then*

$$m(b - a) \leq \int_a^b f \leq M(b - a).$$

Proof. $M(b - a)$ is the upper Riemann sum for the trivial subdivision $a = x_0 < x_1 = b$, where we don't add any new points. The lower Riemann sum is $m(b - a)$, and the lemma is immediate since $l \leq \int_a^b f \leq u$ for *any* upper and lower Riemann sums u and l.   ∎

Theorem 12. *The function*

$$\phi(x) = \int_a^x f(t)\, dt$$

is an antiderivative of f.

Proof. The proof of Theorem 2 in Chapter 8 now applies verbatim.

This theorem is called the Fundamental Theorem of Calculus. It says that $\int_a^b f$, defined analytically as the number calculated by Riemann sums, can also be calculated by the formula

$$\int_a^b f = F(b) - F(a),$$

where F is any antiderivative of f. In other words, the number that is provided by Archimedes' arduous method of exhaustion can sometimes be computed trivially by the great new invention of Newton and Leibniz. We have of course been exploiting this fact all through the book.

PROBLEMS FOR SECTION 7

Once we know that $\int_a^x f(t)\,dt$ is an antiderivative of f (Theorem 12), the remaining basic properties of the definite integral are immediately available from the discussion in Section 8 of Chapter 9, where they were established by differentiation arguments. But it is important for later applications of these ideas to see how certain integral properties follow directly from the Riemann-sum definition of the integral, as in our proofs of Lemmas 4 and 5. Prove the following three properties by the general type of reasoning used in Lemma 4.

1. $\int_a^b (f + g) = \int_a^b f + \int_a^b g$.

2. If $c > 0$, then $\int_a^b cf = c \int_a^b f$.

3. If $g = -f$, then $\int_a^b g = - \int_a^b f$.

4. Prove algebraically that if $a < c < b$, and if m, m_1, and m_2 are, respectively, the minimum values of a function f on $[a,b]$, $[a,c]$, and $[c,b]$, then

$$m(b - a) \leq m_1(c - a) + m_2(b - a).$$

appendixes

appendixes

appendix 1
basic inequality laws

We begin with the following two facts about positive numbers:

P1. If x and y are positive, then so are $x + y$ and xy.

P2. Zero is not positive. If x is not positive, then x is zero or $-x$ is positive.

Starting from P1 and P2 as axioms, it is possible to prove all of the algebraic properties of inequalities. We shall sketch this development below, giving complete details for the proofs of some properties, partial proofs for others, and leaving some as exercises.

We call a number x *negative* if $-x$ is positive. A number x cannot be both positive and negative, for this would imply that $0 = x + (-x)$ is positive (by P1), which contradicts P2. Thus, P2 has the following reformulation:

P2'. A number x is either positive, or zero, or negative, and these possibilities are mutually exclusive.

1. Law of signs. *If x and y have the same sign* (i.e., *if both are positive or both are negative*), *then xy is positive. If x and y have opposite signs, then xy is negative.*

Proof. By hypothesis, x and y are either both positive or both negative. If x and y are both positive, then xy is positive, by P1. If x and y are both negative, then $-x$ and $-y$ are both positive, by P2, so $xy = (-x)(-y)$ is positive, by P1. Thus xy is positive in either case.

The other assertion will be left as an exercise.

2. *If $x \neq 0$, then x^2 is positive. In particular, $1 = 1 \cdot 1$ is positive.*

Proof. This is a corollary of (1), since if $x \neq 0$, then x^2 is a product in which both factors have the same sign.

3. *If $x \neq 0$, then x and $1/x$ have the same sign.*

Proof, by contradiction. Suppose x and $1/x$ have opposite signs. Then $1 = x(1/x)$ is negative, by (1). But 1 is positive, by (2). We therefore have a contradiction, so x and $1/x$ must have the same sign.

Definition. *We say that x is less than y, and write $x < y$, if $y - x$ is positive. Similarly $x > y$ (x is greater than y) if $y < x$, that is, if $x - y$ is positive. In particular, $x > 0$ if and only if x is positive.*

4. *If $x < y$ and $y < z$, then $x < z$.*

Proof. By hypothesis, $x < y$ and $y < z$. Therefore $y - x$ and $z - y$ are both positive, by the definition of inequality. Therefore $z - x = (z - y) + (y - z)$ is positive, by P1. Therefore $x < z$, by the definition of inequality.

5. *If $x < y$, then $x + c < y + c$, no matter what number c is.*

Proof. If $x < y$, then $y - x$ is positive, by the definition of inequality. Then $(y + c) - (x + c) = y - x$ is positive. Then $x + c < y + c$, by the definition of inequality.

6. *If $x < y$ and if c is positive, then $cx < cy$.*

Proof. By hypothesis and the definition of inequality, $y - x$ and c are both positive. Therefore $cy - cx = c(y - x)$ is positive, by P1, and hence $cx < cy$, by the definition of inequality.

7. *If $x < y$ and $a < b$, then $x + a < y + b$.*

8. *If $0 < x < y$ and $0 < a < b$, then $ax < by$.*

9. *If x and y are positive and if $x < y$, then $1/x > 1/y$.*

That is, reciprocals of positive numbers are in opposite order.

Partial proof. We note that

$$\frac{1}{x} - \frac{1}{y} = \frac{y - x}{xy} = (y - x) \cdot \frac{1}{x} \cdot \frac{1}{y}$$

and that all three factors on the right are positive, for various reasons. It follows that $1/x > 1/y$.

10. *If $x < y$, then $-y < -x$.* Changing signs reverses an inequality.

11. *If $x < y$ and c is negative then $cx > cy$.* Multiplying by a negative number reverses an inequality.

Definition. *We say that x is less than or equal to y, and write $x \leq y$ or $x \leqq y$, if either $x < y$ or $x = y$.* This so-called *weak* inequality satisfies a corresponding list of laws that we won't bother to give completely. Here is an example, to show how the two logical alternatives enter in.

12. *If $x \leq y$, then $c + x \leq c + y$, for any number c.*

Proof. If $x \leq y$, then either $x < y$ or $x = y$. If $x < y$, then $x + c < y + c$, by (5). If $x = y$, then $x + c = y + c$ (the sum of a pair of numbers is uniquely defined). Therefore, in either case, $x + c \leq y + c$.

13. $|x| \leq a$ *if and only if* $-a \leq x \leq a$.

Proof. It follows from the definition of absolute value that $|x|$ is the larger of x and $-x$. The inequality $|x| \leq a$ is thus equivalent to the pair of inequalities $-x \leq a, x \leq a$, and hence to the pair $-a \leq x, x \leq a$. But the latter pair is what we mean be the "continued" inequality $-a \leq x \leq a$.

14. $|x + y| \leq |x| + |y|$.

Proof. Since

$$-|x| \leq x \leq |x|,$$
$$-|y| \leq y \leq |y|,$$

it follows that

$$-(|x| + |y|) \leq x + y \leq |x| + |y|,$$

by the weak inequality analogue of (8). But this is equivalent to $|x + y| \leq |x| + |y|$ by (13).

appendix 2
the conic sections

Properties of the parabola, ellipse, and hyperbola are treated in the problem sets at various places in the text, mostly on the basis of parametric and polar-coordinate representations involving the trigonometric functions. Here we shall give a straightforward algebraic treatment of some of these questions, starting with traditional locus definitions of the curves.

1. THE PARABOLA

Definition. *The parabola is the locus traced by a point whose distance from a fixed point* (called the *focus*) *is equal to its distance from a fixed line* (called the *directrix*).

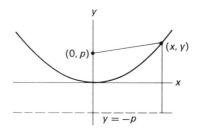

In order to obtain the equation of the parabola in standard form, we take the focus at $(0, p)$ on the y-axis and the directrix parallel to the x-axis with intercept $-p$. Then a point satisfies the locus condition if and only if

$$\sqrt{x^2 + (y - p)^2} = |y + p|.$$

Squaring and simplifying, we obtain first

$$x^2 + y^2 - 2py + p^2 = y^2 + 2py + p^2,$$

and then

$$x^2 = 4py.$$

This is the standard equation of the parabola.

EXAMPLE 1. Identify the graph of the equation $y - 2x^2 = 0$.

. . . .

Solution. We solve for x^2,

$$x^2 = (1/2)y,$$

and interpret the result as the standard form. Thus, $1/2 = 4p$, so $p = 1/8$. The graph is the parabola with focus at $(0,1/8)$ and directrix $y = -1/8$.

Note that we never did write the given equation explicitly in the standard form $x^2 = 4py$. This would be

$$x^2 = 4\left(\frac{1}{8}\right)y.$$

Instead, we found the right value of p by setting $1/2 = 4p$, and that was all we needed to identify the graph.

The parabola $x^2 = 4py$ is symmetric about the y-axis. In terms of the locus definition, the axis of symmetry is the line through the focus perpendicular to the directrix. The **vertex** of a parabola is the point where it intersects its axis of symmetry; i.e., the point halfway from the focus to the directrix. The vertex of the parabola $y^2 = 4px$ is at the origin.

In the derivation above we assumed that p is positive. If p is negative, the algebra is the same, but the picture is different. The focus at $(0, p)$ and directrix $y = -p$ now give us a parabola opening *downward*. Thus, $x^2 = -4y$ is the equation $x^2 = 4py$ with $p = -1$; so its graph is the parabola with focus at $(0, -1)$ and directrix $y = 1$, a downward opening parabola with vertex at the origin.

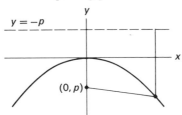

In the same way, the parabola with focus $(p, 0)$ on the x-axis and directrix $x = -p$ parallel to the y-axis has the equation

$$y^2 = 4px.$$

It is symmetric about the x-axis, with vertex at the origin, and opens to the right or to the left depending on whether p is positive or negative.

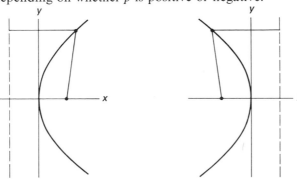

The proof does not have to be repeated. If we interchange the roles of x and y throughout an argument, then we automatically prove a new result, the statement of which is obtained from the old statement by simply interchanging the roles of x and y.

In summary:

Theorem 1. *The equation of a parabola with vertex at the origin is*

$$x^2 = 4py$$

if its axis is the y-axis, and

$$y^2 = 4px$$

if its axis is the x-axis. In each case, p is the coordinate of the focus on the axis.

EXAMPLE 2. Identify the graph of $x + y^2 = 0$.

. . . .

Solution. Solving for y^2,

$$y^2 = -x,$$

we have the form $y^2 = 4px$ with $p = -1/4$. So the graph is the parabola with focus at $(-1/4, 0)$ and directrix at $x = 1/4$. It opens to the left.

Now consider a parabola with vertex at (h, k). If we choose a new XY-axis system, with new origin at the point having old coordinates (h, k), then the vertex of the parabola is at the new origin; so the equation of the parabola in this new system will be

$$X^2 = 4pY$$

if its axis is vertical, and

$$Y^2 = 4pX$$

if its axis is horizontal, by Theorem 1. Since $X = x - h$ and $Y = y - k$ (Chapter 1, Section 8), we have the following generalization of Theorem 1.

Theorem 2. *The equation of a parabola with vertex at the point (h, k) is*

$$(x - h)^2 = 4p(y - k)$$

if its axis is vertical, and

$$(y - k)^2 = 4p(x - h)$$

if its axis is horizontal. In each case p is the signed distance from the vertex to the focus along the axis of the parabola. Thus, the focus is at $(h, k+p)$ in the first case and $(h+p, k)$ in the second case.

EXAMPLE 3. Find the equation of the parabola with focus at $(4,1)$ and directrix $x = 2$.

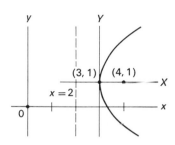

Solution. The vertex is halfway from the focus to the directrix, and hence is at $(h, k) = (3,1)$. The focus at $(4,1)$ is one unit to the right of the vertex at $(3,1)$, so $p = 1$. And the parabola opens in the x-direction, so it has the y-squared equation. Its equation is thus:

$$(y - k)^2 = 4p(x - h)$$

with $(h, k) = (3,1)$ and $p = 1$; that is

$$(y - 1)^2 = 4(x - 3),$$

or

$$y^2 - 2y - 4x + 13 = 0.$$

EXAMPLE 4. Find the equation of the parabola with focus at $(-2,0)$ and with the line $y = 2$ as directrix.

Solution. The vertex is $(-2,1)$, the point halfway between focus and directrix. The focus is one unit *below* the vertex, so $p = -1$, and the axis is vertical. So the equation is

$$(x - h)^2 = 4p(y - k)$$

with $(h, k) = (-2,1)$ and $p = -1$, which works out to

$$x^2 + 4x + 4y = 0.$$

The equation $(x - h)^2 = 4p(y - k)$ can be solved for y, leading to an equivalent equation of the form

$$y = ax^2 + bx + c,$$

with $a \neq 0$. Thus;

Each vertical parabola is the graph of a quadratic function $f(x) = ax^2 + bx + c$.

Conversely, every quadratic function $f(x) = ax^2 + bx + c$, with $a \neq 0$, is the graph of a vertical parabola. To identify the parabola we transform the equation $y = ax^2 + bx + c$ into the standard form $(x - h)^2 = 4p(y - k)$ by completing the square on the x-terms, just as in Section 8 of Chapter 1.

EXAMPLE 5. Identify the graph of the equation $y = 2x^2 - 4x$.

. . . .

Solution. We start by completing the square:

$$y + 2 = 2(x^2 - 2x + 1) = 2(x - 1)^2.$$

This equation can be written

$$(x - 1)^2 = (1/2)(y + 2),$$

which is in the standard form

$$(x - h)^2 = 4p(y - k),$$

with $(h, k) = (1,-2)$ and $p = 1/8$. The graph is thus the parabola with vertex at $(1,-2)$ and focus $1/8$ unit above the vertex, at $(9/8,-2)$.

More generally, any equation in x and y that is quadratic in one variable and linear in the other has a parabolic graph that can be identified by completing a square.

EXAMPLE 6. Identify the graph of $y^2 - 6y - 2x + 7 = 0$.

. . . .

Solution. We convert the equation to standard form, starting by completing the square on the y-terms, as follows:

$$y^2 - 6y + 9 - 2x + 7 = 9,$$

$$(y - 3)^2 = 2x + 2,$$

$$(y - 3)^2 = 2(x + 1).$$

This is of the form

$$(y - k)^2 = 4p(x - h)$$

with $(h, k) = (-1, 3)$ and $p = 1/2$. The graph is thus the parabola with vertex at $(-1, 3)$, opening to the right, with focus $1/2$ unit to the right of the vertex at $(-1/2, 3)$ and with directrix $x = -3/2$.

A unique "vertical" parabola, with equation of the form

$$y = ax^2 + bx + c,$$

can be passed through any given set of three points having distinct x-coordinates, provided they are not collinear, for the three sets of numerical coordinates give us three equations in the three unknowns a, b, and c, and can be solved for these coefficients in the standard manner.

EXAMPLE 7. Find the vertical parabola that contains the points $(-1, 2)$, $(1, 1)$, $(2, 3)$.

. . . .

Solution. The equation of the parabola, of the form

$$y = ax^2 + bx + c,$$

must be satisfied by each of the above three pairs of values. That is,

$$2 = a - b + c$$

$$1 = a + b + c$$

$$3 = 4a + 2b + c$$

The first step in solving three equations in three unknowns is to eliminate one unknown, leaving two equations in two unknowns. For example, we can solve the first equation above for c, substitute in the other two equations and so reduce to two equations in a and b. A simpler way of accomplishing the same result is to subtract the first equation from each of the other two, since c will cancel in the subtraction. We then have

$$-1 = 2b$$

$$1 = 3a + 3b$$

Finally, $b = -1/2$, $a = 5/6$, $c = 2/3$, so the parabola has the equation

$$y = \frac{5}{6}x^2 - \frac{1}{2}x + \frac{2}{3}.$$

The geometry of the parabola is concerned largely with properties of the tangent line. We have to know how to differentiate in order to compute the slope of the tangent line, but from then on it is pure analytic geometry.

Theorem 3. *The tangent line to the parabola*

$$x^2 = 4py$$

at the point (x_0, y_0) on the parabola has the equation

$$x_0 x = 2p(y + y_0).$$

Proof. The parabola $x^2 = 4py$ has the slope

$$\frac{dy}{dx} = \frac{2x}{4p} = \frac{x}{2p}.$$

The tangent line at the point (x_0, y_0) on the parabola thus has the slope $x_0/2p$, and its point–slope equation is

$$y - y_0 = \frac{x_0}{2p}(x - x_0)$$

We multiply out the right side and use the fact that $x_0^2 = 4py_0$ to get the form given in the theorem. ∎

The tangent-line equation and the slope formula $m_0 = x_0/2p$ will be used in the problems. Here we shall check only the optical property of the parabola.

Theorem 4. *The tangent line to the parabola $x^2 = 4py$ at (x_0, y_0) makes equal angles with the focal radius to (x_0, y_0) and the vertical line at that point.*

Proof. The tangent line $x x_0 = 2p(y + y_0)$ has y-intercept $y = -y_0$. The distance from the y-intercept to the focus $(0, p)$ is thus $p + y_0$. But this is also the

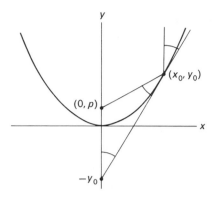

distance from the point (x_0, y_0) to the directrix $y = -p$, and therefore is equal to the distance from (x_0, y_0) to the focus, by the locus definition of the parabola. The triangle in the figure above is thus isosceles and has equal base angles. Since the tangent meets all vertical lines at the same angle, this proves the theorem. ∎

The theorem shows that the light rays emitted by a point source of light placed at the focus will all reflect off the parabola along vertical lines, forming a beam of light shining vertically without any weakening caused by spreading out. This is why a searchlight (or automobile headlight) has a reflector that is parabolic in cross section, and has a small intense light source at the focus. The ideal parallel beam will not quite be realized because the light source can only approximate a point source of light.

Telescopes use the same principle in reverse. The light rays arriving from a distant star form a parallel beam and are therefore focused onto a single point by a parabolic mirror. (Hence the word *focus*.) A camera placed (approximately) at the focus thus gathers all the light intercepted by the mirror, and this huge magnification of the light-gathering power of the camera lens makes it possible to take pictures of objects too faint to be recorded otherwise.

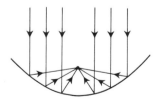

PROBLEMS FOR SECTION 1

Find the equations for each of the following parabolas.

1. Vertex at the origin; focus at (2,0).

2. Vertex at the origin; focus at (0,−1).

3. Directrix $x = 1$; focus $(-1,0)$.

4. Vertex at the origin; directrix $y = 1/4$.

5. Directrix the y-axis; focus $(1,0)$.

6. Directrix the y-axis; focus $(-2,0)$.

7. Directrix the x-axis; vertex $(0,2)$.

8. Directrix $x = 1$; focus at $(3,-1)$.

9. Directrix $x = 1$; focus at the origin.

10. Directrix $y = -2$; focus at the origin.

11. Directrix $y = -2$; Vertex at $(-2,0)$.

Identify each of the following parabolas by finding its vertex and focus.

12. $y = 2x^2$ 13. $y^2 = 2x$

14. $y^2 + 2x = 0$ 15. $y = 2x^2 + 1$

16. $y^2 + 2x + 1 = 0$ 17. $3y + x^2 - 6 = 0$

18. $y = x^2 + 2x + 1$ 19. $y = x^2 + 2x$

20. $y^2 + x + y = 0$ 21. $2y^2 - 4y - x + 2 = 0$

22. $2y^2 - 4y - x = 0$

23. A chord of the parabola $x^2 = 4py$ has endpoints with x-coordinates x_1 and x_2. Show that the slope of the chord is $(x_1 + x_2)/4p$.

24. A focal chord (chord through the focus) of the parabola $x^2 = 4py$ runs from (x_1,y_1) to (x_2,y_2). Show that its length is $y_1 + y_2 + 2p$.

In each of the following problems find the equation of and identify the vertical parabola passing through the three given points.

25. $(1,0)$, $(2,-1)$, $(3,0)$ 26. $(1,0)$, $(2,1)$, $(4,0)$

27. $(-1,1)$, $(0,0)$, $(2,-1)$

28. There can be no parabola having a horizontal axis and passing through the three points in Problem 25. Why?

29. Find the parabola having a horizontal axis and passing through the three points in Problem 27.

30. Find the equation of the parabola with vertical axis, vertex at the origin, and tangent to the line $y = x - 2$.

31. Find the lines tangent to the parabola $y = x^2$ and passing through the point $(2,1)$.

32. Show that the tangent line to the parabola $y^2 = 4px$ at the point (x_0,y_0) has slope $2p/y_0$.

33. The normal to a parabola at a point P on the parabola intersects the axis of the parabola at the point N. Show that P and N are at the same distance from the focus.

34. The tangent to a parabola at a point P intersects the directrix at the point T. Show that the segment PT subtends a right angle at the focus.

35. The tangent to a parabola at a point P intersects the tangent at the vertex at the point Q. If F is the focus, show that the segments FQ and PQ are perpendicular.

36. Prove that the midpoints of all chords of a parabola parallel to a given chord form a straight line that is parallel to the axis of the parabola.

37. Show that the lines tangent to a parabola at the ends of a focal chord (a chord through the focus) meet at right angles.

38. Show that the lines tangent to a parabola at the ends of a focal chord meet on the directrix.

39. Prove that the vertical line $x = x_0$ bisects every chord of the parabola $x^2 = 4py$ that is drawn parallel to the tangent line at (x_0, y_0).

40. Show that

$$x^2 = 4p(y + p)$$

is the standard equation for a parabola with vertical axis and focus at the origin.

41. Show that every upward-opening parabola with focus at the origin intersects at right angles every downward-opening parabola with focus at the origin. (Use the equation in the above problem with two different values of p, one positive and one negative.)

2. THE ELLIPSE

Ellipses and hyperbolas also have focus–directrix locus definitions, but we shall start instead from another locus problem.

Definition. *An ellipse is the locus traced by a point P whose distances from two fixed points F and F' have a constant sum, k,*

$$\overline{PF} + \overline{PF'} = k,$$

where k is greater than the distance $\overline{FF'}$. Each of the fixed points F and F' is called a focus of the ellipse (for reasons having to do with the optical property, which will be discussed later).

In order to obtain the simplest equation for this locus we take the x-axis along the segment FF' and the y-axis as its perpendicular bisector.

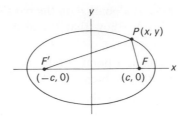

Then $F = (c,0)$ and $F' = (-c,0)$, where c is one-half the segment length. Also, if a is the intercept of the locus on the positive x-axis, then the locus condition says that

$$(a - c) + (a + c) = k,$$

so $k = 2a$. The locus condition for $P = (x,y)$ can thus be rewritten

$$\sqrt{(x - c)^2 + y^2} + \sqrt{(x + c)^2 + y^2} = 2a,$$

where a and c are positive constants such that $a > c$. This, then, is the basic equation of the locus. In order to simplify it, we follow the usual procedure for eliminating radicals: solve for a radical, and square. If we solve for the second radical, square everything, and then simplify, we end up with

$$\sqrt{(x - c)^2 + y^2} = a - \left(\frac{c}{a}\right)x.$$

The ratio c/a occurring here is called the *eccentricity* of the ellipse and is designated e; note that $e = c/a$ is less than 1. The first simplifying step thus results in a formula for the length of the righthand "focal radius" PF. This is an important formula, and we record it as a theorem.

Theorem 5. *Let E be the ellipse having foci $F = (c,0)$ and $F' = (-c,0)$, and x-intercepts at $\pm a$. Then for any point $P = (x,y)$ on E, the focal radii PF and PF' have the lengths*

$$\overline{PF} = a - ex, \qquad \overline{PF'} = a + ex,$$

where $e = c/a$.

Note that the second formula follows from the first, because $\overline{PF} + \overline{PF'} = 2a$. Continuing, we square the equation

$$\sqrt{(x - c)^2 + y^2} = a - \left(\frac{c}{a}\right)x$$

and simplify, this time ending up with

$$\left(\frac{a^2 - c^2}{a^2}\right)x^2 + y^2 = a^2 - c^2.$$

Setting $x = 0$ determines the y-intercepts as the points $\pm\sqrt{a^2 - c^2}$. We call the y-intercepts $\pm b$, as usual. That is, we set

$$b^2 = a^2 - c^2.$$

Then, after dividing by b^2, the above equation turns into the standard equation for the ellipse:

Theorem 6. *An ellipse with foci located on the x-axis and centered at the origin has the equation*

$$\frac{x^2}{a^2} + \frac{y^2}{b^2} = 1.$$

Actually, the reasoning above does only half the job. We have shown that if a point (x,y) is on the locus, then its coordinates satisfy the standard equation. It must also be shown that the argument can be traced backward, so that if a point (x,y) satisfies the standard equation, then it is on the locus. This backwards process involves taking square roots twice, and it must be checked that no sign ambiguity arises. For example, we must check that if (x,y) is on the graph of

$$\frac{x^2}{a^2} + \frac{y^2}{b^2} = 1,$$

then $(a - ex)$ is positive. This will be left as an exercise.

Remember that $a > b$ in the standard equation above, the foci being located by $c^2 = a^2 - b^2$. If we start instead with foci at the points $(0,c)$ and $(0,-c)$ on the y-axis, then we end up with the same standard equation in terms of the intercepts a and b, but this time $b > a$ and $c^2 = b^2 - a^2$. The graph of the standard equation is thus:

1. An ellipse with foci on the x-axis if $a > b$;
2. A circle of radius $r = a = b$ if $a = b$;
3. An ellipse with foci on the y-axis if $a < b$.

We consider that we have identified an ellipse if we have found its standard equation, and hence know a and b. The larger of a and b is (the length of) the *semimajor axis* of the ellipse, and the smaller is (the length of) its *semiminor axis*. We then know the location of the foci, as we noted above, and everything else follows.

EXAMPLE 1. The graph of

$$\frac{x^2}{25} + \frac{y^2}{9} = 1$$

is the ellipse with foci at $\pm\sqrt{25 - 9} = \pm 4$ on the x-axis, and intercepts $\pm a = \pm 5$ on the x-axis and $\pm b = \pm 3$ on the y-axis. However, in view of the symmetry about the axes, we normally record only the positive intercepts $a = 5, b = 3$.

EXAMPLE 2. Identify the graph of

$$x^2 + 4y^2 = 9.$$

. . . .

Solution 1. We throw the equation into the standard form

$$\frac{x^2}{9} + \frac{y^2}{9/4} = 1,$$

and then read off $a^2 = 9$, $b^2 = 9/4$. The graph is an ellipse with foci at

$$\pm c = \pm\sqrt{9 - \frac{9}{4}} = \pm\frac{3}{2}\sqrt{3}$$

on the x-axis, and positive intercepts $a = 3$, $b = 3/2$.

. . . .

Solution 2. We recognize that the graph is an ellipse symmetric in the coordinate axes, so we compute the intercepts $a = 3$, $b = 3/2$ directly from the equation. Since $a > b$, we conclude that the foci are on the x-axis with $c = \sqrt{a^2 - b^2}$ $= 3\sqrt{3}/2$.

EXAMPLE 3. The graph of $4x^2 + y^2 = 1$ is clearly going to be a central ellipse. Its positive intercepts are $a = 1/2$, $b = 1$; so its foci are on the y-axis and $c = \sqrt{1 - (1/2)^2} = \sqrt{3}/2$.

EXAMPLE 4. Compute the lengths of the focal radii to the point(s) on the ellipse

$$\frac{x^2}{4} + \frac{y^2}{2} = 1$$

having x-coordinate 1.

. . . .

Solution. We could solve for the y-coordinates and then use the distance formula, but the formulas for the focal radii are much simpler. We see that $a = 2$, $b = \sqrt{2}$, $c = \sqrt{4 - 2} = \sqrt{2}$, $e = c/a = \sqrt{2}/2$, and

$$\overline{PF} = a - ex = 2 - \frac{\sqrt{2}}{2} \cdot 1 = \frac{4 - \sqrt{2}}{2},$$

$$\overline{PF'} = a + e'x = 2 + \frac{\sqrt{2}}{2} \cdot 1 = \frac{4 + \sqrt{2}}{2}.$$

The standard equation for an ellipse centered at (h,k) and having axes parallel to the coordinate axes is

$$\frac{(x - h)^2}{a^2} + \frac{(y - k)^2}{b^2} = 1.$$

The reasoning goes exactly as it did for the parabola.

EXAMPLE 5. Find the equation of the ellipse whose foci are at $(1,-2)$ and $(5,-2)$, and with $2a = 6$.

. . . .

Solution. The center is midway between the foci, so $(h,k) = (3,-2)$ and $c = 2$. Since we are given that $a = 3$, it follows that $b = \sqrt{a^2 - c^2} = \sqrt{5}$. The standard equation of the ellipse is therefore

$$\frac{(x-3)^2}{9} + \frac{(y+2)^2}{5} = 1.$$

If we expand the squares and collect coefficients, the equation becomes

$$5x^2 + 9y^2 - 30x + 36y + 36 = 0.$$

EXAMPLE 6. Identify the graph of the equation

$$x^2 + 2y^2 + 4x - 4y = 0.$$

. . . .

Solution. After completing the squares the equation is

$$(x+2)^2 + 2(y-1)^2 = 6,$$

from which we get standard equation

$$\frac{(x+2)^2}{6} + \frac{(y-1)^2}{3} = 1.$$

Thus, the graph is an ellipse centered at $(-2,1)$ with $a = \sqrt{6}$ and $b = \sqrt{3}$. Then $c = \sqrt{6-3} = \sqrt{3}$, and since the major axis is parallel to the x-axis, the foci are at $(-2 - \sqrt{3}, 1)$ and $(-2 + \sqrt{3}, 1)$.

Consider again the formula for the right focal radius

$$PF = a - ex.$$

If we factor out e, the resulting expression $(a/e) - x$ can be interpreted as the distance from the point (x,y) to the vertical line $x = a/e$. That is, the distance from P to F is e times its distance from the line $x = a/e$. So:

Theorem 7. *An ellipse can be characterized as the locus traced by a point whose distance from a fixed point F (the focus) is a constant e times its distance from a fixed line (the directrix), the constant e being less than 1.*

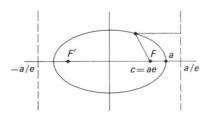

The ellipse $(x^2/a^2) + (y^2/b^2) = 1$ is also the focus–directrix locus defined by the left focus F' and a left directrix at $x = -a/e$. Many geometric properties of an ellipse involve its directrices and foci.

We shall now prove:

Theorem 8. *The tangent line to the ellipse*

$$\frac{x^2}{a^2} + \frac{y^2}{b^2} = 1$$

at the point (x_0, y_0) on the ellipse has the standard equation

$$\frac{x_0 x}{a^2} + \frac{y_0 y}{b^2} = 1.$$

Proof. Calculus is needed to compute the slope of the tangent line. Differentiating the equation of the ellipse implicitly (Chapter 4, Section 6), we get

$$\frac{2x}{a^2} + \frac{2y}{b^2}\frac{dy}{dx} = 0$$

and

$$\frac{dy}{dx} = -\frac{b^2 x}{a^2 y}.$$

The tangent line at (x_0, y_0) thus has the slope $-b^2 x_0/a^2 y_0$, and its point–slope equation is

$$y - y_0 = -\frac{b^2 x_0}{a^2 y_0}(x - x_0).$$

This simplifies to the standard equation when we make use of the fact that

$$\frac{x_0^2}{a^2} + \frac{y_0^2}{b^2} = 1. \quad \blacksquare$$

The optical property of the ellipse is stated in terms of the tangent line.

Theorem 9. *The focal radii to the point P on the ellipse make equal angles with the tangent line at P.*

Thus, if a point source of light is placed at one focus, say F', then all of its reflected rays will pass through the other focus F. The ellipse "focuses" the light on the point F, which is thus one of the two focal points of the optical system.

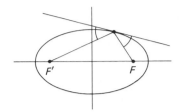

This focusing property of the ellipse explains a startling acoustical phenomenon that has been observed in some buildings. If a hall with walls made of stone, or some other acoustically reflecting material, is in the shape of an ellipse, then a person standing at one focus and speaking quietly can be heard with great clarity at the other focus, but not at points in between.

Proof of theorem. In order to prove the optical property, we make use of the focal radius formulas and the formula

$$d = \frac{|A x_1 + B y_1 + C|}{\sqrt{A^2 + B^2}}$$

for the distance from the point (x_1, y_1) to the line $ax + By + C = 0$ (Problem 37, in Section 4 of Chapter 1).

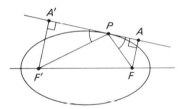

According to this formula, the distances from the foci to the tangent line $x x_0/a^2 + y y_0/b^2 = 1$ have the ratio

$$\frac{F'A'}{FA} = \frac{\left| \dfrac{-C x_0}{a^2} - 1 \right|}{\left| \dfrac{C x_0}{a^2} - 1 \right|}$$

$$= \frac{a^2 + C x_0}{a^2 - C x_0}$$

$$= \frac{a + e x_0}{a - e x_0},$$

which is the ratio $F'P/FP$ of the focal radii. The two right triangles are thus similar, and hence have equal angles at P. ∎

PROBLEMS FOR SECTION 2

In Problems 1 through 9, find the equation of the ellipse determined by the given conditions.

1.   The foci are at $(-1,0)$ and $(1,0)$, and $2a = 3$.

2.   The foci are at $(0,-1)$ and $(0,1)$, and $2b = 3$.

3.   The foci are at $(-1,0)$ and $(1,0)$, and $b = 1$.

4.   The foci are at $(-1,0)$ and $(1,0)$, and the eccentricity $(e = c/a)$ is $1/2$.

5.   The foci are at $(0,1)$ and $(0,3)$, and $e = 1/2$.

6.   The foci are at $(0,1)$ and $(4,1)$, and $e = 3/4$.

7.   The intercepts on the major and minor axes are $(-1,1)$, $(3,1)$, $(1,2)$, and $(1,0)$.

8.   The foci are $(-1,0)$, and $(1,0)$, and the line $x = 2$ is a directrix.

9.   The center is the origin, one focus is at $(0,2)$, and the line $y = 3$ is a directrix.

10.   Find the foci of the ellipse $x^2 + 4x + 2y^2 = 0$.

11.   Same question for $2x^2 + 4x + y^2 = 0$.

12.   Find the equation of the ellipse having the y-axis as a directrix, the point $(1,0)$ as the corresponding focus, and eccentricity $1/2$.

13.   Same question for the point $(1,0)$, line $x = 3$, and $e = 2/3$.

14.   Let P be a point on the ellipse $(x^2/a^2) + (y^2/b^2) = 1$ that is not an intercept. Supposing that $b < a$, show that the distance from P to the origin is greater than b and less than a. (This does not require calculus.)

15.   Show that if $a^2 = b^2 + c^2$ and if the point (x,y) satisfies the equation $(x^2/a^2) + (y^2/b^2) = 1$, then the distance from (x,y) to $(c,0)$ is $(a - ex)$, and the distance from (x,y) to $(-c,0)$ is $(a + ex)$, where $e = c/a$. That is, work out the algebra for Theorem 5.

16.   Complete the proof of Theorem 8 by supplying the missing algebraic calculation at the end of the proof.

17.   Let k be a fixed positive constant. For each value of a greater k, the equation

$$\frac{x^2}{a^2} + \frac{y^2}{a^2 - k^2} = 1$$

represents an ellipse; so, regarding a as a parameter, we thus have a one-parameter family of ellipses. Show that this is a *confocal* one-parameter family; that is, show that all ellipses of this form have the same foci.

18.   The tangent to an ellipse at a point P intersects a directrix at the point D. Prove that the segment PD subtends a right angle at the corresponding focus.

19. Show that the product of the distances from the foci of an ellipse to a tangent is a constant, independent of the tangent.

20. The *vertices* of an ellipse are its intersections with its major axis. The lines from the vertices A' and A of an ellipse, through a point P on the ellipse, meet a directrix at D' and D. Show that the segment $D'D$ subtends a right angle at the corresponding focus.

21. A normal to an ellipse at a point P has intercepts Q_1 and Q_2 on the axes. Show that the product of the lengths of the segments PQ_1 and PQ_2 equals the product of the focal radii to P.

22. The tangent to an ellipse at a point P intersects the tangent at a vertex at the point Q. Show that the line through Q and the other vertex is parallel to the line through P and the center.

23. Show that every ellipse of the form

$$x^2 + 2y^2 = C$$

is orthogonal to every parabola of the form

$$y = kx^2.$$

(Let (x_0, y_0) be a point of intersection. Compute the two slopes at this point, and show that their product is -1.)

3. THE HYPERBOLA

Definition. *A hyperbola is the locus traced by a point P whose distances from two fixed points F and F' have a constant positive difference k, where k is less than the distance between F' and F.*

The locus condition is thus

$$PF' - PF = k \qquad \text{or} \qquad PF - PF' = k,$$

that is,

$$PF' = PF \pm k.$$

The requirement $k < \overline{F'F}$ is necessary if there are to be locus points P that are not on the line through F and F', because any such point P satisfies triangle inequalities like $\overline{PF'} < \overline{PF} + \overline{FF'}$, so

$$k = \overline{PF'} - \overline{PF} < \overline{FF'}.$$

A little thought (or experimentation) will show that the locus has the general features shown on the following page.

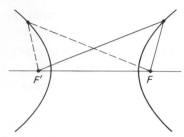

It has two branches, each of which cuts between F' and F. The right branch, bending around F, is the locus of the equation $PF' - PF = k$, while the left branch, bending around F', is the locus of the equation $PF - PF' = k$.

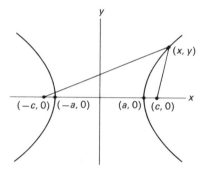

If we take the x-axis along the segment $F'F$ and the origin at its midpoint, then the coordinates of F and F' are $F = (c, 0)$ and $F' = (-c, 0)$, where c is one-half the segment length. If a is the positive x-intercept of the locus, then $a < c$, and for this point the locus condition becomes

$$(a + c) - (c - a) = k.$$

So $k = 2a$, and the locus equation is

$$\sqrt{(x + c)^2 + y^2} = \sqrt{(x - c)^2 + y^2} \pm 2a.$$

The procedure now is exactly as it was in the case of the ellipse. The radicals are eliminated by twice squaring the equation and simplifying.

With $e = c/a$ as before, the first of these two steps yields the focal radius formulas

$$PF = \pm(ex - a),$$

$$PF' = \pm(ex + a),$$

where the plus signs are used for the right branch $(x > 0)$ and the minus sign for the left branch $(x < 0)$. Here the eccentricity e is greater than 1, since $e = c/a$ and $c > a$.

The second step reduces the equation

$$\sqrt{(x-c)^2 + y^2} = \pm\left(\frac{c}{a}x - a\right)$$

to

$$\frac{x^2}{a^2} - \frac{y^2}{b^2} = 1,$$

where $b^2 = c^2 - a^2$. This is the standard equation of the hyperbola.

It can be checked, conversely, that the graph of any equation of this form is a hyperbola with foci on the x-axis. Setting $c = \sqrt{a^2 + b^2}$, the graph is the hyperbola with foci at $F = (c, 0)$ and $F' = (-c, 0)$, and with $k = 2a$.

Certain aspects of the graph can be read directly from the standard equation $(x^2/a^2) - (y^2/b^2) = 1$. It is symmetric with respect to both coordinate axes. It has x-intercepts at $\pm a$, but no y-intercepts. Moreover, $x^2 \geq a^2$ on the graph. This is because the y^2 term is negative, so x^2/a^2 is at least 1. The graph thus consists of two separate pieces (branches), a part to the right of $x = a$ and a part to the left of $x = -a$.

EXAMPLE 1. The hyperbola with foci at $(2,0)$ and $(-2,0)$ and with $k = 2a = 2$ has the equation

$$x^2 - \frac{y^2}{3} = 1$$

because the given data are $a = 1$, $c = 2$, and hence $b = \sqrt{c^2 - a^2} = \sqrt{4-1} = \sqrt{3}$.

EXAMPLE 2. For the hyperbola

$$\frac{x^2}{16} - \frac{y^2}{9} = 1$$

we have $a^2 = 16$ and $b^2 = 9$, so $c^2 = a^2 + b^2 = 25$, and the foci are at $(c, 0) = (5, 0)$ and $(-c, 0) = (-5, 0)$. Its eccentricity e is $5/4$ (since $e = c/a$).

EXAMPLE 3. Find the equation of the hyperbola with foci on the x-axis, x-intercepts at ± 2, and eccentricity $3/2$.

. . . .

Solution. We are given that $a = 2$ and $e = c/a = 3/2$. Therefore, $c = 3$ and $b^2 = c^2 - a^2 = 9 - 4 = 5$. The equation is, therefore,

$$\frac{x^2}{4} - \frac{y^2}{5} = 1.$$

The branching of the hyperbola

$$\frac{x^2}{a^2} - \frac{y^2}{b^2} = 1$$

is related to another distinctive feature of the graph; it is asymptotic to each of the straight lines

$$y = \left(\frac{b}{a}\right)x, \qquad y = -\left(\frac{b}{a}\right)x,$$

in the manner shown in the figure.

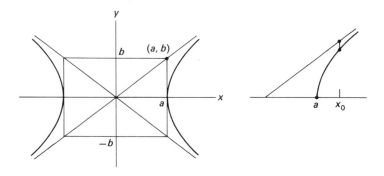

For example, the top of the right branch, whose equation is

$$y = \frac{b}{a}\sqrt{x^2 - a^2},$$

is asymptotic to the line

$$y = \left(\frac{b}{a}\right)x;$$

the two graphs come arbitrarily close together as they run out "to infinity." In order to see this, consider the vertical distance from the hyperbola to the line at a given value of x. It is

$$\frac{b}{a}x - \frac{b}{a}\sqrt{x^2 - a^2} = \frac{b}{a}(x - \sqrt{x^2 - a^2})$$

$$= \frac{b}{a}\frac{(x - \sqrt{x^2 - a^2})(x + \sqrt{x^2 - a^2})}{x + \sqrt{x^2 - a^2}}$$

$$= \frac{ab}{x + \sqrt{x^2 - a^2}},$$

and this is less than ab/x for all x larger than a. The vertical distance between the curves thus approaches zero as x tends to infinity. That is, the curves are asymptotic to each other as $x \to \infty$.

If we interchange the roles of x and y in the preceding discussion, we see that a hyperbola with foci at $(0, c)$ and $(0, -c)$ on the y-axis has the standard equation

$$\frac{y^2}{b^2} - \frac{x^2}{a^2} = 1$$

with $c^2 = a^2 + b^2$.

Note that *the axis containing the foci is not determined by the relative size of a and b*, as it was in the case of the ellipse, *but rather by where the minus sign occurs*. So a and b can be any relative size, and in particular they can be equal. In this case, the hyperbola is called *rectangular* or *equilateral*. Thus, an equilateral hyperbola has the equation

$$x^2 - y^2 = a^2 \qquad \text{or} \qquad y^2 - x^2 = a^2.$$

For given a and b, the equation

$$\frac{x^2}{a^2} - \frac{y^2}{b^2} = \pm 1$$

represents a pair of hyperbolas, one with foci on the x-axis and one with foci on the y-axis. Each of these hyperbolas is said to be the *conjugate* of the other. A pair of conjugate hyperbolas shares the same asymptotes; and they fit together to enclose the origin almost as though they formed a single closed curve.

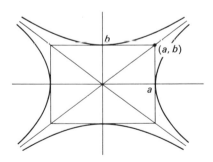

EXAMPLE 4. The hyperbola

$$x^2 - \frac{y^2}{4} = 1$$

has the asymptotes $y = 2x$ and $y = -2x$, because from the standard equation above we see that $a^2 = 1$ and $b^2 = 4$, so the asymptotes $y = \pm(b/a)x$ are the lines $y = \pm 2x$. The conjugate of the above hyperbola is the hyperbola

$$-x^2 + \frac{y^2}{4} = 1.$$

EXAMPLE 5. A hyperbola has asymptotes $y = \pm x$ and passes through the point (2,1). Find its equation.

. . . .

Solution. Since the asymptote $y = (b/a)x$ is the line $y = x$, we have $b/a = 1$ and $a = b$, so the hyperbola is equilateral. Since the hyperbola contains a point (namely (2,1)) "to the right of" the lines $y = \pm x$, its branches are right and left rather than up and down, and its equation has the form

$$\frac{x^2}{a^2} - \frac{y^2}{a^2} = 1.$$

And, since it contains (2,1),

$$\frac{4}{a^2} - \frac{1}{a^2} = 1.$$

So $a = \sqrt{3}$, and the hyperbola has the equation

$$\frac{x^2}{3} - \frac{y^2}{3} = 1,$$

or

$$x^2 - y^2 = 3.$$

PROBLEMS FOR SECTION 3

Find the foci, eccentricity and asymptotes for each of the following hyperbolas.

1. $x^2 - y^2 = 1$.   2. $y^2 - x^2 = 1$.   3. $x^2 - y^2 = 2$.
4. $x^2 - \frac{y^2}{3} = 3$.   5. $3y^2 - x^2 = 3$.   6. $y^2 - 4x^2 = 1$.
7. $2x^2 - 3y^2 = 6$.

Find the equation of the hyperbola for which:

8. The foci are at $(\pm 3, 0)$ and the eccentricity is $3/2$.

9. It is symmetric about the coordinate axes; its x-intercepts are $(\pm 1, 0)$ and its eccentricity is 2.

10. Its foci are at $(\pm, 3, 0)$ and its x-intercepts are $(\pm 2, 0)$.

11. Its foci are at $(0, \pm 3)$ and $a = 2$.

12. Its x-intercepts are $(\pm 1, 0)$ and it contains the two points $(3, \pm 1)$.

Find the equation of the hyperbola with asymptotes $y = \pm 2x$ if:

13. It passes through $(2,0)$.

14. It has y-intercepts $(0, \pm 2)$.

15. Its foci are $(\pm 5, 0)$.

16. It passes through the point $(2,2)$.

17. It passes through the point $(1,1)$.

18. It passes through the point $(1,3)$.

19. Two hyperbolas have the same asymptotes, one with foci on the x-axis and the other with foci on the y-axis. If their eccentrities are e and f, show that

$$\frac{1}{e^2} + \frac{1}{f^2} = 1.$$

20. Carry out the algebra (completing the square twice and simplifying each time) leading from the locus definition of the hyperbola to its standard equation.

21. Show by algebraic computation that if

$$\frac{x^2}{a^2} - \frac{y^2}{b^2} = 1$$

and if $c^2 = a^2 + b^2$, then the distance from (x,y) to $(c,0)$ is $|ex - a|$, where $e = c/a$.

22. Show that the hyperbola

$$\frac{x^2}{2} - \frac{y^2}{2} = 1$$

is the locus of a point whose distance from $(2,0)$ is $\sqrt{2}$ times its distance from line $x = 1$.

23. Show that the hyperbola

$$\frac{x^2}{a^2} - \frac{y^2}{b^2} = 1$$

is the locus of a point whose distance from the point $(c,0)$ is e times its distance from the line $x = a/e$, where $c = \sqrt{a^2 + b^2}$ and $e = c/a$. [*Hint.* Use the focal radius formulas from the text discussion.]

24. Show that the hyperbola

$$\frac{x^2}{a^2} - \frac{y^2}{b^2} = 1$$

is the locus of a point whose distance from the point $(-c, 0)$ is e times its distance from the line $x = -a/e$.

25. Show that a line through a focus perpendicular to an asymptote intersects the asymptote at a point on a directrix.

26. A line through a point P on a hyperbola parallel to an asymptote intersects a directrix at Q. If F is the corresponding focus, show that the triangle FPQ is isosceles.

27. Prove that the tangent line to the hyperbola

$$\frac{x^2}{a^2} - \frac{y^2}{b^2} = 1$$

at the point (x_0, y_0) has the (standard) equation

$$\frac{x x_0}{a^2} - \frac{y y_0}{b^2} = 1.$$

28. *Optical property of the hyperbola.* Prove that the focal radii to a point P on a hyperbola make equal angles with the tangent line at P. (Follow the proof of the optical property of the ellipse, using the above equation of the tangent line.)

29. Prove that each hyperbola having foci F' and F intersects every ellipse having the same foci at right angles. [*Hint.* combine the optical properties of the ellipse and hyperbola.]

30. Illustrate the above problem by taking the foci at $(\pm 3, 0)$ and sketching the hyperbolas whose eccentricities are 3 and 3/2 and the ellipse whose eccentricities are 3/4 and 1/2.

31. Show that the product of the distances from the foci of a hyperbola to the tangent at a point P on the hyperbola is constant; i.e., independent of P.

32. Show that the product of the focal radii to a point P on an equilateral hyperbola is equal to the square of the distance from P to the center of the hyperbola.

33. Show that the product of the distances from a point P on a hyperbola to the asymptotes is constant; i.e., independent of P.

34. The tangent to a hyperbola at a point P intersects the asymptotes at Q_1 and Q_2. Show that P is the midpoint of the segment $Q_1 Q_2$.

tables

Natural logarithms of numbers

n	$\log_e n$	n	$\log_e n$	n	$\log_e n$
0.0	*	4.5	1.5041	9.0	2.1972
0.1	7.6974	4.6	1.5261	9.1	2.2083
0.2	8.3906	4.7	1.5476	9.2	2.2192
0.3	8.7960	4.8	1.5686	9.3	2.2300
0.4	9.0837	4.9	1.5892	9.4	2.2407
0.5	9.3069	5.0	1.6094	9.5	2.2513
0.6	9.4892	5.1	1.6292	9.6	2.2618
0.7	9.6433	5.2	1.6487	9.7	2.2721
0.8	9.7769	5.3	1.6677	9.8	2.2824
0.9	9.8946	5.4	1.6864	9.9	2.2925
1.0	0.0000	5.5	1.7047	10	2.3026
1.1	0.0953	5.6	1.7228	11	2.3979
1.2	0.1823	5.7	1.7405	12	2.4849
1.3	0.2624	5.8	1.7579	13	2.5649
1.4	0.3365	5.9	1.7750	14	2.6391
1.5	0.4055	6.0	1.7918	15	2.7081
1.6	0.4700	6.1	1.8083	16	2.7726
1.7	0.5306	6.2	1.8245	17	2.8332
1.8	0.5878	6.3	1.8405	18	2.8904
1.9	0.6419	6.4	1.8563	19	2.9444
2.0	0.6931	6.5	1.8718	20	2.9957
2.1	0.7419	6.6	1.8871	25	3.2189
2.2	0.7885	6.7	1.9021	30	3.4012
2.3	0.8329	6.8	1.9169	35	3.5553
2.4	0.8755	6.9	1.9315	40	3.6889
2.5	0.9163	7.0	1.9459	45	3.8067
2.6	0.9555	7.1	1.9601	50	3.9120
2.7	0.9933	7.2	1.9741	55	4.0073
2.8	1.0296	7.3	1.9879	60	4.0943
2.9	1.0647	7.4	2.0015	65	4.1744
3.0	1.0986	7.5	2.0149	70	4.2485
3.1	1.1314	7.6	2.0281	75	4.3175
3.2	1.1632	7.7	2.0412	80	4.3820
3.3	1.1939	7.8	2.0541	85	4.4427
3.4	1.2238	7.9	2.0669	90	4.4998
3.5	1.2528	8.0	2.0794	95	4.5539
3.6	1.2809	8.1	2.0919	100	4.6052
3.7	1.3083	8.2	2.1041		
3.8	1.3350	8.3	2.1163		
3.9	1.3610	8.4	2.1282		
4.0	1.3863	8.5	2.1401		
4.1	1.4110	8.6	2.1518		
4.2	1.4351	8.7	2.1633		
4.3	1.4586	8.8	2.1748		
4.4	1.4816	8.9	2.1861		

Exponential functions

x	e^x	e^{-x}	x	e^x	e^{-x}
0.00	1.0000	1.0000	2.5	12.182	0.0821
0.05	1.0513	0.9512	2.6	13.464	0.0743
0.10	1.1052	0.9048	2.7	14.880	0.0672
0.15	1.1618	0.8607	2.8	16.445	0.0608
0.20	1.2214	0.8187	2.9	18.174	0.0550
0.25	1.2840	0.7788	3.0	20.086	0.0498
0.30	1.3499	0.7408	3.1	22.198	0.0450
0.35	1.4191	0.7047	3.2	24.533	0.0408
0.40	1.4918	0.6703	3.3	27.113	0.0369
0.45	1.5683	0.6376	3.4	29.964	0.0334
0.50	1.6487	0.6065	3.5	33.115	0.0302
0.55	1.7333	0.5769	3.6	36.598	0.0273
0.60	1.8221	0.5488	3.7	40.447	0.0247
0.65	1.9155	0.5220	3.8	44.701	0.0224
0.70	2.0138	0.4966	3.9	49.402	0.0202
0.75	2.1170	0.4724	4.0	54.598	0.0183
0.80	2.2255	0.4493	4.1	60.340	0.0166
0.85	2.3396	0.4274	4.2	66.686	0.0150
0.90	2.4596	0.4066	4.3	73.700	0.0136
0.95	2.5857	0.3867	4.4	81.451	0.0123
1.0	2.7183	0.3679	4.5	90.017	0.0111
1.1	3.0042	0.3329	4.6	99.484	0.0101
1.2	3.3201	0.3012	4.7	109.95	0.0091
1.3	3.6693	0.2725	4.8	121.51	0.0082
1.4	4.0552	0.2466	4.9	134.29	0.0074
1.5	4.4817	0.2231	5	148.41	0.0067
1.6	4.9530	0.2019	6	403.43	0.0025
1.7	5.4739	0.1827	7	1096.6	0.0009
1.8	6.0496	0.1653	8	2981.0	0.0003
1.9	6.6859	0.1496	9	8103.1	0.0001
2.0	7.3891	0.1353	10	22026	0.00005
2.1	8.1662	0.1225			
2.2	9.0250	0.1108			
2.3	9.9742	0.1003			
2.4	11.023	0.0907			

Natural trigonometric functions

Angle					Angle				
De-gree	Ra-dian	Sine	Co-sine	Tan-gent	De-gree	Ra-dian	Sine	Co-sine	Tan-gent
0°	0.000	0.000	1.000	0.000					
1°	0.017	0.017	1.000	0.017	46°	0.803	0.719	0.695	1.036
2°	0.035	0.035	0.999	0.035	47°	0.820	0.731	0.682	1.072
3°	0.052	0.052	0.999	0.052	48°	0.838	0.743	0.669	1.111
4°	0.070	0.070	0.998	0.070	49°	0.855	0.755	0.656	1.150
5°	0.087	0.087	0.996	0.087	50°	0.873	0.766	0.643	1.192
6°	0.105	0.105	0.995	0.105	51°	0.890	0.777	0.629	1.235
7°	0.122	0.122	0.993	0.123	52°	0.908	0.788	0.616	1.280
8°	0.140	0.139	0.990	0.141	53°	0.925	0.799	0.602	1.327
9°	0.157	0.156	0.988	0.158	54°	0.942	0.809	0.588	1.376
10°	0.175	0.174	0.985	0.176	55°	0.960	0.819	0.574	1.428
11°	0.192	0.191	0.982	0.194	56°	0.977	0.829	0.559	1.483
12°	0.209	0.208	0.978	0.213	57°	0.995	0.839	0.545	1.540
13°	0.227	0.225	0.974	0.231	58°	1.012	0.848	0.530	1.600
14°	0.244	0.242	0.970	0.249	59°	1.030	0.857	0.515	1.664
15°	0.262	0.259	0.966	0.268	60°	1.047	0.866	0.500	1.732
16°	0.279	0.276	0.961	0.287	61°	1.065	0.875	0.485	1.804
17°	0.297	0.292	0.956	0.306	62°	1.082	0.883	0.469	1.881
18°	0.314	0.309	0.951	0.325	63°	1.100	0.891	0.454	1.963
19°	0.332	0.326	0.946	0.344	64°	1.117	0.899	0.438	2.050
20°	0.349	0.342	0.940	0.364	65°	1.134	0.906	0.423	2.145
21°	0.367	0.358	0.934	0.384	66°	1.152	0.914	0.407	2.246
22°	0.384	0.375	0.927	0.404	67°	1.169	0.921	0.391	2.356
23°	0.401	0.391	0.921	0.424	68°	1.187	0.927	0.375	2.475
24°	0.419	0.407	0.914	0.445	69°	1.204	0.934	0.358	2.605
25°	0.436	0.423	0.906	0.466	70°	1.222	0.940	0.342	2.748
26°	0.454	0.438	0.899	0.488	71°	1.239	0.946	0.326	2.904
27°	0.471	0.454	0.891	0.510	72°	1.257	0.951	0.309	3.078
28°	0.489	0.469	0.883	0.532	73°	1.274	0.956	0.292	3.271
29°	0.506	0.485	0.875	0.554	74°	1.292	0.961	0.276	3.487
30°	0.524	0.500	0.866	0.577	75°	1.309	0.966	0.259	3.732
31°	0.541	0.515	0.857	0.601	76°	1.326	0.970	0.242	4.011
32°	0.559	0.530	0.848	0.625	77°	1.344	0.974	0.225	4.332
33°	0.576	0.545	0.839	0.649	78°	1.361	0.978	0.208	4.705
34°	0.593	0.559	0.829	0.675	79°	1.379	0.982	0.191	5.145
35°	0.611	0.574	0.819	0.700	80°	1.396	0.985	0.174	5.671
36°	0.628	0.588	0.809	0.727	81°	1.414	0.988	0.156	6.314
37°	0.646	0.602	0.799	0.754	82°	1.431	0.990	0.139	7.115
38°	0.663	0.616	0.788	0.781	83°	1.449	0.993	0.122	8.144
39°	0.681	0.629	0.777	0.810	84°	1.466	0.995	0.105	9.514
40°	0.698	0.643	0.766	0.839	85°	1.484	0.996	0.087	11.43
41°	0.716	0.656	0.755	0.869	86°	1.501	0.998	0.070	14.30
42°	0.733	0.669	0.743	0.900	87°	1.518	0.999	0.052	19.08
43°	0.750	0.682	0.731	0.933	88°	1.536	0.999	0.035	28.64
44°	0.768	0.695	0.719	0.966	89°	1.553	1.000	0.017	57.29
45°	0.785	0.707	0.707	1.000	90°	1.571	1.000	0.000	

Powers and roots

No.	Sq.	Sq. Root	Cube	Cube Root	No.	Sq.	Sq. Root	Cube	Cube Root
1	1	1.000	1	1.000	51	2,601	7.141	132,651	3.708
2	4	1.414	8	1.260	52	2,704	7.211	140,608	3.733
3	9	1.732	27	1.442	53	2,809	7.280	148,877	3.756
4	16	2.000	64	1.587	54	2,916	7.348	157,464	3.780
5	25	2.236	125	1.710	55	3,025	7.416	166,375	3.803
6	36	2.449	216	1.817	56	3,136	7.483	175,616	3.826
7	49	2.646	343	1.913	57	3,249	7.550	185,193	3.849
8	64	2.828	512	2.000	58	3,364	7.616	195,112	3.871
9	81	3.000	729	2.080	59	3,481	7.681	205,379	3.893
10	100	3.162	1,000	2.154	60	3,600	7.746	216,000	3.915
11	121	3.317	1,331	2.224	61	3,721	7.810	226,981	3.936
12	144	3.464	1,728	2.289	62	3,844	7.874	238,328	3.958
13	169	3.606	2,197	2.351	63	3,969	7.937	250,047	3.979
14	196	3.742	2,744	2.410	64	4,096	8.000	262,144	4.000
15	225	3.873	3,375	2.466	65	4,225	8.062	274,625	4.021
16	256	4.000	4,096	2.520	66	4,356	8.124	287,496	4.041
17	289	4.123	4,913	2.571	67	4,489	8.185	300,763	4.062
18	324	4.243	5,832	2.621	68	4,624	8.246	314,432	4.082
19	361	4.359	6,859	2.668	69	4,761	8.307	328,509	4.102
20	400	4.472	8,000	2.714	70	4,900	8.367	343,000	4.121
21	441	4.583	9,261	2.759	71	5,041	8.426	357,911	4.141
22	484	4.690	10,648	2.802	72	5,184	8.485	373,248	4.160
23	529	4.796	12,167	2.844	73	5,329	8.544	389,017	4.179
24	576	4.899	13,824	2.884	74	5,476	8.602	405,224	4.198
25	625	5.000	15,625	2.924	75	5,625	8.660	421,875	4.217
26	676	5.099	17,576	2.962	76	5,776	8.718	438,976	4.236
27	729	5.196	19,683	3.000	77	5,929	8.775	456,533	4.254
28	784	5.292	21,952	3.037	78	6,084	8.832	474,552	4.273
29	841	5.385	24,389	3.072	79	6,241	8.888	493,039	4.291
30	900	5.477	27,000	3.107	80	6,400	8.944	512,000	4.309
31	961	5.568	29,791	3.141	81	6,561	9.000	531,441	4.327
32	1,024	5.657	32,768	3.175	82	6,724	9.055	551,368	4.344
33	1,089	5.745	35,937	3.208	83	6,889	9.110	571,787	4.362
34	1,156	5.831	39,304	3.240	84	7,056	9.165	592,704	4.380
35	1,225	5.916	42,875	3.271	85	7,225	9.220	614,125	4.397
36	1,296	6.000	46,656	3.302	86	7,396	9.274	636,056	4.414
37	1,369	6.083	50,653	3.332	87	7,569	9.327	658,503	4.431
38	1,444	6.164	54,872	3.362	88	7,744	9.381	681,472	4.448
39	1,521	6.245	59,319	3.391	89	7,921	9.434	704,969	4.465
40	1,600	6.325	64,000	3.420	90	8,100	9.487	729,000	4.481
41	1,681	6.403	68,921	3.448	91	8,281	9.539	753,571	4.498
42	1,764	6.481	74,088	3.476	92	8,464	9.592	778,688	4.514
43	1,849	6.557	79,507	3.503	93	8,649	9.644	804,357	4.531
44	1,936	6.633	85,184	3.530	94	8,836	9.695	830,584	4.547
45	2,025	6.708	91,125	3.557	95	9,025	9.747	857,375	4.563
46	2,116	6.782	97,336	3.583	96	9,216	9.798	884,736	4.579
47	2,209	6.856	103,823	3.609	97	9,409	9.849	912,673	4.595
48	2,304	6.928	110,592	3.634	98	9,604	9.899	941,192	4.610
49	2,401	7.000	117,649	3.659	99	9,801	9.950	970,299	4.626
50	2,500	7.071	125,000	3.684	100	10,000	10.000	1,000,000	4.642

Table of integrals

1. $\int u\,dv = uv - \int v\,du$

2. $\int a^u\,du = \dfrac{a^u}{\ln a} + C, \qquad a \neq 1, \qquad a > 0$

3. $\int \cos u\,du = \sin u + C$

4. $\int \sin u\,du = -\cos u + C$

5. $\int (ax+b)^n\,dx = \dfrac{(ax+b)^{n+1}}{a(n+1)} + C, \qquad n \neq -1$

6. $\int (ax+b)^{-1}\,dx = \dfrac{1}{a}\ln|ax+b| + C$

7. $\int x(ax+b)^n\,dx = \dfrac{(ax+b)^{n+1}}{a^2}\left[\dfrac{ax+b}{n+2} - \dfrac{b}{n+1}\right] + C, \qquad n \neq -1, -2$

8. $\int x(ax+b)^{-1}\,dx = \dfrac{x}{a} - \dfrac{b}{a^2}\ln|ax+b| + C$

9. $\int x(ax+b)^{-2}\,dx = \dfrac{1}{a^2}\left[\ln|ax+b| + \dfrac{b}{ax+b}\right] + C$

10. $\int \dfrac{dx}{x(ax+b)} = \dfrac{1}{b}\ln\left|\dfrac{x}{ax+b}\right| + C$

11. $\int (\sqrt{ax+b})^n\,dx = \dfrac{2}{a}\dfrac{(\sqrt{ax+b})^{n+2}}{n+2} + C, \qquad n \neq -2$

12. $\int \dfrac{\sqrt{ax+b}}{x}\,dx = 2\sqrt{ax+b} + b\int \dfrac{dx}{x\sqrt{ax+b}}$

13. (a) $\int \dfrac{dx}{x\sqrt{ax+b}} = \dfrac{2}{\sqrt{-b}}\tan^{-1}\sqrt{\dfrac{ax+b}{-b}} + C, \qquad$ if $\ b < 0$

  (b) $\int \dfrac{dx}{x\sqrt{ax+b}} = \dfrac{1}{\sqrt{b}}\ln\left|\dfrac{\sqrt{ax+b}-\sqrt{b}}{\sqrt{ax+b}+\sqrt{b}}\right| + C, \qquad$ if $\ b > 0$

14. $\int \dfrac{\sqrt{ax+b}}{x^2}\,dx = -\dfrac{\sqrt{ax+b}}{x} + \dfrac{a}{2}\int \dfrac{dx}{x\sqrt{ax+b}} + C$

15. $\int \dfrac{dx}{x^2\sqrt{ax+b}} = -\dfrac{\sqrt{ax+b}}{bx} - \dfrac{a}{2b}\int \dfrac{dx}{x\sqrt{ax+b}} + C$

16. $\int \dfrac{dx}{a^2+x^2} = \dfrac{1}{a}\tan^{-1}\dfrac{x}{a} + C$

17. $\int \dfrac{dx}{(a^2+x^2)^2} = \dfrac{x}{2a^2(a^2+x^2)} + \dfrac{1}{2a^3}\tan^{-1}\dfrac{x}{a} + C$

18. $\int \dfrac{dx}{a^2-x^2} = \dfrac{1}{2a}\ln\left|\dfrac{x+a}{x-a}\right| + C$

19. $\int \dfrac{dx}{(a^2-x^2)^2} = \dfrac{x}{2a^2(a^2-x^2)} + \dfrac{1}{2a^2}\int \dfrac{dx}{a^2-x^2}$

20. $\int \dfrac{dx}{\sqrt{a^2+x^2}} = \sinh^{-1}\dfrac{x}{a} + C = \ln|x+\sqrt{a^2+x^2}| + C$

21. $\displaystyle\int \sqrt{a^2 + x^2}\, dx = \frac{x}{2}\sqrt{a^2 + x^2} + \frac{a^2}{2}\sinh^{-1}\frac{x}{a} + C$

22. $\displaystyle\int x^2\sqrt{a^2 + x^2}\, dx = \frac{x(a^2 + 2x^2)\sqrt{a^2 + x^2}}{8} - \frac{a^4}{8}\sinh^{-1}\frac{x}{a} + C$

23. $\displaystyle\int \frac{\sqrt{a^2 + x^2}}{x}\, dx = \sqrt{a^2 + x^2} - a\sinh^{-1}\left|\frac{a}{x}\right| + C$

24. $\displaystyle\int \frac{\sqrt{a^2 + x^2}}{x^2}\, dx = \sinh^{-1}\frac{x}{a} - \frac{\sqrt{a^2 + x^2}}{x} + C$

25. $\displaystyle\int \frac{x^2}{\sqrt{a^2 + x^2}}\, dx = -\frac{a^2}{2}\sinh^{-1}\frac{x}{a} + \frac{x\sqrt{a^2 + x^2}}{2} + C$

26. $\displaystyle\int \frac{dx}{x\sqrt{a^2 + x^2}} = -\frac{1}{a}\ln\left|\frac{a + \sqrt{a^2 + x^2}}{x}\right| + C$

27. $\displaystyle\int \frac{dx}{x^2\sqrt{a^2 + x^2}} = -\frac{\sqrt{a^2 + x^2}}{a^2 x} + C$
    $\qquad$ 28. $\displaystyle\int \frac{dx}{\sqrt{a^2 - x^2}} = \sin^{-1}\frac{x}{a} + C$

29. $\displaystyle\int \sqrt{a^2 - x^2}\, dx = \frac{x}{2}\sqrt{a^2 - x^2} + \frac{a^2}{2}\sin^{-1}\frac{x}{a} + C$

30. $\displaystyle\int x^2\sqrt{a^2 - x^2}\, dx = \frac{a^4}{8}\sin^{-1}\frac{x}{a} - \frac{1}{8}x\sqrt{a^2 - x^2}\,(a^2 - 2x^2) + C$

31. $\displaystyle\int \frac{\sqrt{a^2 - x^2}}{x}\, dx = \sqrt{a^2 - x^2} - a\ln\left|\frac{a + \sqrt{a^2 - x^2}}{x}\right| + C$

32. $\displaystyle\int \frac{\sqrt{a^2 - x^2}}{x^2}\, dx = -\sin^{-1}\frac{x}{a} - \frac{\sqrt{a^2 - x^2}}{x} + C$

33. $\displaystyle\int \frac{x^2}{\sqrt{a^2 - x^2}}\, dx = \frac{a^2}{2}\sin^{-1}\frac{x}{a} - \frac{1}{2}x\sqrt{a^2 - x^2} + C$

34. $\displaystyle\int \frac{dx}{x\sqrt{a^2 - x^2}} = -\frac{1}{a}\ln\left|\frac{a + \sqrt{a^2 - x^2}}{x}\right| + C$
    $\qquad$ 35. $\displaystyle\int \frac{dx}{x^2\sqrt{a^2 - x^2}} = -\frac{\sqrt{a^2 - x^2}}{a^2 x} + C$

36. $\displaystyle\int \frac{dx}{\sqrt{x^2 - a^2}} = \cosh^{-1}\frac{x}{a} + C = \ln\left|x + \sqrt{x^2 - a^2}\right| + C$

37. $\displaystyle\int \sqrt{x^2 - a^2}\, dx = \frac{x}{2}\sqrt{x^2 - a^2} - \frac{a^2}{2}\cosh^{-1}\frac{x}{a} + C$

38. $\displaystyle\int \left(\sqrt{x^2 - a^2}\right)^n dx = \frac{x\left(\sqrt{x^2 - a^2}\right)^n}{n + 1} - \frac{na^2}{n + 1}\int \left(\sqrt{x^2 - a^2}\right)^{n-2} dx, \qquad n \neq -1$

39. $\displaystyle\int \frac{dx}{\left(\sqrt{x^2 - a^2}\right)^n} = \frac{x\left(\sqrt{x^2 - a^2}\right)^{2-n}}{(2 - n)a^2} - \frac{n - 3}{(n - 2)a^2}\int \frac{dx}{\left(\sqrt{x^2 - a^2}\right)^{n-2}}, \qquad n \neq 2$

40. $\displaystyle\int x\left(\sqrt{x^2 - a^2}\right)^n dx = \frac{\left(\sqrt{x^2 - a^2}\right)^{n+2}}{n + 2} + C, \qquad n \neq -2$

41. $\displaystyle\int x^2\sqrt{x^2 - a^2}\, dx = \frac{x}{8}(2x^2 - a^2)\sqrt{x^2 - a^2} - \frac{a^4}{8}\cosh^{-1}\frac{x}{a} + C$

42. $\displaystyle\int \frac{\sqrt{x^2 - a^2}}{x}\, dx = \sqrt{x^2 - a^2} - a\sec^{-1}\left|\frac{x}{a}\right| + C$

43. $\displaystyle\int \frac{\sqrt{x^2 - a^2}}{x^2}\, dx = \cosh^{-1}\frac{x}{a} - \frac{\sqrt{x^2 - a^2}}{x} + C$

44. $\int \dfrac{x^2}{\sqrt{x^2-a^2}}\,dx = \dfrac{a^2}{2}\cosh^{-1}\dfrac{x}{a}+\dfrac{x}{2}\sqrt{x^2-a^2}+C$

45. $\int \dfrac{dx}{x\sqrt{x^2-a^2}} = \dfrac{1}{a}\sec^{-1}\left|\dfrac{x}{a}\right|+C = \dfrac{1}{a}\cos^{-1}\left|\dfrac{a}{x}\right|+C$

46. $\int \dfrac{dx}{x^2\sqrt{x^2-a^2}} = \dfrac{\sqrt{x^2-a^2}}{a^2 x}+C$ 47. $\int \dfrac{dx}{\sqrt{2ax-x^2}} = \sin^{-1}\left(\dfrac{x-a}{a}\right)+C$

48. $\int \sqrt{2ax-x^2}\,dx = \dfrac{x-a}{2}\sqrt{2ax-x^2}+\dfrac{a^2}{2}\sin^{-1}\left(\dfrac{x-a}{a}\right)+C$

49. $\int (\sqrt{2ax-x^2})^n\,dx = \dfrac{(x-a)(\sqrt{2ax-x^2})^n}{n+1}+\dfrac{na^2}{n+1}\int (\sqrt{2ax-x^2})^{n-2}\,dx,$

50. $\int \dfrac{dx}{(\sqrt{2ax-x^2})^n} = \dfrac{(x-a)(\sqrt{2ax-x^2})^{2-n}}{(n-2)a^2}+\dfrac{(n-3)}{(n-2)a^2}\int \dfrac{dx}{(\sqrt{2ax-x^2})^{n-2}}$

51. $\int x\sqrt{2ax-x^2}\,dx = \dfrac{(x+a)(2x-3a)\sqrt{2ax-x^2}}{6}+\dfrac{a^3}{2}\sin^{-1}\dfrac{x-a}{a}+C$

52. $\int \dfrac{\sqrt{2ax-x^2}}{x}\,dx = \sqrt{2ax-x^2}+a\sin^{-1}\dfrac{x-a}{a}+C$

53. $\int \dfrac{\sqrt{2ax-x^2}}{x^2}\,dx = -2\sqrt{\dfrac{2a-x}{x}}-\sin^{-1}\left(\dfrac{x-a}{a}\right)+C$

54. $\int \dfrac{x\,dx}{\sqrt{2ax-x^2}} = a\sin^{-1}\dfrac{x-a}{a}-\sqrt{2ax-x^2}+C$

55. $\int \dfrac{dx}{x\sqrt{2ax-x^2}} = -\dfrac{1}{a}\sqrt{\dfrac{2a-x}{x}}+C$

56. $\int \sin ax\,dx = -\dfrac{1}{a}\cos ax+C$ 57. $\int \cos ax\,dx = \dfrac{1}{a}\sin ax+C$

58. $\int \sin^2 ax\,dx = \dfrac{x}{2}-\dfrac{\sin 2ax}{4a}+C$ 59. $\int \cos^2 ax\,dx = \dfrac{x}{2}+\dfrac{\sin 2ax}{4a}+C$

60. $\int \sin^n ax\,dx = \dfrac{-\sin^{n-1}ax\cos ax}{na}+\dfrac{n-1}{n}\int \sin^{n-2}ax\,dx$

61. $\int \cos^n ax\,dx = \dfrac{\cos^{n-1}ax\sin ax}{na}+\dfrac{n-1}{n}\int \cos^{n-2}ax\,dx$

62. (a) $\int \sin ax\cos bx\,dx = -\dfrac{\cos(a+b)x}{2(a+b)}-\dfrac{\cos(a-b)x}{2(a-b)}+C,\qquad a^2\neq b^2$

(b) $\int \sin ax\sin bx\,dx = \dfrac{\sin(a-b)x}{2(a-b)}-\dfrac{\sin(a+b)x}{2(a+b)},\qquad a^2\neq b^2$

(c) $\int \cos ax\cos bx\,dx = \dfrac{\sin(a-b)x}{2(a-b)}+\dfrac{\sin(a+b)x}{2(a+b)},\qquad a^2\neq b^2$

63. $\int \sin ax\cos ax\,dx = -\dfrac{\cos 2ax}{4a}+C$

64. $\int \sin^n ax\cos ax\,dx = \dfrac{\sin^{n+1}ax}{(n+1)a}+C,\qquad n\neq -1$

65. $\displaystyle\int \frac{\cos ax}{\sin ax}\, dx = \frac{1}{a}\ln |\sin ax| + C$

66. $\displaystyle\int \cos^n ax \sin ax\, dx = -\frac{\cos^{n+1} ax}{(n+1)a} + C, \qquad n \neq -1$

67. $\displaystyle\int \frac{\sin ax}{\cos ax}\, dx = -\frac{1}{a}\ln |\cos ax| + C$

68. $\displaystyle\int \sin^n ax \cos^m ax\, dx = -\frac{\sin^{n-1} ax \cos^{m+1} ax}{a(m+n)} + \frac{n-1}{m+n}\int \sin^{n-2} ax \cos^m ax\, dx,$

$\qquad\qquad\qquad\qquad\qquad\qquad n \neq -m \qquad \text{(If } n = -m, \text{ use No. 86.)}$

69. $\displaystyle\int \sin^n ax \cos^m ax\, dx = \frac{\sin^{n+1} ax \cos^{m-1} ax}{a(m+n)} + \frac{m-1}{m+n}\int \sin^n ax \cos^{m-2} ax\, dx,$

$\qquad\qquad\qquad\qquad\qquad\qquad m \neq -n \qquad \text{(If } m = -n, \text{ use No. 87.)}$

70. $\displaystyle\int \frac{dx}{b+c\sin ax} = \frac{-2}{a\sqrt{b^2-c^2}}\tan^{-1}\left[\sqrt{\frac{b-c}{b+c}}\tan\left(\frac{\pi}{4}-\frac{ax}{2}\right)\right] + C, \qquad b^2 > c^2$

71. $\displaystyle\int \frac{dx}{b+c\sin ax} = \frac{-1}{a\sqrt{c^2-b^2}}\ln\left|\frac{c+b\sin ax + \sqrt{c^2-b^2}\cos ax}{b+c\sin ax}\right| + C, \qquad b^2 < c^2$

72. $\displaystyle\int \frac{dx}{1+\sin ax} = -\frac{1}{a}\tan\left(\frac{\pi}{4}-\frac{ax}{2}\right) + C$

73. $\displaystyle\int \frac{dx}{1-\sin ax} = \frac{1}{a}\tan\left(\frac{\pi}{4}+\frac{ax}{2}\right) + C$

74. $\displaystyle\int \frac{dx}{b+c\cos ax} = \frac{2}{a\sqrt{b^2-c^2}}\tan^{-1}\left[\sqrt{\frac{b-c}{b+c}}\tan\frac{ax}{2}\right] + C, \qquad b^2 > c^2$

75. $\displaystyle\int \frac{dx}{b+c\cos ax} = \frac{1}{a\sqrt{c^2-b^2}}\ln\left|\frac{c+b\cos ax + \sqrt{c^2-b^2}\sin ax}{b+c\cos ax}\right| + C, \qquad b^2 < c^2$

76. $\displaystyle\int \frac{dx}{1+\cos ax} = \frac{1}{a}\tan\frac{ax}{2} + C$
   77. $\displaystyle\int \frac{dx}{1-\cos ax} = -\frac{1}{a}\cot\frac{ax}{2} + C$

78. $\displaystyle\int x\sin ax\, dx = \frac{1}{a^2}\sin ax - \frac{x}{a}\cos ax + C$
   79. $\displaystyle\int x\cos ax\, dx = \frac{1}{a^2}\cos ax + \frac{x}{a}\sin ax + C$

80. $\displaystyle\int x^n \sin ax\, dx = -\frac{x^n}{a}\cos ax + \frac{n}{a}\int x^{n-1}\cos ax\, dx$

81. $\displaystyle\int x^n \cos ax\, dx = \frac{x^n}{a}\sin ax - \frac{n}{a}\int x^{n-1}\sin ax\, dx$

82. $\displaystyle\int \tan ax\, dx = -\frac{1}{a}\ln |\cos ax| + C$
   83. $\displaystyle\int \cot ax\, dx = \frac{1}{a}\ln |\sin ax| + C$

84. $\displaystyle\int \tan^2 ax\, dx = \frac{1}{a}\tan ax - x + C$
   85. $\displaystyle\int \cot^2 ax\, dx = -\frac{1}{a}\cot ax - x + C$

86. $\displaystyle\int \tan^n ax\, dx = \frac{\tan^{n-1} ax}{a(n-1)} - \int \tan^{n-2} ax\, dx, \qquad n \neq 1$

87. $\displaystyle\int \cot^n ax\, dx = -\frac{\cot^{n-1} ax}{a(n-1)} - \int \cot^{n-2} ax\, dx, \qquad n \neq 1$

88. $\displaystyle\int \sec ax\, dx = \frac{1}{a}\ln |\sec ax + \tan ax| + C$
   89. $\displaystyle\int \csc ax\, dx = -\frac{1}{a}\ln |\csc ax + \cot ax| + C$

90. $\displaystyle\int \sec^2 ax\, dx = \frac{1}{a}\tan ax + C$

91. $\displaystyle\int \csc^2 ax\, dx = -\frac{1}{a}\cot ax + C$

92. $\displaystyle\int \sec^n ax\, dx = \frac{\sec^{n-2} ax \tan ax}{a(n-1)} + \frac{n-2}{n-1}\int \sec^{n-2} ax\, dx, \qquad n \neq 1$

93. $\displaystyle\int \csc^n ax\, dx = -\frac{\csc^{n-2} ax \cot ax}{a(n-1)} + \frac{n-2}{n-1}\int \csc^{n-2} ax\, dx, \qquad n \neq 1$

94. $\displaystyle\int \sec^n ax \tan ax\, dx = \frac{\sec^n ax}{na} + C, \qquad n \neq 0$

95. $\displaystyle\int \csc^n ax \cot ax\, dx = -\frac{\csc^n ax}{na} + C, \qquad n \neq 0$

96. $\displaystyle\int \sin^{-1} ax\, dx = x\sin^{-1} ax + \frac{1}{a}\sqrt{1 - a^2x^2} + C$

97. $\displaystyle\int \cos^{-1} ax\, dx = x\cos^{-1} ax - \frac{1}{a}\sqrt{1 - a^2x^2} + C$

98. $\displaystyle\int \tan^{-1} ax\, dx = x\tan^{-1} ax - \frac{1}{2a}\ln(1 + a^2x^2) + C$

99. $\displaystyle\int x^n \sin^{-1} ax\, dx = \frac{x^{n+1}}{n+1}\sin^{-1} ax - \frac{a}{n+1}\int \frac{x^{n+1}\, dx}{\sqrt{1 - a^2x^2}}, \qquad n \neq -1$

100. $\displaystyle\int x^n \cos^{-1} ax\, dx = \frac{x^{n+1}}{n+1}\cos^{-1} ax + \frac{a}{n+1}\int \frac{x^{n+1}\, dx}{\sqrt{1 - a^2x^2}}, \qquad n \neq -1$

101. $\displaystyle\int x^n \tan^{-1} ax\, dx = \frac{x^{n+1}}{n+1}\tan^{-1} ax - \frac{a}{n+1}\int \frac{x^{n+1}\, dx}{1 + a^2x^2}, \qquad n \neq -1$

102. $\displaystyle\int e^{ax}\, dx = \frac{1}{a}e^{ax} + C$

103. $\displaystyle\int b^{ax}\, dx = \frac{1}{a}\frac{b^{ax}}{\ln b} + C, \qquad b > 0, \ b \neq 1$

104. $\displaystyle\int xe^{ax}\, dx = \frac{e^{ax}}{a^2}(ax - 1) + C$

105. $\displaystyle\int x^n e^{ax}\, dx = \frac{1}{a}x^n e^{ax} - \frac{n}{a}\int x^{n-1} e^{ax}\, dx$

106. $\displaystyle\int x^n b^{ax}\, dx = \frac{x^n b^{ax}}{a \ln b} - \frac{n}{a \ln b}\int x^{n-1} b^{ax}\, dx, \qquad b > 0, \ b \neq 1$

107. $\displaystyle\int e^{ax} \sin bx\, dx = \frac{e^{ax}}{a^2 + b^2}(a\sin bx - b\cos bx) + C$

108. $\displaystyle\int e^{ax} \cos bx\, dx = \frac{e^{ax}}{a^2 + b^2}(a\cos bx + b\sin bx) + C$

109. $\displaystyle\int \ln ax\, dx = x\ln ax - x + C$

110. $\displaystyle\int x^n \ln ax\, dx = \frac{x^{n+1}}{n+1}\ln ax - \frac{x^{n+1}}{(n+1)^2} + C, \qquad n \neq -1$

111. $\displaystyle\int x^{-1} \ln ax\, dx = \frac{1}{2}(\ln ax)^2 + C$

112. $\displaystyle\int \frac{dx}{x \ln ax} = \ln|\ln ax| + C$

113. $\displaystyle\int \sinh ax\, dx = \frac{1}{a}\cosh ax + C$

114. $\displaystyle\int \cosh ax\, dx = \frac{1}{a}\sinh ax + C$

115. $\displaystyle\int \sinh^2 ax\, dx = \frac{\sinh 2ax}{4a} - \frac{x}{2} + C$

116. $\displaystyle\int \cosh^2 ax\, dx = \frac{\sinh 2ax}{4a} + \frac{x}{2} + C$

117. $\displaystyle\int \sinh^n ax\, dx = \frac{\sinh^{n-1} ax \cosh ax}{na} - \frac{n-1}{n}\int \sinh^{n-2} ax\, dx, \qquad n \neq 0$

118. $\displaystyle\int \cosh^n ax\, dx = \frac{\cosh^{n-1} ax \sinh ax}{na} + \frac{n-1}{n}\int \cosh^{n-2} ax\, dx, \qquad n \neq 0.$

119. $\displaystyle\int x \sinh ax\, dx = \frac{x}{a}\cosh ax - \frac{1}{a^2}\sinh ax + C$

120. $\displaystyle\int x \cosh ax\, dx = \frac{x}{a}\sinh ax - \frac{1}{a^2}\cosh ax + C$

121. $\displaystyle\int x^n \sinh ax\, dx = \frac{x^n}{a}\cosh ax - \frac{n}{a}\int x^{n-1}\cosh ax\, dx$

122. $\displaystyle\int x^n \cosh ax\, dx = \frac{x^n}{a}\sinh ax - \frac{n}{a}\int x^{n-1}\sinh ax\, dx$

123. $\displaystyle\int \tanh ax\, dx = \frac{1}{a}\ln (\cosh ax) + C$ 124. $\displaystyle\int \coth ax\, dx = \frac{1}{a}\ln |\sinh ax| + C$

125. $\displaystyle\int \tanh^2 ax\, dx = x - \frac{1}{a}\tanh ax + C$ 126. $\displaystyle\int \coth^2 ax\, dx = x - \frac{1}{a}\coth ax + C$

127. $\displaystyle\int \tanh^n ax\, dx = -\frac{\tanh^{n-1} ax}{(n-1)a} + \int \tanh^{n-2} ax\, dx, \qquad n \neq 1$

128. $\displaystyle\int \coth^n ax\, dx = -\frac{\coth^{n-1} ax}{(n-1)a} + \int \coth^{n-2} ax\, dx, \qquad n \neq 1$

129. $\displaystyle\int \text{sech}\, ax\, dx = \frac{1}{a}\sin^{-1} (\tanh ax) + C$

130. $\displaystyle\int \text{csch}\, ax\, dx = \frac{1}{a}\ln \left|\tanh \frac{ax}{2}\right| + C$ 131. $\displaystyle\int \text{sech}^2 ax\, dx = \frac{1}{a}\tanh ax + C$

132. $\displaystyle\int \text{csch}^2 ax\, dx = -\frac{1}{a}\coth ax + C$

133. $\displaystyle\int \text{sech}^n ax\, dx = \frac{\text{sech}^{n-2} ax \tanh ax}{(n-1)a} + \frac{n-2}{n-1}\int \text{sech}^{n-2} ax\, dx, \qquad n \neq 1$

134. $\displaystyle\int \text{csch}^n ax\, dx = -\frac{\text{csch}^{n-2} ax \coth ax}{(n-1)a} - \frac{n-2}{n-1}\int \text{csch}^{n-2} ax\, dx, \qquad n \neq 1$

135. $\displaystyle\int \text{sech}^n ax \tanh ax\, dx = -\frac{\text{sech}^n ax}{na} + C, \qquad n \neq 0$

136. $\displaystyle\int \text{csch}^n ax \coth ax\, dx = -\frac{\text{csch}^n ax}{na} + C, \qquad n \neq 0$

137. $\displaystyle\int e^{ax} \sinh bx\, dx = \frac{e^{ax}}{2}\left[\frac{e^{bx}}{a+b} - \frac{e^{-bx}}{a-b}\right] + C, \qquad a^2 \neq b^2$

138. $\displaystyle\int e^{ax} \cosh bx\, dx = \frac{e^{ax}}{2}\left[\frac{e^{bx}}{a+b} + \frac{e^{-bx}}{a-b}\right] + C, \qquad a^2 \neq b^2$

139. $\displaystyle\int_0^\infty x^{n-1}e^{-x}\, dx = \Gamma(n) = (n-1)!, \qquad n > 0.$

140. $\displaystyle\int_0^\infty e^{-ax^2}\, dx = \frac{1}{2}\sqrt{\frac{\pi}{a}}, \qquad a > 0$

141. $\displaystyle\int_0^{\pi/2} \sin^n x\, dx = \int_0^{\pi/2} \cos^n x\, dx = \begin{cases} \dfrac{1 \cdot 3 \cdot 5 \cdots (n-1)}{2 \cdot 4 \cdot 6 \cdots n} \cdot \dfrac{\pi}{2}, & \text{if } n \text{ is an even integer} \geq 2, \\[2ex] \dfrac{2 \cdot 4 \cdot 6 \cdots (n-1)}{3 \cdot 5 \cdot 7 \cdots n}, & \text{if } n \text{ is an odd integer} \geq 3 \end{cases}$

Greek alphabet

Capital letters	Name of letter	Lower-case letters
A	Alpha	α
B	Beta	β
Γ	Gamma	γ
Δ	Delta	δ
E	Epsilon	ϵ
Z	Zeta	ζ
H	Eta	η
Θ	Theta	θ, ϑ
I	Iota	ι
K	Kappa	κ
Λ	Lambda	λ
M	Mu	μ
N	Nu	ν
Ξ	Xi	ξ
O	Omicron	o
Π	Pi	π
P	Rho	ρ
Σ	Sigma	σ
T	Tau	τ
Υ	Upsilon	υ
Φ	Phi	ϕ, φ
X	Chi	χ
Ψ	Psi	ψ
Ω	Omega	ω

answers to odd-numbered problems

CHAPTER 1

Section 1

1.

3.

5.

7. $6 - 5 = 1.$

9. $-2 - 3 = -5.$

11. $2 - 3 = -1.$

13. $3 - 6 = -3.$

15. O is the midpoint of PQ.

17. O is three-fifths of the way from Q to P.

19. If the points A and B have the coordinates a and b, respectively, then the signed distance from A to B is $b - a$. The sign of $b - a$ thus determines which direction is positive, and $|b - a|$, which is the distance between A and B, determines the unit of length. The origin is then the point whose signed distance from A is ...?

21. Call the common origin O, and let A be the point where $x = 1$, $y = -1$. The system S that assigns to each point P the number $-x$ (minus the x-coordinate of P) is a coordinate system, namely, the coordinate system having the same origin and same unit of distance as the x-coordinate system, but with the opposite direction. The coordinate system S agrees with the y-coordinate system at both points O and A, so S agrees with y everywhere, by Problem 19; that is, $y = -x$ for every point P.

23. Call P_1 the point at which $y = b$, $x = 0$. Call P_2 the point at which $y = b + 1$, $x = 1$. The system S which assigns to each point the coordinate $(x + b)$ is a coordinate system. S agrees with y at P_1 and P_2, so by Problem 19, S agrees with y everywhere; that is, $y = x + b$.

25. At the point P_1 where $x = 0$, y has some value b. At the point P_2 where $x = 1$, $y = a + b$ for some a. The system S which assigns to each point the coordinate $ax + b$ is a coordinate system. S agrees with y at P_1 and P_2, so by Problem 19, S agrees with y everywhere. So $y = ax + b$.

27. $(a + 2b)/3.$

29.

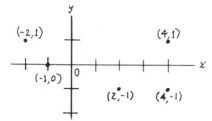

31. This is the straight line parallel to, and three units below, the *x*-axis.

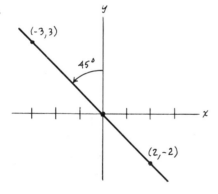

33.

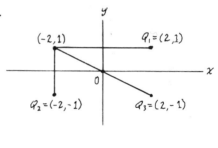

35.

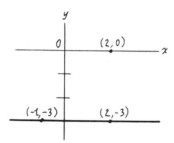

Section 2

1.

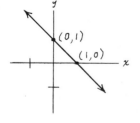

3.

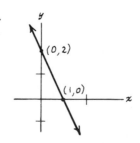

5.

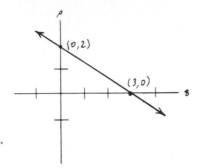

7.

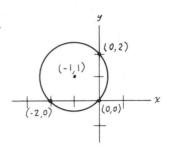

9.

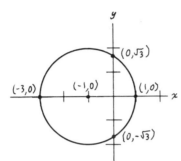

11.

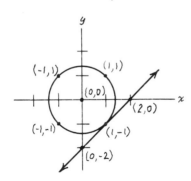

13.

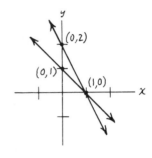

15.

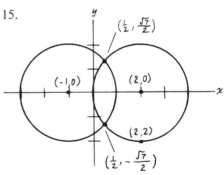

17.

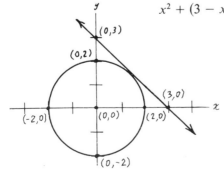

If $y = 3 - x$ and $x^2 + y^2 = 4$, then

$$x^2 + (3 - x)^2 = 4 \Rightarrow 2x^2 - 6x + 5 = 0$$

$$\Rightarrow x^2 - 3x + 5/2 = 0$$

$$\Rightarrow (x - 3/2)^2 + 5/2 - 9/4 = 0$$

$$\Rightarrow (x - 3/2)^2 = -1/4,$$

which is impossible.

19.

x	y
± 2	2
$\pm 3/2$	1/4
± 1	-1
$\pm 1/2$	$-7/4$
0	-2

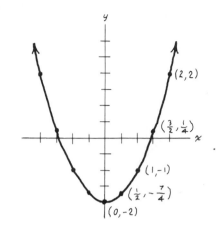

21.

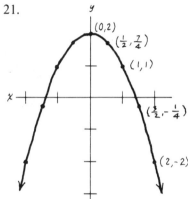

23.

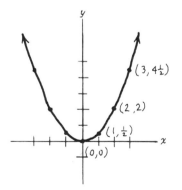

25.

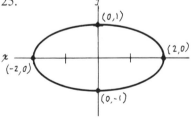

27.

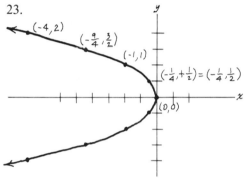

Section 3

1. $y = 3x - 6$; $(2,0)$, $(1,-3)$, $(7/3, 1)$.

3. $m = -1/2$, $b = 2$.

5. $m = -2/3$, $b = 2$.

7. $m = 1, b = 1.$

9.

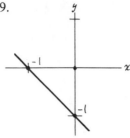

11.

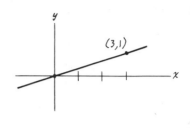

13.

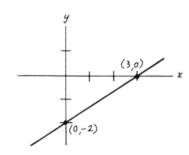

15.

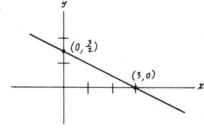

17. $m = 1/2, y = x/2.$

19. No slope, $x = -2.$

21. $m = 0, y = 1.$

23. Point–slope equation: $y - 2 = (-1/2)(x - 1)$, which is equivalent to $2y + x - 5 = 0.$

25. $3y - 2x - 1 = 0.$

27. $y = x + 1.$

29. $(-2, 3).$

31. $y = -x + 1.$

33. $y = (-1/6)x + 4/3.$

35. Say two lines have the same slope m. $l_1: y = mx + b$, $l_2: y = mx + b_2$. If l_1 and l_2 intersect at a point (x_0, y_0), then $y_0 = mx_0 + b_1$, $y_0 = mx_0 + b_2$. Then subtracting: $b_1 - b_2 = 0 \Rightarrow b_1 = b_2 \Rightarrow l_1 = l_2$. Say two lines have different slope, $l_1: y = m_1 x + b_1$; $l_2: y = m_2 x + b_2$. Solve the equations simultaneously to get $x_0 = (b_2 - b_1)/(m_1 - m_2)$, $y_0 = (b_2 m_1 - b_1 m_2)/(m_1 - m_2)$. So l_1 and l_2 are not parallel. (Why?)

37. Yes; the three slopes are all 4/5.

39. No; the slopes are 2/3, 1/2, 3/5.

41. (2,1).

43. Is the given equation the equation of some line? If so, what are its intercepts?

45. Both points are on the side of the line opposite from the origin.

Section 4

1. $\sqrt{5}$.

3. 4.

5. $2\sqrt{26}$.

7. $\sqrt{5}$.

9. $2\sqrt{2}$.

11. $AB = 2\sqrt{2}$, $BC = 3\sqrt{5}$, $CA = \sqrt{41}$.

13. $AB = \sqrt{26}$, $BC = 5$, $CA = \sqrt{13}$.

15. $P_1 P_2 = P_2 P_3 = \sqrt{41}$.

17. (2,1) and (1,3) have slope -2; (2,1) and (8,4) have slope 1/2.

19. $y = 3x - 6$.

21a. Show that the midpoint makes the slope $(b_2 - b_1)/(a_2 - a_1)$ with each endpoint.

23. $3y - 4x + 1 = 0$.

25. $12y - 5x + 59 = 0$.

27. $P_1 P_2$ has midpoint $(7/2, -1/2)$. $P_1 P_3$ has midpoint $(0, 3/2)$. The midpoints have slope $-4/7$. So does $P_2 P_3$.

29. If the two points are (x_1, y_1) and (x_2, y_2), then the locus condition is

$$\sqrt{(x - x_1)^2 + (y - y_1)^2} = \sqrt{(x - x_2)^2 + (y - y_2)^2}.$$

Square; expand the squares, and simplify.

31.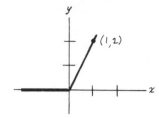

33.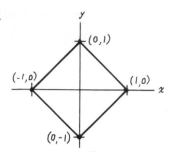

35. Choose the x-axis along one side of the triangle and the y-axis through the opposite vertex. The triangle then has vertices $(a,0)$, $(b,0)$, $(0,c)$. The perpendicular bisectors of the sides meet at the point

$$\left(\frac{a+b}{2},\frac{c}{2}+\frac{ab}{2c}\right).$$

37. $4/\sqrt{5}$.

Section 5

1. $(x-2)^2 + (y-1)^2 = 9$; $x^2 + y^2 - 4x - 2y - 4 = 0$.

3. $(x-2)^2 + (y+3)^2 = 1$; $x^2 + y^2 - 4x + 6y + 12 = 0$.

5. $(x+2)^2 + (y-4)^2 = 25$; $x^2 + y^2 + 4x - 8y - 5 = 0$.

7. $x^2 + y^2 = 16$; $x^2 + y^2 - 16 = 0$.

9. Center $(-3,4)$; radius 5.

11. Center $(-3/2, 5/2)$; radius 3.

13. $C(-1,0)$; $r = 1$.

15. $C(0, -1/2)$; $r = \sqrt{5}/2$.

17. No locus.

19. $(x-5)^2 + (y+2)^2 = 85$.

21. $(x-5/2)^2 + y^2 = 25/4$.

23. $(x-1/2)^2 + (y-3/2)^2 = 10/4$.

25. The circle $(x-4)^2 + y^2 = 4$.

27. The line through the origin and the point (a,b) has the equation $y = bx/a$. Solving this equation simultaneously with $x^2 + y^2 = 1$ (say by substituting $y = bx/a$ into the quadratic equation) gives $x = \pm a/\sqrt{a^2 + b^2}$.

29. If a point lies inside a circle, then its distance from the center is less than the radius of the circle. Write this inequality out for the given point and circle; square, and simplify. The result should be the required inequality.

Section 6

1. Both axes and the origin.

3. The origin.

5. The y-axis.

7. Both axes and the origin.

9. The origin.

11. A line of symmetry l for a triangle must contain a vertex P. (Otherwise, there would be at least two vertices on one side of l, and hence, by symmetry, at least two on the other side also, making at least four vertices altogether.) Then l must be the perpendicular bisector of the side opposite P, by a similar argument, etc.

13. Put coordinate axes along the perpendicular lines of symmetry, and let $P = (x, y)$ by any point in the figure. Show that $Q = (-x, -y)$ must also be in the figure by proceeding in two steps, first to the point of symmetry across the y-axis, and from there across the x-axis.

15. A regular pentagon has five lines of symmetry, each passing through a vertex and the midpoint of the opposite side. (A line of symmetry for any pentagon must contain a vertex: argue as in the answer to Problem 11.)

17. A line of symmetry through a vertex of a quadrilateral must also contain another vertex. (Argue as in the answer to Problem 11.) (The remaining two vertices must be symmetric in this line for the same general reason.) Etc.

19. If a line of symmetry does not contain a vertex, then the four vertices must be two on each side of the line of symmetry (as in Problem 11), and each pair must be the symmetric image of the other pair (same reason). The figure is then an isosceles trapezoid. Now consider the effect of a second line of symmetry.

21. Let C be a bounded configuration with A and B points of symmetry. Call the distance from A to B, $d > 0$. Because C is bounded, there is a smallest circle of radius R about A which completely encloses C. Because R is least, there must be a point X in X at the distance R from A.

Let Y be the reflection of X through A.

Let Z be the reflection of X through B. We may assume that $\sphericalangle BAX$ is at least $90°$; otherwise we just reverse the roles of X and Y. BA and ZY are parallel so $\measuredangle\, ZYA$ is also at least $90°$. Show, therefore, that $\overline{AZ} > R$. But this is a contradiction: Z is a point in the configuration C outside of the circle around A of radius R.

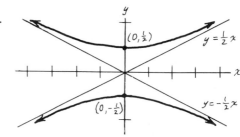

Section 7

1.

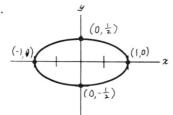

3.

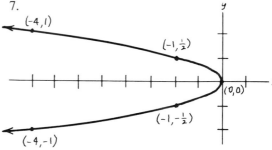

5.

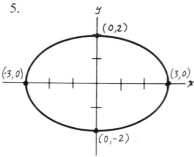

7.

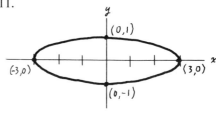

9.

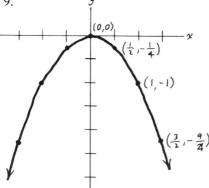

11.

13.

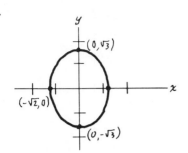

15.

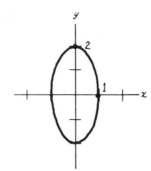

17.

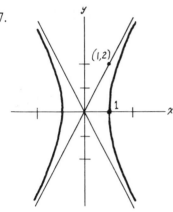

19.

21.

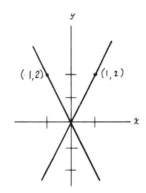

Section 8

1. a) $(2, -1)$	b) $X = x - 1$;	$Y = y - 4.$
3. a) $(3,9)$	b) $X = x - 2$;	$Y = y + 4.$
5. a) $(-2, -1)$	b) $X = x - 3$;	$Y = y - 2.$
7. a) $(2,1)$	b) $X = x + 3$;	$Y = y + 2.$

9. New origin at $(-2, 3)$; $\overline{X}^2 + \overline{Y}^2 = 18$.

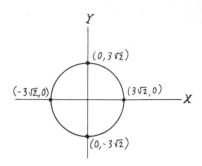

11. New origin at $(0,0)$ (See answer to Problem 11 of Section 7.)

13. New origin at $(-2, 1)$; $\dfrac{\overline{X}^2}{3^2} + \dfrac{\overline{Y}^2}{2^2} = 1$ (See answer to Problem 5 of Section 7.)

15. New origin at $(1,1)$; $\overline{X}^2 - \dfrac{\overline{Y}^2}{2^2} = -1$.

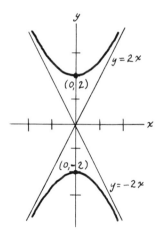

17. New origin at $(1,2)$; $\dfrac{\overline{X}^2}{2^2} + \overline{Y}^2 = 1$.

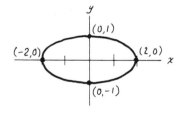

19. $\overline{X}^2 + \overline{Y}^2 + 4\overline{X} - 2\overline{Y} + 1 = 0$.

21. $\overline{X}\,\overline{Y} = 0$.

CHAPTER 2

Section 1

1. $x \geq 0$.

3. $x \neq 2$.

5. $x \neq 2, x \neq -2$.

7. -1.

9. $a^3 + 6a^2 + 11a + 6$.

11. $u - 2$.

13. $H(1/t) = \dfrac{t - 1}{t + 1} = -H(t)$.

15. $H(H(t)) = t$.

17. $1/x - 1/y = (x - y)(-1/xy)$.

19. $\dfrac{1}{x^2 + 1} - \dfrac{1}{y^2 + 1} = (x - y)(x + y)(-1/(x^2 + 1)(y^2 + 1))$.

21. $D = -1 \leq x \leq 1$. For $0 \leq x \leq 1$, $y(y(x)) = \sqrt{1 - (\sqrt{1 - x^2})^2} = \sqrt{x^2} = |x| = x$.

23. $g(y) = y^3$.

25. No; when $x = 0, y = \pm 1/2$.

27. No; when $x = 0, y = \pm 1/2$.

29. $f(x) = x - 1$.

31. No; when $x = -1, y = \pm 1$.

33. $f(x) = \dfrac{x^2 - x}{x + 1}; x \neq -1, x \neq 0$.

35. $f(s) = \dfrac{s^2 \sqrt{3}}{4}$.

Section 2

1. Parabola.

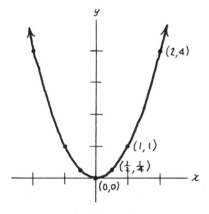

3. The graph is a semicircle, because $(f(x))^2 + x^2 = 1$ and because $f(x) \geq 0$.

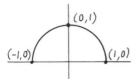

5. 0, 1.

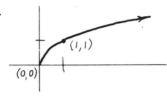

7.

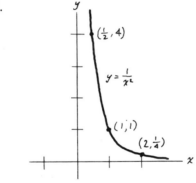

9. f is odd if and only if $f(-x) = -f(x)$, if and only if $(-x, -f(x))$ belongs to the graph of f whenever $(x, f(x))$ does, if and only if the graph of f is symmetric with respect to the origin.

11. $1 + x + x^2 = \underbrace{1 + x^2}_{\text{even}} + \underbrace{x}_{\text{odd}}.$

13. g_1, g_2 odd functions implies $g_1(-x)g_2(-x) = (-g_1(x))(-g_2(x)) = g_1(x)$ $g_2(x)$. So $g_1 g_2$ is an even function; even + even = even; even × odd = odd; even × even = even.

15. Symmetric about the origin.

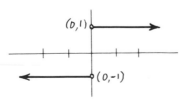

17.

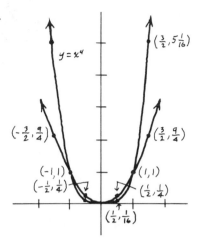

For large n, the graph approaches:

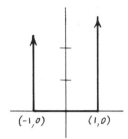

19. $p(x)$ behaves like x^3 for large $|x|$, so for large positive (negative) x, $p(x)$ is large positive (negative). Therefore the graph of $p(x)$ must cross the x-axis at least once. If the graph of $p(x)$ crosses or touches the x-axis at a, then $p(a) = 0$, so $(x - a)$ is a factor of $p(x)$. Since $p(x)$ is third-degree, there can be at most three such a's. In all, the graph crosses the x-axis at least once and crosses or touches at most three times, and $p(x)$ is positive (negative) for large positive (negative) values of x. The only four possibilities are

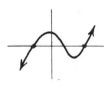

21.

23.

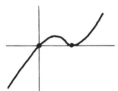

25. $(y - c)/(x - a) = (d - c)/(b - a)$. Solve for y.

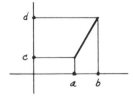

Section 3

1. $(-\infty, \infty)$.

3. The union of $(-\infty, -3)$, $(-3, 2)$ and $(2, \infty)$.

5. $[-6,6]$.

7. The union of $(-\infty, -4]$ and $[4, \infty)$.

9. The union of $(-\infty, 0)$ and $(0, \infty)$.

11. $[0,6]$.

13. $[0, \infty)$.

15. Negative on $(-\infty, 0)$; positive on $(0, \infty)$; increasing on $(-1, 1)$; decreasing on $(-\infty, -1)$ and $(1, \infty)$; turning upward on $(-\sqrt{3}, 0)$ and $(\sqrt{3}, \infty)$; turning downward on $(-\infty, -\sqrt{3})$ and $(0, \sqrt{3})$.

17. $x^3 - 3x + 2 = (x - 1)^2(x + 2)$.

19.

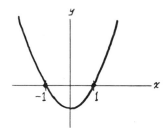

Positive on $(-\infty, -1)$ and $(1, \infty)$; negative on $(-1, 1)$; decreasing on $(-\infty, 0)$; increasing on $(0, \infty)$; turning upward everywhere.

21.

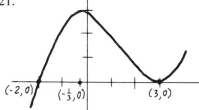

The high point between $x = -2$ and $x = 3$ occurs at $x = -1/3$. (Later on we shall see how to prove this.) Use this fact in writing down the answer.

23. $y = 2x - x^2 = 1 - (x - 1)^2$.

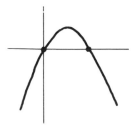

Section 4

1.

3.

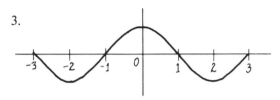

5.

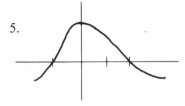

Section 5

1. a) $\dfrac{f(3 + h) - f(3)}{h} = -4 - h;\ -0.41,\ -0.0401,\ -0.004001.$

    b) $\dfrac{f(2 + h) - f(2)}{h} = -2 - h;\ -0.21,\ -0.0201,\ -0.002001.$

3. 3, 5, and x are continuous by I, so by the sum and product rules of II, $3x^2 + 5x$ is continuous.

5. By I, $x^{1/2}$ is continuous ($x \geq 0$); so by II, $1/\sqrt{x}$ is continuous for $x > 0$.

7. $f(x) = x^2$ is continuous by I and II; $g(x) = x^{1/2}$ is continuous by I, so $\sqrt{x^2} = g(f(x))$ is continuous by III.

9. $g(x) = \sqrt{x^2} = |x|$ is continuous
    (by Problem 7); $f(x) = x^{1/3}$ is con-
    tinuous by I, so $g(f(x)) = |x^{1/3}|$
    is continuous by III.

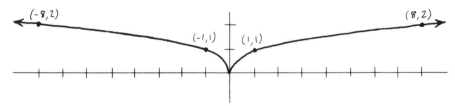

11. Here is one example. Draw a different one.

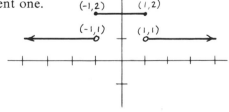

13. $$|x - 2| \le d < 1 \Rightarrow |x - 2| < 1 \Rightarrow -1 < x - 2 < 1$$
$$\Rightarrow 1 < x < 3 \Rightarrow |x + 2| < 5;$$
$$|x^2 - 4| = |(x - 2)(x + 2)| = |x - 2||x + 2| < 5d.$$

15. $f(1) = 1, f(x) = x.$

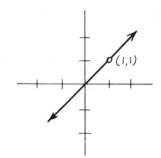

17. $f(1) = 2/3; f(x) = \dfrac{x + 1}{x^2 + x + 1}.$

Section 6

1. $\sqrt{2}.$

3. 2.

5. 1.

7. 12.

9. 3/4.

11. 1/2.

13. −3.

15. 1/7.

17. 1/5.

19. No limit.

21. No limit.

23. −2.

25. 1.

27. $\lim\limits_{x \to 0^-} \dfrac{|x|}{x} = -1; \lim\limits_{x \to 0^+} \dfrac{|x|}{x} = +1.$

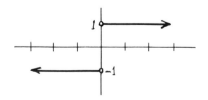

29. -1.

31. 1.

33. -2.

35. $f(x) = [x]$ is continuous from the right everywhere. It is continuous from the left everywhere except at the integer points.

37. $\lim_{x \to 4^-} [x] + p(x) = 7$; $\lim_{x \to 4^+} [x] + p(x) = 9$; but $[4] + p(4) = 8$.

39. $3t + 5$ is continuous, so $\lim_{t \to -1^-} h(t) = 3(-1) + 5 = 2$.
    $t^2 + 1$ is continuous, so $\lim_{t \to -1^+} h(t) = h(-1) = 2$.

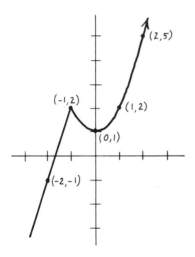

Section 7

1. $f(x) = x$, $g(x) = -4$, $h(x) = 2$ are all continuous. By the product rule gf is continuous, and so is f^2. Applying the rule again to f and f^2, f^3 is continuous. By the sum rule, $f^3 + gf$ is continuous. Then again by the sum rule $(f^3 + gf + h)$ is continuous, and this is p.

3. $\sqrt{x}$ continuous by **L2**

             $1/\sqrt{x}$ continuous by **L3′** $(x > 0)$

             $\sqrt{x} + 1/\sqrt{x}$ continuous by **L3′**

5. $$1, \sqrt{x}, x \text{ continuous by } \mathbf{L2}$$
$$x^2 \text{ continuous by } \mathbf{L3'}$$
$$1 + x^2 \text{ continuous by } \mathbf{L3'}$$
$$\frac{1}{1 + x^2} \text{ continuous by } \mathbf{L3'}$$
$$\frac{\sqrt{x}}{1 + x^2} \text{ continuous by } \mathbf{L3'} \quad (\text{product rule})$$

7. If $\lim_{x \to a^-} f(x) = l$ and $\lim_{x \to a^-} g(x) = m$, then $\lim_{x \to a^-} (f(x) + g(x)) = l + m$, etc.

9. If $g(x) = f^3(x) + 4f(x) + 1 = y^3 - 4y + 1$, then

$$\lim_{x \to a^-} g(x) = \lim_{x \to a^-} (y^3 - 4y + 1) = (-2)^3 - 4(-2) + 1 = 1,$$
$$\lim_{x \to a^+} g(x) = \lim_{x \to a^+} (y^3 - 4y + 1) = 2^3 - 4(2) + 1 = 1,$$
$$g(a) = (y^3 - 4y + 1)_{y=0} = 1.$$

Thus $\lim_{x \to a^-} g(x) = g(a) = \lim_{x \to a^+} g(x) = 1$, and g is continuous at $x = a$.

11. If $f(x) \to l$ as $x \to a^-$ and if $g(y)$ is continuous at $y = l$, then $g(f(x)) \to g(l)$ as $x \to a^-$.

Section 8

1. -7.

3. 1.

5. $-2/3$.

7. 1.

9. $1/2$.

11. $2/3$.

13. $-\infty$.

15. $+\infty$.

17.
$$\lim_{x \to 3} \frac{2x^2 - 5x - 3}{x^2 - x - 6} = \lim_{x \to 3} \frac{(2x + 1)(x - 3)}{(x + 2)(x - 3)} \underset{\text{L5}}{=} \lim_{x \to 3} \frac{2x + 1}{x + 2} \underset{\text{L1}}{=} \frac{7}{5}.$$

This assumes that $(2x + 1)/(x + 2)$ is continuous at $x = 3$, which in turn requires **L3'** and **L2**.

$$\lim_{x \to \infty} \frac{2x^2 - 5x - 3}{x^2 - x - 6} = \lim_{x \to \infty} \frac{2 - \dfrac{5}{x} - \dfrac{3}{x^2}}{1 - \dfrac{1}{x} - \dfrac{6}{x^2}} = \frac{2 - 0 - 0}{1 - 0 - 0} = 2,$$

by **L3** and Example 1.

19. $x > 10$.   21. $x > d^{-1/3}$.

23. $0 < x < 10^{-6}$.   25. $M = d^{-1/2}$.

CHAPTER 3

Section 1

1. $m = 2$.   3. $m = -2,\ y = -2x - 1$.

5. $m = -1,\ y = -x + 2$.   7. $m = 1,\ y = x$.

9. $m = 1 - 2a,\ y = (1 - 2a)x + a^2$. The tangent line is horizontal $(m = 0)$ when $a = 1/2$; its equation is then $y = 1/4$.

Section 2

1. $$\frac{f(r) - f(x)}{r - x} = \frac{r^2 - x^2}{r - x}.$$
    Use the factorization $r^2 - x^2 = (r - x)(r + x)$ to simplify. Then compute the limit.

3. $$\frac{f(r) - f(x)}{r - x} = \frac{\dfrac{1}{r^2} - \dfrac{1}{x^2}}{r - x}.$$
    Cross-multiply in the numerator and then use the factorization $r^2 - x^2 = (r - x)(r + x)$ to simplify. Then compute the limit.

5. You will need to treat $r^{3/2} - t^{3/2}$ as a difference of cubes $(r^{1/2})^3 - (t^{1/2})^3$ and $r - x$ as a difference of squares $(r^{1/2})^2 - (t^{1/2})^2$, and use the formulas $b^3 - a^3 = (b - a)(b^2 + ab + a^2)$, $b^2 - a^2 = (b - a)(b + a)$.   (Another method is to multiply the difference quotient top and bottom by $r^{3/2} + x^{3/2}$.)

7. The difference quotient is

$$\frac{f(r) - f(s)}{r - s} = \frac{-3r - (-3s)}{r - s}.$$

9. Your calculation should show that the difference quotient for the linear function $mx + b$ has the constant value m, regardless of the values of x and r.

11. If $f(x) = x^a$, then $f'(x) = ax^{a-1}$.

13. $2x$.

15. $1/\sqrt{2x}$.

17. $-1/(y - 1)^2$.

19. $1 - 1/x^2$.

21. $x/\sqrt{x^2 + 1}$.

23. $-1/(2(x + 1)^{3/2})$.

25. $(1 - x^2)/(1 + x^2)^2$.

27. Set $y = x^{1/4}$, $s = r^{1/4}$, and show that $(r^{1/4} - x^{1/4})/(r - x)$ simplifies to $1/(s^3 + s^2 y + sy^2 + y^3)$ when $s - y$ is cancelled out. Now take the limit as $s \to y$ and then set $y = x^{1/4}$.

29. $(r^{2/3} - x^{2/3})/(r - x)$ simplifies to $(s + y)/(s^2 + sy + y^2)$, when $s - y$ is cancelled from top and bottom. Now take the limit as $s \to y$, and set $y = x^{1/3}$.

31. This should boil down to $2(a_1 + a_2)x + (b_1 + b_2) = (2a_1 x + b_1) + (2a_2 x + b_2)$.

Section 4

1. $7x^6$.

3. $3x^2 + 6x$.

5. m.

7. $16x^3 - 2x$.

9. $2x - 1$.

11. $2x^{-1/2} + 4x - 3$.

13. $1 - x^{-2}$.

15. $1 - 2x^{-3}$.

19. $dy/dx = 0$ at $x = 1$.

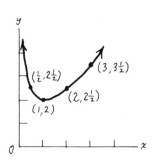

21. $(3, -8)$.

23. $y = -\frac{1}{9}x + 20\frac{1}{3}$.

25. $(-b/2a, c - b^2/4a)$.

27. Your tangent-line equation should be $y - y_0 = 2x_0(x - x_0)$, or an equivalent form using $y_0 = x_0^2$.

29. Picture the light from the point source at $(0, 1/4)$ as shining out along straight-line rays from that point. Then the diagram in Problem 28 contains a typical light ray, and you are to show that it bounces off the parabola as a ray parallel to the y-axis.

31. $y = 6x - 9, y = 2x - 1$.

37. $f'(x) = 1/(3x^{2/3})$.

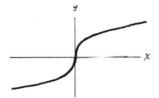

39. We want to show that $[(f + g)(h + k)]' = (f + g)'(h + k) + (f + g)(h + k)'$. We know that $[(f + g)(h + k)]' = [fh + fk + gh + gk]' = (fh)' + (fk)' + (gh)' + (gk)'$ by Theorem 1. By the same theorem, $(f + g)'(h + k) = (f' + g')(h + k) = f'h + f'k + g'h + f'k$, and similarly for the other term on the right in the first line.

41. The problem is to show that

$$\lim_{r \to 0} \frac{f(r) - f(0)}{r - 0} = g(0).$$

This should follow immediately upon setting $f(r) = rg(r)$ in the difference quotient, and simplifying.

43. Show first that the difference quotients for g, f, and h must satisfy a similar inequality, and then apply the squeeze limit law.

Section 5

1. $(x + 2)/\sqrt{x^2 + 4x}$. 3. $-(x + 2)^{-2}$.

5. $-2x/(x^2 + 1)^2$.

7. $-x(1 - x^2)^{-1/2}$.

9. $\frac{8}{5}(2x + 1)^{-1/5}$.

11. $-x^2/(1 + x^3)^{4/3}$.

13. Multiply the difference quotient top and bottom by $\sqrt{g(r)} + \sqrt{g(x)}$; then pass to the limit. Note that $r - x$ does not algebraically cancel out of the difference quotient. Instead, we use the hypothesis that $(g(r) - g(x))/(r - x)$ approaches $g'(x)$ as $r \to x$.

15. Factor $g(r) - g(x)$ out of $[g(r)]^n - [g(x)]^n$ and remember that $(g(r) - g(x))/(r - x) \to g'(x)$ as $r \to x$.

17. $f'(x) = 0$ at $x = 1/8$; $f(1/8) = 8$.

19. The equation for the upper half of the ellipse can be written $y = b\sqrt{1 - x^2/a^2}$. Differentiate and show that the answer can be rewritten

$$\frac{dy}{dx} = -\frac{b^2 x}{a^2 y}.$$

Use this value at the point (x_0, y_0) in the point-slope equation of the tangent line, and simplify.

21. $\dfrac{x x_0}{a^2} - \dfrac{y y_0}{b^2} = 1$.

23. $x^{-1/3}\sqrt{x^{2/3} + a^{2/3}} = \sqrt{1 + (a/x)^{2/3}}$.

25. $\dfrac{-2}{(x + 1)^3}$.

27. $\dfrac{g'(x)}{3(1 + g(x))^{2/3}}$.

29. $\dfrac{-g(x)g'(x)}{\sqrt{1 - g^2(x)}}$.

Section 6

1. $\dfrac{\Delta y}{\Delta x} = 2x + \Delta x$.

3. $\dfrac{\Delta y}{\Delta x} = \dfrac{1}{\sqrt{x + \Delta x} + \sqrt{x}}$.

5. $\dfrac{\Delta y}{\Delta x} = 1 + 2x + \Delta x$.

7. $\dfrac{\Delta y}{\Delta x} = \dfrac{-2x - \Delta x}{[1 + (x + \Delta x)^2][1 + x^2]}$.

9. $\Delta y = 3x^2 \Delta x + 3x(\Delta x)^2 + (\Delta x)^3$, which can be thought of as a sum of seven terms. Your figure should show each of these seven terms as the volume of a rectangular solid.

Section 7

1. $v = 3t^2 - 4t + 1$; $t = \frac{1}{3}$, 1. Forward $(-\infty, \frac{1}{3})$, $(1, \infty)$; backward $(\frac{1}{3}, 1)$.

3. 144 ft.

5. 3/4 sec; 16 ft/sec (up or down).

7. a) -11.73 b) 1.8   c) -10.73

9. -80 ft/sec; $t = 1\frac{1}{4}$ sec.

11. Write $\Delta s = s - s_0$, $\Delta t = t - t_0$ in the increment equation, and solve for s.

Section 8

1. 0.

3. $2/x^3$.

5. $6(x + 1/x^3)$.

7. $12x^{-5}$.

9. 0.

11. $-\frac{1}{4}x^{-3/2}$.

13. 0.

15. $$\frac{d^{m+1}}{dx^{m+1}}x^{m+1} = \frac{d^m}{dx^m}\left(\frac{d}{dx}x^{m+1}\right).$$

Compute the inner derivative and then use the assumed formula.

The formula holds for $n = 1$. Since it holds for $n = 1$ it holds for $n = 2$. Since it holds for $n = 2$ it holds for $n = 3$. So?

17. $f''(x) = \frac{3}{4}x^{-1/2}$.

19. $v_0 = 2$, $a_0 = -2$; $t = 1/3$.

Section 9

1.

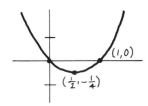

3.

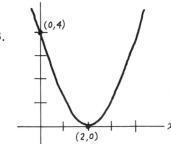

5.

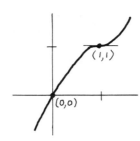

7.

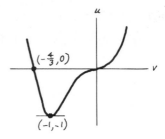

9.

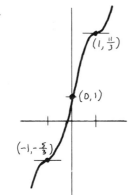

11.

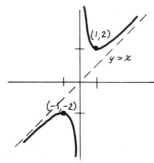

13.

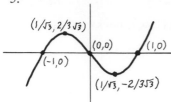

15. $dy/dx = 0$ at $x = -b/2a$. If $a > 0$, then $y \to +\infty$ as $x \to \pm\infty$. Sketch the graph in this case.

Section 10

1. $5/2$.   3. $1/\sqrt{3}$.   5. $2/3^{1/3}$.   7. $\sqrt{6} - 1$.

9. If $g(x) = x^3$, then $(f - g)'(x) = 0$ for all x. By Theorem 7, $(f - g)(x) = c$. $f(x) = g(x) + c = x^3 + c$.

11. Set $g(x) = mx$ and follow the argument of Problem 9 ($b = c$).

13. a) By 11, $v = ds/dt = -32t + b$ for some constant b. Let $g(t) = -16t^2 + bt$; then

$$\frac{d}{dt}(s(t) - g(t)) = v(t) - (-32t + b) = 0.$$

Then by Theorem 7, $s(t) = g(t) + c$ for some constant c. That is, $s = -16t^2 + bt + c$.   b) $s = -16t^2 + 50t$.

15. $X = a^{1/(1-a)}x$.

17. Use Problem 10.

Section 11

1. If $f(x) = x^5$, then $(3.1416)^5 - \pi^5 = f'(X)(3.1416 - \pi)$ for some X between π and 3.1416. Show that the right side above is less than $5(10)^{-3}$, using the assumptions stated in the problem.

3. Here $f(x) = x^{1/5}$, and $(100,001)^{1/5} - 10 = f'(X)(100,001 - 100,000) = f'(X)$ where $100,000 < X < 100,001$. Show that $f'(X) < 10^{-4}/5$.

5. Given $|r - \pi| < 10^{-n}/2$ and $f(x) = x^{1/3}$. Then for some X between r and π,

$$|r^{1/3} - \pi^{1/3}| = |f'(X)(r - \pi)| < \tfrac{1}{3}X^{-2/3}\tfrac{1}{2}(10^{-n}).$$

Show that this is less than $10^{-(n+1)}$, since $X > 3$. (Note: $\sqrt{3} > 5/3$.)

7. $|\pi - r| < 9e$ (assuming $r \geq 3$).

9. $b^n > b^n - 1 > n(b - 1)$. So b^n will be larger than M if $n(b - 1) > M$, and hence if $n > M/(b - 1)$.

11. $\sqrt{1 + x} - 1 = 1/2\sqrt{1 + X} \cdot x$, etc. Here is a direct proof: $1 + x < 1 + x + x^2/4 = (1 + x/2)^2$. Taking the square root gives $\sqrt{1 + x} < 1 + x/2$.

Section 12

1. a) $4\pi r^2 \text{unit}^3/\text{unit}$ b) $(36\pi)^{-1/3} V^{-2/3} \text{unit}/\text{unit}^3$.

3. $-1/\sqrt{15}$ ft/ft.

5. $3r^2 \text{cm}^3/\text{cm}$; volume must be increased by a factor of $2\sqrt{2}$.

7. $(75\pi/4)$ cubic inches per year; $(1875\pi/4)$ cubic inches per year.

9.
$$\frac{1}{l_0} \cdot \frac{dl_t}{dt} = \alpha + 2\beta t.$$

Here β is small compared to α, so β can be ignored if t varies less than $20°$. (See Example 2.)

11. Since M is positive, C is increasing. M is decreasing before x_0 and increasing after x_0, so C should have tangent lines curving down before x_0 and curving up after x_0.

13. Show that the difference quotient $(C(r) - C(x))/(r - x)$ can be rewritten as

$$f(x) + r\left(\frac{f(r) - f(x)}{r - x}\right).$$

Then take the limit.

CHAPTER 4

Section 1

1. $(x - 1)^2(x + 2)^3(7x + 2)$.

3. $3t^5(7t + 12)$.

5. $5x^4 + 36x^3 + 33x^2 - 70x + 13$.

7. $-4(1 + 2v)^3/v^5$.

9. $(x + 1)^3(3x^2 - 1)^{-3/2}(9x^2 - 3x - 4)$.

11. 1.

13. $-1/(z - 1)(z^2 - 1)^{1/2}$.

15. $-(1 - x)^{-1/3}(1 + x)^{-2/3}(x + 1/3)$.

17. First apply the product law to w times (uv), etc.

19. $3x^2 - 7$.

21. $(1 - t^2)/(1 + t^2)^2$.

23. $3x^2(1 + x^2)/(1 + 3x^2)^2$.

25. $-(2x^2 + 4x + 1)(x^2 + 2x + 2)^{-5/2}$.

27. $1/(1 - x)(1 - x^2)^{1/2}$.

29. $\dfrac{1}{(1 + u^2)^{3/2}}$.

31. $2x/(1 - x^2)^2$.

33. $-4(1 - x)/(1 + x)^3$.

35. $\dfrac{-2f'(x)}{(1 + f(x))^2}$.

37. $\dfrac{2(gf' - fg')}{(f + g)^2}$.

39. $\dfrac{2 + 3x - 2x^2 - x^3}{2\sqrt{1 + x}\,(1 + x^2)^2}$.

41. Compute $(f^m f)'$ by the product rule and then use the hypothesis.

43. Write $f = \sqrt{f} \cdot \sqrt{f}$, use the product rule, and solve for $(\sqrt{f}\,)'$.

47. $2(3x^2 - 1)/(1 + x^2)^3$.

49. $\left(\dfrac{x^n + 1}{x^n}\right)' = \left(1 + \dfrac{1}{x^n}\right)' = \dfrac{-n}{x^{n+1}}$ (if $n \neq 0$).

51. $\Delta y = \left(\dfrac{u + \Delta u}{v + \Delta v} - \dfrac{u}{v}\right) = \cdots$

53. $\dfrac{\Delta u}{u} + \dfrac{\Delta v}{v} + \dfrac{\Delta w}{w} + \dfrac{(\Delta u)(\Delta v)}{uv} + \dfrac{(\Delta u)(\Delta w)}{uw} + \dfrac{(\Delta v)(\Delta w)}{vw} + \dfrac{(\Delta u)(\Delta v)(\Delta w)}{uvw}$.

55.

$$\lim_{\Delta x \to 0} \frac{\Delta y}{\Delta x} = \lim_{\Delta x \to 0} \frac{v\dfrac{\Delta u}{\Delta x} - u\dfrac{\Delta v}{\Delta x}}{v(v + \Delta v)} \quad \text{(algebra)}$$

$$= \lim_{\Delta x \to 0}\left[v\frac{\Delta u}{\Delta x} - u\frac{\Delta v}{\Delta x}\right]\lim_{\Delta x \to 0}\left(\frac{1}{v(v + \Delta v)}\right) \qquad \mathbf{L}\ ?$$

$$= \frac{\displaystyle\lim_{\Delta x \to 0}\left[v\frac{\Delta u}{\Delta x} - u\frac{\Delta v}{\Delta x}\right]}{\displaystyle\lim_{\Delta x \to 0}(v(v + \Delta v))} \qquad \mathbf{L}\ ?$$

$$= \frac{\left[\displaystyle\lim_{\Delta x \to 0} v\frac{\Delta u}{\Delta x} - \lim_{\Delta x \to 0} u\frac{\Delta v}{\Delta x}\right]}{\displaystyle\lim_{\Delta x \to 0}(v(v + \Delta v))} \qquad \mathbf{L}\ ?$$

$$= \frac{\left[\left(\displaystyle\lim_{\Delta x \to 0} v\right)\left(\lim_{\Delta x \to 0}\frac{\Delta u}{\Delta x}\right) - \left(\lim_{\Delta x \to 0} u\right)\left(\lim_{\Delta x \to 0}\frac{\Delta v}{\Delta x}\right)\right]}{\left(\displaystyle\lim_{\Delta x \to 0} v\right)\left(\lim_{\Delta x \to 0}(v + \Delta v)\right)} \qquad \mathbf{L}\ ?$$

$$= \frac{(vu' - uv')}{v\lim_{\Delta x \to 0}(v + \Delta v)} \qquad \begin{array}{l}\text{(By Theorem 5, Chapter 3,}\\ \lim_{\Delta x \to 0} v + \Delta v = v.)\end{array}$$

$$= \frac{(vu' - uv')}{v^2}.$$

57. At $x = (mb + na)/(m + n)$. Show that this lies between a and b.

Section 2

1. xe^x.

3. $x^2 e^x$.

5. $e^x/2\sqrt{e^x - 1}$.

7. $(\frac{x}{2}e^{x/2})/\sqrt{x - 1}$.

9. ae^{ax}.

11. $\frac{1}{3}(3e^{3x} + 1)(e^{3x} + x)^{-2/3}$.

13. $e^x - e^{-x}$.

15. $e^{-x}(1 - x - x^2 - x^3)/(1 + x^2)^2$.

17. $1, -2$.

19. $1, -3$.

21. $\dfrac{dy}{dx} = (-2x + 1)e^{-2x}$, $\dfrac{d^2y}{dx^2} = (4x - 4)e^{-2x}$.

23. Show that $(d/dx)(ye^{-kx}) = 0$, by the product law and the assumed equation.

25. $f'(x) = a^x f'(0)$. For $a > 1$, a^x is increasing and $f'(0) > 0$, so $a^x f'(0)$ is increasing. For $0 < a < 1$, the argument is analogous but different in detail. How does it go?

27. The chord over the interval $[0, 1/2]$ has slope $(4^{1/2} - 4^0)/(1/2) = 2$. By MVT there is a number X between 0 and 1/2 such that $g'(X) = 2$. Since g' is increasing, $g'(0) < 2$. Now start from the chord over $[-1/2, 0]$ and show that $g'(0) > 1$.

29. a) $kf(x) + f'(x)$ b) $f(x) = ke^{-3x} + \frac{1}{8}e^{5x}$.

33. $x = e^{-1}$.

Section 3

1.

$P_t = (0, -1)$

$\cos t = 0$, $\sin t = -1$.

3.

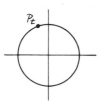

$$\cos t = \frac{\sqrt{2}}{2}, \sin t = \frac{-\sqrt{2}}{2}.$$

5.

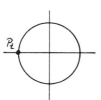

Your estimates should be close to $\cos t = -0.415$, $\sin t = 0.91$, which come from tables.

7.

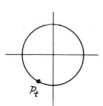

$\cos t = -1, \sin t = 0.$

9.

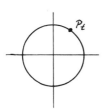

The correct values, to two decimal places, are $\cos t = -0.65$, $\sin t = -0.75$. Your estimates should be close.

11.

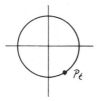

$$\cos t = \frac{\sqrt{2}}{2}, \sin t = \frac{\sqrt{2}}{2}.$$

13. $\frac{5}{4}\pi$.   15. $\frac{2}{3}\pi$.

17. About halfway between $\pi/6$ and $\pi/4$.

19. $\approx \frac{\pi}{2} + \frac{\pi}{7}$.

21. $\sin(\pi + t) = -\sin(\pi - t)$. What is the corresponding cosine identity?

23. The 30°–60°–90° triangle is half an equilateral triangle. This gives $\sin \pi/6 = 1/2$, and then $\cos \pi/6$ is computed by the Pythagorean Theorem.

25. $\cos \theta = 1/\sqrt{5}$, $\sin \theta = 2/\sqrt{5}$.

27. $\pi/4$.   29. $\frac{3}{4}\pi$.

31.   If we guess the angle whose tangent is $1/2$ to be about $\pi/7$, then the answer here is $\approx 6\pi/7$.

33. $13\pi/7$, on the basis of the guess in Problem 31.

49. $\sin(x - y) = \cos((x - y) - \pi/2)$ (by 46)

$$= \cos((x - \pi/2) - y) = \cdots$$

51. $\sin(x + y) = \sin(x - (-y)) = \cdots$ (using Problem 49, etc.)

53.   Divide the answer to Problem 51 by the answer to Problem 52, and then divide top and bottom by $\cos x \cos y$.

55. $\sin 2x = \sin(x + x) = \cdots$

57. $\tan 2x = \tan(x + x) = \cdots$

59.   Set $y = 2x$ in one of the equations in Problem 56.

61. $\sin(x + y) - \sin(x - y) = 2 \sin y \cos x$.

63. $A = \sqrt{a^2 + b^2}$; θ is the angle from the x-axis to the ray through (a, b).

Section 4

1. $\dfrac{d}{dx} \sec x = \dfrac{d}{dx}\left(\dfrac{1}{\cos x}\right) = \cdots$

3. $\dfrac{d}{dx} \cos x = \dfrac{d}{dx}\left(\dfrac{1}{\sin x}\right) = \cdots$

5. $e^x(\cos x - \sin x)$.

7. $\sin 2x$.

9. 0.

11. $1/(1 + \cos x) = \csc x \tan \frac{x}{2} = \frac{1}{2} \sec^2 \frac{x}{2}$.

13. $2 \tan x \sec^2 x$.

15. $-\sin x/2\sqrt{1 + \cos x}$.

17. $3 \sin^2 x \cos^2 x(\cos^2 x - \sin^2 x)$.

19. 0.

21. $e^x \cos x$.

23. $\cos^2 x$.

25. $\dfrac{d}{dx} \cos 2x = \dfrac{d}{dx}(\cos^2 x - \sin^2 x) = \cdots$

27. $\dfrac{d}{dx} \sin(x + a) = \dfrac{d}{dx}(\sin x \cos a + \sin a \cos x) = \cdots$

29. $\dfrac{d}{dx} \tan(x + a) = \dfrac{d}{dx}\left(\dfrac{\sin(x + a)}{\cos(x + a)}\right) = \cdots$

31. By the addition laws, $\sin(x + (\pi/2)) = \cos x$, $\cos(x + (\pi/2)) = -\sin x$.

33. Compute the derivative of $f^2 + (f')^2$ and substitute $f'' = -f$.

35. $\dfrac{d}{dx} e^x \sin x = e^x(\sin x + \cos x);\ \dfrac{d^2}{dx^2} e^x \sin x = 2e^x \cos x$.

37. $\dfrac{d^2 y}{dx^2} = -x \sin x + 2 \cos x;\ \dfrac{d^4 y}{dx^4} = x \sin x - 4 \cos x$.

39. $\dfrac{-2}{(\sin x - \cos x)^2}$.

41. $e^x(x \sin x + x \cos x + \sin x)$.

43. $\cos 2x$.

45. $\cos^5 x$.

47. $\sec^6 x$.

49. $a = -2(v + x)$.

51. For $t > 0, 0 < \sin t < t$, so $\sin^2 x < x^2$ for $x \neq 0$. Manipulate this into the desired inequality.

53. For example $d \cos x/dx = -\sin x = -\cos(\cos x)$.

Section 5

1. $f(x) = ax + b.\ \dfrac{d}{dx} \sin(f(x)) = \cos(f(x))f'(x) = a \cos(ax + b)$.

5. $(\cos \sqrt{x})/2\sqrt{x}$.

7. $-e^x \sin(e^x)$.

9. $e^{(e^x + x)}$.

11. $(3x^2 - 1)\cos(x^3 - x)$.

13. If $e^{f(x)} = x$, then $\dfrac{d}{dx} e^{f(x)} = \dfrac{d}{dx} x = 1$. But $\dfrac{d}{dx} e^{f(x)} = ?$

15. Proceed as in Problems 13 and 14.

17. Let f be even, and set $h(x) = -x$. Then

$$\frac{d}{dx}f(x) = \frac{d}{dx}f(-x) = \frac{d}{dx}f(h(x))$$
$$= f'(h(x))h'(x) = f'(-x)(-1).$$

So $f'(x) = -f'(-x)$: f' is odd. Work out what happens when f is odd in a similar way.

21. $-xe^{\sqrt{1-x^2}}/\sqrt{1-x^2}$.

23. $(-\cos(1/x))/x^2$ $(x \neq 0)$.

25. $2x \sin(1/x) - \cos(1/x)$ $(x \neq 0)$.

27. $(1 - 2x^2 - 2x)e^{-x^2}$.

29. $-[\cos(\cos(\sin x))][\sin(\sin x)]\cos x$.

31. $(1 - 1/x)e^{1/x}$.

33. $-b/2x^2\sqrt{a + b/x}$.

35. $\sin(2\sqrt{t})/4\sqrt{t}(1 + \sin^2(\sqrt{t}))$

37. $2x(\sin(x^2) + \cos(x^2) + 1)/(1 + \cos(x^2))^2$.

39. $2xf'(x^2)$.

41. $2f'(x)e^{2f(x)}$.

43. $f'(e^{g(x)})e^{g(x)}g'(x)$.

45. Show that $f'(1/n\pi) = (-1)^{n+1}$, using the answer to Problem 25.

Section 6

1. $-y/x = 5/2x^2$.

3. $-x/y = -x/\sqrt{16 - x^2}$.

5. $(-3x - 1)/y = (-3x - 1)/\sqrt{10 - 2x - 3x^2}$.

7. $2x/y(x^2 + 1)^2 = 2x/(x^2 - 1)^{1/2}(x^2 + 1)^{3/2}$.

9. $-\sqrt{x/y} = -x^{1/2}/(2 - x^{3/2})^{1/3}$.

11. $(y^2 - 2xy - 2x)/(x^2 - 2xy + 2y)$.

13. $-\csc y = -1/\sqrt{1 - x^2}$.

15. $y/(\cos y - x)$.

17. $\cos x/(3y^2 + 1)$.

19. $-25/16y^3$.

21. $y = (-1/2)x + 3/2$.

23. Slope at $(x_0, y_0) = m = (-x_0 b^2)/(y_0 a^2)$. Solve for k in $y_0 = mx_0 + k$, and recall that $(x_0^2/a^2) + (y_0^2/b^2) = 1$ to reduce $y = mx + k$ to

$$(xx_0/a^2) + (yy_0/b^2) = 1.$$

25. $\dfrac{dy}{dx} = \dfrac{e^{x+y}}{(1 - e^{x+y})} = \dfrac{y}{(1 - y)}, \quad \dfrac{d^2y}{dx^2} = -\dfrac{y}{(1 - y)^3}.$

27. $\dfrac{dy}{dx} = \pm 1, \quad \dfrac{d^2y}{dx^2} = 0.$

Section 7

1. $3x^2 dx + 2dx$.

3. $2(2x + 1)^2(5x^2 + x + 6)dx$.

5. $2xe^{x^2} dx$.

7. $[(x^4 + 7x^2 - 2x + 6)/(x^2 + 3)^2]dx$.

9. $(-1/x^2 + 2)dx$.

11. $dy = 2 \times dx : dy/dx = 2$

13. $dy/dx = x(x + 2)^{-1/3} \times (x - 1)^{-2/3}$

15. $e^y dy = 2x dx$. Solve for dy/dx and manipulate, to get $dy/dx = 2/x$. Similarly, $dx/dy = \frac{1}{2}e^{y/2}$.

17. $(3y^2 + x)dy + (y - 5x^4)dx = 0$. Solve for dy/dx and for dx/dy.

CHAPTER 5

Section 1

1. If $y = e^x$, then $d^2y/dx^2 = e^x$, which is everywhere positive. So the graph of e^x is everywhere concave up, confirming the way we have been drawing it.

3. Critical points: $x = \pm 1/2$, concave down, $(-\infty, 0)$; concave up, $(0, \infty)$;

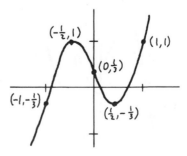

5. No critical points; concave up, $(0, +\infty)$; concave down, $(-\infty, 0)$; slope $= 1$ at point of inflection.

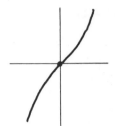

7. Critical point 0; concave up $(-\infty, 4)$, $(4, \infty)$; concave down $(-4, 4)$. Asymptotic to the line $x = -4$, $x = +4$, $y = 2$.

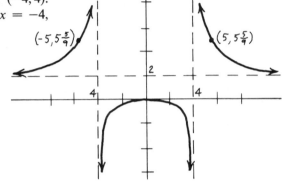

9. Critical points $\pi/6 + n\pi/3$; concave up $((2n + 1)\pi/3, (2n + 2)\pi/3)$; concave down $(2n\pi/3, (2n + 1)\pi/3)$.

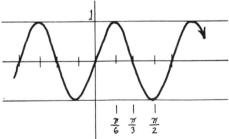

11.

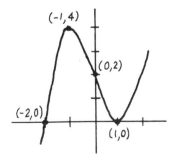

13. Critical point 1; concave up every-
    where.

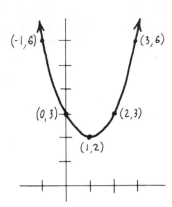

15. Critical points $x = \frac{3}{2}, 2$; concave
    down $(-\infty, 7/4)$; concave up
    $(7/4, \infty)$.

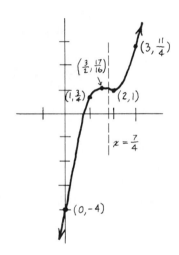

17. Critical points $x = -1, 2$; concave
    up $(-\infty, 0), (2, \infty)$; concave down
    $(0, 2)$.

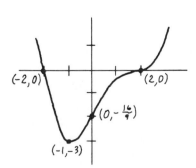

19. Critical points $x = 2, 6$; concave down, $(-\infty, 4)$; concave up, $(4, \infty)$.

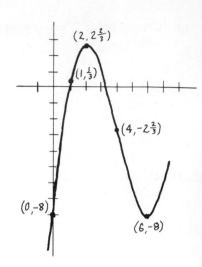

21. Slope at $0 = 1$; slope at $1 = -2$.

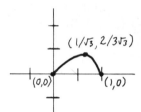

23. Critical points $x = \pi/4,\ \pi,\ 7\pi/4$; concave down $(0, 7\pi/12)$, $(\pi, 17\pi/12)$, approx.; concave up $(7\pi/12, \pi)$, $(17\pi/12, 2\pi)$, approx.

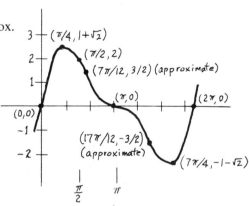

25. Critical point $x = 0$; concave down, $(-1, 1)$; concave up, $(-\infty, -1), (1, \infty)$.

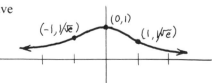

27. Critical point, $x = 0$; concave up, $(-\infty, \sqrt{3})$, $(0, \sqrt{3})$; concave down, $(-\sqrt{3}, 0)$, $(\sqrt{3}, \infty)$.

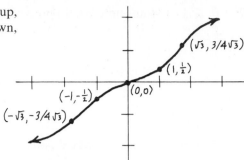

29.

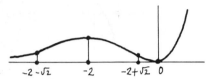

31.

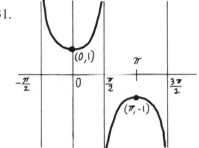

33. Concave up, $(0, \infty)$; concave down, $(-\infty, 0)$.

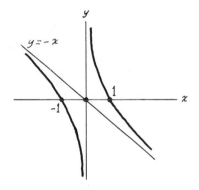

35. Critical point, $2^{2/3}$; concave up, $(0, 4)$; concave down, $(4, +\infty)$.

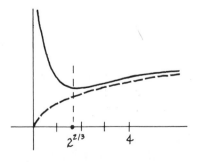

37. $a^2 < 3b$ means no critical points;

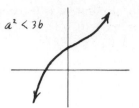

$a^2 = 3b$ means one critical point;

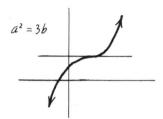

$a^2 > 3b$ means two critical points.

39. Take two nonoverlapping chords C_1 and C_2. By MVT there are tangents T_1 and T_2 with the same slope as C_1 and C_2 (see figure). Since the graph lies above T_1 and T_2, in particular, Y lies above T_1 and X lies above T_2 so $\sphericalangle XzY$ is less than π, in other words, slope T_1 = slope C_1 < slope C_2 = slope T_2.

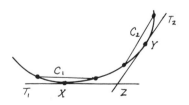

41. Say $x < y$. Let

$$s(u,v) = \frac{f(v) - f(u)}{v - u}.$$

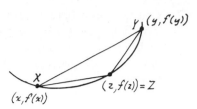

Thus $s(u,v)$ is the slope of the chord from $(u, f(u))$ to $(v, f(v))$. The given is that Z is below XY so that slope $XZ <$ slope $XY <$ slope ZY: $s(x,z) < s(x,y) < s(y,z)$. Now fix z_0 as the midpoint of the interval $[x,y]$, and repeat the above argument for Z between X and z_0. Letting Z approach X, show that $f'(x) \leq s(x,z_0)$. In a similar manner, show that $f'(y) \geq s(z_0,y)$. Conclude that $f'(x) < f'(y)$.

Section 2

1.

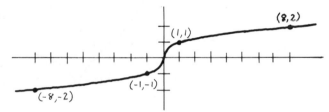

3. Critical points 0, 1; concave up, $(-\infty, -1/2)$; concave down, $(-1/2, 0)$, $(0, \infty)$.

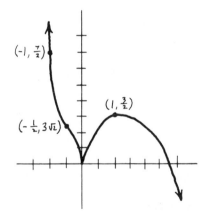

5. Critical points 0, $\pm\pi^3/64$; concave up $(-1.7, 0)$ $(1.7, 4]$ (approx); concave down $[-4, -1.7)$, $(0, 1.7)$.

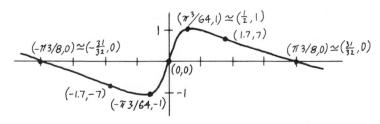

7. Critical points 0, $(2n + 1)\pi/2$; concave up, $((2n + 1)\pi, (2n + 2)\pi)$, $n \geq 0$; $(2n\pi, (2n + 1)\pi)$, $n < 0$; concave down, $(2n\pi, (2n + 1)\pi)$, $n \geq 0$; $((2n + 1)\pi, (2n + 2)\pi)$, $n < 0$.

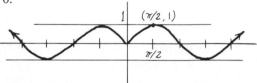

9. Critical points -1, 1; concave up, $(-\infty, -1)$, $(-1, 0)$, concave down, $(0, 1)$, $(1, +\infty)$.

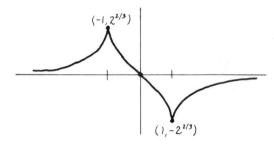

11.

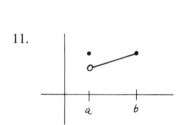

13. a)

(i) $0 \leq f(x) \leq f(y)$ implies $0 \leq f(x)^2 \leq f(y)^2$; multiplying, $0 \leq f(x)^3 \leq f(y)^3$.

(ii) $f(x) \leq f(y) \leq 0$ implies $f(x)^2 \geq f(y)^2 \geq 0$; multiplying, $f(x)^3 \leq f(y)^3$.

(iii) $f(x) \leq 0 \leq f(y)$ implies $f(x)^3 \leq 0 \leq f(y)^3$.

In all three cases $f(x) \leq f(y)$ implies $f(x)^3 \leq f(y)^3$; that is, $x \leq y$. So if $y < x$, then $x \nleq y$ implies $f(x) \nleq f(y)$ implies $f(y) < f(x)$.

b) $f'(x) = \frac{1}{3}x^{-2/3}$, $x \neq 0$. So $f'(x) > 0$ for $x \neq 0$; f is continuous at 0. By Theorem 2, f is increasing on $(-\infty, 0]$, $[0, \infty)$. Now take $x < y$ "straddling" 0. If $x < 0 \leq y$; then $f(x) < f(0) \leq f(y)$. (If $x \leq 0 < y$, then $f(x) \leq f(0) < f(y)$.)

17. The basic fact: $d|x - k|dx = +1$ if $x > k$ and -1 if $x < k$. Thus, if $f(x) = a|x - 2| + bx + c$, then $f'(x) = b - a$ if $x < 2$ and $f'(x) = b + a$ if $x > 2$. So the graph of f is a half-line having slope $b - a$ to the left of $x = 2$, and a half-line with slope $b + a$ to the right of 2: a polygonal graph with vertex over $x = 2$.

Conversely, the polygonal graph with vertex at $(2, h)$ and slopes λ and μ to the left and right of 2 will be the graph of the above function f if $b - a = \lambda$, $b + a = \mu$, and $h = f(2) = 2b + c$; that is, if $a = (\mu + \lambda)/2$, $b = (\mu - \lambda)2$, and $c = h - 2b = h - \mu + \lambda$.

19. If $-1 < x < 1$, then $f'(x) = 1 - 1 = 0$, and the graph runs along a horizontal line over the interval $[-1, 1]$. We can see this directly without using slopes by noting that $|x - 1| = -(x - 1)$ when $x < 1$ and $|x + 1| = x + 1$ when $x > -1$, so

$$|x + 1| + |x - 1| - 1 = (x + 1) - (x - 1) - 1$$

$$= 1 \qquad \text{over } [-1, 1], \text{ etc.}$$

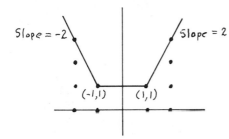

23.

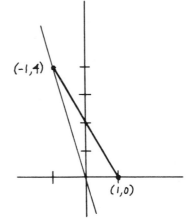

21.

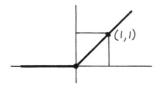

25. $f(x) = \frac{1}{2}|x + 2| - \frac{1}{2}|x + 1| - \frac{1}{2}|x - 1| + \frac{1}{2}|x - 2|$.

27. $\dfrac{d}{dx} \sin x = \cos x \leq 1 = \dfrac{d}{dx} x$, so, by Theorem 3, $\sin x - \sin 0 \leq x - 0$ for all $x \geq 0$; that is, $\sin x \leq x$ for all $x \geq 0$. The second part follows in the same way.

29. Compare the 0 function to f in Theorem 3.

31. By MVT

$$\frac{f(a + h) - f(a)}{h} = f'(X)$$

for some X between a and $(a + h)$, X depending on h. Then $X \to a$ as $h \to 0$ (since X is squeezed between a and $(a + h)$), so $f'(X) \to l$ as $h \to 0$. So?

33. Example: $f(x) = x + x^2\left(\dfrac{\sin 1/x + 1}{2}\right)$.

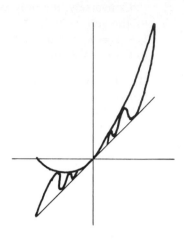

35. $\dfrac{d}{dx}\left(x - \dfrac{x^3}{3!}\right) = 1 - \dfrac{x^2}{2} \leq \cos x = \dfrac{d}{dx}\sin x.$

So, by Theorem 3, $x - x^3/3! - 0 \leq \sin x - 0$ for $x \geq 0.$

Section 3

1. $8\frac{1}{16}$.

3. $2\frac{107}{108}$.

5. 1.02.

7. $3 - \frac{1}{27} \approx 2.96$.

9. $\frac{33}{128} \approx 0.258$.

11. $0.08\pi r^3$.

15. The tangent-line error is less than 0.001. Also, $1/16 \approx 0.06$, with an error of 0.0025. So, $\sqrt{65} \approx 8.06$, with an error less than $0.004 = 4(10)^{-3}$.

17. $|E| \leq 0.002$. Also, $1/108 \approx 0.01$, with an error less than 0.001. So $(80)^{1/4} \approx 2.99$, with an error less than 0.003.

19. $(0.98)^{-1} \approx 1.0200$ with an error less than $4(10)^{-5}$.

21. $\Delta y \approx (\frac{1}{2})x^{1/2}\big|_{x=(81/4)}\Delta x = \frac{1}{2}(81/4)^{-1/2}(-\frac{1}{4}) = -\frac{1}{36} = -0.0277\cdots$, so $\sqrt{20} \approx 4.5 - 0.0277\cdots = 4.4722\cdots$. Since $|f''(x)| = (1/4)x^{3/2} \leq (1/40) \times (20)^{3/2} = B$ for this range of x, we have $|E| \leq B(\Delta x)^2/2 = (1/2^7)20^{3/2}$. You are to show that $|E| \leq 10^{-4}$. This will follow if $2^7 20^{3/2} \geq 10^4$, or (squaring) $2^{17} \geq 10^5$. Check this.

23. The tangent line approximation to $\sin t$ about $a = 0$ is $\sin t = \sin 0 + \cos 0(t - 0) + E = t + E$, where $|E| \leq Bt^2/2$, and B is the maximum

value of $|\sin''(x)| = |\sin(x)|$ on the interval between 0 and t. You are to show that $|E| \le |t|^3/2$, and this will be the case if

$$\frac{B}{2}t^2 \le \frac{|t|^3}{2}.$$

25. The tangent-line approximation to $\sqrt{1+x}$ near $a = 0$ is

$$\sqrt{1+x} \approx \sqrt{1+0} + \frac{1}{2\sqrt{1+0}}x = 1 + \frac{x}{2}$$

so it is only a question of the magnitude of the error estimate $B(\Delta x)^2/2 = Bx^2/2$. What is the largest absolute value of the second derivative of $\sqrt{1+x}$ over the interval $[0, x]$?

31. If $f(x) = \sqrt{x}$, then $f'(x) = 1/2\sqrt{x}$ and $f''(x) = -1/4x^{3/2}$, so f'' is negative and decreasing in magnitude. Therefore,

$$\sqrt{10} = 3 + \frac{1}{2}\left(\frac{1}{3}\right) + E$$

$$= 3.1666\cdots + E,$$

when E is negative, and at most

$$\frac{1}{2}B(\Delta x)^2 = \frac{1}{2}\left(\frac{1}{4}\cdot\left(\frac{1}{3}\right)^3\right)(1)^2$$

in magnitude. Finish the calculation.

Section 4

1. x is the x-intercept of the line of slope $f'(a)$ through $(a, f(a))$. Write down the point–slope equation of this line, set $y = 0$, and solve for x.

3. $1, 3/4$.

7. Let $f(x) = \sin x = 2x/3$, $a = \pi/2$. Then $f'(x) = \cos x - 2/3$, and

$$x_1 = \frac{\pi}{2} - \frac{f(\pi/2)}{f'(\pi/2)} = \frac{3}{2}.$$

Between $3/2$ and $\pi/2$, $\cos x = \sin(\pi/2 - x) < \pi/2 - x < (\pi - 3)/2 < 0.1$, and $|f'(x)| = |\cos x - 2/3| > 1/2 = L$. Also $|f''(x)| = \sin x \le 1 = B$. So

$$|E| \le \frac{B}{2L}(x_1 - a)^2 = (x_1 - a)^2 < 0.01.$$

9. $f(x) = x^2 - 10$, $x_0 = 3$, $x_1 = (19/6)$, $x_2 = (721/228) = 3.162280 \cdots$,
    $|f''(x)| = 2 = B$, $|f'(x)| = 2x \geq 6 = L$,

$$|E| \leq \frac{B}{2L}(x_2 - x_1)^2 = \frac{2}{2 \cdot 6}\left(\frac{1}{19 \cdot 12}\right)^2 < \frac{1}{6(200)^2} = \frac{(0.005)^2}{6} < 5(10)^{-6}.$$

11. $x_1 = 10.050$, $|E| \leq 1.25(10)^{-4}$.

13.
$$x_2 = \frac{17}{12} - \frac{1}{24 \cdot 17} = 1.414215 \cdots,$$

$$|E| < \frac{1}{2}\left(\frac{1}{24 \cdot 17}\right)^2 < 5 \cdot (10)^{-6}.$$

Thus $\sqrt{2} = 1.41421$ to the nearest 5 decimal places.

15.
$$x_2 = \frac{97}{56} - \frac{1}{2 \cdot 56 \cdot 97},$$

$$|E| < \frac{1}{2}\left(\frac{1}{2 \cdot 56 \cdot 97}\right)^2 < 5 \cdot (10)^{-9}.$$

Section 5

1. By Theorem 4, $f'(x)$ changes sign from $-$ to $+$ as x crosses x_0. Thus f' is negative on a small interval (a, x_0), and positive on a small interval (y_0, b). Therefore, ...

3. By Problem 1, f' has a relative minimum at x_0. Thus there is a small interval (a, b) about x_0 on which f' is positive, except at x_0 itself. So, ...

6.   and three similar figures.

7. Relative minimum at $x = 2$; relative maximum at $x = 0$.

9. $x = -1/3$, relative maximum; $x = 1$, relative minimum.

11. $x = -2$, relative maximum; $x = 2$, relative minimum.

13. $x = -3$, relative maximum; $x = -9/5$, relative minimum; $x = 0$, case of Problem 3.

15. Relative minimum at $x = 1$; decreasing through the critical point $x = 0$.

Section 6

1.

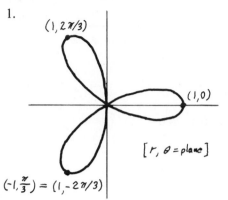

$[r, \theta$-plane

3.

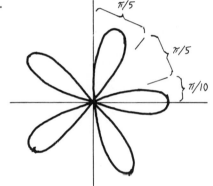

5.

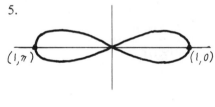

7. Circle of radius 1; center at $(0, 1)$
    (rectangular coordinates).

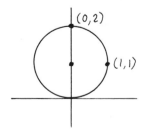

9.   11.

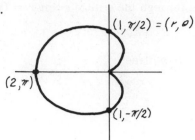

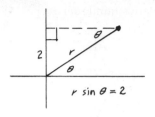

13. The equation in Cartesian coordinates is $y^2 = 1 + 2x$.

CHAPTER 6

Section 1

1. $3\frac{1}{4}$, 1.

3. $22\frac{22}{27}$, -28.

5. 0, -6.

7. $3\sqrt{3}/4$, 0.

9. 3, $-3/8$.

11. $4/e^2$, 0.

13. $5/4$, -1.

15. max $= 3\frac{1}{4}$.

17. max $= 2\frac{5}{27}$.

19. min $= -3/8$.   21. max $= 4/e^2$.

Section 2

1. $\sqrt{2}$.

3. $1/2\sqrt{2}$.

5. $(2, -22)$.

7. -4.

9. $r = \frac{2}{3}R$, $h = \frac{1}{3}H$.

11. 6×8 yards (8 yard length borders the neighbor's lot).

13. $\sqrt{2A} \times \sqrt{A/2}$.

15. Square piece $(96/(4 + \pi))$ in.; circle piece $24\pi/(4 + \pi)$.

17. $\sqrt{3}$.

19. Width: depth $= 1$: $\sqrt{2}$.

21. $8 \times 8 \times 10$ ft.

23. $\overline{AB} = 75/2$ ft.

25. $10/3$ in.

27. $10 - \sqrt{20}$ tons.

29. 100 square yards of land.

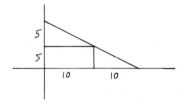

31. Circular cylinder.

33. $h/d = 1$.

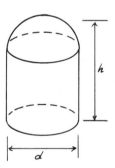

35. $x^2 + f(x)^2$ is minimal, so

$$\frac{d}{dx}(x^2 + f(x)^2) = 0 \Rightarrow 2x + 2f(x)f'(x) = 0 \Rightarrow \frac{f(x)}{x} = \frac{-1}{f'(x)}.$$

In other words, $-1/f'(x)$ is the slope of the line from $(0,0)$ to $(x,f(x))$. Since $-1/f'(x)$ is the negative reciprocal of the slope of the tangent line, the line from $(0,0)$ to $(x,f(x))$ is perpendicular to the graph.

37. $h = 2d$.

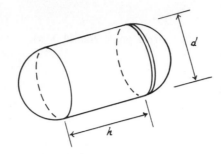

39. 11 thousand dollars per week.

Section 3

1. $-3/4$.

3. $25/12$ ft/sec.

5. $9/5$ ft/sec.

7. 65 mi/hr, 65 mi/hr.

9. $20\sqrt{2}$ ft/min.

11. 0 ft/sec.

13. 10π cu.in./sec.

15. $1/8$ radians/sec.

17. $1/6$ sq.in./sec.

19. -30, 0, 10 (in./sec.).

21. 150π ft./min., -200π ft./min.

25. $3/20$ radians/sec.

Section 4

1. $\log 2/0.01 \approx 70$ hrs.

3. $\log 2/24 \simeq 2.9\%$.

5. 6.6 yrs.

7. $y_0 = ce^{kt_0}$, $y_1 = ce^{kt_1}$; $\log y_0 = \log c + kt_0$; $\log y_1 = \log c + kt_1$. Subtracting: $k(t_1 - t_0) = \log y_1 - \log y_0 = \log(y_1/y_0)$; $k = (\log(y_1/y_0))/(t_1 - t_0)$.

11. The temperature will reach 170° at time $t = 5 \log 1.3/\log(13/12) \approx 16.5$ min.

Section 5

1. $x = 2 \sin(3t - \pi/3)$.

3. $x = 5\sqrt{2} \sin(t + 3\pi/4)$.

5. $x = a \sin(2\pi ft + \alpha)$, so $dx/dt =?$

7. $x = \sqrt{53/2} \sin(\sqrt{2/5}\ t) + \arctan(2\sqrt{2}\ /3\sqrt{5}\ )$.

9. The motion $(d^2x/dt^2) = -kx$ has period $2\pi/\sqrt{k}$. Here $2 = 2\pi\sqrt{l/g}$.

11. 13 ft $\frac{1}{2}$in. (appr.).

13. 13 in. (appr.).

CHAPTER 7

Section 1

3. They are inverses, where defined.

5. $x = (1/3)(y + 5)$.

7. $x = y^{5/3}$.

9. $x = (1 - y)/(1 + y)$.

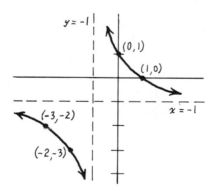

11. $x = y/\sqrt{1 - y^2}$.

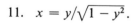

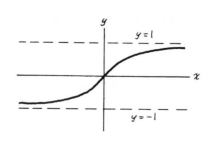

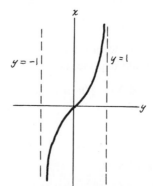

13. Domain $f = [1, \infty)$; domain $g = [0, \infty)$.

15. Domain $f = [0, \infty)$.

Section 2

7. $f'(x) = 3(x - 1)^2$.

9. $f'(x) = 3(x - \frac{1}{3})^2 + \frac{2}{3}$.

11. $f'(x) = 1/(1 + x^2)^{3/2}$.

17. Since $f'(x) = e^x > 0$, f is increasing. Apply Theorem 4. Since $e^x \to \infty$ as $x \to \infty$, and $e^x \to 0$ as $x \to -\infty$, the range of e^x is the interval $(0, \infty)$.

21. Example: $f(x) = \sqrt{1 - x^2}$, on $[0, 1]$.

23. By Theorem 4, $g'(y) = 1/f'(x)$, where $x = g(y)$. This should add up to the required equation, in view of $f'(x) = 1/x$.

Section 3

1. $\log(x^2 - 1)$.

3. $2 \log(x - 1)$.

5. Already shown for $x > 0$. Let $h(x) = -x$. For $x < 0$,

$$\frac{d}{dx} \log|x| = \frac{d}{dx} \log(-x) = \frac{d}{dx} \log(h(x)) = \frac{1}{h(x)} h'(x) \qquad \text{(chain rule)}$$

$$= \frac{-1}{(-x)} = \frac{1}{x}.$$

7. $2x/(x^2 + 1)$.   9. $1 + \log x$.   11. $(1 - \log x)/x^2$.

13. 2.   15. $x^n \log x$.

25. $\log 1 = \log(1 \cdot 1) = \log 1 + \log 1$. So?

27. First show that the derivative of $f(ax) - f(x)$ is the zero function.

29. We require a to be rational because at the moment we have the power rule only for rational exponents.

31. Set $x = y^n$ and $b = 1/n$ in (30).

33. $-1/e$.

Section 4

1. $(\frac{1}{2} \log x + 1)x^{x^{1/2}-1/2}$.

3. $9^x \log 9$.

5. $(a \log x + a)x^{ax}$.

7. $(x \cot x + \log(\sin x))(\sin x)^x$.

9. $(\log 10)(2ax + b)10^{ax^2+bx+c}$.

11. $(\log x \sec^2 x + x^{-1} \tan x)x^{\tan x}$.

13. See Problem 20 of Section 1.

15. $x/(1 + x^2)^{1/2}$.

17. $(m/n)x^{m/n-1}$.

19. $(-1/x^3)4^{1/x}(x + \log 4)$.

21. $(\sec^2 x \log x + x^{-1} \tan x)x^{\tan x}$.

23. $-(1 - x)^{-1/2}(1 + x)^{-3/2}$.

Section 5

1. $1/2\sqrt{x - x^2}$.

3. $2x/\sqrt{1 - x^4}$.

5. ∓ 1 (-1 from 0 to π, $+1$ from π to 2π, etc.).

7. $\arcsin x$.

9. $2\sqrt{1 - x^2}$.

11. $-1/\sqrt{2 \sin x + 2 \sin^2 x}$.

13. 0.

15. The problem is to furnish a value of y such that $\arcsin(\sin y) \neq y$.

17. Show that it must be of the form $2n\pi + \arcsin x$ or $(2n + 1)\pi - \arcsin x$ for some integer n.

19. The identity involved is $\cos 2\theta = 1 - 2 \sin^2\theta$.

Section 6

1. $2x/(1 + x^4)$.

3. $a/(1 + a^2 x^2)$.

5. $1/x\sqrt{x^2 - 1}$.

7. $(1 + x)/(1 + x^2)$.

11. Show that

$$\cot\left(\frac{\pi}{2} - \arctan x\right) = \frac{\cos\left(\frac{\pi}{2} - \arctan x\right)}{\sin\left(\frac{\pi}{2} - \arctan x\right)}$$

reduces to x (by the $\cos(s - t)$ identity, etc.).

13. Show that $\operatorname{arcsec} x = \arccos(1/x)$, and then differentiate.

15. -1.

17. $1/(1 + x^2)$.

19. $-1/(x(\log x)^2 + x)$.

21. Show that the graph looks like this:

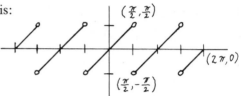

25. $1/\sqrt{2}$.

29. $x = \sqrt{ab + b^2}$.

CHAPTER 8

Section 1

1. $\frac{1}{7}x^7 + c$.

3. $\frac{1}{2}x^2 + c$.

5. $\frac{3}{4}x^{4/3} + c$.

7. $\frac{3}{4}x^4 + \frac{1}{6}x^{-2} + c$.

9. $\frac{2}{3}x^3 - \frac{1}{2}x^2 - 6x + c$.

11. $2\sqrt{x} + c$.

13. $ay^3 + c$.

15. $2x^2 - 4\sqrt{x} + c$.

17. $3t^3 + \frac{25}{2}t^2 + 14t + c$.

19. $3\sqrt[3]{x} + c$.

23. $2(x + 1)^{1/2} + c$.

25. $(3x + 2)^{1/3} + c.$

27. $2x^{3/2}/3 - 4/3.$

Section 2

1. $\int 2x\,dx = x^2 + c.$ When $x = 0,$
   $A = 0,$ so $c = 0;$ $A = 4^2 = 16.$

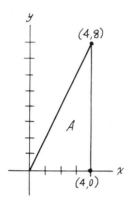

3. 64.

5. 16/3.

7. $15\frac{3}{4}.$

9. 2.

11. 2.

13. The top line has the equation

$$y = \frac{b_2 - b_1}{h}x + b_1;$$

$$A = \frac{b_1 + b_2}{2}h.$$

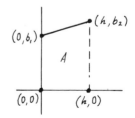

15. 81/4.

17. 32/3.

19. 2.

21. $e - e^{-N};$ $\lim_{n \to \infty} (e - e^{-N}) = e;$ area $= e.$

Section 3

1. The graph being rotated is $y = f(x) = \sqrt{r^2 - x^2},$ so $\pi f^2(x) = \pi(r^2 - x^2).$

3. $\pi/7$.

5. $\pi(6^4/5)$.

7. 3π.

9. $\frac{8}{5}\pi$.

11. π.

17. $\pi h^2(R - (h/3))$.

Section 4

1. $f(x) = x^2/2 - 3x + 13$.

3. $f(x) = (1/4)y^4 - (b^2/2)y^2 + 2b^2 - 4$.

5. $f(t) = (2/3)t^{3/2} + 2t^{1/2} - 28/3$.

7. $8p^2/3$.

9. 80 ft/sec.

11. $126\frac{2}{3}$ ft.

Section 5

1. $1/6$.

3. $-4/15$.

5. $y^3 - x^3$.

7. $-1/2$.

9. 64.

11. $16/3$.

13. $63/4$.

15. 2.

17. 48π.

19. $15\pi/4$.

21. $96\pi/5$.

23. π.

29. $2\cos x$.

31. 1.

33. $36/5$.

35. 2.

37. $\pi/2$.

39. 3.

41. $\pi/2$.

43. Divergent.

45. 6.

Section 6

 1. 8/3.

 3. 18.

 5. 8.

 7. 32/3.

 9. 8.

11. 3/4.

13. 1/12.

15. $2\sqrt{2}/3$.

17. 8 (The sum of two pieces each having area $= 4$).

19. 27/4.

Section 7

 1. $16\pi^5/5$.

 3. $\pi/8$.

 5. Area in the quadrant: $n\pi/2 \leq \theta \leq (n + 1)\pi/2 = \dfrac{e^{n\pi}}{4}(e^{\pi} - 1)$.

Section 8

 1. $2\sqrt{3} - 2/3$.

 3. $(e - e^{-1})/2$.

 5. $2\frac{1}{16}$.

Section 9

 1. $2\pi/3$.

 3. $4\sqrt{3}/3$.

5. 4/3.

7. 14/15.

9. $a^2h/3$.

13. $4r^3/3$.

Section 10

1. $\frac{2}{3}\pi a^3$.

3. $1024\pi/7 = 2^{10}\pi/7$.

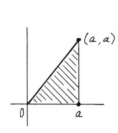

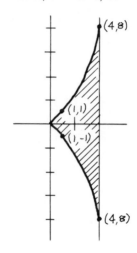

5. $8\pi/15$.

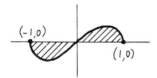

7. a) $-3x(4 - x^2)^{1/2}$ b) $28\pi\sqrt{21}$.

Section 11

1. a) Show that $\rho(x) \geq 0$ at the critical points and endpoints.
    b) 7/12.

3. a) $+1/6$ b) 0.

5. $\sqrt{2}$.

7. $\int_{-\infty}^{0} \rho(x)\,dx = \int_{-\infty}^{0} e^x dx = \lim_{N\to\infty} e^x \Big|_{-N}^{0} = \lim_{N\to\infty}(e^0 - e^{-N}) = 1.$

CHAPTER 9

Section 1

1. $-\frac{1}{2}\cos 2x + c.$

3. $\frac{1}{3}\sin^3 x + c.$

5. $\frac{-1}{3}e^{-t^3} + c.$

7. $-\sin(2 - x) + c.$

9. $-\log(1 + \cos x) + c.$

11. $\frac{1}{2}(\log x)^2 + c.$

13. $\log|x^2 + x - 1| + c.$

15. $\frac{3}{7}(a + y)^{7/3} - \frac{3a}{4}(a + y)^{4/3} + c.$

17. $(4/\sqrt{3})\arctan\sqrt{\frac{1}{3}x - 1}$
    $+ 2\sqrt{x - 3} + c.$

19. $-\frac{4}{3}(x^3 + 2)^{-2} + c.$

21. $2e^{\sqrt{x}} + c.$

23. $\frac{1}{4}(e^x + 1)^4 + c.$

25. $\frac{1}{3}(\log x)^3 + c.$

27. $\frac{1}{2}(\arctan t)^2 + c.$

29. $-2\cos\sqrt{x} + c.$

31. $-2\sqrt{1 - \sin t} + c.$

33. $\frac{1}{3}(x^2 + 1)^{3/2} + c$ (if $x > 0$).

35. $\arctan(e^t) + c.$

37. $\dfrac{1}{n + 1}\tan^{n+1}\theta + c.$

39. $\dfrac{1}{(n + 1)}\sin^{n+1}\theta + c.$

Section 2

1. $\frac{1}{6}\log\left|\dfrac{3 - x}{3 + x}\right| + c$

3. $\frac{1}{9}\log\left|1 - \dfrac{9}{x}\right| + c.$

5. $[1/(2\sqrt{3})]\log\left|\dfrac{x + 1 - \sqrt{3}}{x + 1 + \sqrt{3}}\right| + c$

7. $\log\left|\dfrac{x + 1}{x + 2}\right| + c$

9. $\frac{1}{6}\arctan\frac{2}{3}x + c.$

11. $\frac{1}{6}\log|x - 1| + \frac{5}{6}\log|x + 5| + c.$

13. $x - 2\log|x + 1| + (-1)/(x + 1) + c.$

15. $\frac{2}{3}\log|x + 5| + \frac{1}{3}\log|x - 1| + c.$

17. $(-5)/(x - 1) + 2\log|x| + c.$

19. $\log|\sec\theta + \tan\theta| + c.$

21. $\log\left(\dfrac{e^t + 1}{e^t + 2}\right) + c.$

23. $\frac{1}{8}\log\left(1 - \dfrac{4}{x^2}\right) + c.$

25. $(-1/16)\log|x| + (17/16)\log|x - 4| + (1/4x) + c.$

27. $6x^{1/6} - 6\arctan(x^{1/6}) + c.$

Section 3

1. $x/2 - \dfrac{\sin 2x}{4} + c = \frac{1}{2}(x - \sin x \cos x) + c.$

3. $(-1/4)\sin^3 x \cos x - (3/8)\sin x \cos x + (3/8)x + c.$

5. $(1/24)\sin^4 6x + c.$

7. $(-1/3)\csc^3 x + \csc x + c.$

9. $(1/6)\cos^5\theta \sin\theta + (5/24)\cos^3\theta \sin\theta + (5/16)\cos\theta \sin\theta + (5/16)x + c.$

11. $(-3/4)(\cos 2x)^{2/3} + (3/16)(\cos 2x)^{8/3} + c.$

13. $\log(\tan x) + c.$

15. $(-1/3)\cot^3 x - (1/5)\cot^5 x + c.$

17. $-4 \cot \dfrac{x}{4} - (4/3)\cot^3 \dfrac{x}{4} + c.$

19. $+\tfrac{2}{9} \sec^{9/2} x - \tfrac{2}{3} \sec^{5/2} x + c.$

21. $+\tfrac{1}{2} \sec^2 x + \log(\cos x) + c.$

23. $\tfrac{1}{4} \tan^4 x - \tfrac{1}{2} \sec^2 x + \log(\sec x) + c.$

25. $-x + \tan x - \tfrac{1}{3} \tan^3 x + \tfrac{1}{5} \tan^5 x + c.$

Section 4

1. $\tfrac{1}{3}(4 - x^2)^{3/2} - 4(4 - x^2)^{1/2} + c.$

3. $x/a^2\sqrt{a^2 + x^2} + c.$

5. $-\tfrac{1}{3}(16 - x^2)^{3/2} + c.$

7. $\log\left(\dfrac{\sqrt{1 + x^2} - 1}{|x|}\right) + c = -\log\left(\dfrac{\sqrt{1 + x^2} + 1}{|x|}\right) + c.$

9. $\dfrac{1}{5} \log\left(\dfrac{5 - \sqrt{25 - x^2}}{|x|}\right) + c = -\dfrac{1}{5} \log\left(\dfrac{5 + \sqrt{25 - x^2}}{|x|}\right) + c.$

11. $\dfrac{1}{3} \log\left(\dfrac{\sqrt{9 + 4x^2} - 3}{|2x|}\right) + c.$

13. $x/\sqrt{a^2 - x^2} - \arcsin(x/a) + c.$

15. $-\dfrac{1}{4}(x - 1)(x^2 - 2x - 3)^{-1/2} + c.$

17. $-(1 - x^2)^{1/2} + c.$

19. $(1 + x^2)^{1/2} + c.$

21. $(x^2 - a^2)^{1/2} + c.$

23. $2\sqrt{x} - 2\log(1 + \sqrt{x}) + c.$

25. $\dfrac{6}{7}x^{7/6} - \dfrac{6}{5}x^{5/6} + \dfrac{3}{2}x^{2/3} + 2x^{1/2} - 3x^{1/3} - 6x^{1/6} + 3\log(x^{1/3} + 1)$
$+ 6\arctan(x^{1/6}) + c.$

27. $2\sqrt{1-x} + \sqrt{2}\log\left|\dfrac{\sqrt{1-x} - \sqrt{2}}{\sqrt{1-x} + \sqrt{2}}\right| + c.$

29. $-\arctan(\cos t) + c.$

31. $-(x^2)/2 - (a^2/2)\log|a^2 - x^2| + c.$

33. $2\sqrt{x} - 2\arctan(\sqrt{x}) + c.$

Section 5

1. $\sin x - x \cos x + c.$

3. $\frac{2}{3}x(x + 1)^{3/2} - \frac{4}{15}(x + 1)^{5/2} + c.$

5. $-y^2\cos x + 2y \sin x + 2 \cos x + c.$

7. $-x^2 e^{-x} - 2xe^{-x} - 2e^{-x} + c.$

9. $\frac{1}{2}(x - \sin x \cos x) + c.$

11. $x \arcsin x + \sqrt{1 - x^2} + c.$

13. $\frac{1}{2}(\log x)^2 + c.$

15. $2 \sin \sqrt{x} - 2\sqrt{x} \cos \sqrt{x} + c.$

17. $x \log(1 + x^2) - 2x + 2 \arctan x + c.$

19. $\frac{1}{2}[(x^2 + 1)\arctan x - x] + c.$

21. $(x/(x + 1))\log x - \log(x + 1) + c.$

23. $(x + 1)\arctan \sqrt{x} - \sqrt{x} + c.$

25. $e^{ax}(a \cos bx + b \sin bx)/(a^2 + b^2).$

27. $\dfrac{1}{n - 1}\dfrac{\sin x}{\cos^{n-1} x} + \dfrac{n - 2}{n - 1}\displaystyle\int \sec^{n-2}x\,dx.$

29. $x(\log x)^n - n\displaystyle\int (\log x)^{n-1}dx.$

Section 6

1. $\log\left|\dfrac{x - 1}{x}\right| + \dfrac{1}{x} + \dfrac{1}{2x^2} + c.$

3. $x^{-1} + \dfrac{1}{2} \log \left| \dfrac{x-1}{x+1} \right| + c.$

5. $-x^{-1} - \arctan x + c.$

7. $\dfrac{1}{4} \log \left| \dfrac{1+x}{1-x} \right| + \dfrac{x}{2(1-x^2)} + c$

9. $\dfrac{1}{6} \log \left(\dfrac{x+1}{x-1} \right) + \dfrac{1}{12} \log \left(\dfrac{x-2}{x+2} \right) + c.$

11. $\dfrac{1}{3} \arctan x - \dfrac{1}{6} \arctan \dfrac{x}{2} + c.$

13. a) $a = 1;\ b = 0,\ c = -1,\ d = 0$

   b) $\frac{1}{2} \log(x^2 + 1) + 1/2(x^2 + 1) + c.$

15. $\dfrac{1}{2(n-1)a^2} \dfrac{x}{(x^2+a^2)^{n-1}} + \dfrac{2n-3}{2(n-1)a^2} \displaystyle\int \dfrac{dx}{(x^2+a^2)^{n-1}}.$ Consider $\int u\,dv$, where

   $u = x^{-1}(x^2 + a^2)^{-n-1},\ dv = x\,dx.$

Section 8

1. $2/3.$

3. $9/4.$

5. $(18 - 10\sqrt{3})/27.$

7. $\frac{1}{2} \log 3.$

9. $\log(4/3).$

11. $\frac{3}{2} \arctan(1/\sqrt{2}) - 1/\sqrt{2}.$

13. $(14\sqrt{3}/5)a^5.$

15. $\displaystyle\int_a^{\sqrt{3}a} \dfrac{dx}{x^2\sqrt{a^2+x^2}} = \dfrac{1}{a^2} \left(\dfrac{\sqrt{6}-2}{\sqrt{3}} \right).$

17. $a.$

19. $1/\sqrt{2} + \frac{1}{2} \log(1 + \sqrt{2}).$

21. $\pi/2.$

23. $1.$

25. Apply Theorem 1.

Section 9

3. $2\pi(\log 2 - 1)^2$.

7. $2\pi^2$.   9. π.

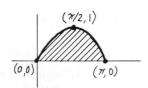

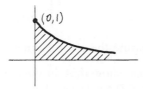

11. $\frac{3}{2} - 2 \log 2$.

13. $\pi[\frac{1}{3} - 2(\log 2 - 1)^2]$ (or $\pi/3 - 2\pi(1 - \log 2)^2$) .

CHAPTER 10

Section 1

3.
$$\sum_{k=0}^{999} \frac{1}{1000 + k}.$$

5. a) $S_4 = 183/256$.
   b) $\int_1^{3/2} x^2 dx = 19/24$. Error estimate of Theorem 1 $= 3/32$. Verify that
   $\frac{19}{24} - \frac{183}{256} < \frac{3}{32}$.

7. a) $S_4 = 1437/1700 = 0.845 \cdots$
   b) Using $K = 2/3$ (from Problem 6), the error estimate of Theorem 1 is
   $1/12$. So

$$0.845 - \tfrac{1}{12} < \frac{\pi}{4} < 0.846 + \tfrac{1}{12}.$$

   Now multiply through by 4 and finish. (Actually, for a decreasing function,
   S_n is larger than the integral. So the $+1/12$ on the right above is not needed,
   and you can conclude that $3.0 < \pi < 3.4$.)

Section 2

1. 0.

3. 4/3.

5. 3.

7. 30 mph.

9. Average of instantaneous velocity $= 1/(t_1 - t_0) \int_{t_0}^{t_1} f'(t)\, dt = \cdots$

Section 3

1. The sum listed is the midpoint evaluation $\overline{S}_8$ for the integral $\log 2 = \int_1^2 dx/x$. A more recognizable form is:

$$\frac{1}{8}\left[\frac{1}{17/16} + \frac{1}{19/16} + \cdots + \frac{1}{31/16}\right].$$

So compute the error estimate for $\overline{S}_8$.

3. One must show that, in the accompanying figure, the shaded region R_1 with corners CDA is larger than the shaded region R_2 with corners DEB, because if that is the case, then the rectangle $ABZX$ misses more of the area under the curve than the extra amount it contains. To show this, use the fact that the graph lies above its tangent line at D, and that the tangent line cuts off congruent triangles.

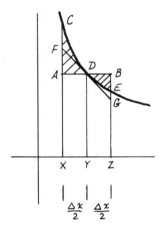

5. The sum is $\overline{S}_5$ in the integral $\pi/4 = \int_1^2 (dx/(1 + x^2))$. So work out the value of the error formula.

Section 4

1. $\log(\sqrt{2} + 1)$.

3. $(1/a^2)((4/9 + a^2)^{3/2} - 8/27)$.

5. $(8/27)(10\sqrt{10} - 1)$.

7. $\sqrt{1 + e^2} - \sqrt{2} - \log[(\sqrt{1 + e^{-2}} + e^{-1})/(\sqrt{2} + 1)]$.

9. Any smooth curve may be approximated arbitrarily closely by an inscribed polygonal path; so it suffices to prove the inequalities for such a polygonal

path P. The curve is decreasing and $0 = x_0 < x_1 < \cdots < x_n = 1$; therefore $1 - y_0 > y_1 > \cdots > y_n = 0$.

$$\sqrt{(x_k - x_{k-1})^2 + (y_k - y_{k-1})^2} \le (x_k - x_{k-1}) + (y_{k-1} - y_k).$$

(Triangle Law)

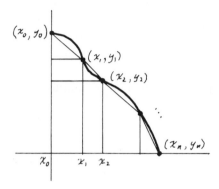

Show, therefore, that

the length of $P \le (x_n - x_0) + (y_0 - y_n) = 2$.

The distance from $(0,1)$ to $(1,0)$ is $\sqrt{2}$, and the shortest path between two points is a straight line, so $\sqrt{2} \le$ length of $P \le 2$.

Section 5

1. $7/6$.

3. 1.

5. $105/62$.

7. $\pi/2$.

9. $(2 \log 2 - 3/4)/(2 \log 2 - 1)$.

11. Since $\bar{x}_1 = \int_a^b x\rho(x)\,dx/m_1$, we have $\int_a^b x\rho(x)\,dx = m_1\bar{x}_1$, etc.

13. $M = \int_a^b (x - p)\rho(x)\,dx = \int_a^b x\rho(x)\,dx - p\int_a^b \rho(x)\,dx = m\bar{x} - pm$.

Section 6

1. $(a/2, b/2)$.   3. $(0, 4/3)$.   5. $(8/5, 0)$.

7. $(1/2, 2/5)$.   9. $(8\sqrt{2}/15,\ 16\sqrt{2}/21)$.   11. $([ae^a/(e^a - 1)] - 1, 0)$.

13. $(4a/3\pi, 0)$.   15. $(1, 0)$.   17. $(1/2, 8/5)$.

19. $(4r/3\pi, 0)$.

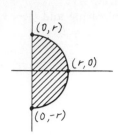

21. $2\pi^2 Rab$.

23. The centroid is at $(a + \sqrt{A/2}, 0)$. Apply the Theorem of Pappus.

27. $(0, \pi/8)$.

29. $(0, 1)$.

Section 7

1. $$\frac{1}{10}\left[\frac{1 + 1/2}{2} + \sum_{k=1}^{9} \frac{10}{10 + k}\right] = \frac{3}{40} + \frac{1}{11} + \frac{1}{12} + \frac{1}{13} + \cdots + \frac{1}{19};$$

$$K\frac{(b - a)(\Delta x)^2}{12} = \frac{1}{600}.$$

3.

$$S = \frac{1}{10}\left[\frac{3}{4} + \frac{100}{101} + \frac{25}{26} + \frac{100}{109} + \frac{25}{29} + \frac{4}{5} + \frac{25}{34} + \frac{100}{199} + \frac{25}{41} + \frac{100}{181}\right].$$

The main problem in the error estimate is finding a reasonable bound K for the second derivative of $f(x) = 1/(1 + x^2)$. Compute $f''(x)$ and show that $|f''(x)| \le 2$ (because $|1 - 3x^2| \le 1 + 3x^2$).

9. $K = 6$ and

$$\frac{K(b - a)(\Delta x)^3}{24} = \frac{1}{4096}.$$

CHAPTER 11

Section 1

1. $y = x^2$; parabola $dy/dx = 2t$; critical point $t = 0$.

3. $y = (x + 1)^3$; $dy/dt = 3t^2$,
   critical point at $t = 0$.

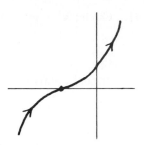

5. $y = 3x/2$; straight line;
   $$\frac{dy}{dx} = \frac{dy/dt}{dx/dt} = \frac{3}{2}\frac{t^{-2/3}}{t^{-2/3}} = \frac{3}{2},$$
   except at $t = 0$, where undefined.

(2,3)

7. $\dfrac{x^2}{4} + \dfrac{y^2}{9} = 1$;   ellipse;
   $$\frac{dy}{dx} = -\frac{3}{2}\tan t;$$
   critical points at $t = n\pi/2$.

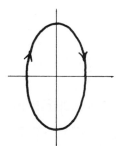

9. Ellipse; $\dfrac{x^2}{4} + \dfrac{y^2}{16} = 1$;
   $$\frac{dy}{dx} = -2\cot t;$$
   critical points $t = n\pi/2$.

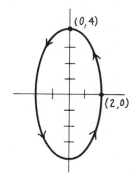

(0,4)

(2,0)

11. Circle; $x^2 + y^2 = 4$;

$$\frac{dy}{dx} = -\cot t;$$

critical points $t = n\pi/2$.

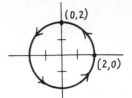

13.

15.

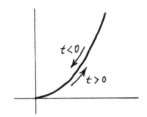

17. 5 sec; 300 ft; 100 ft.

19.

t	$-\infty$	0	$+\infty$
x	$+\infty$	0	$+\infty$
y	$+\infty$	0	$+\infty$

Singular point $t = 0$.

21.

t	$-\infty$	-1	0	1	2	$+\infty$
x	$-\infty$	2	0	-2	2	$+\infty$
y	$-\infty$	-4	0	-2	-4	$+\infty$

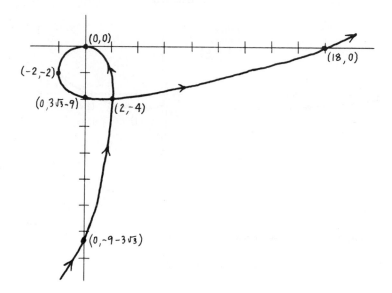

23.

t	$-\infty$	0	1	2	$+\infty$
x	$-\infty$	0	1	2	$+\infty$
y	$-\infty$	0	-2	-4	$+\infty$

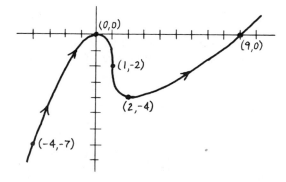

25. Singular point at $t = 0$.

t	$-\infty$	-1	0	$1/2$	1	$+\infty$
x	$-\infty$	-7	0	$-1/4$	1	$+\infty$
y	$+\infty$	-1	0	$-7/16$	-1	$+\infty$

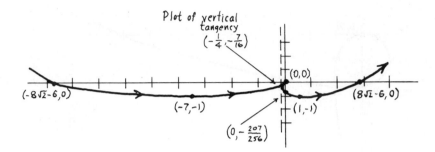

Plot of vertical tangency

Section 3

1. $(1 + 4t^2)^{1/2}$.

3. $(9\cos^2 t + \sin^2 t)^{1/2} = \sqrt{1 + 8\cos^2 t}$.

5. $a^2/2 + a$.

7. a) $5\sqrt{10} - 8\sqrt{2}$ b) $32 - 8\sqrt{2}$.

9. $\sqrt{2}(e - 1)$.   11. $1/2$.

Section 4

1. $(8/3)((\pi^2 + 1)^{3/2} - 1)$.

3. 1.

5. $\log(2[\sqrt{1 + \pi^2} + \pi]/[\sqrt{4 + \pi^2} + \pi]) + \dfrac{1}{\pi}[\sqrt{4 + \pi^2} - \sqrt{1 + \pi^2}]$.

9. $\cot\psi = 1$. So ψ has the constant value $\pi/4$. The curve is turning counterclockwise at a constant rate equal to the turning rate of a ray from the origin. $\phi = \theta + \pi/4$.

11. $\cot\psi = -\tan\theta$, so $\psi = \theta + \pi/2$.
    Erect the perpendicular to the tangent line T at the point of tangency P, intersecting the axis $\theta = 0$ at C. The fact that $\psi = \theta + \pi/2$ means that $\triangle OCP$ is isosceles consistent with the fact that the graph is a circle centered at C.

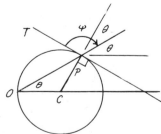

Section 5

1. $2/(1 + 4x^2)^{3/2}$.

3. $-x/(1 + x^2)^{3/2}$.

5. 0.

7. The curvature takes on the value -1 at $x = 0$ and $+1$ at $x = \pi$. It is periodic, of period 2π, and is bounded by 1 in absolute value.

9. $2/3|a|$.

11. If a function graph has 0 curvature, then $d^2y/dx^2 = 0$. Integrate this equation twice.

13. $K = ab/(a^2 \sin^2 t + b^2 \cos^2 t)^{3/2}$. Max and min: a/b^2, b/a^2 (supposing $0 < b < a$).

17. $K = 3/2|a|\sqrt{2 + 2 \cos \theta}$.

19. $K = (\theta^2 + 2)/|a|(\theta^2 + 1)^{3/2}$.

Section 6

1. 0.   3. 1.   5. 1/2.

7. 1/2.   9. $-1/2$.   11. $-1/4$.

13. 1.

15. 0 (but not by l'Hôpital; see page 283).

17. 1.

19. 0.

21. The definition of $f'(0)$ leads to the following direct proof.

$$\lim_{x \to 0} \frac{f(x) - f(-x)}{x} = \lim_{x \to 0} \frac{f(x) - f(0) - [f(-x) - f(0)]}{x}$$

$$= \lim_{x \to 0} \frac{f(x) - f(0)}{x} + \lim_{x \to 0} \frac{f(-x) - f(0)}{-x}$$

$$= f'(0) + f'(0) = 2f'(0).$$

But work the problem by l'Hôpital's rule, assuming that $f'(x)$ is continuous.

23. l'Hôpital gives
$$\lim_{x \to 0} \frac{\log(1 + x)}{x} = 1.$$

25. If we define $H(0) = 0$, then H becomes continuous at the origin, and $H'(y)/G'(y)$ turns out to be $h'(1/y)/g'(1/y)$.

Section 7

3. Let

$$g(x) = f(x) - \left[\frac{f(b) - f(a)}{b - a}\right](x - a).$$

Assume there is no point X as in (D); then g' is not 0 in (a, b), so, by (A), either $g' > 0$ on all (a, b) or $g' < 0$ on all (a, b). In the former case, g is increasing, by (B). In the latter case, g is decreasing, by (B'). But $g(a) = g(b) = f(a)$, so g can be neither decreasing or increasing. Contradiction. So the point X exists after all.

7. Say p has roots $x_1, \ldots, x_n$ with $x_1 < x_2 < \cdots < x_n$. By Rolle's theorem, p' is 0 in each interval (x_k, x_{k+1}) $k = 1, \ldots, n - 1$. p' is of degree $n - 1$, so this accounts for all its roots and forces them to be distinct.

CHAPTER 12

Section 1

1. $0.\overline{3}$.

3. $0.\overline{142857}$.

5. $0.\overline{714285}$.

7. $0.\overline{2}$.

9. $0.0\overline{9}$.

11. $.\overline{076923}$.

13. $.\overline{230769}$.

15. $.\overline{012987}$.

17. $.0\overline{27}$.

19. $.004\overline{1841}$.

21. $5/9$.

23. $1/33$.

25. $1/9$.

27. $37/99$.

29. $1/18$.

31. $16/45$.

33. $109/330$.

35. $8/11$.

37. $31/55$.

39. $0.03 < b - a < 0.05$, $2.46 \le b + a < 2.48$.

41. $e < 2.8 = 2(1.4) \le 2\sqrt{2}$, so $e^2 < (2\sqrt{2})^2 = 8$.

43. $3.111 < x < 3.171$.

45. $1.985 < x < 1.995$.

47. $2.108 < x < 2.118$.

49. $|x - 1.416| < 0.012$.

51. $|x - 2.718| < 0.001$.

53. $|x - 2.236| < 0.05$.

55. We are assuming that

$$\frac{1}{n} = \frac{m}{10^k},$$

where m is a positive integer. So $mn = 10^k$ and n divides 10^k. Therefore, ...

Section 2

1. 2.

3. 0.

5. -5.

7. 2.

9. $1/2$.

11. $\lim n \sin \frac{1}{n} = \lim \dfrac{\sin \frac{1}{n}}{\frac{1}{n}} = \lim\limits_{x \to 0} \dfrac{\sin x}{x} = 1.$

    The last limit was fundamental in Chapter 4, but it also follows from l'Hôpital's rule.

13. 1.

15. -1.

Section 3

1. $\{a_n\}$ decreasing and bounded below by B implies $\{-a_n\}$ increasing and bounded above by $-B$. So?

3. $b - a = \lim(b_n - a_n)$; therefore, ...

5. Since each m-place decimal is also an $(m + 1)$-place decimal (with zero in the last place), it follows that b_{m+1} is the smallest of a collection of numbers that includes b_m. So, ...

7. This just repeats an argument given in the text.

Section 4

1. $4/3$.   3. $3/5$.   5. 3.

7. $1/6$.   9. $3/4$.   11. $27/16$.

13. 12.

15. $1/6$.

19. Multiply by 2, and compare to $\sum 1/n$.

21. Multiply by 4 and compare to $\sum 1/n$.

23. $\displaystyle\sum_{n=1}^{\infty} \frac{23}{(100)^n} = \frac{23}{99}$.

25. $\displaystyle\sum_{n=1}^{\infty} \frac{315}{(1000)^n} = \frac{35}{111}$.

27. $\displaystyle 2 + \sum_{n=1}^{\infty} \frac{1}{100^n} = 2\frac{1}{99}$.

31.
$$2s_n = \frac{2}{1 \cdot 3} + \frac{2}{3 \cdot 5} + \frac{2}{5 \cdot 7} + \cdots + \frac{2}{(2n-1)(2n+1)}$$
$$= \left[\frac{1}{1} - \frac{1}{3}\right] + \left[\frac{1}{3} - \frac{1}{5}\right] + \left[\frac{1}{5} - \frac{1}{7}\right] + \cdots + \left[\frac{1}{2n-1} - \frac{1}{2n+1}\right]$$
$$= 1 - \frac{1}{2n+1} = \frac{2n}{2n+1}.$$

Therefore $s = \frac{1}{2}$.

33. This series can be obtained as the sum of the series in Problem 31 and the series in Problem 32. Its sum is $3/4$.

35. 390 ft.

Section 5

3. Diverges.

5. Converges.

7. Diverges.

9. Converges   (because $\log n < n$).

11. Converges   ($\sin x < x$ for positive x).

13. Converges   ($n!/n^n \leq 2/n^2$).

Section 6

1. Converges

3. Converges

5. Test fails   (but the series converges by comparison with $\sum 1/n^2$)

7. Test fails   (but have divergence by comparison with harmonic series)

9. Converges

13. Diverges

15. Converges   ($\sin x/x \to 1$ as $x \to 0$)

17. Converges

19. Converges

21. Converges

23. Diverges

25. Converges

27. $n!/n^n < 1/n$

Section 7

1. Converges

3. Converges

5. Diverges

7. $1/x^2$ is decreasing and continuous, so, by Theorem 7,

$$\sum_{n=2}^{\infty} \frac{1}{n^2} < \int_1^{\infty} \frac{dx}{x^2} = 1.$$

9. $\sum_2^{\infty} 1/n^3 < 1/2$, as in Problem 7.

11. $\displaystyle\int_2^{\infty} \frac{x\,dx}{x^3+1} < \int_2^{\infty} \frac{dx}{x^2} = \frac{1}{2}.$

13. $\displaystyle\int_1^{\infty} e^{-x^2}dx < \int_1^{\infty} xe^{-x^2}dx = \frac{1}{2e}.$

15. The trapezoidal approximation to $\int_a^b f$ is less than the integral if f is concave down. Therefore,

$$\frac{\log 1}{2} + \log 2 + \cdots + \log(n-1) + \frac{\log n}{2} < \int_1^n \log x\,dx$$

or

$$\log[(n-1)!\,\sqrt{n}] < n\log n - n + 1.$$

CHAPTER 13

Section 1

1. $(-1, 1)$ 3. $(-\infty, \infty)$ 5. $(-1, 1)$

7. $(-1, 1)$ 9. $(0, 4)$ 11. $(\frac{1}{2}, \frac{3}{2})$

13. $(0, 6)$

15. 6

17. $f(x) = x - \dfrac{x^3}{3} + \dfrac{x^5}{5} + \cdots + (-1)^n \dfrac{x^{2n+1}}{2n+1} + \cdots$

converges on $(-1, 1)$ by the ratio test. Show that the series for $f'(x)$ has the sum

$$\frac{1}{1 - (-x^2)} = \frac{1}{1 + x^2} = \frac{d}{dx}(\arctan x), \quad \text{etc.}$$

19. $-\log(1 - x)/x$

21. $(2 - x)/(1 - x)^2$

23. $(1 - x)\log(1 - x) + 2x + 1$

27. 0.40546510

Section 2

1. $1 + x + x^2 + x^3 + \cdots$

3. $0 + 0 + 0 + x^3 + 0 + \cdots$

5. $5 - 2x + 0 + x^3 + 0 + \cdots$

7. $\displaystyle\sum_{n=0}^{\infty} \frac{(-1)^n x^{2n}}{n!}$

9. $\displaystyle\sum_{n=0}^{\infty} \frac{(-1)^n x^{6n+3}}{(2n + 1)}$

11. $\displaystyle\sum_{n=0}^{\infty} \frac{(-1)^n x^{2n+1}}{2^n(2n + 1)n!}$

Section 4

1. $(-1, 1]$ 3. $[-1, 1)$ 5. $(0, 4)$

7. $(-1, 3)$

9. $[-2, 4]$

11. $\sin 0.2 \simeq 0.19866933$ (The next digit is between 0 and 3.)

13. Integrate $e^{-t^2/2} = \displaystyle\sum_{n=0}^{\infty} \frac{(-t^2/2)^n}{n!}$ termwise; 0.86.

15. Error $< (1/200)^3/3 = (1/2.4)10^{-7}$ must be multiplied by 4, since we are approximating $4 \arctan 1/239$, so

$$\text{Error} < \left(\frac{4}{2.4}\right)10^{-7} < 2(10^{-7}).$$

Section 5

1. $p_3(x) = x + x^3/3!$ 3. $p_3(x) = 1 + x^2/2$

5. $p_4(x) = (\sqrt{2}/2)(1 + x - (x^2/2!) - (x^3/3!) + (x^4/4!))$

7. $p_4(x) = x + (x^3/3)$

9.

$$f(x) = 1 + \frac{x^2}{2} + \frac{(5 \sec^3 X \tan X + \sec X \tan^2 X)}{6} x^3.$$

We can take $K = 1$. (Use the fact that $1/2 < \pi/6$.)

11. $f(x) = 1 - x + (x^2/2) - (x^3/3!) + (x^4/4!) - (e^{-X}/5!)x^5$ can take $K = 1/60$ (because $e^{1/2} < 2$).

13. $f(x) = 1 + (3/2)x + (3/8)x^2 - (1/16)x^3 + (\frac{9}{16}(1 + X)^{-5/2}/4!)x^4$; $K = 3\sqrt{2}/32$.

15. We proved earlier that the derivative of an odd function is even and vice versa, so if f is odd, then $f^{(2n)}$ is odd for all $n = 0, 1, \ldots$. So f has no even terms in its expansion. The argument for f even is similar.

17. $A = 1/2, B = -1/8, C = 1/16$

19. $x + x^3/3$

21. For $-1 < x \le 0$, the remainder is less than $|x|^{n+1}$. For $0 < x < 1/2$, the remainder is less than $(2x)^{n+1}$.

23. Let $g(x) = f(x) - p_n(x)$. Then g has derivatives up through order $n - 1$ on an interval about the origin, $g''(0)$ exists, and $g(0) = g'(0) = \cdots = g''(0) = 0$.

25. Suppose f is even and suppose p_n is one of its Maclaurin polynomials.

$$\lim_{x \to 0} \frac{f(x) - p_n(x)}{x^n} = 0; \qquad \lim_{x \to 0} \frac{f(-x) - p_n(-x)}{(-x)^n} = 0.$$

$$f(x) = f(-x), \text{ so } \lim_{x \to 0} \frac{f(x) - p_n(-x)}{x^n} = 0.$$

Therefore, by Problem 24, $p_n(-x)$ is also the Maclaurin polynomial of f of degree n; that is, $p_n(-x) = p_n(x)$. So p_n is even. The proof for f odd is similar.

Section 6

1. $(x - 1) - ((x - 1)^2/2) + ((x - 1)^3/3)$

3. $1/2 + (\sqrt{3}/2)(x - \pi/6) - (1/4)(x - \pi/6)^2 - (\sqrt{3}/12)(x - \pi/6)^3$
    $+ (1/48)(x - \pi/6)^4$

5. $e^a + e^a(x - a) + (e^a/2)(x - a)^2 + (e^a/6)(x - a)^3$

7. $\dfrac{\sqrt{2}}{2}\left[1 - (x - \pi/4) - \dfrac{(x - \pi/4)^2}{2} + \dfrac{(x - \pi/4)^3}{3!} + \dfrac{(x - \pi/4)^4}{4!}\right]$

    $-\dfrac{\sin X}{5!}(x - \pi/4)^5$

9. p_4 as in 3. $r_4(x) = (\cos X/5!)(x - \pi/6)^5$

11. $1 + (1/3)(x - 1) - (1/9)(x - 1)^2 + (5/81)(x - 1)^3$
    $-(10/243)X^{-11/3}(x - 1)^4$

13.
$$1 + r(x - 1) + \frac{r(r - 1)}{2}(x - 1)^2 + \frac{r(r - 1)(r - 2)}{3!}(x - 1)^3$$
$$+ \frac{r(r - 1)(r - 2)(r - 3)}{4!}X^{r-4}(x - 1)^4$$

15.
$$-\tfrac{1}{2}\log 2 - (x - \pi/4) - (x - \pi/4)^2 - \tfrac{2}{3}(x - \pi/4)^3$$
$$- \frac{4\sec^2 X \tan^2 X + 2\sec^4 X}{4!}(x - \pi/4)^4$$

17. $1 + \tfrac{1}{2}(x - 1) - \tfrac{1}{8}(x - 1)^2 + \cdots$
    $+ \dfrac{\tfrac{1}{2}(-\tfrac{1}{2})(-\tfrac{3}{2}) \cdots ((3 - 2n)/2)}{n!}(x - 1)^n + \cdots$

19. $\displaystyle\sum_{n=0}^{\infty} \frac{(-1)^n(x - \pi/2)^{2n}}{(2n)!}$

21. $\displaystyle\sum_{n=0}^{\infty} (-1)^{n+1}\frac{(x - \pi)^{2n+1}}{(2n + 1)!}$

25. $4 + 7(x - 1) + 5(x - 1)^2 + (x - 1)^3$

27. $|r_n(x)| = \left| \frac{\left(\begin{smallmatrix} \sin \\ \cos \end{smallmatrix}\right)(X)(x - \pi/2)^{n+1}}{(n+1)!} \right| \leq \frac{(x - \pi/2)^{n+1}}{(n+1)!}$

Since x is fixed, you must show that $(K^n/n!) \to 0$ as $n \to \infty$.

29. If $x > a$, then $|r_n(x)| \leq e^x \left| \frac{(x-a)^{n+1}}{(n+1)!} \right| = \frac{K_1(K_2)^{n+1}}{(n+1)!}$

Section 7

1. $\frac{1}{3} \times 10^{-5}$; $\frac{1}{48} \times 10^{-5}$; $10^{-5}/3$; $10^{-5}/48$

5. Say we want $(p(a), p'(a), p(b), p'(b)) = (A, B, C, D)$. Let $p_1(x) = B(x - a) + A$; then $p_1(a) = A$ and $p_1'(a) = B$. Choose p_2 by (4) such that $p_2(a) = p_2'(a) = 0$ and $p_2(b) = C - p_1(b)$ and $p_2'(b) = D - p_1'(b)$. Then it is easy to check that $p = p_1 + p_2$ satisfies the correct equalities. For uniqueness, say there were another polynomial q that satisfied the same requirements. Show that $p - q = 0$ by the uniqueness in Problem 4.

7. If f is a third-degree polynomial, then $f''' \equiv 0$, so one may set $K = 0$ in the error formula.

9. p a polynomial of degree at most 5;

$$p(b) - p(a) = \frac{p'(b) + p'(a)}{2}(b - a) - \frac{p''(b) - p''(a)}{12}(b - a)^2 + \frac{p^{(5)}(0)(b - a)^5}{720}.$$

Section 8

1. A polynomial of degree three or less has fourth derivative 0, so one can set $K = 0$ in the error estimate.

CHAPTER 14

Section 1

1. $(-1, 3)$

3. $(23, 5/2)$

5. $A + X = (a, b) + (x, y) = (a + x, b + y)$. Thus,

$$A + X = A \leftrightarrow (a + x, b + y) = (a + b)$$
$$\leftrightarrow a + x = a \quad \text{and} \quad b + y = b$$
$$\leftrightarrow x = 0 \quad \text{and} \quad y = 0.$$

7.

$$(U + X) + A = ((u,v) + (x,y)) + (a + b)$$
$$= (u + x, v + y) + (a + b)$$
$$= ((u + x) + a, (v + y) + b)$$
$$= (u + (x + a), v + (y + b)), \quad \text{etc.}$$

9. $X' - A' = (-1, -2) = -(1, 2) = -(X - A)$

11. $\overrightarrow{A'X'}$ is half as long as $\overrightarrow{AX}$, and in the opposite direction.

13. $\overrightarrow{A'X'}$ is one-third as long as $\overrightarrow{AX}$ and in the opposite direction.

Section 2

1. $X = (1, 4) + t(1, -6);\ x = 1 + t,\ y = 4 - 6t.$

3. $X = (1, 1) + t(3, 0);\ x = 1 + 3t,\ y = 1.$

5. $X = (-1, 1/2) + t(3/2, 1/4);\ x = -1 + (3/2)t,\ y = 1/2 + (1/4)t$

7. $(5/2, 7/4)$

9. $(-13/8, -15/4)$

11. Say the parallelogram has vertices x_1, x_2, x_3, x_4 (in clockwise order); then the diagonals go from x_1 to x_3 and from x_2 to x_4. We know that $x_2 - x_1 = x_3 - x_4$. Show that, therefore, $\frac{1}{2}(x_2 + x_4) = \frac{1}{2}(x_1 + x_3)$, and interpret this result.

13. The new vertices are $A = \frac{1}{2}(x_1 + x_2)$, $B = \frac{1}{2}(x_2 + x_3)$, $C = \frac{1}{2}(x_3 + x_4)$, and $D = \frac{1}{2}(x_4 + x_1)$. Check that $A - B = D - C$, and interpret.

15. A is on the line from 0 to Y, so $A = t_1 Y$. B is on the line from X to Y, so $B = X + t_2\ (Y - X)$. $\overrightarrow{AB}$ is parallel to $\overrightarrow{OX}$, so $B - A = tX$. Thus $X + t_2(Y - X) - t_1 Y = tX$. Collect coefficients and show that either $t_1 = t_2$ or the vectors X and Y are parallel.

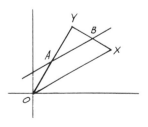

Section 3

3. $X = (-2, 3) + t(1, 1)$; $x = -2 + t$, $y = 3 + t$

5. $X = t(a, b)$; $x = ta$, $y = tb$

7. $X = (2, -1) + t(1, 4)$; $x = 2 + t$, $y = -1 + 4t$

9. $(5/4, 9/4)$

11. $(-6, -13)$

13. $(-5, -11)$

15. $(4, 4)$

17. $(1, 1)$

19. $(1, 0)$ has components $(-1-2)$;
     $(0, 1)$ has components $(1, 1)$;
     So $(7,1)$ has components

$$7(-1, -2) + 1(1, 1) = (-6, -13)$$

and $(-3,4)$ has components

$$-3(-1, -2) + 4(1, 1) = (7, 10),$$

and so on.

Section 4

1. $|V| = 5$; $\dfrac{V}{|V|} = (3/5, 4/5)$

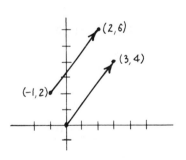

3. $|A| = \sqrt{10}$, $\dfrac{A}{|A|} = \left(\dfrac{-3}{\sqrt{10}}, \dfrac{1}{\sqrt{10}} \right)$

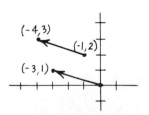

5. $|V| = \sqrt{5}, \quad \dfrac{V}{|B|} = \left(\dfrac{1}{\sqrt{5}}, \dfrac{-2}{\sqrt{5}} \right)$

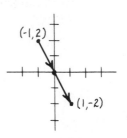

7. $A \cdot X = 5/2; \phi = \arccos(5/\sqrt{34})$

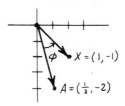

9. $A \cdot X = 0; \phi = \pi/2.$

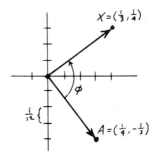

11. $\pm(2/\sqrt{5}, -1\sqrt{5})$

13. $\left(\dfrac{\sqrt{2} - \sqrt{6}}{4}, \dfrac{\sqrt{6} + \sqrt{2}}{4} \right)$ or $\left(\dfrac{\sqrt{6} + \sqrt{2}}{4}, \dfrac{\sqrt{2} - \sqrt{6}}{4} \right)$

15. $t(1, \sqrt{3}), s(1, -\sqrt{3})$

17. $-2x + 3y = 0$

19. The equation of l is $A \cdot (x - x_0) = 0$. The point x_p closest to the origin is on the line through the origin perpendicular to l, so $x_p = tA$. So

$$A \cdot (tA - x_0) = 0 \Rightarrow tA \cdot A - A \cdot x_0 = 0$$

$$\Rightarrow t = A \cdot x_0/A \cdot A.$$

21. $(2\sqrt{5}, \sqrt{5})$

23. $(1/\sqrt{5}, 2/\sqrt{5})$

25. $(2/\sqrt{5}, -1/\sqrt{5})$

27. $(2, -2)$

29. $(3/5, -4/5)$

31. Existence: Let $x_i = X \cdot B_i$, $i = 1, 2$, and let $U = X - (x_1 B_1 + x_2 B_2)$. Then

$$U \cdot B_1 = X \cdot B_1 - x_1 B_1 \cdot B_1 - x_2 B_2 \cdot B_1 = x_1 - x_1 - 0 = 0.$$

Similarly, $U \cdot B_2 = 0$. Therefore $U = 0$, and $X = x_1 B_1 + x_2 B_2$. What about uniqueness?

33. $X \cdot U = X \cdot (A/|A|) = (X \cdot A)/|A| = 3/\sqrt{2}$

35. $-3/\sqrt{5}$

37. 10.

39. The collection of all vectors having the component $k/|A|$ in the direction of A.

Section 5

1. $\dfrac{dF}{dt} = (1, -1/t^2)$; $t = \pm 1$ $(F(1) = \pm(1, 1))$

3 $\dfrac{d\Phi}{dt} = (2t, 1)$; $t = -1$ $(\Phi = (1, 2))$

5. $\dfrac{dX}{dt} = (-\sin t, \cos t)$; everywhere

7. $(dG/dt) = (-3 \sin t, 2 \cos t)$; $t = (n\pi/2)$ $(G = (0, \pm2), (\pm3, 0))$

9. $(40, -2)$; $X = (40, 14) + t(40, -2) = (40 + 40t, 14 - 2t)$, or $x = 40 + 40t$, $y = 14 - 2t$

11. $(1, 1)$; $X = (1, 0) + u(1, 1) = (1 + u, u)$, or $x = 1 + u$, $y = u$

13. $(0, 2)$; $X = (2, 0) + u(0, 2) = (2, 2u)$

15. $(1, -2)$; $x = u$, $y = 2 - 2u$

17. $x = x_0 - at \sin \theta_0$, $y = y_0 + bt \cos \theta_0$

19. The increment arrows are drawn to points on the other side of X_0, but when divided by the *negative* Δt, the adjusted arrows approach the same limiting arrow (representing the tangent vector).

21. Let $h(t) = \int_0^t f(s)\, ds$; then $h'(t) = f(t)$. By hypothesis, $x'(t) = uf(t)$ $y'(t) = vf(t)$, so $x(t) = uh(t) + k$, $y(t) = vh(t) + l$, and $X(t) = K + h(t)U$. So $X(t)$ is a straight line.

23. If d is the shortest distance between the two paths, then it is certainly the shortest distance from x_2 to path $x = G(t)$. Therefore, by a slight generalization of Example 5, $x_1 x_2$ is perpendicular to $X = G(t)$. By a similar argument, $x_1 x_2$ is perpendicular to the other curve.

Section 6

1. By rule (5), $dX/dt = A$, a constant vector. Then by the example in this section $x = At + C$.

5. $X \cdot dX/dt = 0$ implies

$$0 = X \cdot \frac{dX}{dt} + \frac{dX}{dt} \cdot X = \frac{d}{dt} X \cdot X = \frac{d}{dt} |X|^2.$$

So $|X|^2 = c$, a constant. So $X = F(t)$ runs along the circle of radius $\sqrt{c}$ about O.

7. Assume that the path does not contain the origin (since the zero vector does not specify a direction). Solve for $|X|dU/dt$ in the identity

$$\frac{dX}{dt} = \frac{d|X|}{dt} U + |X| \frac{dU}{dt}$$

and use the hypothesis of the problem, to conclude that dU/dt must be zero. Therefore, . . .

9. Your differentiated equation should be equivalent to $\cos \theta_1 + \cos \theta_2 = 0$, where θ_1 and θ_2 are the angles that $X - F$ and $X - G$ make with the tangent vector. Therefore $\theta_2 = \pi - \theta_1$, so the angle of incidence $=$ angle of reflection $= \theta_1$.

Section 7

1. a) $(x, y) = \left(\frac{s}{\sqrt{2}} \cos\left(\log \frac{s}{\sqrt{2}}\right), \frac{s}{\sqrt{2}} \sin\left(\log \frac{s}{\sqrt{2}}\right) \right)$

   b) $T = \left(\frac{1}{\sqrt{2}} \left(\cos\left(\log \frac{s}{\sqrt{2}}\right) - \sin\left(\log \frac{s}{\sqrt{2}}\right) \right), \frac{1}{\sqrt{2}} \left(\sin\left(\log \frac{s}{\sqrt{2}}\right) \right. \right.$
   $\left. \left. + \cos\left(\log \frac{s}{\sqrt{2}}\right) \right) \right)$

   $\frac{dT}{ds} = \frac{1}{s\sqrt{2}} \left(-\sin\left(\log \frac{s}{\sqrt{2}}\right) - \cos\left(\log \frac{s}{\sqrt{2}}\right), \cos\left(\log \frac{s}{\sqrt{2}}\right) - \sin\left(\log \frac{s}{\sqrt{2}}\right) \right)$

   c) $K = 1/s$ (Note that in all the above, $0 < s < \infty$.)

3. Note: $0 \le t < \infty, 0 \le s < \infty$.

   a) $(x,y) = (\frac{4}{3}((1 + 2s)^{1/2} - 1)^{3/2}, 2 - 2(1 + 2s)^{1/2} + s)$

   b) $T = (2(1 + 2s)^{1/2} - 1)^{1/2}(1 + 2s)^{-1/2}, -2(1 + 2s)^{-1/2} + 1); (dT/ds) =$
   $(((1 + 2s)^{1/2} - 1)^{-1/2}(1 + 2s)^{-1} - 2((1 + 2s)^{1/2} - 1)^{1/2}(1 + 2s)^{-3/2},$
   $2(1 + 2s)^{-3/2})$

   c) $K = 1/(1 + 2s)\sqrt{\sqrt{1 + 2s} - 1}$.

Section 8

1. $V = (-3 \sin 3t, 3 \cos 3t)$;
   $A = (-9 \cos 3t, -9 \sin 3t)$;
   $v = 3, dv/dt = 0$. We know that the unit circle has curvature $K = 1$
   (However, see the next answer); so $Kv^2 = 9$.

3. $V = (-3 \sin t, 2 \cos t)$;
   $A = (-3 \cos t, -2 \sin t)$;
   $v = (4 + 5 \sin^2 t)^{1/2}, dv/dt = \frac{5}{2} \sin 2t/(4 + 5 \sin^2 t)^{1/2}$ The absolute curvature
   K can be computed from the formula $K = |\dot{x}\ddot{y} - \ddot{y}\dot{x}|/v^3$ (Chapter 11,
   Section 5); so $Kv^2 = |\dot{x}\ddot{y} - \ddot{y}\dot{x}|/v$. Here, $Kv^2 = 6/(4 + 5 \sin^2 t)^{1/2}$.

5. $V = (1, 2t); \quad A = (0, 2); \quad v = (1 + 4t^2)^{1/2}, \quad dv/dt = 4t(1 + 4t^2)^{-1/2}. \quad Kv^2$
   $= |\dot{x}\ddot{y} - \ddot{y}\dot{x}|/v = 2(1 + 4t^2)^{-1/2}$.

7. $V = (-6 \sin 3t, 6 \cos 2t)$;
   $A = (-18 \cos 3t, -12 \sin 2t)$;
   $v = 6(\sin^2 3t + \cos^2 2t)^{1/2}$.
   $dv/dt = (9 \sin 6t - 6 \sin 4t)(\sin^2 3t + \cos^2 2t)^{-1/2}$.
   $Kv^2 = (18 \cos 3t \cos 2t + 12 \sin 3t \sin 2t)(\sin^2 3t + \cos^2 2t)^{-1/2}$

9. $v \le 25\sqrt{2} \quad (K = |d^2y/dx^2|/(1 + (dy/dx)^2)^{3/2}$
   $1/50$ and $Kv^2 = $ must be ≤ 25.)

11. If the particle is moving ($v \ne 0$) and the path is curved, $K \ne 0$, then the
    normal component $Kv^2 \ne 0$.

13. $50\sqrt{2}$ mph ≈ 70 mph

CHAPTER 15

Section 1

1. $\dfrac{\partial z}{\partial y} = \lim\limits_{\Delta y \to 0} \dfrac{f(x, y + \Delta y) - f(x, y)}{\Delta y}$

3. $\dfrac{\partial z}{\partial x} = 6x - 2y; \dfrac{\partial z}{\partial y} = -2x + 1$

5. $\dfrac{\partial z}{\partial x} = ae^{ax+by}; \dfrac{\partial z}{\partial y} = bc^{ax+by}$

7. $\dfrac{\partial z}{\partial x} = 2\cos(2x + 3y); \dfrac{\partial z}{\partial y} = 3\cos(2x + 3y)$

9. $\dfrac{\partial z}{\partial x} = ye^{-x^2/2} - x^2ye^{-x^2/2}; \dfrac{\partial z}{\partial y} = xe^{-x^2/2} = (1 - x^2)ye^{-x^2/2}$

11. $\dfrac{\partial z}{\partial x} = \dfrac{y}{x}; \dfrac{\partial z}{\partial y} = \log(xy) + 1$

13. $\dfrac{\partial z}{\partial x} = a\cos ax \cos by; \dfrac{\partial z}{\partial y} = -b\sin ax \sin by$

15. $\dfrac{\partial z}{\partial x} = \dfrac{y}{x^2 + y_2}; \dfrac{\partial z}{\partial y} = \dfrac{-x}{x^2 + y_2}$

17. $\dfrac{\partial z}{\partial x} = \dfrac{-2y}{(x - y)^2}; \dfrac{\partial z}{\partial y} = \dfrac{2x}{(x - y)^2}$

19. $D_1 f(s, t) = (1 + st)e^{st}, D_2 f(s, t) = s^2 e^{st}; D_1(2, 3) = 7e^6, D_2(-2, 1) = 4e^{-2}$

21. $D_1 f(x, y) = y^2 z^3, D_2 f(x, y) = 2xyz^3, D_3 f(x, y) = 3xy^2 z^2; -32; 0; 48.$

23. $2a \sin b; a^2$

25. 4.

27. -1

29. The boundary consists of just one point, the origin.

31. One example is

$$\log\left(\dfrac{1}{x^2 + y^2} - 1\right).$$

Find another (possibly by modifying this one).

Section 2

1. $(-3, 3); -9$ (min)

3. $(0, \pm\sqrt{3}); -9$ (min)

5. $\pm(\sqrt{\tfrac{2}{3}}, \sqrt{\tfrac{2}{3}}); -4/27$ (min)

7. $(\tfrac{1}{2}, \tfrac{1}{4}); -1/8$ (min)

9. $\tfrac{3}{8}\sqrt{2}$

11. $4, -\tfrac{1}{3}$

13. $2/\sqrt{5}$

15. $x = 6, y = 2, z = 3.$

Section 3

1.

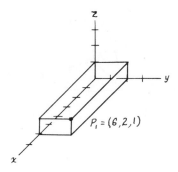

$P_1 = (6,2,1)$

3.

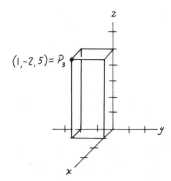

$(1,-2,5) = P_3$

5.

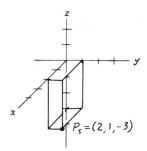

$P_5 = (2,1,-3)$

7.

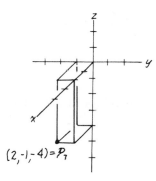

$(2,-1,-4) = P_7$

9.

$(-6,-2,-1) = P_9$

11. $\sqrt{26}$

13. $\sqrt{42}$

15. $3\sqrt{10}$

17. The plane perpendicular to the y-axis (parallel to the xz-coordinate plane) and cutting the y-axis at $y = 3$.

Section 4

1. The surface of a sphere of radius 6.

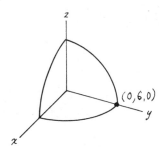

3. The plane parallel to the x-axis intersecting the yz-plane in the line $z = 1 - y$.

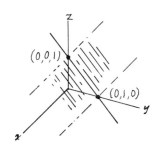

5. A parabola in the plane $x = 2$.

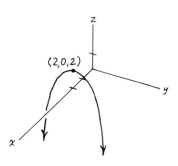

7. A circle of radius 2 in the plane $z = 1$, with center at $(0, 0, 1)$.

9. The cylinder of radius 2 with axis the x-axis.

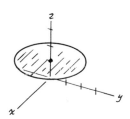

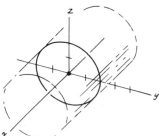

11. A cylinder with an ellipse as perpendicular cross section, and axis the z-axis.

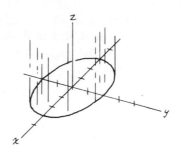

13. The sphere with center $(2, 1, 3)$ and radius $5\sqrt{2}$.

15.

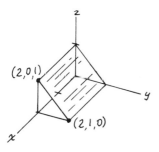

17.

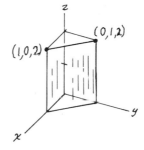

19.

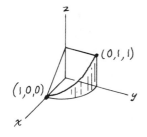

21. The intersection of the cylinders of radius 4 with axes the x- and z-axes. The curve lies in the plane $x = z$.

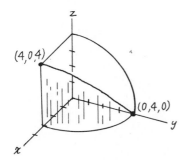

23. The intersection of the ellipsoid with semiaxes 4, 4, and 8, the first octant, and the region under the plane $y + z = 4$.

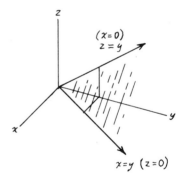

Section 5

1. (Plane)

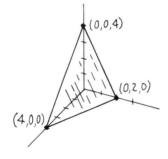

3. (Plane)

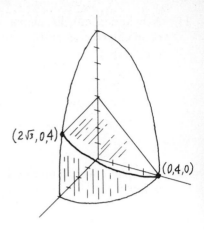

5.

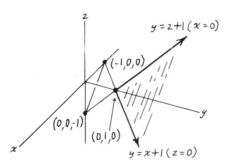

7. $x = -2$

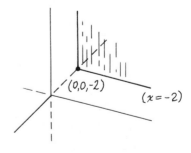

9.

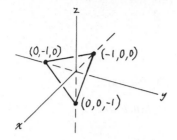

11. If $A = a$ and $B = b$, then there can be a point (x, y, z) on both planes only
     if $C = c$, in which case the planes are identical. Therefore, if $A = a$ and
     $B = b$ the planes are parallel. On the other hand, if $A \neq a$. then the two
     planes intersect the xz-coordinate plane $y = 0$ in the nonparallel lines
     $z = ax + c$, $z = Ax + C$, so the two planes are not parallel. Similarly, if
     $B \neq b$.

13. $z = 1$

15. $5x - 4y + 3z = 15$

17. $9x - 8y + 13z = 33$

19. $5x + 4y + 3z = 22$

21. $3x - 4y + 25z = 25(\log 5 - 1)$

23. $x - 2y + z = 2$

25. $(2/7, 4/7, 6/7)$

CHAPTER 16

Section 1

1. $32t$

3. $-\sec t$

5. $2(e^{2t} - e^{-2t})/(e^{2t} + e^{-2t})$

7. $\dfrac{\partial z}{\partial u} = 4u;\ \dfrac{\partial z}{\partial v} = 4v$

9.
$$\frac{\partial z}{\partial u} = 2[v^3 + 5v^2u + 3vu^2]e^{(u^2 + 2uv)(2uv + v^2)};$$

$$\frac{\partial z}{\partial v} = 2[u^3 + 5u^2v + 3uv^2]e^{(u^2 + 2uv)(2uv + v^2)}.$$

11. $\dfrac{\partial z}{\partial x} = f'\left(\dfrac{y}{x}\right)\left(\dfrac{-y}{x^2}\right);\ \dfrac{\partial z}{\partial y} = f'\left(\dfrac{y}{x}\right)\left(\dfrac{1}{x}\right).$

15. If $x = g(u,v),\ y = h(u,v)$, then

$$dx = D_1 g(u,v)du + D_2 g(u,v)dv,$$

$$dy = D_1 h(u,v)du + D_2 h(u,v)dv.$$

Also

$$z = f(x,y) = f(g(u,v),\quad h(u,v)) = F(u,v),$$

and

$$dz = D_1 F(u,v)du + D_2 F(u,v)dv.$$

But

$$D_1 F(u,v) = D_1 f(x,y)D_1 g(u,v) + D_2 f(x,y)D_1 h(u,v).$$

Section 2

1. If the functions $x = g(t),\ y = h(t)$, and $z = k(t)$ are differentiable at $t = t_0$, and if $f(x,y,z)$ is smooth (continuously differentiable) inside a small sphere about $(x_0,y_0,z_0) = (g(t_o),h(t_0),k(t_0))$, then $F(t) = f(g(t),h(t),k(t))$ is a differentiable function of t at $t = t_0$, and

$$F'(t_0) = D_1 f(x_0,y_0,z_0)g'(t_0) + D_2 f(x_0,y_0,z_0)h'(t_0) + D_3 f(x_0,y_0,z_0)k'(t_0).$$

3. If w is a smooth function of $x_1,\ x_2,\ x_3,\ x_4$ which in turn are differentiable functions of t, then w is a differentiable function of t and

$$\frac{dw}{dt} = \frac{\partial w}{\partial x_1}\frac{dx_1}{dt} + \frac{\partial w}{\partial x_2}\frac{dx_2}{dt} + \frac{\partial w}{\partial x_3}\frac{dx_3}{dt} + \frac{\partial w}{\partial x_4}\frac{dx_4}{dt}.$$

Section 3

1. Yes, everywhere.

3. Everywhere but $(-3/4^{1/3}, 2^{1/3})$

5. Everywhere but $(0,0)$

7. Let $g(t) = f(t) - mt$; then $g'(t) = f'(t) - m > 0$ on $[c,d]$, so g is increasing on $[c,d]$. Therefore, $g(d) > g(c)$, etc.

9. a) $0 < a < b \Rightarrow (1/b) < (1/a)$
   b) $0 < a < 0$ and $0 < c < d \Rightarrow ac < bd.$

Section 4

1. $(6, 8)$

3. $(-\pi/12, -1/2)$

5. $(8, 4)$

7. $(2, -1)$

9.

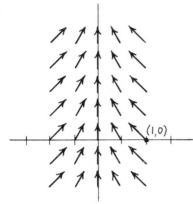

11. $(\sqrt{2}, -1)$

13. $(1, \sqrt{3})$

15. $2x + y = 4$

17. $3x + 4y = 25$

19. $3x + y = 5$

21. $-x + 2y = 4$

23. $(5/8 + \pi/4)(x - 2) + (5/2 + \pi/4)(y - 1/2) = 0$

25. grad $f = (2x, 2y)$,   grad $g = (-y/x^2, 1/x)$
    So grad $f \cdot$ grad $g = 0$.

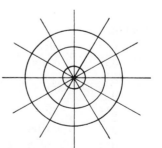

27. The vector X_0 is perpendicular to the level curve at the point X_0, by Example 5 of Section 5 of Chapter 14. So is grad $f(X_0)$.

29. (Consequences of the corresponding laws for derivatives in one variable.)

31. $f(x,y) = \frac{1}{2}(x^2 + y^2) + c$

Section 5

1. Set $U = A/|A|$; then apply Theorem 7.

3. $\sqrt{3}/2 + 1$

5. $2\sqrt{10}/75$

7. $e\sqrt{5}$

9. $54/169$

11. 2.

13. $(11\sqrt{2}\,; (1/\sqrt{2})(-1,-1)$

15. $32\sqrt{5}/15; (1/\sqrt{5})(1,-2)$

17. $\sqrt{10}/2; (1/\sqrt{10})(3,-1)$

19. $2/\sqrt{5}\,; (1/\sqrt{5})(1,-2)$

21. Let $A = \dfrac{dX}{dt}$.

$$\frac{dz}{dt} = \operatorname{grad} f \cdot \frac{dX}{dt} = \frac{\operatorname{grad} f \cdot A}{|A|}|A| = (D_{A/|A|}f)|A|.$$

23. $\operatorname{grad} f(X_0) = (4, 9/2)$.

25. $u + v$

27. $bu + av$

Section 6

1. $\dfrac{\partial^2 z}{\partial x^2} = 6;\ \dfrac{\partial^2 z}{\partial x \partial y} = -2 = \dfrac{\partial^2 z}{\partial y \partial x};\ \dfrac{\partial^2 z}{\partial y^2} = 0.$

3. $\dfrac{\partial^2 z}{\partial x^2} = (4x^2y + 2y)e^{x^2y};\ \dfrac{\partial^2 z}{\partial x \partial y} = (2x^3y + 2x)e^{x^2y} = \dfrac{\partial^2 z}{\partial y \partial x};\ \dfrac{\partial^2 z}{\partial y^2} = x^4 e^{x^2y}.$

5. $\dfrac{\partial^2 z}{\partial x^2} = 2;\ \dfrac{\partial^2 z}{\partial x \partial y} = 3 = \dfrac{\partial^2 z}{\partial y \partial x};\ \dfrac{\partial^2 z}{\partial y^2} = 2$

7. $\dfrac{\partial^2 z}{\partial x^2} = 0 = \dfrac{\partial^2 z}{\partial y^2};\ \dfrac{\partial^2 z}{\partial x \partial y} = \dfrac{\partial^2 z}{\partial y \partial x} = 1.$

9. $\dfrac{\partial^2 z}{\partial x^2} = -\dfrac{6y}{x^4};\ \dfrac{\partial^2 z}{\partial x \partial y} = \dfrac{2}{x^3} - \dfrac{2}{y^3} = \dfrac{\partial^2 z}{\partial y \partial x};\ \dfrac{\partial^2 z}{\partial y^2} = \dfrac{6x}{y^4}.$

11. Set $g(x) = x - ct$ (t fixed), $h(t) = x - ct$ (x fixed). Then:

$$\frac{\partial z}{\partial x} = f'(g(x))g'(x) = f'(g(x))$$

$$\frac{\partial^2 t}{\partial x^2} = f''(g(x))g'(x) = f''(g(x))$$

$$\frac{\partial z}{\partial t} = f'(h(t))h'(t) = -cf'(h(t))$$

$$\frac{\partial^2 z}{\partial t^2} = -cf''(h(t))h'(t) = c^2 f''(h(t))$$

$$c^2 \frac{\partial^2 z}{\partial x^2} = \frac{\partial^2 z}{\partial t^2}$$

13. $\dfrac{\partial^2 z}{\partial y^2} = \dfrac{2x^2 - 2y^2}{(x^2 + y^2)^2}; \dfrac{\partial^2 z}{\partial x^2} = \dfrac{2y^2 - 2x^2}{(x^2 + y^2)^2}.$

15. $\dfrac{\partial z}{\partial y} = \dfrac{\partial^2 z}{\partial x^2} = (x^2 - 2y)e^{-x^2/4y}/4y^2 \sqrt{y}.$

Section 7

1. Rel. min. at $(-1, 1/4)$

3. Saddle point at $(0,0)$; rel. min. at $(1/6, 1/12)$

5. Saddle point at $(0,0)$

7. Saddle point at $(1, -1/2)$

9. Saddle point at $(1/3, -1/3)$; rel. min. at $(3/4, 1/2)$

11. Rel. min. at $(-1, 1)$

13. Rel. min. at $(1, 1)$ and $(-1, -1)$

15. Saddle point at $(0,0)$

CHAPTER 17

Section 1

1. $(2/3, 1/3, 2/3)$

3. $(7/11, -6/11, -6/11)$

5. $(2/3, 2/3, 1/3)$

7. $0, \phi = \pi/2$

9. $0, \phi = \pi/2$

11. $X = (1 + t, -2 + 3t, 1 - 2t)$

13. $X = (-2 + 3t, -1 + 4t, 2 + 2t)$

15. $X = (-1 + 3t, -2 + 2t, 1 + t)$

17. $x = 1, y = -1 + \frac{3}{2}t, z = 4 - 4t$

19. $x = -2 + \frac{5}{2}t, y = -1, z = -2t$

21. $X = (t, -2t, -t)$

23. $X = (1 + 2t, 1, 1 - t)$

25. $A = (\cos \alpha, \cos \beta, \cos \gamma)$
    $L = (\cos \lambda, \cos \mu, \cos \nu)$
    $A \cdot L = \cos \phi$

27. $(t, -t, -t)$

31. Note that reversing the roles of A_1 and A_2 reverses the sign of each expression in the formula for $A_1 \times A_2$.

33.
$$(A_1 \times A_2) \cdot A_2 = A_1 \cdot (A_2 \times A_2) \qquad (32)$$
$$= A_1 \cdot 0 \qquad (30)$$
$$= 0.$$

35. Expand both sides in coordinate notation to verify the formula.

37. The parallelogram with base $|A_2|$ has altitude $|A_1|\sin \phi$. So its area is $|A_1||A_2|\sin \phi = |A_1 \times A_2|$

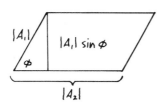

Section 2

1. $x + 2y + z = 0$

3. $x + 2y = 6$

5. $4x + 3y - 2z = 48$

7. $(1, 0, 1), (2, -1, 0)$

9. $(1, 5, 0), (0, 3, 1)$

11. $tA_1 + sA_2$ is in the xy-plane if and only if $t + s = 0$. So $s = -t$, and points
$$tA_1 + sA_2 = t(A_1 - A_2) = t(1, -4, 0)$$
form the intersection.

13. $A_1 \times A_2 \cdot X = (-4, -1, 7) \cdot X = 0$

15. $(3t, -5t, -7t)$

17. $x + 9y + 7z = 12$

Section 3

1. $\{(2/\sqrt{6}, 1/\sqrt{6}, 1/\sqrt{6}), (1/\sqrt{3}, -1/\sqrt{3}, -1/\sqrt{3}), (0, 1/\sqrt{2}, -1/\sqrt{2})\}$

3. $\{(1/\sqrt{3}, 1/\sqrt{3}, 1/\sqrt{3}), (1/\sqrt{2}, -1/\sqrt{2}, 0), (1/\sqrt{6}, 1/\sqrt{6}, -2/\sqrt{6})\}$

5. $(2/3, 2/3, 1/3)$.

$$x_1 B_1 + x_2 B_2 + x_3 B_3 = \tfrac{2}{3}(2, 1, 2) + \tfrac{2}{3}(2, -2 - 1) + \tfrac{1}{3}(1, 2, -2)$$
$$= \tfrac{1}{3}(4 + 4 + 1, 2 - 4 + 2, 4 - 2 - 2) = \tfrac{1}{3}(9, 0, 0) = (1, 0, 0).$$

7. $(2/3, -1/3, -2/3)$

9. $(4/3, 10/3, 8/3)$;

$$x_1 B_1 + x_2 B_2 + x_3 B_3 = \tfrac{4}{9}(2, 1, 2) + \tfrac{10}{9}(2, -2, -1) + \tfrac{8}{9}(1, 2, -2)$$
$$= \tfrac{1}{9}(8 + 20 + 8, 4 - 20 + 16, 8 - 10 - 16)$$
$$= \tfrac{1}{9}(36, 0, -18) = (4, 0, -2)$$

11. $(-4/3, -1/3, -8/3)$

13. This boils down to the fact that O is the only scalar multiple of A that is perpendicular to A.

15. $X = x_1 B_1 + x_2 B_2 + x_3 B_3$, so

$$X \cdot A = x_1 B_1 \cdot A + x_2 B_2 \cdot A + x_3 B_3 \cdot A$$
$$= x_1 a_1 + x_2 a_2 + x_3 a_3 = \sum_1^3 a_i x_i$$

Section 4

1. $A \times X = (a_2 x_3 - x_2 a_3, \ a_3 x_1 - x_3 a_1, \ a_1 x_2 - x_1 a_2)$

Section 5

1. $(1, -2, 3)$

3. $(3, 1, 1/2)$

5. $(1/4, 1/4, -1/16)$

7. (a, b, c)

9. $(0, 1 - \pi/2, 0)$

11. $\sqrt{2/7}$

13. $(18 - 12e^2)/7$

15. $4e^{-4}$

17. $(-3/\sqrt{2}, 1/\sqrt{2}, 5/\sqrt{2})$

19. $\sqrt{14}/6;\ 1/\sqrt{14}\,(-1, 2, -3)$

21.
$$\frac{d}{dt}wX = \left(\frac{d}{dt}wx_1,\ \frac{d}{dt}wx_2,\ \frac{d}{dt}wx_3 \right)$$
$$= \left(w\frac{dx_1}{dt} + \frac{dw}{dt}x_1,\ w\frac{dx_2}{dt} + \frac{dw}{dt}x_2,\ w\frac{dx_3}{dt} + \frac{dw}{dt}x_3 \right)$$
$$= w\frac{dX}{dt} + \frac{dw}{dt}X.$$

23. Change to coordinate notation and use the ordinary chain rule, as in (21).

25. The function $|U - x_0|^2$ attains a minimum at U_0, so
$$\frac{d}{dt}|U - x_0|^2 = 0$$
at this point. Apply Problem 22, and conclude that
$$2\frac{dU}{dt} \cdot [U_0 - x_0] = 0.$$
(dU/dt) is the vector tangent to the curve U at U_0, so $U_0 - x_0$ is perpendicular to U. Similarly, $U_0 - x_0$ is perpendicular to x at x_0.

27. Exactly similar to Chapter 14, Section 6, Problems 1 and 3.

29. Prove that $|X|^2 = X \cdot X$ is a constant.
$$0 = \frac{dX}{dt} \cdot X$$
implies
$$0 = \frac{dX}{dt} \cdot X + X \cdot \frac{dX}{dt} = \frac{d}{dt}[X \cdot X]$$
$$= \frac{d}{dt}(|X|^2).$$
So $|X|^2$ is constant.

31. $(dX/dt) = Xf(t)$, where f is a scalar function. Let $U|X| = X$; then follow the argument of Problem 7, Chapter 14, Section 6.

Section 6

1. $(-1, 3, -3)$

3. $(-1, 0, 1)$

5. a) $(2, 3, 1/6)$ if $(1/2, 1/3, 6)$
   b) $(2, 3, 1/6)$

7. $X = (1 + et, \pi/2, e - t)$

9. $X = (2 + 2t, 3 + t, -1 + 5t)$

11. $-x + 3y + 3z = 2$

13. $x = z$

15. $x - y - z = 1$

17. Choose any positive r less than $\min(1/|a|^3, 1/|b|^3)$. Show that if $x^2 + y^2 < r^2$, then, in particular, $x^2 < 1/a^{2/3}$ and $|x|^{1/3}|a| < 1$. Similarly, $|y|^{1/3}|b| < 1$. Show that, therefore, $|ax + by| < x^{2/3} + y^{2/3}$. Thus, within the circle of radius r, $ax + by > -(x^{2/3} + y^{2/3})$.

19. Separate into components, and apply the usual product formula, viewing the functions as *functions of a single variable* for *each* of their variables.

23. $1/30$.

25. $16/\sqrt{3}$

27. $\pm(98)^{3/4}/2$

Section 8

1. Differentiate $|X|^2 = X \cdot X$

3. $4/5$.

5. $\sqrt{2}/(\cos^2 t + 1)^{3/2}$

CHAPTER 18

Section 1

1. $1/4$

3. 1

5. 0

7. 0

9. $3/20$

11. 1

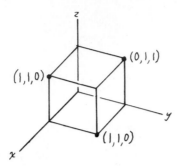

13. 1

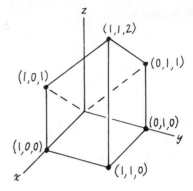

15. 5/12

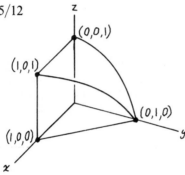

17. 1/3

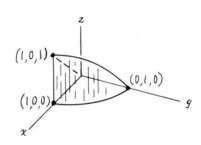

19. $\frac{4}{3}\pi abc$

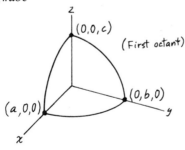

21. 27/4

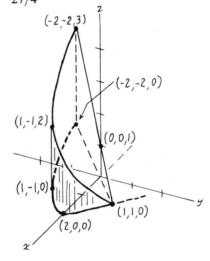

23. 1/4

25. 4/15

27. 1/3

29. 27/4

Section 2

1. 469/1800

7. $\int_0^1 \int_y^1 f(x,y)\, dx\, dy$

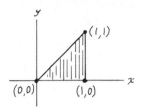

9. $\int_0^1 \int_y^{\sqrt{y}} \cdots dx\, dy$

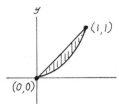

11. $\int_0^1 \int_{\sqrt{x}}^1 \cdots dy\, dx$
    $+ \int_0^1 \int_{-1}^{\sqrt{x}} \cdots dy\, dx$

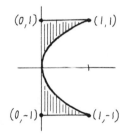

13. $1/2 - 1/2e$

15. $1/3$

17. a) $1/4$ b) $1/4$

19. a) $2/3$ b) $5/6$

21. $3/4$

23. a) $4 \cdot 1 + 8 \cdot 2 + 4 \cdot 4 = 36$
    b) $(1/36)(19/9)^2 = (19/18)^2(1/9)$
    c) $1/9$

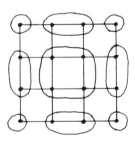

Section 3

1. Take a curve $H(s)$ from $(x_1 y_1)$ to (x_2, y_2) that stays in G. We may assume that H is parametrized by arc-length s, so that $H(0) = (x_1, y_1)$, $H(L) = (x_2, y_2)$, where L is the length of the path. Note $|dH/ds| = 1$. Use the dot-product form of the chain rule to show that

$$\left| \frac{d}{ds} f(H(s)) \right| = |\text{grad } f| \le K\sqrt{2}.$$

Show, therefore, that $|f(H(L)) - f(H(0))| \le K\sqrt{2}|L - 0|$, etc.

3. Let $M_i = f(a_i, b_i)$ be the maximum value of f in ΔG_i. Let $m_i = f(c_i, d_i)$ be the minimum value of f in ΔG_i. Now, ΔG_i has length less than e, so by (1),

$$|M_i - m_i| \leq \sqrt{2}\,Ke.$$

Now reason exactly as in the text.

Section 4

1. $1/4$

3. $4/35$

5. $8k$

7. $4k/3$

9. $8ka^3/3$

11. πk

13. 1

Section 5

1. $(2/3, 2/3)$

3. $(5/9, 5/11)$

5. $(0, 4/3)$

7. $(3/8, 0)$

9. $(0, 0)$

11. $((\pi^2 - 4)/\pi, \pi/8)$

13. $(2, 1/8)$

Section 6

1. $122/3\pi$

3. $\pi/3$

5. $4\pi/3$

7. $\pi/3 - 4/9$

9. $\int_0^{2\pi} \int_0^R (kr)r\,dr\,d\theta = 2\pi k R^3/3$

11. Mass $k\pi/12$; center of mass $(0, 6/5\pi)$ (rectangular coordinates)

13. $16a^3/3$

Section 7

1. $(\pi/6)(17\sqrt{17} - 1)$

3. 20π

5. $\pi - 2$

7. $3\sqrt{14}$

9. $\pi a\sqrt{a^2 + h^2}$

Section 8

1. 6

3. $abc/6$

5. $18/35$

7. $1/3$

9. $16/3$

11. $64/15$

13. $8/3$

15. $(3/8)(a, b, c)$

Section 9

1. 8π

3. $\dfrac{16\pi}{3}\left(1 - \dfrac{\sqrt{3}}{2}\right) \approx \dfrac{4\pi}{5}$

5. $\frac{2}{3}\pi a^3(1 - \cos \alpha)$

7. $\frac{2}{3}\pi[6^{3/2} - 11]$

9. $2\pi a^2 k$, where k is the constant of proportionality.

11. $\frac{1}{2}\pi a^4 k$, where a is the radius, and k is the proportionality constant.

13. $\frac{1}{3}\pi a^3 b^2$

15. $\bar{x} = \bar{y} = 0; \quad \bar{z} = 3/(16(1 - \sqrt{3}/2)) \approx 5/4$

17. $(3a/8)(1, 1, 1)$

Section 10

3. If Σ_1 and Σ_2 are Riemann sums for f over G_1 and G_2. respectively, then $\Sigma = \Sigma_1 + \Sigma_2$ is a Riemann sum for f over G. Each of these Riemann sums approaches the corresponding triple integral as the maximum subregion diameter approaches zero.

9. We can suppose that f is positive. Since ΔG lies between the cylinders C' and C'', we have

$$0 \le \frac{1}{\Delta z}\left[\iiint\limits_{\Delta G} f - \iiint\limits_{C'} f\right] \le \frac{1}{\Delta z}\iiint\limits_{C''-C'} f \le B(A'' - A'),$$

where B is a bound for f. Since $A'' - A' \to 0$ as $\Delta z \to 0$, the above difference of triple integers approaches 0. Let C be the cylinder of altitude Δz based on the cross section R_{z_0}. Then, by the same argument,

$$\frac{1}{\Delta z}\left[\iiint\limits_{C} f - \iiint\limits_{C'} f\right] \to 0.$$

Finally,

$$\frac{1}{\Delta z}\iiint\limits_{C} f \to \iint\limits_{R_{z_0}} f(x, y, z_0),$$

by Problem 8. Now combine these three limits.

CHAPTER 19

Section 1

1. $y = Ce^{2x}$

3. $y = Ce^{kx}$

5. $y = -2/(x^2 + C)$

7. $y = \log(2/(C - x^2))$

9. $y = \sin(x + C)$

11. $y^2 - 2x^2 = C$

13. $y = x^2/(Cx^2 + x - 2)$

17. $y = -\arctan(\log cx)$ or $y = \pi/2 + n\pi$ for fixed n.

19. $y = \arctan(c - \cos x)$ or $y = \pi/2 + n\pi$ for fixed n.

21. $\dfrac{dv}{f(v) - v} = \dfrac{dx}{x}$

23. $P = Ct^k$ $\left(\dfrac{dP}{dt} = k\dfrac{P}{t}\right)$

25. $t = (\sqrt{10} - \sqrt{x})/(\sqrt{10} - 3)$ hours

Section 2

1. $dy = dx$

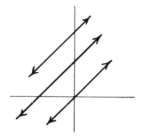

3. $dy/dx = -x/y$

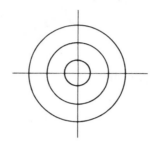

5. $\dfrac{dy}{dx} = \dfrac{-x}{\sqrt{1 - x^2}}$

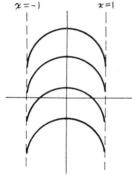

7. $\dfrac{dy}{dx} = \pm 2\sqrt{y}\,;\ \left(\dfrac{dy}{dx}\right)^2 = 4y$

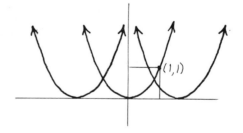

9. Circles: $x^2 + y^2 = C$

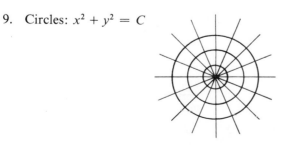

11. $y = Ce^{-2x}$

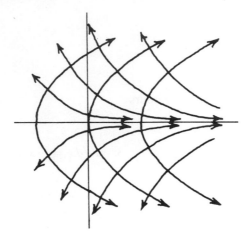

13. $y = Cx$ (See (9))

15. $x^2 + Cy + y^2 = 0$

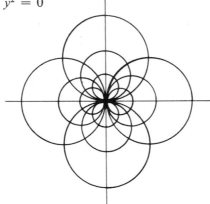

Section 3

1.

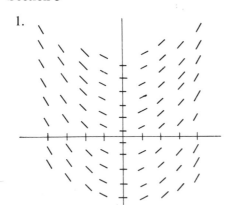

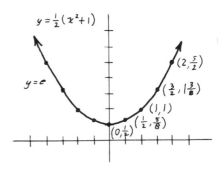

3.

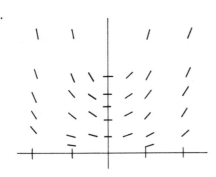

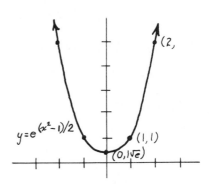

$y = e^{(x^2 - 1)/2}$

$(2,$

$(1, 1)$

$(0, 1\sqrt{e})$

Section 4

1. $y = -e^x + Ce^{+2x}$

3. $y = e^x/(1 + a) + Ce^{-ax}$, for $a \neq -1$; $y = xe^x + Ce^x$ for $a = -1$.

5. $y = x - 1 + Ce^{-x}$

7. $y = x^2/3 + C/x$

9. $y = 1 + C/x$

11. $y = \left(1/(1 + a^2)\right)(a \sin x - \cos x) + Ce^{-ax}$

13. $y = (1/2)(x^2 - 1) + Ce^{-x^2}$

15. $y = \tan x + C \sec x$

Section 5

1. $y = -e^x + Ce^{2x}$

3. $y = x - 1 + Ce^{-x}$

5. $y = (1/a) + Ce^{-ax}$ $(a \neq 0)$; $y = x + C$ $(a = 0)$.

7. $y = (1/3)xe^x - (1/9)e^x + Ce^{-2x}$

9. $y = (1/2)(x \sin x - x \cos x + \cos x) + Ce^{-x}$

11. $\dfrac{d}{dx}(e^{ax}y) = e^{ax}r(x) \Leftrightarrow e^{ax}\dfrac{dy}{dx} + ae^{ax}y = e^{ax}r(x)$ (Product formula)

13. Use the derivative product rule to expand $L(xy)$.

15. $y = xe^{-4x} + Ce^{-4x}$

17. $y = (1/2)e^x + 1 + Ce^{-x}$

19. $y = -x - 1 + xe^x + Ce^x$

21. a) $3x^3$ b) x c) 0 d) xe^x

23. Use the derivative product rule to expand $L(u, v)$.

Section 6

1. $y = x$

3. $y = -\frac{1}{3}e^x$

5. $y = \frac{1}{4}\cos x$

7. $y = \frac{1}{3}x^3 - x^2 + 2x$

9. $y = \frac{1}{3}e^x + \frac{2}{3}e^{-x}$

11. $y = \frac{1}{2}(xe^x - e^x)$

13. $y = -\frac{1}{2}e^{-x}$

15. $y = -\frac{3}{10}\sin x - \frac{1}{10}\cos x$

17. $y = \frac{1}{2}x^2 - x - \frac{1}{2}\sin x - \frac{1}{2}\cos x$

19. $y = \frac{1}{12}x^4 e^x$

21. $y = -\frac{1}{4}e^x(\sin x + \cos x)$

23. $y = e^x \sin x$

Section 7

1. $y = c_1 \sin \sqrt{2} + c_2 \cos \sqrt{2}x$

3. $y = \frac{1}{2}(x^2 - 1) + c_1 \sin \sqrt{2}x + c_2 \cos \sqrt{2}x$

5. $y = -\frac{1}{4}x \cos 2x + c_1 \sin 2x + c_2 \cos 2x = c_1 \sin 2x + (c_2 - (x/4))\cos 2x$

7. $y = \frac{1}{2}xe^x + c_1 e^x + c_2 e^{-x} = (c_1 + (x/2))e^x + c_2 e^{-x}$

9. $y = -\frac{1}{5}e^x \sin x - \frac{2}{5}e^x \cos x + c_1 e^x + c_2 e^{-x}$

11. $y = c_1 e^x + c_2 e^{-2x}$

13. $-\frac{3}{10}\sin x - \frac{1}{10}\cos x + c_1 e^x + c_2 e^{-2x}$

15. $y = +c_1 e^x + c_2 xe^x$

17. $y = \frac{1}{2}\cos x + c_1 e^x + c_2 xe^x$

19. $y = c_1 e^{-x}\sin x + c_2 e^{-x}\cos x$

21. $y = \frac{1}{3}e^x + c_1 e^{-x} \sin x + c_2 e^{-x} \cos x$

23. $(D - k_1)[(D - k_2)f] = (D - k_1)[f' - k_2 f]$,   etc.

25. $y = c_1 e^{kx} + c_2 x e^{kx}$

27. $y = \frac{1}{3}x e^{2x} + c_1 e^{-x} + c_2 e^{2x}$

Section 8

1. For $m^2 > 4ms$, $m\ddot{x} + r\dot{x} + sx = 0$ has solutions $x = c_1 e^{-kt} + c_2 e^{-lt}$, where

$$k = \frac{r + \sqrt{r^2 - 4ms}}{2m} \quad \text{and} \quad l = \frac{r - \sqrt{r^2 - 4ms}}{2m}, \quad (0 < l < k).$$

$0 = \dot{x}(0) = -kc_1 e^{-k \cdot 0} - lc_2 e^{-l \cdot 0}$, so $c_2 = (-k/l)c_1$; $0 < x(0) = c_1 e^{-k \cdot 0} - (k/l)c_1 e^{-l \cdot 0} = (1 - (k/l))c_1$; so $c_1 < 0$. Let $c = -c_1/l$; then $c > 0$ and $x = c[ke^{-lt} - le^{-kt}]$.

3. Similarly, as in Problem 1, $x = c_1 e^{-kt} + c_2 e^{-lt}$ with $0 < l < k$.

$0 = x(0) = c_1 e^{-k \cdot 0} + c_2 e^{-l \cdot 0} = c_1 + c_2$, so $c_1 = -c_2$ and $x = c_2[e^{-lt} - e^{-kt}]$. $0 < \dot{x}(0) = c_2[-le^{-l \cdot 0} + ke^{-k \cdot 0}] = c_2(k - l)$, so $c_2 = c > 0$.

5. As in (1) and (3), $x = c_1 e^{-kt} + c_2 e^{-lt}$, $0 < l < k$. $1 = x(0) = c_1 + c_2$, so $c_1 = 1 - c_2$. Let $a = c_2$; then $x = ae^{-lt} + (1 - a)e^{-kt}$. $\dot{x}(0) = a(-l)e^{-l \cdot 0} + (1 - a)(-k)e^{-k \cdot 0} = a(k - l) - k$. So $a < 0$ implies $\dot{x}(0) = v_0 = a(k - l) - k < -k$, and $v_0 = a(k - l) - k < -k$ implies $a < 0$.

7. (This is just like the case of Problem 25 in Section 7.) $x = c_1 e^{-kt} + c_2 t e^{-kt}$; $x_0 = 1 = c_1 + (c_2)(0) = c_1$,   so $x = e^{-kt} + c_2 t e^{-kt}$. $0 = v_0 = \dot{x}(0) = -ke^{-k \cdot 0} + c_2(-k)(0)e^{-k(0)} + c_2 e^{-k \cdot 0} = -k + c_2$, so $c_2 = k$ and $x = (1 + kt)e^{-kt}$.

9. This is a result of Theorem 9 (Section 7) and Problem 35, Chapter 4, Section 3.

11. See Problem 35, Chapter 4, Section 3.

13. Differentiate (∗) to get (∗∗)

Section 9

1. $c_1 = 0$, $c_n = (n + 2)c_{n+2}$; converges for all x;

$$y = c_0\left(1 + \frac{x^2}{2} + \frac{x^4}{2 \cdot 4} + \frac{x^6}{2 \cdot 4 \cdot 6} + \cdots + \frac{x^{2n}}{2^n \cdot n!} + \cdots\right)$$

3. $c_2 = 0,\ 2c_2 = 1,\ (n + 3)c_{n+3} = c_n$;

$$y = c_0\left(1 + \frac{x^3}{3} + \frac{x^6}{3 \cdot 6} + \cdots + \frac{x^{3n}}{3^n n!} + \cdots\right)$$

$$+ \left(\frac{x^2}{2} + \frac{x^5}{2 \cdot 5} + \frac{x^8}{2 \cdot 5 \cdot 8} + \cdots + \frac{x^{3n-1}}{2 \cdot 5 \cdot 8 \cdots (3n - 1)} + \cdots\right)$$

$$= c_0 f(x) = g(x).$$

Each series converges for all x.

5. $c_0,\ (n + 1)c_{n+1} = nc_{n-1}$;

$$y = c_1\left(1 + \frac{x^2}{2} + \frac{3}{2 \cdot 4}x^4 + \frac{3 \cdot 5}{2 \cdot 4 \cdot 6}x^6 + \cdots\right);\ \text{converges on } (-1, 1).$$

7. $c_3 = c_2 = 0,\ (n + 2)(n + 1)c_{n+2} = c_{n-2}$;

$$y = c_0\left[1 + \frac{x^4}{4 \cdot 3} + \frac{x^8}{8 \cdot 7 \cdot 4 \cdot 3} + \cdots\right]$$

$$+ c_1\left[x + \frac{x^5}{5 \cdot 4} + \frac{x^9}{9 \cdot 8 \cdot 5 \cdot 4} + \cdots\right].$$

Each series converges for all x.

9. $(n + 2)(n + 1)c_{n+2} = [n(n - 1) + 1]c_n$;

$$y = c_0\left[1 + \frac{x^2}{2} + \frac{x^4}{8} + \cdots\right] + c_1\left[x + \frac{x^3}{6} + \frac{7x^5}{120} + \cdots\right].$$

Converges on $(-1, 1)$, by the ratio test.

CHAPTER 20

Section 1

1. $\dfrac{1}{7} - \dfrac{10}{71} = \dfrac{1}{497}$

3. $\dfrac{385}{536} - \dfrac{334}{465} = \dfrac{1}{249{,}240}$

5. 3.8   (correct expansion: next place between 2 and 7)

7. 1.4   (closest; within 0.02)

9. 3.1   (closest; within 0.025)

11. 1.0 (correct); 1.1 (closest, within 0.032)

13. within 0.006

15. within 0.005; within 0.011

19. First, $p(1) > \sin(1)$. (Why?) Let d be the 10-place truncation of the expansion of $p(1)$, and let k be the digit in the eleventh place (we suppose we have computed d and k).

   a) If $k \geq 4$, then d is the 10-place truncation in the expansion of $\sin 1$.

   b) If $k \leq 4$, then d is the 10-place decimal closest to $\sin 1$.

   (If $k = 4$, then *both* conclusions are valid.)

21. $s - s_{n-1} = \sum_n^\infty a_j < k \sum_n^\infty r^j = (k/(1-r))r^n$. Go on from here.

Section 2

1. 3 places

3. 5 places

5. $x^3 = (3.1415)^3 = 31.003 \cdots < \pi^3;\ (\pi^3 - x^3) < 0.005$, so $\pi^3 = 31.00 \cdots$

7. There are two cases, $f'(x) \geq b$ everywhere, and $f'(x) \leq -b$ everywhere. In the first case, apply the mean-value theorem to $g(x) = f(x) - bx$. In the second base, apply it to $h(x) = -f(x) - bx$.

9. Use the factorization

$$a^4 - x^4 = (a - x)(a^3 + a^2 x + ax^2 + x^3).$$

11. Use the factorization

$$a^n - x^n = (a - x)(a^{n-1} + a^{n-2}x + \cdots + x^{n-1}) = (a - x)\sum_{k=0}^{n-1} a^{n-1-k}x^k.$$

15. From Problem 9 (relabeled), we have

$$|\Delta y| \leq 4a^3 |\Delta x|.$$

Thus we will have $|\Delta y| \leq \epsilon$ if $4a^3|\Delta x| \leq \epsilon$; that is, if $|\Delta x| \leq \epsilon/4a^3$.

17. $\sqrt{1 - x^2} = \sqrt{1 - x}\sqrt{1 + x}$. Also, $1 + x \leq 2$ (because $\sqrt{1 - x}$ doesn't make sense unless $x \leq 1$).

19. If $\epsilon \leq 1$ it is sufficient to take $|\Delta x| \leq \epsilon^{5/4}/5$.

21. $\epsilon^2/3$ (supposing $\epsilon \leq 1$).

23. If $z = x + y$, then $\Delta z = \Delta x + \Delta y$, and $|\Delta z| \leq |\Delta x| + |\Delta y|$. So $|\Delta z| < \epsilon$ if $|\Delta x|$ and $|\Delta y|$ are both less than $\epsilon/2$.

25. Suppose that $x > a > 0$ and that r is a rational number approximating x, that is also greater than a. Then,

$$\left| \frac{1}{x} - \frac{1}{r} \right| < \frac{|r - x|}{|xr|} < \frac{|r - x|}{a^2}.$$

Thus, $1/r$ is a rational number approximating $1/x$ and, according to the above inequality, the error will be less than ϵ in magnitude if $(|r - x|/a^2) < \epsilon$; that is, if $|r - x| < a^2\epsilon$. Since x is calculable, we can find such an r, so $1/x$ is also calculable. Note that the requirement on r is a double one:

$$|r - x| < a^2\epsilon \qquad \text{and} \qquad r > a.$$

27. Using Problem 26(a) repeatedly, we have that $x_1 + x_2$, $(x_1 + x_2) + x_3$, ..., $x_1 + x_2 + \cdots + x_n$ are successively all calculable numbers.

Section 3

1. $\sqrt{10}$ is the solution of $f(x) = x^2 - 10 = 0$ and $f'(x) = 2x \geq 6 = m$ when $x \geq 3$. So 3.16 approximates $\sqrt{10}$ with an error at most

$$\frac{|f(3.16)|}{6} = \frac{|(3.16)^2 - 10|}{6} = \frac{10 - 9.9856}{6}$$

$$= \frac{.0144}{6} < 0.003.$$

3. $30^{1/3}$ is the solution of $x^3 - 30 = 0$. Also, $d(x^3 - 30)/dx = 3x^2 \geq 27$ when $x \geq 3$. Therefore, 3.11 approximates $30^{1/3}$ with an error at most

$$\frac{|f(3.11)|}{27} = \frac{30.080 \cdots - 30}{27} = \frac{0.080 \cdots}{27} < 0.003.$$

Section 4

1. This is an initial-value problem. Show that $dL(ax)/dx = L'(x)$, so that $L(ax) - L(x) = c$ for some constant c. Then evaluate c by giving x a special value.

5. Use the integral test and the known fact that the harmonic series diverges.

7. $E(x + y) = E(L(E(x)) + L(E(y))) = \cdots$ A similar proof was given in Chapter 7, Section 3.

9. Problem 24, Section 1, Chapter 13, provides the formula

$$\log \frac{1 + y}{1 - y} = F(y).$$

Set $y = (x - 1)/(x + 1)$ and reduce, to obtain

$$\log x = F\left(\frac{x - 1}{x + 1}\right).$$

11. The differentiated power series should be twice the geometric series expansion of $1/(1 - x^2)$. The second part is a chain-rule calculation.

Section 5

1. Show that $dF(\tan x)/dx = 1$ by the chain rule. Then $F(\tan x) = x + C$. At $x = 0$, $F(0) = 0$; so $C = 0$. In all x, $F(\tan x) = x$.

3. $\int_0^a \frac{dt}{\sqrt{1 - t^2}}$ is increasing as a increases to 1. Also

$$\int_0^a \frac{dt}{\sqrt{1 - t^2}} = \int_0^a \frac{dt}{\sqrt{1 - t}\sqrt{1 + t}} < \int_0^a \frac{dt}{\sqrt{1 - t}} = 2(1 - \sqrt{1 - a}) < 2.$$

Section 7

1. Show that, on a given closed interval, $\max(f + g) \leq \max f + \max g$. This leads to an inequality for the upper Riemann sums of f, g, and $(f + g)$. Similarly for $\min(f + g)$ and lower Riemann sums. The final result is that $\int_a^b (f + g)$ and $\int_a^b f + \int_a^b g$ both lie between $l_f + l_g$ and $u_f + u_g$, where l and u are the lower and upper Riemann sums. The identity follows.

3. By (1), $\int_a^b f + g = \int_a^b f + \int_a^b g$. But $f + g = 0$.

index